Foundations of Optimal Control Theory

The SIAM Series in Applied Mathematics
R. F. Drenick, Harry Hochstadt, Dean Gillette, ***Editors***

RIORDAN, *Stochastic Service Systems*
ROSENBLATT, *Time Series Analysis*
HENRICI, *Error Propagation for Difference Methods*
MILLER, *Multidimensional Gaussian Distributions*
BARLOW AND PROSCHAN, *Mathematical Theory of Reliability*
ISAACS, *Differential Games*
NOETHER, *Elements of Nonparametric Statistics*
LEE AND MARKUS, *Foundations of Optimal Control Theory*
HART, *Handbook of Computer Approximations*

In preparation (*titles tentative*):
BAREISS, *Transport Theory*
DORN AND LEMKE, *Mathematical Programming*
HEADING, *The Stokes Phenomenon*
KELLEY, *Critical-Path Scheduling*
MACCOLL, *Modern Dynamical Theory*
MARSAGLIA, *Monte Carlo Methods*
SNEDDON AND LOWENGRUB, *Crack Problems*

Foundations of Optimal Control Theory

E. B. Lee

L. Markus

Center for Control Sciences
Institute of Technology
University of Minnesota

John Wiley & Sons, Inc., New York · London · Sydney

10 9 8 7 6

Library of Congress Catalog Card Number: 67-22414
Printed in the United States of America

ISBN 0471 52263 5

To Judy and Lois

Preface

The mathematical theory of optimal control began about twenty years ago as a special topic within the discipline of differential equations. After the maximal principle and the method of dynamic programming were discovered, it was recognized that optimal control theory could be treated within the framework of the calculus of variations. However, many of the basic concepts of control theory still demanded an approach through the qualitative theory of differential systems, and it is from this viewpoint that this text has been prepared.

In the past three or four years the theory of control of deterministic processes with several degrees of freedom has reached a satisfying stage of completeness. As interpreted by the theory of nonlinear ordinary differential equations, the fundamental problems of control theory have been mathematically posed and answered.

Because the theory had reached a certain degree of stability and perfection, the authors believed that a thorough and careful presentation of the current status of control theory would serve the useful purpose of offering a foundation on which later researches could be based. It was with this intent that the *Foundations of Optimal Control Theory* was written. Our goal was to present an organized treatment of control theory that would be thorough and complete within the limitations set by the restriction to deterministic (nonstochastic) problems definable in terms of ordinary differential systems.

In general the exposition is in the mathematical style of definitions, theorems, and proofs. The hypotheses for each analytical or geometrical conclusion are clearly displayed. However, occasionally certain standing hypotheses on continuity or boundedness are listed at the beginning of the sections and some care is needed not to overlook these assertions. Exercises follow most sections of the text. Some of these exercises are simple problems to make the general theory more concrete, but some exercises give refinements and extensions of the text material; very rarely an exercise is needed to complete the details of some technical calculation in one of the theorems.

The prerequisite for studying this text is a good undergraduate course in differential equations and advanced calculus. Of course, a student knowing real analysis and some engineering linear control methods would understand and digest the material much more easily.

ACKNOWLEDGMENTS

During the several years of writing this treatise Dr. E. B. Lee received financial support from the Air Force Office of Scientific Research and Dr. L. Markus received similar support from the Office of Naval Research. Both authors express their gratitude for the cooperation of these U.S. governmental agencies.

Dr. Markus thanks the Guggenheim Foundation for a fellowship that enabled him to spend an uninterrupted year in writing this book. Dr. Lee has received much valuable advice and many scientific insights in his consulting duties at the Honeywell Corporation, and he thanks Dr. H. Schuck, Dr. C. Harvey, and Mr. C. R. Stone in particular.

Portions of the text were read and much-appreciated suggestions and corrections were submitted by Dr. H. K. Wilson and Mr. H. Gollwitzer. However each author firmly states that the final responsibility for all errors in fact and judgment lies with his co-author.

E. B. Lee
L. Markus

Minneapolis, Minnesota
August 1967

Contents

Foundations of Optimal Control Theory

CHAPTER 1

Methods, Theory, and Practice in Optimal Control Synthesis

The general theory of optimal control for linear and nonlinear processes will be described in this chapter and illustrated by the application of these basic principles in the synthesis of optimal controllers. Thus a statement of all the fundamental theory and a definite procedure for application of this theory are contained in this chapter. The systematic mathematical development of these ideas will be presented in the later chapters. Only continuous deterministic processes are investigated here, although many of the concepts and results are significant for stochastic control processes.

1.1 EXAMPLES OF OPTIMAL CONTROL PROBLEMS

Optimal design of automatic control processes usually leads to nonlinear phenomena and so differs markedly from elementary linear feedback analysis. We shall here introduce the concepts and methods of optimal control theory by a study of several particular examples.

Example* 1. *Control of the Angular Speed of a Rotor. Consider a disk or rotor R free to rotate about a fixed axis through the center of R and perpendicular to its plane. Let $\omega(t)$ be the angular speed of the rotor at time t, as measured around the axis in appropriate coordinates. Let the initial angular speed $\omega(0) = \omega_0$ and we seek to stop the rotor. That is, the problem is to control the output or response $\omega(t)$ from $\omega = \omega_0$ to $\omega = 0$ by the application of some external torque $L(t)$ about the axis of rotation.

Here the dynamical equation of motion is

$$I \frac{d\omega}{dt} = L(t),$$

where I is the positive constant moment of inertia of R about its axis and $L(t)$ is the input torque or controller. Mathematically, the problem is to choose $L(t)$, in a manner compatible with the particular mechanics of the system or process, so that the response $\omega(t)$, which is the solution of the dynamical differential equation with prescribed initial state $\omega(0) = \omega_0$,

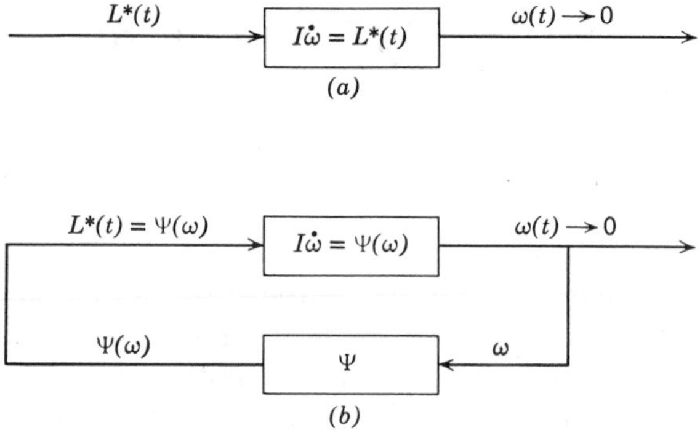

Figure 1.1 Scheme of a control process (*a*) open-loop control, (*b*) closed-loop control.

tends to zero as t increases. Moreover, we shall seek an optimal controller $L^*(t)$ such that the corresponding optimal response $\omega^*(t)$ will reach zero in a most efficient manner, for example, in minimal possible time.

Such a control problem might arise if the rotor R is a roller in some manufacturing process or if R is the cross section of a rocket missile. In the first situation the controlling torque could be effected by some electromechanical device and in the second instance by a couple from auxiliary steering jets. The problem of steering $\omega = \omega_0$ to $\omega = 0$ can also arise if there is some ideal constant angular speed for R, for then ω can indicate the error between the actual and the ideal angular speed. Thus our problem can be viewed in the general framework of controlling or regulating an error to zero.

If the initial angular speed ω_0 is known beforehand, the control signal $L(t)$ might well be specified as an open-loop input (see Figure 1.1*a*) and we require $L^*(t)$ to be optimal for the desired purpose. However, if we wish to construct a self-correcting control device that operates satisfactorily for all possible initial values ω_0 and even corrects for repeated perturbations of the response $\omega(t)$, we must synthesize the optimal controller $L^*(t)$ by an appropriately designed feedback loop (see Figure 1.1*b*). That is, we must calculate a function $\Psi(\omega)$ of the current state ω of the rotor and use $\Psi(\omega)$ as a feedback controlling signal. Then the solution $\omega(t)$ of the process

$$I\frac{d\omega}{dt} = \Psi(\omega),$$

for each initial state ω_0, will be an optimal response. That is, $\omega(t)$ will coincide with the optimal response $\omega^*(t)$, which would result from the optimal controller $L^*(t)$ in the open-loop process.

Consider the linear feedback control

$$\Psi(\omega) = -k\omega$$

where $k > 0$ is a constant gain. Then

$$I\dot{\omega} = -k\omega, \qquad \omega(0) = \omega_0,$$

has the solution

$$\omega(t) = \omega_0 e^{-(k/I)t},$$

which tends toward zero as $t \to +\infty$. If we wish to increase the rate of retardation of $\omega(t)$ we can augment the gain constant k, but however large we fix k in this mathematical model, the rotor never actually stops; it only tends towards rest. Furthermore, the problem of selecting an optimal linear feedback control is not mathematically sensible, since each such controller can be improved by increasing the gain. Also the physical problem is not well posed since there is a limit to how much amplification in the feedback loop is practical, or even possible, because nonlinearities such as saturation phenomena are then significant for the physical apparatus.

A reasonable consideration for the optimal control of the rotor is the requirement that the controlling torque remain within certain prescribed bounds. For convenience of notation, let us demand that

$$-1 \leq L(t) \leq +1.$$

The control torque $L^*(t)$, which is allowed to vary discontinuously to account for sudden switching, must satisfy the restriction $|L^*(t)| \leq 1$ and must steer the initial state $\omega = \omega_0$ to the target state $\omega = 0$ in a minimal possible time.

The solution for the open-loop time-optimal controller $L^*(t)$ is obvious from physical considerations. If $\omega_0 > 0$, take

$$L^*(t) = -1 \quad \text{and} \quad \omega^*(t) = \frac{I\omega_0 - t}{I}$$

until $T = I\omega_0$ when $\omega^*(T) = 0$. If $\omega_0 < 0$, take

$$L^*(t) = +1 \quad \text{and} \quad \omega^*(t) = \frac{I\omega_0 + t}{I}$$

until $T = -I\omega_0$ when $\omega^*(T) = 0$. Because of the constancy of the sign of the optimal response $\omega^*(T)$ in each instance, it is easy to construct the desired synthesizing function $\Psi(\omega)$ for the feedback optimal control. Take

$$\Psi(\omega) = -\text{sgn}\ \omega,$$

where the signum function is defined by

$$\text{sgn}\ \omega = \begin{cases} +1 & \text{if} \quad \omega > 0 \\ 0 & \text{if} \quad \omega = 0. \\ -1 & \text{if} \quad \omega < 0 \end{cases}$$

Then the nonlinear differential equation

$$I\dot{\omega} = -\text{sgn}\ \omega$$

will have, for each initial state ω_0, a solution that is the same as the time-optimal response $\omega^*(t)$ of the corresponding open loop circuit.

Example* 2. *Control of a Mechanism Along a Smooth Track. Consider a mechanism, such as a crane or trolley, of mass m, which moves along a horizontal track with negligible friction. The position coordinate x of the trolley at time t is determined by Newton's law of motion

$$m\ddot{x} = u(t),$$

where $u(t)$ is the external controlling force applied to the trolley, measured in appropriate coordinates and units. Suppose that the initial position and velocity of the trolley along the track are $x = x_0$ and $\dot{x} = y = y_0$ respectively. We consider the problem of stopping the trolley at a prescribed target position—say, $x = 0$, $y = 0$—in minimal possible time by means of a (possibly discontinuous) controlling force $u(t)$ subject to the restraint

$$|u(t)| \leq 1.$$

Here the synthesis of the optimal controller is not obvious and the final results obtained here are surprising. The methods we sketch in connection with this example form the main part of the theory examined in detail in Chapter 2. The geometrical properties in this analysis, although quite plausible intuitively, will be rigorously proved in Chapter 2. Hence our treatment is rather lengthy since it is meant to illustrate a general theoretical approach.

For convenience choose the unit of mass so $m = 1$ and introduce the velocity $y = \dot{x}$ to write the Newton's equation of motion as a first-order differential system

$$(\mathcal{S}) \qquad \begin{aligned} \dot{x} &= y \\ \dot{y} &= u(t), \end{aligned}$$

or in matrix notation (as explained in the appendices of this chapter)

$$\begin{pmatrix} \dot{x} \\ \dot{y} \end{pmatrix} = \begin{pmatrix} 0 & 1 \\ 0 & 0 \end{pmatrix} \begin{pmatrix} x \\ y \end{pmatrix} + \begin{pmatrix} 0 \\ 1 \end{pmatrix} u(t)$$

or

$$\dot{\mathbf{x}} = A\mathbf{x} + bu.$$

Here the vector $\mathbf{x} = \begin{pmatrix} x \\ y \end{pmatrix}$ and the matrices $A = \begin{pmatrix} 0 & 1 \\ 0 & 0 \end{pmatrix}$ and $b = \begin{pmatrix} 0 \\ 1 \end{pmatrix}$ are related as above. We shall write all significant formulas in this example in both coordinate and in matrix notation for comparison.

It is useful to consider a solution curve

$$\mathbf{x}(t) = \begin{pmatrix} x(t) \\ y(t) \end{pmatrix}$$

as a parametrized curve in the xy-plane, which is known as the phase plane. Thus we select some controller $u(t)$ with values on the restraint interval $|u(t)| \le 1$ and then examine the corresponding solution $\mathbf{x}(t)$ with initial data $\mathbf{x}_0 = \begin{pmatrix} x_0 \\ y_0 \end{pmatrix}$. In this manner we seek to steer $\mathbf{x}_0$ to the origin $\mathbf{x} = 0$ in a minimal possible time.

Fix a time $t_1 > 0$ and consider all the various possible controllers $u(t)$ on $0 \le t \le t_1$ with $|u(t)| \le 1$. Each of these controllers determines a corresponding response $\mathbf{x}(t)$ initiating at the given point $\mathbf{x}_0$. By direct substitution we easily verify the formula for the solution is

$$\begin{aligned} x(t) &= x_0 + y_0 t + \int_0^t \left[\int_0^s u(\sigma)\, d\sigma \right] ds \\ y(t) &= y_0 + \int_0^t u(\sigma)\, d\sigma \end{aligned}$$

or

$$\mathbf{x}(t) = e^{At}\mathbf{x}_0 + e^{At} \int_0^t e^{-As} b\, u(s)\, ds.$$

Define the set $K(t_1)$ in the phase plane to be the totality of all endpoints of all the above responses which initiate at $\mathbf{x}_0$ at $t = 0$. In other words,

$K(t_1)$ is the set of all points that can be attained at time t_1 when starting from $\mathbf{x}_0$ at time $t_0 = 0$, and using controllers satisfying the required restraints. It is not difficult to verify in this special example, and it will be proved later in the general theory, that $K(t_1)$ is a closed, bounded, convex set and, moreover, that $K(t_1)$ varies continuously with the end time t_1.

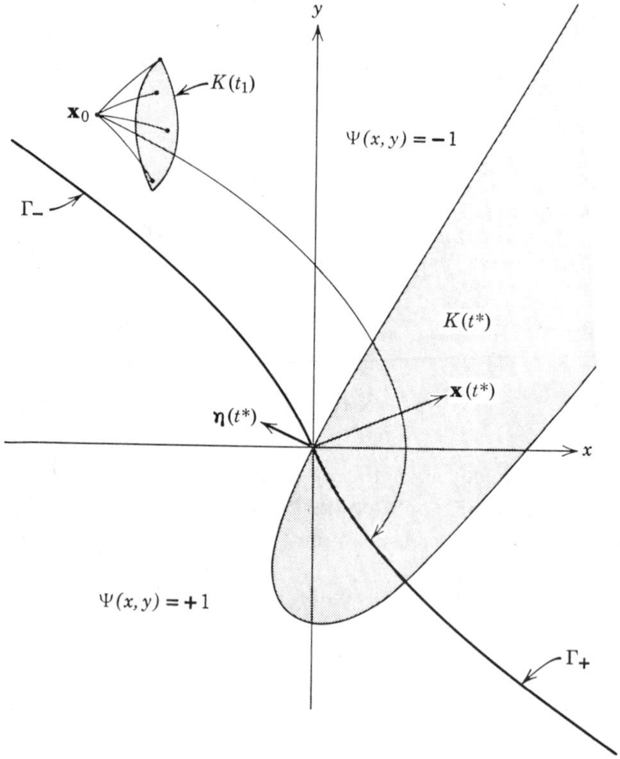

Figure 1.2 Set of attainability and switching locus for $\ddot{x} = u$, $|u(t)| \leq 1$.

The minimal optimal time $t = t^*$ is determined as the first time at which $K(t)$ contains the target (0, 0). Since $K(t)$ varies continuously with t, (0, 0) can be proved to lie on the boundary of the set $K(t^*)$. The optimal response $\mathbf{x}^*(t) = \begin{pmatrix} x^*(t) \\ y^*(t) \end{pmatrix}$ leads to the origin at $t = t^*$ and the optimal controller $u^*(t)$, $0 \leq t \leq t^*$, is that controller that produces the optimal response.

Let $\boldsymbol{\eta}(t^*) = (\eta_1(t^*), \eta_2(t^*))$ be a constant unit vector at the origin that is an outward normal for the convex set $K(t^*)$ (see Figure 1.2). Then for each response $\mathbf{x}(t) = \begin{pmatrix} x(t) \\ y(t) \end{pmatrix}$ leading to a point $\mathbf{x}(t^*)$ in $K(t^*)$ we must have

$$\eta_1(t^*)\, x(t^*) + \eta_2(t^*)\, y(t^*) \leq 0$$

or

$$\boldsymbol{\eta}(t^*)\, \mathbf{x}(t^*) \leq 0;$$

that is, the vector $\mathbf{x}(t^*)$, from the origin to the point $\mathbf{x}(t^*)$, has no positive component along the direction designated by $\boldsymbol{\eta}(t^*)$. We write this arithmetical statement, which is equivalent to the geometrical observation that $\boldsymbol{\eta}(t^*)$ is an outward normal to $K(t^*)$ at the boundary point $\mathbf{x}^*(t^*) = 0$:

$$\eta_1(t^*)\, x^*(t^*) + \eta_2(t^*)\, y^*(t^*) = \max_{(x,y)\in K(t^*)} [\eta_1(t^*)\, x + \eta_2(t^*) y]$$

or

$$\boldsymbol{\eta}(t^*)\, \mathbf{x}^*(t^*) = \max \boldsymbol{\eta}(t^*)\, \mathbf{x}(t^*).$$

In this last equality, which is called the maximal principle, the maximum is taken over all responses $\mathbf{x}(t)$ leading to points $\mathbf{x}(t^*)$ in $K(t^*)$. From the maximal principle we shall deduce certain extremal properties of the optimal controller $u^*(t)$ and so construct the desired synthesis function $\Psi(x, y)$. Since the maximal principle involves the optimal time t^* and the normal vector $\boldsymbol{\eta}(t^*)$, which are not directly known, the use of the maximal principle will be rather indirect.

We use the explicit integral expressions found for the response $\mathbf{x}(t)$ and we consider the maximal principle for

$$\eta_1(t^*)\left[x_0 + y_0 t^* + \int_0^{t^*}\int_0^s u(\sigma)\, d\sigma\, ds\right] + \eta_2(t^*)\left[y_0 + \int_0^{t^*} u(\sigma)\, d\sigma\right].$$

Thus when we disregard all terms not involving $u(t)$ we note that

$$\eta_1(t^*)\int_0^{t^*}\int_0^s u(\sigma)\, d\sigma\, ds + \eta_2(t^*)\int_0^{t^*} u(\sigma)\, d\sigma$$

must be maximized by the optimal controller $u^*(t)$. Use the identity, proved by differentiating both terms,

$$\int_0^t\int_0^s u(\sigma)\, d\sigma\, ds = \int_0^t (t - \sigma)\, u(\sigma)\, d\sigma$$

and define $\eta_1(s) = \eta_1(t^*)$, $\eta_2(s) = \eta_1(t^*)(t^* - s) + \eta_2(t^*)$ on the interval $0 \leq s \leq t^*$, to find that $u^*(t)$ maximizes

$$\int_0^{t^*} \eta_2(s)\, u(s)\, ds.$$

In matrix notation the above calculation reads that $u^*(t)$ maximizes

$$\boldsymbol{\eta}(t^*)e^{At^*}\mathbf{x}_0 + \boldsymbol{\eta}(t^*)e^{At^*}\int_0^{t^*} e^{-As}b\, u(s)\, ds$$

so $u^*(t)$ also maximizes the second term

$$\int_0^{t^*} \boldsymbol{\eta}(s)b\, u(s)\, ds = \int_0^{t^*} \eta_2(s)\, u(s)\, ds,$$

where we define

$$\boldsymbol{\eta}(t^*)e^{At^*}e^{-As} = \boldsymbol{\eta}(s) = (\eta_1(s), \eta_2(s)).$$

It is clear that the maximum for the integral (recalling that $|u(t)| \leq 1$)

$$\int_0^{t^*} \eta_2(s)\, u(s)\, ds$$

is achieved only by the controller

$$u^*(t) = \operatorname{sgn} \eta_2(t) \quad \text{on} \quad 0 \leq t \leq t^*.$$

Therefore the optimal controller $u^*(t)$ is a relay controller in that it assumes only the values $+1$ and -1 except where it switches between these, precisely at the zeros of the unknown function $\eta_2(t)$.

However, note from the definition of $\boldsymbol{\eta}(t)$ that

$$\dot{\eta}_1 = 0, \qquad \dot{\eta}_2 = -\eta_1$$

or in matrix notation,

$$\dot{\boldsymbol{\eta}}(t) = \boldsymbol{\eta}(t^*)e^{At^*}e^{-At}(-A) = -\boldsymbol{\eta}(t)\, A.$$

Therefore

$$\ddot{\eta}_2 = 0$$

and $\eta_2(t)$ is a linear polynomial in t. The conclusion is that $\eta_2(t)$, hence $u^*(t)$, can have at most one zero. So we find that the optimal control $u^*(t)$ is a relay control with values $+1$ and -1 and it has at most one switch or discontinuity. Using this information we shall now construct the synthesis function $\Psi(x, y)$ for our problem.

The optimal response from $\mathbf{x}_0$ to the origin must follow an arc of a solution of the extremal differential system ($u \equiv -1$)

$$(\mathcal{S}_-) \qquad \begin{aligned} \dot{x} &= y \\ \dot{y} &= -1 \end{aligned}$$

and then an arc of a solution of the extremal system ($u \equiv +1$)

$$(\mathcal{S}_+) \qquad \begin{aligned} \dot{x} &= y \\ \dot{y} &= +1 \end{aligned}$$

or vice versa. Since the extremal differential systems $\mathcal{S}_-$ and $\mathcal{S}_+$ are autonomous (the coefficients do not depend on time t) we can construct extremal responses that terminate at the origin by a process of backing out of the origin as $-t$ increases. That is, we start the extremal response at $t = 0$ at the origin and follow the appropriate solution curves of $\mathcal{S}_-$ and $\mathcal{S}_+$ backwards in time to reach the point $\mathbf{x}_0$ at some negative value of $t = -t^*$. Then reverse the time sense and start from $\mathbf{x}_0$ at $t = 0$ to arrive at the origin at $t = t^*$, thus obtaining the optimal response $\mathbf{x}^*(t)$, the optimal time t^*, and the optimal control $u^*(t)$.

Let us construct all possible extremal responses from all possible initial points to the origin. Choose a unit vector $\boldsymbol{\eta}(0) = (\eta_1(0), \eta_2(0))$ and use this as initial data to determine a solution of

$$\dot{\eta}_1 = 0, \qquad \dot{\eta}_2 = -\eta_1.$$

Use

$$u(t) = \operatorname{sgn} \eta_2(t)$$

and

$$\dot{x} = y$$
$$\dot{y} = \operatorname{sgn} \eta_2(t),$$

with $x(0) = 0, y(0) = 0$, to construct an extremal response $\mathbf{x}(t)$ terminating at the origin at $t = 0$.

In this way we can construct all possible extremal responses that lead to the origin as t increases, and one of these will steer $\mathbf{x}_0$ to the origin. For example, if we take $\eta_1(0) = 0, \eta_2(0) = +1$, then $\eta_2(t) \equiv +1$ on $t \leq 0$ and we trace the response along

$$\dot{x} = y$$
$$\dot{y} = +1$$

or

$$\frac{dy}{dx} = \frac{1}{y}$$

which yields the curve through the origin

$$\Gamma_+ \colon 2x = y^2 \quad \text{for} \quad y \leq 0,$$

(see Figure 1.2). Similarly if we take $\eta_1(0) = 0, \eta_2(0) = -1$ we trace out the response along

$$\Gamma_- \colon -2x = y^2 \quad \text{for} \quad y \geq 0.$$

For any other values of $\eta_1(0), \eta_2(0)$ with $\eta_2(0) > 0$ we trace out a response along Γ_+ until $\eta_2(t) = 0$ and then we follow backwards along the appropriate solution of $\mathcal{S}_-$. A similar process occurs for $\eta_2(0) < 0$. An easy study of the geometry of the solution-curve families of $\mathcal{S}_-$ and $\mathcal{S}_+$ shows that there is one and only one extremal response that leads from any prescribed initial point $\mathbf{x}_0$ to the origin. Thus this extremal response must be the optimal response, which is known to exist by the general theory developed later.

The curve consisting of Γ_- and Γ_+ is called the switching locus W. In this example the switching locus has the explicit form

$$y = W(x) = \begin{cases} -\sqrt{2x} & \text{for} \quad x \geq 0 \\ +\sqrt{-2x} & \text{for} \quad x < 0. \end{cases}$$

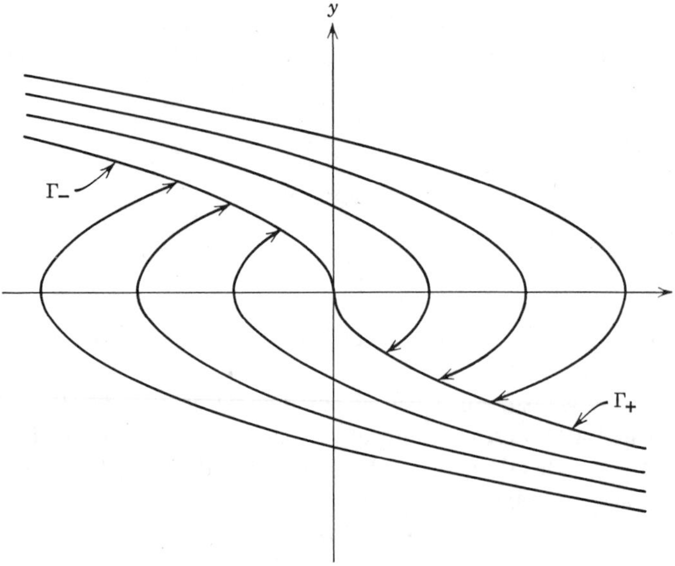

Figure 1.3 Minimal time-optimal responses for $\dot{x} = y$ and $\dot{y} = u(t)$, $|u(t)| \leq 1$ as constructed by analogue computer.

We define the synthesizer by

$$\Psi(x, y) = \begin{cases} -1 & \text{if} \quad y > W(x) \quad \text{or if} \quad (x, y) \neq (0, 0) \quad \text{lies on} \quad \Gamma_- \\ \ \ 0 & \text{if} \quad x = y = 0 \\ +1 & \text{if} \quad y < W(x) \quad \text{or if} \quad (x, y) \neq (0, 0) \quad \text{lies on} \quad \Gamma_+. \end{cases}$$

Then the optimal response from any initial state $\begin{pmatrix} x_0 \\ y_0 \end{pmatrix}$ to the origin is just the solution of

$$\ddot{x} = \Psi(x, \dot{x})$$

initiating at $x(0) = x_0$, $\dot{x}(0) = y_0$. It is clear from the geometry of the phase plane that the solutions of

$$\ddot{x} = \Psi(x, \dot{x})$$

are well defined, even though $\Psi(x, y)$ is discontinuous on $y = W(x)$. The synthesis function $\Psi(x, y)$ for this problem can be stored in the electronic structure of the feedback loop in an effective way. Figure 1.3 is an actual reproduction of the optimal responses of the process

$$\ddot{x} = u(t), \qquad |u(t)| \leq 1$$

with the synthesized optimal controller

$$u^*(t) = \Psi(x(t), \dot{x}(t)).$$

The optimal control $u^*(t)$ for the mechanical trolley can be understood as a maximal accelerating force followed by a maximal braking deceleration until the mechanism stops just at the required position $x = 0$. The critical time for the switch from acceleration to deceleration (or vice versa) is computed from our graphical analysis.

Example* 3. *Control of a Harmonic Oscillator. Consider a particle of mass m with position coordinate x at time t, attracted towards the origin by a linear restoring force $-k^2x$, with the spring constant $k^2 > 0$. The Newton equation of motion is (in appropriate units and coordinates)

$$m\ddot{x} + k^2x = u(t),$$

where the external controlling force $u(t)$ is required to be bounded in magnitude, say

$$|u(t)| \leq 1.$$

For simplicity of exposition assume $m = 1$ and $k^2 = 1$. Again we wish to steer the initial state $x(0) = x_0$, $\dot{x}(0) = y_0$ to the origin in minimal time.

In the phase plane the corresponding differential system is

$$\dot{x} = y$$

$$\dot{y} = -x + u(t),$$

or, in matrix notation,

$$\dot{\mathbf{x}} = A\mathbf{x} + bu(t)$$

with the vector response

$$\mathbf{x}(t) = \begin{pmatrix} x(t) \\ y(t) \end{pmatrix}$$

and the coefficient matrices

$$A = \begin{pmatrix} 0 & 1 \\ -1 & 0 \end{pmatrix}, \qquad b = \begin{pmatrix} 0 \\ 1 \end{pmatrix}.$$

Using the same arguments concerning the convex set $K(t_1)$ of attainability as in Example 2, we are led to the maximal principle and the formula for the optimal controller

$$u^*(t) = \operatorname{sgn} \eta_2(t),$$

where $\boldsymbol{\eta}(t) = (\eta_1(t), \eta_2(t))$ is a solution of

$$\dot{\eta}_1 = \eta_2, \qquad \dot{\eta}_2 = -\eta_1$$

or

$$\dot{\boldsymbol{\eta}}(t) = -\boldsymbol{\eta} A.$$

Thus

$$\ddot{\eta}_2 + \eta_2 = 0$$

and $\eta_2(t)$ is a harmonic oscillation. The time duration between consecutive zeros of $\eta_2(t)$ is exactly π.

Again we construct the switching locus W and the synthesizer $\Psi(x, y)$ by considering all possible extremal responses that terminate at the origin. We must consider the solution-curve families in the phase plane of the extremal differential systems

$$(\mathcal{S}_-) \qquad \begin{aligned} \dot{x} &= y \\ \dot{y} &= -x - 1 \end{aligned}$$

and

$$(\mathcal{S}_+) \qquad \begin{aligned} \dot{x} &= y \\ \dot{y} &= -x + 1. \end{aligned}$$

The solution curves of $\mathcal{S}_-$ are circles concentric about $x = -1, y = 0$ and traced with a period of 2π. The solutions of $\mathcal{S}_+$ are circles concentric about $x = +1, y = 0$, again with period 2π.

If we take the unit vector $\boldsymbol{\eta}(0)$ so that $\eta_1(0) = 1$, $\eta_2(0) = 0$, then $\eta_2(t) = -\sin t$ and, on the interval $-\pi < t < 0$, $\operatorname{sgn} \eta_2(t) = +1$. The corresponding extremal response traces out the solution curve of $\mathcal{S}_+$ through the origin, that is,

$$\Gamma_+ : x = -\cos t + 1 \qquad \text{on } -\pi < t < 0$$
$$y = \sin t$$

or

$$(x - 1)^2 + y^2 = 1, \qquad y < 0.$$

If $\eta_1(0) = -1$, $\eta_2(0) = 0$, then $\eta_2(t) = \sin t$ and, on the interval $-\pi < t < 0$, we have $\operatorname{sgn} \eta_2(t) = -1$. The corresponding extremal response traces out the solution curve of $\mathcal{S}_-$ through the origin, that is,

$$\Gamma_- : x = \cos t - 1 \quad \text{on} \quad -\pi < t < 0$$
$$y = -\sin t$$

or

$$(x + 1)^2 + y^2 = 1, \qquad y > 0.$$

However, for any other choice of $\boldsymbol{\eta}(0)$ with $\eta_2(0) > 0$ the extremal response traces back from the origin along Γ_+ until $\eta_2(t) = 0$. At this point the extremal response switches to a solution of $\mathcal{S}_-$ which it follows for a duration of length π before switching again to a solution of $\mathcal{S}_+$ (see Figure 1.4). A similar process occurs for initial data with $\eta_2(0) < 0$ but here the extremal response starts back from the origin along Γ_-.

The switching locus W, which consists of the points where the above extremal responses switch between the solution families of $\mathcal{S}_-$ and $\mathcal{S}_+$,

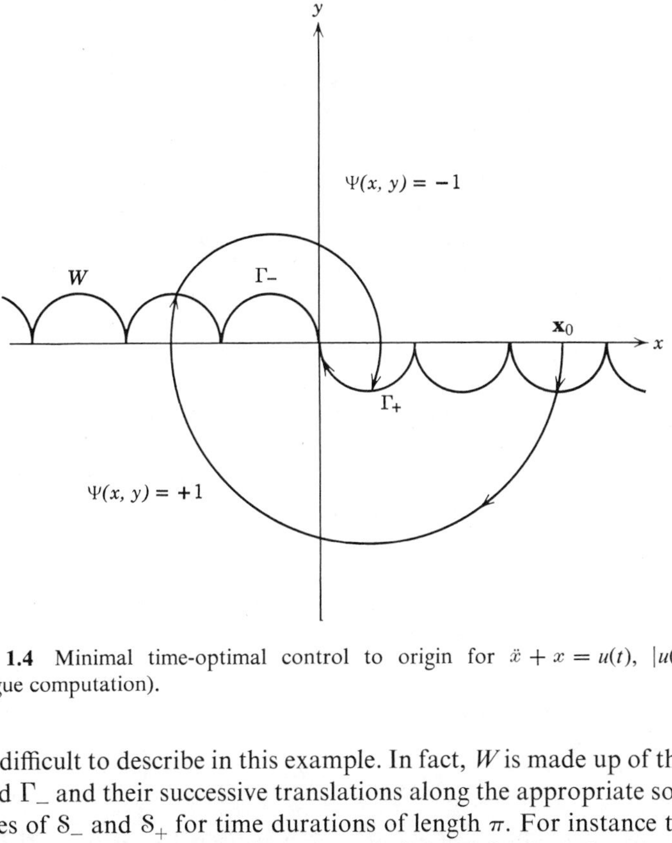

Figure 1.4 Minimal time-optimal control to origin for $\ddot{x} + x = u(t)$, $|u(t)| \leq 1$ (analogue computation).

is not difficult to describe in this example. In fact, W is made up of the arcs Γ_+ and Γ_- and their successive translations along the appropriate solution families of $\mathcal{S}_-$ and $\mathcal{S}_+$ for time durations of length π. For instance the arc Γ_+ is to be translated back along the solutions of $\mathcal{S}_-$ for a duration of π; this image of Γ_+ is then translated back along the solutions of $\mathcal{S}_+$ for a duration of π, and so forth. Note that such a translation along the solution family of $\mathcal{S}_+$ or of $\mathcal{S}_-$ for a duration of π is just a rigid rotation of the phase plane through π radians about the corresponding center $x = 1$, $y = 0$ or $x = -1$, $y = 0$. This process of successive switching leads to a final construction of W as a collection of semicircles of radius one, as indicated in Figure 1.4.

Define the synthesizer for real $(x, y) \neq (0, 0)$

$$\Psi(x, y) = \begin{cases} -1 & \text{if } (x, y) \text{ lies above } W \text{ or on } \Gamma_- \\ 0 & \text{if } (x, y) \text{ lies on } W \text{ otherwise} \\ +1 & \text{if } (x, y) \text{ lies below } W \text{ or on } \Gamma_+. \end{cases}$$

Then the optimal responses for the control of the harmonic oscillator are just the solutions of

$$\ddot{x} + x = \Psi(x, \dot{x})$$

for an arbitrary choice of initial phase point (x_0, y_0). Figure 1.4 is an actual reproduction of the optimal responses of a controlled harmonic oscillator, as described above.

While it may be possible to guess the qualitative shape of W from a physical description of the control process, the exact quantitative calculation of W and of the synthesizer $\Psi(x, y)$ requires a theoretical investigation such as has been presented here.

Example* 4. *Control of Chemical Reaction with Nonlinear Cost. Consider a chemical mixture A which is added to a tank at a constant rate for a fixed time $0 \leq t \leq T$. Assume that the pH value x at which the reaction occurs determines the quality of the final product and that this pH value can be controlled by the strength u of some component of A.

Suppose that the reaction takes place so that the rate of change of pH value x is proportional to the sum of the current pH value and the strength u of the controlling ingredient, that is,

$$\frac{dx}{dt} = \alpha x + \beta u$$

where α and β are known positive constants. Suppose further that the decrease in yield due to variations in the pH value is

$$\int_0^T x^2 \, dt$$

and that the rate of cost of maintaining the strength u is proportional to u^2. Then the total cost associated with the control $u(t)$ on $0 \leq t \leq T$ is

$$C(u) = \int_0^T (ax^2 + u^2) \, dt$$

where $a > 0$ converts the two costs to common units.

The problem is now well set mathematically. The initial pH value is specified, say $x(0)$, and we seek a control function $u^*(t)$ on $0 \leq t \leq T$, which determines a response $x^*(t)$ so that

$$C(u(t)) = \int_0^T [ax(t)^2 + u(t)^2]\, dt$$

is a minimum. The controller $u(t)$ is not restricted by any preassigned bound, but the positive nature of the cost integrand indicates that there will exist an optimal controller $u^*(t)$. We seek to synthesize $u^*(t)$, that is, to determine the optimal controller in terms of the state $x^*(t)$.

The method of convex sets and the maximal principle are available, and will be developed in detail in Chapter 3, however the nonlinearity of the cost functional $C(u)$ introduces new technical difficulties. Here we shall obtain the optimal controller using certain plausible arguments of the theory of dynamic programming. Our methods follow the principle of optimality, which states that an optimal controller $u^*(t)$ on $0 \leq t \leq T$ must also yield an optimal controller on each subinterval of $0 \leq t \leq T$. The rigorous treatment of these methods is based on convex sets of attainability and is similar to the analysis of the maximal principle we shall present later.

Consider the chemical reaction at some state x_0 at an instant t_0 in the interval $[0, T]$. There is an optimal controller $u^*(t)$ for the interval $[t_0, T]$, with response starting at x_0, and this produces the minimal cost $V(x_0, t_0) = C(u^*)$. We shall assume that the real-valued function $V(x, t)$ is sufficiently smooth and differentiable so that the succeeding analysis is valid.

For each controller $u(t)$ on $[t_0, T]$, with corresponding response $x_u(t)$ initiating at x_0, the cost is

$$\int_{t_0}^{t_0+\delta} [ax_u(t)^2 + u(t)^2]\, dt + \int_{t_0+\delta}^{T} [ax_u(t)^2 + u(t)^2]\, dt$$

where $\delta > 0$ is an arbitrary small number. By modifying $u(t)$ to become optimal on $[t_0 + \delta, T]$ one can obtain the cost

$$\int_{t_0}^{t_0+\delta} [ax_u(t)^2 + u(t)^2]\, dt + V(x_u(t_0 + \delta), t_0 + \delta).$$

But the minimal cost from x_0 at time t_0 does not exceed this and so we obtain

$$V(x_0, t_0) = \min_{u(t)} \left\{ \int_{t_0}^{t_0+\delta} [ax_u(t)^2 + u(t)^2]\, dt + V(x_u(t_0 + \delta), t_0 + \delta) \right\}$$

where the minimum is taken over all controllers $u(t)$ on $[t_0, T]$. This equation illustrates the idea of dynamic programming in that we think of the optimal-control program decomposed into the sum of two programs, on $[t_0, t_0 + \delta]$ and on $[t_0 + \delta, T]$, with the dynamics indicated by the possibility of varying δ.

Using a Taylor series expansion in terms of the small number $\delta > 0$ we obtain

$$V(x_0, t_0) = \min_{u(t)} \left\{ \delta[a{x_0}^2 + u(t_0)^2] + V(x_0, t_0) + \left[\frac{\partial V}{\partial x}(x_0, t_0)\frac{dx_u}{dt}(t_0) + \frac{\partial V}{\partial t}(x_0, t_0)\right]\delta + o(\delta)\right\}$$

where the remainder $o(\delta)$ tends to zero with a higher order than does δ. Write

$$\frac{dx_u}{dt}(t_0) = \alpha x_0 + \beta\, u(t_0)$$

and let δ approach zero to obtain

$$-\frac{\partial V}{\partial t}(x, t) = \min_u \left\{ax^2 + u^2 + \frac{\partial V}{\partial x}(x, t)(\alpha x + \beta u)\right\},$$

where we have written the generic point (x_0, t_0) as (x, t). Here the minimum is computed for the real function

$$h(u) = ax^2 + u^2 + \frac{\partial V}{\partial x}\cdot(\alpha x + \beta u)$$

for each fixed value of (x, t). Set

$$\frac{\partial h}{\partial u} = 2u + \beta\frac{\partial V}{\partial x} = 0,$$

and we find that the minimum is attained at

$$u = -\frac{\beta}{2}\frac{\partial V}{\partial x}.$$

Thus $V(x, t)$ is a solution of the nonlinear partial differential equation

$$-\frac{\partial V}{\partial t} = ax^2 - \frac{\beta^2}{4}\left(\frac{\partial V}{\partial x}\right)^2 + \alpha x\frac{\partial V}{\partial x},$$

and we have the required condition $V(x, T) = 0$. This partial differential equation for the minimal cost $V(x, t)$ is the main result of the techniques of dynamic programming as applied to our problem.

Since $V(x, t)$ is prescribed on the line $t = T$ and the only time derivative entering the partial differential equations is $\partial V/\partial t$, it is known that there exists a unique solution $V(x, t)$. We try to compute the solution in the form

$$V(x, t) = c(t)x^2$$

where the function $c(t)$ is to be determined. Direct substitution into the partial differential equation yields the requirement

$$\frac{dc}{dt} = \beta^2 c^2 - 2\alpha c - a,$$

with $c(T) = 0$. This first-order ordinary differential equation uniquely defines the required function $c(t)$ that produces the correct cost function

$$V(x, t) = c(t)x^2.$$

We can obtain a formula for $c(t)$ in terms of elementary functions by using the substitution

$$\zeta(t) = e^{-\beta^2 \int_0^t c(t)\,dt}$$

or

$$\frac{\dot{\zeta}}{\zeta} = -\beta^2 c.$$

For then we compute easily

$$\ddot{\zeta} + 2\alpha\dot{\zeta} - a\beta^2\zeta = 0$$

with $\dot{\zeta}(T) = 0$ and we can also set $\zeta(T) = 1$ since only the ratio $\dot{\zeta}/\zeta$ is of interest. The solution here is

$$\zeta(t) = e^{-\alpha(t-T)}\left[\cosh\sqrt{\alpha^2 + a\beta^2}(t - T) + \frac{\alpha}{\sqrt{\alpha^2 + a\beta^2}} \sinh\sqrt{\alpha^2 + a\beta^2}(t - T)\right].$$

Then $c(t) = -\dot{\zeta}/\zeta\beta^2$ and $V(x, t) = c(t)x^2$ can be computed explicitly.

Now the optimal controller $u^*(t)$ on $[0, T]$ with a response $x^*(t)$ that initiates from the given $x(0)$ satisfies the minimizing condition

$$V(x^*(t), t) = \int_t^{t+\delta} [ax^*(s)^2 + u^*(s)^2]\,ds + V(x^*(t + \delta), t + \delta)$$

for all t on $[0, T]$. When we follow the earlier analysis we obtain

$$-\frac{\partial V}{\partial t}(x^*(t), t) = ax^*(t)^2 + u^*(t)^2 + \frac{\partial V}{\partial x}(x^*(t), t) \cdot (\alpha x^*(t) + \beta u^*(t)).$$

Thus for fixed values of t, $u^*(t)$ must be the value of u that minimizes

$$h(u) = ax^*(t)^2 + u^2 + \frac{\partial V}{\partial x}(x^*(t), t) \cdot (\alpha x^*(t) + \beta u).$$

That is,

$$u^*(t) = -\frac{\beta}{2}\frac{\partial V}{\partial x}(x^*(t), t) = -\frac{\beta}{2} \cdot [2c(t)\,x^*(t)]$$

or

$$u^*(t) = -\beta c(t)x^*(t).$$

Moreover this equality characterizes the optimal controller.

Therefore we synthesize the optimal controller $u^*(t)$ by using the feedback loop

$$u = -\beta c(t)x,$$

which is a linear control system with variable gain $c(t)$. This is the promised solution and it is easily mechanized (see Figure 1.5).

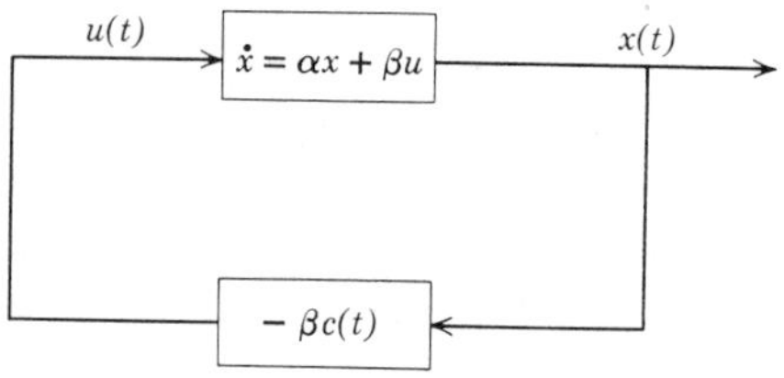

Figure 1.5 Schematic synthesis of optimal control $u(t) = -\beta c(t)x$.

As we shall show in Chapter 3, there is a whole class of problems that can be solved in the manner indicated, namely problems that involve a quadratic cost function of the output and control variables while the basic process is linear.

Example 5. A Classical Variational Approach. In this example we examine the optimal control problem from the viewpoint of the classical calculus of variations. Because of the complicated dependence of the response $x(t)$ on the controller $u(t)$, as determined through the dynamical differential equations, this variational problem leads to the rather difficult problem of Bolza or Mayer. We consider this problem here without any discussion of continuity or differentiability and we follow the classical notation of the calculus of variations.

Consider the control process in R^n, that is, x is a real n-vector,

$$(\mathcal{S}) \qquad \dot{x} = f(x, u), \qquad x(0) = x_0,$$

with controllers $u(t) \subset R^m$ on $0 \leq t \leq 1$. For each controller $u(t)$ there is a corresponding response $x(t)$, with $x(0)$ at the fixed initial state x_0, and with cost

$$C(u) = \int_0^1 h(x, u)\, dt.$$

We impose no restraints on $u(t)$ or on $x(1)$. Let $u^*(t)$ be an optimal controller minimizing $C(u)$ and let $x^*(t)$ be the corresponding optimal response (see Appendix for vector and matrix notation).

Let

$$u(t, \epsilon) = u^*(t) + \epsilon\, \delta u(t)$$

be a one-parameter family of perturbations or variations with corresponding responses

$$x(t, \epsilon) = x^*(t) + \epsilon\, \delta x(t) + o(\epsilon), \qquad \delta x(0) = 0.$$

Note that

$$u(t, 0) = u^*(t) \quad \text{and} \quad \frac{\partial u}{\partial \epsilon}(t, 0) = \delta u(t)$$

$$x(t, 0) = x^*(t) \quad \text{and} \quad \frac{\partial x}{\partial \epsilon}(t, 0) = \delta x(t).$$

Consider the variation in cost

$$\left.\frac{\partial C}{\partial \epsilon}\right|_{\epsilon=0} = \delta C = \int_0^1 \left[\frac{\partial h(t)}{\partial x}\, \delta x(t) + \frac{\partial h(t)}{\partial u}\, \delta u(t)\right] dt.$$

Here $\partial h(t)/\partial x$ means $(\partial h/\partial x)(x^*(t), u^*(t))$. Similarly all other such functions are evaluated at $x = x^*(t)$, $u = u^*(t)$. Since $C(u(\cdot, \epsilon))$ is minimized at $\epsilon = 0$, we must have

$$\delta C \equiv 0$$

for all variations $\delta u(t)$. This necessary condition for the optimal controller will now be clarified.

The variation $\delta u(t)$ yields the response variation $\delta x(t)$, which satisfies the variational differential equation

$$\delta \dot{x} = \frac{\partial f(t)}{\partial x}\, \delta x + \frac{\partial f(t)}{\partial u}\, \delta u, \qquad \delta x(0) = 0.$$

Hence

$$\delta x(t) = \int_0^t \Phi(t)\, \Phi^{-1}(s)\, \frac{\partial f(s)}{\partial u}\, \delta u(s)\, ds$$

where the fundamental matrix $\Phi(t)$ satisfies

$$\dot{\Phi} = \frac{\partial f(t)}{\partial x}\, \Phi \quad \text{and} \quad \Phi(0) = I.$$

Then

$$\delta C = \int_0^1 \left[\frac{\partial h(t)}{\partial x} \int_0^t \Phi(t)\, \Phi^{-1}(s)\, \frac{\partial f(s)}{\partial u}\, \delta u(s)\, ds + \frac{\partial h(t)}{\partial u}\, \delta u(t)\right] dt.$$

To simplify the notation we introduce the vector function

$$\eta^*(t) = -\eta_0\, \Phi^{-1}(t) + \int_0^t \frac{\partial h(s)}{\partial x}\, \Phi(s)\, \Phi^{-1}(t)\, ds$$

with the constant vector η_0 so chosen that

$$\eta^*(1) = -\eta_0\,\Phi^{-1}(1) + \int_0^1 \frac{\partial h(s)}{\partial x}\,\Phi(s)\,\Phi^{-1}(1)\,ds = 0.$$

This means that $\eta^*(t)$ is the unique solution of the adjoint variational differential equation

$$\dot{\eta} = -\eta\,\frac{\partial f}{\partial x} + \frac{\partial h}{\partial x}, \qquad \eta(1) = 0.$$

We further simplify the notation by defining the Hamiltonian function of $2n + m$ real variables

$$H(\eta, x, u) = \eta f(x, u) - h(x, u).$$

Then

$$\dot{\eta} = -\frac{\partial H}{\partial x}, \qquad \eta(1) = 0$$

and

$$\dot{x} = \frac{\partial H}{\partial \eta}, \qquad x(0) = x_0,$$

are satisfied by $\eta^*(t)$ and $x^*(t)$ when $u = u^*(t)$.

Next we use these notations to clarify the necessary condition $\delta C \equiv 0$. First verify the integration by parts

$$\int_0^1 \frac{\partial h(t)}{\partial x}\,\Phi(t)\left(\int_0^t \Phi^{-1}(s)\,\frac{\partial f(s)}{\partial u}\,\delta u(s)\,ds\right) dt$$

$$= \left(\int_0^1 \frac{\partial h(t)}{\partial x}\,\Phi(t)\,dt\right)\left(\int_0^1 \Phi^{-1}(s)\,\frac{\partial f(s)}{\partial u}\,\delta u(s)\,ds\right) - \int_0^1\left(\int_0^t \frac{\partial h(s)}{\partial x}\,\Phi(s)\,ds\right)\Phi^{-1}(t)\,\frac{\partial f(t)}{\partial u}\,\delta u(t)\,dt.$$

Using this formula we obtain

$$\delta C = \left(\int_0^1 \frac{\partial h(t)}{\partial x}\,\Phi(t)\,dt\right)\left(\int_0^1 \Phi^{-1}(s)\,\frac{\partial f(s)}{\partial u}\,\delta u(s)\,ds\right) - \int_0^1\left[\left(\int_0^t \frac{\partial h(s)}{\partial x}\,\Phi(s)\,ds\right)\Phi^{-1}(t)\,\frac{\partial f(t)}{\partial u} - \frac{\partial h(t)}{\partial u}\right]\delta u(t)\,dt.$$

But this means that

$$\delta C = \int_0^1\left[(\eta^*(1)\,\Phi(1) + \eta_0)\,\Phi^{-1}(t)\,\frac{\partial f(t)}{\partial u} - \left(\int_0^t \frac{\partial h(s)}{\partial x}\,\Phi(s)\,\Phi^{-1}(t)\,ds\right)\frac{\partial f(t)}{\partial u} + \frac{\partial h(t)}{\partial u}\right]\delta u\,dt,$$

or

$$\delta C = \int_0^1 - \left[\eta^* \frac{\partial f(t)}{\partial u} + \frac{\partial h(t)}{\partial u} \right] \delta u(t)\, dt.$$

Since

$$\delta C \equiv 0$$

for all variations $\delta u(t)$ about the optimal $u^*(t)$ we find that

$$-\eta^*(t) \frac{\partial f(t)}{\partial u} + \frac{\partial h(t)}{\partial u} = 0$$

or

$$\frac{\partial H}{\partial u}(\eta^*(t), x^*(t), u^*(t)) \equiv 0.$$

A more detailed study of the variations about the minimizing controller $u^*(t)$ shows that $u = u^*(t)$ is a maximum, rather than an arbitrary critical point of $H(\eta^*(t), x^*(t), u)$. That is

$$H(\eta^*(t), x^*(t), u^*(t)) = \max_{u \in R^m} H(\eta^*(t), x^*(t), u).$$

This is the maximal principle which plays such an important role in the theory of optimal control.

The system of equations

$$\dot{x} = \frac{\partial H}{\partial \eta}, \qquad \dot{\eta} = -\frac{\partial H}{\partial x}, \qquad \frac{\partial H}{\partial u} = 0$$

are the Euler-Lagrange equations of our variational problem—but they are in Hamiltonian form. In classical literature, in which there are no boundary restraints on the controllers, these conditions are usually termed *the Weierstrass necessary conditions for an extremal.* To see this clearly consider the special case where the dynamics are described by the scalar equation

$$\dot{x} = u, \qquad x(0) = x_0, \qquad x(T) = x,$$

and the cost is

$$C(u) = \int_0^T h(x, u)\, dt = \int_0^T h(x, \dot{x})\, dt.$$

Here the Lagrangian function is $h(x, \dot{x})$ and Lagrange's necessary condition for a minimizing smooth curve $x^*(t)$ is

$$\frac{d}{dt}\left(\frac{\partial h}{\partial \dot{x}}\right) - \frac{\partial h}{\partial x} = 0.$$

Using $H = \eta u - h$, we write

$$\frac{d}{dt}\left[\eta - \frac{\partial H}{\partial u}\right] + \frac{\partial H}{\partial x} = 0.$$

Since $\partial H/\partial u = 0$, this means that

$$\dot{\eta} = -\frac{\partial H}{\partial x}.$$

The functions $\eta^*(t)$ are called the Lagrange multipliers for the variational problem (and are often denoted by $\lambda(t)$ in classical texts). The Hamiltonian function $H(\eta, x, u)$ is often replaced by the negative of the expression we have used; but we prefer the notation established here since it agrees with the usage in current optimal control literature.

If the controller $u(t)$ is restricted in magnitude, or the target $x(1)$ is specified, the variational techniques become much more complicated both from a formal viewpoint and from a technical viewpoint. For these reasons we shall not follow the classical calculus of variations in our presentation, but shall rely on a geometrical development that will be completely correct in all mathematical detail.

EXERCISES

1. Consider a control process described by $\dot{x} + bx = u$ for real constant b, with the restraint $|u(t)| \leq 1$. Verify that the response $x(t)$ with $x(0) = x_0$ to a controller $u(t)$ is

$$x(t) = e^{-bt}x_0 + e^{-bt}\int_0^t e^{bs}\, u(s)\, ds.$$

(a) If $b \geq 0$, show that every initial point x_0 can be controlled to $x_1 = 0$.
(b) If $b < 0$, describe precisely those points x_0 which can be steered to $x_1 = 0$.

2. Consider a control process

$$\dot{x} + bx = u \quad \text{for real constant } b$$

and the restraint $|u(t)| \leq 1$. Let x_0 be an initial point that can be steered to the target $x_1 = 0$. Show that the minimal time-optimal controller is synthesized by

$$u^* = -\operatorname{sgn} x.$$

Compute the minimal time t^* in terms of x_0 and b.

3. Consider a control process

$$\ddot{x} + 2b\dot{x} = u \quad \text{for real constant} \quad b \neq 0,$$

and the restraint $|u(t)| \leq 1$. Show that a change of scale of $x = z/b^2$ and in time $t = \tau/|b|$ reduces the general problem to the two cases $2b = +1$ and $2b = -1$. Show that the minimal time-optimal controller steering (x_0, y_0) to $(0, 0)$ is $u^*(t) = \operatorname{sgn} \eta_2(t)$, where $\eta_2(t)$ has at most one zero. Construct the switching locus and describe the optimal control and optimal response in terms of this locus and the extremal systems for which $u(t) = +1$ and $u(t) = -1$. Distinguish carefully between the two cases $2b = +1$ and $2b = -1$.

4. Consider the control process

$$\ddot{x} + 2b\dot{x} + k^2x = u \qquad \text{for real constants } b \text{ and } k^2 > 0,$$

with restraint $|u(t)| \leq c$ for $c > 0$. Verify that an appropriate change of scale reduces the problem to the case where $k^2 = 1$ and $c = 1$.

5. What is the shortest time in which a passenger can be transported from New York to Los Angeles? Assume that a missile with ultimate mechanical and thermodynamical properties is available but that the passenger imposes the restraint that the maximal acceleration is 100 ft/sec². (The missile starts from rest in New York and stops at Los Angeles. Assume the path is a straight line of length 2400 miles and neglect the rotation and curvature of the earth.)

6. Consider the control process

$$\ddot{x} + x = u$$

with the restraint $|u(t)| \leq 1$. Let (x_0, y_0) be an initial state that lies above the switching locus $y = W(x)$ for the minimal time-optimal control to the origin. Let l be a positive integer such that

$$2l - 1 < [(x_0 + 1)^2 + y_0^2]^{1/2} < 2l + 1.$$

Show that the optimal controller $u^*(t)$ has precisely l switches. Make a corresponding statement for the case where $y_0 < W(x_0)$.

7. Consider the control process

$$\dot{x} = \alpha x + \beta u,$$

as in Example 4. However let the cost be

$$C(u) = \int_0^T (ax^2 + e^u)\, dt.$$

Use the technique of dynamic programming to obtain a partial differential equation for the minimal cost function $V(x, t)$.

1.2 STATEMENT OF A GENERAL OPTIMAL CONTROL PROBLEM

The most general control problem considered here is described by four types of data: (1) the plant or process, (2) the initial state and the target state of the physical system, (3) the class of admissible controllers, and (4) the cost-functional or performance index, which quantitatively compares the effectiveness of the various controllers. We shall discuss each of these four factors before making a precise statement of the optimal control problem

1. *The plant or process* relates the state or response $x(t)$ to the input or control $u(t)$ by a vector ordinary differential system

$$(\mathcal{S}) \qquad \dot{x}^i = f^i(t, x^1, x^2, \ldots, x^n, u^1, \ldots, u^m) \qquad i = 1, 2, \ldots, n.$$

Sometimes $x(t)$ is called the output, but we shall later define the output as some function of x determined by a further condition of observability or measurement. The process is autonomous, or linear, or nth order, according to the nature of the differential system $\mathcal{S}$ (see appendices to this chapter).

Nonlinearities can occur even in elementary physical theories because of the effects of nonlinear friction, amplification, or saturation. However, even for linear processes we shall deliberately introduce nonlinear feedback, as in relay control, in order to synthesize the optimal controller. Moreover many physical systems have major nonlinear aspects that cannot be ignored or successfully treated by the method of perturbations or linear approximations. The two examples offered below in this section illustrate nonlinear systems that are qualitatively different from any linear approximation. Because of the presence of these nonlinearities we shall have limited reference to the classical techniques of linear control theory such as integral transforms or transfer functions.

In each optimal control problem our ultimate goal is to synthesize the optimal controller by an appropriately designed closed or feedback loop. The advantage of such a closed-loop control, as against an open-loop control, is that the process then becomes self-adjusting and self-correcting. A feedback control can often correct for unpredictable variations in the environment of the plant or for repeated perturbations or irregularities in the process.

***Example* 1.** Consider the damped nonlinear oscillator with one degree of freedom $x(t)$ and with control $u(t)$ as defined by the process

$$\ddot{x} + f(x, \dot{x})\dot{x} + g(x) = u(t).$$

The frictional coefficient $f(x, y) \in C^1$ in R^2, the elastic restoring force $g(x) \in C^1$ in R^1; and the controlling force $u(t)$ is bounded and measurable on $0 \leq t < \infty$. Because of the physical nature of the mechanism we are led to the hypotheses

$$f(x, y) \geq 0$$

$$xg(x) \geq 0$$

$$|u(t)| \leq B, \qquad \text{for a constant } B > 0.$$

We first show that the solution $S(t) = (x(t), y(t))$, with $x(0) = x_0$, $y(0) = y_0$, of the differential system

$$\dot{x} = y$$
$$\dot{y} = -g(x) - f(x, y)y + u(t)$$

is defined in the phase plane R^2 for all times $0 \leq t < +\infty$.

If $S(t)$ were only defined on some maximal time interval $0 \leq t < \tau_+ < \infty$, then $r(t)^2 = x(t)^2 + y(t)^2$ would necessarily assume arbitrarily large values as $t \to \tau_+$. To prove the contrary define the energy function

$$V(x, y) = \frac{y^2}{2} + \int_0^x g(s)\, ds.$$

Note that $V(x, y) \geq 0$ in R^2 and $V = 0$ only on that segment of the x-axis that contains the origin and whereon $g(x) = 0$. Along the solution $S(t)$ we define $V(t) = V(x(t), y(t))$, and compute

$$\frac{dV}{dt} = y\dot{y} + g(x)\dot{x} = -f(x, y)y^2 + y\, u(t).$$

Then, using elementary inequalities,

$$\frac{dV}{dt} \leq B\,|y| \leq B\left(\frac{y^2}{2} + 1\right)$$

and

$$\frac{d}{dt}[V(t) + 1] \leq B[V(t) + 1].$$

Integrating this inequality we obtain

$$V(t) + 1 \leq [V(0) + 1]e^{B\tau}$$

and so

$$V(t) \leq [V(0) + 1]e^{B\tau_+}$$

on $0 \leq t < \tau_+$. But since $G(x) = \int_0^x g(s)\, ds \geq 0$,

$$y^2 \leq 2[V(0) + 1]e^{B\tau_+} = C^2$$

and

$$|y(t)| \leq C, \quad \text{for some constant } C.$$

Now

$$x(t) = x_0 + \int_0^t y(s)\, ds$$

and

$$|x(t)| \leq x_0 + C\tau_+.$$

But this proves that $V(t)$ is bounded on the finite interval $0 \le t < \tau_+$, which is impossible. Thus $S(t)$ is defined on all $0 \le t < +\infty$.

We next show that for each initial state $(x_0, y_0) \in R^2$ at $t = 0$, we can choose the control $u(t)$, satisfying $|u(t)| \le B$, so that the corresponding response steers (x_0, y_0) to an arbitrarily prescribed neighborhood of the origin. The controllability of (x_0, y_0) to the exact origin in a finite time is discussed in the next section of this chapter.

For each constant $V_0 \ge 0$ consider the phase plane locus

$$V(x, y) = \frac{y^2}{2} + G(x) = V_0.$$

This curve has two branches

$$y = \pm\sqrt{2(V_0 - G(x))}$$

which join wherever $G(x) = V_0$. Thus the above locus may consist of two separate curves, a simple closed curve around the origin, or possibly a figure shaped like $\supset$ or $\subset$ in the phase plane.

Consider the free oscillator where $u(t) \equiv 0$ and note that because

$$\dot{x} = y$$

the solutions must move toward increasing x where $y > 0$ and toward decreasing x where $y < 0$. Also the vertical velocity of the differential system on the x-axis is positive or zero on $x \le 0$ and is negative or zero on $x \ge 0$, since here $\dot{y} = -g(x)$. Now prescribe a small disk D centered at the origin of the phase plane and we shall prove that the solution $S(t) = (x(t), y(t))$ can be steered into D in a finite time by a suitable controller $u(t)$ (see Figure 1.6). We discuss the details of the case where $g(x) \ne 0$ for $x \ne 0$, that is, the origin is the unique critical point. The more general case can be treated in the same manner using a small amount of control $u(t)$ to keep the solution $S(t)$ from stagnating at any point of the x-axis other than the origin.

Suppose, at first, that (x_0, y_0) lies in the second quadrant $x_0 < 0$, $y_0 > 0$. Take $u(t) = 0$ and follow the solution $S(t)$ until it enters either D or the first quadrant of R^2. One of these two alternatives must arise since $x(t)$ is increasing in $y > 0$, and $S(t)$ cannot approach the negative x-axis [where $\dot{y} = -g(x) > 0$] and $S(t)$ cannot become unbounded—it cannot rise above the level curve $V(x, y) = V(x_0, y_0)$ because $\dot{V} = -f(x, y)y^2 \le 0$.

If (x_0, y_0) lies in the first quadrant, or when $S(t)$ arrives in the first quadrant as t increases, we impose the control $u(t) = -B < 0$. Then $\dot{y} \le -B$ and eventually $S(t)$ must cross the positive x-axis and enter the fourth quadrant. In the fourth quadrant we set $u(t) = 0$ and $S(t)$ must then move into D or else into the third quadrant. In the third quadrant we impose the control $u(t) = +B$.

In this manner we control $S(t)$ in a tightening spiral clockwise around the origin with controller $u(t)$ as determined above. Since

$$\dot{V} = -f(x, y)y^2 + y\,u(t) \leq 0$$

(and $\dot{V} < 0$ where $y \neq 0$), it is easy to see that

$$\lim_{t \to +\infty} V(t) = 0.$$

But the region $V(x, y) < \epsilon$, for sufficiently small $\epsilon > 0$, intersects D in such a way that the spiral $S(t)$ must enter D.

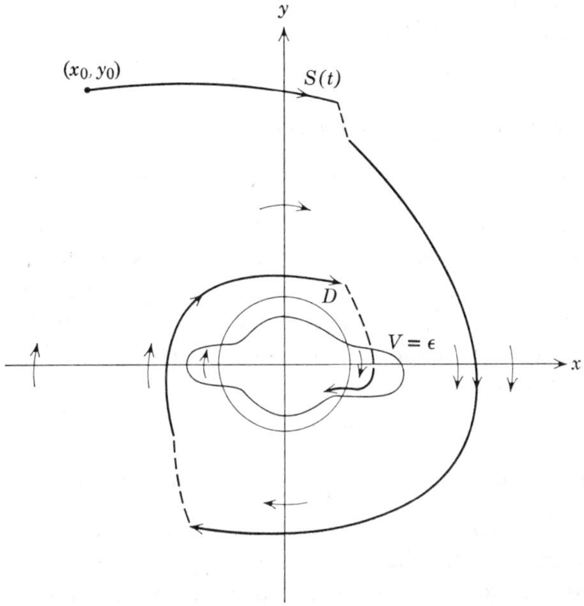

Figure 1.6 Controllability of the nonlinear process $\ddot{x} + f(x, \dot{x})\dot{x} + g(x) = u(t)$.

Thus $S(t)$ can be controlled to an arbitrarily small neighborhood of the origin.

***Example* 2.** Consider a rigid body, say a space capsule, rotating in inertial space about its centroid with angular velocity $\boldsymbol{\omega}(t)$ and subjected to the controlling torque $\mathbf{u}(t)$. At each instant t let ω_1, ω_2, ω_3 be the components of $\boldsymbol{\omega}$ analyzed along the principal axes fixed in the body B. Then the Euler equations of motion in the $(\omega_1, \omega_2, \omega_3)$ phase space R^3 are

$$I_1\dot{\omega}_1 = (I_2 - I_3)\omega_2\omega_3 + u_1(t)$$
$$I_2\dot{\omega}_2 = (I_3 - I_1)\omega_3\omega_1 + u_2(t)$$
$$I_3\dot{\omega}_3 = (I_1 - I_2)\omega_1\omega_2 + u_3(t).$$

Here I_1, I_2, I_3 are the principal moments of inertia of B and u_1, u_2, u_3 are the components of $\mathbf{u}(t)$ along the principal axes.

Let us assume that the controlling torque $\mathbf{u}(t)$ is effected by steering jets that are affixed to the body B and that operate with the same maximum thrust regardless of the state or motion of B, that is,

$$|u_i(t)| \leq 1 \quad \text{for} \quad i = 1, 2, 3.$$

We wish to regulate $\boldsymbol{\omega}(t)$ so that it approaches zero. In the next section we shall discuss the existence of a control $\mathbf{u}(t)$ that steers $\boldsymbol{\omega}(t)$ precisely to zero in a finite time duration.

Define the energy function

$$E = \tfrac{1}{2}(I_1\omega_1^2 + I_2\omega_2^2 + I_3\omega_3^2)$$

and compute, along any solution $\boldsymbol{\omega}(t)$,

$$\dot{E} = I_1\omega_1\dot{\omega}_1 + I_2\omega_2\dot{\omega}_2 + I_3\omega_3\dot{\omega}_3 = \omega_1 u_1 + \omega_2 u_2 + \omega_3 u_3.$$

Then choose $\mathbf{u}(t)$ so that $\dot{E}(t) < 0$. Let $E(\omega_1, \omega_2, \omega_3) = E_0$, for a constant E_0, be the smallest ellipsoid in phase space towards which we can control the solution $\boldsymbol{\omega}(t)$. If $E_0 > 0$, the solution $\boldsymbol{\omega}(t)$ must come arbitrarily near to some point on the ellipsoid $E(\omega_1, \omega_2, \omega_3) = E_0$ and then we can modify the control $\mathbf{u}(t)$ so that the response $\boldsymbol{\omega}(t)$ penetrates the interior of the indicated ellipsoid. Thus

$$\lim_{t \to +\infty} E(t) = 0,$$

and we can steer $\boldsymbol{\omega}(t)$ arbitrarily near to zero in finite time.

2. *The initial point or state* x_0 is a known vector in phase space. In a real physical process x_0 and the response $x(t)$ might have components that describe position, velocities, angular velocities, temperatures, or other data that are measured and recorded by appropriate sensing instruments (see discussion of observability in Chapters 2 and 6). *Also a target set G is prescribed* for the control problem.

Sometimes the target will be a continuously moving compact set $G(t)$ on $\tau_0 \leq t \leq \tau_1$. That is, for each such instant t we designate a nonempty compact set $G(t)$ in R^n. The continuity of $G(t)$ as a function of the real variable t is defined in terms of the following concept of the distance from $G(t)$ to $G(t')$:

$$\text{dist}\,(G(t), G(t')) = \max \left[\max_{P \in G(t)} \text{dist}\,(P, G(t')),\ \max_{P' \in G(t')} \text{dist}\,(P', G(t)) \right].$$

Thus for instant t and given $\epsilon > 0$ there is $\delta > 0$ such that

$$\text{dist}\,(G(t), G(t') < \epsilon \quad \text{whenever} \quad |t - t'| < \delta.$$

If $G(t)$ is a continuously moving point tracing a smooth curve $\xi(t)$ in R^n, we often study the error or deviation of $x(t)$ from the target,

$$e(t) = x(t) - \xi(t).$$

Here $x(t)$ is the response for the control process

$$(\mathcal{S}) \qquad \dot{x} = f(t, x, u)$$

and the error $e(t)$ satisfies the control process

$$\dot{e} = f(t, e + \xi(t), u) - \dot{\xi}(t) = \tilde{f}(t, e, u).$$

In this interpretation we wish to control $e(t)$ to zero, that is, to regulate the error to zero.

3. *The class* Δ *of admissible controllers* usually consists of measurable functions $u(t)$, on various time intervals $t_0 \leq t \leq t_1$, each of which steers x_0 to the target $G(t)$. That is, the solution $x(t)$ of

$$(\mathcal{S}) \qquad \dot{x} = f(t, x, u) \qquad \text{with } x(t_0) = x_0$$

satisfies $x(t_1) \in G(t_1)$.

Suppose $\mathcal{S}$ is autonomous and x_0 is steered to x_1 by $u_1(t)$ on $t_0 \leq t \leq t_1$. If $u_2(t)$ on $t_0 \leq t \leq t_2$ steers x_1 to x_2, then the total control

$$u(t) = \begin{cases} u_1(t) & \text{on} \quad t_0 \leq t \leq t_1 \\ u_2(t - t_1 + t_0) & \text{on} \quad t_1 \leq t \leq t_1 + t_2 - t_0 \end{cases}$$

steers x_0 to x_2. A similar observation shows that for autonomous systems we can always begin the control at $t = 0$ with no loss of generality.

Various additional restrictions are often imposed on the functions comprising Δ. For example the condition $u(t) \subset \Omega$, where Ω is a fixed compact convex restraint set in R^m. Also the initial or final time for the duration of the controllers is sometimes prescribed.

4. *The cost functional* (or performance index, or objective functional, or effort) is an accepted quantitative criterion for the efficiency of each controller $u(t)$ on $t_0 \leq t \leq t_1$ in the class Δ. If Δ consists of controllers on various time intervals that steer x_0 to the target, one often defines the cost of $u(t)$ to be

$$C(u) = \int_{t_0}^{t_1} f^0(t, x(t), u(t))\, dt$$

where $f^0(t, x, u)$ is a given continuous function. If $f^0(t, x, u) \equiv 1$, then $C(u) = t_1 - t_0$ and we have the minimal time problem.

Sometimes Δ consists of controllers on a fixed specified time duration, say $t_0 \leq t \leq T$, without the requirement of reaching the target, say the curve $\xi(t)$. Then one often has cost functional

$$C(u) = |x(T) - \xi(T)| + \int_{t_0}^{T} f^0(t, x(t), u(t))\, dt.$$

In particular, there are certain important quadratic cost functionals, referring to the mean error of the response $x(t)$ and the energy expended by the controller $u(t)$,

$$C(u) = g(x(T)) + \int_{t_0}^{T} [x'(t)\, W(t)\, x(t) + u'(t)\, U(t)\, u(t)]\, dt.$$

Here $g(x)$ is specified as are the symmetric matrices $W(t)$ and $U(t)$; these are positive (semi-) definite so that $x'Wx \geq 0$ and $u'Uu > 0$ for nonzero vectors x and u.

The minimal time-optimal control problem for linear systems is treated in Chapter 2 and the quadratic cost functional is studied in Chapter 3.

Now consider a control problem with (1) process $\mathcal{S}$, (2) initial point x_0 and target $G(t)$, (3) admissible controllers Δ, and (4) cost functional $C(u)$ defined on all controllers u in the nonempty set Δ.

Definition. A controller $u^*(t)$ in the admissible class Δ is called (minimal) optimal, with respect to the real cost functional $C(u)$, in case

$$C(u^*) \leq C(u)$$

for all $u(t)$ in Δ.

In Chapters 2 and 3 we prove the existence of the optimal controller for linear processes with various cost functionals. In Chapter 4 we prove rather general theorems for the existence of optimal controllers for nonlinear processes, such as the following.

THEOREM 1. *Let be given the control problem consisting of*

1. *A differential system*

   $$(\mathcal{S})\quad \dot{x}^i = g^i(t, x) + h_j^{\,i}(t, x)u^j \quad \text{for } i = 1, \ldots, n \quad \text{and} \quad j = 1, \ldots, m$$

   with

   $$g^i(t, x),\ h_j^{\,i}(t, x) \quad \text{and} \quad \frac{\partial g^i}{\partial x^k}(t, x),\ \frac{\partial h_j^{\,i}}{\partial x^k}(t, x),$$

   for $k = 1, \ldots, n$, all continuous in $R^1 \times R^n$
2. *A nonempty, convex, compact restraint set $\Omega \subset R^m$*
3. *The initial point $x_0 \in R^n$ and the continuously moving compact target set $G(t) \subset R^n$ on the finite interval $\tau_0 \leq t \leq \tau_1$*
4. *The cost functional*

   $$C(u) = \int_{t_0}^{t_1} [g^0(t, x(t)) + h_j^{\,0}(t, x(t))\, u^j(t)]\, dt$$

   where $g^0(t, x)$ and $h_j^{\,0}(t, x)$ are continuous on $R^1 \times R^n$.

Let $\Delta = \Delta(\mathcal{S}, \Omega, x_0, G)$ *be the class of measurable controllers* $u(t) \subset \Omega$ on *subintervals* $t_0 \le t \le t_1$ of $\tau_0 \le t \le \tau_1$ *with responses* $x(t)$ *steering* $x(t_0) = x_0$ *to* $x(t_1) \in G(t_1)$.

Assume

(a) Δ *is nonempty;*
(b) *There exists a uniform bound* $B < \infty$ *for all responses* $x(t)$ *corresponding to controls in* Δ, *that is,* $|x(t)| \le B$.

Then there exists an optimal control $u^*(t)$ *in* Δ.

It can also be proved that if $\Delta(\alpha) \subset \Delta$ (those controllers with $t_0 = \alpha$ fixed) or if $\Delta(\alpha, \beta) \subset \Delta(\alpha)$ (controllers with $t_0 = \alpha$ and $t_1 = \beta$ fixed) is nonempty, then an optimal controller will exist for the corresponding admissible class of controllers.

The proof of this theorem and other existence theorems, as well as examples in which there is no optimal controller, will be presented in Chapter 4. All such existence proofs involve three observations: (1) Δ is nonempty, (2) Δ is weakly compact, so that there exists a limit controller $u^*(t)$ to an appropriate sequence $u_k(t)$ with decreasing costs, and (3) there is a continuity property of the cost functional so that $\lim_{n\to\infty} C(u_n) = C(u^*)$. Unfortunately all these existence theorems are nonconstructive. Further analysis is thus required before the optimal controller can be computed effectively.

For the case of a linear control process

$$\dot{x} = A(t)x + B(t)u$$

with integrable coefficients, it is easy to see that hypothesis (b) of the above theorem is always fulfilled automatically. We investigate hypothesis (a) of the theorem in the next section, which deals with the concept of controllability.

1.3 RESULTS ON CONTROLLABILITY

In this section we discuss the possibility of steering an initial state x_0 precisely to a target state x_1 in a finite time duration.

Definition. An autonomous control process

$$\dot{x}^i = f^i(x^1, \ldots, x^n, u^1, \ldots, u^m), \qquad i = 1, \ldots, n,$$

with $f(x, u) \in C^1$ in $R^n \times R^m$, is called (completely) controllable in case: for each pair of points x_0 and x_1 in R^n there exists a bounded measurable controller $u(t)$ on some finite interval $0 \le t \le t_1$ such that the corresponding response $x(t)$ steers $x(0) = x_0$ to $x(t_1) = x_1$.

Remarks. For a nonautonomous process

$$\dot{x} = f(t, x, u)$$

we define a corresponding concept of controllability for each initial time t_0 in case: for each initial point x_0 and each target point x_1 there exists a bounded measurable controller $u(t)$ on $t_0 \leq t \leq t_1$ with response $x(t)$ steering $x(t_0) = x_0$ to $x(t_1) = x_1$.

In Chapter 2 we prove the following theorem on the controllability of linear processes:

THEOREM 2. *The linear control process*

$$\dot{x} = Ax + Bu,$$

where A is a real constant $n \times n$ matrix and B is a real constant $n \times m$ matrix, is controllable if and only if $n \times nm$ matrix

$$[B, AB, A^2B, \ldots, A^{n-1}B]$$

has rank n.

In the examples of Section 1.2 we considered the problem of controlling or regulating a response to zero. We demonstrated cases in which every initial state could be controlled to a neighborhood of the origin. For the precise control to the origin we then apply an appropriate control, as is described in the following results.

Definition. For the control process

$$\dot{x}^i = f^i(x^1, \ldots, x^n, u^1, \ldots, u^m) \qquad i = 1, \ldots, n,$$

with $f(x, u) \in C^1$ in $R^n \times R^m$, the domain $\mathcal{C}$ of (null) controllability consists of all those points $x_0 \in R^n$ that can be steered to the origin by an admissible controller $u(t)$ on some finite time interval $0 \leq t \leq t_1$.

In Chapter 6 we shall prove the following basic theorem on regulation to rest.

THEOREM 3. *Consider the control process*

$$\dot{x}^i = f^i(x^1, \ldots, x^n, u^1, \ldots, u^m) \qquad i = 1, \ldots, n,$$

with $f(x, u) \in C^1$ in $R^n \times R^m$.

Assume

(a) $f(0, 0) = 0$;
(b) *the class Δ of admissible controllers includes all measurable controllers $u(t)$ on finite time intervals satisfying $|u(t)| \leq \epsilon$ for some $\epsilon > 0$;*

(c) *the linear differential system*

$$\dot{x} = Ax + Bu,$$

with constant coefficient matrices

$$A = \left(\frac{\partial f^i}{\partial x^j}(0, 0)\right) \quad \text{and} \quad B = \left(\frac{\partial f^j}{\partial u^k}(0, 0)\right)$$

is controllable, that is

$$\text{rank}\,[B, AB, A^2B, \ldots, A^{n-1}B] = n.$$

Then the domain $\mathcal{C}$ *of controllability contains an open neighborhood of the origin in* R^n.

To illustrate the force of this last theorem we note the following immediate corollary, which also will be proved independently in Chapter 2.

COROLLARY. *Consider the linear control process*

$$\dot{x} = Ax + Bu,$$

where A *is a real constant* $n \times n$ *matrix and* B *is a real constant* $n \times m$ *matrix.*

Assume

(a) A *is stable, that is, each eigenvalue* λ *of* A *has* $\text{Re}\,\lambda < 0$;
(b) *controllability, that is,* $\text{rank}\,[B, AB, A^2B, \ldots, A^{n-1}B] = n$.
Then each initial state x_0 *can be steered to* $x_1 = 0$ *by some measurable controller* $u(t)$ *on a finite interval* $0 \leq t \leq t_1$. *Furthermore we can require that* $|u(t)| \leq \epsilon$ *for an arbitrarily prescribed* $\epsilon > 0$.

The applications of the above theorem to Examples 1 and 2 of Section 1.2, and other interesting special cases, are left to the exercises.

During the remainder of this section we introduce certain control problems that usually arise from a single high-order linear differential equation rather than from a matrix system. Consider a linear control process

$$x^{(n)} + a_1(t)x^{(n-1)} + \cdots + a_n(t)x = u,$$

where $u(t)$ is a scalar control restricted in magnitude to some bounded interval $\mathfrak{J}$. We may wish to steer an initial state $(x_0, \dot{x}_0, \ddot{x}_0, \ldots, x_0^{(n-1)})$ to a target G, say $x = 0$, and furthermore hold the response in G forever afterwards by a control in $\mathfrak{J}$. This type of problem arises in the theory of single- and multiple-component control illustrated by the next example and treated in detail in Chapter 2.

***Example* 1.** Consider the linear control problem

$$\dddot{x} + 2\ddot{x} + 2\dot{x} + x = u, \quad \text{with} \quad |u(t)| \leq 1.$$

Suppose that we wish to regulate to rest the velocity $\dot{x}$ and acceleration $\ddot{x}$, but that the displacement x is not significant. That is, we wish to steer each initial condition $(x_0, \dot{x}_0, \ddot{x}_0)$ to the target $\dot{x} = 0$, $\ddot{x} = 0$ and hold the velocity and acceleration at zero forever after by a control satisfying the restraint $|u(t)| \leq 1$. As a differential system this control problem reads, setting $x = x^1$,

$$\dot{x}^1 = x^2$$

$$\dot{x}^2 = x^3$$

$$\dot{x}^3 = -x^1 - 2x^2 - 2x^3 + u(t).$$

The target G is the line $x^2 = 0$, $x^3 = 0$ in R^3. We shall define the core of G to be those points in G that can be held forever in G by admissible controls. For a point in core (G) there is a controller $u(t)$ on $0 \leq t < \infty$ with $|u(t)| \leq 1$ and response with $x^2(t) = 0$, $x^3(t) = 0$ so that $\dot{x}^2(t) = 0$ and $\dot{x}^3(t) = 0$. But this means that in core (G) we have

$$x^1(t) = u(t) \quad \text{so} \quad |x^1| \leq 1.$$

On the other hand an initial point $(x_0{}^1, 0, 0)$ with $|x_0{}^1| \leq 1$ can be held fixed in G forever with the constant control $u(t) = x_0{}^1$. Thus

$$\text{core}\,(G) = \{|x^1| \leq 1,\ x^2 = 0,\ x^3 = 0\},$$

which is a segment on the x^1-axis.

Thus the problem of controlling a point to G and holding thereafter in G is entirely equivalent to controlling the point to core (G). We have reduced the problem of controlling to the target G, with the supplementary requirement of holding in G, to the more standard problem of controlling to the new target core (G), with no further mention of additional holding. Note that the set core (G) is a compact convex target in R^3.

It is also interesting to note that control, with further holding, to the plane

$$G' = \{x^2 = 0\}$$

requires that $\dot{x}^2 = x^3 = 0$ and so

$$\text{core}\,(G') = \text{core}\,(G).$$

Therefore the original problem of 2-component control to $G = \{x^2 = 0,$ $x^3 = 0\}$ could be restated as a problem of single component control to $G' = \{x^2 = 0\}$. This illustrates a general result obtained later.

Another type of linear control problem, in which the derivatives of the control function occur, is referred to as *numerator dynamics*. This term arises because the usual transfer function is a rational function whose numerator relates to the control function and its derivatives. The next example indicates the nature of such problems. A detailed study will appear in Chapter 2.

***Example* 2.** Consider the linear control problem with numerator dynamics

$$(\mathfrak{L}) \qquad \ddot{x} + 3\dot{x} + 2x = 2\dot{u} + u,$$

with controllers $u(t)$ in C^1 and satisfying $|u(t)| \leq 1$. The transfer function for the open loop circuit is

$$\frac{2p + 1}{p^2 + 3p + 2}$$

and we note that $(2p + 1)$ arises from the terms $(2\dot{u} + u)$.

We wish to steer an initial state $(x_0, \dot{x}_0)$ to $(0, 0)$ in minimal time. In order to describe the control problem by a linear system in phase space we could write $\dot{x} = y$ and obtain, as usual,

$$\dot{x} = y$$
$$\dot{y} = -2x - 3y + 2\dot{u} + u.$$

Later theory will show that there is no optimal controller $u(t)$ in class C^1 and we must extend our class of admissible controllers to include discontinuous, and hence nondifferentiable, functions before an optimal can be achieved.

For this purpose write the given control problem $\mathfrak{L}$ as a linear system in the following new way (see Chapter 2, Exercise 4.5):

$$(\mathcal{S}) \qquad \begin{aligned} \dot{x}^1 &= x^2 + 2u \\ \dot{x}^2 &= -2x^1 - 3x^2 - 5u. \end{aligned}$$

Then the transfer function from the input u to the output x^1 is computed by

$$p\bar{x}^1 = \bar{x}^2 + 2\bar{u}$$
$$p\bar{x}^2 = -2\bar{x}^1 - 3\bar{x}^2 - 5\bar{u}$$

using the Laplace transforms $\bar{x}^1$, $\bar{x}^2$, $\bar{u}$; so

$$\bar{x}^1 = \frac{2p + 1}{p^2 + 3p + 2}\,\bar{u}.$$

Moreover the system $\mathcal{S}$ no longer involves the derivatives of u and hence it can be studied within the standard control theory as expounded in Chapter 2.

It should be noted that the phase plane coordinates x, $y = \dot{x}$ now enter the problem as

$$x^1 = x, \qquad x^2 = y - 2u$$

and the initial state $x^1(0) = x_0$, $x^2(0) = \dot{x}_0 - 2u(0)$ depends on the controller $u(t)$. This difficulty can be met either by using controllers with $u(0) = 0$, or else by replacing the initial point by the initial segment $x^1 = x_0$, $y_0 - 2 \leq x^2 \leq y_0 + 2$, or else by discarding $\mathcal{L}$ completely and considering $\mathcal{S}$ as the true description of the control problem with an initial state (x_0^1, x_0^2) prescribed.

Actually the final system $\mathcal{S}$ is often the physically significant one and the single equation $\mathcal{L}$, which displays numerator dynamics, has been previously derived from $\mathcal{S}$ by operations of differentiation and elimination —operations that are not strictly permissible when discontinuous control functions are allowed in $\mathcal{S}$.

1.4 EXTREMAL AND MAXIMAL PROPERTIES OF OPTIMAL CONTROL AND SYNTHESIS

In the differential calculus the minimum of a function of a real variable is located by examining the critical points, those points at which the derivative is zero. An analogous procedure is followed in the theory of optimal control.

In this section we state the maximal principle that every optimal controller $u(t)$ is a maximal controller, that is, $u(t)$ is critical for the control problem. We present here the concepts for the autonomous case only; the more general nonautonomous theory is proved in detail in Chapter 5. We consider the autonomous control problem:

1. ($\mathcal{S}$) $\qquad \dot{x}^i = f^i(x^1, \ldots, x^n, u^1, \ldots, u^m), \qquad i = 1, \ldots, n,$

   with $f(x, u) \in C^1$ in $R^n \times \Omega$.
2. The initial point x_0 and the target set G, a nonempty compact set in R^n.
3. The class Δ of all measurable controllers $u(t)$ on various finite intervals $0 \leq t \leq t_1$, which steer x_0 to G, with $u(t) \subset \Omega$, the nonempty compact restraint set in R^m.
4. The cost functional $C(u) = \int_0^{t_1} f^0(x(t), u(t))\, dt$, where $f^0(x, u) \in C^1$ in $R^n \times \Omega$.

Definition. Consider the autonomous control process $\{\mathcal{S}, x_0, G, \Omega, \Delta, C\}$ as described above. Let $u(t)$ on $0 \leq t \leq t_1$ be a controller in Δ with corresponding response $x(t) = (x^i(t))$, $i = 1, \ldots, n$. We extend the response to an $(n + 1)$ vector

$$\hat{x}(t) = (x^\alpha(t)); \qquad \alpha = 0, 1, \ldots, n,$$

by defining

$$x^0(t) = \int_0^t f^0(x(t), u(t))\, dt.$$

An $(n + 1)$ vector $\hat{\eta}(t) = (\eta_\alpha(t))$ on $0 \leq t \leq t_1$ is called an adjoint response in case $\hat{\eta}(t)$ is a nonvanishing solution of the Hamiltonian differential system,

$$\dot{x}^\alpha = \frac{\partial \hat{H}}{\partial \eta_\alpha} = f^\alpha(x, u(t)) \qquad \alpha = 0, 1, \ldots, n.$$

$$\dot{\eta}_\alpha = -\frac{\partial \hat{H}}{\partial x^\alpha} = -\eta_0 \frac{\partial f^0}{\partial x^\alpha}(x, u(t)) - \cdots - \eta_n \frac{\partial f^n}{\partial x^\alpha}(x, u(t))$$

Here the Hamiltonian function is defined by

$$\hat{H}(\hat{\eta}, \hat{x}, u) = \eta_0 f^0(x, u) + \eta_1 f^1(x, u) + \cdots + \eta_n f^n(x, u)$$

and we define

$$\hat{M}(\hat{\eta}, \hat{x}) = \max_{u \in \Omega} \hat{H}(\hat{\eta}, \hat{x}, u).$$

Then we define the controller $u(t)$ on $0 \leq t \leq t_1$ to be maximal in case there exists an adjoint response $\hat{\eta}(t)$ so that

1. $\hat{H}(\hat{\eta}(t), \hat{x}(t), u(t)) = \hat{M}(\hat{\eta}(t), \hat{x}(t))$ almost everywhere on $0 \leq t \leq t_1$, and
2. $\hat{M}(\hat{\eta}(t), \hat{x}(t)) = 0$ everywhere, and $\eta_0 \leq 0$.

The next theorem is called the maximal principle for autonomous systems.

THEOREM 4. *Consider the autonomous control problem* $\{\mathcal{S}, x_0, G, \Omega, \Delta, C\}$ *as described above. Let* $u(t)$ *on* $0 \leq t \leq t_1$ *be an optimal controller in* Δ. *Then* $u(t)$ *is a maximal controller.*

Note that $u(t)$ is called a maximal controller even though it yields a minimum cost. We prefer this confusion, which enters only in the later chapters, to the alternative of changing the traditional notation used throughout the literature of control theory.

In order to clarify the nature of the maximal principle let us consider the statement of the principle for an autonomous linear control problem:

1. $\mathcal{S}: \dot{x} = Ax + Bu$
   for real constant $n \times n$ matrix A and $n \times m$ matrix B;
2. Initial point x_0 in the domain of null controllability and target G as the origin;

3. Compact convex restraint set $\Omega \subset R^m$
4. $C(u) = \int_0^{t_1} dt = t_1$, the time duration of control.

In this case the Hamiltonian function is

$$\hat{H}(\hat{\eta}, \hat{x}, u) = \eta_0 + \eta[Ax + Bu] = \eta_0 + H(\eta, x, u)$$

where $\eta = (\eta_i)$, $i = 1, \ldots, n$ is an n-row vector, and $\eta(Ax + Bu) = H$. Then

$$\hat{M}(\hat{\eta}, \hat{x}) = \eta_0 + \eta Ax + \max_{u \in \Omega} \eta Bu = \eta_0 + M(\eta, x),$$

where $M = \max_{u \in \Omega} H$. If $u(t)$ on $0 \leq t \leq t_1$ is maximal, then the response $x(t) = (x^i(t))$, $i = 1, \ldots, n$, and an adjoint response $\eta(t) = (\eta_i(t))$, $i = 1, \ldots, n$ satisfy

$$\dot{x} = Ax + Bu(t)$$

$$\dot{\eta} = -\eta A,$$

and $x^0 = t$, $\eta_0 =$ constant.

The maximal principle requires that

$$\eta_0 + \eta(t)A\, x(t) + \eta(t)\, B\, u(t) = \eta_0 + \eta(t)A\, x(t) + \max_{u \in \Omega} \eta(t)Bu$$

or

$$\eta(t)B\, u(t) = \max_{u \in \Omega} \eta(t)Bu,$$

almost everywhere on $0 \leq t \leq t_1$.

The second condition of the maximal principle asserts that

$$\eta_0 + \eta(t)A\, x(t) + \max_{u \in \Omega} \eta(t)B\, u = 0$$

everywhere. If $\eta(t)$ vanished at one point on $0 \leq t \leq t_1$, then it vanishes identically since it is a solution of the homogeneous linear system

$$\dot{\eta} = -\eta A.$$

But if $\eta(t) \equiv 0$, then $\eta_0 = 0$, and this contradicts the nonvanishing of the $(n + 1)$ vector $\hat{\eta}(t)$. Therefore $\eta(t)$ is nowhere zero.

Thus, in this case, we can ignore the new response components $x^0 = t$ and $\eta_0 =$ const., and refer only to the n-vectors $x(t)$ and $\eta(t)$, and find a maximal controller $u(t)$ in terms of $H(\eta, x, u)$ and $M(\eta, x)$. It is important to notice that the adjoint response satisfies the fixed differential system

$$\dot{\eta} = -\eta A$$

whose coefficients do not depend on the control $u(t)$ or the response $x(t)$. Thus $\eta(t)$ is entirely determined by its initial conditions, which we can take as a unit row vector since the conditions of the maximal principle are homogeneous and remain satisfied if $\eta(t)$ is multiplied by a positive constant.

Consider the important case where Ω is the m-cube $|u^j| \leq 1$ for the above autonomous linear control process. Then

$$\eta(t)B\, u(t) = \max_{u \in \Omega} \eta(t)Bu$$

means that each component of the maximal controller $u(t)$ should be chosen as $+1$ or -1, according to the sign of the corresponding component of the vector $\eta(t)B$. That is, a maximal controller must satisfy

$$u(t) = [\operatorname{sgn} \eta(t)B]'$$

almost everywhere, provided the components of $\eta(t)B$ do not vanish on a set of positive measure on $0 \leq t \leq t_1$. Note that this is the same extremal condition deduced for an optimal controller in the examples of Section 1.1 by means of the geometry of convex sets of attainability.

Here we can attempt to synthesize the maximal controller

$$u(t) = [\operatorname{sgn} \eta(t)B]'$$

by determining first an adjoint response $\eta(t)$, and then integrating

$$\dot{x} = Ax + B[\operatorname{sgn} \eta(t)B]'$$

backwards in time, beginning at the origin $x = 0$ and reaching the required state x_0. Thus we try various beginning values for the vector $\eta(t)$, e.g., give a unit vector at $t = 0$, then construct $u(t)$ and $x(t)$ for $t \leq 0$. If we construct all such possible maximal controls and their responses, one of these will be the optimal control (if such exists) steering x_0 to the origin. Afterwards we revert to increasing t and translate the time scale so that x_0 corresponds to $t = 0$. This procedure was followed in constructing the switching locus and synthesizing the optimal controller in the examples of Section 1.1.

An important logical consideration in the synthesis of the maximal, and hence optimal controller is that there should be just one maximal controller steering x_0 to the target. A control process having this uniqueness property will be called *normal* and a theory of normal control processes will be developed later. In particular it will be shown that the linear autonomous control process

$$x^{(n)} + a_1 x^{(n-1)} + \cdots + a_n x = u$$

with restraint $|u(t)| \leq 1$ is normal, with respect to minimal time-optimal control to the origin. Hence the synthesis of the optimal controller as a relay controller is possible through the techniques utilizing the maximal principle, maximal controllers, and the switching locus.

1.5 SYNTHESIS OF OPTIMAL CONTROLLERS FOR SECOND ORDER LINEAR PROCESSES

In this section we shall complete the synthesis for the minimal time optimal controller for regulation of $(x, \dot{x})$ to $(0, 0)$ for the most general second order linear process

$$\ddot{x} \pm 2b\dot{x} \pm k^2 x = u,$$

where $b \geq 0$ and $k^2 > 0$ are constants and the control restraint is $|u(t)| \leq 1$.

In Section 1.1 the most general first order linear process

$$\dot{x} \pm bx = u, \qquad |u(t)| \leq 1,$$

was treated. Certain special second order cases were also analyzed, in particular,

$$\ddot{x} \pm b\dot{x} = u, \qquad |u(t)| \leq 1,$$

and

$$\ddot{x} + k^2 x = u, \qquad |u(t)| \leq 1.$$

It was also stated in the exercises that an apparently more general restraint $|u(t)| \leq c$ for $c > 0$, reduces to the standard problem where $|u(t)| \leq 1$ under appropriate changes of scale and units. The cases treated in this section will complete the synthesis for the time-optimal controller for all second order autonomous linear processes.

Consider the synthesis of the minimal time-optimal controller for the regulation to zero of the linear process

$$(\mathcal{L}) \qquad \ddot{x} \pm 2b\dot{x} \pm k^2 x = u,$$

with constants $b \geq 0$ and $k^2 > 0$ and control restraint $|u(t)| \leq 1$. We note that the corresponding linear system

$$(\mathcal{S}) \qquad \begin{pmatrix} \dot{x} \\ \dot{y} \end{pmatrix} = \begin{pmatrix} 0 & 1 \\ \mp k^2 & \mp 2b \end{pmatrix} \begin{pmatrix} x \\ y \end{pmatrix} + \begin{pmatrix} 0 \\ 1 \end{pmatrix} u$$

is controllable and normal. Hence by the theorems stated in Sections 1.2 and 1.3 there exists an optimal controller $u^*(t)$ on $0 \leq t \leq t^*$ steering a given initial state (x_0, y_0), which lies in the domain $\mathcal{C}$ of controllability, to $(0, 0)$. Also in this case $\mathcal{C}$ is an open connected set in the phase plane R^2.

By the maximal principle of Section 1.4 the optimal controller $u^*(t)$ is maximal and has the extremal-value

$$u^*(t) = \operatorname{sgn} \eta_2(t)$$

almost everywhere on $0 \leq t \leq t^*$. Here the adjoint response $\eta(t) = (\eta_1(t), \eta_2(t))$ satisfies

$$\dot{\eta}_1 = \pm k^2 \eta_2$$
$$\dot{\eta}_2 = -\eta_1 \pm 2b\eta_2$$

or

$$\ddot{\eta}_2 \mp 2b\dot{\eta}_2 \pm k^2\eta_2 = 0.$$

Note that $\eta_2(t)$ is not identically zero, for in such a case

$$\eta_1(t) = \pm 2b\eta_2(t) - \dot{\eta}_2(t) = 0,$$

which contradicts the requirement that $\eta(t)$ vanish nowhere on $0 \leq t \leq t^*$. Since $\eta_2(t)$ is analytic, it has only a finite number of zeros on $0 \leq t \leq t^*$.

Since $\mathcal{L}$ is normal, there is just one maximal controller steering (x_0, y_0) to $(0, 0)$ and hence this one must be the optimal controller $u^*(t)$. We shall construct the appropriate switching locus $y = W(x)$ where the extremal response switches between the systems

$$(\mathcal{S}_-) \qquad \dot{x} = y \qquad \dot{y} = \mp k^2 x \mp 2by - 1$$

and

$$(\mathcal{S}_+) \qquad \dot{x} = y \qquad \dot{y} = \mp k^2 x \mp 2by + 1.$$

Since $\mathcal{S}_-$ becomes $\mathcal{S}_+$ under the substitution $x \to -x$, $y \to -y$, we shall find that $W(-x) = -W(x)$ and so the switching locus need only be found for $x > 0$.

Thus the synthesis of the optimal controller $u^*(t)$ is reduced to a determination of the domain $\mathcal{C}$ of controllability and the construction of the switching locus $y = W(x)$.

Example* 1. *Damped Linear Oscillator. Consider the minimal time-optimal control to the origin for

$$(\mathcal{L}) \qquad \ddot{x} + 2b\dot{x} + k^2 x = u,$$

Here $b > 0$ and $k^2 > 0$ are constants and $|u(t)| \leq 1$. Since the matrix

$$A = \begin{pmatrix} 0 & 1 \\ -k^2 & -2b \end{pmatrix}$$

is stable, the corollary to a theorem in Section 1.3 states that the domain $\mathcal{C}$ of controllability is all R^2.

There are two qualitatively different cases:

1. underdamping, $b^2 - k^2 < 0$
2. critical and overdamping, $b^2 - k^2 \geq 0$.

Consider first the case, $b^2 - k^2 < 0$. Here each solution of the extremal system

$$(\mathcal{S}_-) \qquad \begin{pmatrix} \dot{x} \\ \dot{y} \end{pmatrix} = \begin{pmatrix} 0 & 1 \\ -k^2 & -2b \end{pmatrix} \begin{pmatrix} x \\ y \end{pmatrix} - \begin{pmatrix} 0 \\ 1 \end{pmatrix}$$

is a spiral approaching the critical (equilibrium) point $0_-: x = -1/k^2, y = 0$; and each solution of

$$(\mathcal{S}_+) \qquad \begin{pmatrix} \dot{x} \\ \dot{y} \end{pmatrix} = \begin{pmatrix} 0 & 1 \\ -k^2 & -2b \end{pmatrix} \begin{pmatrix} x \\ y \end{pmatrix} + \begin{pmatrix} 0 \\ 1 \end{pmatrix}$$

is a spiral approaching the critical point $0_+ : x = 1/k^2, y = 0$, as t increases towards $+\infty$. Each extremal response reaching the origin must be composed of a finite number of solution segments switching alternately between

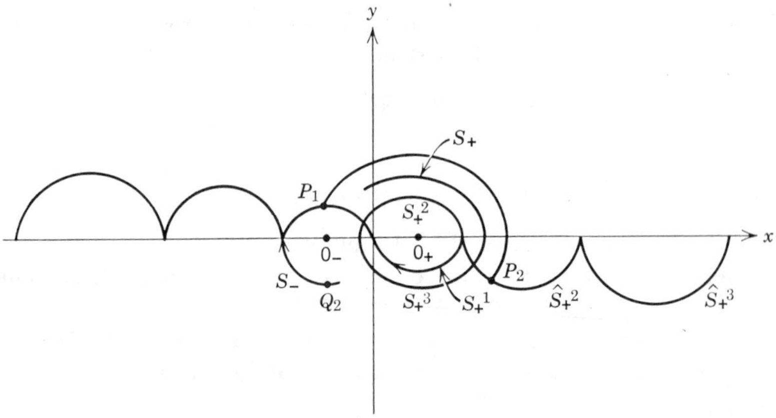

Figure 1.7 Minimal time-optimal control to origin. Switching locus diagram for $\ddot{x} + 2b\dot{x} + k^2x = u$, $|u(t)| \leq 1$. Case 1, Example 1: $b^2 - k^2 < 0$, $b > 0$, $k^2 > 0$.

$\mathcal{S}_-$ and $\mathcal{S}_+$. An examination of the most general adjoint response

$$\eta_2(t) = \alpha e^{bt} \sin(\omega t + \beta),$$

for constants $\alpha \neq 0$, β, and $\omega = \sqrt{k^2 - b^2}$, shows that the time duration between switches is exactly $T = \pi/\omega$.

Let S_+ be the solution of $\mathcal{S}_+$ leading backwards in time from the origin (see Figure 1.7). We shall describe a construction for the switching locus $y = W(x)$ on $x \geq 0$ in terms of the solution S_+. A proof of the validity of the construction is then presented.

We shall construct the switching locus by unfolding S_+ with kinks only at the points of intersection with the x-axis. Let S_+^1 be the arc of S_+ leading backwards from (0, 0) for a half revolution to the preceding intercept with the x-axis.

Clearly S_+^1 and its reflection S_-^1 in the origin are part of the switching locus since each point of S_+^1 is a switch point for a maximal or extremal response from a solution of $\mathcal{S}_-$ that meets S_+^1 and thence leads into (0, 0).

Let $S_+^1, S_+^2, S_+^3, \ldots$ be the successive arcs of S_+, each leading backwards for one-half revolution of the spiral S_+ and each terminating at x-intercepts of S_+. Form the switching locus $y = W(x)$ on $x \geq 0$ from the arc S_+^1 and arcs $\hat{S}_+^2, \hat{S}_+^3, \ldots$ rigidly congruent to $S_+^2, S_+^3, \ldots$, respectively. We fit these together as a continuous curve $y = W(x)$, single valued along the positive x-axis, so that tracing the switching locus with increasing x describes the same ordering of the points of the $\hat{S}_+^1, \hat{S}_+^2, \hat{S}_+^3, \ldots$ as does tracing $S_+^1, S_+^2, S_+^3, \ldots$ along S_+ for increasing $(-t)$.

Then on $x < 0$ define $W(x) = -W(-x)$. To complete the synthesis of the optimal controller through a feedback circuit we define the synthesizing function $\Psi(x, y)$ so that solutions of

$$\ddot{x} + 2b\dot{x} + k^2x = \Psi(x, \dot{x})$$

always yield the optimal-response steering any initial point (x_0, y_0) to $(0, 0)$. To obtain this define [for all real $(x, y) \neq (0, 0)$],

$$\Psi(x, y) = \begin{cases} -1 & \text{for} \quad y > W(x) \quad \text{and on} \quad S_-^1 \\ 0 & \text{for} \quad y = W(x) \quad \text{otherwise} \\ +1 & \text{for} \quad y < W(x) \quad \text{and on} \quad S_+^1. \end{cases}$$

In order to verify the correctness of our construction of the switching locus we begin with the arc S_+^1 of the solution S_+ of $\mathcal{S}_+$ and the arc S_-^1 of the solution S_- of $\mathcal{S}_-$ leading backwards from the origin. The switching locus to the right of S_+^1 is defined by following the solutions of $\mathcal{S}_+$ backwards in time for a duration $T = \pi/\omega$, starting at points of S_-^1.

Take a point $P_1 = (x^1, y^1)$ on S_-^1 and call $P_2 = (x^2, y^2)$ the image after a duration of $T = \pi/\omega$ backwards in time along a solution of $\mathcal{S}_+$. Then $P_1, 0_+$, and P_2 are colinear and the ratio of the distance P_10_+ to 0_+P_2 is e^{-bT}. This is easily computed since P_1 and P_2 lie on the same spiral solution around 0_+ at an interval of one-half revolution, and the damping factor is e^{-bt}. But the point $Q_2 = (x_-^2, y_-^2)$ lying on S_- opposite from P_1 through 0_- also has the same ratio e^{-bT} for the corresponding distances from the center 0_-. By similarity of triangles we find that $y_-^2 = y^2$ and an easy calculation shows that $x^2 = x_-^2 + (e^{b\pi/\omega} + 1)(2/k^2)$. This means that the switching locus arc $\hat{S}_+^2$ is merely a rigid translation of the arc S_-^2, which is the continuation of S_-^1 backwards along S_- for one-half revolution. But S_-^2 is rigidly congruent to S_+^2 because the differential systems $\mathcal{S}_-$ and $\mathcal{S}_+$ are interchanged by the substitution $(x, y) \rightarrow (-x, -y)$. Thus the continuation of $y = W(x)$ on $x > 0$ from S_+^1 is the arc $\hat{S}_+^2$ which is congruent to S_+^2, as required. A repetition of this argument yields the complete description of $y = W(x)$.

We next turn to case (2), $b^2 - k^2 \geq 0$. Here each solution of $\mathcal{S}_{\pm}$ approaches $0_{\pm}$, but on each such solution x can vanish at most once. The general solution of the adjoint equation is

$$\eta_2(t) = e^{bt}(\alpha + \beta t) \quad \text{if} \quad b^2 - k^2 = 0$$

or

$$\eta_2(t) = \alpha e^{bt} \sinh(\mu t + \beta) \quad \text{if} \quad b^2 - k^2 > 0,$$

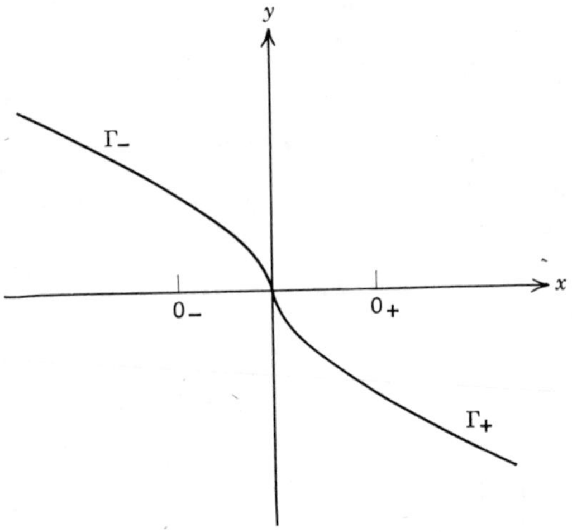

Figure 1.8 Minimal time-optimal control to origin. Switching locus diagram for $\ddot{x} + 2b\dot{x} + k^2x = u$, $|u(t)| \leq 1$. Case 2, Example 1: $b^2 - k^2 \geq 0$, $b > 0$, $k^2 > 0$.

where α and β are constants and $\mu = \sqrt{b^2 - k^2}$. In any case, $\eta_2(t)$ has at most one zero and the appropriate optimal controller is

$$u^*(t) = \operatorname{sgn} \eta_2(t)$$

In case (2) the switching locus $y = W(x)$ consists of exactly two solution arcs, Γ_- and Γ_+. Here Γ_- is the solution of $\mathcal{S}_-$ passing through (0, 0) and Γ_+ is the solution of $\mathcal{S}_+$ leading backwards from (0, 0) in the fourth quadrant (see Figure 1.8). Thus it is easy to construct this switching locus $y = W(x)$, which is single valued over the entire x-axis, and the corresponding optimal control synthesizer

$$\Psi(x, y) = \begin{cases} -1 & \text{for} \quad y > W(x) \quad \text{and on} \quad \Gamma_- \\ +1 & \text{for} \quad y < W(x) \quad \text{and on} \quad \Gamma_+. \end{cases}$$

Example* 2. *Linear Oscillator with Negative Friction. Consider the minimal time-optimal control to the origin for

$$(\mathcal{L}) \qquad \ddot{x} - 2b\dot{x} + k^2x = u.$$

Here $b > 0$ and $k^2 > 0$ are constants and $|u(t)| \leq 1$. We note that the differential system

$$(\mathcal{S}) \qquad \begin{pmatrix} \dot{x} \\ \dot{y} \end{pmatrix} = \begin{pmatrix} 0 & 1 \\ -k^2 & 2b \end{pmatrix} \begin{pmatrix} x \\ y \end{pmatrix} + \begin{pmatrix} 0 \\ 1 \end{pmatrix} u$$

is not stable at the origin (for $u \equiv 0$), but it is controllable and normal.

To construct the switching locus and to synthesize the optimal controller we proceed as in Example 1 and consider the extremal systems

$$(\mathcal{S}_-) \qquad \dot{x} = y \qquad \dot{y} = -k^2x + 2by - 1$$

and

$$(\mathcal{S}_+) \qquad \dot{x} = y \qquad \dot{y} = -k^2x + 2by + 1.$$

Again there are two cases:

1. underdamping, $b^2 - k^2 < 0$
2. critical and overdamping, $b^2 - k^2 \geq 0$.

Consider first case (1), $b^2 - k^2 < 0$. Here each solution curve of $\mathcal{S}_-$ spirals outwards from 0_-: $x = -(1/k^2)$, $y = 0$; and each solution curve of $\mathcal{S}_+$ spirals outwards from 0_+: $x = 1/k^2$, $y = 0$ as t increases. The most general adjoint response is

$$\eta_2(t) = \alpha e^{-bt} \sin(\omega t + \beta)$$

for constants α, β, and $\omega = \sqrt{k^2 - b^2}$. Thus there can be many switches between arcs of $\mathcal{S}_-$ and $\mathcal{S}_+$ for an extremal response leading to (0, 0), however, the time duration between switches is exactly $T = \pi/\omega$.

The switching locus $y = W(x)$ is constructed in a manner similar to that used in Example 1, above. Let S_+ be the solution of $\mathcal{S}_+$ leading backwards from 0 and let S_+^1 be the arc of S_+ from (0, 0) backwards in time for a duration $T = \pi/\omega$. Let $S_+^2, S_+^3, \ldots$ be arcs of S_+, each leading backwards along S_+ for a duration of $T = \pi/\omega$ and each terminating on the x-axis.

Then $y = W(x)$, for $x > 0$, is composed of $S_+{}^1$ followed by $S_+{}^2$ and continued by a rigid unfolding of S_+ with hinges inserted at the x-intercepts of S_+ (see Figure 1.9).

Since the diameter of $S_+{}^1$ is $(e^{-b\pi/\omega} + 1)(1/k^2)$, and since the diameter of

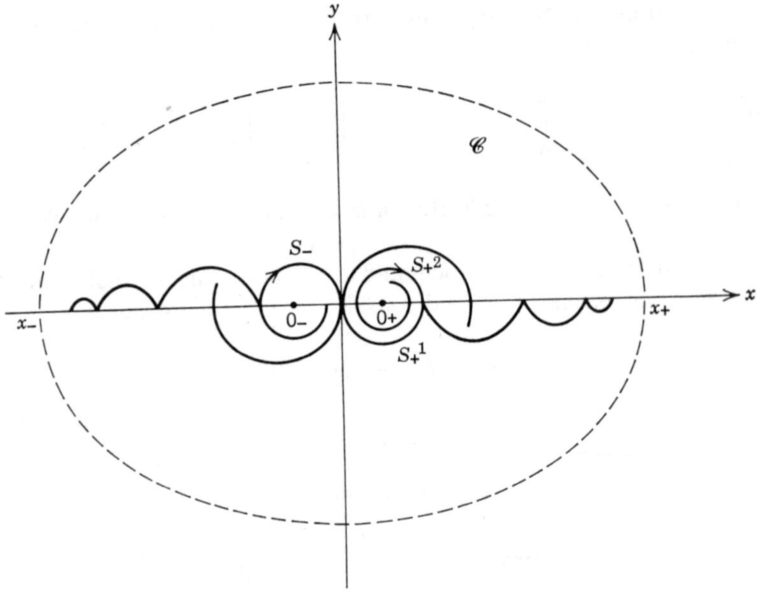

Figure 1.9 Minimal time-optimal control to origin. Switching locus diagram for $\ddot{x} - 2b\dot{x} + k^2x = u$, $|u(t)| \leq 1$. Case 1, Example 2: $b^2 - k^2 < 0$, $b > 0$, $k^2 > 0$.

each succeeding loop along S_+ is decreased by a factor of $e^{-b\pi/\omega}$, we compute that $y = W(x)$ is defined on

$$0 \leq x < x_+ = \frac{1 + e^{-b\pi/\omega}}{(1 - e^{-b/\pi\omega})k^2}.$$

Using the oddness property $W(x) = -W(-x)$, we define $W(x)$ on $x_- < x \leq 0$ where $x_- = -x_+$.

It is easy to see that the domain $\mathcal{C}$ of controllability is the open region bounded by the solution of $\mathcal{S}_-$ that passes from $(x_-, 0)$ to $(x_+, 0)$ in $y \geq 0$, and by the solution of $\mathcal{S}_+$ that passes from $(x_+, 0)$ to $(x_-, 0)$ in $y \leq 0$. Thus the synthesizer is defined in $\mathcal{C}$ as follows

$$\Psi(x, y) = \begin{cases} -1 & \text{for} \quad y > W(x) \quad \text{and on} \quad S_-{}^1 \\ 0 & \text{for} \quad y = W(x) \quad \text{otherwise} \\ +1 & \text{for} \quad y < W(x) \quad \text{and on} \quad S_+{}^1. \end{cases}$$

Now consider case (2), $b^2 - k^2 \geq 0$. Every solution of $\mathcal{S}_\pm$ leaves the critical point $0_\pm$ and crosses the x-axis once at most. Also each extremal response leading to $(0, 0)$ will have at most one switch. The solutions of $\mathcal{S}_+$ are easily obtained from the solutions of $\mathcal{S}_-$ of Example 1, case (2), upon the substitution $x \to -x$, $y \to y$, $t \to -t$; and the solutions of $\mathcal{S}_-$ are similarly obtained from $\mathcal{S}_+$ of Example 1, case (2).

Let S_+^1 be the arc of the solution of $\mathcal{S}_+$ leading from 0_+ to $(0, 0)$. Then the switching locus $y = W(x)$ is just S_+^1 on $x \geq 0$ and the corresponding

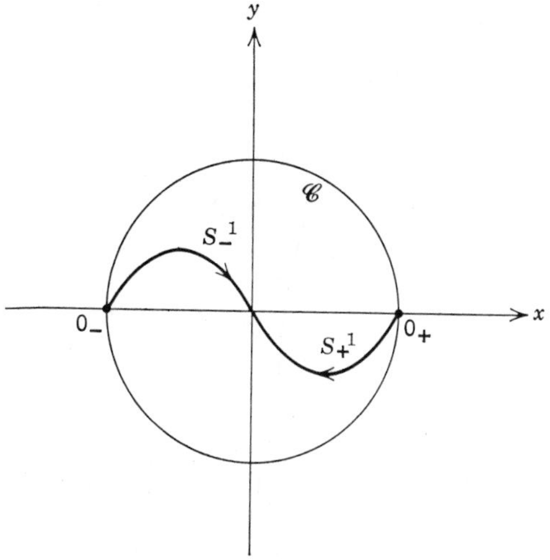

Figure 1.10 Minimal time-optimal control to origin. Switching locus diagram for $\ddot{x} - 2b\dot{x} + k^2x = u$, $|u(t)| \leq 1$. Case 2, Example 2: $b^2 - k^2 \geq 0$, $b > 0$, $k^2 > 0$.

arc S_-^1 of $\mathcal{S}_-$ on $x \leq 0$. The domain $\mathcal{C}$ of controllability is the open region bounded by the solution of $\mathcal{S}_+$ from 0_+ to 0_- in $y \leq 0$ and the solution of $\mathcal{S}_-$ from 0_- to 0_+ in $y \geq 0$ (see Figure 1.10). The synthesizer is defined in $\mathcal{C}$ as usual:

$$\Psi(x, y) = \begin{cases} -1 & \text{for} \quad y > W(x) \quad \text{and on} \quad S_-^1 \\ +1 & \text{for} \quad y < W(x) \quad \text{and on} \quad S_+^1. \end{cases}$$

Example* 3. *Control Under a Repulsive Force. Consider the synthesis for the minimal time-optimal controller for regulation to zero for the linear process

$$(\mathcal{L}) \qquad \ddot{x} + 2b\dot{x} - k^2x = u,$$

with constants b and $k^2 > 0$, and $|u(t)| \leq 1$. Again the optimal controller $u^*(t)$ on $0 \leq t \leq t^*$, steering any point (x_0, y_0) in the domain of controllability to $(0, 0)$, is the unique maximal controller steering (x_0, y_0) to $(0, 0)$,

and

$$u^*(t) = \operatorname{sgn} \eta_2(t).$$

The adjoint response $\eta_2(t) \not\equiv 0$ satisfies

$$\ddot{\eta}_2 - 2b\dot{\eta}_2 - k^2\eta_2 = 0,$$

and has at most one zero; for the most general solution is

$$\eta_2(t) = \alpha e^{bt} \sinh(\nu t + \beta)$$

where α and β are constants and $\nu = \sqrt{b^2 + k^2}$.
Thus $u^*(t)$ has at most one switch.

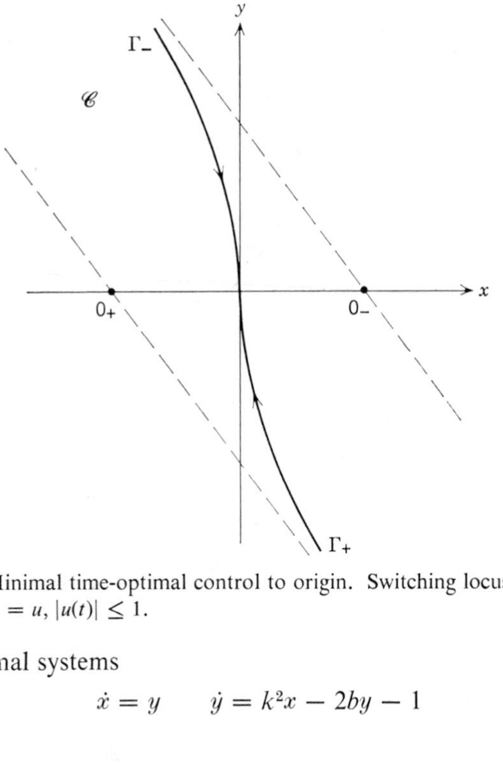

Figure 1.11 Minimal time-optimal control to origin. Switching locus for Example 3: $\ddot{x} + 2b\dot{x} - k^2x = u$, $|u(t)| \leq 1$.

The extremal systems

$(\mathcal{S}_-)$ $\qquad \dot{x} = y \qquad \dot{y} = k^2x - 2by - 1$

and

$(\mathcal{S}_+)$ $\qquad \dot{x} = y \qquad \dot{y} = k^2x - 2by + 1$

have corresponding critical points $0_-: x = 1/k^2$, $y = 0$ and $0_+: x = -1/k^2$, $y = 0$. The nature of $\mathcal{S}_-$ at 0_-, and of $\mathcal{S}_+$ at 0_+, is that of a saddle-point singularity.

The switching locus $y = W(x)$ is formed from exactly two arcs Γ_+ and Γ_-. Here Γ_+ is the solution of $\mathcal{S}_+$ through (0, 0) and lies in the fourth quadrant of the phase plane. Similarly Γ_- is the solution of $\mathcal{S}_-$ through (0, 0) and lies in the second quadrant (see Figure 1.11).

The domain $\mathcal{C}$ of controllability is the open band between the lines

$$y = (-b - \nu)\left(x - \frac{1}{k^2}\right)$$

and

$$y = (-b - \nu)\left(x + \frac{1}{k^2}\right).$$

The qualitative situation is the same for $b > 0$, $b = 0$, and $b < 0$.
The synthesizer is defined in $\mathcal{C}$ as usual

$$\Psi(x, y) = \begin{cases} -1 & \text{in} \quad y > W(x) \quad \text{and on} \quad \Gamma_- \\ +1 & \text{in} \quad y < W(x) \quad \text{and on} \quad \Gamma_+. \end{cases}$$

EXERCISES

1. Consider the control of a Hamiltonian system

$$\dot{x}^i = \frac{\partial H}{\partial y^i} + u^i$$

$$\dot{y}^i = -\frac{\partial H}{\partial x^i} + v^i \qquad i = 1, \ldots, n.$$

Here $(x^1, \ldots, x^n, y^1, \ldots, y^n) = (x, y)$ is a point in phase space R^{2n} and the Hamiltonian function $H(x, y)$ is given in class C^2 in R^{2n}. The control vector (u, v) satisfies the restraint

$$|u^i| \leq 1, \qquad |v^i| \leq 1.$$

Assume $H(x, y) > 0$ and $|\text{grad } H(x, y)| > 0$ in $R^{2n} - (0, 0)$; $H(0, 0) = 0$ and $\lim_{|x|+|y| \to \infty} H(x, y) = +\infty$. Show that every initial state (x_0, y_0) can be steered to a prescribed neighborhood of the origin.

2. Consider the nonlinear differential equation

$$\ddot{x} + f(x, \dot{x})\dot{x} + g(x) = 0$$

with $f(x, y)$ and $g(x)$ in C^1 in the phase plane R^2. Assume $f(x, y) \geq 0$ and $xg(x) > 0$ for $x \neq 0$.

(a) Show that if $G(x) = \int_0^x g(x)\, dx$ satisfies $\lim_{|x| \to \infty} G(x) = \infty$, then each energy locus

$$V(x, y) = \frac{y^2}{2} + G(x) = V_0 > 0$$

is a simple closed curve encircling the origin.

(b) Let $f(x, y) \equiv 0$ and $g(x) = xe^{-x^2}$ so that the nonlinear equation

$$\ddot{x} + xe^{-x^2} = 0$$

has the linear approximation

$$\ddot{x} + x = 0$$

near the origin. Show that the two differential equations are qualitatively different with respect to the global behavior of the solutions in the phase plane.

3. Consider the following control processes with the indicated targets and cost functionals. Show that, in each case, it is possible to steer the initial point to the target but that there does not exist any optimal controller. Contrast each example with the hypotheses of the existence theorem stated in Section 1.2.

(a) $\dot{x} = \sin 2\pi u$, $\dot{y} = \cos 2\pi u$, $\dot{z} = -1$ in R^3 with restraint $|u(t)| \leq 1$.

Steer $(0, 0, 1)$ to $(0, 0, 0)$ minimizing $C(u) = \int_0^t (x^2 + y^2)\, dt$.

Hint: construct controllers $u_k(t)$ on $0 \leq t \leq 1$ so that $\sin 2\pi u_k(t) = \sin 2\pi kt$, $\cos 2\pi u_k(t) = \cos 2\pi kt$ for $k = 1, 2, 3, \ldots$.

(b) $\dot{x} = u$, $\dot{y} = v$, $\dot{z} = -1$ in R^3 with restraint $u(t)^2 + v(t)^2 = 1$.

Steer $(0, 0, 1)$ to $(0, 0, 0)$ minimizing $C(u, v) = \int_0^t (x^2 + y^2)\, dt$.

(c) $\dot{x} = 1$, $\dot{y} = -xe^y u$ in R^2 with restraint $0 \leq u(t) \leq 2$. Steer $(-1, 0)$ to $(1, 0)$ minimizing $C(u) = \int_0^t (2 - y)\, dt = \int_{-1}^1 (2 - y)\, dx$.

[Hint: for each response $x(t) = t - 1$, $y(t)$ show $0 \leq y(x) \leq -\ln x^2$ for $x \neq 0$. Then $C(u) > \int_{-1}^1 (2 + \ln x^2)\, dx = 0$. Try controllers $u_\epsilon(t) = 2 - \epsilon$ for small $\epsilon > 0$.]

4. Consider the nonlinear control process

$$\ddot{x} - \dot{x}^2 - x^2 = u^2 \quad \text{with} \quad |u(t)| \leq 1$$

in the phase plane $x, y = \dot{x}$. Show that the domain of null controllability lies entirely within the fourth quadrant and hence fails to contain a neighborhood of the origin.

5. Write the linear control process with constant coefficients

$$x^{(n)} + a_1 x^{(n-1)} + \cdots + a_n x = u$$

as a first-order matrix system

$$(\mathcal{S}) \qquad \dot{x} = Ax + bu$$

by setting $x^1 = x$, $x^2 = \dot{x}, \ldots, x^n = x^{(n-1)}$. Show that $\mathcal{S}$ is controllable by using the theorems stated in Section 1.3.

6. Consider the nonlinear control process

$$x^{(n)} + f(x, \dot{x}, \ddot{x}, \ldots, x^{(n-1)}, u) = 0$$

with $f(x^1, x^2, x^3, \ldots, x^n, u)$ in C^1 in R^{n+1} and control restraint $|u(t)| \leq 1$. Assume $f(0, 0, \ldots, 0) = 0$ and $(\partial f/\partial u)(0, 0, \ldots, 0) \neq 0$. Using the theorems stated in Section 1.3 show that the domain of null controllability, for the corresponding first-order vector system in R^n, contains an open neighborhood of the origin.

7. Show that the process

$$(\mathcal{S}) \qquad \dot{x} = -x + u, \qquad \dot{y} = -2y$$

is not controllable in R^2 by a direct examination of the solution curves in the phase plane. Make the coordinate transformation

$$\bar{x} = 2x - y, \qquad \bar{y} = x - y$$

to obtain the corresponding system $\bar{\mathcal{S}}$. Check the algebraic condition of Section 1.3 for the controllability of the system $\bar{\mathcal{S}}$.

8. Consider Examples 1 and 2 of Section 1.2.

(a) Show that every initial point can be steered to the origin.

(b) In each case verify all the hypotheses of the existence theorem of Section 1.2 to show that minimal time-optimal control exists.

9. (a) Consider an autonomous nonlinear control process,

$$(\mathcal{S}) \qquad \dot{x} = f(x, u). \qquad \text{with } u(t) \in \Omega$$

steering x_0 to $x_1 = 0$ minimizing $C(u) = \int_0^t f^0(x(t), u(t))\,dt$, as in Section 1.4. State the maximal principle for an optimal controller $u(t)$ entirely in terms of the controller $u(t)$ and the responses $x(t)$ and $\eta(t)$.

(b) State the corresponding principle for a control $u(t)$ that maximizes the cost $C(u)$.

10. Consider the control process

$$\dot{x} = u - u^2$$

with restraint $|u(t)| \leq 1$. Show that the optimal controller $u(t)$ steering $x_0 = -1$ to $x_1 = 0$ in minimal time is $u(t) \equiv \frac{1}{2}$. Note that this optimal controller is not a relay that switches between $+1$ and -1.

APPENDIX 1. GEOMETRIC THEORY OF ORDINARY DIFFERENTIAL EQUATIONS

In Examples 2 and 3 of Section 1.1 we noted how to reduce the study of a second order differential equation, say,

$$\ddot{x} = f(t, x, \dot{x})$$

to the theory of two simultaneous first order differential equations, say

$$\dot{x} = y$$
$$\dot{y} = f(t, x, y).$$

In the same manner, higher-order real scalar differential equations can be studied by a corresponding first-order real vector differential equation, which is a special case of the general first-order vector differential system.

$$\begin{aligned} \dot{x}^1 &= f^1(t, x^1, \ldots, x^n) \\ \dot{x}^2 &= f^2(t, x^1, \ldots, x^n) \\ &\cdot \\ &\cdot \\ &\cdot \\ \dot{x}^n &= f^n(t, x^1, \ldots, x^n). \end{aligned}$$

We write such a vector differential system as

$$\dot{x}^i = f^i(t, x^1, \ldots, x^n) = f^i(t, x) \qquad i = 1, \ldots, n$$

or just

$$(\mathcal{S}) \qquad \dot{x} = f(t, x), \text{ or } \frac{dx}{dt} = f(t, x).$$

A solution

$$x(t) = \begin{pmatrix} x^1(t) \\ \cdot \\ \cdot \\ \cdot \\ x^n(t) \end{pmatrix}$$

is a column vector of real differentiable functions of t (called time) defined on some open interval $\mathfrak{J}$ and there satisfying the differential system

$$(\mathcal{S}) \qquad \dot{x}^i(t) = f^i(t, x(t)), \qquad i = 1, \ldots, n.$$

In this section we give a geometric interpretation of the vector differential system $\mathcal{S}$ as a vector field in the space R^n, the space of n real variables $(x^1, \ldots, x^n)$. We shall also list terminology and notation concerning scalar and vector functions and use these concepts to state general theorems on vector differential equations. The reader may wish to glance lightly through these definitions and summaries and then to return for more detailed inspection as these ideas arise in the theory of optimal control.

Notations for Sets in R^n, $n = 1, 2, 3, \ldots$

The space R^n is the space of all real n-tuples of coordinates $(x^1, \ldots, x^n)$ with $-\infty < x^i < \infty$ for $i = 1, \ldots n$. Thus R^1 is the real line and R^2 is the real plane. If a point or vector $x_0 = (x_0^1, \ldots, x_0^n)$ belongs to a subset A of R^n we write $x_0 \in A$. If every point of A lies in the subset B, that is, A is contained in B, we write $A \subset B$. The set of points in set A_1 but not in A_2 is the difference $A_1 - A_2$; if $A_1 \subset A_2$ this is the empty set that contains no points. The intersection $A_1 \cap A_2$ is the set of points in both A_1 and A_2, and the union $A_1 \cup A_2$ is the set of points that lie in either A_1 or A_2 (or both); the intersection and union of many sets is defined similarily.

For sets $A \subset R^n$ and $B \subset R^m$ we define the product set $A \times B \subset R^{n+m}$ as the set of all points (x, y) with $x \in A$ and $y \in B$.

The concepts and notations $P_0 \in S$, $A \subset B$, $A_1 - A_2$, $A_1 \cap A_2$, $A_1 \cup A_2$ and $A \times B$ are defined for subsets of more general spaces in the same way.

Notations for geometry in R^n

For the point or vector $x_0 = (x_0^1, \ldots, x^n) \in R^n$ we shall use the norm (different than Euclidean length)

$$|x_0| = |x_0^1| + |x_0^2| + \cdots + |x_0^n|.$$

With the distance $d(x_0, y_0) = |x_0 - y_0|$ the set R^n becomes a metric space, that is, a space with prescribed real valued distance function satisfying the axioms

1. $d(x_0, y_0) > 0$ if $y_0 \neq x_0$ and $d(x_0, x_0) = 0$;
2. $d(x_0, y_0) = d(y_0, x_0)$;
3. $d(x_0, z_0) \leq d(x_0, y_0) + d(y_0, z_0)$.

A set $\mathcal{O} \subset R^n$ is called open if, for each point $x_0 \in \mathcal{O}$, there is a number $r > 0$ such that the set $\{x \mid x \in R^n \text{ and } |x - x_0| < r\}$ lies entirely in $\mathcal{O}$. That is, $\mathcal{O}$ contains the set of points $x \in R^n$ satisfying the indicated condition (a replacement of the norm $|x - x_0|$ by the Euclidean length of $x - x_0$ would define precisely the same sets of R^n to be open). A set $C \subset R^n$ is called closed if $R^n - C$ is open in R^n. An open set $\mathcal{O} \subset R^n$ contains none of its boundary points, whereas a closed set $C \subset R^n$ contains all its boundary points. Any union of open sets is itself open in R^n, and any intersection of closed sets is closed in R^n. The union of all open sets contained in a given set $A \subset R^n$ is called the interior of A. A set $N \subset R^n$ that contains A in its interior is called a neighborhood of A. The intersection of all closed sets of R^n, each of which contains A, is called $\bar{A}$, the closure of A in R^n. A point P belongs to the boundary ∂A of a set A just in case every neighborhood of P meets A and the complement $R^n - A$. Similar definitions and properties of open and closed sets holds in any metric space.

A set $K \subset R^n$ is called compact in case K is closed and bounded in R^n (that is, K is closed and $|x|$ is bounded for $x \in K$). The distance of a point

$P \in R^n$ to a compact set $K \subset R^n$ is the shortest Euclidean distance from P to a point of K. A set $A \subset R^n$ is called convex if, for each pair of points x_0 and x_1 in A, the entire segment $\mu x_0 + (1 - \mu)x_1$, for $0 \le \mu \le 1$, lies in A (here the linear combination of vectors is computed componentwise). An open set $\mathcal{O} \subset R^n$ is called connected if any two points in $\mathcal{O}$ can be joined by a continuous curve that lies in $\mathcal{O}$. Note that R^n itself is open, closed, convex, and connected in R^n, but not compact.

The vector differential system

$$(\mathcal{S}) \qquad \dot{x}^i = f^i(t, x^1, \ldots, x^n) \qquad i = 1, 2, \ldots, n$$

can be interpreted geometrically as a vector field, with components $f^i(t, x)$ at each instant t, in the space R^n. We call R^n, or the subset of R^n where $\mathcal{S}$ is defined, the phase space of $\mathcal{S}$. A solution curve of $\mathcal{S}$ in R^n is a time-parametrized curve $x(t) = (x^i(t))$ whose tangent vector or velocity vector $\dot{x}(t) = (\dot{x}^i(t))$ coincides with the vector $f(t, x(t))$ of the field $\mathcal{S}$.

If $f(t, x) = f(x)$ is independent of t, the differential system $\mathcal{S}$ is called autonomous. In this case the vector field of $\mathcal{S}$ can be interpreted as the velocity field of a steady-state fluid flow in R^n. If $f(x_0) = 0$, then x_0 is a critical point (or equilibrium point) of the autonomous system $\mathcal{S}$ and $x(t) = x_0$ is a solution whose graph in phase space is just one point. A periodic solution $x(t) = x(t + P)$ for some constant period $P > 0$ describes a simple closed curve in phase space. This geometric language is often useful for analyzing the qualitative behavior of the solutions of $\mathcal{S}$.

In order to state precisely the fundamental theorems on the existence, uniqueness, and regularity of solutions of $\mathcal{S}$, we need to define certain ideas of continuity and differentiability.

Notations for Continuity and Differentiability in R^n

The vector function $f(x)$ with values in R^n is called continuous in $A \subset R^n$ in case each real component $f^i(x^1, \ldots, x^n)$ is continuous in A. Also $f(x)$ is in class C^k for $k = 1, 2, 3, \ldots$, in an open set $\mathcal{O} \subset R^n$ if each real component $f^i(x^1, \ldots, x^n)$ is continuous and has continuous partial derivatives of all orders $\le k$ in $\mathcal{O}$. If $f(x)$ is in all C^k we write $f(x) \in C^\infty$ in $\mathcal{O}$. If $f(x)$ is analytic in $\mathcal{O}$, that is, each $f^i(x^1, \ldots, x)$ has an absolutely convergent power-series expansion about each point $x \in \mathcal{O}$, then it follows that $f(x) \in C^\infty$. As noted earlier a vector function $x(t)$ with components $(x^1(t), \ldots, x^n(t))$ for $t \in R^1$ has the derivative $\dot{x}(t)$ with components $(\dot{x}^1(t), \ldots, \dot{x}^n(t))$.

Consider the real vector differential system

$$(\mathcal{S}) \qquad \dot{x}^i = f^i(t, x^1, \ldots, x^n) \qquad i = 1, 2, \ldots, n$$

where $f(t, x)$ is defined and continuous in $\mathfrak{J} \times \mathcal{O} \subset R^{1+n}$ where $\mathfrak{J} \subset R^1$ is an open interval and $\mathcal{O}$ is open in R^n. Assume the functions $(\partial f^i/\partial x^j)(t, x)$ are continuous in $\mathcal{I} \times \mathcal{O}$. Then for each initial point $(t_0, x_0) \in \mathcal{I} \times \mathcal{O}$ there exists a unique solution of $\mathcal{S}$

$$x = \phi(t, t_0, x_0)$$

through the given initial point at time t_0,

$$\phi(t_0, t_0, x_0) = x_0,$$

and the solution is defined in $\mathcal{I} \times \mathcal{O}$ for some maximal time duration $\tau_-(t_0, x_0) < t < \tau_+(t_0, x_0)$ whereon it satisfies the differential system $\mathcal{S}$. This fundamental existence and uniqueness theorem is proved in texts on the theory of ordinary differential equations.

We shall utilize a stronger existence theorem than that described above since we shall have occasion to study differential systems in which the coefficients $f(t, x)$ are discontinuous functions of t. These discontinuities are important in optimal-control theory since they describe the instantaneous switches that occur in optimal relay controllers. We shall always assume that $f^i(t, x)$ is measurable in t for each fixed x. The measurable functions of a real variable t form an extremely general class of functions that includes all continuous functions, piecewise continuous functions, and even all limits of such functions. It is difficult to give an example of a nonmeasurable function and we shall encounter only measurable functions (usually piecewise continuous) in this work. We give here a brief summary of the basic concepts of measurable and Lebesgue integrable functions.

Notations for Measurable and Integrable Functions on R^n, and Weak Compactness

A subset N of R^n is called a null set (or set of measure zero) in case N can be covered by (is contained in) a countable union of n-cubes whose total n-volume is less than an arbitrarily prescribed number $\epsilon > 0$. For example, any finite or countable infinite set of points in R^n has measure zero. Two functions $f_1(x)$ and $f_2(x)$ defined on $A \subset R^n$ that differ in value only on a null set are said to be equal almost everywhere on A. The measurable sets of R^n are defined as the members of the smallest family of sets of R^n that contains all open sets, all closed sets, all null sets of R^n, and also every difference, and countable union, and countable intersection of its members. A real-valued function $h(t)$ on a real interval $\mathcal{I}$ is called measurable in case for all real α and β, the set $\{t \mid t \in \mathcal{I} \text{ and } \alpha < h(t) < \beta\}$ is measurable in R^1. If $h(t)$ is measurable on $\mathcal{I}$, there exists a closed subset $\mathcal{C}$ on which $h(t)$ is continuous, and we can require that the measure of $\mathcal{I}$-$\mathcal{C}$ be arbitrarily small.

If $h(t)$ is measurable on $\mathfrak{J}$ then we can define the Lebesgue integral

$$\int_{\mathfrak{J}} h(t)\, dt$$

by considering appropriate limits of approximating sums. The function $h(t)$ is called integrable on $\mathfrak{J}$ in case the above integral, as well as the integral of $|h(t)|$, is a finite real number. A change of the values of $h(t)$ on a null set does not effect the value of the integral. If $h(t)$ is piecewise continuous and $\mathfrak{J}$ is compact the value of the Lebesgue integral of $h(t)$ is the same as that of the usual Riemann integral.

Let $h(t)$ be integrable on the interval $\mathfrak{J} = (t_0, t_1)$ and consider

$$H(t) = \int_{t_0}^{t} h(s)\, ds \qquad \text{for} \quad t_0 \le t \le t_1.$$

Such an indefinite integral $H(t)$ defines an absolutely continuous function. An absolutely continuous function can be proved to be continuous and to have a derivative almost everywhere (that is, everywhere on $\mathfrak{J}$ excepting a null set) and there

$$\frac{d}{dt} H(t) = h(t).$$

Every Lipschitz continuous function $H(t)$ is absolutely continuous.

Thus the main novelty arising from the discontinuities in $f(t, x)$ is that while the usual formulas of differential and integral calculus are valid, there is an occassional warning that certain equalities may hold only "almost everywhere." The reader who is willing to grant the correctness of this extension of the calculus can proceed with the motto, "Damn the null sets, full differentiability ahead."

In proving the existence of certain optimal controllers we shall later require certain sequences of controls to converge. A sequence $\{u_n(t)\}$ $n = 1, 2, 3, \ldots$ of real (or vector valued) integrable functions on a real interval $\mathfrak{J}$ is called weakly convergent to $u^*(t)$ in case, for each bounded measurable test function $g(t)$, we have

$$\lim_{n\to\infty} \int_{\mathfrak{J}} g(t)\, u_n(t)\, dt = \int_{\mathfrak{J}} g(t)\, u^*(t)\, dt.$$

The collection of all vector functions $u(t)$ which are measurable on a given finite interval $\mathfrak{J}$ and have values in a given compact convex set $\Omega \subset R^m$ is known to be (sequentially) weakly compact. That is, each such sequence of functions has a subsequence that converges weakly on $\mathfrak{J}$ to a function of the given collection. Of course, the limit function $u^*(t)$ is only determined almost everywhere on $\mathfrak{J}$.

The set of all real functions $u(t)$ on an interval $\mathfrak{J}$ for which

$$\int_{\mathfrak{J}} |u(t)|^p \, dt < \infty, \text{ for a given } 1 \leq p < \infty,$$

defines the space L_p. With the norm $\|u\|_p = \left(\int_{\mathfrak{J}} |u(t)|^p \, dt\right)^{1/p}$ the vector space L_p acquires a complete metric and L_p is a Banach space (that is, upon identifying functions differing only on null sets, L_p becomes a complete normed vector space). For $p = \infty$ we define the Banach space L_∞ of all essentially (on $\mathfrak{J}$ minus a null set) bounded measurable functions with the norm $\|u\|_\infty = \text{ess. sup } |u(t)|$. If $\mathfrak{J}$ is compact, then $L_p \subset L_1$ for $1 \leq p \leq \infty$.

A closed ball in L_p for $1 < p \leq \infty$ (that is, the set $|u|_p \leq B$) is weakly (often called weak*) sequentially compact; and, in fact, the appropriate limits of integrals are valid for each test function $g(t) \in L_q$ on $\mathfrak{J}$, where $1/p + 1/q = 1$ and $1/\infty = 0$. The case $p = q = 2$ is of special interest, and L_2 is called Hilbert space.

A vector (or matrix) $u(t)$ is in L_p, $1 \leq p \leq \infty$, on $\mathfrak{J}$ just in case each component $u^i(t)$ is in L_p. This obtains if and only if $\int_{\mathfrak{J}} |u(t)|^p \, dt < \infty$.

We now state the basic existence, uniqueness, and regularity theorem for real vector differential systems. For proofs and related discussion see the text on integration [McShane].

THEOREM 1A. *Consider the real vector differential system*

$$(\mathcal{S}) \qquad \dot{x}^i = f^i(t, x^1, \ldots, x^n) \qquad i = 1, \ldots, n$$

where $f(t, x)$ is defined in an open set $\mathfrak{J} \times \mathcal{O} \subset R^{1+n}$.

Assume

(a) *For each fixed $t \in \mathfrak{J}$ the functions $f^i(t, x)$ are in class C^1 for $x \in \mathcal{O}$;*
(b) *For each fixed $x \in \mathcal{O}$ the functions $f^i(t, x)$ are measurable in t in $\mathfrak{J}$;*
(c) *Given any compact sets $\mathfrak{J}_c \subset \mathfrak{J}$ and $K \subset \mathcal{O}$ there exists an integrable function $m(t)$ on $\mathfrak{J}_c$ such that*

$$|f(t, x)| \leq m(t) \qquad \text{and} \qquad \left|\frac{\partial f}{\partial x}(t, x)\right| \leq m(t)$$

for all $(t, x) \in \mathfrak{J}_c \times K$.

Then for each initial point (t_0, x_0) in $\mathfrak{J} \times \mathcal{O}$ there exists a unique solution of $\mathcal{S}$

$$x = \phi(t, t_0, x_0)$$

with

$$\phi(t_0, t_0, x_0) = x_0,$$

defined for a maximal time duration

$$\tau_-(t_0, x_0) < t < \tau_+(t_0, x_0).$$

Moreover the vector function $\phi(t, t_0, x_0)$ *is then defined and continuous in an open set* $D \subset R^{1+1+n}$. *For each fixed* (t_0, x_0) *the function* $\phi(t, t_0, x_0)$ *is absolutely continuous in t and satisfies the vector differential equation*

$$\frac{d\phi(t, t_0, x_0)}{dt} = f(t, \phi(t, t_0, x_0))$$

almost everywhere on $\tau_- < t < \tau_+$. *For each fixed* (t, t_0) *the function* $\phi(t, t_0, x_0)$ *is in class* C^1 *in* x_0 *and the vector*

$$\frac{\partial\phi(t, t_0, x_0)}{\partial x_0{}^j}, \quad \textit{for each} \quad j = 1, 2, \ldots n,$$

is absolutely continuous in t and satisfies the linear differential system

$$\frac{d}{dt}\left(\frac{\partial \phi^i}{\partial x_0{}^j}\right) = \sum_{k=1}^{n} \frac{\partial f^i}{\partial x^k}(t, \phi(t, t_0, x_0))\left(\frac{\partial \phi^k}{\partial x_0{}^j}\right).$$

This basic theorem has a number of extensions and modifications that occasionally arise. We list these further results as a series of remarks.

Remarks. 1. Assume the coefficients of $\mathcal{S}$ are $f(t, x, \lambda)$, where λ or $(\lambda^1, \ldots, \lambda^m)$ is a real vector parameter in R^m. If $f(t, x, \lambda)$ is defined in an open set $\mathfrak{I} \times \mathcal{O} \times \Lambda \subset R^{1+n+m}$ and if for each $\lambda_0 \in \Lambda$ hypotheses (a), (b), and (c) of the theorem hold, then there exists a solution of $\mathcal{S}$

$$x = \phi(t, t_0, x_0, \lambda_0)$$

through the initial point (t_0, x_0) for the parameter value λ_0. Further if hypothesis (a) and (c) are strengthened to state that

(a′) For each fixed $t \in \mathfrak{I}$, the functions $f^i(t, x, \lambda)$ are in class C^k $(k = 1, 2, 3, \ldots)$ in (x, λ) in $\mathcal{O} \times \Lambda$ and

(c′) Given compact sets $\mathfrak{I}_c \subset \mathfrak{I}$, $K \subset \mathcal{O}$, $L \subset \Lambda$ there exists an integrable function $m(t)$ on $\mathfrak{I}_c$ such that

$$|Df(t, x, \lambda)| \leq m(t) \quad \text{for all} \quad (t, x, \lambda) \in \mathfrak{I}_c \times K \times L$$

for every partial differentiation D of order $\leq k$ with respect to (x, λ), then $\phi(t, t_0, x_0, \lambda)$ is continuous in an open set $\mathcal{O}' \subset R^{1+1+n+m}$ and is in class C^k in (x_0, λ).

2. If $f(t, x, \lambda)$ is continuous in $\mathfrak{J} \times \mathcal{O} \times \Lambda \subset R^{1+n+m}$ and satsifies hypotheses (a′) and (c′), then hypothesis (b) automatically holds. In this case the solution $\phi(t, t_0, x_0, \lambda)$ is in class C^1 in $\mathcal{O}' \subset R^{1+1+n+m}$ and satisfies the differential system at every point on the interval $\tau_- < t < \tau_+$.

3. In many problems we can conclude that the maximal duration for solutions is given by $\tau_+ = +\infty$. If $\mathfrak{J} = (t_0, +\infty)$ and if the solution through (t_0, x_0) lies in a compact subset $K \subset \mathcal{O} \subset R^n$ for $t \geq t_0$, then it can be shown that $\tau_+(t_0, x_0) = +\infty$.

APPENDIX 2. ALGEBRAIC THEORY OF LINEAR DIFFERENTIAL EQUATIONS

In Examples 2 and 3 of Section 1 we studied linear control problems described by second order differential equations and we reduced the problems to linear first order vector differential systems in the phase plane. Higher order linear processes lead to more extensive linear vector differential systems for which the general process is

$$
\begin{aligned}
\dot{x}^1 &= a_1{}^1(t)x^1 + a_2{}^1(t)x^2 + \cdots + a_n{}^1(t)x^n + b_1{}^1(t)u^1 + \cdots + b_m{}^1(t)u^m \\
\dot{x}^2 &= a_1{}^2(t)x^1 + a_2{}^2(t)x^2 + \cdots + a_n{}^2(t)x^n + b_1{}^2(t)u^1 + \cdots + b_m{}^2(t)u^m \\
&\;\;\vdots \\
\dot{x}^n &= a_1{}^n(t)x^1 + a_2{}^n(t)x^2 + \cdots + a_n{}^n(t)x^n + b_1{}^n(t)u^1 + \cdots + b_m{}^n(t)u^m.
\end{aligned}
$$

This linear differential system corresponds to a physical system with m input or control variables

$$
u = \begin{pmatrix} u^1 \\ u^2 \\ \cdot \\ \cdot \\ \cdot \\ u^m \end{pmatrix}
$$

and n output or response variables

$$
x = \begin{pmatrix} x^1 \\ x^2 \\ \cdot \\ \cdot \\ \cdot \\ x^n \end{pmatrix}.
$$

The plant or process is described by the given coefficients, possibly time dependent, which we write as matrix arrays

$$A(t) = \begin{pmatrix} a_1^{\ 1}(t) & \cdots & a_n^{\ 1}(t) \\ a_1^{\ 2}(t) & \cdots & a_n^{\ 2}(t) \\ \cdot & & \\ \cdot & & \\ \cdot & & \\ a_1^{\ n}(t) & \cdots & a_n^{\ n}(t) \end{pmatrix} \quad \text{and} \quad B(t) = \begin{pmatrix} b_1^{\ 1}(t) & \cdots & b_m^{\ 1}(t) \\ b_1^{\ 2}(t) & \cdots & b_m^{\ 2}(t) \\ \cdot & & \\ \cdot & & \\ \cdot & & \\ b_1^{\ n}(t) & \cdots & b_m^{\ n}(t) \end{pmatrix}$$

The matrix $A(t) = (a_j^{\ i}(t))$, with $a_j^{\ i}(t)$ in the ith row and the jth column, is an $n \times n$ matrix and $B(t) = (b_k^{\ j}(t))$ is an $n \times m$ matrix having n rows and m columns.

The linear control process

$$\dot{x}^i = \sum_{j=1}^{n} a_j^{\ i}(t)x^j + \sum_{k=1}^{m} b_k^{\ i}(t)u^k \quad \text{for} \quad i = 1, 2, \ldots, n$$

(or, designating summation by repeating indices,

$$\dot{x}^i = a_j^{\ i}(t)x^j + b_k^{\ i}(t)u^k \qquad i = 1, 2, \ldots, n)$$

can be written in matrix notation as

$$(\mathfrak{L}) \qquad \dot{x} = A(t)x + B(t)u.$$

The sums and products of matrices and vectors are defined so that $\mathfrak{L}$ can be written in this convenient notation. We tabulate here the main facts and formulas of matrix algebra and matrix calculus that will be used in the theory of optimal control of linear systems. The reader may wish to refer back to this tabulation as these concepts arise in the subsequent analysis.

Algebra of Constant Matrices

A matrix is a rectangular array of components that can be numbers or functions. Vectors and scalars are special types of matrices.

The rank of a constant $n \times m$ matrix is the maximal number of linearly independent rows (which is same as the number of linearly independent columns). Let $F = (f_j^{\ i})$ and $G = (g_j^{\ i})$ be $n \times m$ matrices with real constant entries or components. For given real numbers α and β define the linear combination $\alpha F + \beta G$ to be the $n \times m$ matrix with entries $(\alpha f_j^{\ i} + \beta g_j^{\ i})$ in the ith row and the jth column. Let $H = (h_k^{\ j})$ be an $m \times r$ matrix and define the product GH to be the $n \times r$ matrix with entry $\sum_{j=1}^{m} g_j^{\ i} h_k^{\ j}$ in the ith row and the kth column. Note that in general $GH \neq HG$ even when both G and H are $n \times n$ matrices so the products can be defined.

This "row by column" multiplication rule allows us to write

$$(a_j^{\ i}x^j) = Ax \quad \text{and} \quad (b_k^{\ i}u^k) = Bu,$$

as in our linear control problem $\mathfrak{L}$.

Let $A = (a_j^{\ i})$ be a square $n \times n$ matrix. The transpose A' of A is the $n \times n$ matrix with entry $a_j^{\ i}$ in the jth row and ith column. That is, A' is obtained from A by interchanging the components by a reflection in the principal diagonal. Clearly $(A')' = A$. If $A' = A$ then A is called symmetric. If x is a column vector, then x' denotes a row vector. If A is a real symmetric matrix and $x'Ax > 0$ (or ≥ 0) for all real vectors $x \neq 0$, then A is called positive definite (or semidefinite).

Let A be a square $n \times n$ matrix and assume that for every vector y there exists exactly one vector x such that $y = Ax$. Then A is called nonsingular and we write the solution for x in terms of y by $x = A^{-1}y$. The inverse matrix A^{-1}, which exists just in case the determinant of A is not zero, can be computed by solving for x in terms of y in $y = Ax$ by the usual processes of elimination. Clearly $(A^{-1})^{-1} = A$.

The following rules of matrix algebra can be proved for all real constant $n \times m$ matrices F, G, and H:

1. $(F + G) + H = F + (G + H)$
2. $F + G = G + F$
3. $F + (-F) = 0$ or $F - F = 0$, where 0 is the $n \times m$ matrix with all entries zero and $-F = (-1)F$.
4. $F + 0 = F$
5. $(\alpha\beta)F = \alpha(\beta F)$ for numbers α, β
6. $\alpha(F + G) = \alpha F + \alpha G$
7. $(\alpha + \beta)F = \alpha F + \beta F$
8. $1F = F$, $0F = 0$ for the number 0 and the zero matrix 0.

For matrices of the correct sizes, so that the products are defined, we have the following laws:

9. $F(\alpha G + \beta H) = \alpha FG + \beta FH$ and $(\alpha G + \beta H)F = \alpha GF + \beta HF$
10. $(FG)H = F(GH)$

For square $n \times n$ real matrices we have the laws:

11. $IA = AI = A$, $0A = A0 = 0$
    where I is the identity matrix with entires $+1$ on the principal diagonal and zeros elsewhere, that is $I = (\delta_j^{\ i})$ where

$$\delta_j^{\ i} = \begin{cases} 1 & \text{if} \quad i = j \\ 0 & \text{if} \quad i \neq j. \end{cases}$$

12. $(A + B)' = A' + B'$
13. $(AB)' = B'A'$

For $n \times n$ nonsingular matrices we have

14. $(AB)^{-1} = B^{-1}A^{-1}$ and $AA^{-1} = A^{-1}A = I$
15. $(A^{-1})' = (A')^{-1}$

Let A be an $n \times n$ matrix and define the n eigenvalues or characteristic values $\{\lambda_1, \lambda_2, \ldots, \lambda_n\}$ of A to be the roots, counting multiplicities, of the determinantal or characteristic polynomial equation

$$\det (A - \lambda I) = 0.$$

If A is real symmetric, all eigenvalues of A are real and there exists a real nonsingular matrix P so that

$$PAP^{-1} = \operatorname{diag} \{\lambda_1, \lambda_2, \ldots, \lambda_n\} = \begin{pmatrix} \lambda_1 & & & & \mathbf{0} \\ & \lambda_2 & & & \\ & & \cdot & & \\ & & & \cdot & \\ & & & & \cdot \\ \mathbf{0} & & & & \lambda_n \end{pmatrix}$$

If A has distinct complex eigenvalues, then there exists a nonsingular complex matrix P so that PAP^{-1} is diagonal, as above.

For matrices with complex entries, or with entries which are complex valued functions of time evaluated at some fixed time instant t, the same algebraic laws (1) to (15) still hold.

Calculus of Matrices

For a real or complex $n \times m$ matrix $A = (a_j{}^i)$ define the norm $|A| = \sum_{i,j} |a_j{}^i|$. Then it is easy to verify

1. $|\alpha A| \leq |\alpha|\, |A|$ for numbers α
2. $|A + B| \leq |A| + |B|$
3. $|AB| \leq |A|\, |B|$

where the matrices are of the correct sizes so that the sums and products are defined. In particular, for a vector x, with $|x| = \sum_{i=1}^{m} |x^i|$ we have,

4. $|Ax| \leq |A|\, |x|$.

If $A(t) = (a_j{}^i(t))$ has entries that are functions of t on some interval $\mathfrak{J}$, then define

$$\int_{\mathfrak{J}} A(t)\, dt \qquad \text{and} \qquad \frac{d}{dt} A(t)$$

to be the matrices with entries $\int_{\mathfrak{J}} a_j^i(t)\,dt$ and $\dot{a}_j^i(t)$, respectively. Thus $A(t)$ is measurable, integrable, continuous, absolutely continuous, or in C^k just in case each $a_j^i(t)$ has the corresponding property. Also

5. $\left|\int_{\mathfrak{J}} A(t)\,dt\right| \le \int_{\mathfrak{J}} |A(t)|\,dt$

6. $\left|\dfrac{d}{dt}|A(t)|\right| \le \left|\dfrac{dA}{dt}\right|,$

wherever these exist.

For a square real or complex $n \times n$ matrix A we define the $n \times n$ matrix

$$\exp A = e^A = I + A + \frac{A^2}{2!} + \frac{A^3}{3!} + \cdots + \frac{A^k}{k!} + \cdots$$

Here the convergence of a sequence or a series of matrices is defined by the convergence in each component. For every $n \times n$ matrix A the series for the exponential function converges and defines e^A. Using power series techniques one can prove the following laws:

7. $e^{-A} = (e^A)^{-1}$ and $e^0 = I$
8. $e^A \cdot e^B = e^{A+B}$ whenever $AB = BA$
9. $\dfrac{d}{dt}e^{At} = Ae^{At} = e^{At}A$
10. $\exp(PAP^{-1}) = Pe^AP^{-1}$ and

$$\exp\{P \operatorname{diag}(\lambda_1, \ldots, \lambda_n)P^{-1}\} = P \operatorname{diag}\{e^{\lambda_1}, \ldots, e^{\lambda_n}\}P^{-1}.$$

If each eigenvalue λ_i of A satisfies the bound $\operatorname{Re}\lambda_i < \lambda$, then the following important estimate holds

11. $|e^{At}| \le ce^{\lambda t}$ on $0 \le t < \infty$

for some constant $c > 0$. If all eigenvalues of A satisfy $\operatorname{Re}\lambda_i < 0$, then A is called a stability matrix and $|e^{At}| \to 0$ as $t \to +\infty$.

We now use these laws of matrix algebra and calculus to describe the general theory of linear differential systems. Consider the first-order, linear, homogeneous, differential system

$$\dot{x}^i = a_j^i(t)x^j \qquad i = 1, \ldots, n$$

or, in matrix notation,

$$\dot{x} = A(t)x.$$

Here $A(t) = (a_j^i(t))$ is a $n \times n$ real or complex matrix function of time t on some real interval $\mathfrak{J}$. A set of n solutions, or column vectors,

$$x = \phi_1(t) = \begin{pmatrix} \phi_1^1(t) \\ \cdot \\ \cdot \\ \cdot \\ \phi_1^n(t) \end{pmatrix}, \ldots, x = \phi_n(t) = \begin{pmatrix} \phi_n^1(t) \\ \cdot \\ \cdot \\ \cdot \\ \phi_n^n(t) \end{pmatrix}$$

defines a fundamental set or basis of solutions on $\mathfrak{J}$ in case the $n \times n$ matrix

$$\Phi(t) = \begin{pmatrix} \phi_1^1(t) & \cdots & \phi_n^1(t) \\ \cdot & & \\ \cdot & & \\ \cdot & & \\ \phi_1^n(t) & & \phi_n^n(t) \end{pmatrix}$$

is nonsingular everywhere on $\mathfrak{J}$. Thus the columns of an $n \times n$ matrix $X(t)$ form a basis of solutions for

$$\dot{x} = A(t)x \quad \text{on} \quad \mathfrak{J}$$

just in case $X(t)$ is a fundamental matrix solution, that is,

$$\frac{d}{dt} X(t) = A(t)X(t),$$

and

$$\det X(t) \neq 0 \quad \text{on} \quad \mathfrak{J}.$$

If $A(t)$ is integrable on each compact subinterval of $\mathfrak{J}$, then for given initial data $t_0 \in \mathfrak{J}$ and x_0 there exists a unique, absolutely continuous, fundamental solution matrix $\Phi(t)$ (or $\Phi(t, t_0)$) on $\mathfrak{J}$ with $\Phi(t_0) = I$. The solution of

$$\dot{x} = A(t)x \quad \text{with} \quad x(t_0) = x_0$$

is

$$x(t) = \Phi(t)x_0,$$

and this is real if $A(t)$ and x_0 are real. If $x_0 = 0$, then $x(t) \equiv 0$ on $\mathfrak{J}$ and hence a *nontrivial* solution $x(t)$ vanishes nowhere on $\mathfrak{J}$.

If $A(t) = A$ is constant, then the fundamental solution matrix, which is the identity at t_0, is

$$\Phi(t) = e^{A(t-t_0)}.$$

Now consider the nonhomogeneous linear differential system

$$\dot{x} = A(t)x + b(t)$$

where the $n \times n$ matrix $A(t)$ and the n-column vector $b(t)$ are integrable on each compact subinterval of a given real interval $\mathfrak{J}$. Let $t_0 \in \mathfrak{J}$ and x_0 be prescribed initial data. Then there exists a unique solution $x(t)$ on $\mathfrak{J}$ with $x(t_0) = x_0$. This solution $x(t)$ is given by the fundamental formula (called the *variation of parameters* formula for historical reasons)

$$x(t) = \Phi(t)x_0 + \Phi(t)\int_{t_0}^{t} \Phi(s)^{-1}\, b(s)\, ds.$$

Here $\Phi(t)$ is the fundamental matrix solution of the corresponding homogeneous system

$$\dot{x} = A(t)x$$

with $\Phi(t_0) = I$. A direct calculation verifies that the variation of parameters formula yields the required solution. For compute

$$\begin{aligned} \frac{d}{dt}\, x(t) &= A(t)\left[\Phi(t)x_0 + \Phi(t)\int_{t_0}^{t} \Phi(s)^{-1}\, b(s)\, ds\right] + \Phi(t)\,\Phi(t)^{-1}\, b(t) \\ &= A(t)\, x(t) + b(t). \end{aligned}$$

The initial value is also as prescribed,

$$x(t_0) = \Phi(t_0)x_0 = Ix_0 = x_0.$$

If $A(t) = A$ is constant, and say $t_0 = 0$, then

$$\Phi(t) = e^{At}$$

and the fundamental variation of parameter formula is

$$x(t) = e^{At}x_0 + \int_0^t e^{A(t-s)}\, b(s)\, ds.$$

Most estimates of the qualitative behavior of the solution $x(t)$, as in control theory or stability theory, start with an analysis of the fundamental formula of variation of parameters.

The following simple example illustrates the use of matrix theory in solving linear differential systems.

Example

Consider the damped harmonic oscillator

$$\ddot{x} + 2b\dot{x} + k^2x = f(t)$$

where b, k are real constants, and $f(t)$ is a real function which is integrable on every compact interval. Assume $k > 0$ and $k^2 > b^2$.

Consider the homogeneous system

$$\begin{pmatrix} \dot{x} \\ \dot{y} \end{pmatrix} = A \begin{pmatrix} x \\ y \end{pmatrix}$$

where

$$A = \begin{pmatrix} 0 & 1 \\ -k^2 & -2b \end{pmatrix}.$$

A fundamental matrix solution of the homogeneous system can be verified to be

$$e^{At} =$$

$$\frac{k}{\omega} e^{-bt} \begin{pmatrix} \sin(\omega t + \alpha) & \frac{1}{k} \sin \omega t \\ -b \sin(\omega t + \alpha) + \omega \cos(\omega t + \alpha), & \frac{-b}{k} \sin \omega t + \frac{\omega}{k} \cos \omega t \end{pmatrix}$$

where $\omega = \sqrt{k^2 - b^2}$, $\sin \alpha = \omega/k$, $\cos \alpha = b/k$. This matrix solution is most easily computed by utilizing elementary methods of solving

$$\ddot{x} + 2b\dot{x} + k^2 x = 0$$

with initial conditions $x(0) = 1$, $\dot{x}(0) = 0$ or $x(0) = 0$, $\dot{x}(0) = 1$, rather than by power-series methods.

The solution of the nonhomogeneous equation with initial data (x_0, y_0) at $t = 0$ is then

$$\begin{pmatrix} x(t) \\ \dot{x}(t) \end{pmatrix} = e^{At} \begin{pmatrix} x_0 \\ y_0 \end{pmatrix} + \int_0^t e^{A(t-s)} \begin{pmatrix} 0 \\ f(s) \end{pmatrix} ds.$$

The first component of this vector solution is

$$x(t) = \frac{k}{\omega} e^{-bt} \left[x_0 \sin(\omega t + \alpha) + \frac{y_0}{k} \sin \omega t \right] + \frac{1}{\omega} \int_0^t f(s) e^{-b(t-s)} \sin \omega(t - s)\, ds,$$

which is the required solution of the given scalar nonhomogeneous linear differential equation.

CHAPTER 2

Optimal Control of Linear Systems

In this chapter we shall be primarily concerned with the minimal time-optimal control problem for the linear process $\mathcal{L}$ defined in Section 2.1 and we shall develop this theory thoroughly.

In Sections 2.2 and 2.3 we discuss the qualitative aspect of control theory. Here we define the set of attainability to consist of all states to which x_0 can be steered by an admissible control function $u(t)$. We develop the basic theory of controllability by a study of the geometry of the set of attainability.

In the later sections we turn to the quantitative aspects of control theory. Here we prove the existence of optimal controllers, demonstrate the maximal principle that optimal controllers have certain maximal and extremal properties, and synthesize the desired optimal controllers through closed-loop design. In each situation we present the concepts and principles for general nonautonomous linear processes and then turn to a detailed analysis of the special results and formulas for the case of constant coefficients.

The appendix to this chapter summarizes the basic definitions and properties of convex sets. Also we there prove some mathematically difficult theorems that extend the results on controllability given in Section 2.2.

2.1 THE LINEAR CONTROL PROCESS

Consider a linear control process described by a linear differential system

$$(\mathcal{L}) \qquad \dot{x} = A(t)x + B(t)u + v(t).$$

The coefficients $A(t)$, $B(t)$, $v(t)$ are known matrices or vectors, as stated below, and we choose the controller $u(t)$ to steer or control the response $x(t)$ from an initial state x_0 to some desired target state in R^n.

We assume throughout this chapter that the coefficients of $\mathfrak{L}$ satisfy the following standing hypotheses:

1. $A(t)$ is an $n \times n$ matrix, $B(t)$ is an $n \times m$ matrix and $v(t)$ is a column n-vector, each of which is real and measurable on the line $-\infty < t < \infty$.
2. The norms $|A(t)|$, $|B(t)|$, and $|v(t)|$ are integrable on each compact subinterval of the time t.
3. A controller $u(t)$ is a real bounded measurable m-vector on some interval $\mathfrak{J}: t_0 \leq t \leq t_1$ (usually $t_0 < t_1 < \infty$) with values in a non-empty restraint set $\Omega \subset R^m$, and possibly satisfying certain other restrictions mentioned later.

A response or solution $x(t)$ is a real absolutely continuous n-vector on $\mathfrak{J}$ which satisfies the corresponding differential system

$$(\mathfrak{L}_u) \qquad \dot{x} = A(t)x + B(t)\,u(t) + v(t),$$

as described in the background theory of linear differential systems in Appendix 2 of Chapter 1.

On occasion we shall assume further hypotheses on the coefficients of $\mathfrak{L}$ or the restraint set Ω. But the indicated standing hypotheses, which guarantee the existence of solutions and the standard theory of linear differential systems, will be maintained throughout Chapter 2. In particular, the variation of parameters formula holds for the solution $x(t)$, with $x(t_0) = x_0$,

$$x(t) = \Phi(t)x_0 + \Phi(t)\int_{t_0}^{t} \Phi(s)^{-1}[B(s)\,u(s) + v(s)]\,ds.$$

Here $\Phi(t)$ is the fundamental matrix solution of the homogeneous system

$$\dot{x} = A(t)x$$

with $\Phi(t_0) = I$. If $A(t) = A$ is constant, $\Phi(t) = e^{A(t-t_0)}$.

2.2 CONTROLLABILITY: THE SET OF ATTAINABILITY

Consider the linear control process

$$(\mathfrak{L}) \qquad \dot{x} = A(t)x + B(t)u + v(t).$$

Given the initial state x_0 we study the set $K(t_1)$ of all points of R^n to which x_0 can be steered by controllers $u(t) \subset \Omega$ on $t_0 \leq t \leq t_1$.

Definition. Consider the control system $\mathcal{L}$ with restraint set Ω, initial state x_0, and all controllers $u(t) \subset \Omega$ on $t_0 \le t \le t_1$. Let $x(t)$ denote the corresponding responses initiating at $x(t_0) = x_0$. The set of attainability $K(\mathcal{L}, \Omega, x_0, t_0, t_1)$ is the set of all endpoints $x(t_1)$ in R^n. Usually we suppress mention of all quantities but the final time t_1 and write $K(t_1)$ for the set of attainability. For notational purposes set $K(t_0) = x_0$.

We note that $K(\mathcal{L}, \Omega, x_0, t_0, t_1)$ is merely a rigid translate of $K(\mathcal{L}, \Omega, 0, t_0, t_1)$ by the vector $\Phi(t_1)x_0$. Thus the geometry of the set of attainability $K(\mathcal{L}, \Omega, x_0, t_0, t_1)$ does not depend on the initial point x_0, except for the location of this set in R^n. For autonomous linear processes $\mathcal{L}$, only the difference $t_1 - t_0$ is significant and we usually take $t_0 = 0$.

The restraint set Ω will usually be compact and convex. Such a restraint might be the m-cube

$$|u^i| \le 1 \qquad i = 1, \ldots, m.$$

In the appendix we show that the following, Theorem 1, holds for an arbitrary compact Ω, which need not be convex, but here we present a somewhat easier proof using the assumption that Ω is convex.

THEOREM 1. *Consider the linear control process in* R^n

$$(\mathcal{L}) \qquad \dot{x} = A(t)x + B(t)u + v(t)$$

with compact convex restraint set Ω, *initial state* x_0, *and controllers* $u(t)$ *on* $t_0 \le t \le t_1$. *Then the set of attainability* $K(t_1)$ *is compact, convex, and varies continuously with* t_1 *on* $t_1 \ge t_0$.

Proof. To prove that $K(t_1)$ is compact—that is, closed and bounded in R^n—we prove that every sequence of points $x_1(t_1), x_2(t_1), \ldots, x_r(t_1), \ldots$ in $K(t_1)$ has a subsequence which converges to some limit point $\bar{x}(t_1)$ in $K(t_1)$. Consider the corresponding solutions $x_r(t)$ and controllers $u_r(t) \subset \Omega$ on $t_0 \le t \le t_1$ for $r = 1, 2, 3, \ldots$. The variation of parameters formula states

$$x_r(t) = \Phi(t)x_0 + \Phi(t)\int_{t_0}^{t} \Phi(s)^{-1}\,[B(s)\,u_r(s) + v(s)]\,ds,$$

where $\Phi(t)$ is the appropriate fundamental matrix solution with $\Phi(t_0) = I$.

The set of all controllers $u(t) \subset \Omega$ on $t_0 \le t \le t_1$ is weakly compact (in fact, the set of all functions $B(t)\,u(t)$ is weakly compact, see Lemma 1A of the appendix to Chapter 2) and so there is a subsequence $u_{r_i}(t)$ which converges weakly to a controller $\bar{u}(t) \subset \Omega$ on $t_0 \le t \le t_1$, that is,

$$\lim_{i\to\infty} \int_{t_0}^{t} \Phi(s)^{-1}\,B(s)\,u_{r_i}(s)\,ds = \int_{t_0}^{t} \Phi(s)^{-1}\,B(s)\,\bar{u}(s)\,ds.$$

Let $\bar{x}(t)$ be the response to the control $\bar{u}(t)$ so that on $t_0 \le t \le t_1$ we have

$$\bar{x}(t) = \Phi(t)x_0 + \Phi(t)\int_{t_0}^{t} \Phi(s)^{-1}\,[B(s)\,\bar{u}(s) + v(s)]\,ds = \lim_{i\to\infty} x_{r_i}(t).$$

Thus

$$\lim_{i\to\infty} x_{r_i}(t_1) = \bar{x}(t_1) \in K(t_1),$$

and so $K(t_1)$ is compact.

To prove that $K(t_1)$ is convex we show that the line segment

$$(1-\lambda)\,x_0(t_1) + \lambda\,x_1(t_1), \qquad 0 \le \lambda \le 1,$$

joining two points $x_0(t_1)$ and $x_1(t_1)$ in $K(t_1)$, lies entirely in $K(t_1)$. Let $u_0(t)$ and $u_1(t)$ be controllers with corresponding responses $x_0(t)$ and $x_1(t)$. Define the controllers $u_\lambda(t) \subset \Omega$ on $t_0 \le t \le t_1$ by

$$u_\lambda(t) = (1-\lambda)\,u_0(t) + \lambda\,u_1(t).$$

The response $x_\lambda(t)$ to $u_\lambda(t)$ is

$$x_\lambda(t) = \Phi(t)x_0 + \Phi(t)\int_{t_0}^{t} \Phi(s)^{-1}\,[B(s)\,u_\lambda(s) + v(s)]\,ds.$$

Thus

$$x_\lambda(t) = (1-\lambda)\left\{\Phi(t)x_0 + \Phi(t)\int_{t_0}^{t} \Phi(s)^{-1}\,[B(s)\,u_0(s) + v(s)]\,ds\right\}$$
$$+ \lambda\left\{\Phi(t)x_0 + \Phi(t)\int_{t_0}^{t} \Phi(s)^{-1}\,[B(s)\,u_1(s) + v(s)]\,ds\right\},$$

and so

$$x_\lambda(t_1) = (1-\lambda)\,x_0(t_1) + \lambda\,x_1(t_1).$$

Hence $K(t_1)$ is convex.

We now fix all the data $(\mathfrak{L}, \Omega, x_0, t_0)$ and study the dependence of the set $K(t_1)$ on the variable end time $t_1 > t_0$. Given $\epsilon > 0$ we show that there exists $\delta > 0$ so that the distance between $K(t_1)$ and $K(t_2)$ is less than ϵ whenever $|t_1 - t_2| < \delta$; that is, we show that the correspondence

$$t \to K(t) \quad \text{for} \quad t > t_0$$

is a continuous map of the real half-line into the metric space of all nonempty compact subsets of R^n (see Appendix 1 of Chapter 1). Here the distance between $K(t_1)$ and $K(t_2)$ is the infimum of all $\epsilon > 0$ such that each point of $K(t_2)$ is closer than ϵ to some point of $K(t_1)$ and also each point of $K(t_1)$ is closer than ϵ to some point of $K(t_2)$.

Let $\hat{u}(t) \subset \Omega$ be a controller with response $\hat{x}(t)$ on $t_0 \le t \le t_1 + 1$. Then, for $t_0 < t_1$, $t_2 < t_1 + 1$ we have

$$\hat{x}(t_2) - \hat{x}(t_1) = \Phi(t_2)\int_{t_0}^{t_2} \Phi(s)^{-1}[B(s)\,\hat{u}(s) + v(s)]\,ds$$
$$- \Phi(t_2)\int_{t_0}^{t_1} \Phi(s)^{-1}[B(s)\,\hat{u}(s) + v(s)]\,ds$$
$$+ [\Phi(t_2) - \Phi(t_1)]\left\{\int_{t_0}^{t_1} \Phi(s)^{-1}[B(s)\,\hat{u}(s) + v(s)]\,ds + x_0\right\}.$$

On the interval $t_0 \le t \le t_1 + 1$ the continuous functions $\Phi(t)$ and $\Phi(t)^{-1}$ satisfy the bounds

$$|\Phi(t)| < C_1, \qquad |\Phi(t)^{-1}| < C_1,$$

for some constant C_1. Because of the integrability of $|B(t)|$ and $|v(t)|$ and the boundedness of $|\hat{u}(t)|$, we obtain the bound

$$|x_0| + \int_{t_0}^{t_1+1} |\Phi(s)^{-1}| \cdot |B(s)\,\hat{u}(s) + v(s)|\,ds < C_2.$$

Since an integral is a continuous function

$$\left|\int_{t_1}^{t} \Phi(s)^{-1}[B(s)\,\hat{u}(s) + v(s)]\,ds\right| < \frac{\epsilon}{2C_1}$$

and

$$|\Phi(t) - \Phi(t_1)| \le \left|\int_{t_1}^{t} A(s)\,\Phi(s)\,ds\right| < \frac{\epsilon}{2C_2}$$

for a prescribed $\epsilon > 0$ and $|t - t_1| < \delta$. Thus for $|t_2 - t_1| < \delta$ we have

$$|\hat{x}(t_2) - \hat{x}(t_1)| \le |\Phi(t_2)| \cdot \left|\int_{t_1}^{t_2} \Phi(s)^{-1}[B(s)\,\hat{u}(s) + v(s)]\,ds\right|$$
$$+ |\Phi(t_2) - \Phi(t_1)|\left[\int_{t_0}^{t_1+1} |\Phi(s)^{-1}| \cdot |B(s)\,\hat{u}(s) + v(s)|\,ds + |x_0|\right]$$

and

$$|\hat{x}(t_2) - \hat{x}(t_1)| < C_1 \cdot \frac{\epsilon}{2C_1} + C_2 \cdot \frac{\epsilon}{2C_2} = \epsilon.$$

Now let $\hat{x}(t_1) \in K(t_1)$ correspond to a control $\hat{u}(t)$ on $t_0 \le t \le t_1$. Define $\hat{u}(t) \subset \Omega$ on $t_0 \le t \le t_1 + 1$ by assigning the values $\hat{u}(t) = \hat{u}(t_1)$ on $t_1 \le t \le t_1 + 1$, and let $\hat{x}(t)$ be the corresponding response. Then $\hat{x}(t_2) \in K(t_2)$ and $|\hat{x}(t_2) - \hat{x}(t_1)| < \epsilon$. On the other hand if $\tilde{x}(t_2)$ is a point of $K(t_2)$ corresponding to the controller $\tilde{u}(t)$ on $t_0 \le t \le t_2$, again extend $\tilde{u}(t) \subset \Omega$ to the interval $t_0 \le t \le t_1 + 1$ to obtain $|\tilde{x}(t_2) - \tilde{x}(t_1)| < \epsilon$.

The above construction shows that the distance between $K(t_1)$ and $K(t_2)$

is less than ϵ whenever $|t_1 - t_2| < \delta$, where $0 < \delta < 1$ depends on ϵ and t_1. The same procedure shows that the distance between $K(t_0) = x_0$ and $K(t_1)$ is less than ϵ when $|t_1 - t_0|$ is sufficiently small. Thus $K(t_1)$ varies continuously with $t_1 \geq t_0$. Q.E.D.

COROLLARY. *If P is a point in the interior of the set $K(t_1)$, then there exists a neighborhood N of P and $\delta > 0$ such that each set $K(t_2)$, with $|t_2 - t_1| < \delta$, contains N in its interior.*

Proof. Let

$$x_0(t_1), x_1(t_1), \ldots, x_n(t_1)$$

be the vertices of an n-simplex S (the convex hull of $n + 1$ independent points) that lies in the interior of $K(t_1)$ and that has P at its centroid. Let N be the interior of a similar simplex but with all lengths halved. Let the corresponding controllers be $u_0(t), u_1(t), \ldots, u_n(t) \subset \Omega$ for the extended interval $t_0 \leq t \leq t_1 + 1$.

Let $\epsilon > 0$ be so small that every n-simplex $Q_0, Q_1, \ldots, Q_n$ with $|Q_i - x_i(t_1)| < \epsilon$ still contains N in its interior. By Theorem 1 there exists $\delta > 0$ such that for $|t_2 - t_1| < \delta$ we have $|x_i(t_2) - x_i(t_1)| < \epsilon$, $i = 0, 1, \ldots, n$. Since $K(t_2)$ contains $x_0(t_2) \ldots, x_n(t_2)$ and since $K(t_2)$ is convex, $K(t_2)$ contains N in its interior. Q.E.D.

Remarks. Theorem 1A of the appendix to this chapter proves the conclusion of Theorem 1 and its corollary without the hypothesis that Ω should be convex. We shall use the statement of Theorem 1A henceforth since the convexity of Ω does not enter into the subsequent theory. In fact, in the general *bang-bang* principle of Theorem 4 we necessarily must discuss compact restraint sets that are definitely not convex.

It is often important to choose controllers $u(t)$ that lie always on the boundary $\partial\Omega$ of the restraint set $\Omega \subset R^m$. For instance, in general the optimal controller has such extremal properties, as we shall show later in this chapter. This will follow from the geometric fact that the optimal response $x^*(t)$ on $t_0 \leq t \leq t^*$ has its endpoint $x^*(t^*)$ in the boundary (or frontier) $\partial K(t^*)$ of the set of attainability $K(t^*)$. In fact, we shall note that $x^*(t^*)$ lies in that part of $\partial K(t^*)$ that belongs to no earlier set $K(t)$ for $t_0 \leq t \leq t^*$.

Definition. Let $K(t_1)$ be the set of attainability for the control process $\mathfrak{L}$ with initial state x_0 and all controllers $u(t) \subset \Omega$ on $t_0 \leq t \leq t_1$. A point $P \in \partial K(t_1)$ is defined to lie in the new frontier of $K(t_1)$ just in case P does not lie in any set $K(t)$ for $t_0 \leq t < t_1$, that is,

$$P \in [\partial K(t_1) - \bigcup_{[t_0 \leq t < t_1]} K(t)].$$

We next investigate controllers $u(t)$ on $t_0 \leq t \leq t_1$, which steer x_0 to $\partial K(t_1)$. Such controllers will be called extremal, and these play a basic role in the geometry of $K(t)$ and in the solution of the time-optimal control problem for the linear process $\mathfrak{L}$.

Definition. Let $u(t) \subset \Omega$ on $t_0 \leq t \leq t_1$ be a controller for the linear process

$$(\mathfrak{L}) \qquad \dot{x} = A(t)x + B(t)u + v(t)$$

with restraint set Ω and initial state x_0 at time t_0. If the corresponding response $x(t)$ has an endpoint $x(t_1)$ in the boundary $\partial K(t_1)$ of the set of attainability $K(t_1)$, then $u(t)$ is called an extremal controller and $x(t)$ is an extremal response on $t_0 \leq t \leq t_1$.

In order to express the extremal condition analytically we shall refer to the linear differential system

$$\dot{x} = A(t)x$$

and its adjoint linear differential system

$$\dot{\eta} = -\eta A(t).$$

Here $\eta(t)$ is an n-row vector and every such solution is of the form $\eta(t) = \eta_0\, \Phi(t)^{-1}$, where η_0 is a constant vector and $\Phi(t)$ is the fundamental matrix solution of

$$\dot{x} = A(t)x$$

with $\Phi(t_0) = I$. This formula for $\eta(t)$ can be verified to be a solution with $\eta(t_0) = \eta_0$ by direct differentiation. If $\eta_0 \neq 0$, then $\eta(t)$ is a nontrivial solution of the adjoint equation, that is, $\eta(t)$ vanishes nowwhere on the interval $t_0 \leq t \leq t_1$. If $A(t) = A$ is constant, then $\eta(t) = \eta_0 e^{-(t-t_0)A}$.

The following, Theorem 2, is the principal analytical device in the theory of time-optimal control problems for the linear process $\mathfrak{L}$, and it is equivalent to the Pontryagin maximal principle for this case. Theorem 2 asserts essentially that a control is extremal if it is maximal (in the sense of Chapter 1), and so the confusing phrase "maximal controller" can be omitted in the theory of linear processes.

THEOREM 2. *Consider the linear control process in* R^n

$$(\mathfrak{L}) \qquad \dot{x} = A(t)x + B(t)u + v(t)$$

with compact restraint set Ω *and initial state* x_0 *at time* t_0. *A controller* $u(t) \subset \Omega$ *on* $t_0 \leq t \leq t_1$ *is extremal if and only if there exists a nontrivial solution* $\eta(t)$ *of*

$$\dot{\eta} = -\eta\, A(t)$$

such that

$$\eta(t)\, B(t)\, u(t) = \max_{u \in \Omega} \eta(t)\, B(t) u,$$

for almost all t on $t_0 \leq t \leq t_1$.

Proof. Assume that $u(t)$ on $t_0 \leq t \leq t_1$ is extremal and so steers x_0 to $x(t_1) \in \partial K(t_1)$ by the response

$$x(t) = \Phi(t)x_0 + \Phi(t) \int_{t_0}^{t} \Phi(s)^{-1}[B(s)\, u(s) + v(s)]\, ds.$$

Since $K(t_1)$ is compact and convex, there is a supporting hyperplane π to $K(t_1)$ at the boundary point $x(t_1)$. Let $\eta(t_1)$ denote the unit normal vector to π at $x(t_1)$, with $\eta(t_1)$ directed out of $K(t_1)$, that is, $\eta(t_1)$ points into an open half-space defined by π, which does not meet $K(t_1)$. Define the nontrivial adjoint response

$$\eta(t) = \eta_0\, \Phi(t)^{-1} \quad \text{with} \quad \eta(t_1) = \eta_0\, \Phi(t_1)^{-1}.$$

Then compute the inner product of $\eta(t)$ and $x(t)$,

$$\eta(t)\, x(t) = \eta_0 x_0 + \int_{t_0}^{t} \eta(s)[B(s)\, u(s) + v(s)]\, ds.$$

Now suppose that

$$\eta(t)\, B(t)\, u(t) < \max_{u \in \Omega} \eta(t)\, B(t)\, u$$

on some set of positive duration in $t_0 \leq t \leq t_1$. Define a controller $\hat{u}(t) \subset \Omega$ on $t_0 \leq t \leq t_1$ so

$$\eta(t)\, B(t)\, \hat{u}(t) = \max_{u \in \Omega} \eta(t)\, B(t)\, u$$

(for measurability of $\hat{u}(t)$ see Lemmas 2A and 3A in the appendix to this chapter).

But then, for the corresponding response $\hat{x}(t)$ in R^n, we compute

$$\eta(t_1)\, \hat{x}(t_1) = \eta_0 x_0 + \int_{t_0}^{t_1} \eta(s)\, B(s)\, \hat{u}(s)\, ds + \int_{t_0}^{t_1} \eta(s)\, v(s)\, ds.$$

Since

$$\int_{t_0}^{t_1} \eta(s)\, B(s)\, u(s)\, ds < \int_{t_0}^{t_1} \eta(s)\, B(s)\, \hat{u}(s)\, ds,$$

we find that

$$\eta(t_1)\, x(t_1) < \eta(t_1)\, \hat{x}(t_1).$$

But this inequality contradicts the construction of $\eta(t_1)$ as the outward normal to π at $x(t_1)$. For the inequality indicates that point $\hat{x}(t_1)$ is separated by π from $K(t_1)$, which is impossible since $\hat{x}(t_1) \in K(t_1)$. Therefore we conclude that

$$\eta(t)\, B(t)\, u(t) = \max_{u \in \Omega} \eta(t)\, B(t)\, u$$

almost everywhere in $t_0 \leq t \leq t_1$.

Conversely, assume that, for some nontrivial adjoint response $\eta(t) = \eta_0\, \Phi(t)^{-1}$, the control $u(t) \subset \Omega$ satisfies

$$\eta(t)\, B(t)\, u(t) = \max_{u \in \Omega} \eta(t)\, B(t)\, u$$

almost everywhere on $t_0 \leq t \leq t_1$. We must show that the corresponding response $x(t)$ terminates at a boundary point of $K(t_1)$, not at an interior point of $K(t_1)$.

Suppose that $x(t_1)$ lies in the interior of $K(t_1)$. For the specified adjoint response $\eta(t)$ we consider a point $\tilde{x}(t_1)$ in $K(t_1)$ with

$$\eta(t_1)\, x(t_1) < \eta(t_1)\, \tilde{x}(t_1).$$

Let $\tilde{u}(t) \subset \Omega$ be the control which yields the response $\tilde{x}(t)$. Our hypothesis states

$$\eta(t)\, B(t)\, \tilde{u}(t) \leq \eta(t)\, B(t)\, u(t) = \max_{u \in \Omega} \eta(t)\, B(t)\, u$$

almost always on $t_0 \leq t \leq t_1$. As above, we compute

$$\eta(t_1)\, \tilde{x}(t_1) \leq \eta(t_1)\, x(t_1),$$

which is a contradiction. Hence $x(t_1) \in \partial K(t_1)$, as required. Q.E.D.

Theorem 2 merely states that the response $x(t)$ leading to a boundary point of $K(t_1)$, say to the "south-east corner" of $K(t_1)$, must almost always go in the appropriate "southeasterly" direction at the greatest possible speed compatible with the restraint Ω. However the flow along the solutions of $\mathfrak{L}$ has its own peculiar local geometry, and the appropriate "southeasterly" direction at each point on $x(t)$ is designated by the changing vector $\eta(t)$. We formalize these remarks as a corollary.

COROLLARY 1. *Let $u(t) \subset \Omega$ on $t_0 \leq t \leq t_1$ be an extremal controller for the process $\mathfrak{L}$, with corresponding responses $x(t)$ and adjoint response $\eta(t)$ so*

$$\eta(t)\, B(t)\, u(t) = \max_{u \in \Omega} \eta(t)\, B(t)\, u$$

almost everywhere on $t_0 \leq t \leq t_1$. Then on each subinterval $t_0 \leq t \leq \tau < t_1$, $u(t)$ is also an extremal controller with $x(\tau) \in \partial K(\tau)$, Furthermore, $\eta(\tau)$ is an exterior normal to a supporting hyperplane π_τ to $K(\tau)$ at $x(\tau)$.

Proof. On the subinterval $t_0 \leq t \leq \tau$ we have

$$\eta(t)\, B(t)\, u(t) = \max_{u \in \Omega} \eta(t)\, B(t)\, u$$

almost always, and hence $u(t)$ is extremal on this interval, and so $x(\tau) \in \partial K(\tau)$.

In the theorem we prove that

$$\eta(\tau)\, \tilde{x}(\tau) \leq \eta(\tau)\, x(\tau)$$

for every response $\tilde{x}(\tau)$ on $t_0 \leq t \leq \tau$. Let π_τ be the hyperplane through $x(\tau)$ having normal vector $\eta(\tau)$. The above inequality shows that $K(\tau)$ does not meet the open half-space into which $\eta(\tau)$ points. Thus π_τ is a supporting hyperplane to $K(\tau)$ at the point $x(\tau)$. Q.E.D.

COROLLARY 2. *Let $K(t)$ on $t_0 \leq t \leq t_1$ be the set of attainability for the control process $\mathfrak{L}$. If for some time τ on $t_0 < \tau < t_1$ the set $K(\tau)$ has a nonempty interior, then $K(t)$ has a nonempty interior for all $\tau \leq t \leq t_1$.*

Proof. Let $x(t)$ on $t_0 \leq t \leq t_1$ be a response with $x(\tau)$ an interior point of $K(\tau)$. If $x(t)$ on $\tau \leq t \leq t_1$ were a boundary point of $K(t)$, then $x(t)$ would be extremal and hence $x(\tau)$ would be in $\partial K(\tau)$. Thus $x(t)$ on $\tau \leq t \leq t_1$ must be an interior point of $K(t)$. Q.E.D.

In Theorem 2 we observe that a controller $u(t)$ on $t_0 \leq t \leq t_1$ steers x_0 to the boundary $\partial K(t_1)$ just in case

$$\eta(t)\, B(t)\, u(t) = \max_{u \in \Omega} \eta(t)\, B(t)\, u \quad \text{a.e. (almost everywhere)}$$

for an appropriate adjoint response $\eta(t)$. It often happens that for each boundary point $P_1 \in \partial K(t_1)$ there is just one extremal controller $u(t) \subset \Omega$ on $t_0 \leq t \leq t_1$, which steers x_0 to P_1.

Definition. Consider the linear control process

$$(\mathfrak{L}) \qquad \dot{x} = A(t)x + B(t)u + v(t)$$

with restraint set Ω and initial state x_0 at time t_0. The problem defined by the data $(\mathfrak{L}, \Omega, x_0, t_0, t_1)$ is called normal in case any two controllers $u_1(t)$ and $u_2(t)$ on $t_0 \leq t \leq t_1$, which steer x_0 to the same boundary point $P_1 \in \partial K(t_1)$, must be equal almost everywhere.

THEOREM 3. *Consider the linear control process in R^n*

$$(\mathfrak{L}) \qquad \dot{x} = A(t)x + B(t)u + v(t)$$

with compact restraint set Ω and initial state x_0 at time t_0. The problem $(\mathfrak{L}, \Omega, x_0, t_0, t_1)$ is normal if and only if the following uniqueness property *obtains: for each nontrivial solution* $\eta(t)$ *of* $\dot{\eta} = -\eta A(t)$ *and for any two controllers* $u_1(t)$ *and* $u_2(t) \subset \Omega$ *satisfying*

$$\eta(t) B(t) u_1(t) = \eta(t) B(t) u_2(t) = \max_{u \in \Omega} \eta(t) B(t) u, \qquad \text{a.e.,}$$

the controllers $u_1(t)$ *and* $u_2(t)$ *are the same, that is,*

$$u_1(t) = u_2(t), \quad \text{a.e. on} \quad t_0 \leq t \leq t_1.$$

If the problem is normal, and if Ω *contains more than one point, the set of attainability* $K(t_1)$ *is strictly convex; thus* $K(t_1)$ *is a compact convex set with nonempty interior.*

Proof. If Ω contains just one point, then all controllers are equal; thus $K(t_1)$ contains just one point and the theorem holds. Now assume that Ω contains more than one point.

Let the problem $(\mathfrak{L}, \Omega, x_0, t_0, t_1)$ be normal and we show that $K(t_1)$ is strictly convex. Suppose, to the contrary, that there is a supporting hyperplane π such that $\pi \cap K(t_1)$ contains more than one point and hence a compact line segment L. Let $u_a(t)$ and $u_b(t) \subset \Omega$ steer x_0 to the distinct endpoints P_a and P_b of L.

For each measurable subset $D \subset \mathfrak{J}$ consider the real $2n$-vector

$$w(D) = \begin{bmatrix} \int_D \Phi(s)^{-1} B(s) u_a(s)\, ds \\ \int_D \Phi(s)^{-1} B(s) u_b(s)\, ds \end{bmatrix}$$

where $\Phi(s)$ is the usual fundamental matrix solution. The vector valued set function $w(D)$ has some values

$$w(\mathfrak{J}) = \begin{pmatrix} r_a \\ r_b \end{pmatrix} \quad \text{and} \quad w(\varnothing) = \begin{pmatrix} 0 \\ 0 \end{pmatrix}$$

for the empty set $\varnothing$. By Liapunov's result (Lemma 4A of the appendix to this chapter) there is a set $D_{.5} \subset \mathfrak{J}$ for which

$$w(D_{.5}) = \begin{pmatrix} r_{a/2} \\ r_{b/2} \end{pmatrix} \quad \text{and} \quad w(\mathfrak{J} - D_{.5}) = \begin{pmatrix} r_{a/2} \\ r_{b/2} \end{pmatrix}.$$

Since $P_a \neq P_b$, we know $r_a \neq r_b$ and so neither $D_{.5}$ nor $\mathfrak{J} - D_{.5}$ is a null set. Define the controllers

$$u_1(t) = \begin{cases} u_a(t) & \text{for} \quad t \in D_{.5} \\ u_b(t) & \text{for} \quad t \in \mathfrak{J} - D_{.5} \end{cases}$$

and

$$u_2(t) = \begin{cases} u_a(t) & \text{for} \quad t \in \mathfrak{J} - D_{.5} \\ u_b(t) & \text{for} \quad t \in D_{.5} \end{cases}.$$

Then the response $x_1(t)$ to $u_1(t)$ yields

$$x_1(t_1) = \Phi(t_1)x_0 + \Phi(t_1)\int_{D_{.5}} \Phi(s)^{-1}[B(s)\, u_a(s) + v(s)]\, ds$$
$$+ \Phi(t_1)\int_{\mathfrak{J}-D_{.5}} \Phi(s)^{-1}[B(s)\, u_b(s) + v(s)]\, ds.$$

A short computation shows that $x_1(t_1) = \frac{1}{2}P_a + \frac{1}{2}P_b$ is the midpoint of L. The response to $u_2(t)$ also yields the midpoint $x_2(t_1) = x_1(t_1) = \frac{1}{2}P_a + \frac{1}{2}P_b$. By normality $u_1(t) = u_2(t)$ almost everywhere on $\mathfrak{J}$. However this means that $u_a(t) = u_b(t)$ almost everywhere on $D_{.5}$ and also on $\mathfrak{J} - D_{.5}$. But this contradicts the supposition that P_a and P_b are distinct endpoints of L. Therefore $K(t_1)$ is strictly convex.

Let $\eta(t)$ be a nontrivial adjoint response and let π be the supporting hyperplane to the strictly convex set $K(t_1)$, where $\eta(t_1)$ is an outward orthogonal vector to π. If $\hat{u}_1(t)$ and $\hat{u}_2(t)$ are any controllers satisfying

$$\eta(t)\, B(t)\, \hat{u}_1(t) = \eta(t)\, B(t)\, \hat{u}_2(t) = \max_{u \in \Omega} \eta(t)\, B(t)\, u, \quad \text{a.e.},$$

then both $\hat{u}_1(t)$ and $\hat{u}_2(t)$ steer x_0 to the unique point P_1 in $\pi \cap K(t_1)$. By normality

$$\hat{u}_1(t) = \hat{u}_2(t) \quad \text{a.e. on} \quad \mathfrak{J}.$$

Conversely assume the uniqueness property of Theorem 3. Let $P \in \partial K(t_1)$ and let $\eta(t)$ be a nontrivial adjoint response with $\eta(t_1)$ an outward pointing orthogonal vector to a supporting hyperplane π to $K(t_1)$ at P_1. Let $u_1(t)$ and $u_2(t) \subset \Omega$ be controllers steering x_0 to P_1. Then by Theorem 2

$$\eta(t)\, B(t)\, u_1(t) = \eta(t)\, B(t)\, u_2(t) = \max_{u \in \Omega} \eta(t)\, B(t)\, u, \quad \text{a.e.}$$

The uniqueness hypothesis then yields the desired result

$$u_1(t) = u_2(t) \quad \text{a.e.}$$

Hence the problem $(\mathfrak{L}, \Omega, x_0, t_0, t_1)$ is normal. Q.E.D.

COROLLARY. *If the problem* $(\mathfrak{L}, \Omega, x_0, t_0, t_1)$ *is normal, then for each* τ *on* $t_0 < \tau < t_1$ *the problem* $(\mathfrak{L}, \Omega, x_0, t_0, \tau)$ *is normal.*

Proof. Suppose $u_1(t)$ and $u_2(t)$ on $t_0 \le t \le \tau$ steer x_0 to the same point

$P_0 \in \partial K(\tau)$. Then using an appropriate nontrivial adjoint response $\eta(t)$ we obtain

$$\eta(t)\, B(t)\, u_1(t) = \eta(t)\, B(t)\, u_2(t) = \max_{u \in \Omega} \eta(t)\, B(t)u$$

almost everywhere on $t_0 \leq t \leq \tau$. Continue the solution $\eta(t)$ on $t_0 \leq t \leq t_1$ as the outward normal to a supporting hyperplane π_t to $K(t)$. Choose $u(t) \leq \Omega$ on $\tau < t \leq t_1$ so

$$\eta(t)\, B(t)\, u(t) = \max_{u \in \Omega} \eta(t)\, B(t)u$$

and then continue the $u_1(t)$ and $u_2(t)$ as $u(t)$ on $\tau < t \leq t_1$.

Then $u_1(t)$ and $u_2(t)$ steer x_0 to the same point $P_1 \in K(t_1)$. By Theorem 2 both $u_1(t)$ and $u_2(t)$ are extremal controllers and so $P_1 \in \partial K(t_1)$. However $(\mathfrak{L}, \Omega, x_0, t_0, t_1)$ is normal and so $u_1(t) = u_2(t)$ almost everywhere on $t_0 \leq t \leq t_1$. Therefore $(\mathfrak{L}, \Omega, x_0, t_0, \tau)$ is normal. Q.E.D.

Remark. Theorem 2 shows that an extremal controller $u(t)$ on $t_0 \leq t \leq t_1$ for initial point x_0 is also extremal for every other initial point $\bar{x}_0$, that is, $u(t)$ steers $\bar{x}_0$ to the boundary of $K(t_1, \bar{x}_0)$. Similarly Theorem 3 shows that if $(\mathfrak{L}, \Omega, x_0, t_0, t_1)$ is normal, then so is $(\mathfrak{L}, \Omega, \bar{x}_0, t_0, t_1)$ for every initial point $\bar{x}_0$ in R^n.

We shall investigate the hypothesis of normality later in this chapter in connection with the synthesis of optimal controllers.

In exploiting the first three theorems of this chapter we shall find that the properties of $K(t_1)$ expressed in Theorem 1 will relate to the existence of an optimal controller, Theorem 2 will describe and characterize optimal controllers as extremal controllers, and Theorem 3 will yield uniqueness theorems for the synthesis of optimal controllers.

The next theorem shows that whatever steering can be accomplished by using controls $u(t) \subset \Omega$, can already be done by restricting attention to controls $u(t)$ lying always on the boundary $\partial\Omega$ of the restraint set. To prove this we shall refer to Theorem 1A in the appendix of this chapter since we must consider the restricted restraint set $\partial\Omega$, which is not convex.

Since a controller $u(t) \subset \partial\Omega$ is often represented physically by a mechanism that switches abruptly from one extreme value to another, Theorem 4 is referred to as the *general bang-bang* principle.

THEOREM 4. *Consider the linear control process in* R^n

$$(\mathfrak{L}) \qquad \dot{x} = A(t)x + B(t)u + v(t)$$

with compact restraint set Ω *and initial state* x_0 *at time* t_0. *Let* Ω_0 *be a compact subset of* Ω *with convex hull*

$$H(\Omega_0) = H(\Omega).$$

Let $K(t_1)$ be the set of attainability for all controllers $u(t) \subset \Omega$ on $t_0 \leq t \leq t_1$ and let $K_0(t_1)$ be the corresponding set for controllers $u(t) \subset \Omega_0$. Then $K(t)$ is compact and convex and

$$K_0(t_1) = K(t_1).$$

Proof. Theorem 1A of the appendix to Chapter 2 shows that $K(t_1)$ is compact and convex. Also the sets of attainability corresponding to the restraint sets Ω_0, Ω, and $H(\Omega)$ are all $K(t_1)$, as shown in Theorem 1A.

Q.E.D.

COROLLARY 1. *Let $\Omega_0 = \partial\Omega$ for the compact restraint set Ω. Then the corresponding sets of attainability are $K_0(t_1) = K(t_1)$.*

COROLLARY 2. *If Ω is a convex polytope with vertices Ω_0, then the corresponding sets of attainability are $K_0(t_1) = K(t_1)$.*

Example For the linear process $\mathcal{L}$ let Ω be the m-cube $|u^i| \leq 1$, $i = 1, \ldots, m$. Take Ω_0 to be the set of vertices. Then every point in $K(t_1)$ can be attained using a bang-bang controller $u(t)$ with

$$|u^i(t)| = 1 \quad \text{on} \quad t_0 \leq t \leq t_1.$$

However take $m = 1$ and consider the scalar control problem

$$\dot{x} = b(t)u, \qquad -1 \leq u \leq 1,$$

where $b(t) = t^3 \sin 1/t$ and $b(0) = 0$ so $b(t) \in C^1$. Take $x_0 = 0$, $t_0 = 0$ and $t_1 = 1$. Then $K(t_1)$ is a compact interval $-\alpha \leq x \leq \alpha = \int_0^1 |b(t)|\, dt$. But note that the point $x_1 = \alpha$ can be attained only by a controller $u(t)$, which has an infinite number of switches on $0 \leq t \leq 1$, namely, at the zeros of $b(t)$. Thus the general bang-bang principle does not assure that the switches of the bang-bang controller should be finite in number or a particularly simple set.

2.3 CONTROLLABILITY AND STABILITY FOR AUTONOMOUS SYSTEMS

Consider the autonomous linear control process

$$(\mathcal{L}) \qquad \dot{x} = Ax + Bu$$

for real constant matrices A and B of dimensions $n \times n$ and $n \times m$ respectively. Here we have assumed that the origin $x = 0$ is an equilibrium point for the free or uncontrolled system where $u = 0$. More general autonomous linear systems can often be brought to this form by a translation of coordinates in both x and u space. In this and the next section we shall study controllability, stability, and observability properties of autonomous linear processes; for nonautonomous processes see the exercises and also Theorem 6, Chapter 3.

Let us impose no restraint on the magnitude of the controller $u(t)$, that is, take the restraint set $\Omega = R^m$ and seek to steer an arbitrary initial point x_0 to an arbitrary target point x_1 in a finite time duration.

Definition. The autonomous linear control process

$$(\mathcal{L}) \qquad \dot{x} = Ax + Bu,$$

with $\Omega = R^m$, is (completely) controllable in case: for each pair of points x_0 and x_1 in R^n, there exists a bounded measurable controller $u(t)$ on some finite interval $0 \leq t \leq t_1$, which steers x_0 to x_1.

The next theorem asserts a computable criterion for the controllability of autonomous linear processes.

THEOREM 5. *The autonomous linear process in* R^n

$$(\mathcal{L}) \qquad \dot{x} = Ax + Bu$$

is controllable if and only if the $n \times nm$ *matrix*

$$[B, AB, A^2B, \ldots, A^{n-1}B]$$

has rank n.

Proof. Assume that $\mathcal{L}$ is controllable so that x_0 can be steered to an arbitrary target point x_1 in R^n. Suppose, contrary to the theorem, that

$$\text{rank } [B, AB, \ldots, A^{n-1}B] < n.$$

Then the rows of this matrix are linearly dependent and there exists a nonzero constant row vector v such that

$$v[B, AB, \ldots, A^{n-1}B] = 0$$

or

$$vB = vAB = vA^2B = \cdots = vA^{n-1}B = 0.$$

By the Cayley-Hamilton theorem matrix A satisfies its own characteristic equation

$$A^n = c_1A^{n-1} + c_2A^{n-2} + \cdots + c_nI$$

for certain real numbers $c_1, c_2, \ldots, c_n$. Thus

$$vA^nB = c_1vA^{n-1}B + \cdots + c_nvB = 0,$$

and by induction,

$$vA^{n+k}B = 0 \quad \text{for all} \quad k = 0, 1, 2, 3, \ldots.$$

Therefore

$$ve^{At}B = v\left[I + At + \frac{1}{2!}A^2t^2 + \cdots\right]B = 0$$

for all real t.

Now the response initiating at $x_0 = 0$ with controller $u(t)$ is

$$x(t) = e^{At}\int_0^t e^{-As}B\,u(s)\,ds.$$

Then

$$vx(t) = \int_0^t ve^{A(t-s)}B\,u(s)\,ds = 0$$

for all controllers. Thus every response $x(t)$ lies in the hyperplane of R^n orthogonal to v. This contradicts the controllability of $\mathcal{L}$. Hence we conclude that rank $[B, AB, \ldots, A^{n-1}B] = n$.

Conversely, assume that $[B, AB, \ldots, A^{n-1}B]$ has rank n and we shall show that $\mathcal{L}$ is controllable. Let K_0' be the set of all points to which the origin can be steered in a prescribed duration $0 \leq t \leq 1$ with controllers restrained by $|u^i| \leq 1$ for $i = 1, \ldots, m$. Then K_0' is compact and convex in R^n.

Suppose the dimension of K_0' is less than n. Then there exists a unit vector v such that

(1) $$\int_0^1 ve^{A(1-s)}B\,u(s)\,ds = 0$$

for all such controllers. Since $u(s)$ is arbitrary, within the magnitude restraint, the integrand in (1) vanishes, and

(2) $$ve^{A(1-s)}B \equiv 0 \quad \text{for} \quad 0 \leq s \leq 1.$$

Set $s = 1$ to get $vB = 0$. Next differentiate Eq. (2) with respect to s and again set $s = 1$ to obtain $vAB = 0$. By repeated differentiation of Eq. (2) with respect to s, we obtain

$$vB = vAB = vA^2B = \cdots = vA^{n-1}B = 0.$$

But this means that $[B, AB, \ldots, A^{n-1}B]$ has dependent rows, contrary to our assumption, and hence dimension $K_0' = n$.

Since the controller $u(t)$ can be replaced by $-u(t)$, K_0' is symmetric about the origin. Since K_0' contains an open set and K_0' is convex, K_0' must contain the origin in its interior. If the control restraint is relaxed to $|u^i| \leq l$ for $l = 1, 2, 3, \ldots$, then the corresponding set of attainability of the origin is lK_0'. Therefore the set K_0 of attainability of the origin for all controllers $u(t)$ on $0 \leq t \leq 1$ must be the entire space R^n.

For an arbitrary point x_0 the set K of attainability for controllers $u(t)$ on $0 \leq t \leq 1$ is $K = e^A x_0 + K_0 = R^n$. Thus $\mathcal{L}$ is controllable. Q.E.D.

Since the concept of controllability for an autonomous linear process in R^n

$$(\mathfrak{L}) \qquad \dot{x} = Ax + Bu$$

is defined geometrically, the property of controllability must be independent of the linear coordinates employed in R^n. If we define new coordinates $\bar{x} = Px$, for a real constant nonsingular matrix P, then $\mathfrak{L}$ assumes the form

$$(\mathfrak{L}) \qquad \dot{\bar{x}} = \bar{A}\bar{x} + \bar{B}u$$

where $\bar{A} = PAP^{-1}$ and $\bar{B} = PB$. For this reason we define an autonomous linear process in R^n

$$(\bar{\mathfrak{L}}) \qquad \dot{x} = \bar{A}x + \bar{B}u$$

to be *linearly equivalent* to $\mathfrak{L}$ just in case there exists a real constant nonsingular matrix P such that $\bar{A} = PAP^{-1}$ and $\bar{B} = PB$. Thus linearly equivalent processes represent the same geometrical or physical system referred to different coordinates in R^n.

It is easy to verify algebraically that the property of controllability is intrinsic, that is,

$$[\bar{B}, \bar{A}\bar{B}, \bar{A}^2\bar{B}, \ldots, \bar{A}^{n-1}\bar{B}] = P[B, AB, A^2B, \ldots, A^{n-1}B]$$

and

$$\operatorname{rank}\,[\bar{B}, \bar{A}\bar{B}, \ldots, \bar{A}^{n-1}\bar{B}] = \operatorname{rank}\,[B, AB, \ldots, A^{n-1}B].$$

The matrix $[B, AB, \ldots, A^{n-1}B]$ is called the *controllability matrix* of $\mathfrak{L}$ and we note the invariance of its rank under linear equivalence.

For controllable processes the control function can be made continuous and arbitrarily smooth without losing the basic steering properties. The following corollary emphasizes this fact, which will be of importance later in Theorem 8.

COROLLARY 1. *Let*

$$(\mathfrak{L}) \qquad \dot{x} = Ax + Bu$$

be controllable in R^n. Then for each pair of points x_0 and $x_1 \in R^n$, and each arbitrary short positive time duration $0 \leq t \leq t_1$, there exists a smooth controller $u(t) \in C^\infty$ on $0 \leq t \leq t_1$ that steers x_0 to x_1. Moreover, under restraints $|u^i(t)| \leq \epsilon$, $|\dot{u}^i(t)| \leq \epsilon, \ldots, |u^{i(r)}(t)| \leq \epsilon$ for $\epsilon > 0$, the initial point x_0 can be steered to all points in a neighborhood of $e^{At_1}x_0$.

Proof. Let K_∞ be the set of attainability from x_0 in time $0 \leq t \leq t_1$ (the use of $t_1 > 0$ instead of $t_1 = 1$ is unessential) using controllers $u(t) \in C^\infty$. Clearly K_∞ is convex. Since each bounded measurable controller on $0 \leq t \leq t_1$ can be uniformly approximated by a C^∞ controller (except on a duration of small measure where a uniform bound can be maintained) K_∞ is dense in $K = R^n$. Thus $K_\infty = R^n$ and x_0 can be steered to x_1 by a smooth controller.

Now consider only smooth controllers $u(t) \in C^\infty$ on $0 \le t \le t_1$ restrained by $|u^i(t)| \le \epsilon, \ldots, |u^{i(r)}(t)| \le \epsilon$. Let $K_\infty{}^r$ be the corresponding set of attainability from x_0. Then $K_\infty{}^r$ is convex and is symmetric about $e^{At_1}x_0$. If the dimension of $K_\infty{}^r$ were less than n there would exist a unit vector v so that

$$v \int_0^{t_1} e^{A(t_1-s)} B\, u(s)\, ds = 0$$

for all admissible controllers. But this still implies that $ve^{A(t_1-s)}B = 0$ on $0 \le s \le t_1$. However, as in the theorem, this condition contradicts the controllability of $\mathcal{L}$. Thus $K_\infty{}^r$ has dimension n and must contain a neighborhood of $e^{At_1}x_0$. Q.E.D.

The concept of controllability of $\mathcal{L}$ is of great importance in studying the domain $\mathcal{C}$ of (null) controllability—those points of R^n that can be steered to the origin in a finite time by admissible controllers $u(t)$ lying in a restraint set $\Omega \subset R^m$. The domain $\mathcal{C}$ is always connected, and $\mathcal{C}$ is open in R^n if and only if $\mathcal{C}$ contains a neighborhood of the origin. This last assertion follows immediately from the continuous dependence of the solutions of $\mathcal{L}$ upon the initial point $x_0 \in R^n$.

COROLLARY 2. *Consider the autonomous linear process in R^n*

$$(\mathcal{L}) \qquad \dot{x} = Ax + Bu$$

with control restraint set $\Omega \subset R^m$ containing $u = 0$ in its interior. Then the domain $\mathcal{C}$ of null controllability is open in R^n if and only if $\mathcal{L}$ is controllable.

Proof. First note that $\mathcal{L}$ is controllable if and only if

$$(\mathcal{L}_-) \qquad \dot{x} = -Ax - Bu$$

is also controllable, since the columns of $[B, AB, A^2B, \ldots, A^{n-1}B]$ and of $[-B, AB, -A^2B, \ldots, (-1)^n A^{n-1}B]$ span the same space. If $u(t)$ steers $x_0 = 0$ to x_1 along a solution of $\mathcal{L}_-$ on $[0, t_1]$, then $u(t_1 - t)$ steers x_1 to the origin along a solution of $\mathcal{L}$ on $[0, t_1]$. Thus the set K_- of attainability from the origin along solutions of $\mathcal{L}_-$ with controllers $u(t) \subset \Omega$ is precisely the set $\mathcal{C}$ of null controllability of $\mathcal{L}$.

Let $\mathcal{L}$ and hence $\mathcal{L}_-$ be controllable. Then, with the further restraint $|u^i(t)| \le \epsilon$ for some $\epsilon > 0$, the corresponding set $K_-{}^\epsilon$ contains a neighborhood of $x_0 = 0$. Since $K_-{}^\epsilon \subset K_- = \mathcal{C}$, the set $\mathcal{C}$ contains a neighborhood of $x_0 = 0$ and hence $\mathcal{C}$ is open in R^n.

Conversely assume that $\mathcal{C}$ is open in R^n. Then each point $x_1 \in \mathcal{C}$ is attainable from $x_0 = 0$ by a solution of $\mathcal{L}_-$ with a control in Ω. Hence $\mathcal{C}$ is precisely the set of all points

$$x = -\int_0^{t_1} e^{A(s-t_1)}B\, u(s)\, ds,$$

for $u(t) \subset \Omega$ and all time durations $0 \leq t_1 < \infty$. Suppose $\mathcal{L}_-$ is not controllable. Then there exists a unit vector v such that

$$vB = vAB = vA^2B = \cdots = vA^{n-1}B = 0$$

and hence

$$ve^{At}B = 0 \quad \text{for all real } t.$$

But this implies that $\mathcal{C}$ lies in the hyperplane orthogonal to v, which is impossible because $\mathcal{C}$ has a nonempty interior. Thus $\mathcal{L}_-$ and hence $\mathcal{L}$ must be controllable. Q.E.D.

COROLLARY 3. *Consider the autonomous linear process in* R^n

$$(\mathcal{L}) \qquad \dot{x} = Ax + Bu$$

with control restraint $\Omega \subset R^m$.

Assume

(a) $u = 0$ *lies in the interior of* Ω;
(b) $\mathcal{L}$ *is controllable;*
(c) A *is stable, that is, each eigenvalue* λ *of* A *satisfies* $\operatorname{Re} \lambda < 0$.

Then the domain of null controllability $\mathcal{C} = R^n$.

Proof. Let an initial point $x_1 \in R^n$ be steered by the null control $u(t) \equiv 0$ until the response $x(t)$ approaches $x_0 = 0$ and enters $\mathcal{C}$. But then $x(t)$ can be steered to the origin in a finite time. Hence $x_1 \in \mathcal{C}$ and $\mathcal{C} = R^n$. Q.E.D.

If $m = 1$, that is, B is a column vector b, then for

$$(\mathcal{L}) \qquad \dot{x} = Ax + bu$$

the following statements are equivalent:

1. $\mathcal{L}$ is controllable;
2. rank $[b, Ab, A^2b, \ldots, A^{n-1}b] = n$;
3. det $[b, Ab, A^2b, \ldots, A^{n-1}b] \neq 0$;
4. the vectors $b, Ab, A^2b, \ldots, A^{n-1}b$ are linearly independent.

Some of these elementary criteria for controllability fail for $m \geq 2$. For example, take $A = 0$ and $B = I$ to define a controllable system even though all the columns of AB are zero. Thus the theory of controllability is simpler for the case $m = 1$, a single scalar controller. The next four theorems relate to processes with single scalar controllers.

Often a controllable linear process in R^n,

$$(\mathfrak{L}) \qquad \dot{x} = Ax + Bu$$

with a vector controller $u(t) \subset R^m$, can be effectively controlled with a scalar controller $\mu(t)$, where $u(t) = c\,\mu(t)$ for a constant vector c. Here we define the constant column vector $b = Bc$ and consider

$$(\mathfrak{L}_1) \qquad \dot{x} = Ax + b\mu$$

for scalar controllers $\mu(t) \subset R^1$. This reduction of the control space from R^m to R^1, with the preservation of controllability, is always possible except for some rather exceptional cases.

The possibility of the reduction of the control space from R^m to R^1 depends only on the Jordan canonical structure of the matrix A. For every complex $n \times n$ matrix A there is a nonsingular complex matrix P such that

$$PAP^{-1} = \text{diag}\,\{A_1, A_2, \ldots, A_k\}$$

where each block

$$A_j = \begin{pmatrix} \lambda_j & 1 & 0 & 0 & \ldots & 0 & 0 \\ 0 & \lambda_j & 1 & 0 & \ldots & 0 & 0 \\ \cdot & & & & & & \\ \cdot & & & & & & \\ \cdot & & & & & & \\ 0 & 0 & & & \ldots & \lambda_j & 1 \\ 0 & 0 & & & \ldots & 0 & \lambda_j \end{pmatrix}, \qquad j = 1, 2, \ldots, k.$$

involves an eigenvalue λ_j of A. This Jordan canonical form for A is unique, except for the order of the blocks A_j on the diagonal. If the $n \times n$ matrix A has n distinct eigenvalues, then each block $A_j = (\lambda_j)$ is a 1×1 matrix. Matrix A defines a linear transformation of the complex n-dimensional vector space X onto itself, and the coordinates corresponding to each block A_j determine an invariant subspace X_j, on which A_j acts as a transformation.

THEOREM 6. *Consider the autonomous linear process in* R^n

$$(\mathfrak{L}) \qquad \dot{x} = Ax + Bu,$$

which is controllable with $u(t) \subset R^m$. *There exists a real constant vector* c *such that*

$$(\mathfrak{L}_1) \qquad \dot{x} = Ax + (Bc)\mu$$

is controllable for $\mu(t) \subset R^1$ *if and only if each two distinct Jordan canonical blocks of* A *contain unequal eigenvalues of* A.

Proof. If the real matrices A, B satisfy the controllability condition

$$\text{rank}\,[B, AB, A^2B, \ldots, A^{n-1}B] = n,$$

then so do the complex matrices $\bar{A} = PAP^{-1}$, $\bar{B} = PB$ for a nonsingular complex matrix P.

Let

$$\bar{A} = \text{diag}\,\{A_1, A_2, \ldots, A_k\}$$

be the Jordan canonical form for A where each A_j is a square $n_j \times n_j$ complex matrix

$$A_j = \begin{pmatrix} \lambda_j & 1 & 0 & 0 & \ldots & 0 \\ 0 & \lambda_j & 1 & 0 & \ldots & 0 \\ \cdot & & & & & \\ \cdot & & & & & \\ \cdot & & & & & \\ 0 & 0 & 0 & & \ldots & \lambda_j \end{pmatrix}.$$

We label the rows β of $\bar{B}$ accordingly

$$\bar{B} = \begin{pmatrix} \beta_{11} \\ \cdot \\ \cdot \\ \cdot \\ \beta_{1n_1} \\ \text{-----} \\ \beta_{21} \\ \cdot \\ \cdot \\ \cdot \\ \beta_{2n_2} \\ \text{------} \\ \cdot \\ \cdot \\ \cdot \end{pmatrix}.$$

Note that the n_1-row of $\bar{A}^l\bar{B}$ is $(\lambda_1)^l\beta_{1n_1}$ and, since $\mathfrak{L}$ is controllable, $\beta_{1n_1} \neq 0$. Similarly $\beta_{2n_2} \neq 0, \ldots, \beta_{kn_k} \neq 0$.

Assume two blocks, say A_1 and A_2, have the same eigenvalue $\lambda_1 = \lambda_2$. For such a matrix $\bar{A}$ and any complex vector $\bar{b}$, the system

$$\dot{x} = \bar{A}x + \bar{b}\,\mu(t)$$

is not controllable with $\mu(t) \subset R^1$. This obtains because the n_1 and $n_1 + n_2$ rows in the matrix $[\bar{b}, \bar{A}\bar{b}, \ldots, \bar{A}^{n-1}\bar{b}]$ are just $(b_{1n_1}, \lambda_1 b_{1n_1}, \ldots, \lambda_1^{n-1}b_{1n_1})$ and $(b_{2n_2}, \lambda_2 b_{2n_2}, \ldots, \lambda_2^{n-1}b_{2n_2})$, which are linearly dependent when $\lambda_1 = \lambda_2$. Therefore, in this case, there exists no reduction of $\mathfrak{L}$ to a controllable system $\mathfrak{L}_1$.

Conversely, assume all the blocks $A_1, \ldots, A_k$ contain unequal eigenvalues $\lambda_1, \ldots, \lambda_k$. Choose a real constant vector c so that

$$\bar{B}c = \bar{b} = \begin{pmatrix} b_{11} \\ \cdot \\ \cdot \\ \cdot \\ b_{1n_1} \\ \text{------} \\ b_{21} \\ \cdot \\ \cdot \\ \cdot \\ b_{2n_2} \\ \text{------} \\ \cdot \\ \cdot \\ \cdot \end{pmatrix}$$

has $b_{1n_1} \neq 0, b_{2n_2} \neq 0, \ldots, b_{kn_k} \neq 0$. (This can be done, for instance, by choosing the entries of c to be algebraically independent of all the entries of $\bar{B}$). We shall show that the columns of the matrix $[\bar{b}, \bar{A}\bar{b}, \ldots, \bar{A}^{n-1}\bar{b}]$ are linearly independent.

First note that the vectors

$$\bar{b}, (\bar{A} - \lambda_k I)\bar{b}, (\bar{A} - \lambda_k I)^2\bar{b}, \ldots, (\bar{A} - \lambda_k I)^{n_k-1}\bar{b}$$

span the same space as do the vectors

$$\bar{b}, \bar{A}\bar{b}, \bar{A}^2\bar{b}, \ldots, \bar{A}^{n_k-1}\bar{b}.$$

Now write

$$h = (\bar{A} - \lambda_k I)^{n_k}\bar{b} = \begin{pmatrix} h_{11} \\ \cdot \\ \cdot \\ \cdot \\ h_{1n_1} \\ \text{-----} \\ h_{21} \\ \cdot \\ \cdot \\ \cdot \\ h_{2n_2} \\ \text{-----} \\ \cdot \\ \cdot \\ \cdot \end{pmatrix}$$

so

$$h = \bar{A}^{n_k}\bar{b} + \text{linear combination of } \{\bar{b}, \bar{A}\bar{b}, \ldots, \bar{A}^{n_k-1}\bar{b}\}.$$

Then the vectors

$$h, (\bar{A} - \lambda_{k-1}I)h, \ldots, (\bar{A} - \lambda_{k-1}I)^{n_{k-1}-1}h$$

span the same space as do the vectors

$$h, \bar{A}h, \bar{A}^2h, \ldots, \bar{A}^{n_{k-1}-1}h.$$

As we continue in this way we compute that the vectors

$$\bar{b}, (\bar{A} - \lambda_k I)\bar{b}, (\bar{A} - \lambda_k I)^2\bar{b}, \ldots, (\bar{A} - \lambda_k I)^{n_k-1}\bar{b}$$

$$(\bar{A} - \lambda_k I)^{n_k}\bar{b},\ (\bar{A} - \lambda_{k-1}I)(\bar{A} - \lambda_k)^{n_k}\bar{b}, \ldots, (\bar{A} - \lambda_{k-1}I)^{n_{k-1}-1}(\bar{A} - \lambda_k)^{n_k}\bar{b},$$

$$\vdots$$

$$(\bar{A} - \lambda_2 I)^{n_2} \cdots (\bar{A} - \lambda_k I)^{n_k}\bar{b}, (\bar{A} - \lambda_1 I)(\bar{A} - \lambda_2 I)^{n_2} \cdots (\bar{A} - \lambda_k I)^{n_k}\bar{b}, \ldots,$$
$$(\bar{A} - \lambda_1 I)^{n_1-1}(\bar{A} - \lambda_2 I)^{n_2} \cdots (\bar{A} - \lambda_k I)^{n_k}\bar{b}$$

span the same space as do the columns of

$$[\bar{b}, \bar{A}\bar{b}, \bar{A}^2\bar{b}, \ldots, \bar{A}^{n-1}\bar{b}].$$

The last n_k rows of the vectors

$$(\bar{A} - \lambda_k I)^{n_k-1}\bar{b}, \ldots, (\bar{A} - \lambda_k I)^2\bar{b}, (\bar{A} - \lambda_k I)\bar{b}, \bar{b}$$

form the matrix

$$\begin{pmatrix} b_{kn_k} & & & & * & \\ 0 & b_{kn_k} & & & & \\ 0 & 0 & \cdot & & & \\ \cdot & & & \cdot & & \\ \cdot & & & & \cdot & \\ \cdot & & & & & \\ 0 & 0 & & & b_{kn_k} & \\ 0 & 0 & & & 0 & b_{kn_k} \end{pmatrix},$$

which has $b_{kn_k} \neq 0$ on the main diagonal and zeros below. Note that the vector h has zeros as the last n_k entries and that

$$h_{k-1,n_{k-1}} = (\lambda_{k-1} - \lambda_k)^{n_k} b_{k-1,n_{k-1}} \neq 0.$$

Then it is easy to compute that the column vectors

$$(\bar{A} - \lambda_1 I)^{n_1-1}(\bar{A} - \lambda_2 I)^{n_2} \cdots (\bar{A} - \lambda_k I)^{n_k}\bar{b}, \ldots, (\bar{A} - \lambda_k I)\bar{b}, \bar{b}$$

describe a triangular matrix with nonzero diagonal elements

$$(\lambda_1 - \lambda_2)^{n_2} \cdots (\lambda_1 - \lambda_k)^{n_k} b_{1n_1} \neq 0, \ldots b_{kn_k} \neq 0.$$

Since the determinant of such a triangular matrix is nonzero,

$$\det [\bar{b}, \bar{A}\bar{b}, \bar{A}^2\bar{b}, \ldots, \bar{A}^{n-1}\bar{b}] \neq 0.$$

Since $\bar{A} = PAP^{-1}$ and $\bar{b} = \bar{B}c = PBc$, we find that

$$\det [Bc, ABc, A^2Bc, \ldots, A^{n-1}Bc] \neq 0 \quad \text{so that}$$

$$(\mathfrak{L}_1) \qquad \dot{x} = Ax + (Bc)\mu$$

is controllable for $\mu(t) \subset R^1$. Q.E.D.

The next theorem obtains a physically significant and mathematically useful canonical form for controllable processes with a single scalar controller.

THEOREM 7. *The autonomous linear process*

$$x^{(n)} + a_1 x^{(n-1)} + \cdots + a_n x = u,$$

or the corresponding system in phase space R^n

$$(\mathfrak{D}) \qquad \begin{aligned} \dot{x}^1 &= x^2 \\ \dot{x}^2 &= x^3 \\ &\vdots \\ \dot{x}^n &= -a_n x^1 - a_{n-1} x^2 - \cdots - a_1 x^n + u, \end{aligned}$$

is controllable with $u \in R^1$. *Moreover every controllable process in* R^n,

$$(\mathfrak{L}) \qquad \dot{x} = Ax + bu$$

with $u \in R^1$, *is linearly equivalent to such a control process* $\mathfrak{D}$.

Proof. For the matrices

$$A_1 = \begin{pmatrix} 0 & 1 & 0 & 0 & \ldots & 0 \\ 0 & 0 & 1 & 0 & \ldots & 0 \\ \cdot & & & & & \\ \cdot & & & & & \\ \cdot & & & & & \\ 0 & 0 & 0 & 0 & \ldots & 1 \\ -a_n & -a_{n-1} & & & \ldots & -a_1 \end{pmatrix} \quad \text{and} \quad b_1 = \begin{pmatrix} 0 \\ 0 \\ \cdot \\ \cdot \\ \cdot \\ 0 \\ 1 \end{pmatrix}$$

the controllability condition of Theorem 5 is easily verified.

Now consider the controllable process $\mathfrak{L}$. Define the real nonsingular $n \times n$ matrix

$$P = [A^{n-1}b, A^{n-2}b, \ldots, A^2b, Ab, b].$$

Introduce new coordinates in R^n by $\bar{x} = P^{-1}x$ so $\mathfrak{L}$ becomes

$$\dot{\bar{x}} = P^{-1}AP\bar{x} + P^{-1}bu.$$

By direct matrix multiplication verify that

$$b = [A^{n-1}b, \ldots, Ab, b]\begin{pmatrix} 0 \\ 0 \\ \cdot \\ \cdot \\ \cdot \\ 0 \\ 1 \end{pmatrix} \quad \text{or} \quad P^{-1}b = \begin{pmatrix} 0 \\ 0 \\ \cdot \\ \cdot \\ \cdot \\ 0 \\ 1 \end{pmatrix} = b_1$$

and $AP = PN$ or $P^{-1}AP = N$ where

$$N = \begin{pmatrix} \alpha_1 & 1 & 0 & 0 & \ldots & 0 \\ \alpha_2 & 0 & 1 & 0 & \cdots & 0 \\ \cdot & & & & & \\ \cdot & & & & & \\ \cdot & & & & & \\ \alpha_{n-1} & 0 & 0 & 0 & \cdots & 1 \\ \alpha_n & 0 & 0 & 0 & \cdots & 0 \end{pmatrix}.$$

The constants $\alpha_1, \alpha_2, \ldots, \alpha_n$ are uniquely specified by the expansion

$$A^nb = \alpha_1 A^{n-1}b + \alpha_2 A^{n-2}b + \cdots + \alpha_n b.$$

For the process $\mathfrak{D}$ the characteristic equation of the corresponding matrix A_1 is

$$A_1{}^n = -a_1 A_1^{n-1} - a_2 A_1^{n-2} \cdots - a_n I.$$

Thus the analogous coordinate change applied to $\mathfrak{D}$, in place of $\mathfrak{L}$, reduces the coefficient matrices of $\mathfrak{D}$ to N and $\begin{pmatrix} 0 \\ 0 \\ \cdot \\ \cdot \\ \cdot \\ 1 \end{pmatrix}$, provided we take

$-a_1 = \alpha_1, -a_2 = \alpha_2, \ldots, -a_n = \alpha_n$. Therefore $\mathcal{L}$ is linearly equivalent to the process

$$x^{(n)} - \alpha_1 x^{(n-1)} + \cdots - \alpha_n x = u,$$

as required. Q.E.D.

In the corollaries to Theorem 5 we studied the domain $\mathcal{C}$ of null controllability of an autonomous linear process $\mathcal{L}$ in R^n when the controllers satisfy a restraint $u(t) \subset \Omega \subset R^m$. The next theorem yields a definitive description of the important case when each initial point can be steered to the origin, that is, $\mathcal{C} = R^n$. Since this is a global rather than local analysis, appropriate hypotheses must be placed on the global properties of $\mathcal{L}$ and Ω.

THEOREM 8. *Consider the autonomous linear process in* R^n

$$(\mathcal{L}) \qquad \dot{x} = Ax + bu$$

with a compact restraint interval $u \subset \Omega \subset R^1$ *having* $u = 0$ *in its interior. Then the domain of null controllability* $\mathcal{C} = R^n$ *if and only if the two conditions hold:*

(a) $\mathcal{L}$ *is controllable;*
(b) *every eigenvalue* λ *of* A *satisfies* $\operatorname{Re} \lambda \leq 0$.

Proof. If $\mathcal{L}$ is not controllable, then, by Corollary 2 of Theorem 5, there are points in R^n that cannot be steered to the origin.

Assume that $\mathcal{L}$, controllable or not, has an eigenvalue λ of A with $\operatorname{Re} \lambda > 0$. There exists a real nonsingular coordinate transformation in R^n, $y = Px$, so $\mathcal{L}$ becomes

$$\dot{y} = PAP^{-1}y + Pbu,$$

where

$$\dot{y}^1 = \lambda y^1 + b_1 u \quad \text{if} \quad \lambda > 0 \text{ is real},$$

or otherwise

$$\dot{y}^1 = \alpha y^1 + \beta y^2 + b_1 u$$

$$\dot{y}^2 = -\beta y^1 + \alpha y^2 + b_2 u$$

with $\lambda = \alpha + i\beta$ and $\alpha > 0$. Take initial conditions $y_0 \in R^n$ with $y_0{}^1 > 0$ very large (or $(y_0{}^1)^2 + (y_0{}^2)^2 > 0$ very large in the second case) and compute $dy^1/dt > 0$ (or $d/dt/(y^{1^2} + y^{2^2}) > 0$) for $t > 0$ for all choices of u in the compact set Ω. Thus the initial point y_0 cannot be steered to the origin by a controller $u(t) \subset \Omega$. Therefore $\mathcal{C} = R^n$ implies conditions (a) and (b).

Now assume that $\mathcal{L}$ is controllable and that every eigenvalue λ of A satisfies $\operatorname{Re} \lambda \leq 0$. We first give a preliminary discussion that shows that we need consider only the case where all eigenvalues of A are pure imaginary.

After a nonsingular coordinate transformation in R^n we can assume that $\mathcal{L}$ has the form

$$\begin{pmatrix} \dot{x}_p \\ \dot{x}_q \end{pmatrix} = \begin{pmatrix} A_p & 0 \\ 0 & A_q \end{pmatrix} \begin{pmatrix} x_p \\ x_q \end{pmatrix} + \begin{pmatrix} b_p \\ b_q \end{pmatrix} u$$

where each eigenvalue λ_p of A_p satisfies $\operatorname{Re} \lambda_p = 0$, and each eigenvalue λ_q of A_q satisfies $\operatorname{Re} \lambda_q < 0$. With the null control $u(t) = 0$ the solutions of the asymptotically stable system

$$\dot{x}_q = A_q x_q$$

decay towards $x_q = 0$ as $t \to +\infty$. Moreover, provided the coordinates x_q are suitably chosen, the radial velocity is negative

$$x_q' \dot{x}_q = x_q' A_q x_q \leq -\zeta x_q' x_q \quad \text{for} \quad 0 < \zeta < -\operatorname{Re} \lambda_q,$$

and further use of controllers $u(t)$ with sufficiently small norms $|u(t)|$ will retain a solution $x_q(t)$ within a prescribed neighborhood N_q of $x_q = 0$. Thus, if we can steer an arbitrary initial point of (the controllable system)

$$\dot{x}_p = A_p x_p + b_p u$$

to a suitably small neighborhood N_p of $x_p = 0$ by the use of controllers $u(t)$ with arbitrarily small $|u(t)|$, then the controllability of $\mathcal{L}$ implies that $\mathcal{C} = R^n$. Therefore we can reduce our problem to the study of a controllable system, which we shall again write as

$$(\mathcal{L}) \qquad \dot{x} = Ax + bu,$$

such that each eigenvalue λ of A satisfies $\operatorname{Re} \lambda = 0$. We must prove that each initial point $x_0 \in R^n$ can be steered to a prescribed neighborhood N of $x = 0$ by a controller $u(t)$ having $|u(t)| < \epsilon$, for a prescribed $\epsilon > 0$. Such a demonstration will complete the proof of the theorem.

By Theorem 7 the process $\mathcal{L}$ is linearly equivalent in R^n to a process defined by

$$(\mathcal{D}) \qquad D^s(D^2 + \gamma_1)(D^2 + \gamma_2) \cdots (D^2 + \gamma_r)x = u,$$

where $s \geq 0$, $r \geq 0$ are integral orders of differentiation and $\gamma_1, \ldots, \gamma_r$ are positive constants; and we write the state $(x, \dot{x}, \ldots, x^{(n-1)})$, or simply $(x, x', \ldots, x^{n-1})$.

Consider first the case $r = 0$, $s \geq 1$. If $s = 1$ the required control $u(t)$ is easily constructed (see Example 1 of Chapter 1). Moreover we can also choose $u(t) \in C^\infty$ on $t_0 \leq t \leq t_1$, steering x_0 to $x_1 = 0$ such that:

1. $u(t) = 0$ for t in some neighborhoods of $t = t_0$ and $t = t_1$;
2. $x(t) = 0$ for t in some neighborhood of t_1.

These technical requirements on the controller $u(t)$ are made so that several such controllers can be fitted in sequence to form a differentiable function. For the remainder of this proof we shall call such a controller *acceptable*. We introduce the σ-norm of a C^∞ controller

$$|u(t)|_\sigma = |u(t)| + |\dot{u}(t)| + \cdots + |u^{(\sigma)}(t)|,$$

and we leave as an exercise the construction of an acceptable controller $u(t)$ with $|u(t)|_\sigma \leq \epsilon$, for preassigned $\sigma \geq 0$ and $\epsilon > 0$, steering an arbitrary x_0 to $x_1 = 0$ for the process $Dx = u$. We proceed by mathematical induction and assume that such acceptable controllers have been constructed for all processes

$$D^j x = u, \qquad j = 1, 2, \ldots, s-1.$$

Consider the process

$$D^s x = u$$

which we relate to the differential systems

$$D^{s-1} z = \xi \quad \text{and} \quad D\xi = u.$$

Take $u(t) \in C^\infty$ and let $x(t)$, $\xi(t)$, and $z(t)$ be solutions with matching initial data $\xi_0 = x_0^{(s-1)}$ and $z_0 = x_0$, $z_0^{(1)} = x_0^{(1)}, \ldots, z_0^{(s-2)} = x_0^{(s-2)}$. Then note that $z(t) \equiv x(t)$.

First choose an acceptable controller $u(t)$ on $0 \leq t \leq t_1$ with $|u(t)|_\sigma < \epsilon$ steering $\xi(t)$ from $\xi_0 = x_0^{s-1}$ to $\xi_1 = 0$ by $D\xi = u$. This control $u(t)$ defines a response $x(t)$ of $D^s x = u$ steering $(x_0, x_0{}^1, \ldots, x_0^{s-1})$ to some state $(x_1, x_1{}^1, \ldots, x_1^{s-2}, 0)$. Use the induction hypothesis to determine an acceptable controller $\xi(t)$ on $t_1 \leq t \leq t_2$ with $|\xi(t)|_{\sigma+1} < \epsilon$ steering $z(t)$ from $(x_1, x_1{}^1, \ldots, x_1^{s-2})$ to $(0, 0, \ldots, 0)$ by $D^{s-1} z = \xi$. Define $u(t) = D\xi(t)$ on $t_1 \leq t \leq t_2$; then $u(t)$ is an acceptable controller on $0 \leq t \leq t_2$ steering $x(t)$ from $(x_0, x_0{}^1, \ldots, x_0^{s-1})$ to $(0, 0, \ldots, 0)$ and $|u(t)|_\sigma < \epsilon$. The induction conclusion is that for the process

$$D^s x = u \qquad s \geq 1,$$

each initial state can be steered to the origin by an acceptable controller $u(t)$ with $|u(t)|_\sigma < \epsilon$, for preassigned $\sigma \geq 0$ and $\epsilon > 0$.

Next consider the case $r \geq 1$, $s = 0$ so

$$(D^2 + \gamma_1)(D^2 + \gamma_2) + \cdots + (D^2 + \gamma_r)x = u.$$

For $r = 1$ we use the techniques of Example 1 of Section 1.2 and of Corollary 1 of Theorem 5 (Chapter 2) to construct an acceptable controller $u(t)$ with $|u(t)|_\sigma < \epsilon$ steering a given initial state to rest. We leave the explicit description of such an acceptable controller to the exercises. Let $r > 1$, $s = 0$, and proceed by induction as before.

Write the process $\mathfrak{D}$ as

$$(D^2 + \gamma_2) \cdots (D^2 + \gamma_r)z = \xi$$
$$(D^2 + \gamma_1)\xi = u$$

and use matching initial data for a given $u(t) \in C^\infty$ and corresponding $x(t)$, $z(t)$, $\xi(t)$,

$$z_0 = x_0, z_0^{\ 1} = x_0^{\ 1}, \ldots, z_0^{2r-3} = x_0^{2r-3}$$

$$\xi_0 = (D^2 + \gamma_2) \cdots (D^2 + \gamma_r)x(0), \qquad \xi_0^{\ 1} = (D^2 + \gamma_2) \cdots (D^2 + \gamma_r)x^1(0)$$

so $z(t) = x(t)$.

First choose an acceptable controller $u(t)$ on $0 \leq t \leq t_1$ with $|u(t)|_\sigma < \epsilon$ steering $\xi(t)$ from $(\xi_0, \xi_0^{\ 1})$ to $(0, 0)$ by $(D^2 + \gamma_1)\xi = u$. This control $u(t)$ defines a response $x(t)$ of $\mathfrak{D}$ steering $(x_0, x_0^{\ 1}, \ldots, x_0^{2r-1})$ to some state $(x_1, x_1^{\ 1}, \ldots, x_1^{2r-1})$. Use the induction hypothesis to determine an acceptable controller $\xi(t)$ on $t_1 \leq t \leq t_2$ with $|\xi(t)|_{\sigma+2} < \epsilon/(1 + \gamma_1)$ steering $z(t)$ from $(x_1, x_1^{\ 1}, \ldots, x_1^{2r-3})$ to $(0, 0, \ldots, 0)$. Define $u(t) = (D^2 + \gamma_1)\, \xi(t)$ on $t_1 \leq t \leq t_2$; then $u(t)$ is an acceptable controller on $0 \leq t \leq t_2$ steering $x(t)$ from $(x_0, x_0^{\ 1}, \ldots, x_0^{2r-1})$ to $(0, 0, \ldots, 0)$. Also $|\xi(t)|_{\sigma+2} + \gamma_1 |\, \xi(t)|_\sigma < \epsilon$, from which it follows easily that $|u(t)|_\sigma < \epsilon$. Thus there exists an acceptable controller $u(t)$ for $\mathfrak{D}$ in all cases $r \geq 1$, $s = 0$.

Finally consider the general case of $\mathfrak{D}$ for $r \geq 1$, $s \geq 1$. Here set

$$(D^2 + \gamma_1) \cdots (D^2 + \gamma_r)z = \xi$$
$$D^s\xi = u$$

and use matching initial data for given $u(t) \in C^\infty$ and corresponding solutions $x(t)$, $z(t)$, $\xi(t)$,

$$z_0 = x_0, z_0^{\ 1} = x_0^{\ 1}, \ldots, z_0^{2r-1} = x_0^{2r-1}$$
$$\xi_0 = (D^2 + \gamma_1) \cdots (D^2 + \gamma_r)\, x(0), \ldots,$$
$$\xi_0^{s-1} = D^{s-1}(D^2 + \gamma_1) \cdots (D^2 + \gamma_r)\, x(0),$$

so $z(t) = x(t)$.

First choose an acceptable controller $u(t)$ on $0 \leq t \leq t_1$ with $|u(t)|_\sigma < \epsilon$ steering $\xi(t)$ from $(\xi_0, \ldots, \xi_0^{s-1})$ to $(0, 0, \ldots, 0)$. This controller $u(t)$ steers $x(t)$ from $(x_0, x_0^{\ 1}, \ldots, x_0^{2r+s-1})$ to some state $(x_1, x_1^{\ 1}, \ldots, x_1^{2r+s-1})$. Take $\xi(t)$ as an acceptable controller on $t_1 \leq t \leq t_2$ steering $z(t)$ from $(x_1, x_1^{\ 1}, \ldots, x_1^{2r-1})$ to $(0, 0, \ldots, 0)$ with $|\xi(t)|_{\sigma+s} < \epsilon$. Define $u(t) = D^s\, \xi(t)$ on $t_1 \leq t \leq t_2$. Then $u(t)$ is an acceptable controller on $0 \leq t \leq t_2$ with $|u(t)|_\sigma < \epsilon$ steering $x(t)$ in $\mathfrak{D}$ from $(x_0, x_0^{\ 1}, \ldots, x_0^{2r+s-1})$ to $(0, 0, \ldots, 0)$.

Therefore the theorem is proved in all cases. Q.E.D.

COROLLARY. *Consider the autonomous linear process in* R^n

$$(\mathcal{L}) \qquad \dot{x} = Ax + Bu$$

with a compact restraint set $\Omega \subset R^m$ *containing* $u = 0$ *in its interior. Assume that no two Jordan canonical blocks of* A *contain equal eigenvalues of* A. *Then the domain of null controllability* $\mathcal{C} = R^n$ *if and only if:*

(a) $\mathcal{L}$ *is controllable*
(b) *every eigenvalue* λ *of* A *satisfies* $\operatorname{Re} \lambda \leq 0$.

Proof. If $\mathcal{C} = R^n$, then the proof of the theorem also applies if $m \geq 1$ to show that conditions (a) and (b) must hold. Conversely, if $\mathcal{L}$ is controllable and $\operatorname{Re} \lambda \leq 0$, then Theorem 6 shows that we can effectively replace the restraint set Ω by a compact interval that contains $u = 0$ in its interior. Then Theorem 8 applies to this case and so $\mathcal{C} = R^n$. Q.E.D.

For a controllable system in R^n

$$(\mathcal{L}) \qquad \dot{x} = Ax + Bu$$

every initial state x_0 can be regulated to zero in a finite time by some controller $u(t) \subset R^m$. This behavior is in contrast with that of a completely uncontrolled process ($B = 0$) which is stable (each eigenvalue λ of A satisfies $\operatorname{Re} \lambda < 0$) where each solution $x(t)$ approaches the origin as $t \to +\infty$.

Definition. The autonomous linear process in R^n

$$(\mathcal{L}) \qquad \dot{x} = Ax + Bu$$

is called stabilizable if there exists a constant linear feedback synthesis

$$u = Dx$$

such that

$$\dot{x} = Ax + BDx = (A + BD)x$$

is stable. That is, D is a real $m \times n$ matrix such that each eigenvalue of $(A + BD)$ has negative real part.

If $\mathcal{L}$ and $\bar{\mathcal{L}}$ are linearly equivalent processes, with $\bar{A} = PAP^{-1}$ and $\bar{B} = PB$, and if $\mathcal{L}$ is stabilizable, then $\bar{\mathcal{L}}$ is also stabilizable. For if $(A + BD)$ is a stability matrix, then

$$P(A + BD)P^{-1} = \bar{A} + \bar{B}\bar{D}, \qquad (\bar{D} = DP^{-1}),$$

is also a stability matrix.

THEOREM 9. *Consider the autonomous linear process in* R^n

$$(\mathfrak{L}) \qquad \dot{x} = Ax + bu$$

with controllers $u(t) \subset R^1$. *If* $\mathfrak{L}$ *is controllable, then* $\mathfrak{L}$ *is stabilizable by some linear feedback synthesis* $u = Dx$.

Proof. By Theorem 7 we can replace $\mathfrak{L}$ by the linearly equivalent system determined by

$$(\mathfrak{D}) \qquad x^{(n)} + a_1 x^{(n-1)} + \cdots + a_n x = u,$$

or

$$\dot{x}^1 = x^2$$

$$\dot{x}^2 = x^3$$

$$\vdots$$

$$\dot{x}^n = -a_n x^1 - a_{n-1} x^2 - \cdots - a_1 x^n + u.$$

For each choice of the real constant vector

$$D = (d_n, d_{n-1}, \ldots, d_1)$$

so

$$u = d_n x^1 + d_{n-1} x^2 + \cdots + d_1 x^n$$

the feedback process $\mathfrak{D}$ becomes

$$x^{(n)} + (a_1 - d_1)x^{(n-1)} + \cdots + (a_n - d_n)x = 0.$$

In particular, a stable process can be achieved by determining D so that the characteristic polynomial is

$$\lambda^n + (a_1 - d_1)\lambda^{n-1} + \cdots + (a_n - d_n) = (\lambda + 1)^n,$$

or

$$(a_i - d_i) = \binom{n}{i} \quad \text{for} \quad 1 \le i \le n. \qquad \text{Q.E.D.}$$

For the final two theorems of this section we return to the general case of a multidimensional controller $u(t) \subset R^m$, for $m \geq 1$. Here we study processes that are not controllable and seek to select a controllable part or else to approximate the general process by a controllable process.

THEOREM 10. *Consider an autonomous linear process in* R^n

$$(\mathfrak{L}) \qquad \dot{x} = Ax + Bu.$$

Then there exists a unique linear subspace C *in* R^n *such that*

(a) *C is invariant under $\mathfrak{L}$, that is, no point in C can be steered out of C, and no point outside of C can be steered into C,*

and

(b) $\mathcal{L}$ *restricted to C is controllable.*

Proof. Let C be the set of all points in R^n to which the origin can be steered in some finite time by controllers $u(t) \subset R^m$. We first show that C is a linear space.

Let $0 < t_1 < t_2$ and consider points in C

$$x_1(t_1) = \int_0^{t_1} e^{A(t_1-s)} B\, u_1(s)\, ds$$

$$x_2(t_2) = \int_0^{t_2} e^{A(t_2-s)} B\, u_2(s)\, ds.$$

In the first integral set $\sigma = s + t_2 - t_1$ and define the controller

$$U_1(\sigma) = \begin{cases} 0 & \text{on} \quad 0 \le \sigma \le t_2 - t_1 \\ u_1(\sigma - t_2 + t_1) & \text{on} \quad t_2 - t_1 < \sigma \le t_2. \end{cases}$$

Then

$$x_1(t_1) = \int_0^{t_2} e^{A(t_2-\sigma)} B\, U_1(\sigma)\, d\sigma$$

and so $x_1(t_1)$ can be attained from the origin at time t_2. Thus a linear combination of the controllers $U_1(t)$ and $u_2(t)$ on $0 \le t \le t_2$ determines the same linear combination of the points $x_1(t_1)$ and $x_2(t_2)$. Hence C is a linear space.

Note that C is just the origin if and only if $B = 0$, that is, $\mathcal{L}$ is completely uncontrollable. In this case the theorem holds and henceforth we assume that the dimension of C is $k \ge 1$.

By our construction no point of C can be steered out of C. Hence there exist coordinates $(\bar{x}^1, \ldots, \bar{x}^n)$ in R^n such that C is precisely $\bar{x}^{k+1} = 0, \ldots, \bar{x}^n = 0$ and $\mathcal{L}$ is written

$$\dot{\bar{x}}_1 = A_{11}\bar{x}_1 + A_{12}\bar{x}_2 + B_1 u$$

$$\dot{\bar{x}}_2 = \qquad\qquad A_{22}\bar{x}_2$$

where

$$\bar{x}_1 = \begin{pmatrix} \bar{x}^1 \\ \cdot \\ \cdot \\ \cdot \\ \bar{x}^k \end{pmatrix} \quad \text{and} \quad \bar{x}_2 = \begin{pmatrix} \bar{x}^{k+1} \\ \cdot \\ \cdot \\ \cdot \\ \bar{x}^n \end{pmatrix}.$$

This form for $\mathcal{L}$ obtains since on C we have $\dot{\bar{x}}_2 = 0$. We now observe that no point where $\bar{x}_2 \neq 0$ can be steered into C, and hence C is invariant.

Now restrict $\mathcal{L}$ to C to obtain

$$(\mathcal{L}_c) \qquad \dot{\bar{x}}_1 = A_{11}\bar{x}_1 + B_1 u \quad \text{on} \quad \bar{x}_2 = 0.$$

Since the origin can be steered to every point of C, Corollary 2 of Theorem 5 proves that $\mathcal{L}_c$ is controllable on C.

Finally let C' be any invariant linear subspace of R^n on which $\mathcal{L}$ is controllable. Since C' is invariant, it includes all points to which the origin can be steered, so $C \subset C'$. Since $\mathcal{L}$ is controllable on C', each point of C' can be steered to the origin and so $C' \subset C$. Thus $C' = C$, which proves the uniqueness of the subspace C satisfying properties (a) and (b) of the theorem. Q.E.D.

The invariant subspace C is called the *controllability space* for $\mathcal{L}$ and the restriction $\mathcal{L}_c$ of $\mathcal{L}$ to C is called the controllable part of $\mathcal{L}$.

COROLLARY. *Let C be the controllability space in R^n for*

$$(\mathcal{L}) \qquad \dot{x} = Ax + Bu.$$

Then there exist coordinates $\bar{x} = \begin{pmatrix} \bar{x}_1 \\ \bar{x}_2 \end{pmatrix}$ *in R^n so C is just $\bar{x}_2 = 0$ and $\mathcal{L}$ is*

$$\dot{\bar{x}}_1 = A_{11}\bar{x}_1 + A_{12}\bar{x}_2 + B_1 u$$
$$\dot{\bar{x}}_2 = A_{22}\bar{x}_2.$$

Also

$$\textit{dimension } C = \textit{rank } [B, AB, A^2B, \ldots, A^{n-1}B].$$

Proof. In the proof of the theorem we established the existence of the required coordinates $\bar{x} = \begin{pmatrix} \bar{x}_1 \\ \bar{x}_2 \end{pmatrix}$. (Note that $\bar{x}_1$ is vacuous if $C = 0$ so $B = 0$ and $\mathcal{L}$ is completely uncontrollable—in any case the space $\bar{x}_1 = 0$ is not determined intrinsically but is any complement to the space C). Since $\mathcal{L}_c$ is controllable on C,

$$\text{dimension } C = \text{rank } [B_1, A_{11}B_1, A_{11}{}^2, B_1, \ldots, A_{11}^{k-1}B_1] = k.$$

Also

$$\text{rank } [B_1, A_{11}B_1, \ldots, A_{11}^{k-1}B_1] = \text{rank } [B_1, A_{11}B_1, \ldots, A_{11}^{n-1}B_1].$$

But

$$\bar{A}^l\bar{B} = \begin{pmatrix} A_{11}{}^l B_1 \\ 0 \end{pmatrix} \quad \text{for} \quad 0 \leq l \leq n,$$

where

$$\bar{A} = \begin{pmatrix} A_{11} & A_{12} \\ 0 & A_{22} \end{pmatrix}, \qquad \bar{B} = \begin{pmatrix} B_1 \\ 0 \end{pmatrix}.$$

Therefore

$$\text{dimension } C = \text{rank } [\bar{B}, \bar{A}\bar{B}, \bar{A}^2\bar{B}, \ldots, \bar{A}^{n-1}\bar{B}]$$

and the result follows from the invariance of the rank of the controllability matrix under linear equivalence. Q.E.D.

THEOREM 11. *Consider an autonomous linear process in* R^n

$$(\mathfrak{L}_0) \qquad \dot{x} = A_0 x + B_0 u.$$

If $\mathfrak{L}_0$ *is controllable, then there exists an* $\epsilon_1 > 0$ *such that every autonomous linear process*

$$(\mathfrak{L}) \qquad \dot{x} = Ax + Bu \quad \text{with} \quad |A - A_0| < \epsilon_1 \quad \text{and} \quad |B - B_0| < \epsilon_1$$

is also controllable.

If $\mathfrak{L}_0$ *is not controllable, then for each* $\epsilon > 0$, *there exists a controllable process*

$$(\mathfrak{L}_1) \qquad \dot{x} = A_1 x + B_1 u \quad \text{with} \quad |A_1 - A_0| < \epsilon, \qquad |B_1 - B_0| < \epsilon.$$

That is, the set of all controllable processes is open and dense in the metric space of all autonomous linear processes in R^n, *the distance from* $\mathfrak{L}_1$ *to* $\mathfrak{L}_0$ *being* $|A_1 - A_0| + |B_1 - B_0|$.

Proof. If $\mathfrak{L}_0$ is controllable in R^n, the rows of $[B_0, A_0B_0, \ldots, A_0^{n-1}B_0]$ describe n linearly independent vectors in R^{nm}. If $|A - A_0| < \epsilon_1$ and $|B - B_0| < \epsilon_1$ for a sufficiently small $\epsilon_1 > 0$, then the rows of $[B, AB, \ldots, A^{n-1}B]$ must approximate these n vectors of R^{nm} and hence must also be linearly independent. In this case

$$(\mathfrak{L}) \qquad \dot{x} = Ax + Bu$$

is also controllable.

On the other hand assume that $\mathfrak{L}_0$ is not controllable. For a given $\epsilon > 0$ choose matrices A_1 and B_1 with $|A_1 - A_0| < \epsilon$, $|B_1 - B_0| < \epsilon$ such that all entries of A_1 and B_1 are algebraically independent over the rational numbers (that is, no nontrivial rational polynomial relations hold between the entries of A_1 and B_1—the existence of such A_1 and B_1 is a standard property of the arithmetic of real numbers). Then

$$\text{rank } [B_1, A_1B_1, \ldots, A_1^{n-1}B_1] = n$$

since no $n \times n$ subdeterminant can be zero because each such determinant is a polynomial in the entries of A_1 and B_1. Thus $\mathfrak{L}_1$ is controllable. Q.E.D.

Despite the artificiality of the constructions in the proof of Theorem 11, the final result is of physical interest. It assures us that, in the typical or *generic* case, an autonomous linear process $\mathfrak{L}$ is controllable. In particular,

9. The Routh-Hurwitz criteria for stability state that every root λ of the real polynomial

$$\lambda^n + a_1\lambda^{n-1} + \cdots + a_n = 0$$

satisfies Re $\lambda < 0$ if and only if $D_k > 0$ for $k = 1, 2, \ldots, n$. Here

$$D_1 = a_1, \qquad D_2 = \begin{vmatrix} a_1 & a_3 \\ 1 & a_2 \end{vmatrix}, \qquad D_k = \begin{vmatrix} a_1 & a_3 & a_5 & \cdots & a_{2k-1} \\ 1 & a_2 & a_4 & \cdots & a_{2k-2} \\ 0 & a_1 & a_3 & \cdots & a_{2k-3} \\ 0 & 1 & a_2 & \cdots & a_{2k-4} \\ 0 & 0 & a_1 & \cdots & a_{2k-5} \\ \cdot & & & & \\ \cdot & & & & \\ \cdot & & & & \\ 0 & 0 & 0 & \cdots & a_k \end{vmatrix}$$

with $a_j = 0$ for $j > n$.

(a) Show that if all roots λ satisfy Re $\lambda \leq 0$, then $a_k \geq 0$ and $D_k \geq 0$ for $k = 1, \ldots, n$.

(b) Show that if $a_k \geq 0$ and $D_k \geq 0$ for $k = 1, \ldots, n$, then all roots λ satisfy Re $\lambda \leq 0$, provided $n \leq 3$. The example $\lambda^4 + \lambda^2 + 1 = 0$ shows that the criterion fails for $n = 4$.

10. Show that the processes

$$Dx = u \quad \text{and} \quad (D^2 + 1)x = u$$

have acceptable controllers $u(t)$ with $|u(t)|_\sigma < \epsilon$ (in the sense of Theorem 8) steering an arbitrary initial state to rest.

11. Show that the autonomous linear process in R^n

$$(\mathcal{L}) \qquad \dot{x} = Ax + Bu,$$

which is generic (in the sense of Theorem 11), is controllable by a single scalar controller.

12. The autonomous linear process in R^n

$$(\mathcal{L}) \qquad \dot{x} = Ax + bu, \qquad u \in R^1,$$

is called controllable with arbitrarily restrained control in case: for each pair of points x_0 and x_1 in R^n and $\epsilon > 0$, there exists a controller $u(t)$ with $|u(t)| \leq \epsilon$ steering x_0 to x_1 in a finite time. Show that $\mathcal{L}$ is controllable with arbitrarily restrained control if and only if

(a) $\mathcal{L}$ is controllable;

(b) every eigenvalue λ of A satisfies Re $\lambda = 0$.

13. Two springs are connected in series to a fixed wall and oscillate along a horizontal frictionless track. The equations of motion are

$$\ddot{x} = -k_1 x + k_2(y - x)$$

$$\ddot{y} = -k_2(y - x) + u$$

where x and y are the extensions of the free ends beyond their equilibrium positions, $k_1 > 0$ and $k_2 > 0$ are spring constants, and $u(t)$ is a controlling force applied to the end of the second spring. Show that this system is controllable with arbitrarily restrained control. (Assume that all motions occur near the equilibrium state $x = \dot{x} = 0$, $y = \dot{y} = 0$, so that the wall support does not interfere with the control.)

14. Consider the autonomous linear process in R^n

$$(\mathcal{L}) \qquad \dot{x} = Ax + Bu$$

with initial state x_0 at $t_0 = 0$ and compact restraint set Ω. Assume that $\mathcal{L}$ is controllable, $H(\Omega)$ has nonempty interior in R^m, and rank $B = m$. Then prove that a compact subset $Z \subset \Omega$ has the bang-bang property, that is,

$$K_Z(t_1) \equiv K_\Omega(t_1) \quad \text{for all} \quad t_1 \geq 0,$$

if and only if

$$H(Z) = H(\Omega).$$

[Hint: Theorem 4 asserts that $H(Z) = H(\Omega)$ implies that Z has the bang-bang property. For the converse suppose $H(Z) \neq H(\Omega)$ and take a supporting hyperplane π to $H(\Omega)$, which does not meet $H(Z)$. Let $\eta_1 B$ be the outward unit normal to π for some η_1. Take a point $P_0 \in \partial K_\Omega(t_1)$ at which the outward normal is η_1, which is possible since $K_\Omega = K_{H(\Omega)}$ is a convex body and rank $B = m$. Now any controller $u_0(t) \subset H(\Omega)$ steering x_0 to P_0 must satisfy the maximal principle, and this shows that $u_0(t)$ lies outside $H(Z)$ for all t near t_1.]

15. Let $\dot{x} = Ax + bu$ be controllable, as in Theorem 9. Show that a feedback synthesis $u = Dx$ can be selected so that the eigenvalues of $A + bD$ have arbitrarily prescribed values (compatible with the reality of $A + bD$).

2.4 CONTROLLABILITY AND OBSERVABILITY

Consider the real autonomous linear control process

$$(\mathcal{L}) \qquad \dot{x} = Ax + Bu$$

with input or control vector $u \in R^m$ and response or state vector $x \in R^n$.

It may happen that only certain components, or linear combination of components, of the state vector x are physically significant or even observable. In these cases we augment the description of the process $\mathcal{L}$ by an observability equation

$$(\mathcal{O}) \qquad \omega = Hx.$$

Here H is a real $r \times n$ constant matrix that defines the observable output r-vector ω in terms of the state n-vector x. The complete relation of input to output, as described by the equations $\mathcal{L}$ and $\mathcal{O}$, is called an autonomous linear observed process.

Example. Consider the control process

$$x^{(n)} + a_1 x^{(n-1)} + \cdots + a_n x = u$$

where only the output x, and none of the higher derivatives, is observable. Then we write the observed process

$$\begin{pmatrix} \dot{x}^1 \\ \dot{x}^2 \\ \cdot \\ \cdot \\ \cdot \\ \cdot \\ \dot{x}^n \end{pmatrix} = \begin{pmatrix} 0 & 1 & 0 & 0 & \cdots & 0 \\ 0 & 0 & 1 & 0 & \cdots & 0 \\ \cdot & & & & & \\ \cdot & & & & & \\ \cdot & & & & & \\ 0 & 0 & 0 & 0 & \cdots & 1 \\ -a_n & -a_{n-1} & & & \cdots & -a_1 \end{pmatrix} \begin{pmatrix} x^1 \\ x^2 \\ \cdot \\ \cdot \\ \cdot \\ \cdot \\ x^n \end{pmatrix} + \begin{pmatrix} 0 \\ 0 \\ \cdot \\ \cdot \\ \cdot \\ 0 \\ 1 \end{pmatrix} u$$

with output

$$\omega = (1 \quad 0 \quad 0 \quad \cdots \quad 0) \begin{pmatrix} x^1 \\ x^2 \\ \cdot \\ \cdot \\ \cdot \\ x^n \end{pmatrix}.$$

For the linear observed process

$$\dot{x} = Ax + Bu, \qquad \omega = Hx$$

the *input-output relation* is

$$\omega(t) = He^{At} \int_0^t e^{-As} B\, u(s)\, ds,$$

for the initial data $x_0 = 0$ at $t = 0$. If all components of the input are zero except $u^j(t)$, and

$$u(t) = \frac{1}{\epsilon}\begin{pmatrix} 0 \\ \cdot \\ \cdot \\ \cdot \\ 0 \\ 1 \\ 0 \\ \cdot \\ \cdot \\ \cdot \\ 0 \end{pmatrix} = \frac{1}{\epsilon} e_j \quad \text{on} \quad 0 \leq t \leq \epsilon \quad \text{and} \quad u(t) = 0 \quad \text{otherwise},$$

then

$$\omega(t) = He^{At}\frac{1}{\epsilon}\int_0^\epsilon e^{-As}Be_j\, ds \quad \text{for} \quad t \geq \epsilon > 0.$$

The limiting case as $\epsilon \to 0$ defines the output to a unit impulse input $u(t) = \delta(t)e_j$ (the $\delta(t)$-function can be defined as the idealized limiting step function input or, more technically, as a measure that assigns a weight of $+1$ at the instant $t = 0$). The limiting output is

$$\omega(t) = He^{At}Be_j \quad \text{for} \quad t \geq 0.$$

In other words the (i, j) entry of

$$W(t) = He^{At}B \quad \text{for} \quad t \geq 0$$

defines the output component $\omega^i(t)$ for the impulse input $u(t) = \delta(t)e_j$. The matrix $W(t)$ determines the total input-output relation of the observed process since, for arbitrary input $u(t)$, we have

$$\omega(t) = \int_0^t W(t - s)\, u(s)\, ds \quad \text{for} \quad t \geq 0$$

where $x_0 = 0$ at $t = 0$.

Because of the convolution integral relating $u(t)$ to $\omega(t)$, it is often convenient to use the Laplace transforms $U(p)$ and $\Omega(p)$, respectively, and the transfer-function matrix

$$Z(p) = L(W(t)) = \int_0^\infty W(t)e^{-pt}\, dt.$$

Then the input-output relation becomes

$$\Omega(p) = Z(p)\, U(p).$$

Definition. For the autonomous linear observed process

$$(\mathfrak{L}) \qquad \dot{x} = Ax + Bu \quad \text{with} \quad \omega = Hx,$$

the impulse-response or weighting-function matrix is

$$W(t) = \begin{cases} H\, e^{At}B & \text{for} \quad t \geq 0 \\ 0 & \text{for} \quad t < 0. \end{cases}$$

Also the transfer-function matrix, from the input u to the output ω, is

$$Z(p) = L(W(t)) = \int_0^\infty W(t)e^{-pt}\, dt.$$

In this section we shall take the view that the impulse-response matrix, or equally well the transfer-function matrix, completely characterizes the physically observable aspects of an observed process. The main Theorem 14 is a precise statement of the exact correspondence between matrices $Z(p)$ of rational functions and observed processes. We first discover which matrices $Z(p)$ can serve as transfer-functions, then we define the physically significant class of controllable and observable linear processes, and then we demonstrate the basic equivalence of these two approaches to linear control theory.

By the known structure of the matrix e^{At}, for a real constant $n \times n$ matrix A, it is clear that $He^{At}B$ is an $r \times m$ matrix with entries of the form $t^\sigma e^{\alpha t} \cos \beta t$ or $t^\sigma e^{\alpha t} \sin \beta t$ for $\sigma = 0, 1, 2, 3, \ldots$ and real α, β, or else a finite real linear combination of such terms. Call an $r \times m$ matrix with such entries an exponential-polynomial matrix.

THEOREM 12. *A real $r \times m$ matrix*

$$W(t) = \begin{cases} W_0(t) & \textit{for} \quad t \geq 0 \\ 0 & \textit{for} \quad t < 0 \end{cases}$$

is the impulse-response matrix of some real autonomous linear observed process if and only if $W_0(t)$ is an exponential-polynomial matrix.

Also, an $r \times m$ matrix $Z(p)$ is the transfer-function matrix of some real autonomous linear observed process if and only if each entry of $Z(p)$ is a real rational function of p, with a numerator of lesser degree than the denominator.

Proof. The elementary Laplace transform formulas

$$L(t^\sigma e^{\alpha t} \cos \beta t) = (-1)^\sigma \frac{d^\sigma}{dp^\sigma}\left[\frac{p - \alpha}{(p-\alpha)^2 + \beta^2}\right]$$

$$L(t^\sigma e^{\alpha t} \sin \beta t) = (-1)^\sigma \frac{d^\sigma}{dp^\sigma}\left[\frac{\beta}{(p-\alpha)^2 + \beta^2}\right],$$

and the inverses

$$L^{-1}\left(\frac{1}{p}\right) = 1$$

$$L^{-1}\left(\frac{p^\rho}{(p-a)^\sigma}\right) = \frac{d^\rho}{dt^\rho}\left[\frac{t^{\sigma-1}}{(\sigma-1)!}e^{at}\right] \quad \text{for} \quad \sigma \geq 1, \qquad 0 \leq \rho < \sigma,$$

and the techniques of partial fraction decomposition show that $W_0(t)$ is an exponential-polynomial matrix if and only if $Z(p)$ is a matrix of rational functions, each with a numerator (which might be identically zero) of degree less than the degree of the denominator. Thus we need only prove the theorem for the matrices $W_0(t)$.

Since a real autonomous observed process has an impulse-response matrix $W(t)$ of the form $He^{At}B$ for $t \geq 0$, both $W_0(t)$ and its Laplace transform $Z(p)$ must have the special forms required in the theorem.

Now let $W_0(t)$ be an exponential-polynomial matrix. We must construct a real autonomous observed process with the appropriate impulse-response matrix.

Let $W_0(t) = (l_{ij}(t))$ where each $l_{ij}(t)$, for $1 \leq i \leq r$, $1 \leq j \leq m$, is a real finite linear combination of terms of the form $t^\sigma e^{\alpha t} \cos \beta t$ and $t^\sigma e^{\alpha t} \sin \beta t$. Thus each $l_{ij}(t)$ is a solution of some homogeneous linear differential equation with constant coefficients; say a differential equation of some high order N. Hence each $l_{ij}(t)$ appears in a fundamental $N \times N$ matrix $e^{A_{ij}t}C_{ij}$; and, if the constant matrices A_{ij} and C_{ij} are correctly chosen, the term $l_{ij}(t)$ appears in the upper left corner.

Next construct the matrix differential system of order Nrm, $\dot{x} = Ax$, where

$$A = \text{diag}\,\{A_{11}, A_{12}, \ldots, A_{1m}, A_{21}, \ldots, A_{2m}, \ldots, A_{r1}, \ldots, A_{rm}\}.$$

Let

$$C = \text{diag}\,\{C_{11}, \ldots, C_{1m}, \ldots, C_{r1}, \ldots, C_{rm}\}$$

and consider the matrix

$$e^{At}C = \text{diag}\,\{e^{A_{11}t}C_{11}, \ldots, e^{A_{rm}t}C_{rm}\},$$

which contains each $l_{ij}(t)$ in the upper left corner of the corresponding block $e^{A_{ij}t}C_{ij}$. We must now choose constant matrices H and B_1 so that

$$W_0(t) = He^{At}CB_1 = He^{At}B,$$

as required for the exponential-polynomial matrix $W_0(t)$.

Let H and B_1 be matrices with entries 1 and 0 placed so as to select the appropriate elements for $W_0(t)$. We illustrate the general technique by the case $r = 2$, $m = 3$. Define

$$H = \left(\begin{array}{c|c|c|c|c|c} 10 \cdots 0 & 10 \cdots 0 & 10 \cdots 0 & 00 \cdots 0 & 00 \cdots 0 & 00 \cdots 0 \\ 00 \cdots 0 & 00 \cdots 0 & 00 \cdots 0 & 10 \cdots 0 & 10 \cdots 0 & 10 \cdots 0 \end{array}\right)$$

so

$He^{At}C =$

$$\left(\begin{array}{c|c|c|c|c|c} l_{11}^* \cdots * & l_{12}^* \cdots * & l_{13}^* \cdots * & 0 \cdots 0 & 0 \cdots 0 & 0 \cdots 0 \\ 0 \cdots 0 & 0 \cdots 0 & 0 \cdots 0 & l_{21}^* \cdots * & l_{22}^* \cdots * & l_{23}^* \cdots * \end{array}\right)$$

Then define

$$B_1 = \left(\begin{array}{c|c|c}
1 & 0 & 0 \\
0 & \vdots & \vdots \\
\vdots & & \\
0 & 0 & 0 \\
\hline
0 & 1 & 0 \\
\vdots & 0 & \vdots \\
 & \vdots & \\
0 & 0 & 0 \\
\hline
0 & 0 & 1 \\
\vdots & \vdots & 0 \\
 & & \vdots \\
0 & 0 & 0 \\
\hline
1 & 0 & 0 \\
0 & & \vdots \\
\vdots & \vdots & \\
0 & 0 & 0 \\
\hline
0 & 1 & 0 \\
\vdots & 0 & \vdots \\
 & \vdots & \\
0 & 0 & 0 \\
\hline
0 & 0 & 1 \\
\vdots & \vdots & 0 \\
 & & \vdots \\
0 & 0 & 0
\end{array}\right)$$

so

$$H\, e^{At} C B_1 = \begin{pmatrix} l_{11} & l_{12} & l_{13} \\ l_{21} & l_{22} & l_{23} \end{pmatrix}$$

as required. Q.E.D.

Consider an autonomous linear observed process in R^n

$$(\mathcal{L}) \qquad \dot{x} = Ax + Bu \quad \text{and} \quad \omega = Hx.$$

A constant nonsingular linear transformation in R^n

$$\bar{x} = Px$$

brings $\mathcal{L}$ to the form

$$(\bar{\mathcal{L}}) \qquad \dot{\bar{x}} = \bar{A}\bar{x} + \bar{B}u \quad \text{and} \quad \omega = \bar{H}\bar{x}$$

where $\bar{A} = PAP^{-1}$, $\bar{B} = PB$, and $\bar{H} = HP^{-1}$. We define $\mathcal{L}$ and $\bar{\mathcal{L}}$ to be *linearly equivalent* under the coordinate transformation $\bar{x} = Px$, and linearly equivalent observed processes have the same intrinsic properties. For example, the impulse-response matrices are the same since these relate u to ω and do not depend on the coordinates for x in R^n, thus

$$W(t) = He^{tA}B = HP^{-1}e^{tPAP^{-1}}PB = \bar{H}e^{t\bar{A}}\bar{B}, \quad \text{for} \quad t \geq 0.$$

In Theorem 10 we noted that every autonomous linear observed process in R^n

$$(\mathcal{L}) \qquad \dot{x} = Ax + Bu \quad \text{and} \quad \omega = Hx$$

is linearly equivalent to an observed process of the canonical form

$$\dot{\bar{x}}_1 = \bar{A}_{11}\bar{x}_1 + \bar{A}_{12}\bar{x}_2 + \bar{B}_1 u$$

$$\dot{\bar{x}}_2 = \bar{A}_{22}\bar{x}_2$$

and

$$\omega = \bar{H}_1\bar{x}_1 + \bar{H}_2\bar{x}_2.$$

Here $\bar{x} = \begin{pmatrix}\bar{x}_1 \\ \bar{x}_2\end{pmatrix}$ are coordinates in R^n and $\bar{x}_2 = 0$ defines the intrinsic *controllable part* of $\mathcal{L}$. Furthermore $\mathcal{L}$ is controllable if and only if the coordinate set $\bar{x}_2$ is vacuous, that is, $\bar{x}_1$ spans all of R^n. The observed process $\mathcal{L}$ is called completely uncontrolled (or free) if the coordinate set $\bar{x}_1$ is vacuous, that is, $B = 0$. The decomposition of $\mathcal{L}$ shows that on $\bar{x}_2 = 0$ the restriction of $\mathcal{L}$ is $\dot{\bar{x}}_1 = \bar{A}_{11}\bar{x}_1 + \bar{B}_1 u$, which is controllable; and the projection of $\mathcal{L}$ onto $\bar{x}_1 = 0$ is $\dot{\bar{x}}_2 = \bar{A}_{22}\bar{x}_2$, which is completely uncontrollable. Thus we could define a controllable observed process as one that has no completely uncontrollable part.

In an analogous manner we shall define the observed process

$$\dot{x} = Ax + Bu \quad \text{and} \quad \omega = 0$$

to be completely unobservable, that is, $H = 0$. An observed process which has no such completely unobservable part will be called (completely) observable.

Definition. The autonomous linear observed process in R^n

$$(\mathfrak{L}) \qquad \dot{x} = Ax + Bu \quad \text{and} \quad \omega = Hx$$

is observable if it is not linearly equivalent to a dynamical system of the form

$$\dot{\bar{x}}_1 = \bar{A}_{11}\bar{x}_1 + \bar{A}_{12}\bar{x}_2 + \bar{B}_1 u$$

$$\dot{\bar{x}}_2 = \bar{A}_{22}\bar{x}_2 + \bar{B}_2 u$$

and

$$\omega = \bar{H}_2\bar{x}_2,$$

with $\bar{x} = \begin{pmatrix} \bar{x}_1 \\ \bar{x}_2 \end{pmatrix}$ and the coordinate set $\bar{x}_1$ nonvacuous.

Note that if $\mathfrak{L}$ does admit such a representation with $\bar{x}_1$ nonvacuous, then the restriction of $\mathfrak{L}$ to the subspace $\bar{x}_2 = 0$ (say for $u = 0$) is

$$\dot{\bar{x}}_1 = \bar{A}_{11}\bar{x}_1 \quad \text{and} \quad \omega = 0,$$

which is completely unobservable. The intrinsic decomposition of an arbitrary autonomous linear observed process into observable and completely unobservable parts will be determined later, and we show that $\mathfrak{L}$ is observable just in case the free system with $u \equiv 0$ has no nontrivial response $x(t)$ for which the output $\omega(t) \equiv 0$.

THEOREM 13. *The autonomous linear observed process in* R^n

$$(\mathfrak{L}) \qquad \dot{x} = Ax + Bu \quad \text{and} \quad \omega = Hx$$

is observable if and only if the dual dynamical system

$$(\mathfrak{L}') \qquad \dot{x} = A'x + H'u \quad \text{and} \quad \omega = B'x$$

is controllable. This obtains if and only if

$$\text{rank}\,[H' \quad A'H' \quad A'^2H' \quad \cdots \quad A'^{n-1}H'] = n.$$

Proof. The observed process $\mathfrak{L}$ fails to be observable just in case there is a coordinate transformation $\bar{x} = \begin{pmatrix} \bar{x}_1 \\ \bar{x}_2 \end{pmatrix} = Px$ (with $\bar{x}_1$ nonvacuous) such that

$$\bar{A} = PAP^{-1} = \begin{pmatrix} \bar{A}_{11} & \bar{A}_{12} \\ 0 & \bar{A}_{22} \end{pmatrix},$$

$$\bar{B} = PB = \begin{pmatrix} \bar{B}_1 \\ \bar{B}_2 \end{pmatrix} \quad \text{and} \quad \bar{H} = HP^{-1} = (0 \quad \bar{H}_2).$$

But in this case the transformation $\tilde{x} = \begin{pmatrix} \tilde{x}_1 \\ \tilde{x}_2 \end{pmatrix} = (P^{-1})'x$ acting on $\mathfrak{L}'$ yields the coefficients

$$P^{-1'}A'P' = \begin{pmatrix} \bar{A}'_{11} & 0 \\ \bar{A}'_{12} & \bar{A}'_{22} \end{pmatrix},$$

$$P^{-1'}H' = \begin{pmatrix} 0 \\ \bar{H}'_2 \end{pmatrix} \quad \text{and} \quad B'P' = (\bar{B}'_1 \quad \bar{B}'_2),$$

so that $\mathcal{L}'$ fails to be controllable. The same calculation shows that if $\mathcal{L}'$ fails to be controllable, then $\mathcal{L}$ fails to be observable.

Therefore $\mathcal{L}$ is observable if and only if $\mathcal{L}'$ is controllable, that is,

$$\text{rank}\,[H' \quad A'H' \quad A'^2H' \quad \cdots \quad A'^{n-1}H'] = n. \qquad \text{Q.E.D.}$$

Note that the dual of $\mathcal{L}'$ is $\mathcal{L}$ and hence $\mathcal{L}$ is controllable if and only if $\mathcal{L}'$ is observable. The duality mentioned above indicates that the theorems on controllability must have dually related theorems on observability (see Exercise 4 below). For example, the next result defines the *unobservable part* of a free observed process.

LEMMA 1. *Consider the autonomous linear observed process*

$$(\mathcal{L}) \qquad \dot{x} = Ax \quad \text{and} \quad \omega = Hx$$

which is completely uncontrollable in R^n. *Then there exists a unique linear subspace* $\mathcal{U}$, *of maximal dimension* $0 \leq l \leq n$, *such that*

(a) $\mathcal{U}$ *is invariant;*
(b) $\mathcal{L}$ *restricted to* $\mathcal{U}$ *is completely unobservable.*

Moreover $\mathcal{L}$ *is observable if and only if* $\mathcal{U} = 0$. *In appropriate coordinates* $\bar{x} = \begin{pmatrix} \bar{x}_1 \\ \bar{x}_2 \end{pmatrix}$, *where* $\mathcal{U}$ *is* $\bar{x}_2 = 0$, *the observed process is*

$$\dot{\bar{x}}_1 = \bar{A}_{11}\bar{x}_1 + \bar{A}_{12}\bar{x}_2$$

$$\dot{\bar{x}}_2 = \qquad\qquad \bar{A}_{22}\bar{x}_2$$

and

$$\omega = \qquad\qquad \bar{H}_2\bar{x}_2.$$

The projected system on $\bar{x}_1 = 0$, *which is vacuous if* $l = n$,

$$\dot{\bar{x}}_2 = \bar{A}_{22}\bar{x}_2 \quad \text{and} \quad \omega = \bar{H}_2\bar{x}_2$$

is observable.

Proof. If $\mathcal{U}_1$ and $\mathcal{U}_2$ are invariant subspaces on which $\mathcal{L}$ is completely unobservable, then $\omega = Hx$ is identically zero on the linear space $\mathcal{U}_1 + \mathcal{U}_2$ spanned by $\mathcal{U}_1$ and $\mathcal{U}_2$. Hence $\mathcal{U}_1 + \mathcal{U}_2$ is an invariant, completely unobservable subspace of $\mathcal{L}$. Define $\mathcal{U}$ as the linear space spanned by all invariant subspaces on which $\mathfrak{L}$ is completely unobservable. This construction yields $\mathfrak{U}$ as an invariant, completely unobservable subspace for $\mathfrak{L}$; and every other such subspace has a dimension less than that of $\mathfrak{U}$.

In appropriate coordinates $\bar{x} = \begin{pmatrix} \bar{x}_1 \\ \bar{x}_2 \end{pmatrix}$, where $\mathfrak{U}$ is $\bar{x}_2 = 0$, the observed process has the form stated in the lemma. If $l < n$, then $\bar{x}_2$ is nonvacuous and the projected system

$$\dot{\bar{x}}_2 = \bar{A}_{22}\bar{x}_2 \quad \text{and} \quad \omega = \bar{H}_2\bar{x}_2$$

is observable; for otherwise a further decomposition $\bar{x}_2 = \begin{pmatrix} \bar{x}_3 \\ \bar{x}_4 \end{pmatrix}$ violates the maximality property of $\mathfrak{U}$.

Finally $l = n$ just in case $H = 0$ so that $\mathfrak{L}$ is completely unobservable; and $l = 0$ just in case $\bar{x}_2$ spans R^n so that $\mathfrak{L}$ is observable.

Thus $\mathfrak{U} = 0$ implies that $\mathfrak{L}$ is observable. On the other hand, if $\mathfrak{L}$ is observable it admits a decomposition as in the lemma only when $\bar{x}_1$ is vacuous, so $\mathfrak{U} = 0$. Q.E.D.

In order to construct examples of controllable and observable linear processes we provide a canonical form for such systems. We consider here only the case of a single scalar controller $u(t)$, that is $m = 1$, since then Theorem 7 offers a basic canonical form for controllable processes.

LEMMA 2. *Consider an autonomous observed process in R^n*

$$(\mathfrak{L}) \qquad \dot{x} = Ax + bu, \quad \textit{with} \quad u \in R^1$$
$$\omega = Hx.$$

Assume $\mathfrak{L}$ is controllable with the form

$$A = \begin{pmatrix} 0 & 1 & 0 & 0 & \cdots & 0 & 0 \\ 0 & 0 & 1 & 0 & \cdots & 0 & 0 \\ \cdot & & & & & & \\ \cdot & & & & & & \\ \cdot & & & & & & \\ 0 & 0 & 0 & 0 & \cdots & 0 & 1 \\ -a_n & -a_{n-1} & & & \cdots & -a_2 & -a_1 \end{pmatrix}, \qquad b = \begin{pmatrix} 0 \\ 0 \\ \cdot \\ \cdot \\ \cdot \\ 0 \\ 1 \end{pmatrix}$$

and

$$H = \begin{pmatrix} b_{1n} & b_{1,n-1} & \cdots & b_{11} \\ b_{2n} & b_{2,n-1} & \cdots & b_{21} \\ \cdot & & & \\ \cdot & & & \\ \cdot & & & \\ b_{rn} & b_{r,n-1} & \cdots & b_{r1} \end{pmatrix}.$$

Then $\mathfrak{L}$ *is observable if and only if the polynomials*

$$D(p) = p^n + a_1 p^{n-1} + \cdots + a_n$$

and

$$N_1(p) = b_{11} p^{n-1} + b_{12} p^{n-2} + \cdots + b_{1n}$$
$$\vdots$$
$$N_r(p) = b_{r1} p^{n-1} + b_{r2} p^{n-2} + \cdots + b_{rn}$$

have no common root.

Proof. By duality $\mathfrak{L}$ is observable if and only if

$$(\mathfrak{L}') \qquad \dot{x} = Fx + Gu$$

is controllable, where

$$F = \begin{pmatrix} 0 & 0 & \cdots & 0 & -a_n \\ 1 & 0 & \cdots & 0 & -a_{n-1} \\ 0 & 1 & \cdots & 0 & \\ 0 & 0 & & 0 & \cdot \\ \cdot & & & & \cdot \\ \cdot & & & & \cdot \\ \cdot & & & & \\ 0 & 0 & \cdots & 1 & -a_1 \end{pmatrix} = A'$$

$$G = \begin{pmatrix} b_{1n} & b_{2n} & \cdots & b_{rn} \\ b_{1,n-1} & b_{2,n-1} & \cdots & b_{r,n-1} \\ \cdot & & & \\ \cdot & & & \\ \cdot & & & \\ b_{11} & b_{21} & \cdots & b_{r1} \end{pmatrix} = H'.$$

Thus $\mathfrak{L}$ is observable if and only if

$$\text{rank}\,[G \quad FG \quad F^2G \quad \cdots \quad F^{n-1}G] = n.$$

Introduce the notation

$$\Delta[P, Q] = [Q \quad PQ \quad P^2Q \quad \cdots \quad P^{n-1}Q]$$

for an arbitrary $n \times n$ matrix P and column vector Q. Note that $\Delta[P, Q]$ is a matrix valued function which is linear in Q. We shall evaluate this function for $P = F$ and for Q a column of G.

First let $e_1, e_2, \ldots, e_n$ be the consecutive column vectors of the identity matrix I, so $e_{i+1} = Fe_i$ for $1 \leq i \leq n - 1$. Next compute

$$\Delta[F, e_1] = [e_1 e_2 \cdots e_n] = I$$

and

$$\Delta[F, e_i] = \Delta[F, F^{i-1}e_1] = F^{i-1}\,\Delta[F, e_1] = F^{i-1},$$

for $1 \leq i \leq n$. Write the jth column of G as

$$\beta_j = b_{jn}e_1 + b_{jn-1}e_2 + \cdots + b_{j1}e_n$$

and compute

$$\Delta[F, \beta_j] = b_{jn}I + b_{jn-1}F + \cdots + b_{j1}F^{n-1} = N_j(F).$$

Thus

$$\text{rank}\,[G \quad FG \quad \cdots \quad F^{n-1}G] = \text{rank}\,[N_1(F), N_2(F), \ldots, N_r(F)],$$

since these two $n \times nr$ matrices have precisely the same columns except for arrangement.

To compute the rank of $[N_1(F), \ldots, N_r(F)]$ for the given polynomials $N_1, \ldots, N_r$ we can use convenient (perhaps complex) coordinates wherein F becomes triangular, say,

$$\begin{pmatrix} \lambda_1 & 0 & 0 & \cdots & 0 \\ * & \lambda_2 & 0 & \cdots & 0 \\ \cdot & & \cdot & & \\ \cdot & & & \cdot & \\ \cdot & & & & \cdot \\ * & & & * & \lambda_n \end{pmatrix}.$$

Here $\lambda_1, \ldots, \lambda_n$ are the eigenvalues of A and hence the roots of the polynomial $D(p)$. Then $[N_1(F), \ldots, N_r(F)]$ has the same rank as has [M]:

$$\left[\begin{pmatrix} N_1(\lambda_1) & & & 0 \\ & \cdot & & \\ & & \cdot & \\ * & & & N_1(\lambda_n) \end{pmatrix} \begin{pmatrix} N_2(\lambda_1) & & & 0 \\ & \cdot & & \\ & & \cdot & \\ * & & & N_2(\lambda_n) \end{pmatrix} \cdots \begin{pmatrix} N_r(\lambda_1) & & & 0 \\ & \cdot & & \\ & & \cdot & \\ * & & & N_r(\lambda_n) \end{pmatrix}\right].$$

Suppose among the roots $\lambda_1, \lambda_2, \ldots, \lambda_n$ of $D(p)$ there is a root, say λ_1, which is also a root of each of the polynomials $N_1(p), N_2(p), \ldots, N_r(p)$. Then the first row of the matrix in [M] is zero, and so the rank is less than n, and $\mathcal{L}$ is not observable. Thus, if $\mathcal{L}$ is observable, then $D, N_1, N_2, \ldots, N_r$ have no common root.

Conversely assume there is no common root for the polynomials $D, N_1, N_2, \ldots, N_r$. Then λ_1 fails to annihilate some N_{j_1}, and we select the column of [M] that contains $N_{j_1}(\lambda_1) \neq 0$ on the main diagonal of the corresponding $n \times n$ submatrix. For each $\lambda_1, \lambda_2, \ldots, \lambda_n$ we select such a column from matrix [M], and these n columns are clearly linearly independent. In this case matrix [M] has rank n, and hence $\mathfrak{L}$ is observable. Q.E.D.

The next theorem, the principal result of this section, unites the two basic approaches to linear control theory: the transfer-function or input-output approach, and the approach of observed processes as linear differential equations.

THEOREM 14. *Let $Z(p)$ be an $r \times m$ nonzero matrix of real rational functions, each with numerator of lesser degree than the denominator. Then there exists a real autonomous linear observed process*

$$(\mathfrak{L}) \qquad \dot{x} = Ax + Bu \quad \textit{and} \quad \omega = Hx$$

that is controllable and observable and that has $Z(p)$ for its transfer-function matrix. Moreover, $\mathfrak{L}$ is unique up to a linear equivalence, when $m = 1$.

Proof. By Theorem 12 we can realize $Z(p)$ as the transfer-function matrix of some autonomous linear observed process in R^N

$$(\tilde{\mathfrak{L}}) \qquad \dot{x} = Ax + Bu \quad \text{and} \quad \omega = Hx.$$

If we use Theorem 10, and assume that appropriate coordinates $x = \begin{pmatrix} x_a \\ x_b \end{pmatrix}$ have been chosen in R^N, we can denote the controllable part of $\tilde{\mathfrak{L}}$ by $x_b = 0$ and write the observed process

$$\dot{x}_a = A_{aa}x_a + A_{ab}x_b + B_a u$$

$$\dot{x}_b = \qquad A_{bb}x_b$$

and

$$\omega = H_a x_a + H_b x_b.$$

Note that x_a is not vacuous, for this would imply that $B = 0$ and that $\tilde{\mathfrak{L}}$ is completely uncontrollable and has a zero transfer-function matrix. Also the restriction of $\tilde{\mathfrak{L}}$ to $x_b = 0$ is the controllable system

$$(\mathfrak{L}_a) \qquad \dot{x}_a = A_{aa}x_a + B_a u \quad \text{and} \quad \omega = H_a x_a.$$

The impulse-response matrix of $\mathfrak{L}_a$ is the same as that of $\tilde{\mathfrak{L}}$, since

$$He^{tA}B = (H_a \; H_b)\begin{pmatrix} e^{tA_{aa}} & * \\ 0 & e^{tA_{bb}} \end{pmatrix}\begin{pmatrix} B_a \\ 0 \end{pmatrix} = H_a e^{tA_{aa}} B_a.$$

Thus $\mathfrak{L}_a$ has the transfer-function matrix $Z(p)$.

Next use Lemma 1 above to define coordinates $x_a = \begin{pmatrix} x_1 \\ x_2 \end{pmatrix}$ in order to obtain an observable projection of $\mathcal{L}_a$. That is, write

$$(\mathcal{L}_a) \qquad \begin{aligned} \dot{x}_1 &= A_{11}x_1 + A_{12}x_2 + B_1u \\ \dot{x}_2 &= \qquad\quad A_{22}x_2 + B_2u \end{aligned}$$

and

$$\omega = \qquad H_2x_2.$$

Here x_2 is nonvacuous, for this would imply that $H_a = 0$ and that $\mathcal{L}_a$ is completely unobservable with a zero transfer-function matrix. Also the projection of $\mathcal{L}_a$ on $x_1 = 0$ is the observable process (in Lemma 1 we have $B_a = 0$, but observability does not depend on B_a)

$$(\mathcal{L}) \qquad \begin{aligned} \dot{x}_2 &= A_{22}x_2 + B_2u \\ \omega &= H_2x_2. \end{aligned}$$

We verify the controllability of $\mathcal{L}$. The system $\mathcal{L}_a$, with dimension n_a is controllable and so

$$\operatorname{rank} \begin{bmatrix} * & * & * & & * \\ B_2 & A_{22}B_2 & A_{22}{}^2B_2 & \cdots & A_{22}^{n_a-1}B_2 \end{bmatrix} = n_a.$$

Thus the rows of the matrix

$$[B_2 \quad A_{22}B_2 \quad A_{22}{}^2B_2 \quad \cdots \quad A_{22}^{n_a-1}B_2]$$

are linearly independent. From the Cayley-Hamilton theorem we conclude that

$$\operatorname{rank}\,[B_2 \quad A_{22}B_2 \quad A_{22}{}^2B_2 \quad \cdots \quad A_{22}^{n-1}B_2] = n,$$

where n is the dimension of $\mathcal{L}$. Hence $\mathcal{L}$ is controllable and observable. The impulse-response matrix of $\mathcal{L}$ is that of $\mathcal{L}_a$ since

$$H_ae^{tA_{aa}}B_a = (0 \quad H_2)\begin{pmatrix} e^{tA_{11}} & * \\ 0 & e^{tA_{22}} \end{pmatrix}\begin{pmatrix} B_1 \\ B_2 \end{pmatrix} = H_2e^{tA_{22}}B_2.$$

Thus $\mathcal{L}$ has the required transfer-function matrix $Z(p)$.

Finally we prove that $\mathcal{L}$ is unique up to a linear equivalence in R^n. We restrict our proof to the case $m = 1$ for here we can assume that $\mathcal{L}$ is equivalent to an observed process

$$(\mathfrak{L}') \qquad \dot{x} = \begin{pmatrix} 0 & 1 & 0 & 0 & \cdots & 0 & 0 \\ 0 & 0 & 1 & 0 & \cdots & 0 & 0 \\ \cdot & & & & & & \\ \cdot & & & & & & \\ \cdot & & & & & & \\ 0 & 0 & 0 & 0 & \cdots & 0 & 1 \\ -a_n' & -a_{n-1}' & & & \cdots & -a_2' & -a_1' \end{pmatrix} x + \begin{pmatrix} 0 \\ 0 \\ \cdot \\ \cdot \\ \cdot \\ 0 \\ 1 \end{pmatrix} u$$

and

$$\omega = \begin{pmatrix} b_{1n}' & b_{1,n-1}' & \cdots & b_{11}' \\ b_{2n}' & b_{2,n-1}' & \cdots & b_{21}' \\ \cdot & & & \\ \cdot & & & \\ \cdot & & & \\ b_{rn}' & b_{r,n-1}' & \cdots & b_{r1}' \end{pmatrix} x.$$

Also the transfer-function matrix of $\mathfrak{L}'$ is

$$Z(p) = \frac{1}{D'(p)} \begin{pmatrix} N_1'(p) \\ \cdot \\ \cdot \\ \cdot \\ N_r'(p) \end{pmatrix}$$

where the polynomials in this matrix are

$$D'(p) = p^n + a_1' p^{n-1} + \cdots + a_n'$$
$$N_1'(p) = b_{11}' p^{n-1} + \cdots + b_{1n}'$$
$$\cdot$$
$$\cdot$$
$$\cdot$$
$$N_r'(p) = b_{r1}' p^{n-1} + \cdots + b_{rn}'.$$

Since $\mathfrak{L}'$ is observable the polynomials $D', N_1', \ldots, N_r'$ have no common root (note that the prime $'$ here does not signify differentiation).

We prove that $\mathfrak{L}'$ is uniquely determined by the known transfer-function matrix

$$Z(p) = \frac{1}{D(p)} \begin{pmatrix} N_1(p) \\ \cdot \\ \cdot \\ \cdot \\ N_r(p) \end{pmatrix},$$

where we have cancelled common factors to obtain the relatively prime polynomials

$$D(p) = p^n + a_1 p^{n-1} + \cdots + a_n$$

$$N_1(p) = b_{11} p^{n-1} + \cdots + b_{1n}$$

$$\cdot$$
$$\cdot$$
$$\cdot$$

$$N_r(p) = b_{r1} p^{n-1} + \cdots + b_{rn}.$$

In this way we show that two controllable, observable processes with the same transfer-function matrix $Z(p)$ must both be equivalent to $\mathfrak{L}'$.

We have

$$\frac{N_j(p)}{D(p)} = \frac{N_j'(p)}{D'(p)} \quad \text{for} \quad j = 1, 2, \ldots, r$$

and so

$$D(p)\, N_j'(p) = D'(p)\, N_j(p).$$

If $p = \lambda_1$ is a root of $D(p)$, then there is a polynomial $N_{j_1}(p)$ which does not have the root λ_1. Hence $D'(p)$ must have the root λ_1. Hence $D(p)$ and $D'(p)$ have the same roots and, since they each have leading coefficient $+1$, $D(p) \equiv D'(p)$ (and they have the same degree n, as already indicated in the notation). Also $N_j(p) \equiv N_j'(p)$ for $1 \leq j \leq r$.

Thus the dimensions r and n in $\mathfrak{L}'$ are fixed by the dimensions of $Z(p)$ and the degree of the common denominator $D(p)$ in $Z(p)$. Also the exact coefficients of the observed process $\mathfrak{L}'$ are completely fixed. Hence $\mathfrak{L}'$ serves as the unique canonical form for any controllable, observable, autonomous observed process with transfer-function matrix $Z(p)$. Q.E.D.

Example. Let us consider the case $r = m = 1$ and construct the controllable, observable process for the transfer function

$$\frac{N(p)}{D(p)} = \frac{b_1 p^{n-1} + b_2 p^{n-2} + \cdots + b_n}{p^n + a_1 p^{n-1} + \cdots + a_n}$$

where the real polynomials $N(p)$ and $D(p)$ are relatively prime (in particular, $N(p) \not\equiv 0$).

On occasion this transfer function is represented by the differential equation with *numerator dynamics*

$$x^{(n)} + a_1 x^{(n-1)} + \cdots + a_n x = b_1 u^{(n-1)} + b_2 u^{(n-2)} + \cdots + b_n u.$$

However this differential equation is not an observed process since it involves derivatives of the control function $u(t)$. Hence discontinuous relay controllers cannot be discussed within this framework (unless an extensive theory of generalized functions and their derivatives is developed). We therefore abandon the techniques of numerator dynamics and return to the construction of an appropriate observed process.

We obtain the required realization of $N(p)/D(p)$ with

$$x^{(n)} + a_1 x^{(n-1)} + \cdots + a_n x = u$$

and

$$\omega = b_1 x^{(n-1)} + \cdots + b_n x.$$

That is, take

$$A = \begin{pmatrix} 0 & 1 & 0 & 0 & \cdots & 0 & 0 \\ 0 & 0 & 1 & 0 & \cdots & 0 & 0 \\ \cdot & & & & & & \\ \cdot & & & & & & \\ \cdot & & & & & & \\ 0 & 0 & 0 & 0 & \cdots & 0 & 1 \\ -a_n & -a_{n-1} & & & \cdots & -a_2 & -a_1 \end{pmatrix} \qquad B = \begin{pmatrix} 0 \\ 0 \\ \cdot \\ \cdot \\ \cdot \\ 0 \\ 1 \end{pmatrix}$$

and

$$H = (b_n \quad b_{n-1} \quad \cdots \quad b_2 \quad b_1).$$

This observed process is controllable and observable and it realizes the transfer function $N(p)/D(p)$ from the input u to the output ω.

The proof of the uniqueness of the controllable, observable process with a prescribed transfer-function matrix is omitted for the more complicated situations where $m > 1$. It should also be remarked that the theory of controllability and observability can be extended to nonautonomous linear processes (see exercises below) but in these cases the practical computable criteria listed in Theorems 5 and 13 are not available.

For the last topic in this section we study the problem of control where the state vector x is to be steered in a finite time to a prescribed nonempty target G, and then x is to be held within G forever afterwards. While we shall not write formal observability equations, the most interesting case is that of multicomponent control, where $x^1, x^2, \ldots, x^r$ are the observed components that are to be regulated to zero and held there. In the case of multicomponent control the target G is the linear space $x^1 = 0,\ x^2 = 0, \ldots, x^r = 0$ in R^n.

Definition. Consider the autonomous linear control process in R^n

$$(\mathfrak{L}) \qquad \dot{x} = Ax + Bu$$

with compact restraint set $\Omega \subset R^m$, and target set $G \subset R^n$. The core of G consists of all points $x_1 \in G$ for which there exists a control function

$u(t) \subset \Omega$ on $0 \leq t < \infty$, which steers x_1 along a response $x_1(t) \subset G$ on $0 \leq t < \infty$.

If an initial state $x_0 \in R^n$ is to be steered to G and then held in G forever afterwards, we must steer x_0 to the set core (G), which is defined above. Thus the control process of steering x_0 to G, with the subsequent condition of holding the response within G, can be reduced to the standard problem of steering x_0 to core (G), with no further discussion of the response after it meets the set core (G).

THEOREM 15. *Consider the autonomous linear control process in* R^n

$$(\mathfrak{L}) \qquad \dot{x} = Ax + Bu$$

with compact convex restraint set $\Omega \subset R^m$ *and closed convex target set* G. *Then core* (G) *is a closed convex subset of* G. *Furthermore*

$$\text{core}\,(\text{core}\,(G)) = \text{core}\,(G).$$

Proof. Let x_1 and x_2 be controlled in G by $u_1(t)$ and $u_2(t) \subset \Omega$ on $0 \leq t < \infty$, that is,

$$x_i(t) = e^{At}x_i + e^{At}\int_0^t e^{-As}B\, u_i(s)\, ds \quad \text{for} \quad i = 1, 2.$$

Then for $0 \leq \lambda \leq 1$, the convex combination

$$\lambda\, x_1(t) + (1 - \lambda)\, x_2(t) = e^{At}(\lambda x_1 + (1 - \lambda)x_2) + e^{At}\int_0^t e^{-As}\, B(\lambda u_1(s) + (1 - \lambda)\, u_2(s))\, ds$$

and so $[\lambda\, u_1(t) + (1 - \lambda)\, u_2(t)] \subset \Omega$ on $0 \leq t < \infty$ steers $[\lambda x_1 + (1 - \lambda)x_2]$ in G. Thus core (G) is convex.

Let $x_1, x_2, \ldots$ be points in core (G) with $\lim_{k\to\infty} x_k = \bar{x} \in G$. Let $u_1(t), u_2(t), \ldots$ be the corresponding controllers that hold the points $x_1, x_2, \ldots$ in G. Select a subsequence, still written $x_1, x_2, \ldots$ such that $\lim_{k\to\infty} u_k(t) = \bar{u}(t) \subset \Omega$ with weak convergence on each finite time interval $0 \leq t \leq t_1$. Then the limiting response is

$$\bar{x}(t) = e^{At}\bar{x} + e^{At}\int_0^t e^{-As}B\, \bar{u}(s)\, ds$$

and

$$\bar{x}(t) = \lim_{k\to\infty} x_k(t) \quad \text{for each fixed} \quad t \geq 0.$$

Since G is closed, $\bar{x}(t) \subset G$ for all $t \geq 0$. Hence $\bar{x} \in$ core (G) and core (G) is closed.

If $x_0 \in \text{core}\,(G)$, then there is a response $x_0(t) \subset G$ for some controller $u(t) \subset \Omega$ on $0 \leq t < \infty$. But then for each fixed $t \geq 0$, $x_0(t)$ serves as the initial point of a response in G with a controller in Ω. Hence $x_0(t) \in \text{core}\,(G)$ for each $t \geq 0$. Thus $x_0 \in \text{core}\,(\text{core}\,(G))$ and so $\text{core}(\text{core}\,(G)) = \text{core}\,(G)$.

Q.E.D.

It is often difficult to determine whether core (G) is compact, even if G is a linear subspace of R^n (see example in Section 1.3), However, for time-optimal control problems the extremely remote part of G is usually eliminated as an appropriate target. Thus we shall often find problems where the target is compact.

The case where G is a linear subspace of R^n is called multicomponent regulation. The next theorem shows that this problem reduces, in the generic case, to single-component regulation.

THEOREM 16. *Consider the autonomous linear control process in R^n*

$$(\mathfrak{L}) \qquad \dot{x} = Ax + bu$$

with compact restraint interval $\Omega \subset R^1$ containing $u = 0$. Assume $\mathfrak{L}$ is controllable and let π be a proper linear subspace of R^n. Then there exists a hyperplane $\hat{\pi}$ of dimension $n - 1$ such that

$$\text{core}\,(\hat{\pi}) = \text{core}\,(\pi)$$

Proof. The set core (π) is a closed convex subset of π, nonempty since it contains $x = 0$, and so core (π) has a nonempty interior relative to some $(n-r)$-dimensional subspace $\pi_1 \subset \pi$. Also $\text{core}\,(\pi_1) = \text{core}\,(\pi)$, and we choose coordinates $\bar{x} = \begin{pmatrix} \bar{x}_1 \\ \bar{x}_2 \end{pmatrix}$ in R^n so that π_1 is just $\bar{x}_1 = \begin{pmatrix} \bar{x}^1 \\ \cdot \\ \cdot \\ \cdot \\ \bar{x}^r \end{pmatrix} = 0$ (if π_1 is just the origin, $\bar{x}_2$ is vacuous).

Write $\mathfrak{L}$ as

$$\dot{\bar{x}}_1 = \bar{A}_{11}\bar{x}_1 + \bar{A}_{12}\bar{x}_2 + \bar{b}_1 u$$

$$\dot{\bar{x}}_2 = \bar{A}_{21}\bar{x}_1 + \bar{A}_{22}\bar{x}_2 + \bar{b}_2 u.$$

Suppose $\bar{b}_1 = 0$. Then at points of core (π_1) where $\bar{x}_1 = 0$, we have $\dot{\bar{x}}_1 = 0$ and $\bar{A}_{12}\bar{x}_2 = 0$. Since core (π_1) has an interior in π_1, we find that $\bar{A}_{12} = 0$. But this implies that

$$\dot{\bar{x}}_1 = \bar{A}_{11}\bar{x}_1,$$

which contradicts the controllability of $\mathfrak{L}$. Therefore $\bar{b}_1 \neq 0$.

Define new coordinates in R^n, still called $\bar{x}$, so that π_1 remains $\bar{x}_1 = 0$ but

$$\bar{b}_1 = \begin{pmatrix} 0 \\ 0 \\ \cdot \\ \cdot \\ \cdot \\ 0 \\ 1 \end{pmatrix},$$

and write

$$\bar{A}_{12} = \begin{pmatrix} a_{r+1}^1 & \cdots & a_n{}^1 \\ \cdot & & \\ \cdot & & \\ \cdot & & \\ a_{r+1}^r & \cdots & a_n{}^r \end{pmatrix}.$$

Suppose the kth row of $\bar{A}_{12}$, for $1 \leq k < r$, fails to vanish. Then

$$\dot{\bar{x}}^k = a_{r+1}^k \bar{x}^{r+1} + \cdots + a_n{}^k \bar{x}^n = 0$$

everywhere in core (π_1). But this implies that core (π_1) lies within the intersection of

$$\bar{x}^1 = \bar{x}^2 = \cdots = \bar{x}^r = 0 \quad \text{and} \quad a_{r+1}^k \bar{x}^{r+1} + \cdots + a_n{}^k \bar{x}^n = 0,$$

which contradicts the assertion that dimension core (π_1) $= n - r$. Thus

$$\bar{A}_{12} = \begin{pmatrix} 0 & 0 & \cdots & 0 \\ 0 & 0 & \cdots & 0 \\ \cdot & & & \\ \cdot & & & \\ \cdot & & & \\ 0 & 0 & \cdots & 0 \\ a_{r+1}^r & & \cdots & a_n{}^r \end{pmatrix}.$$

For any two points $(\bar{x}_{1a}, 0)$ and $(\bar{x}_{1b}, 0)$ in R^n there exists a controller $u(t) \subset R^1$ steering the first point to the second along a response $\bar{x}(t)$. But then

$$\bar{u}(t) = a_{r+1}^r \bar{x}^{r+1}(t) + \cdots + a_n{}^r \bar{x}^n(t) + u(t)$$

steers $\bar{x}_1(t)$ from $\bar{x}_{1a}$ to $\bar{x}_{1b}$ in R^r along a solution of

$$\dot{\bar{x}}_1 = \bar{A}_{11}\bar{x}_1 + \bar{b}_1 \, \bar{u}(t).$$

Thus the process in R^r is controllable and

$$P = [\bar{A}_{11}^{r-1}\bar{b}_1 \quad \bar{A}_{11}^{r-2}\bar{b}_1 \quad \cdots \quad \bar{A}_{11}\bar{b}_1 \quad \bar{b}_1]$$

is a nonsingular $r \times r$ matrix.

Define new coordinates in $\bar{x}_2 = 0$ by

$$\bar{x}_1 = Py$$

to obtain the process $\mathfrak{L}$ in the form

$$\begin{aligned} \dot{y} &= P^{-1}\bar{A}_{11}Py + P^{-1}\bar{A}_{12}\bar{x}_2 + P^{-1}\bar{b}_1 u \\ \dot{\bar{x}}_2 &= \bar{A}_{21}Py + \bar{A}_{22}\bar{x}_2 + \bar{b}_2 u. \end{aligned}$$

But now a direct calculation, as in Theorem 6, shows that the process $\mathfrak{L}$ assumes the simple form

$$\begin{aligned} \dot{y}^1 &= \alpha_1 y^1 + y^2 \\ \dot{y}^2 &= \alpha_2 y^1 + y^3 \\ \dot{y}^3 &= \alpha_3 y^1 + y^4 \\ &\cdot \\ &\cdot \\ &\cdot \\ \dot{y}^{r-1} &= \alpha_{r-1} y^1 + y^r \\ \dot{y}^r &= \alpha_r y^1 + a_{r+1}^r \bar{x}^{r+1} + \cdots + a_n{}^r \bar{x}^n + u \end{aligned}$$

and

$$\dot{\bar{x}}_2 = \bar{A}_{21}Py + \bar{A}_{22}\bar{x}_2 + \bar{b}_2 u,$$

where $\alpha_1, \alpha_2, \ldots, \alpha_r$ are certain real constants.

Now consider the hyperplane $\hat{\pi}$ defined by $y^1 = 0$ in R^n. Then $\pi_1 \subset \hat{\pi}$ so core $(\pi_1) \subset$ core $(\hat{\pi})$. Take a point $Q \in$ core $(\hat{\pi})$. There is a response from Q, by a controller $u(t) \subset \Omega$, whereon

$$\dot{y}^1 = 0 \quad \text{so} \quad y^2 = 0.$$

But then

$$\dot{y}^2 = 0 \quad \text{so} \quad y^3 = 0,$$

and, continuing this analysis, we find that $y^1 = 0$, $y^2 = 0$, $y^3 = 0, \ldots,$ $y^r = 0$ on the response from Q. Thus $Q \in$ core (π_1) and

$$\text{core}\,(\hat{\pi}) = \text{core}\,(\pi_1) = \text{core}\,(\pi)$$

as required. Q.E.D.

EXERCISES

1. For an autonomous observed process with impulse-response matrix $W(t)$ show that the input $u(t)$, where $u(t) = 0$ for $t < 0$, forces the initial

state $x_0 = 0$ to produce the output

$$\omega(t) = \int_{-\infty}^{\infty} W(s)\, u(t - s)\, ds \quad \text{for} \quad t \geq 0.$$

2. Construct the autonomous linear process that is observable, controllable, and that realizes the transfer-function matrix

$$Z(p) = \begin{pmatrix} \dfrac{1}{p^2 - 1} \\[2ex] \dfrac{p}{p^2 + p} \\[2ex] \dfrac{p}{p^2 - p} \end{pmatrix}.$$

3. Interpret the transfer-function matrix $Z(p)$, for an autonomous observed process, as the (complex) amplitude of the periodic output to a unit harmonic input.

4. Consider the linear process in R^n

$$(\mathfrak{L}) \qquad \dot{x} = A(t)x + B(t)u$$

with controllers $u(t) \subset R^m$ and continuous coefficient matrices $A(t)$, $B(t)$ for all real t. Define $\mathfrak{L}$ to be completely controllable if, for each pair of points x_0, x_1 in R^n and for each initial time t_0, there exists a controller $u(t)$ on some interval $t_0 \leq t \leq t_1$ steering x_0 to x_1.

The following exercises develop the theory of controllability for nonautonomous systems $\mathfrak{L}$. For notational purposes denote the initial state x_0 at time t_0 by $\{x_0, t_0\}$.

(a) Let $C(t_0)$ be the collection of all points $x_0 \in R^n$ such that $\{x_0, t_0\}$ can be steered to the origin. Show that $C(t_0)$ is a linear subspace of R^n and that there exists a time $\bar{t}_0 > t_0$ such that each $\{x_0, t_0\} \in \{C(t_0), t_0\}$ can be steered to $\{0, \bar{t}_0\}$.

(b) Each $\{x_0, t_0\} \in \{C(t_0), t_0\}$ can be steered to each $\{x_1, t_1\} \in \{C(t_1), t_1\}$ for $t_1 \geq \bar{t}_0$.

(c) $\mathfrak{L}$ is completely controllable if and only if every $C(t_0) = R^n$.

(d) Define the symmetric positive semidefinite matrix

$$W(t_0, t_1) = \int_{t_0}^{t_1} \Phi(t_0, t)\, B(t)\, B'(t)\, \Phi'(t_0, t)\, dt$$

where $\Phi(t, t_0)$ is the fundamental solution matrix of $\dot{x} = A(t)x$ with $\Phi(t_0, t_0) = I$. Show that the range of the transformation W on R^n, for $t_1 = \bar{t}_0$, is $R[W(t_0, \bar{t}_0)] = C(t_0)$. Hence $\mathfrak{L}$ is completely controllable if and only if $W(t_0, \bar{t}_0)$ is nonsingular for each $t_0 \in R^1$.

(e) The observed process in R^n.

$$(\mathfrak{L}) \qquad \dot{x} = A(t)x + B(t)u \quad \text{and} \quad \omega = H(t)x$$

is called observable on $t \leq t_0$ in case the dual process

$$(\mathfrak{L}') \qquad \dot{x} = -A'(t_0 - t)x - H'(t_0 - t)u \quad \text{and} \quad \omega = B'(t_0 - t)x$$

is controllable on $t \geq t_0$ (that is, $C(t_0) = R^n$). Define $\mathfrak{L}$ to be completely observable if it is observable for each $t_0 \in R^1$. Using these definitions show that an autonomous process is completely observable if and only if

$$\operatorname{rank}\,[H', A'H', \ldots, A'^{n-1}H'] = n.$$

5. Consider the linear system with numerator dynamics

$$(\mathfrak{D}) \qquad x^{(n)} + a_1(t)x^{(n-1)} + \cdots + a_n(t)x = b_1(t)u^{(n-1)} + \cdots + b_n(t)u$$

where the coefficients $a_1(t), \ldots, b_n(t)$ are smooth C^∞ functions for all t. Then, with initial data $x = \dot{x} = \cdots = x^{(n-1)} = 0$ at $t = 0$, and with a smooth input $u(t)$ on $t \geq 0$, there is a well-determined response or output $x(t)$. Consider the observed process

$$(\mathfrak{L}) \qquad \begin{aligned} \dot{x}^1 &= x^2 + G_1(t)u \\ \dot{x}^2 &= x^3 + G_2(t)u \\ &\;\;\vdots \\ \dot{x}^{n-1} &= x^n + G_{n-1}(t)u \\ \dot{x}^n &= -a_n(t)x^1 - a_{n-1}(t)x^2 - \cdots - a_1(t)x^n + G_n(t)u \end{aligned}$$

and

$$\omega = x^1.$$

Here we define the coefficients $a_0 \equiv 1$, $G_0 \equiv 0$,

$$G_1(t) = b_1(t)$$

and for $2 \leq i \leq n$

$$G_i(t) = b_i(t) - \sum_{k=0}^{i-1} \sum_{s=0}^{i-k} \binom{n+s-i}{n-i} a_{i-k-s}(t) \frac{d^s G_k}{dt^s}.$$

(a) Show that the coefficients $G_i(t)$ can be determined formally by successively eliminating $x^2, x^3, \ldots, x^n$ from $\mathfrak{L}$ and demanding that the resulting differential equation for x^1 be precisely $\mathfrak{D}$.

(b) Show that the output $\omega = x^1(t)$ for $\mathfrak{L}$, with initial state $x_0 = 0$ at $t = 0$ and input function with $u(0) = \dot{u}(0) = \cdots = u^{(n-2)}(0) = 0$, is precisely the same as the corresponding output $x(t)$ of $\mathfrak{D}$.

(c) It is commonly asserted that the process $\mathfrak{L}$ is completely controllable provided $G_n(t) \not\equiv 0$. Discuss the controllability of $\dot{x}^1 = x^2 + tu$, $\dot{x}^2 = u$ in this context.

6. Consider the control process in R^2 defined by $\ddot{x} = u$ with restraint $|u(t)| \leq 1$. Let the target G be the line $x^1 + x^2 = 0$ and compute core (G).

7. Consider the set of all autonomous observed processes in R^n with $u \in R^m$ and $\omega \in R^r$,

$$\dot{x} = Ax + Bu \quad \text{and} \quad \omega = Hx.$$

Show that the generic process (in the sense of Theorem 11) is controllable and observable.

2.5 TIME OPTIMAL CONTROL FOR LINEAR PROCESSES

In this section we shall prove the fundamental existence and uniqueness theorems for optimal controllers of linear processes. We shall establish the maximal principle, which characterizes the optimal controller as an extremal controller, and use this to synthesize the optimal controller through the technique of the switching locus. In each case we shall first state the general theory for time-dependent processes and then give specific, detailed, and computable results for autonomous processes.

We shall study the minimal time-optimal control problem for the linear process in R^n

$$(\mathfrak{L}) \qquad \dot{x} = A(t)x + B(t)u + v(t)$$

where the coefficient matrices $A(t)$, $B(t)$, and $v(t)$ are integrable on every finite interval, as postulated in the first section of this chapter. The restraint set Ω is a nonempty compact set in R^m and the target $G(t)$ is a continuously varying nonempty compact set on $\tau_0 \leq t \leq \tau_1$. The set Δ of admissible controllers consists of all measurable vectors $u(t) \subset \Omega$, on various finite time intervals $\tau_0 \leq t \leq t_1 \leq \tau_1$, steering the initial point x_0 at $t = \tau_0$ to the target $G(t_1)$ at $t = t_1$.

THEOREM 17. *Consider the linear control process in* R^n

$$(\mathfrak{L}) \qquad \dot{x} = A(t)x + B(t)u + v(t)$$

with compact restraint set $\Omega \subset R^m$, *initial state* $x_0 \in R^n$, *and continuously varying compact target set* $G(t)$ *on* $\tau_0 \leq t \leq \tau_1$. *If there exists a controller*

$u(t) \subset \Omega$ on $\tau_0 \leq t \leq t_1 \leq \tau_1$ steering x_0 to $G(t_1)$, then there exists a minimal time-optimal controller $u^*(t) \subset \Omega$ on $\tau_0 \leq t \leq t^* \leq \tau_1$ steering x_0 to $G(t^*)$.

Proof. If $x_0 \in G(\tau_0)$, then we agree that the minimal time duration is zero, that is $t^* = \tau_0$. Now assume $x_0 \notin G(\tau_0)$ and consider controllers $u(t)$ on $\tau_0 \leq t \leq t_1$ with $\tau_0 < t_1 \leq \tau_1$. Consider the set of attainability $K(t_1)$ initiating from x_0 at τ_0.

Define t^* as the greatest lower bound of all times t_1 such that $K(t_1)$ meets $G(t_1)$. By the continuous dependence of $K(t_1)$ and $G(t_1)$ on t_1, the set of times for which $K(t_1)$ meets $G(t_1)$ is a closed set in R^1. Hence t^* on $\tau_0 < t^* \leq \tau_1$ is the first or minimal time at which $K(t)$ meets $G(t)$. Let $u^*(t) \subset \Omega$ on $\tau_0 \leq t \leq t^*$ be any controller steering x_0 to $K(t^*) \cap G(t^*)$. Then $u^*(t)$ is an optimal controller, as required. Q.E.D.

In the existence theorem (Theorem 17) we seek a minimal time-optimal controller $u^*(t) \subset \Omega$ on $\tau_0 \leq t \leq t^*$, steering the initial state x_0 at $t = \tau_0$ to $G(t^*)$. If we free the initial time instant and merely seek an optimal controller on some finite interval $\tau_0 \leq t_0^* \leq t \leq t_1^* \leq \tau_1$, then the corresponding existence can be demonstrated by using the limit t_0^* of a sequence of starting times $t_0^{(\nu)}$ with appropriately decreasing control durations $t^{*(\nu)} - t_0^{(\nu)}$.

In the following corollary we give explicit criteria for the existence of an optimal controller for an autonomous process.

COROLLARY. *Consider the autonomous linear process in R^n*

$$(\mathcal{L}) \qquad \dot{x} = Ax + Bu$$

with compact restraint set $\Omega \subset R^m$, initial state x_0, and the origin as the fixed target in R^n.

Assume

(a) *$u = 0$ lies in the interior of Ω;*

(b) *$\mathcal{L}$ is controllable;*

(c) *A is stable, that is each eigenvalue λ of A satisfies* Re $\lambda < 0$. *Then there exists a minimal time-optimal controller $u^*(t) \subset \Omega$ on $0 \leq t \leq t^*$ steering x_0 to the origin.*

Proof. By Corollary 3 of Theorem 5 the domain of null controllability of $\mathcal{L}$ is all R^n. Thus there exists a controller $u(t) \subset \Omega$ on $0 \leq t \leq t_1$ steering x_0 to the origin. By the theorem there exists an optimal controller $u^*(t) \subset \Omega$ on $0 \leq t \leq t^* \leq t_1$ steering x_0 to the origin. Q.E.D.

If $m = 1$, that is $\Omega \subset R^1$ in the above corollary, then we can relax the third hypothesis to

(c′) each eigenvalue λ of A satisfies Re $\lambda \leq 0$, according to Theorem 8.

For nonautonomous linear processes there is a useful global stability criterion (see Exercise 6) but it is difficult to test the required controllability in any direct manner.

The next theorem, known as the maximal principle for linear processes, establishes important extremal properties for an optimal controller. In fact, under rather general normality conditions described below, these extremal properties completely characterize the optimal controller. In the remaining theorems of this chapter it is assumed that the target set is compact and moves continuously with time. It is only necessary to assume that the target set is closed and moves continuously in the sense that its intersection with each fixed compact set moves continuously.

THEOREM 18. *Consider the linear control process in* R^n

$$(\mathcal{L}) \qquad \dot{x} = A(t)x + B(t)u + v(t)$$

with compact restraint set $\Omega \subset R^m$, *initial state* $x_0 \in R^n$, *and continuously varying compact target set* $G(t)$ *on* $\tau_0 \leq t \leq \tau_1$. *Let* $u^*(t) \subset \Omega$ *on* $\tau_0 \leq t \leq t^*$ *be a minimal time-optimal controller with response* $x^*(t)$ *steering* x_0 *to* $G(t^*)$. *Then* $u^*(t)$ *is extremal, that is,*

$$m(t) \equiv \max_{u \in \Omega} \eta(t)\, B(t)u = \eta(t)\, B(t)\, u^*(t)$$

and so

$$M(t) \equiv \max_{u \in \Omega} \eta(t)[A(t)\, x^*(t) + B(t)u + v(t)]$$
$$= \eta(t)[A(t)\, x^*(t) + B(t)\, u^*(t) + v(t)]$$

almost always on $\tau_0 \leq t \leq t^*$. *Here* $\eta(t)$ *is a nontrivial solution of the adjoint system*

$$\dot{\eta} = -\eta\, A(t)$$

and $\eta(t^*)$ *is an outwards unit normal to a supporting hyperplane to the set of attainability* $K(t^*)$ *at* $x^*(t^*)$ *in* $\partial K(t^*)$.

Furthermore if $G(t) = G$ *is constant, then* $x^*(t^*)$ *lies in the new frontier of* $K(t^*)$. *In this case, provided* $A(t)$, $B(t)$, *and* $v(t)$ *are continuous, the normal* $\eta(t^*)$ *can be selected so that*

$$M(t^*) \geq 0.$$

If, in addition, G *is convex, then* $\eta(t^*)$ *can also be selected to satisfy the transversality condition, namely that* $\eta(t^*)$ *is normal to a common supporting hyperplane separating* $K(t^*)$ *and* G.

Proof. The response endpoint $x^*(t^*)$ must lie on the boundary $\partial K(t^*)$. For if $x^*(t^*)$ were in the interior of $K(t^*)$, then, by Theorem 1, some open neighborhood N of $x^*(t^*)$ lies in the interior of $K(t)$ for all t sufficiently near t^*. But then the continuity of $G(t)$ implies that $G(t_1)$ meets N for some $t_1 < t^*$, which contradicts the optimality of $u^*(t)$. Hence $x^*(t^*) \in \partial K(t^*)$, that is, $u^*(t)$ is an extremal controller.

By Theorem 2, for the extremal controller $u^*(t)$, there exists a nontrivial adjoint response $\eta(t)$ so that

$$m(t) = \eta(t)\, B(t)\, u^*(t)$$

and

$$M(t) = \eta(t)[A(t)\, x^*(t) + B(t)\, u^*(t) + v(t)]$$

almost always on $\tau_0 \leq t \leq t^*$. Also $\eta(t)$ can be selected as any solution of

$$\dot{\eta} = -\eta\, A(t)$$

with $\eta(t^*)$ an outwards normal to any supporting hyperplane to $K(t^*)$ at the point $x^*(t^*)$.

Now let $G(t) = G$ be a constant nonempty compact set in R^n. Then, from the minimal time optimality of $u^*(t)$, we note that $x^*(t^*) \in K(t^*) \cap G$ lies in the new frontier of $K(t^*)$. Since $x^*(t)$ may not have a derivative at $t = t^*$, we must use a limiting process to prove that $M(t^*) \geq 0$.

At each instant t there is a hyperplane $\hat{\pi}(t)$ midway between $K(t)$ and the endpoint $x^*(t^*)$, that is, $\hat{\pi}(t)$ is the perpendicular bisector of the shortest chord from $x^*(t^*)$ to $K(t)$. Choose t_1 on $\tau_0 \leq t_1 < t^*$ and then $\hat{\pi}(t_1)$ separates $x^*(t_1)$ from $x^*(t^*)$. Thus, at some instant $\hat{t}_1 > t_1$, the velocity $\dot{x}^*(\hat{t}_1) = A(\hat{t}_1)x^*(\hat{t}_1) + B(\hat{t}_1)\, u^*(\hat{t}_1) + v(\hat{t}_1)$ must have a positive component along the unit normal $\hat{\eta}(t_1)$ to $\hat{\pi}(t_1)$, directed out of the half-space containing $K(t_1)$. Now choose t_2 on $\hat{t}_1 < t_2 < t^*$ and let $\hat{t}_2 > t_2$ be such that $\hat{\eta}(t_2)\, \dot{x}^*(\hat{t}_2) > 0$. In this way we define a sequence of times

$$\tau_0 \leq t_1 < \hat{t}_1 < t_2 < \hat{t}_2 < \cdots < t_\nu < \hat{t}_\nu < \cdots < t^*$$

with

$$\hat{\eta}(t_\nu)[A(\hat{t}_\nu)\, x^*(\hat{t}_\nu) + B(\hat{t}_\nu)\, u^*(\hat{t}_\nu) + v(\hat{t}_\nu)] > 0.$$

Now use the compactness of Ω and of the sphere of unit directions to select a subsequence, still denoted by t_ν, such that the following limits exist:

$$\lim_{\nu\to\infty} u^*(\hat{t}_\nu) = \bar{u} \in \Omega$$

$$\lim_{\nu\to\infty} \hat{\pi}(t_\nu) = \pi(t^*) \quad \text{and} \quad \lim_{\nu\to\infty} \hat{\eta}(t_\nu) = \eta(t^*).$$

Then $\pi(t^*)$ is a supporting hyperplane to $K(t^*)$ at $x^*(t^*)$, with outward unit normal $\eta(t^*)$. From the continuity of $A(t)$, $B(t)$, $v(t)$, and $x^*(t)$ we obtain

$$\eta(t^*)[A(t^*)\, x^*(t^*) + B(t^*)\bar{u} + v(t^*)] \geq 0$$

so

$$M(t^*) \geq 0, \quad \text{as required.}$$

If G is a convex target, then we repeat the above argument with $\hat{\pi}(t)$ as the perpendicular bisector of any (and hence all) shortest chord from G to $K(t)$. Then the limiting hyperplane $\pi(t^*)$ and unit normal $\eta(t^*)$ satisfy the required transversality condition. Q.E.D.

For the autonomous linear process we can augment the maximal principle by the following corollary.

COROLLARY. *Consider the autonomous linear process in* R^n

$$(\mathfrak{L}) \qquad \dot{x} = Ax + Bu + v$$

with compact restraint set $\Omega \subset R^m$. *Let* $u(t) \subset \Omega$ *on* $\tau_0 \leq t \leq \tau_1$ *be any extremal controller so*

$$M(t) \equiv \max_{u \in \Omega} \eta(t)[A\, x(t) + Bu + v] = \eta(t)[A\, x(t) + B\, u(t) + v]$$

almost always, for corresponding responses $x(t)$ *and* $\eta(t)$. *Then* $M(t)$ *is constant on* $\tau_0 \leq t \leq \tau_1$.

Proof. By Lemma 2A of the appendix to this chapter, $M(t)$ is absolutely continuous and has a derivative almost always. We compute the derivative $\dot{M}(t)$ at an instant $t = t_1$ when it exists. Thus for $t_2 > t_1$ we have

$$\frac{M(t_2) - M(t_1)}{t_2 - t_1}$$
$$\geq \frac{\eta(t_2)[A\, x(t_2) + B\, u(t_1) + v] - \eta(t_1)[A\, x(t_1) + B\, u(t_1) + v]}{t_2 - t_1},$$

assuming $\dot{x}(t)$ satisfies $\mathfrak{L}$ and $M(t)$ satisfies the maximal principle at time t_1. Upon adding and subtracting $\eta(t_2)A\, x(t_1)$ to the numerator and computing the limit as $t_2 \to t_1$, we find

$$\frac{dM}{dt}(t_1) \geq \eta(t_1)A\, \dot{x}(t_1) + \dot{\eta}(t_1)A\, x(t_1) + \dot{\eta}(t_1)[B\, u(t_1) + v]$$

and

$$\dot{M}(t_1) \geq \eta A[Ax + Bu + v] - \eta AAx - \eta A[Bu + v] = 0.$$

A similar calculation shows that $\dot{M}(t_1) \leq 0$ and hence $\dot{M}(t) = 0$ almost everywhere on $\tau_0 \leq t \leq \tau_1$. Thus $M(t)$ is constant. Q.E.D.

The next theorem proves that, under conditions of normality, that maximal principle is sufficient, as well as necessary, for optimality. This will be accomplished by proving that the optimal controller is the unique extremal controller steering x_0 to the fixed convex target set G, and satisfying the transversality conditions. Using the results of the theorem we shall be able to synthesize the optimal controller as function of the state x in R^n.

THEOREM 19. *Consider the linear control process in* R^n

$$(\mathfrak{L}) \qquad \dot{x} = A(t)x + B(t)u + v(t)$$

with compact restraint set $\Omega \subset R^m$, *initial state* $x_0 \in R^n$, *and constant target set* G. *Assume that* $A(t)$, $B(t)$, *and* $v(t)$ *are continuous on* $t \geq \tau_0$ *and*

(a) $(\mathfrak{L}, \Omega, x_0, \tau_0, t)$ *is normal for* $t > \tau_0$;
(b) G *is a compact convex set in* R^n;
(c) *for each* $\bar{x}(\bar{t}) \in G$ *and time* $\bar{t} \geq \tau_0$ *there is a controller* $\bar{u}(t) \subset \Omega$ on $\bar{t} \leq t < \infty$, *with response* $\bar{x}(t) \subset G$, *and* $\bar{u}(t)$ *is nonextremal on every open interval* $\bar{t} < t < \bar{t}_1$.

Let $u_1(t) \subset \Omega$ *on* $\tau_0 \leq t \leq t_1$ *and* $u_2(t) \subset \Omega$ *on* $\tau_0 \leq t \leq t_2$ *be extremal controllers satisfying the transversality condition, namely, for the corresponding adjoint responses* $\eta_1(t)$ *and* $\eta_2(t)$ *the vectors* $\eta_1(t_1)$ *and* $\eta_2(t_2)$ *are inward unit normals to supporting hyperplanes to* G. *Then*

$$t_1 = t_2 = t^*$$

and

$$u_1(t) = u_2(t), \quad \text{almost always on} \quad \tau_0 \leq t \leq t^*,$$

so $u_1(t) = u^*(t)$ *is the unique minimal time-optimal controller steering* x_0 *to* G.

Proof. If $t_1 = t_2$ then the set of attainability $K(t_1)$ meets the convex target G at points $x_1(t_1)$ and $x_2(t_1)$, the terminal ends of the responses to $u_1(t)$ and $u_2(t)$, respectively. Since the control problem is normal, $K(t_1)$ is strictly convex and contains no line segment in its boundary. (If $t_1 = \tau_0$ or if Ω is just a single point the theorem follows trivially and these cases will not be mentioned further).

But, by the transversality condition, there is a supporting hyperplane π separating $K(t_1)$ and G. If $x_1(t_1) \neq x_2(t_1)$ then the segment joining them would lie in $K(t_1) \cap G$. Hence the segment between $x_1(t_1)$ and $x_2(t_1)$ would lie in π and so in $\partial K(t_1)$. But this contradicts the strict convexity of $K(t_1)$ and therefore $x_1(t_1) = x_2(t_1)$. In this case the normality implies that $u_1(t) = u_2(t)$ almost always on $\tau_0 \leq t \leq t_1$.

Suppose $t_1 < t_2$. Then the strictly convex set $K(t_2)$ is separated from G by a common supporting hyperplane. But hypothesis (c) implies that the interior of $K(t)$ meets G for all $t > t_1$. If the interior of $K(t_2)$ meets G then $K(t_2)$ cannot be separated from G. Thus this case is impossible, so $t_1 = t_2$.

Thus any extremal controller satisfying the transversality condition, in particular the optimal controller $u^*(t)$ on $\tau_0 \leq t \leq t^*$, must equal $u_1(t)$ almost everywhere on $\tau_0 \leq t \leq t_1 = t^*$. Q.E.D.

In the following corollaries we give explicitly verifiable hypotheses, to replace conditions (a) and (c) of the theorem, for autonomous linear processes. We shall discuss the unique optimal controller $u^*(t)$ where it is recognized that the uniqueness obtains only when the initial time is fixed, for example, on $0 \leq t \leq t^*$, and even there the values of $u^*(t)$ may be modified on a null set.

COROLLARY 1. *Consider the autonomous linear process in* R^n

$$(\mathcal{L}) \qquad \dot{x} = Ax + Bu + v$$

with convex polyhedral restraint set $\Omega \subset R^m$ *and initial state* $x_0 \in R^n$. *Assume the normality condition*:

(a) $Bw, ABw, \ldots, A^{n-1}Bw$ *are linearly independent for each nonzero vector* w *along an edge of* Ω, *or along* Ω *itself if it is a segment.*

Then the problem $(\mathcal{L}, \Omega, x_0, 0, t)$ *is normal for all* $t > 0$. *If* $u_1(t) \subset \Omega$ *on* $0 \leq t \leq t_1$ *is an extremal controller, then* $u_1(t)$ *is (almost always) piecewise constant with values at the vertices of* Ω *and with only a finite number of discontinuities, called switches.*

If the normality condition (a) *holds and*

(b) G *is a fixed compact convex target set;*

(c) *for each point* $\bar{x} \in G$ *there is a controller* $\bar{u}(t) \subset \Omega$ *on* $0 \leq t < \infty$, *with response* $\bar{x}(t) \subset G$, *and* $\bar{u}(t)$ *is nonextremal on an open interval* $0 < t < \bar{t}_1$,

then any extremal controller $u_1(t) \subset \Omega$ *on* $0 \leq t \leq t_1$, *steering* x_0 *to* G *and satisfying the transversality condition, must equal the unique optimal controller* $u^*(t)$ *almost everywhere on* $0 \leq t \leq t_1 = t^*$.

Proof. We first show that hypothesis (a) implies that the control problem is normal on every interval $0 \leq t \leq \tau_1$. Suppose $(\mathcal{L}, \Omega, x_0, 0, \tau_1)$ is not normal. Then there exist distinct extremal controllers $u_1(t)$ and $u_2(t)$ so that

$$\eta(t)B\, u_1(t) = \eta(t)B\, u_2(t) = \max_{u \in \Omega} \eta(t)Bu$$

almost everywhere on $0 \leq t \leq \tau_1$, where $\eta(t) = \eta_0 e^{-At}$ and $u_1(t) \neq u_2(t)$ on some set S of positive duration in $0 \leq t \leq \tau_1$.

At each fixed instant t examine the real valued linear function of u,

$$F_t(u) = \eta(t)Bu.$$

Since Ω is a convex polyhedron, $F_t(u)$ assumes its maximum for $u \in \Omega$ on precisely one face of Ω where it is constant (a face of Ω is a vertex, edge, or intersection of a hyperplane with $\partial\Omega$, or Ω itself). Thus at each instant $t \in S$, the linear function $F_t(u)$ assumes its maximal value on some (possibly many) edge e_t. Since Ω has only a finite number of edges, there is a positive time duration $S_1 \subset S$ wherein $F_t(u)$ assumes its maximal value on a fixed edge e_1. Let $w \neq 0$ be a vector parallel to the edge e_1 and then

$$\eta_0 e^{-tA} Bw = 0 \quad \text{for} \quad t \in S_1.$$

Discarding the countable number of isolated points of S_1 we can compute the derivative

$$-\eta_0 e^{-tA} ABw = 0 \quad \text{almost everywhere on } S_1.$$

Repeating this process, we find that

$$\eta_0 e^{-tA} Bw = 0, \quad \eta_0 e^{-tA} ABw = 0, \ldots, \eta_0 e^{-tA} A^{n-1} Bw = 0.$$

But this implies that the vectors

$$Bw, ABw, A^2Bw, \ldots, A^{n-1}Bw$$

are all orthogonal to $\eta_0 e^{-tA} \neq 0$ and hence are linearly dependent. This statement contradicts the normality condition (a) and hence we conclude that the control problem is normal.

Redefine the extremal controller $u_1(t)$ on a null set so

$$\eta(t)B\, u_1(t) = \max_{u \in \Omega} \eta(t)\, Bu$$

everywhere on $0 \leq t \leq t_1$. Then $u_1(t)$ lies almost always at a vertex of Ω since $F_t(u)$ assumes its maximum only at a vertex because of the normality condition.

Those times t at which $F_t(u)$ assumes its maximum at just a vertex form an open set in R^1, and the complement, which includes the set of switches, is a closed set. If there were an infinite number of discontinuities of $u_1(t)$, then $\eta(t_\nu)Bu$ attains its maximum all along some fixed edge e of Ω for an infinite set of times $\{t_\nu\}$. This implies that $\eta(t_\nu)Bw = 0$, where w is a unit vector parallel to the edge e. Since $\eta(t)Bw$ is a real analytic function with an infinite set of zeros $\{t_\nu\}$, it follows that $\eta(t)Bw \equiv 0$ for all t in the interval $0 \leq t \leq t_1$. But this implies that

$$\eta_0 e^{-tA} Bw = 0, \quad \eta_0 e^{-tA} ABw = 0, \ldots, \eta_0 e^{-tA} A^{n-1} Bw = 0,$$

which contradicts the normality condition. Therefore there are only a finite number of switches for the extremal controller $u_1(t)$ on $0 \leq t \leq t_1$.

Hypotheses (b) and (c) imply that $t_1 = t^*$ and $u_1(t) = u^*(t)$ almost everywhere on $0 \leq t \leq t_1 = t^*$, just as in the theorem. Q.E.D.

The next result follows immediately from Corollary 1, but we state it separately because of its great importance in applications.

COROLLARY 2. *Consider the autonomous linear process in* R^n

$$(\mathcal{L}) \qquad \dot{x} = Ax + Bu + v$$

with cubical restraint set Ω: $|u^j| \le 1$ *in* R^m. *Assume the normality condition:* $Bw, ABw, \ldots, A^{n-1}Bw$ *are linearly independent for each unit vector* w *along an edge of* Ω, *or along the segment* Ω *if* $m = 1$. *Then each extremal controller* $u(t) \subset \Omega$ *on* $0 \le t \le t_1$ *is determined by a nontrivial adjoint response* $\eta(t) = \eta_0 e^{-tA}$,

$$u(t) = \operatorname{sgn}(\eta(t)B)' \quad (\textit{almost everywhere}).$$

Thus $u(t)$ *is a relay controller, that is,* $u(t)$ *is piecewise constant with each component assuming only the values* ± 1 *and having only a finite number of switches.*

The requirement that

$$Bw, ABw, A^2Bw, \ldots, A^{n-1}Bw$$

be linearly independent for each unit vector w parallel to an edge of the convex polyhedron Ω (or along Ω if it is a segment) is called the *normality condition*. If

$$(\mathcal{L}) \qquad \dot{x} = Ax + Bu \quad \text{with} \quad u \subset \Omega$$

satisfies the normality condition, then $\mathcal{L}$ is controllable. For the existence of a single unit vector w, such that $Bw, ABw, \ldots, A^{n-1}Bw$ are independent, implies that the vectors

$$[B, AB, \ldots, A^{n-1}B]w^{(i)} \qquad i = 1, \ldots, n$$

are linearly independent, where $w^{(i)}$ is a column nm-vector with w filling the components $(i-1)m + 1$ through im and zeros elsewhere. If $m = 1$, that is, if Ω is a segment in R^1, then $\mathcal{L}$ is controllable if and only if the normality condition holds.

COROLLARY 3. *Consider the autonomous linear process in* R^n

$$(\mathcal{L}) \qquad \dot{x} = Ax + Bu$$

with convex polyhedral restraint set $\Omega \subset R^m$ *containing* $u = 0$ *in its interior, and target* G *as the origin* $x = 0$.

Assume that $\mathcal{L}$ *satisfies the normality condition. Then for each point* x_0 *in the domain of null controllability* $\mathcal{C}$ *there exists a unique extremal controller* $u^*(t)$ *steering* x_0 *to the origin, and* $u^*(t)$ *is the optimal controller.*

If A is stable then $\mathcal{C} = R^n$ and so each point $x_0 \in R^n$ can be steered to the origin by just one extremal controller, namely, the optimal controller.

Proof. The existence of the unique extremal controller steering $x_0 \in \mathcal{C}$ to the origin follows from Theorem 17 and Corollary 1. The assertion that $\mathcal{C} = R^n$ if A is stable follows from Corollary 3 of Theorem 5. Q.E.D.

We can now utilize Theorem 19 to synthesize the optimal controller by means of the switching locus and the technique of *backing out of the target.* For an autonomous control problem such as is discussed in Corollary 1 above we can synthesize the optimal controller by the following steps:

1. Use the process differential system and the adjoint system with time reversed

$$\dot{x} = -Ax - Bu(t) - v$$

$$\dot{\eta} = \eta A$$

with initial conditions $x(0) \subset \partial G$ and with $\eta(0)$ an inward unit normal to a supporting hyperplane to G at $x(0)$. Use only data for which $M \equiv \max_{u \in \Omega} \eta(0)[Ax(0) + Bu + v] \geq 0$. Define $u(t)$ by the maximal principle

$$\eta(t)B\,u(t) = \max_{u \in \Omega} \eta(t)Bu.$$

2. Find the unique datum $(x(0), \eta(0))$ which produces a solution $x(t)$, $\eta(t)$ passing through the prescribed initial state x_0 at some time $t^* > 0$.
3. Reverse the time sense again and define

$$x^*(t) = x(t^* - t) \quad \text{and} \quad \eta^*(t) = \eta(t^* - t)$$

on $0 \leq t \leq t^*$. Then $u^*(t)$ defined by

$$\eta^*(t)B\,u^*(t) = \max_{u \in \Omega} \eta^*(t)Bu \quad \text{on} \quad 0 \leq t \leq t^*$$

is the optimal controller, and $x^*(t)$ is the corresponding optimal-response steering x_0 to G.

Steps (1) and (2) can be carried out by a computer (analogue or digital) once the control process equations and restraints are given. Then for each initial state $x_0 \in R^n$ the corresponding optimal controller can be stored in some computer memory for future reference. A very useful method of storing the information concerning the optimal controller is by a description of the switching locus.

The switching locus W in $R^n - G$ consists in all points $x(t)$ at the instants when $u(t)$ is discontinuous; where $\eta(t)B\,u(t) = \max_{u \in \Omega} \eta(t)Bu$. Here

$x(t)$ and $\eta(t)$ are extremal responses satisfying the appropriate transversality conditions at G as defined in step (1) above. The switching locus W, for $m = 1$ where Ω is a segment $|u| \leq 1$, is usually a rather simple curve in the phase plane (as seen in the examples of Chapter 1) or a hypersurface in higher-dimensional space R^n. In this case W separates $R^n - G$ into two open sets: M_+, in which we store the control value $u = +1$, and M_-, in which we store $u = -1$. The synthesis function

$$\Psi(x) = \begin{cases} +1 & \text{for} \quad x \in M_+ \\ -1 & \text{for} \quad x \in M_- \end{cases}$$

then achieves the required optimal-control synthesis by the solutions of $\dot{x} = Ax + B\Psi(x) + v$.

If $m > 1$ and Ω is a cube $|u^j| \leq 1$ in R^m, it is often convenient to consider the switching locus independently for each component of the extremal controller

$$u(t) = \operatorname{sgn}\,(\eta(t)B)'.$$

It seems awkward to formulate general properties of such switching loci. However the following two examples illustrate this important technique of synthesis.

***Example* 1.** Consider the autonomous control process in R^2

$$(\mathfrak{L}) \qquad \dot{x}^1 = x^2 + u$$

$$\dot{x}^2 = -x^2 + u$$

with restraint Ω: $|u| \leq 1$ in R^1. We wish to synthesize the minimal time-optimal controller steering to the line $x^1 = 0$, with the additional requirement that the response can thereafter be held on this line. Thus the target is

$$G = \text{core}\,\{x^1 = 0\}.$$

If a response lies in $x^1 = 0$ then $\dot{x}^1(t) \equiv 0$, $x^2(t) = -u(t)$, and so $|x^2| \leq 1$. Conversely each point $x_0^1 = 0$, $|x_0^2| \leq 1$ can be controlled in the set $|x^2| \leq 1$ by $u(t) = -x_0^2e^{-2t}$ for $t \geq 0$. Thus G is the set $\{x^1 = 0, |x^2| \leq 1\}$. We note that G is a compact convex set in R^2 and also that each point $(x_0^1, x_0^2) \in G$ can be steered in G by the nonextremal controller $u(t) = -x_0^2e^{-2t}$ on $t \geq 0$.

Using the coefficient matrices

$$A = \begin{pmatrix} 0 & 1 \\ 0 & -1 \end{pmatrix}, \qquad B = \begin{pmatrix} 1 \\ 1 \end{pmatrix},$$

and vector $w = 1$ along Ω, we verify the normality condition. Thus $\mathfrak{L}$ is controllable and Theorem 8 asserts that the domain of null controllability is all R^2. Then Theorem 17 and Corollary 1 of Theorem 19 assert that each initial state in R^2 can be steered to G by a unique extremal controller satisfying the transversality condition, and that this is the optimal controller.

We proceed by the method of backing out of the target. Write the system $\mathcal{L}$ and the adjoint system with time reversed,

$$\begin{aligned} \dot{x}^1 &= -x^2 - u \quad \text{and} \quad u = \operatorname{sgn}(\eta_1 + \eta_2) \\ \dot{x}^2 &= x^2 - u \\ \dot{\eta}_1 &= 0 \\ \dot{\eta}_2 &= \eta_1 - \eta_2. \end{aligned}$$

Note that along a solution where $\dot{\eta}_1 = 0$,

$$\ddot{\eta}_1 + \ddot{\eta}_2 = \dot{\eta}_1 - \dot{\eta}_2 = -\dot{\eta}_1 - \dot{\eta}_2 \quad \text{so} \quad \eta_1 + \eta_2 = c_1 + c_2 e^{-t}.$$

Thus each extremal controller $u(t)$ has at most one switch on $0 \le t < \infty$. We examine all extremal controllers satisfying the transversality condition to obtain the switching locus W in $R^2 - G$.

Take initial data $x_0^1 = 0$, $|x_0^2| < 1$, $\eta_{10} = \pm 1$, $\eta_{20} = 0$. Then $\eta_2 + \eta_1 = \pm 2 \mp e^{-t}$ so there are no switches for these controllers. Use the value $u = -1$ and start from $x_0^1 = 0$, $x_0^2 = +1$ to define the curve

$$\Gamma_- = \{x^1 = -2e^t + 2t + 2, \qquad x^2 = 2e^t - 1 \quad \text{for} \quad t \ge 0\}.$$

We show that every point of Γ_- lies on W. The extremal with data $x_0^1 = 0$, $x_0^2 = +1$, $\eta_{10} = \cos\theta$, $\eta_{20} = \sin\theta$ for each fixed θ on $\pi \le \theta \le 2\pi$ follows Γ_- as long as $\eta_1(t) + \eta_2(t) < 0$. But

$$\eta_1(t) + \eta_2(t) = (\sin\theta - \cos\theta)e^{-t} + 2\cos\theta \quad \text{for} \quad t \ge 0.$$

Thus for each θ on $\pi \le \theta \le 3\pi/2$, we find

$$u(t) = \operatorname{sgn}(\eta_1(t) + \eta_2(t)) = -1 \quad \text{for} \quad t \ge 0.$$

For each θ on $3\pi/2 < \theta < 7\pi/4$ there is just one positive zero $t_1(\theta) > 0$ of $(\sin\theta - \cos\theta)e^{-t} + 2\cos\theta$. An easy calculation shows that $t_1(\theta)$ decreases monotonically from $+\infty$ to 0 as θ increases. Thus there exist extremal controllers satisfying the transversality condition at G, which switch from $u = +1$ to $u = -1$ at an arbitrarily prescribed point of Γ_- and then follow Γ_- into the target G.

Define Γ_+ by reflecting Γ_- through the origin to obtain the complete switching locus $W = \Gamma_+ \cup \Gamma_-$, and note that the corresponding curve $x^2 = W(x^1)$ separates $R^2 - G$. Define the synthesizing function

$$\Psi(x^1, x^2) = \begin{cases} -1 & \text{for} \quad x^2 > W(x^1) \quad \text{and on} \quad x^2 = \Gamma_-(x^1) \\ +1 & \text{for} \quad x^2 < W(x^1) \quad \text{and on} \quad x^2 = \Gamma_+(x^1). \end{cases}$$

The optimal responses, for various initial states in $R^2 - G$, are sketched in Figure 2.1.

***Example* 2.** Consider the autonomous control process in R^3

$$\dddot{x} = u$$

or

$$(\mathfrak{L}) \qquad \dot{x}^1 = x^2, \qquad \dot{x}^2 = x^3, \qquad \dot{x}^3 = u$$

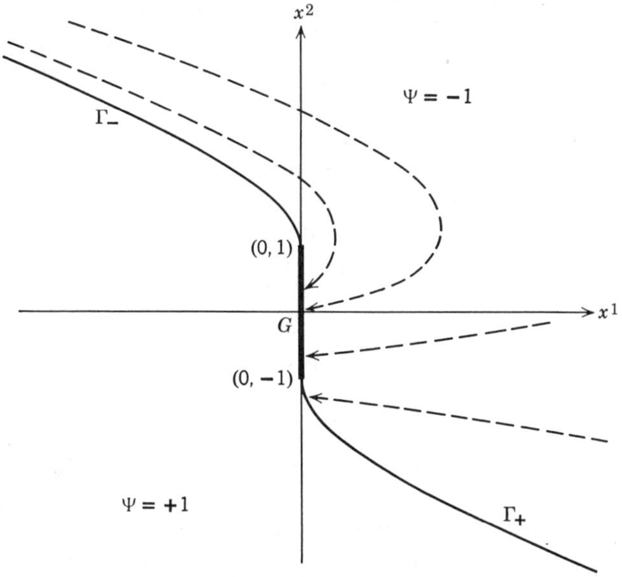

Figure 2.1 Switching locus and synthesis of optimal control $\dot{x}^1 = x^2 + u$, $\dot{x}^2 = -x^2 + u$, $|u| \leq 1$. Target G: $x^1 = 0$, $|x^2| \leq 1$.

with restraint Ω: $|u| \leq 1$ in R^1. We seek to synthesize the minimal time optimal controller to the origin. Theorems 8 and 17 guarantee the existence of an optimal controller from each initial state in R^3, and Corollary 3 of Theorem 19 asserts that this optimal controller is the unique extremal controller steering the response to the origin.

We proceed to construct the switching locus by backing out of the origin. Write the system $\mathfrak{L}$ and the adjoint system with time reversed,

$$\dot{x}^1 = -x^2, \qquad \dot{x}^2 = -x^3, \qquad \dot{x}^3 = -u \quad \text{with} \quad u = \operatorname{sgn} \eta_3(t)$$

$$\dot{\eta}_1 = 0, \qquad \dot{\eta}_2 = \eta_1, \qquad \dot{\eta}_3 = \eta_2.$$

Note that $\dddot{\eta}_3 = 0$ so $\eta_3(t) = \eta_{30} + \eta_{20}t + \eta_{10}t^2/2$ so that each extremal controller has at most two switches corresponding to the positive zeros of $\eta_3(t)$.

Define the curve Γ_+ as the response leading out of the origin with $u = +1$, that is,

$$x^1 = \frac{-t^3}{6}, \qquad x^2 = \frac{t^2}{2}, \qquad x^3 = -t \quad \text{for} \quad t > 0.$$

Since we can select the initial data $\eta_{10}, \eta_{20}, \eta_{30}$ so $\eta_3(t) > 0$ on $0 < t < t_1$, $\eta_3(t) < 0$ on $t_1 < t < t_2 \leq +\infty$, and $\eta_3(t) > 0$ on $t > t_2$ for arbitrary $0 < t_1 < t_2 \leq +\infty$, each point on Γ_+ can occur as a switch from $u = -1$ to $u = +1$ in an extremal control steering the response into the origin along Γ_+.

From each point on Γ_+ defined by a value of $t > 0$ we compute the responses (in the backwards-process differential equations) with the control value $u = -1$. Using the independent variable $s \geq 0$ we compute the responses

$$x^1 = \frac{s^3}{6} - \frac{s^2 t}{2} - \frac{st^2}{2} - \frac{t^3}{6}$$

$$x^2 = \frac{-s^2}{2} + st + \frac{t^2}{2}$$

$$x^3 = s - t.$$

For $t > 0$, $s \geq 0$ these equations define the switching surface W_-, which contains the switching curve Γ_+.

Next we repeat this integration procedure by starting at the origin with control value $u = -1$ to define the curve Γ_-,

$$x^1 = \frac{t^3}{6}, \qquad x^2 = \frac{-t^2}{2}, \qquad x^3 = t \quad \text{for} \quad t > 0.$$

Now integrate backwards from each point of Γ_-, using the control value $u = +1$, to define the switching surface W_+,

$$x^1 = \frac{t^3}{6} + \frac{t^2 s}{2} + \frac{ts^2}{2} - \frac{s^3}{6}$$

$$x^2 = \frac{-t^2}{2} - ts + \frac{s^2}{2}$$

$$x^3 = t - s$$

for $s \geq 0$, $t > 0$.

The total switching surface is $W = W_- \cup W_+$, which contains the switching curve $\Gamma = \Gamma_+ \cup \Gamma_-$. We shall show that W separates R^3 and that Γ separates W (with origin added to Γ). In fact, W is a single-valued surface over the entire (x^1, x^3) coordinate plane. To prove this let $(\bar{x}^1, \bar{x}^3)$ be given and we seek a unique parameter pair (s, t) defining a corresponding point $(\bar{x}^1, \bar{x}^2, \bar{x}^3)$ on W.

On W_- we have

$$x^1 = \frac{(x^3)^3}{6} - st^2 \qquad s \geq 0, t > 0$$

and on W_+

$$x^1 = \frac{(x^3)^3}{6} + st^2 \qquad s \geq 0, t > 0.$$

Thus if $\bar{x}^1 = (\bar{x}^3)^3/6$ we require $s = 0$; take the point where $\bar{x}^2 = (\bar{x}^3)^2/2$ on Γ_- if $\bar{x}^1 < 0$; otherwise take the point where $\bar{x}^2 = -(\bar{x}^3)^2/2$ on Γ_+. If $\bar{x}^1 < (\bar{x}^3)^3/6$ we seek a point on W_-, but if $\bar{x}^1 > (\bar{x}^3)^3/6$ we seek a point on W_+. For example, suppose $\bar{x}^1 < (\bar{x}^3)^3/6$ we seek a root $t > 0$ of

$$st^2 + \left[\bar{x}^1 - \frac{(\bar{x}^3)^3}{6}\right] = 0$$

or

$$(\bar{x}^3 + t)t^2 + \left[\bar{x}^1 - \frac{(\bar{x}^3)^3}{6}\right] = 0.$$

Since this cubic polynomial in t has a horizontal tangent at $t = 0$ it is easy to see that there is exactly one positive root. A similar analysis shows that there is just one point on W_+ where $\bar{x}^1 > (\bar{x}^3)^3/6$.

Thus W separates R^3 into two regions called M_+ where $x^2 \to +\infty$, and M_- where $x^2 \to -\infty$. Since the curve Γ_- matches the boundary of W_- corresponding to the parameter values $t = 0$, $s > 0$ and, similarly, Γ_+ matches the boundary of W_+ we find that Γ separates the surface W into two pieces.

If an initial point (x_0^1, x_0^2, x_0^3) lies in M_+, use the control value $u = +1$ until the trajectory hits W_-. Then switch to $u = -1$ until the trajectory hits Γ_+ and then switch to $u = +1$ and follow Γ_+ into the origin. An opposite pattern of switches holds for an initial point in M_-. The required synthesis function is thus

$$\Psi(x^1, x^2, x^3) = \begin{cases} +1 & \text{in} \quad M_+ \\ -1 & \text{in} \quad M_- \\ -1 & \text{on} \quad W_- - \Gamma_+. \\ +1 & \text{on} \quad W_+ - \Gamma_- \\ +1 & \text{on} \quad \Gamma_+ \\ -1 & \text{on} \quad \Gamma_- \end{cases}$$

Example 3. Consider the autonomous control process in R^n defined by

$$x^{(n)} + a_1 x^{(n-1)} + a_2 x^{(n-2)} + \cdots + a_n x = u$$

with restraint $|u| \leq 1$ in R^1. We wish to steer an initial state $(x_0, x_0^{(1)}, \ldots, x_0^{(n-1)})$ to $x = 0$ in minimal time and hold it there forever after. The corresponding system in R^n is

$$(\mathfrak{L}) \qquad \dot{\mathbf{x}} = A\mathbf{x} + bu$$

where

$$A = \begin{pmatrix} 0 & 1 & 0 & 0 & \cdots & 0 \\ 0 & 0 & 1 & 0 & \cdots & 0 \\ \cdot \\ \cdot \\ \cdot \\ 0 & 0 & 0 & 0 & \cdots & 1 \\ -a_n & -a_{n-1} & & & \cdots & -a_1 \end{pmatrix}, \qquad b = \begin{pmatrix} 0 \\ 0 \\ \cdot \\ \cdot \\ \cdot \\ 0 \\ 1 \end{pmatrix}.$$

We easily compute that core $\{x^1 = 0\}$ is the origin in R^n, which is thereby the target. The process $\mathfrak{L}$ is controllable and hence normal. Thus the domain of null controllability $\mathfrak{C}$ is an open connected subset of R^n and we assume that the initial state $\mathbf{x}_0 \in \mathfrak{C}$. There is therefore a unique extremal controller steering $\mathbf{x}_0$ to the origin, and this is the optimal controller $u^*(t)$ on $0 \leq t \leq t^*$ with corresponding response $\mathbf{x}^*(t)$ and adjoint response $\boldsymbol{\eta}^*(t)$.

Here $\boldsymbol{\eta}^*(t)$ is a nontrivial solution of

$$\dot{\boldsymbol{\eta}} = -\boldsymbol{\eta}A \quad \text{or} \quad \dot{\boldsymbol{\eta}}' = -A'\boldsymbol{\eta}' \quad \text{where} \quad \boldsymbol{\eta} = (\eta_1, \eta_2, \ldots, \eta_n).$$

Then we compute

$$\begin{aligned} \dot{\eta}_1 &= \qquad\qquad a_n\eta_n \\ \dot{\eta}_2 &= -\eta_1 \quad + a_{n-1}\eta_n \\ &\cdot \\ &\cdot \\ &\cdot \\ \dot{\eta}_{n-1} &= -\eta_{n-2} + a_2\eta_n \\ \dot{\eta}_n &= -\eta_{n-1} + a_1\eta_n. \end{aligned}$$

Elimination leads to the differential equation for $\eta_n(t)$,

$$\eta^{(n-1)} - a_1\eta^{(n-1)} + a_2\eta^{(n-2)} - a_3\eta^{(n-3)} + \cdots + (-1)^n a_n\eta = 0.$$

The optimal controller satisfies the maximal principle

$$\eta^*(t)b\,u^*(t) = \max_{|u|\leq 1} \eta^*(t)bu$$

so

$$u^*(t) = \operatorname{sgn} \eta_n(t) \quad \text{almost always on} \quad 0 \leq t \leq t^*.$$

However, the geometry of $\mathcal{C}$ and the description of the switching locus W is too complicated for any practical use in the generality considered in this example.

The difficulties of analyzing the geometry of the switching locus W and storing its description in the memory of an automatic computer indicate that this method is usually impractical for control processes of an order higher than three. In Appendix A we present a direct digital computation for the optimal controller without any intricate geometric subtleties.

While the complete description of the switching locus is extremely awkward in higher dimensions, there are two important cases for which certain definite switching properties can be easily established.

THEOREM 20. *Consider the autonomous linear process in R^n*

$$(\mathcal{L}) \qquad \dot{x} = Ax + bu$$

with restraint Ω: $|u| \leq 1$ in R^1. Assume $\mathcal{L}$ is controllable and hence normal.

If every eigenvalue of A is real, then each extremal controller has at most $n - 1$ switches on $0 \leq t < \infty$.

If every eigenvalue of A has a nonzero imaginary part, then each extremal controller has an infinite number of switches on $0 \leq t < \infty$. Thus, for each number $N > 0$, there exists an initial state $x_0 \in R^n$ for which the corresponding optimal controller steering x_0 to the origin has more than N switches.

Proof. Since $\mathcal{L}$ is controllable we can assume that new coordinates in R^n are introduced as in Theorem 7 so that

$$A = \begin{pmatrix} 0 & 1 & 0 & 0 & \cdots & 0 \\ 0 & 0 & 1 & 0 & \cdots & 0 \\ \cdot & & & & & \\ \cdot & & & & & \\ \cdot & & & & & \\ 0 & 0 & 0 & 0 & \cdots & 1 \\ -a_n & -a_{n-1} & & & \cdots & -a_1 \end{pmatrix} \quad \text{and} \quad b = \begin{pmatrix} 0 \\ 0 \\ \cdot \\ \cdot \\ \cdot \\ 0 \\ 1 \end{pmatrix}.$$

Then $\mathcal{L}$ can be described by a scalar differential equation of nth order

$$x^{(n)} + a_1 x^{(n-1)} + \cdots + a_n x = u, \qquad |u| \leq 1.$$

An extremal controller $u(t)$ has the form

$$u(t) = \operatorname{sgn} \eta(t)$$

where $\eta(t)$ is the last component of a nontrivial solution of

$$\begin{pmatrix} \dot{\eta}_1 \\ \cdot \\ \cdot \\ \cdot \\ \dot{\eta}_n \end{pmatrix} = -A' \begin{pmatrix} \eta_1 \\ \cdot \\ \cdot \\ \cdot \\ \eta_n \end{pmatrix}.$$

An easy calculation shows that $\eta_n(t)$ is a solution of

$$\eta^{(n)} - a_1\eta^{(n-1)} + a_2\eta^{(n-2)} - \cdots + (-1)^n a_n \eta = 0.$$

The distinct eigenvalues $\{\lambda_1, \lambda_2, \ldots, \lambda_r\}$ of $-A'$ are the negatives of the eigenvalues of A and hence these are all real or have nonzero imaginary parts according to the hypotheses on A.

Assume that all eigenvalues of A are real. Then

$$\eta(t) = P_1(t)e^{\lambda_1 t} + \cdots + P_r(t)e^{\lambda_r t}$$

where the real polynomials $P_j(t)$ have degrees $\leq n_j - 1$, and n_j is the multiplicity of eigenvalue λ_j for $1 \leq j \leq r$. But $n_1 + n_2 + \cdots + n_r = n$ and a known result on real exponential polynomials (see Exercise 13, below) asserts that $\eta(t)$ can have at most $n - 1$ real zeros on $-\infty < t < \infty$. Thus an extremal controller $u(t)$ has at most $n - 1$ switches on $0 \leq t < \infty$.

Next assume that every eigenvalue of A has a nonzero imaginary part and hence the same condition holds for the eigenvalues $\lambda_j = \alpha_j + i\beta_j$ of $-A'$. In this case

$$\eta(t) = e^{\alpha_1 t}[P_1(t) \cos \beta_1 t + Q_1(t) \sin \beta_1 t] + \cdots$$
$$+ e^{\alpha_r t}[P_r(t) \cos \beta_r t + Q_r(t) \sin \beta_r t]$$

where $P_1(t), Q_1(t), \ldots, P_r(t), Q_r(t)$ are real polynomial that are not all zero. For notational simplicity take α_1 to be the largest of the $\alpha_1, \alpha_2, \ldots, \alpha_r$ that appear with nonzero coefficients in $\eta(t)$. Then

$$\eta(t) = e^{\alpha_1 t} t^k \sum_{j=1}^{r} (a_j \cos \beta_j t + b_j \sin \beta_j t) + R(t).$$

Here $k \geq 0$, and the trigonometric sum

$$T(t) = \sum_{j=1}^{r} (a_j \cos \beta_j t + b_j \sin \beta_j t)$$

is not identically zero. The remainder term $R(t)$ is such that

$$\lim_{t\to\infty} e^{-\alpha_1 t}t^{-k} R(t) = 0.$$

Note that $T(t)$ is a finite trigonometric sum that has mean zero on $0 \leq t < \infty$. Also there exists $L > 0$ such that $T(t)$ assumes values greater than some $\epsilon > 0$ and less than $-\epsilon < 0$ on each interval of length L. [This follows from the theory of almost periodic functions or else from an elementary analysis of the explicit expression for $T(t)$.] Let $\bar{t} > 0$ be such that

$$|e^{-\alpha_1 t}t^{-k} R(t)| < \frac{\epsilon}{2}$$

for $t \geq \bar{t}$. Then

$$e^{-\alpha_1 t}t^{-k}\,\eta(t) = T(t) + e^{-\alpha_1 t}t^{-k} R(t)$$

must have a zero on each interval $t_1 \leq t \leq t_1 + L$ for $t_1 > \bar{t}$. Thus $\eta(t)$ has an infinite number of zeros and $u(t)$ has an infinite number of switches on $0 \leq t < \infty$.

Since the optimal controller for a given initial state x_0 is obtained by integrating an extremal controller backwards out of the origin, there exist points $x_0 \in R^n$ for which the optimal controller $u^*(t)$ on $0 \leq t \leq t^*$ has more than the prescribed number N switches. Q.E.D.

Another technique for synthesizing the optimal controller utilizes the isochronal hypersurfaces in R^n. Let $T(x)$ be the minimal time for steering the initial point x to the target; then the locus in R^n

$$T(x) = t \quad \text{for} \quad t \geq 0$$

is the isochronal hypersurface with parameter value t. Along an optimal response $x^*(t)$ on $t_0 \leq t \leq 0$ we have

$$T(x^*(t)) = -t$$

and

$$\nabla T(x^*(t))\,\dot{x}^*(t) = -1$$

wherever the row vector $\nabla T = \operatorname{grad} T$ and the derivative $\dot{x}^*(t)$ exist.

We shall find that $-\nabla T(x)$ can be used in place of the adjoint response $\eta(t)$ in synthesizing the optimal controller $u^*(t)$. In order to simplify the statement of the result we assume that the target G is just the origin and that there exists exactly one extremal controller steering an initial state to the origin, as in Corollary 3 of Theorem 19.

THEOREM 21. *Consider the autonomous linear process in* R^n

$$(\mathfrak{L}) \qquad \dot{x} = Ax + Bu$$

with compact restraint set $\Omega \subset R^m$ containing $u = 0$ in its interior. Assume that the process $\{\mathfrak{L}, \Omega\}$ is normal on every interval and that the domain of null controllability is all R^n. Let $T(x)$ be the minimal time for steering an initial state $x \in R^n$ to the origin. Then $T(x)$ is continuous in R^n and the isochronal hypersurfaces

$$T(x) = t \quad \textit{for each} \quad t > 0$$

form a family of closed convex hypersurfaces that increase monotonically from $x = 0$ to ∞ as t increases from $t = 0$ to ∞.

Proof. Consider $K(t_1)$ for $t_1 > 0$, the set of attainability from $x_0 = 0$ with controllers in Ω. Each $K(t_1)$ is a compact, strictly convex set and $K(t_1) \subset K(t_2)$ for $0 < t_1 \leq t_2$ (see the remarks following Theorem 3 and also Exercise 4 of Section 3). We shall prove that the locus $T(x) = t_1$ is precisely the boundary of $K(t_1)$.

Let $x_1 \in \partial K(t_1)$ so only one extremal controller $u_1(t)$ on $0 \leq t \leq t_1$ steers x_0 to x_1. Since $u_1(-t)$ is the optimal controller steering x_1 to x_0 we note that $T(x_1) = t_1$. Conversely a point x^* on the locus $T(x) = t_1$ is the terminus of an optimal response $x^*(t)$ steering x_0 to $x^*(t_1) = x^*$. Thus x^* lies on the boundary of $K(t_1)$. Therefore the isochronal locus $T(x) = t_1$ is just the closed convex hypersurface that is the boundary of $K(t_1)$ in R^n. We note that the isochronal hypersurfaces of the family

$$T(x) = t_1 \quad \text{for} \quad t_1 > 0$$

are nonintersecting and each encloses the origin. Also the isochronal hypersurfaces increase monotonically from $x_0 = 0$ towards ∞ as t_1 increases from 0 to ∞ since the sets $K(t_1)$ likewise increase.

To prove the continuity of $T(x)$ in R^n let $T(x_1) = t_1$. Take $\epsilon > 0$ and consider the shell bounded by the hypersurfaces $T(x) = t_1 - \epsilon$ and $T(x) = t_1 + \epsilon$. [If $x_1 = 0$ so $T(x_1) = 0$; the corresponding argument is similar.] Then for a sufficiently small $\delta > 0$ the neighborhood $|x - x_1| < \delta$ lies within the shell and herein $|T(x) - t_1| < \epsilon$. Thus $T(x)$ is continuous at x_1 and hence at every point of R^n. Q.E.D.

COROLLARY. *Assume that $T(x) \in C^1$ in an open set $\mathcal{O} \subset R^n$ that does not meet the switching locus of the autonomous process*

$$(\mathfrak{L}) \qquad \dot{x} = Ax + Bu.$$

Then

$$\max_{u \in \Omega} [-\nabla T(x)][Ax + Bu] = 1 \quad \text{in} \quad \mathcal{O}.$$

If, in addition, the restraint set Ω is a cube $|u^j| \leq 1$ in R^m, then at each point $x \in \mathcal{O}$ the optimal controller at x has the value $\Psi(x)$ where

$$\Psi(x) = -\text{sgn}\,[\nabla T(x)B].$$

Proof. Let $x^*(t)$ on $0 \leq t \leq t^*$ be the optimal response steering a point $x_1 \in \mathcal{O}$ to the origin by the controller $u^*(t)$. Then

$$T(x^*(t)) = -t$$

and so

$$\nabla T(x^*(t))\dot{x}^*(t) = \nabla T(x^*(t))[A\,x^*(t) + B\,u^*(t)] = -1$$

for $0 \leq t \leq t_1$ when $x^*(t) \subset \mathcal{O}$.

Thus $\nabla T(x_1)$ does not vanish in $\mathcal{O}$ and so it defines the outwards normal vector to the tangent hyperplane to the isochronal hypersurface through x_1. Hence $\nabla T(x_1)$ is a positive scalar multiple of $\eta(t^*)$, the adjoint response corresponding to the optimal controller $u^*(t^* - t)$ steering $x_0 = 0$ to x_1. Then $-B\,u^*(0)$ has the maximal possible component along $\nabla T(x_1)$, or

$$-\nabla T(x_1)B\,u^*(0) = \max_{u \in \Omega} [-\nabla T(x_1)Bu].$$

Hence

$$\max_{u \in \Omega} \{-\nabla T(x_1)[Ax_1 + Bu]\} = -\nabla T(x_1)[Ax \;\; + B\,u^*(0)] = 1.$$

Therefore at each point $x \subset \mathcal{O}$ we have

$$\max_{u \in \Omega} [-\nabla T(x)][Ax + Bu] = 1.$$

Finally, let Ω be the cube $|u^j| \leq 1$ in R^m. Then the optimal controller from $x_0 = 0$ to $x_1 \in \mathcal{O}$ is

$$u^*(t^* - t) = \operatorname{sgn}[-\eta(t)B]$$

and so, at each point, $x_1 \in \mathcal{O}$,

$$\Psi(x_1) = \operatorname{sgn}[-\eta(t^*)B] = -\operatorname{sgn}[\nabla T(x_1)B]. \qquad \text{Q.E.D.}$$

The corollary indicates a method of synthesizing the optimal controller $u^*(t)$ by use of the isochronal function $T(x)$ as follows:

1. Find $T(x)$ explicitly by solving the partial differential equations

$$\nabla T[Ax + B] = -1 \quad \text{in} \quad M_+$$

$$\nabla T[Ax - B] = -1 \quad \text{in} \quad M_-.$$

   Here we assume that Ω is the interval $|u| \leq 1$ and M_+ and M_- are the regions where the optimal control values are $+1$ and -1, respectively.
2. Define the synthesizing function $\Psi(x)$ by

$$\Psi(x) = -\operatorname{sgn}[\nabla T(x)B] \quad \text{for} \quad x \in M_+ \cup M_-.$$

This method, however, involves all the complications of the previous method depending on the switching locus. In fact, one must first locate the switching locus in order to specify the regions M_+ and M_-. Then the partial differential equations for $T(x)$ are to be solved as Cauchy initial value problems with the values of $T(x)$ precomputed on the switching locus. The optimal responses are the characteristics for these partial differential equations; hence the optimal responses must be known in order to compute $T(x)$.

The method of isochronal surfaces is conceptually interesting and occasionally useful numerically, but it is difficult to give a precise and general analysis of this technique.

We conclude this chapter on the control of linear processes by proving that the minimal optimal time t^* and the optimal controller $u^*(t)$ depend on all the data $\{\mathfrak{L}, \Omega, x_0, t_0, G\}$ in an appropriately continuous manner. This continuity permits one to approximate a complicated physical problem by a simpler mathematical idealization and to obtain a useful approximation for the optimal controller. To simplify the statement of the result we consider autonomous processes, with target G as the origin, and a polyhedral restraint set Ω. Since the edges of Ω are important for the normality condition, we designate by E_Ω the set of all unit vectors in R^m that are parallel to the edges of Ω (or along Ω if it is a segment).

THEOREM 22. *Consider an autonomous linear process in* R^n

$$(\mathfrak{L}) \qquad \dot{x} = Ax + Bu$$

with convex polyhedron restraint $\Omega \subset R^m$ *containing* $u = 0$ *in its interior. Assume the normality condition:*

$$Bw, ABw, \ldots, A^{n-1}Bw$$

are independent for $w \in E_\Omega$. *Let* x_0 *be an initial state in the domain* $\mathcal{C}$ *of null controllability and let* $u^*(t) \subset \Omega$ *on* $0 \leq t \leq t^*$ *be the optimal controller steering* x_0 *to the origin by the response* $x^*(t)$.

Consider a perturbed autonomous process in R^n

$$(\hat{\mathfrak{L}}) \qquad \dot{x} = \hat{A}x + \hat{B}u$$

with convex polyhedral restraint $\hat{\Omega} \subset R^m$ *and initial point* $\hat{x}_0$. *Then for each* $\epsilon > 0$ *there exists* $\delta > 0$ *such that*

$$|A - \hat{A}| + |B - \hat{B}| + |x_0 - \hat{x}_0| + \text{dist}(\Omega, \hat{\Omega}) + \text{dist}(E_\Omega, E_{\hat{\Omega}}) < \delta$$

implies that $\hat{x}_0$ *lies in the domain* $\hat{\mathcal{C}}$ *of null controllability of* $\hat{\mathfrak{L}}$, *that the problem* $\{\hat{\mathfrak{L}}, \hat{\Omega}, \hat{x}_0\}$ *is normal, and that there is a unique optimal controller* $\hat{u}^*(t) \subset \hat{\Omega}$ *on* $0 \leq t \leq \hat{t}^*$ *steering* $\hat{x}_0$ *to the origin with the response* $\hat{x}^*(t)$.

Furthermore

$$|t^* - \hat{t}^*| < \epsilon$$

and

$$|x^*(t) - \hat{x}^*(t)| < \epsilon \quad \text{on} \quad 0 \leq t \leq \tau^* = \min [t^*, \hat{t}^*]$$

and

$$\int_0^{\tau^*} |u^*(t) - \hat{u}^*(t)|\, dt < \varepsilon.$$

Proof. Since a determinant varies continuously with its entries, there exists $\delta_1 > 0$ such that:

$$|\mathfrak{L} - \hat{\mathfrak{L}}| \equiv |A - \hat{A}| + |B - \hat{B}| + |x_0 - \hat{x}_0| + \text{dist}\,(\Omega, \hat{\Omega}) + \text{dist}\,(E_\Omega, E_{\hat{\Omega}}) < \delta_1$$

implies that

$$\det [\hat{B}\hat{w}, \hat{A}\hat{B}\hat{w}, \ldots, \hat{A}^{n-1}\hat{B}\hat{w}] \neq 0 \quad \text{for} \quad \hat{w} \in E_{\hat{\Omega}}.$$

We further restrict $\delta_1 > 0$ so that there exists a compact cubical neighborhood N of $u = 0$ that lies interior to all $\hat{\Omega}$ with dist $(\Omega, \hat{\Omega}) \leq \delta_1$. We consider only these problems $(\hat{\mathfrak{L}}, \hat{\Omega}, \hat{x}_0\}$, with $|\mathfrak{L} - \hat{\mathfrak{L}}| < \delta_1$; each of which is normal, controllable, and has an open set $\hat{\mathfrak{C}} \subset R^n$ for the domain of null controllability.

Take any $0 < \epsilon < 1$ and let $\hat{\mathfrak{C}}(\epsilon/2, N) \subset R^n$ be the set of points attainable from the origin in time duration $\epsilon/2$ by controllers $u(t) \subset N$, using responses of $\hat{\mathfrak{L}}$. Since $\hat{\mathfrak{L}}$ is controllable each $\hat{\mathfrak{C}}(\epsilon/2, N)$ contains an inscribed ball of radius $\hat{r} > 0$ and centered at the origin. A slight change in the coefficients of $\hat{\mathfrak{L}}$ produces a corresponding change in $\hat{\mathfrak{C}}(\epsilon/2, N)$ to a new convex set for which the inscribed radius is greater than $\hat{r} - \xi$, for a prescribed $\xi > 0$. Thus $\hat{r}$ is a lower semicontinuous function of the matrices $\hat{A}$ and $\hat{B}$, and hence $\hat{r}$ has a positive minimum $r_\epsilon > 0$ when $|A - \hat{A}| + |B - \hat{B}| \leq \delta_1$.

Let $u^*(t) \subset \Omega$ on $0 \leq t \leq t^*$ be the optimal controller steering x_0 to the origin by a response $x^*(t)$ of $\mathfrak{L}$, and let $\hat{u}(t) \subset \hat{\Omega}$ be any controller with response $\hat{x}(t)$ of $\hat{\mathfrak{L}}$ and $|u^*(t) - \hat{u}(t)| < \delta_2$ on $0 \leq t \leq t^* + 1$ (use $u^*(t) \equiv 0$ for $t > t^*$). Then if $|\mathfrak{L} - \hat{\mathfrak{L}}| < \delta_2 < \delta_1$, where $\delta_2 > 0$ is sufficiently small, we find that $|x^*(t) - \hat{x}(t)| < r_\epsilon 2^{-1} |e^{\hat{A}\epsilon/2}|^{-1}$ on $0 \leq t \leq t^* + 1$. This follows from the variation of parameters formula for the response $\hat{x}(t)$ as a continuous function of $\hat{A}$, $\hat{B}$, $\hat{x}_0$, and $\hat{u}(t)$. But then $|\hat{x}(t^*)| < r_\epsilon 2^{-1} |e^{\hat{A}\epsilon/2}|^{-1}$ and so $e^{\hat{A}\epsilon/2}\, \hat{x}(t^*) \in \hat{\mathfrak{C}}(\epsilon/2, N)$. Thus $\hat{x}(t^*)$ can be steered to the origin in duration $\epsilon/2$ with a controller in $N \subset \hat{\Omega}$ and along a response of $\hat{\mathfrak{L}}$. Therefore $\hat{x}_0 \in \hat{\mathfrak{C}}$ and the optimal time for steering $\hat{x}_0$ to the origin is $\hat{t}^* < t^* + \epsilon/2$. By defining the restrictions on $\delta_2 > 0$ symmetrically in terms of $\{\mathfrak{L}, \Omega, x_0, t^*\}$ and $\{\hat{\mathfrak{L}}, \hat{\Omega}, \hat{x}_0, \hat{t}^*\}$ we can conclude that $t^* < \hat{t}^* + \epsilon/2$ and so

$$|t^* - \hat{t}^*| < \epsilon \quad \text{for} \quad |\mathfrak{L} - \hat{\mathfrak{L}}| < \delta_2.$$

Next take $|\mathfrak{L} - \hat{\mathfrak{L}}| < \delta_2$ and let $u^*(t) \subset \Omega$ on $0 \leq t \leq t^*$ and $\hat{u}^*(t) \subset \hat{\Omega}$ on $0 \leq t \leq \hat{t}^*$ be the corresponding optimal controllers with responses $x^*(t)$ and $\hat{x}^*(t)$. By the continuous dependence of the response on $\hat{A}$, $\hat{B}$, $\hat{x}_0$, and $\hat{u}(t)$ we note that there exists $\epsilon_1 > 0$ such that

$$\int_0^{\tau^*} |u^*(t) - \hat{u}^*(t)|\, dt < \varepsilon_1 < \varepsilon \quad \text{for} \quad \tau^* = \min(t^*, \hat{t}^*)$$

implies that (with a possible further restriction of $\delta_2 > 0$)

$$|x^*(t) - \hat{x}^*(t)| < \epsilon \quad \text{on} \quad 0 \leq t \leq \tau^*.$$

We shall prove that for each $\epsilon_2 > 0$ there exists a positive $\delta < \delta_2$ such that $|\mathfrak{L} - \hat{\mathfrak{L}}| < \delta$ implies that $|u^*(t) - \hat{u}^*(t)| < \epsilon_2$ except for some duration of ϵ_2 on $0 \leq t \leq \tau^*$. Such a demonstration will complete the proof of the theorem.

Recall the maximal principle that

$$\eta(t)B\, u^*(t) = \max_{u \in \Omega} \eta(t)Bu$$

for some adjoint response $\eta(t)$, so that $u^*(t)$ is piecewise constant at the vertices of Ω. From this follows the existence of a constant $\theta > 0$ such that responses $x^*(t)$ and $\bar{x}(t)$ of $\mathfrak{L}$, for controllers $u^*(t)$ and $\bar{u}(t)$ in Ω, can satisfy $|x^*(t^*) - \bar{x}(t^*)| < \theta$ only if $|u^*(t) - \bar{u}(t)| < \epsilon_2/2$ outside of a set of duration ϵ_2 on $0 \leq t \leq t^*$.

Use the continuity of the optimal time t^*, as proved above, to obtain a positive $\delta_3 < \delta_2$ such that $|x^*(t^*) - \hat{x}^*(t^*)| < \theta/2$ for $|\mathfrak{L} - \hat{\mathfrak{L}}| < \delta_3$. Choose a positive $\delta_4 < \delta_3$ so that there exists a controller $\bar{u}(t) \subset \Omega$, when $|\mathfrak{L} - \hat{\mathfrak{L}}| < \delta_4$, with

$$|\hat{u}^*(t) - \bar{u}(t)| < \delta_4 \quad \text{on} \quad 0 \leq t \leq \hat{t}^* + 1$$

and

$$|\hat{x}^*(t^*) - \bar{x}(t^*)| < \theta/2.$$

Then

$$|u^*(t) - \hat{u}^*(t)| \leq |u^*(t) - \bar{u}(t)| + |\hat{u}^*(t) - \bar{u}(t)| < \frac{\varepsilon_2}{2} + \delta_4$$

except on a set of duration ϵ_2 on $0 \leq t \leq t^*$. Finally define $\delta = \min(\epsilon_2/2, \delta_4)$. Then $|\mathfrak{L} - \hat{\mathfrak{L}}| < \delta$ implies that $|u^*(t) - \hat{u}^*(t)| < \epsilon_2$ except on a set of duration of ϵ_2 in $0 \leq t \leq t^*$.

For an appropriate choice of $\epsilon_2 > 0$, and corresponding determination of $\delta > 0$, we have

$$\int_0^{\tau^*} |u^*(t) - \hat{u}^*(t)|\, dt < \epsilon_1 < \epsilon$$

and

$$|x^*(t) - \hat{x}^*(t)| < \epsilon \quad \text{on} \quad 0 \le t \le \tau^*,$$

as required. Q.E.D.

If we perturb the normal problem $\{\mathfrak{L}, \Omega, x_0\}$ to $\{\hat{\mathfrak{L}}, \hat{\Omega}, \hat{x}_0\}$ so that

$$|A - \hat{A}_0| + |B - \hat{B}| + |x_0 - \hat{x}_0| + \text{dist}\,(\Omega, \hat{\Omega}) < \delta$$

then we cannot conclude that the perturbed problem has a unique optimal controller $\hat{u}^*(t) \subset \hat{\Omega}$ steering $\hat{x}_0$ to the origin on $0 \le t \le \hat{t}^*$. However, one can show that every optimal controller $\hat{u}^*(t)$ of $\{\hat{\mathfrak{L}}, \hat{\Omega}, \hat{x}_0\}$ approximates $u^*(t)$ in the sense of Theorem 22.

EXERCISES

1. Consider the control process

$$\dot{x} = A(t)x + B(t)u$$

with $u \subset \Omega$, where the initial point x_0 at $t = 0$ is to be steered in minimal time to a moving target point $x = G(t)$, which traces a smooth curve. Show that the substitution $y = x - G(t)$ reduces the control problem to the regulation of y from $y_0 = x_0 - G(0)$ to $y = 0$ along a response of $\dot{y} = A(t)y + B(t)u + v(t)$. Compute $v(t)$.

2. Consider the minimal time controller to the origin for the system

$$\begin{aligned}\dot{x}^1 &= -\alpha_1 x^1 + b_1 u \\ \dot{x}^2 &= -\alpha_2 x^2 + b_2 u \\ &\cdot \\ &\cdot \\ &\cdot \\ \dot{x}^n &= -\alpha_n x^n + b_n u\end{aligned}$$

with all $b_j \ne 0$, $\alpha_j > 0$, and $|u| \le 1$. Show that each initial point x_0 satisfies the n transcendental equations

$$\pm \frac{x_0{}^j}{\alpha_j} = \frac{2}{\alpha_j}\left[\frac{1}{2} - e^{b_j t_1} + e^{b_j t_2} - e^{b_j t_3} + \cdots + (-1)^{r-1} e^{b_j t_{r-1}} + \frac{(-1)^r e^{b_j t_r}}{2}\right]$$

where $t_1 < t_2 < \cdots < t_{r-1}$ are the switching times for an extremal controller steering x_0 to the origin in time $0 \le t \le t_r$, and the initial sign $(\pm)$ is determined by the first value of the control as $u = \pm 1$. Show that we can require $r \le n$ and characterize the optimal time t^* as the minimal t_r for which such a solution $(0 < t_1 < t_2 < \cdots < t_r)$ of the transcendental equations exists.

3. For which values of the real parameter ρ does the control system

$$\begin{pmatrix} \dot{x}^1 \\ \dot{x}^2 \end{pmatrix} = \begin{pmatrix} -2 & \rho \\ 0 & -1 \end{pmatrix} \begin{pmatrix} x^1 \\ x^2 \end{pmatrix} + \begin{pmatrix} 1 & 0 \\ \rho & \rho \end{pmatrix} \begin{pmatrix} u^1 \\ u^2 \end{pmatrix}$$

satisfy the normality condition, with restraint Ω: $|u^1| \leq 1$, $|u^2| \leq 1$.

4. Compute the switching locus W and sketch the synthesis for the minimal time optimal controllers for $\ddot{x} = u$ with restraint $|u| \leq 1$ and target G as the disk $(x^1)^2 + (x^2)^2 \leq 1$ in the phase plane. Verify the required stability, normality, and transversality conditions.

5. Consider the minimal time-optimal control problem in R^2

$$\dot{x}^1 = -x^1 - x^2 + u^1$$
$$\dot{x}^2 = \quad - x^2 + u^1 + u^2$$

with restraint $|u^1| \leq 1$, $|u^2| \leq 1$, and target as the origin. Compute the switching loci W_1 and W_2 for the control components u^1 and u^2 and sketch the synthesis of the optimal controllers. Verify the required stability, normality, and transversality conditions.

6. Let $A(t)$ be a real continuous $n \times n$ matrix on $t \geq 0$. Assume that there exists $\epsilon > 0$ such that the eigenvalues of the symmetric matrix $A(t) + A(t)'$, at each instant $t \geq 0$, are less than $-\varepsilon$. Prove that the differential system $\dot{x} = A(t)x$ is stable at the origin, that is, for each solution $x(t)$

$$\lim_{t \to +\infty} x(t) = 0.$$

[Hint: $(d/dt)(x'x) = x'(A + A')x \leq -\varepsilon(x'x)$].

7. Consider the linear control problem in R^n

$$(\mathcal{L}) \qquad \dot{x} = Ax + Bu + v(t)$$

where A and B are constant and satisfy the controllability condition

$$\text{rank}\,[B, AB, A^2B, \ldots, A^{n-1}B] = n,$$

and $v(t)$ is continuous on R^1. Assume the restraint set $\Omega \subset R^m$ is compact and strictly convex and that the target G satisfies hypotheses (b) and (c) of Theorem 19. Then prove that the control process is normal and that the uniqueness of extremal controls transversal to G holds as in Theorem 19. Comment on the case where $v \equiv 0$, G is the origin, and $u = 0$ lies interior to Ω.

8. Consider the set of all autonomous control processes in R^n

$$(\mathcal{L}) \qquad \dot{x} = Ax + Bu + v$$

with a fixed convex polyhedral restraint set $\Omega \subset R^m$. Note that the normality condition does not involve v. Show that a generic process $\mathcal{L}$ satisfies the normality condition in the sense that those matrices (A, B) satisfying the normality condition form an open dense set in the metric space of all pairs of matrices (see Theorem 11).

9. Consider $\dot{x} = Ax + bu$ where A is a real 2×2 matrix with complex eigenvalues $\alpha \pm i\beta$ (and $\beta > 0$), with restraint Ω: $|u| \leq 1$ in R^1. Assume that $\mathcal{L}$ is controllable. Then for $t > 2\pi/\beta$ the set $K(t)$ has no corners, that is, at each boundary point of $K(t)$ there exists a unique supporting hyperplane (see Exercise 2 of Section 3).

10. Consider the linear control process in R^n

$$(\mathcal{L}) \qquad \dot{x} = A(t)x + B(t)u + v(t)$$

with $A(t)$, $B(t)$, and $v(t)$ continuous on R^1, initial point x_0, and compact restraint set $\Omega \subset R^m$. Show that $K(t)$ is Lipschitz continuous on a compact interval $\tau_0 \leq t \leq \tau_1$, that is,

$$\operatorname{dist}(K(t_1), K(t_2)) \leq k\,|t_1 - t_2| \quad \text{for} \quad \tau_0 \leq t_1 < t_2 \leq \tau_2$$

for some constant $k > 0$.

11. Consider the linear control process in R^n

$$(\mathcal{L}) \qquad \dot{x} = A(t)x + B(t)u + v(t)$$

with continuously varying, nonempty, compact convex restraint set $\Omega(t) \subset R^m$ on $\tau_0 \leq t \leq \tau_1$. Let x_0 be the initial state and $G(t)$ the compact continuously varying target set for $\tau_0 \leq t \leq \tau_1$. Prove that $K(t)$ is a compact, convex, continuously moving set, and hence obtain the analogies of the existence theorems 1 and 17. Prove the analogue of Theorem 2 and assert the maximal principle for extremal controls

$$\eta(t)\, B(t)\, u(t) = \max_{u \in \Omega(t)} \eta(t)\, B(t)u,$$

almost always.

12. Consider the linear control process in R^n

$$\dot{x} = A(t)x + B(t)u + v(t)$$

with initial state x_0 at τ_0 and fixed compact target G. Consider controls $u(t)$ on various $\tau_0 \leq t \leq t_1$ restrained by $\|u\|_2 = \int_{\tau_0}^{t_1} u(s)^2\, ds \leq 1$ and also $u(t) \subset \Omega$ where Ω is a closed convex set containing $u = 0$ in R^m. Prove that $K(t)$ is a compact, convex, continuously moving set, and hence obtain the analogues of the existence theorems 1 and 17.

By defining the new function $x^0(t) = \int_{\tau_0}^{t} u(s)^2\, ds$ we have the corresponding control process in R^{n+1}

$$\begin{aligned} \dot{x} &= A(t)x + B(t)u + v(t) \\ \dot{x}^0 &= u(t)^2 \end{aligned}$$

with initial state $(x_0, 0)$ and target as the compact cylinder $G \times [0 \le x^0 \le 1]$. The only restraint is $u(t) \subset \Omega$. The maximal principle for this nonlinear process in R^{n+1} will be discussed later.

13. Let $P_j(t)$ be a real polynomial of degree $\le n_j - 1$ for $1 \le j \le r$ and let $\lambda_1 < \lambda_2 < \cdots < \lambda_r$ be distinct real numbers. Prove that

$$\eta(t) = P_1(t)e^{\lambda_1 t} + \cdots + P_r(t)e^{\lambda_r t}$$

has at most $n_1 + n_2 + \cdots + n_r - 1$ real zeros. [Hint: Use induction on r. If $\eta(t)e^{-\lambda_r t}$ has $n_1 + \cdots + n_r$ real roots, then its n_rth derivative must have $n_1 + n_2 + \cdots + n_{r-1}$ real roots.]

14. Consider the controllable autonomous process in R^n

$$(\mathfrak{L}) \qquad \dot{x} = Ax + Bu$$

with compact convex restraint set $\Omega \subset R^m$ containing $u = 0$ in its interior.

(a) If $u_1(t)$ on $0 \le t \le t_1$ and $u_2(t)$ on $0 \le t \le t_2$ are extremal controllers steering x_0 to the origin, prove that $t_1 = t_2 = t^*$ is the minimal optimal time of regulation.

(b) If $\{\hat{\mathfrak{L}}, \hat{x}_0, \hat{\Omega}\}$ is any sufficiently nearby control problem of the same type, then $\hat{t}^*$ approximates t^*.

15. Consider the linear process in R^n

$$(\mathfrak{L}) \qquad \dot{x} = A(t)x + B(t)u + v(t)$$

with convex polyhedral restraint set $\Omega \subset R^m$, initial state x_0 at time $\tau_0 = 0$, and constant compact target set G. Assume that

$$\begin{aligned} A(t) &= A_0 + tA_1 + t^2A_2 + \cdots \\ B(t) &= B_0 + tB_1 + t^2B_2 + \cdots \\ v(t) &= v_0 + tv_1 + t^2v_2 + \cdots \end{aligned}$$

are real analytic on $t \ge 0$.

Show that if the problem is normal, an optimal controller $u^*(t)$ on $0 \le t \le t^*$ is piecewise constant (after redefinition on a null set) with a finite number of switches between the vertices of Ω. Moreover, if $n = 2$,

$$\det |B_0 w, (-A_0B_0 + B_1)w| \ne 0$$

for every edge w of Ω implies the normality of the control problem.

16. Consider the linear control process in R^n

$$(\mathcal{L}) \qquad \dot{x} = A(t)x + B(t)u$$

where $A(t) \in L_1$ and $B(t) \in L_q$, $1 \leq q < \infty$, on a finite interval $0 \leq t \leq T$. The admissible controllers are m-vectors $u(t)$ on various intervals $0 \leq t \leq t_1 \leq T$, with the restraint

$$\left(\int_0^{t_1} \sum_{i=1}^{m} |u^i(t)|^p \, dt\right)^{1/p} \leq 1,$$

where $1/p + 1/q = 1$, and for $q = 1$, $p = \infty$ use

$$\underset{0 \leq t \leq t_1, 1 \leq i \leq m}{\text{ess. sup.}} |u^i(t)| \leq 1.$$

Show that the corresponding set $K(t_1)$ of attainability from an initial state x_0 is compact, convex, and varies continuously with the time t_1. If a compact convex target set G is assigned, state the appropriate existence theorem for a minimal time-optimal controller $u^*(t)$ on $0 \leq t \leq t^*$.

Assume A and B are constants and the controllability condition obtains

$$\text{rank } [B, AB, A^2B, \ldots, A^{n-1}B] = n,$$

and assume $1 < p < \infty$ so the unit ball in L_p is strictly convex. Then prove that the time-optimal control $u^*(t)$ on $0 \leq t \leq t^*$ is unique and satisfies the maximal principle

$$u^{i*}(t) = |v^{i*}(t)|^{q/p} \operatorname{sgn} v^{i*}(t) \qquad i = 1, \ldots m$$

where $v^*(t) = \eta^*(t)B$ and $\eta^*(t) = \eta_0 e^{-At}$ is some nontrivial adjoint response. Also $u^*(t)$ lies on the boundary of the unit ball in L_p. [Hint: Use the weak compactness and the convexity of the unit ball in L_p to prove the properties of $K(t_1)$ and the existence theorem. The maximal principle follows from the condition for equality in Hölder's inequality.]

APPENDIX: CONVEX SETS

A subset K of a real vector space V is convex in case the segment $\lambda P_1 + (1 - \lambda)P_2$, $0 \leq \lambda \leq 1$, joining each pair of points P_1, P_2 in K lies entirely in K. The empty set, a single point $P \in V$, a segment joining two points $P_1, P_2 \in V$, and the entire space V itself are examples of convex sets. The intersection of convex subsets of V is a convex set.

We shall deal mainly with convex subsets of the real n-dimensional vector space R^n. Convex subsets of R^n are always connected but may be open, or closed, or neither as the following examples indicate:

1. a hyperplane π: $\sum_{i=1}^{n} a_i x^i + b = 0$ with $a \neq 0$ in cartesian coordinates $(x^1, \ldots, x^n)$ of R^n;
2. a closed half-space $\sum_{i=1}^{n} a_i x^i + b \geq 0$ (or ≤ 0);
3. an open half-space $\sum_{i=1}^{n} a_i x^i + b > 0$ (or < 0);
4. an open (or closed) ball $\sum_{i=1}^{n} (x^i - x_0{}^i)^2 < r^2$ (or $\leq r^2$) for a radius $r > 0$ and center x_0;
5. an n-cube $|x^i| \leq a$, $i = 1, \ldots, n$ and side length $2a > 0$, or the n-cube with some of its boundary points deleted.

The closure $\bar{K}$ and the interior int K of a convex set $K \subset R^n$ are convex and moreover $\overline{\text{int } K} = \bar{K}$ and int $\bar{K} =$ int K. A convex set $K \subset R^n$ has a dimension $r \leq n$, the dimension of the unique smallest linear manifold $L(K) \subset R^n$ that contains K. Relative to $L(K)$ the (nonempty) convex set K has a nonempty interior; furthermore if K is compact then K is topologically equivalent to a closed r-dimensional ball.

For an arbitrary subset $M \subset R^n$ define the convex hull $H(M)$ as the intersection of all convex sets containing M, so $H(M)$ is the smallest convex set that contains M. Thus M is convex if and only if $M = H(M)$. If M is compact, then $H(M)$ is also compact, and each point in $H(M)$ is a convex combination of some $n + 1$ points of M. The convex hull $H(P_0, P_1, \ldots, P_k)$ of a finite set of points in R^n is called a convex polytope. If the points $P_0, P_1, \ldots, P_k$ are linearly independent in R^n (that is, the vectors $P_1 - P_0, P_2 - P_0, \ldots, P_k - P_0$ are linearly independent), then $H(P_0, P_1, \ldots, P_k)$ is called a k-simplex. In particular a 1-simplex is a segment, a 2-simplex is a triangle, and a 3-simplex is a tetrahedron. It can be proved that a compact subset M of R^n is a convex polytope if and only if M is the intersection of a finite number of closed half-spaces. An arbitrary closed convex subset $K \subset R^n$ is the intersection of a countable number of closed half-spaces.

A hyperplane π separates two sets M_1 and M_2 of R^n in case M_1 lies in one closed half-space bounded by π and M_2 in the other closed half-space. Two disjoint convex sets K_1 and K_2 can be separated by a hyperplane in R^n if K_1 has a nonempty interior or if K_1 is closed and K_2 is compact.

Let K be a closed convex set in R^n. A hyperplane π that meets K and that contains K in one closed half-space is called a supporting hyperplane to K. Through each point of ∂K there passes a supporting hyperplane to the closed convex set $K \subset R^n$.

A point P is called an extreme point of the convex set $K \subset R^n$ if P does not lie on any segment $H(P_1, P_2)$ spanned by points $P_1 \neq P$ and $P_2 \neq P$ in K. Each supporting hyperplane to a compact convex set $K \subset R^n$ contains at least one extreme point of K. Furthermore K is the convex hull of the set of its extreme points.

A closed convex set K, which contains more than one point in R^n, is called strictly convex if each supporting hyperplane meets K in exactly one point. A strictly convex set $K \subset R^n$ has a nonempty interior and also each boundary point of K is an extreme point.

We next present several lemmas that are needed for Theorems 1 and 2 and that lead up to the stronger result of Theorem 1A. All these results will be applied to the linear control process

$$(\mathfrak{L}) \qquad \dot{x} = A(t)x + B(t)u + v(t)$$

with controllers $u(t)$ on $\mathfrak{J}$: $t_0 \leq t \leq t_1$ having values in the restraint set $\Omega \subset R^m$. Here $A(t)$, $B(t)$, and $v(t)$ are integrable.

LEMMA 1A. *Let Ω be a compact convex set in R^m and let $\mathcal{F}$ be the family of all measurable vector functions $u(t) \subset \Omega$ on the compact real interval $\mathfrak{J}$. Then $\mathfrak{F}$ is weakly compact (sequentially).*

Proof. Let $u_1(t), u_2(t), \ldots, u_k(t), \ldots$ be a sequence of functions in $\mathfrak{F}$, and we wish to select a subsequence $u_{k_i}(t)$ that converges weakly to a limit function $\bar{u}(t)$ in $\mathfrak{F}$, that is,

$$\lim_{i\to\infty} \int_{\mathfrak{J}} h(t)\, u_{k_i}(t)\, dt = \int_{\mathfrak{J}} h(t)\, \bar{u}(t)\, dt$$

for every bounded measurable n-vector function $h(t)$ on $\mathfrak{J}$. We need only prove the weak convergence for each component of $u_{k_i}(t)$. Thus consider a sequence of real scalar functions $w_k(t)$ with a uniform bound

$$|w_k(t)| \leq C \quad \text{on} \quad \mathfrak{J}: t_0 \leq t \leq t_1.$$

Hence $w_k(t)$ lie in the Hilbert space $L_2(t_0, t_1)$.

Let $\phi_1(t), \phi_2(t), \ldots, \phi_k(t), \ldots$ be a complete orthonormal set of real functions (for example, trigonometric functions) so that $w_k(t)$ has the expansion

$$w_k(t) \sim \alpha_k{}^1\phi_1(t) + \alpha_k{}^2\phi_2(t) + \cdots$$

with generalized Fourier coefficients $\alpha_k{}^j$ uniformly bounded since

$$\sum_{j=1}^{\infty} |\alpha_k{}^j|^2 = \int_{\mathfrak{J}} w_k(t)^2\, dt \leq C_1{}^2 = C^2\, |\mathfrak{J}|^2.$$

Let $w_{k1}(t)$ be a subsequence of $w_k(t)$ so

$$\lim_{k1\to\infty} \alpha_{k1}{}^1 = \gamma^1.$$

Let $w_{k2}(t)$ be a subsequence of $w_{k1}(t)$ so

$$\lim_{k2\to\infty} \alpha_{k2}^{\ 2} = \gamma^2.$$

Continue to extract subsequences with the successive generalized Fourier coefficients converging to γ^j. Now pick a diagonal subsequence

$$w_{1'}(t) = w_{11}(t),\ w_{2'}(t) = w_{22}(t), \ldots,\ w_{k'}(t) = w_{kk}(t), \ldots,$$

with the expansions

$$w_{k'}(t) \sim \beta_k^{\ 1}\phi_1(t) + \beta_k^{\ 2}\phi_2(t) + \cdots$$

and

$$\lim_{k\to\infty} \beta_k^{\ j} = \gamma^j \quad \text{for} \quad j = 1, 2, 3, \ldots.$$

For each finite integer k and real $\delta > 0$ we have

$$(\gamma^1)^2 + (\gamma^2)^2 + \cdots + (\gamma^k)^2 \le C_1^{\ 2} + \delta.$$

Thus

$$\sum_{j=1}^{\infty} |\gamma^j|^2 \le C_1^{\ 2}$$

and so by the Riesz-Fischer theorem there exists a measurable function $\bar{w}(t)$ on $\mathfrak{J}$ with the expansion

$$\bar{w}(t) \sim \gamma^1\,\phi_1(t) + \gamma^2\,\phi_2(t) + \cdots.$$

We claim that

$$\lim_{k'\to\infty} w_{k'}(t) = \bar{w}(t)$$

weakly on $\mathfrak{J}$. For let $\psi(t)$ be a real bounded measurable function with

$$|\psi(t)| \le C_2 \quad \text{on} \quad \mathfrak{J}.$$

Then there is a finite sum (say, a trigonometric polynomial)

$$P(t) = b^1\,\phi_1(t) + \cdots + b^l\,\phi_l(t)$$

with good approximation

$$\int_{\mathfrak{J}} |\psi(t) - P(t)|^2\,dt < \epsilon^2,$$

for a prescribed $\epsilon > 0$. Note that

$$\lim_{k'\to\infty} \int_{\mathfrak{J}} P(t)\,w_{k'}(t)\,dt = \lim_{k\to\infty} (b^1\beta_k^{\ 1} + \cdots + b^l\beta_k^{\ l})$$

which equals

$$b^1\gamma^1 + \cdots + b^l\gamma^l = \int_{\mathfrak{J}} P(t)\,\bar{w}(t)\,dt.$$

Now estimate

$$\left| \int_{\mathfrak{J}} \psi(t)\, w_{k'}(t)\, dt - \int_{\mathfrak{J}} \psi(t)\, \bar{w}(t)\, dt \right| \leq \left| \int_{\mathfrak{J}} P(w_{k'} - \bar{w})\, dt \right|$$

$$+ \left| \int_{\mathfrak{J}} (\psi - P)\, w_{k'}\, dt - \int_{\mathfrak{J}} (\psi - P)\bar{w}\, dt \right| .$$

Using the Schwarz inequality we obtain

$$\left| \int_{\mathfrak{J}} \psi(w_{k'} - \bar{w})\, dt \right| \leq \epsilon + C_1\epsilon + C_1\epsilon$$

for all sufficiently large k'. Thus the weak convergence of $w_{k'}(t)$ is established.

From the vector sequence $u_k(t)$ we extract subsequences to obtain $u_{k_i}(t)$ for which each component converges weakly on $\mathfrak{J}$, that is, there is a limit vector function $\bar{u}(t)$ so

$$\lim_{i\to\infty} \int_{\mathfrak{J}} h(t)\, u_{k_i}(t)\, dt = \int_{\mathfrak{J}} h(t)\, \bar{u}(t)\, dt$$

for each bounded measurable vector $h(t)$ on $\mathfrak{J}$.

Finally we must prove that $\bar{u}(t) \subset \Omega$ on $\mathfrak{J}$ (note that $\bar{u}(t)$ can be modified on a null set without changing the integral $\int_{\mathfrak{J}} h(t)\, \bar{u}(t)\, dt$). Let

$$(\pi) \qquad a_1x^1 + \cdots + a_nx^n + b = 0 \quad \text{or} \quad ax + b = 0$$

be a supporting hyperplane to Ω, so Ω lies in the closed half-space $ax + b \leq 0$. Let E be the subset of $\mathfrak{J}$ where

$$a\, \bar{u}(t) + b > 0.$$

Then by weak convergence

$$\lim_{i\to\infty} \int_{\mathfrak{J}} \chi_E(t)(a\, u_{k_i}(t) + b)\, dt = \int_{\mathfrak{J}} \chi_E(t)(a\, \bar{u}(t) + b)\, dt$$

where $\chi_E(t)$ is $+1$ on E and 0 on $\mathfrak{J} - E$. But

$$\int_{\mathfrak{J}} \chi_E(t)(a\, u_{k_i}(t) + b)\, dt \leq 0$$

and, if E has positive measure,

$$\int_{\mathfrak{J}} \chi_E(t)\, (a\, \bar{u}(t) + b)\, dt > 0.$$

This contradiction shows that E is a null set and so $\bar{u}(t)$ lies on the appropriate side of π almost always on $\mathfrak{J}$.

Now Ω is precisely the intersection of a countable number of closed half-spaces, and so $\bar{u}(t) \subset \Omega$ except on the union of a countable number of null sets. Therefore $\bar{u}(t) \subset \Omega$ almost everywhere on $\mathfrak{J}$, as required. Q.E.D.

Remark. A slightly stronger form of the lemma follows from the uniform bound on $|u|$, namely

$$\lim_{i\to\infty} \int_{\mathfrak{J}} h(t)\, u_{k_i}(t)\, dt = \int_{\mathfrak{J}} h(t)\, \bar{u}(t)\, dt$$

for each integrable vector $h(t)$ on $\mathfrak{J}$. To prove this it is only necessary to obtain the corresponding result for a sequence of scalar functions $w_k(t)$ converging weakly to $\bar{w}(t)$ on $\mathfrak{J}$. From the theorem there is a uniform bound

$$|w_k(t)| \le C \quad \text{and} \quad |\bar{w}(t)| \le C \quad \text{on} \quad \mathfrak{J}.$$

Let $\psi(t)$ be integrable on $\mathfrak{J}$ and choose

$$P(t) = b^1\phi_1(t) + \cdots + b^l\phi_l(t)$$

so

$$\int_{\mathfrak{J}} |\psi(t) - P(t)|\, dt < \epsilon.$$

For a subsequence $w_{k'}(t)$ we obtain

$$\lim_{k'\to\infty} \int_{\mathfrak{J}} P(t)(w_{k'}(t) - \bar{w}(t))\, dt = 0.$$

Then the estimate in the theorem shows that

$$\lim_{k'\to\infty} \int_{\mathfrak{J}} \psi(t)(w_{k'}(t) - \bar{w}(t))\, dt = 0,$$

and the corresponding result holds for the sequence of vectors $u_{k_i}(t)$.

In particular let

$$h(s) = \chi_t(s)\, \Phi(s)^{-1}\, B(s) \quad \text{for} \quad t_0 \le s \le t_1$$

where $\chi_t(s) = 1$ on $t_0 \le s \le t$ and 0 elsewhere on $\mathfrak{J}$,
$\Phi(s)$ is continuous and $B(s)$ integrable on $\mathfrak{J}$.
Then for each fixed t on $\mathfrak{J}$.

$$\lim_{i\to\infty} \int_{t_0}^{t} \Phi(s)^{-1}\, B(s)\, u_{k_i}(s)\, ds = \int_{t_0}^{t} \Phi(s)^{-1}\, B(s)\, \bar{u}(s)\, ds.$$

LEMMA 2A. *Let $\Omega \subset R^m$ be compact and let $\eta(t)$ be an absolutely continuous vector on $\mathfrak{J}$. For each $t \in \mathfrak{J}$ define*

$$m(t) = \max_{u\in\Omega} \eta(t)\, B(t) u.$$

Then $m(t)$ is integrable on $\mathfrak{J}$. If $B(t)$ is continuous (or absolutely continuous), then $m(t)$ is continuous (or absolutely continuous).

Proof. Let E be a closed subset of $\mathfrak{J}$ on which $B(t)$ is continuous and we show that $m(t)$ is measurable on E. Take any real number α and consider the set $E_\alpha \subset E$ where $m(t) \geq \alpha$. We show that every such E_α is closed and hence that $m(t)$ is measurable on E.

If E_α is not closed there is a sequence

$$t_k \to \bar{t}$$

with t_k and $\bar{t}$ in E, and

$$m(t_k) \geq \alpha \quad \text{but} \quad m(\bar{t}) < \alpha.$$

For a corresponding sequence of points $u_k \in \Omega$,

$$m(t_k) = \eta(t_k)\, B(t_k)u_k \geq \alpha$$

and we select a subsequence, still called u_k, so

$$u_k \to \bar{u} \in \Omega.$$

Then

$$m(\bar{t}) \geq \lim_{k\to\infty} m(t_k) = \eta(\bar{t})\, B(\bar{t})\bar{u} \geq \alpha.$$

This contradiction shows that E_α is closed.

Now let $E_1, E_2, \ldots, E_l, \ldots$ be a sequence of closed subsets of $\mathfrak{J}$ such that the measure

$$(\mathfrak{J} - E_l) \leq 2^{-l} \qquad l = 1, 2, 3, \ldots$$

and $B(t)$ is continuous on E_l (the existence of E_l is a general property of measurable functions). On each E_l the function $m(t)$ is measurable, and so $m(t)$ is measurable on their union, which differs from $\mathfrak{J}$ by a null set. Thus $m(t)$ is measurable on $\mathfrak{J}$. Since $|\eta(t)|$ on $\mathfrak{J}$ and $|u|$ are bounded, $m(t)$ is integrable on $\mathfrak{J}$.

Now assume that $B(t)$ is continuous, or absolutely continuous, on $\mathfrak{J}$. At times $t_1, t_2 \in \mathfrak{J}$ let $m(t)$ be achieved at corresponding points u_1 and $u_2 \in \Omega$. Then

$$m(t_2) - m(t_1) \leq \eta(t_2)\, B(t_2)u_2 - \eta(t_1)\, B(t_1)u_2 = [\eta(t_2)\, B(t_2) - \eta(t_1)\, B(t_1)]u_2$$

$$m(t_2) - m(t_1) \geq \eta(t_2)\, B(t_2)u_1 - \eta(t_1)\, B(t_1)u_1 = [\eta(t_2)\, B(t_2) - \eta(t_1)\, B(t_1)]u_1$$

From these estimates the continuity or absolute continuity of $m(t)$ follows immediately. Q.E.D.

LEMMA 3A. *Let $\Omega \subset R^m$ be compact and let $\phi(t, u)$ be a real m-vector function which is continuous in (t, u) for $u \in \Omega$ and all real t. For* *each fixed t, the set*

$$B(t)\,\phi(t, \Omega) = \{x \in R^n \mid x = B(t)\,\phi(t, u) \quad \text{for} \quad u \in \Omega\}$$

is compact in R^n. Let $g(t)$ be a measurable n-vector function with

$$g(t) \in B(t)\,\phi(t, \Omega) \quad \text{for each real} \quad t.$$

Then there exists a measurable m-vector function $u(t) \subset \Omega$ such that

$$g(t) = B(t)\,\phi(t, u(t)) \quad \text{for all real} \quad t.$$

Proof. For each fixed t_0 consider all points $u \subset \Omega$ for which

$$B(t_0)\,\phi(t_0, u) = g(t_0).$$

Select $u(t_0)$ so that the first component $u^1(t_0)$ is the smallest possible. If there is more than one such point u, required that $u^2(t_0)$ is the smallest among these, then further that $u^3(t_0)$ is smallest among these, and so forth, to define a unique vector $u(t_0) \subset \Omega$. We shall prove that $u(t)$ is measurable, and this need only be demonstrated for a compact interval $\mathfrak{J}$.

Suppose that $u^1(t), \ldots, u^{s-1}(t)$ are measurable on $\mathfrak{J}$ (if $s = 1$ nothing is assumed here), and we prove that $u^s(t)$ is measurable on $\mathfrak{J}$.

There exists a closed set $E_l \subset \mathfrak{J}$ with

$$\text{measure}\,(\mathfrak{J} - E_l) \leq 2^{-l} \qquad l = 1, 2, 3, \ldots$$

and such that on E_l the functions $u^1(t), \ldots, u^{s-1}(t)$, $B(t)$, and $g(t)$ are continuous. Pick a real number α and we show that the subset of E_l where $u^s(t) \leq \alpha$ is closed.

Suppose the contrary so that there is a sequence

$$t_k \to \bar{t} \quad \text{with} \quad t_k \quad \text{and} \quad \bar{t} \quad \text{in} \quad E_l$$

and

$$u^s(t_k) \leq \alpha < u^s(\bar{t}).$$

Choose a subsequence, still called t_k, so

$$\lim_{k \to \infty} u(t_k) = \bar{u} \in \Omega.$$

By the given continuity on E_l,

$$u^i(t_k) \to u^i(\bar{t}) = \bar{u}^i \quad \text{for} \quad i = 1, 2, \ldots, s-1,$$
$$g(t_k) \to g(\bar{t})$$
$$B(t_k) \to B(\bar{t}).$$

Thus

$$B(\bar{t})\,\phi(\bar{t}, \bar{u}) = g(\bar{t}).$$

But

$$\bar{u}^s \leq \alpha < u^s(\bar{t}),$$

which contradicts the definition of $u^s(t)$. Therefore $u^s(t)$ is measurable on E_l. As in Lemma 2*A*, $u^s(t)$ is measurable on $\mathfrak{J}$ and so the induction shows that $u(t)$ is measurable on $\mathfrak{J}$. Hence $u(t)$ is measurable on the entire real line. Q.E.D.

LEMMA 4A. *Let $y(t)$ be an integrable m-vector on the compact interval $\mathfrak{J}$. For each measurable subset $E \subset \mathfrak{J}$ consider the m-vector*

$$x_E = \int_E y(t)\, dt.$$

Let K be the set of all such points x_E as E varies over the collection of all measurable subsets of $\mathfrak{J}$. Then K is convex in R^m. If $y(t)$ is also bounded, then K is compact.

Proof. We here offer a compressed proof of this important measure-theoretic result. We shall be concerned with the interval $\mathfrak{J}$ and the σ-algebra $\mathfrak{B}$ of all Lebesgue measurable subsets of $\mathfrak{J}$ (a σ-algebra is a collection of subsets of $\mathfrak{J}$ closed under the operations of taking countable unions and intersections and complementation, in particular $\mathfrak{J}$ and the empty set ϕ are in the algebra). We shall also discuss certain σ-subalgebras $\mathfrak{A} \subset \mathfrak{B}$ [and all such algebras will be nonatomic, that is, if the Lebesgue measure $\mu(E) > 0$ for $E \in \mathfrak{A}$ there exists a subset $E_1 \subset E$ in $\mathfrak{A}$ with $0 < \mu(E_1) < \mu(E)$].

We first remark that for each σ-algebra $\mathfrak{A} \subset \mathfrak{B}$ there exists a continuous family of sets D_α, $0 \leq \alpha \leq 1$, with $D_\alpha \in \mathfrak{A}$, $D_{\alpha_1} \subset D_{\alpha_2}$ if and only if $\alpha_1 \leq \alpha_2$, and $\mu(D_\alpha) = \alpha\mu(\mathfrak{J})$. For convenience we henceforth assume $\mu(\mathfrak{J}) = 1$ so $\mu(D_\alpha) = \alpha$ on $0 \leq \alpha \leq 1$. Such a continuous family is easily constructed as a maximal, linearly ordered (by inclusion) chain of sets in $\mathfrak{A}$, with the aid of the axiom of choice.

Now let $f(t)$ be a real integrable function on $\mathfrak{J}$ and $\mathfrak{A}$ a σ-algebra. Then there exists a σ-algebra $\mathfrak{A}_1 \subset \mathfrak{A}$ whereon $\int_E f\, dt = \mu(E) \int_{\mathfrak{J}} f\, dt$ (we simplify this calculation by normalizing $\mu(\mathfrak{J}) = 1$ and $\int_{\mathfrak{J}} f\, dt = 1$). To prove this we first construct a set $E_1 \in \mathfrak{A}$ whereon $\int_{E_1} f\, dt = \frac{1}{2}$ and $\mu(E_1) = \frac{1}{2}$. The existence of E_1 is found by using a continuous family D_α, $0 \leq \alpha \leq 1$, in $\mathfrak{A}$. Note that $\mu(D_\alpha - D_{\alpha-1/2}) = \frac{1}{2}$ on $\frac{1}{2} \leq \alpha \leq 1$ and the integral of f over $D_\alpha - D_{\alpha-1/2}$ is a real continuous function $\phi(\alpha)$ such that $[\phi(1) + \phi(\frac{1}{2})]/2 = \frac{1}{2}$. Thus for some intermediate α_1 on $\frac{1}{2} \leq \alpha \leq 1$ we obtain $\phi(\alpha_1) = \frac{1}{2}$. Next, partition E_1 and $E_2 = \mathfrak{J} - E_1$ into two similar subsets E_3, E_4 and E_5, E_6, on which $\int_{E_i} f\,dt = \frac{1}{4} = \mu(E_i)$. Continue this partitioning to obtain a countable collection of such sets E and then let $\mathfrak{A}_1$ be the σ-algebra generated by all these sets. Since $\int_E f\,dt$ and $\mu(E)$ are each measures defined on $\mathfrak{A}_1$ and they are equal on the algebra generated by the above countable family of sets $E_1, E_2, \ldots$, we have $\int_E f\,dt = \mu(E)$ for all $E \in \mathfrak{A}_1$.

From a finite repetition of the above construction of the σ-algebra $\mathfrak{A}_1$ we obtain the following result:

Let $f = (f^1, \ldots, f^k)$ be a real k-vector of integrable functions on $\mathfrak{J}$. Then there exists a σ-algebra $\mathfrak{A} \subset \mathfrak{B}$ on which

$$\int_E f\,dt = \mu(E) \int_{\mathfrak{J}} f\,dt, \quad \text{for all} \quad E \in \mathfrak{A}.$$

Then it is easy to prove the convexity of $K = \left\{x_E = \int_E y(t)\,dt \mid E \in \mathfrak{B}\right\}$. Suppose

$$\int_{F_1} y(t)\,dt = a_1, \qquad \int_{F_2} y(t)\,dt = a_2,$$

and consider the intermediate point

$$\lambda a_1 + (1 - \lambda) a_2 \quad \text{for some} \quad 0 < \lambda < 1.$$

Consider the $2m$-vector $y^*(t) = (y(t)\,\chi_1(t), y(t)\,\chi_2(t))$ with the characteristic functions $\chi_1(t)$, $\chi_2(t)$ of F_1 and F_2. Let $\mathfrak{A} \subset \mathfrak{B}$ be a σ-algebra whereon

$$\int_E y^*(t)\,dt = \mu(E) \int_{\mathfrak{J}} y^*(t)\,dt = \mu(E) \begin{pmatrix} a_1 \\ a_2 \end{pmatrix}.$$

Let D_α, with $\mu(D_\alpha) = \alpha$, be a continuous family of $\mathfrak{A}$ and define $F = (D_\lambda \cap F_1) \cup [(\mathfrak{J} - D_\lambda) \cap F_2]$. Then

$$\int_F y(t)\,dt = \int_{D\lambda} y(t)\,\chi_1(t)\,dt + \int_{\mathfrak{J}-D\lambda} y(t)\,\chi_2(t)\,dt = \lambda a_1 + (1 - \lambda) a_2.$$

Hence K is convex.

The compactness of K is not proved here but follows from the argument of Theorem 1A. Q.E.D.

THEOREM 1A. *Consider the linear control process in R^n*

$$(\mathfrak{L}) \qquad \dot{x} = A(t)x + B(t)u + v(t)$$

with compact restraint set Ω, initial state x_0, and controllers $u(t) \subset \Omega$ on

$\mathfrak{J}: t_0 \leq t \leq t_1$. *Then the set of attainability* $K(t_1)$ *is compact, convex, and varies continuously with* t_1 *on* $t_1 \geq t_0$. *Moreover, if* Ω *is relaxed to its convex hull* $H(\Omega)$ *and if* $K_H(t_1)$ *is the corresponding set of attainability for all controllers* $u(t) \subset H(\Omega)$ *on* $t_0 \leq t \leq t_1$, *then*

$$K(t_1) = K_H(t_1).$$

Proof. Using the result of Liapunov as in the proof of Theorem 3, we find that $K(t_1)$ is convex.

The variation of parameter formula for a response $x(t)$ to a controller $u(t) \subset \Omega$ is

$$x(t) = \Phi(t)x_0 + \Phi(t)\int_{t_0}^{t} \Phi(s)^{-1}[B(s)\,u(s) + v(s)]\,ds.$$

Since Ω is compact, and $\Phi(t)$ is continuous on $\mathfrak{J}$, and $B(t)$ and $v(t)$ are integrable on $\mathfrak{J}$, we note that $K(t_1)$ is bounded. Thus the closure $\overline{K(t_1)}$ is a compact convex set in R^n. We shall prove that $K(t_1) = \overline{K(t_1)}$ or briefly $K = \bar{K}$.

Let P_0 be a point in $\bar{K}$. Since the interior of $\bar{K}$ is just the interior of K, we take $P_0 \in \partial\bar{K}$. At first suppose there is a supporting hyperplane π to $\bar{K}$ such that $\pi \cap \bar{K} = P_0$. Let $\eta(t_1)$ be a unit vector orthogonal to π and pointing into an open half-space not containing $\bar{K}$. Use the adjoint response

$$\eta(t) = \eta_0\,\Phi(t)^{-1} \quad \text{with} \quad \eta(t_1) = \eta_0\,\Phi(t_1)^{-1}.$$

Then, by Lemmas 2A and 3A, there is a controller $\hat{u}(t) \subset \Omega$ with

$$\eta(t)\,B(t)\,\hat{u}(t) = \max_{u\in\Omega} \eta(t)\,B(t)u \equiv m(t).$$

For the corresponding response $\hat{x}(t)$ in R^n we have

$$\eta(t_1)\,\hat{x}(t_1) = \max_{x\in K} \eta(t_1)x = \max_{x\in\bar{K}} \eta(t_1)x.$$

Thus $\hat{x}(t_1) \in \pi \cap \bar{K}$ and $\hat{x}(t_1) = P_0 \in K$.

In case $P_0 \in \partial\bar{K}$ fails to be the intersection of a supporting hyperplane with $\bar{K}$, we take a supporting hyperplane π at P_0 so that $\pi \cap \bar{K}$ is a compact convex set S_1 of lowest possible dimension. We demonstrate that $S_1 \subset K$ in the case where S_1 is a compact line segment and indicate the modifications in the argument for the higher-dimensional cases.

Define $\eta(t)$ as above for the hyperplane π. For each $t \in \mathfrak{J}$ let Ω_t be the compact subset of Ω where

$$\eta(t)\,B(t)u = m(t).$$

A control $u(t)$ will steer x_0 to the segment S_1 if and only if $u(t) \subset \Omega_t$ almost always.

Let $\eta_1(t_1)$ be an external vector to S_1 at the endpoint P_1 of S_1 and lying in the line generated by S_1. Define the adjoint response $\eta_1(t)$ accordingly and let

$$m_1(t) = \max_{u \in \Omega_t} \eta_1(t)\, B(t)u.$$

Then, by a slight extension of Lemmas 2A and 3A, we show that $m_1(t)$ is measurable and there exists a measurable controller

$$\hat{u}_1(t) \subset \Omega_t$$

such that

$$\eta_1(t)\, B(t)\, \hat{u}_1(t) = m_1(t).$$

We shall prove that $\hat{u}_1(t)$ steers x_0 to the endpoint P_1 of S_1.

Since $P_1 \in \bar{K}$, there is a sequence of controllers $u_j(t) \subset \Omega$ with responses $x_j(t)$ so

$$\lim_{j\to\infty} x_j(t_1) = P_1.$$

Since

$$\eta(t_1)x_j(t_1) \to \eta(t_1)P_1 = \eta(t_1)P_0,$$

$$\lim_{j\to\infty} \eta(t)\, B(t)\, u_j(t) = m(t), \quad \text{in measure,}$$

that is, for each $\epsilon > 0$ there is a subset of $\mathfrak{J}$ with measure ϵ off which

$$|m(t) - \eta(t)\, B(t)\, u_j(t)| < \epsilon$$

for all sufficiently large j.

For each $\epsilon > 0$ define

$$m_\epsilon(t) = \max_{u \in \Omega_{t,\epsilon}} \eta_1(t)\, B(t)u$$

where $\Omega_{t,\epsilon}$ is the subset of Ω where $\eta(t)\, B(t)u \geq m(t) - \epsilon$. Note that $\Omega_{t,\epsilon}$ is compact and $m_\epsilon(t)$ is measurable. Also for each fixed $t \in \mathfrak{J}$

$$\lim_{\epsilon\to 0} \Omega_{t,\epsilon} = \Omega_t$$

$$\lim_{\epsilon\to 0} m_\epsilon(t) = m(t),$$

both sequences converging monotonically nonincreasing. The interpretation is that a controller $u(t)$ must lie in $\Omega_{t,\epsilon}$ except for some set of small time duration if it is to steer x_0 near to the segment S_1 in $\bar{K}$.

Fix a small $\epsilon > 0$. Then for sufficiently large j we have

$$u_j(t) \subset \Omega_{t,\epsilon}$$

except on a subset of $\mathfrak{J}$ with measure ϵ. Because $\pi \cap \bar{K} = S_1$

$$\lim_{j\to\infty} \eta_1(t_1)\, x_j(t_1) = \lim_{j\to\infty} \sup_{x} \eta_1(t)x = \eta_1(t_1)P_1$$

where the supremum is taken over all $x \in \bar{K}$ such that

$$\eta(t_1)x \geq \eta(t_1)P_0 - \frac{1}{j}.$$

Thus for a prescribed $\epsilon > 0$ there exists $0 < \epsilon_1 < \epsilon$ so that

$$|m_{\epsilon_1}(t) - \eta_1(t)\, B(t)\, u_j(t)| < \epsilon,$$

except on a set of measure ϵ, for all sufficiently large j.

By Egoroff's theorem

$$\lim_{\epsilon \to 0} m_\epsilon(t) = m_1(t)$$

almost uniformly on $\mathfrak{J}$, and therefore

$$\lim_{j \to \infty} \eta_1(t)\, B(t)\, u_j(t) = m_1(t),$$

in measure on $\mathfrak{J}$. It follows that there is a subsequence, still called $u_j(t)$, such that at almost every point $t \in \mathfrak{J}$

$$\lim_{j \to \infty} \eta(t)\, B(t)\, u_j(t) = m(t) = \eta(t)\, B(t)\, \hat{u}_1(t)$$

$$\lim_{j \to \infty} \eta_1(t)\, B(t)\, u_j(t) = m_1(t) = \eta_1(t)\, B(t)\, \hat{u}_1(t).$$

Since $u_j(t)$ steers x_0 to $x_j(t_1) \to P_1$, the limit controller $\hat{u}_1(t)$ steers x_0 to P_1 and hence $P_1 \in K$. In the same way we construct a controller $\hat{u}_2(t) \subset \Omega$, which steers x_0 to the other endpoint P_2 of the segment S_1. Because K is convex, the entire segment $S_1 \subset K$.

If $P_0 \in \partial\bar{K}$ lies in no supporting hyperplane meeting $\bar{K}$ in a segment, we take a supporting hyperplane π at P_0 so that $\pi \cap \bar{K}$ is a compact convex set S of lowest possible dimension. If S has dimension two then we consider the boundary of S relative to the plane $L(S)$ spanned by S. Each such boundary point is isolated by a supporting line to S in $L(S)$ or else lies on a segment that is the intersection of S with such a line. In either case a repetition of the above argument shows that the boundary of S in $L(S)$ lies in K and hence S lies in the convex set K. If S has dimension three or more then consider the boundary of S relative to the linear manifold $L(S)$ spanned by S, and a repetition of our argument proves that $S \subset K$.

Thus every point $P_0 \in \partial\bar{K}$ lies in K, and so $K = \bar{K}$.

Finally we prove that $K(t_1) = K_H(t_1)$, or $K = K_H$. Both K and K_H are compact convex sets and $K \subset K_H$. We shall show that K is dense in K_H, from which it follows that $K = K_H$.

First let $u_H(t) \subset H(\Omega)$ be a step function with a finite number of values on the disjoint intervals $\mathfrak{J}_1, \mathfrak{J}_2, \ldots, \mathfrak{J}_s$ that comprise $\mathfrak{J}$. Write

$$u_H(t) = u_{H1} + \cdots + u_{Hs}$$

where u_{Hj} is constant on the jth interval $\mathfrak{J}_j \subset \mathfrak{J}$ and is zero elsewhere on $\mathfrak{J}$. On the first interval $\mathfrak{J}_1$: $t_0 \leq t \leq \tau_1$ write

$$u_{H1} = \lambda_0 u_{01} + \cdots + \lambda_n u_{n1},$$

a constant convex combination of the vectors $u_{01}, \ldots, u_{n1}$ in Ω. By the convexity of K there is a controller $u_1(t) \subset \Omega$ on $\mathfrak{J}_1$, which steers x_0 to the same point $x_H(\tau_1)$ as does u_{H1}. Now start with $x_H(\tau_1)$ as initial point and use the controller u_{H2} on $\mathfrak{J}_2$: $\tau_1 \leq t \leq \tau_2$ to find a controller $u_2(t) \subset \Omega$ on $\mathfrak{J}_2$ that steers $x_H(\tau_1)$ to the same point $x_H(\tau_2)$ as does u_{H2}. Continuing in this way we construct the controller

$$u(t) = u_1(t) + u_2(t) + \cdots + u_s(t) \subset \Omega$$

where $u_j(t) \equiv 0$ on $\mathfrak{J}_i$ for $i \neq j$, which steers x_0 to the same point $x_H(t_1)$ as does $u_H(t)$ on $\mathfrak{J}$.

Each controller $\tilde{u}(t) \subset H(\Omega)$ on $\mathfrak{J}$ is continuous on a closed subset $E \subset \mathfrak{J}$ with measure of $(\mathfrak{J} - E)$ arbitrarily small. Since $H(\Omega)$ is convex we can modify $\tilde{u}(t)$ on the open intervals comprising $\mathfrak{J} - E$ by using line segments in $H(\Omega)$ so that the resulting function $\bar{u}(t) \subset H(\Omega)$ is continuous on $\mathfrak{J}$. Next choose points on $\bar{u}(t)$ to construct a step function $u_H(t) \subset H(\Omega)$ so that $u_H(t)$ uniformly approximates $\tilde{u}(t)$ off some set of small measure. Thus the response $x_H(t)$ of $u_H(t)$ uniformly approximates the response $\tilde{x}(t)$ of $\tilde{u}(t)$ on $\mathfrak{J}$. Therefore K is dense in K_H and so $K = K_H$. Q.E.D.

CHAPTER 3

Optimal Control for Linear Processes with Integral Convex Cost Criteria

In the present chapter we develop the theoretical foundation for the optimalization of the integral square error criterion and for generalizations of this criterion over a fixed time duration. The first part of the chapter deals only with quadratic error criteria and applications of the corresponding theory. The second part involves general convex integral criteria and also processes in which the control function is further restrained. The results, necessary and sufficient conditions for optimal control, are achieved by exploring the geometric properties of the set of attainability.

3.1 SIGNIFICANCE OF INTEGRAL COST CRITERIA

An integral cost criterion is used whenever the average performance over the chosen time duration is of primary importance and brief disturbances from the ideal can be disregarded. For example, a frequently used criterion in the design of controllers is the integral of the error squared. This criterion has been studied in great detail and the optimal controller has been tabulated as an explicit function of certain linear control parameters depending on the coefficients and initial conditions of the linear system (see examples in Section 3.3 below).

Even though the systems studied in this chapter are linear they are important for the analysis of nonlinear guidance and control decisions such as are involved in space maneuvers, since the linear processes may arise as variational equations obtained by linearization near a known solution of the nonlinear system.

In many physical problems the selection of the cost functional, the specified integral criterion, is not easily determined. The correct choice is not obvious and, in practice, the best policy often indicates the selection of a cost functional that leads to an easily constructed optimal controller and that seems reasonably to approximate the ideal goal. After the design of the optimal-control synthesizer we must experiment to note if the so controlled system meets all other physical requirements. Until more is known about the properties of various optimal controllers this method of approximation and successive adjustment appears to be the most effective way to proceed. It thus appears that the motivation for studying certain specific optimal control systems is that such investigations can be formulated as definite mathematical problems that lead to useful techniques and methods for control synthesis.

3.2 INTEGRAL QUADRATIC COST CRITERIA

Measurement of the quality of a control system by integrating the system error squared over a fixed time interval provides a criterion of performance for which the minimizing control is easily calculated. We first investigate the general properties of this class of linear control systems. Optimal controllers are shown to be extremal controllers, satisfying the maximal principle and corresponding to the boundary of the set of attainability (the necessary condition), and a one-to-one correspondence between such boundary points and extremal controllers is established (the sufficiency condition). A number of special problems are solved in the next section using the general theory developed here.

In this section we consider a real linear control process

$$(\mathcal{L}) \qquad \dot{x} = A(t)x + B(t)u$$

where $A(t)$ and $B(t)$ are continuous matrices, of size $n \times n$ and $n \times m$ respectively, on a given finite time interval $t_0 \leq t \leq T$. The state n-vector $x(t)$ starts from the fixed initial state $x(t_0) = x_0$ and is steered by the control m-vector $u(t)$ to the endpoint $x(T)$. The cost functional of control is prescribed by

$$C(u) = g(x(T)) + \int_{t_0}^{T} [x(s)'\, W(s)\, x(s) + u(s)'U(s)\, u(s)]\, ds.$$

Here $g(x)$ is a given real continuous function on R^n and $W(s)$ and $U(s)$ are real square matrices which are continuous and symmetric on $t_0 \leq s \leq T$. Also we assume that $W(s)$ is positive semidefinite and $U(s)$ is positive definite for all s, that is,

$$W(s) = W(s)' \geq 0 \quad \text{and} \quad U(s) = U(s)' > 0 \quad \text{so}$$

$$x(s)'\, W(s)\, x(s) = \|x(s)\|_W^2 \geq 0 \quad \text{and}$$

$$u(s)'\, U(s)\, u(s) = \|u(s)\|_U^2 > 0 \quad \text{if} \quad u(s) \neq 0.$$

The problem of optimal control is to minimize the cost functional $C(u)$, among all measurable controllers $u(s)$ for which

$$\int_{t_0}^{T} \|u(s)\|_U^{\,2}\, ds < \infty.$$

All the above hypotheses and notations hold throughout this section. We may also require that the controller steer the response to a prescribed target in R^n.

Since $U(s) > 0$ is continuous and bounded, it is easy to see that

$$\int_{t_0}^{T} \|u(s)\|_U^{\,2}\, ds < \infty$$

if and only if $u(t)$ lies in the Hilbert space $L_2(t_0, T)$, that is,

$$\int_{t_0}^{T} u(s)'\, u(s)\, ds = \int_{t_0}^{T} \|u(s)\|^2\, ds < \infty.$$

Such admissible controllers are certainly integrable and the continuous response $x(t)$ is bounded on $t_0 \leq t \leq T$. Because of the nonnegative nature of the (semidefinite) norms

$$x(s)'\, W(s)\, x(s) = \|x(s)\|_W^2 \geq 0$$
$$u(s)'\, U(s)\, u(s) = \|u(s)\|_U^2 > 0, \quad \text{for} \quad u(s) \neq 0,$$

we can expect that the minimum for the cost $C(u)$ can be realized, at least under reasonable restrictions on $g(x(T))$ such as are discussed in Theorem 2.

For notational convenience we define

$$x_u^{\,0}(t) = \int_{t_0}^{t} [\|x_u(s)\|_W^{\,2} + \|u(s)\|_U^{\,2}]\, ds$$

and consider the response $\hat{x}_u(t) = (x_u^{\,0}(t), x_u(t))$ in R^{n+1} for each control vector $u(t)$. We first consider the case $g(x) \equiv 0$ and we shall find that the results obtained are basic for the analysis of the general case.

Definition. Consider the control problem in R^n

$$(\mathfrak{L}) \qquad \dot{x} = A(t)x + B(t)u$$

with cost functional

$$C_0(u) = \int_{t_0}^{T} [\|x(s)\|_W^{\,2} + \|u(s)\|_U^{\,2}]\, ds.$$

The set $\hat{K} = \hat{K}(T, x_0)$ of attainability is the totality of all response endpoints

$$\hat{x}_u(T) = (x_u^{\,0}(T), x_u(T)) \quad \text{in} \quad R^{n+1},$$

for all admissible control vectors $u(t)$ on $t_0 \leq t \leq T$.

In this section $\hat{K}$ will always refer to the process $\mathfrak{L}$ with cost $C_0(u)$. Because of the nonlinearities in $C_0(u)$ the set $\hat{K}(T, x_0)$ depends significantly on the point x_0. It is clear that $\hat{K}$ lies in the half-space $x^0 > 0$ (except possibly for a single point corresponding to the null control $u(t) \equiv 0$). The convexity of $\hat{K}$ will be shown to be a consequence of the norm convexity relations

$$\begin{aligned}\|\lambda\, u_1(s) + (1-\lambda)\, u_2(s)\|_U{}^2 &= \lambda^2\, \|u_1\|_U{}^2 \\ &\quad + 2\lambda(1-\lambda)u_1'Uu_2 + (1-\lambda)^2\, \|u_2\|_U{}^2 \\ &\leq \lambda^2\, \|u_1\|_U{}^2 + \lambda(1-\lambda)[\|u_1\|_U{}^2 + \|u_2\|_U{}^2] \\ &\quad + (1-\lambda)^2\, \|u_2\|_U{}^2 \\ &= \lambda\, \|u_1(s)\|_U{}^2 + (1-\lambda)\, \|u_2(s)\|_U{}^2\end{aligned}$$

and similarly

$$\|x_{\lambda u_1 + (1-\lambda) u_2}(s)\|_W{}^2 \leq \lambda\, \|x_{u_1}(s)\|_W{}^2 + (1-\lambda)\, \|x_{u_2}(s)\|_W{}^2,$$

for $0 \leq \lambda \leq 1$.

LEMMA. *Consider the control process in R^n*

$$(\mathfrak{L}) \qquad \dot{x} = A(t)x + B(t)u$$

with cost functional

$$C_0(u) = \int_{t_0}^{T} [\|x(s)\|_W{}^2 + \|u(s)\|_U{}^2]\, ds$$

and set of attainability $\hat{K} \subset R^{n+1}$. Then the orthogonal projection of $\hat{K}$ on the hyperplane $x^0 = 0$ is an entire linear manifold. Also if the point $\hat{y} = (y^0, y) \in \hat{K}$, then the entire half-line $x^0 \geq y^0$, $x = y$ lies in $\hat{K}$.

Proof. The variations of parameter formula

$$x_u(T) = \Phi(T)x_0 + \Phi(T) \int_{t_0}^{T} \Phi(s)^{-1}\, B(s)\, u(s)\, ds,$$

where $\Phi(t)$ is the solution of $\dot{x} = A(t)x$ with $\Phi(t_0) = I$, shows that the points $x_u(T) - \Phi(T)x_0$ fill an entire linear subspace of $x^0 = 0$ when u varies over the linear space $L_2(t_0, T)$ of all admissible controllers.

Now let $\bar{u}(t)$ steer the initial state $(0, x_0)$ to the point (y^0, y) in $\hat{K}$. We shall construct a control $u(t) = \bar{u}(t) + u_\beta(t)$ such that

$$(1) \qquad \int_{t_0}^{T} \Phi(s)^{-1}\, B(s)\, u_\beta(s)\, ds = 0$$

$$(2) \qquad \int_{t_0}^{T} \|x_u(s)\|_W{}^2 + \|u(s)\|_U{}^2\, ds = y^0 + b,$$

for a prescribed $b \geq 0$. Let

$$u_\beta^1(s) = X(s, T)\beta_1 + X\left(s, \frac{T + t_0}{2}\right)\beta_2 + \cdots + X\left(s, \frac{T + nt_0}{n + 1}\right)\beta_{n+1}$$

and

$$u_\beta^j(s) = 0 \quad \text{for} \quad j = 2, 3, \ldots, m \quad \text{on} \quad t_0 \leq s \leq T.$$

Here define the function

$$X(s, h) = \begin{cases} 1 & \text{if} \quad s \leq h \\ 0 & \text{if} \quad s > h \end{cases}$$

and the constants $\beta_1, \beta_2, \ldots, \beta_{n+1}$ are determined below.

To satisfy property (1) we require

$$\beta_1 \int_{t_0}^{T} \Phi(s)^{-1}\, b_1(s)\, ds + \beta_2 \int_{t_0}^{\frac{T+t_0}{2}} \Phi(s)^{-1}\, b_1(s)\, ds + \cdots$$
$$+ \beta_{n+1} \int_{t_0}^{\frac{T+nt_0}{n+1}} \Phi(s)^{-1}\, b_1(s)\, ds = 0$$

where $b_1(s)$ is the first column of $B(s)$. Thus property (1) obtains if we fix the $n + 1$ real numbers $\beta_1, \beta_2, \ldots, \beta_{n+1}$ as a nontrivial solution of the above n linear homogeneous scalar equations. For each real $\rho \neq 0$ the values $\rho\beta = (\rho\beta_1, \rho\beta_2, \ldots, \rho\beta_{n+1})$ with the corresponding control

$$u(s) = \bar{u}(s) + u_{\rho\beta}(s) = \bar{u}(s) + \rho\, u_\beta(s)$$

satisfies condition (1).

We shall choose $\rho \neq 0$ so that $u(s)$ satisfies property (2). Now

$$C_0(u) = \int_{t_0}^{T} [\|x_u\|_W^{\;2} + \|u\|_U^{\;2}]\, ds = \int_{t_0}^{T} [\|x_{\bar{u}+\rho u_\beta}\|_W^{\;2} + \|\bar{u} + \rho u_\beta\|_U^{\;2}]\, ds.$$

For abbreviation write $x_u(s) = \Phi(s)x_0 + P_u(s)$ so

$$C_0(u) = \int_{t_0}^{T} [\|x_{\bar{u}} + \rho P_{u_\beta}\|_W^{\;2} + \|\bar{u} + \rho u_\beta\|_U^{\;2}]\, ds$$
$$= \int_{t_0}^{T} [\|x_{\bar{u}}\|_W^{\;2} + \|\bar{u}\|_U^{\;2}]\, ds + 2\rho \int_{t_0}^{T} [x_{\bar{u}}' W P_{u_\beta} + \bar{u}' U u_\beta]\, ds$$
$$+ \rho^2 \int_{t_0}^{T} [\|P_{u_\beta}\|_W^{\;2} + \|u_\beta\|_U^{\;2}]\, ds.$$

Since $u_\beta(s) \not\equiv 0$, the coefficient of ρ^2 is positive and hence, by appropriate choice of ρ, we can demand that the last two terms total any preassigned number $b \geq 0$. In this case

$$C_0(u) = y^0 + b, \quad \text{as required.} \qquad \text{Q.E.D.}$$

THEOREM 1. *Consider the control process in* R^n

$$(\mathfrak{L}) \qquad \dot{x} = A(t)x + B(t)u$$

with cost functional

$$C_0(u) = \int_{t_0}^{T} [\|x\|_W^2 + \|u\|_U^2]\, ds.$$

Then the set of attainability $\hat{K} \subset R^{n+1}$ *is convex and closed.*

Proof. Let $\hat{x}_1 = (x_1^0, x_1)$ and $\hat{x}_2 = (x_2^0, x_2)$ be two points in $\hat{K}$ corresponding to the controllers $u_1(s)$ and $u_2(s)$ on $t_0 \leq s \leq T$. Let

$$\hat{y} = (y^0, y) = \lambda\hat{x}_1 + (1 - \lambda)\hat{x}_2 \quad \text{for} \quad 0 < \lambda < 1.$$

To prove that $\hat{K}$ is convex we must construct a controller steering $(0, x_0)$ to $\hat{y}$.

Define

$$\bar{u}(s) = \lambda\, u_1(s) + (1 - \lambda)\, u_2(s)$$

with corresponding responses

$$x_{\bar{u}}(s) = \lambda\, x_1(s) + (1 - \lambda)\, x_2(s)$$

so

$$x_{\bar{u}}(T) = \lambda x_1 + (1 - \lambda)\, x_2 = y.$$

By the convexity properties of the norms we obtain

$$x_{\bar{u}}^0(T) = \int_{t_0}^{T} [\|x_{\bar{u}}\|_W^2 + \|\bar{u}\|_U^2]\, ds \leq \lambda \int_{t_0}^{T} [\|x_1\|_W^2 + \|u_1\|_U^2]\, ds + (1 - \lambda) \int_{t_0}^{T} [\|x_2\|_W^2 + \|u_2\|_U^2]\, ds$$

and

$$x_{\bar{u}}^0(T) \leq \lambda x_1^0 + (1 - \lambda)\, x_2^0 = y^0.$$

But $\hat{K}$ contains the entire half-line $x^0 \geq x_{\bar{u}}^0(T)$, $x = y$ and hence $\hat{K}$ contains $\hat{y}$. Thus $\hat{K}$ is convex.

It is convenient to show that even relative to the nonlinear coordinates $(\sqrt{x^0}, x^1, \ldots, x^n)$ in $x^0 \geq 0$ in R^{n+1} the set $\hat{K}$ is convex. For this we must construct a control steering $(0, x_0)$ to

$$\hat{z} = (z^0, z) = \lambda(\sqrt{x_1^0}, x_1) + (1 - \lambda)(\sqrt{x_2^0}, x_2).$$

Again

$$\bar{u}(s) = \lambda\, u_1(s) + (1 - \lambda)\, u_2(s)$$

steers x_0 to

$$x_{\bar{u}}(T) = \lambda\, x_1 + (1 - \lambda)x_2 = z.$$

Now introduce the notation

$$\xi(s) = \begin{pmatrix} x(s) \\ u(s) \end{pmatrix}, \qquad V(s) = \begin{pmatrix} W(s) & 0 \\ 0 & U(s) \end{pmatrix}$$

and define the norm

$$|||\xi|||^2 = \int_{t_0}^{T} \|\xi(s)\|_V{}^2 \, ds.$$

Then the triangle inequality

$$|||\lambda\xi_1 + (1-\lambda)\xi_2||| \le \lambda\, |||\xi_1||| + (1-\lambda)\, |||\xi_2|||$$

proves that

$$\sqrt{x_{\bar{u}}{}^0(T)} \le \lambda\sqrt{x_1{}^0} + (1-\lambda)\sqrt{x_2{}^0} = z^0.$$

The above lemma again proves that $\hat{K}$ contains $\hat{z}$ and hence $\hat{K}$ is convex relative to the coordinates $(\sqrt{x^0}, x^1\, x^2, \ldots, x^n)$ as required.

In the remainder of the proof we use the coordinates $(\sqrt{x^0}, x^1, \ldots, x^n)$ in $x^0 \ge 0$ and $\hat{K}$ is closed with reference to these coordinates if and only if $\hat{K}$ is closed in R^{n+1} using the usual coordinates $(x^0, x^1, \ldots, x^n)$. We shall also assume that $\hat{K}$ has a nonempty interior, for otherwise we could define our subsequent constructions within the linear manifold spanned by $\hat{K}$.

Each boundary point $\hat{p} = (\sqrt{p^0}, p)$ of $\bar{\hat{K}}$ has a supporting hyperplane with exterior normal extending towards the hyperplane $x^0 = 0$, as follows from the lemma. Therefore there exists a point $\hat{q} = (0, q)$ such that $\hat{p}$ is the unique point in $\bar{\hat{K}}$ that is closest to $\hat{q}$. That is, $\hat{p}$ is characterized as the unique point in $\bar{\hat{K}}$ satisfying the condition

$$|p^0| + \|p - q\|^2 = \inf_{\hat{r}\in\hat{K}} \{|r^0| + \|r - q\|^2\}$$

We complete the proof of the theorem by showing that for each given $(0, q)$ there exists a point $\hat{p}$ in $\hat{K}$ which satisfies this condition. Consider a sequence of controllers $u_i(s)$ such that

$$\lim_{i\to\infty}\left\{\int_{t_0}^{T} \|x_i(s)\|_W{}^2 + \|u_i(s)\|_U{}^2 \, ds + \|x_i(T) - q\|^2\right\} = \alpha$$

where

$$\alpha = \inf_{\hat{r}\in\hat{K}} \{|r^0| + \|r - q\|^2\}.$$

For each controller $u_i(s)$ write the response

$$x_i(s) = H_i(s) + P_i(s)$$

where

$$H(s) = \Phi(s)x_0,$$

$$P_i(s) = \Phi(s)\int_{t_0}^{s} \Phi(\sigma)^{-1}\, B(\sigma)\, u_i(\sigma)\, d\sigma$$

and define the functional

$$J(u) = \int_{t_0}^{T} \|x(s)\|_W^2 + \|u(s)\|_U^2 \, ds + \|x(T) - q\|^2 - \int_{t_0}^{T} \|H(s)\|_W^2 \, ds - \|H(T) - q\|^2$$

or

$$J(u) = 2P(T)'(H(T) - q) + \|P(T)\|^2 + \int_{t_0}^{T} \|P(s)\|_W^2 + 2H(s)' \, W(s) \, P(s) + \|u(s)\|_U^2 \, ds.$$

It is straightforward to compute

$$J\left(\frac{u_i - u_j}{2}\right) + J\left(\frac{u_i + u_j}{2}\right) = \frac{1}{2} J(u_i) + \frac{1}{2} J(u_j) + (H(T) - q)'(P_i(T) - P_j(T)) + \int_{t_0}^{T} H(s)' \, W(s)(P_i(s) - P_j(s)) \, ds.$$

Further compute

$$\begin{aligned} \frac{1}{2}\left[J(u_i) + J(u_j) - 2J\left(\frac{u_i + u_j}{2}\right)\right] &= J\left(\frac{u_i - u_j}{2}\right) - (H(T) - q)' \, (P_i(T) - P_j(T)) \\ &\quad - \int_{t_0}^{T} H(s)' \, W(s)(P_i(s) - P_j(s)) \, ds \\ &= \left\| \frac{P_i(T) - P_j(T)}{2} \right\|^2 + \int_{t_0}^{T} \left\| \frac{P_i(s) - P_j(s)}{2} \right\|_W^2 + \left\| \frac{u_i(s) - u_j(s)}{2} \right\|_U^2 ds. \end{aligned}$$

Now

$$J(u_i) \to \inf_u J(u) = \beta \quad \text{and} \quad J\left(\frac{u_i + u_j}{2}\right) \geq \beta \quad \text{so}$$

$$\tfrac{1}{2}[J(u_i) + J(u_j) - 2\beta] \geq \left\| \frac{P_i(T) - P_j(T)}{2} \right\|^2 + \int_{t_0}^{T} \left\| \frac{P_i(s) - P_j(s)}{2} \right\|_W^2 + \left\| \frac{u_i(s) - u_j(s)}{2} \right\|_U^2 ds.$$

Since $J(u_i) + J(u_j) - 2\beta$ is positive, yet approaches zero, we must have

$$\lim_{i,j\to\infty} \int_{t_0}^{T} \|u_i(s) - u_j(s)\|_U^2 \, ds = 0.$$

By the Riesz-Fischer theorem the sequence $\{u_i\}$ converges in $L_2(t_0, T)$ to some limit controller $u^*(s)$ with response $x^*(s)$. Thus

$$\int_{t_0}^{T} \|x^*(s)\|_W^2 + \|u^*(s)\|_U^2 \, ds + \|x^*(T) - q\|^2 = \alpha$$

and hence the point $(\sqrt{p^0}, p) = (\sqrt{x^{0*}(T)}, x^*(T))$ lies in $\hat{K}$. Thus $\hat{K} = \bar{\hat{K}}$ and $\hat{K}$ is closed. Q.E.D.

For the control process in R^n

$$(\mathfrak{L}) \qquad \dot{x} = A(t)x + B(t)u$$

we could assign a cost functional

$$\tilde{C}_0(u) = \left[\int_{t_0}^{T} \|x\|_W^2 + \|u\|_U^2 \, ds\right]^{1/2}$$

and so determine a corresponding set of attainability $\tilde{K} \subset R^{n+1}$ consisting of all points $(\tilde{C}_0(u), x(T))$. We note that the proof of the following corollary is contained in the proof of Theorem 1.

COROLLARY. *Consider the control process in* R^n

$$(\mathfrak{L}) \qquad \dot{x} = A(t)x + B(t)u$$

with cost functional

$$\tilde{C}_0(u) = \left[\int_{t_0}^{T} \|x\|_W^2 + \|u\|_U^2 \, ds\right]^{1/2}.$$

Then the corresponding set of attainability $\tilde{K} \subset R^{n+1}$ *is convex and closed.*

An analogue of the following existence theorem, and other results of this chapter, are valid for the cost functionals $g(x(T)) + \tilde{C}_0(u)$ as well as $g(x(T)) + C_0(u)$, but we give the proofs only for the second type and leave the first to the exercises.

THEOREM 2. *Consider the control process in* R^n

$$(\mathfrak{L}) \qquad \dot{x} = A(t)x + B(t)u$$

with cost functional

$$C(u) = g(x(T)) + \int_{t_0}^{T} \|x\|_W^2 + \|u\|_U^2 \, ds.$$

If either

(a) $g(x) > a$, *that is*, $g(x)$ *is bounded below*

or

(b) $g(\lambda x_1 + (1 - \lambda)x_2) \leq \lambda\, g(x_1) + (1 - \lambda)g(x_2), 0 \leq \lambda \leq 1,$ *that is* $g(x)$ *is a convex function,*

then there exists a (minimal cost) optimal controller.

Proof. Consider the set of attainability $\hat{K} \subset R^{n+1}$ corresponding to the conrol process $\mathfrak{L}$ with the cost coordinate $x^0(T) = \int_{t_0}^{T} \|x\|_W^2 + \|u\|_U^2\, ds.$ Then $\hat{K}$ is convex and closed as proved in Theorem 1. Since each allowable

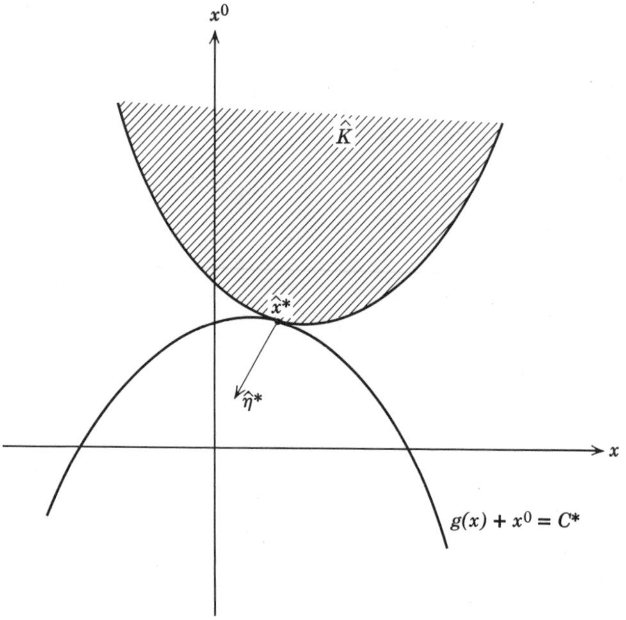

Figure 3.1 Optimal response in the convex set of attainability.

control $u(t)$ defines a point $(x_u{}^0(T), x_u(T)) \in \hat{K}$ we need only prove that the real function $g(x) + x^0$ assumes its minimum in $\hat{K}$ (see Figure 3.1)

If $g(x) > a$, then

$$\lim_{x^0 \to \infty} [g(x) + x^0] = +\infty$$

uniformly on $\hat{K}$. Thus there exists a bound $\alpha > 0$ such that the minimum of $[g(x) + x^0]$ on $\hat{K}$ is assumed on the compact set $\hat{K} \cap [x^0 \leq \alpha]$.

Next assume that $g(x)$ is a convex function. For each real number c_1 the set in R^{n+1} defined by

$$g(x) + x^0 \leq c_1$$

is closed and has nonempty interior. This set is also convex since

$$g(x_1) + x_1{}^0 \leq c_1 \quad \text{and} \quad g(x_2) + x_2{}^0 \leq c_1$$

implies that

$$g(\lambda x_1 + (1 - \lambda)x_2) + \lambda x_1{}^0 + (1 - \lambda)x_2{}^0 \leq c_1.$$

Consider a constant c_1 such that the corresponding set meets $\hat{K}$ and we prove that the intersection is bounded and hence compact. From this assertion the existence of the optimal control follows immediately.

Let π be a hyperplane in R^{n+1} such that $g(x) + x^0 \leq c_1$ for (x^0, x) below π; for instance take π as a supporting hyperplane to this convex set. We show that for points $(x^0, x) \in \hat{K}$, with sufficiently large $|x|$, we have $x^0 > k\,|x|$ for a prescribed constant $k > 0$. Such points (x^0, x) in $\hat{K}$ must lie above π and hence satisfy $g(x) + x^0 > c_1$. Upon establishing this calculation, we obtain the required compactness and the theorem is proved.

For points (x^0, x) in $\hat{K}$ we have

$$|x(T)| \leq |\Phi(T)x_0| + \int_{t_0}^{T} |\Phi(T)\,\Phi(s)^{-1}\,B(s)|\,|u(s)|\,ds.$$

If $|x(T)| \geq 2\,|\Phi(T)x_0|$ and we write

$$|\Phi(T)\,\Phi(s)^{-1}\,B(s)| \leq M \quad \text{on} \quad t_0 \leq s \leq T,$$

then

$$\int_{t_0}^{T} |u(s)|\,ds \geq \frac{1}{2M}\,|x(T)|.$$

By Schwarz' inequality

$$\int_{t_0}^{T} |u(s)|\,ds \leq c_2\left[\int_{t_0}^{T} \|u(s)\|^2\,ds\right]^{1/2}$$

for a constant $c_2 > 0$. Thus for $|x(T)| \geq 2\,|\Phi(T)x_0|$ we have

$$|x(T)|^2 \leq 4M^2c_2{}^2\int_{t_0}^{T} \|u(s)\|^2\,ds \leq c_3\,x^0(T).$$

Hence, for sufficiently large $|x(T)|$, we have

$$x^0(T) \geq k\,|x(T)|,$$

and $(x^0(T), x(T))$ in $\hat{K}$ lies above the hyperplane π.

Therefore the closed intersection of

$$g(x) + x^0 \leq c_1 \quad \text{and} \quad \hat{K}$$

is bounded, hence compact, and the existence theorem is proved. Q.E.D.

Because the real function $g(x) + x^0$ decreases monotonically as x^0 decreases, the optimal control must steer to a point on the boundary of $\hat{K}$

in R^{n+1}. In fact, we consider the convex set $\hat{K}$ within the linear manifold $L(\hat{K})$, which it spans, and the optimal control must steer to the relative boundary of $\hat{K}$ in $L(\hat{K})$. Thus the controllers that steer to the relative boundary of $\hat{K}$ in $L(\hat{K})$ are of special importance.

Definition. Consider the control problem in R^n

$$(\mathfrak{L}) \qquad \dot{x} = A(t)x + B(t)u$$

with set of attainability $\hat{K} \subset R^{n+1}$ corresponding to the cost functional $C_0(u)$. A control $\bar{u}(t)$ on $t_0 \leq t \leq T$, which steers $(0, x_0)$ to a relative boundary point of $\hat{K}$ [relative to the linear manifold $L(\hat{K})$], is called an extremal control and the corresponding response $\bar{x}(t)$ is also extremal.

The next theorem, the maximal principle for our control problem, asserts that, for an extremal controller $\bar{u}(t)$,

$$\eta_0 \, \|u\|_U{}^2 + \eta(t)\, B(t)u$$

attains its maximal value for $u = \bar{u}(t)$. Here $\hat{\eta}(t) = (\eta_0, \eta(t))$ is a certain $(n+1)$ row vector with a constant component $\eta_0 < 0$. Since

$$\|u\|_U{}^2 - \frac{\eta}{|\eta_0|} Bu = \left\| u - \frac{U^{-1}B'\eta'}{2\,|\eta_0|} \right\|_U^2 - \left\| \frac{U^{-1}B'\eta'}{2\,|\eta_0|} \right\|_U^2,$$

we find the maximum of $\eta_0 \, \|u\|_U{}^2 + \eta Bu$ only at

$$u(t) = \frac{U(t)^{-1}\, B(t)'\, \eta(t)'}{2\,|\eta_0|}.$$

THEOREM 3. *Consider the control process in* R^n

$$(\mathfrak{L}) \qquad \dot{x} = A(t)x + B(t)u.$$

A control $\bar{u}(t)$ with response $\bar{x}(t)$ on $t_0 \leq t \leq T$ is extremal if and only if there exists an $(n+1)$ vector $\hat{\eta}(t) = (\eta_0, \eta(t))$ satisfying

$$\dot{\eta} = -2\eta_0\, \bar{x}(t)'\, W(t) - \eta\, A(t), \qquad \eta_0 < 0 \quad \textit{constant}$$

such that

$$\eta_0 \, \|\bar{u}(t)\|_U{}^2 + \eta(t)\, B(t)\, \bar{u}(t) = \max_{u \in R^m} \{\eta_0 \, \|u\|_U{}^2 + \eta(t)\, B(t)u\}$$

or

$$\bar{u}(t) = -\frac{1}{2\eta_0}\, U(t)^{-1}\, B(t)'\, \eta(t)', \quad \textit{almost everywhere.}$$

Proof. Let $\bar{x}(t)$ correspond to the controller

$$\bar{u}(t) = -\frac{1}{2\eta_0}\, U(t)^{-1}\, B(t)'\, \eta(t)'$$

where $\hat{\eta}(t) = (\eta_0, \eta(t))$ satisfies the differential equation

$$\dot{\eta} = -2\eta_0 \bar{x}(t)' W(t) - \eta A(t)$$

and $\eta_0 < 0$ is constant. Only the ratio η/η_0 enters the hypothesis and so we can choose $\eta_0 = -\frac{1}{2}$ for convenience. We shall prove that

$$\hat{\eta}(T)\hat{x}(T) > \hat{\eta}(T)\hat{\omega}$$

for all points $\hat{\omega} = (\omega^0, \omega)$ of $\hat{K}$ different from $\hat{x}(T) = (\bar{x}^0(T), \bar{x}(T))$. From this inequality we shall deduce that $\hat{x}(T)$ lies on the relative boundary of $\hat{K}$. Here $\hat{\omega}(t)$ is the general response to a controller $u(t)$ as determined by

$$\omega^0(t) = \int_{t_0}^{t} \|\omega(s)\|_W^2 + \|u(s)\|_U^2 \, ds$$

$$\omega(t) = \Phi(t)x_0 + \int_{t_0}^{t} \Phi(t)\,\Phi(s)^{-1} B(s)\, u(s)\, ds.$$

Consider

$$\frac{d}{dt}(\hat{\eta}(t)\,\hat{\omega}(t)) = -\tfrac{1}{2}\dot{\omega}^0 + \dot{\eta}\omega + \eta\dot{\omega}$$
$$= -\tfrac{1}{2}[\|\omega\|_W^2 + \|u\|_U^2] + \bar{x}'W\omega - \eta A\omega + \eta A\omega + \eta Bu.$$

Integration from t_0 to t yields

$$-\tfrac{1}{2}\omega^0(t) + \eta(t)\,\omega(t) - \eta(t_0)x_0$$
$$= \int_{t_0}^{t} -\tfrac{1}{2}[\|\omega\|_W^2 + \|u\|_U^2] + [\omega'\, W(s)\, \bar{x}(s) + \eta\, B(s)\, u(s)]\, ds.$$

For the case where $u(s) = \bar{u}(s) = U(s)^{-1} B(s)' \eta(s)'$ and $\hat{\omega}(s) = \hat{x}(s)$ this simplifies to

$$-\tfrac{1}{2}\bar{x}^0(t) + \eta(t)\,\bar{x}(t) - \eta(t_0)x_0 = \tfrac{1}{2}\int_{t_0}^{t} \|\bar{x}(s)\|_W^2 + \|\eta(s)'\|^2_{BU^{-1}B'}\, ds.$$

Now the maximum of $-\frac{1}{2}\|u\|_U^2 + \eta Bu$ is attained only for

$$u = U(s)^{-1} B(s)' \eta(s)'.$$

So if $u(s) \neq \bar{u}(s)$,

$$-\tfrac{1}{2}\|u(s)\|_U^2 + \eta\, B(s)\, u(s) < \tfrac{1}{2}\|\eta(s)'\|^2_{BU^{-1}B'}.$$

Also $\|\bar{x}(s) - \omega(s)\|_W^2 \geq 0$ implies that

$$\tfrac{1}{2}\|\bar{x}(s)\|_W^2 \geq \omega'\, W(s)\bar{x} - \tfrac{1}{2}\|\omega(s)\|_W^2.$$

Thus

$$-\tfrac{1}{2}\bar{x}^0(t) + \eta(t)\,\bar{x}(t) - \eta(t_0)x_0 > -\tfrac{1}{2}\omega^0(t) + \eta(t)\,\omega(t) - \eta(t_0)x_0$$

unless $u(t) = \bar{u}(t)$ almost everywhere on $0 \le t \le T$. Therefore

$$\hat{\eta}(T)\,\hat{x}(T) > \hat{\eta}(T)\,\hat{\omega}(T)$$

for all $\hat{\omega}(T) \neq \hat{x}(T)$ in $\hat{K}$.

But this inequality means that there is a supporting hyperplane to $\hat{K}$ at $\hat{x}(T)$ having the exterior normal vector $\hat{\eta}(T)$. Since $\eta_0 < 0$, the supporting hyperplane cannot meet $\hat{K}$ in its relative interior, but must meet only the relative boundary of $\hat{K}$. Thus $\bar{u}(t)$ and $\bar{x}(t)$ are extremal.

Conversely assume that $\bar{u}(t)$ has a response $\hat{x}(t) = (\bar{x}^0(t), \bar{x}(t))$ leading to the relative boundary point $\hat{x}(T) \in \partial\hat{K}$. Let $\hat{\eta}(T) = (-\frac{1}{2}, \bar{\eta}(T))$ be an external normal to $\hat{K}$ at $\hat{x}(T)$ and define $\bar{\eta}(t)$ as the corresponding solution of the adjoint system

$$\dot{\eta} = \bar{x}(t)'\, W(t) - \eta A(t).$$

We must show that $\bar{u}(t) = U(t)^{-1}\, B(t)'\, \bar{\eta}(t)$ almost everywhere on $t_0 \le t \le T$.

Suppose $\bar{u}(t)$ fails to satisfy the maximal principle on a subset Δ of positive duration (we can take Δ compact and $\bar{u}(t)$ bounded on Δ), where $-\frac{1}{2}\,\|\bar{u}\|_U^{\,2} + \bar{\eta}(t)\,B(t)\,\bar{u}(t) + \delta \le \max_u\,[-\frac{1}{2}\,\|u\|_U^{\,2} + \bar{\eta}(t)\,B(t)u]$, for some $\delta > 0$. For each small $\epsilon > 0$ define a perturbed controller

$$u_\epsilon(t) = \begin{cases} U(t)^{-1}\, B(t)'\, \bar{\eta}(t) & \text{on a subset } \Delta_\epsilon \text{ of measure } \epsilon \text{ in } \Delta \\ \bar{u}(t) & \text{elsewhere on } t_0 \le t \le T. \end{cases}$$

Let the corresponding response be $\hat{x}_\epsilon(t)$ so

$$|\hat{x}_\epsilon(t) - \hat{x}(t)| \le c_1\epsilon \quad \text{for some constant } c_1.$$

Now compute, as above,

$$\hat{\eta}(T)\,\hat{x}(T) - \hat{\eta}(T)\,\hat{x}_\epsilon(T) \le \int_{t_0}^{T} \tfrac{1}{2}\,\|\bar{x} - x_\epsilon\|_W^{\,2}\,dt - \int_{\Delta\epsilon} \delta\,dt \le c_2\epsilon^2 - \delta\epsilon$$

for a constant c_2. Thus, for sufficiently small $\epsilon > 0$,

$$\hat{\eta}(T)\,\hat{x}_\epsilon(T) > \hat{\eta}(T)\,\hat{x}(T),$$

which is impossible when $\hat{\eta}(T)$ is an exterior normal to $\hat{K}$ at $\hat{x}(T)$. Therefore $\bar{u}(t)$ satisfies the required maximal principle. Q.E.D.

Theorem 3 asserts that a controller $\bar{u}(t)$ is extremal if and only if $\bar{u}(t)$ is maximal (in the sense of Chapter 1). Thus we shall avoid the term *maximal controller* in the remainder of this chapter.

COROLLARY. *Consider the control process in* R^n

$$(\mathfrak{L}) \qquad \dot{x} = A(t)x + B(t)u$$

with cost functional

$$C(u) = g(x(T)) + \int_{t_0}^{T} \|x(s)\|_W{}^2 + \|u(s)\|_U{}^2\, ds.$$

Let $u^*(t)$ be an optimal controller with response $x^*(t)$ on $t_0 \leq t \leq T$. Then $u^*(t)$ is extremal, that is, there exists an n-vector $\eta(t)$ satisfying

$$\dot{\eta} = x^*(t)'\, W(t) - \eta\, A(t)$$

such that

$$u^*(t) = U(t)^{-1}\, B(t)'\, \eta(t)', \text{ almost everywhere.}$$

The normality conditions, which guarantee the uniqueness of the extremal control steering $(0, x_0)$ to a boundary point of $\hat{K}$, are automatically satisfied for linear control processes with integral quadratic cost criteria. Thus the maximal condition of Theorem 3 is a sufficient as well as necessary condition for the optimality of a chosen controller. We prove the required uniqueness in Theorem 4 and apply this to optimal control theory in Theorem 5 and in the examples of the next section.

THEOREM 4. *Consider the control process in R^n*

$$(\mathfrak{L}) \qquad \dot{x} = A(t)x + B(t)u$$

with the set of attainability $\hat{K} \subset R^{n+1}$ corresponding to the cost functional $C_0(u)$. Let $u_1(t)$ and $u_2(t)$ be extremal controls with responses $\hat{x}_1(t)$ and $\hat{x}_2(t)$ in R^{n+1} for $t_0 \leq t \leq T$. If

$$\hat{x}_1(T) = \hat{x}_2(T), \quad \text{then} \quad u_1(t) = u_2(t) \quad \text{almost everywhere.}$$

Proof. Let $\hat{\eta}(T) = (-\frac{1}{2}, \eta(T))$ be an exterior normal to $\hat{K}$ at $\hat{x}_1(T) = \hat{x}_2(T)$ and let $\eta(t)$ be the corresponding solution of

$$\dot{\eta} = x_1(t)'\, W(t) - \eta\, A(t).$$

Then, as proved in Theorem 3,

$$u_1(t) = u_2(t) = U(t)^{-1}\, B(t)'\, \eta(t)' \quad \text{almost everywhere.}$$

For otherwise $\hat{\eta}(T)\,\hat{x}_1(T) = \hat{\eta}(T)\,\hat{x}_2(T)$ would be less than some $\hat{\eta}(T)\hat{\omega}$ with $\hat{\omega} \in \hat{K}$. Q.E.D.

THEOREM 5. *Consider the control process in R^n*

$$(\mathfrak{L}) \qquad \dot{x} = A(t)x + B(t)u$$

with cost functional

$$C(u) = g(x(T)) + \int_{t_0}^{T} \|x\|_W^2 + \|u\|_U^2\, ds = g(x(T)) + C_0(u)$$

where $g(x)$ is a C^1 convex function. Then there exists a unique hypersurface

S_m among the family

$$S_c\colon\ g(x) + x^0 = c$$

such that S_m is tangent to $\hat{K}$; hence m is the optimal cost. Also there exists a unique extremal controller, namely the optimal $u^*(t)$, which attains the single point at which S_m touches $\hat{K}$.

Furthermore there is a unique solution of the differential system

$$\dot{x} = A(t)x + B(t)\,U(t)^{-1}\,B(t)'\eta'$$

$$\dot{\eta} = x'\,W(t) - \eta\,A(t),$$

satisfying the boundary conditions

$$x(t_0) = x_0 \quad \textit{and} \quad \eta(T) = -\tfrac{1}{2}\,\mathrm{grad}\,g(x(T)),$$

namely, the optimal response $x^*(t)$ and $\eta^*(t)$ such that

$$u^*(t) = U(t)^{-1}\,B(t)'\,\eta^*(t)'$$

is the optimal controller on $t_0 \leq t \leq T$.

Proof. We first show that there exists a unique constant m such that S_m is tangent to $\hat{K}$ (the set of attainability corresponding to $C_0(u)$), that is, the convex set $g(x) + x^0 \leq m$ meets $\hat{K}$ but is separated from the relative interior of $\hat{K}$ by a common supporting hyperplane π^*, which is the tangent hyperplane to S_m. From this it follows easily that m is the minimal cost. By Theorem 2 the intersection of $\hat{K}$ with $g(x) + x^0 \leq c$ is compact for all large c. Hence we define m as the infimum of all such numbers c such that the corresponding intersection is nonempty. For $c > m$ the hypersurface S_c meets the relative interior of $\hat{K}$; and for $c < m$ the hypersurface S_c does not meet $\hat{K}$. Thus only for $c = m$ can S_c be tangent to $\hat{K}$.

Let P be a point in $\hat{K} \cap S_m$ and let π^* be the tangent hyperplane to S_m at P. Then π^* fails to separate $\hat{K}$ and S_m only in the case where π^* meets the relative interior of $\hat{K}$. But suppose that a (relative) open set N of the interior of $\hat{K}$ lies below π^*. Then the cone with base N and vertex P lies interior to $\hat{K}$ and also below π^*. However S_m is tangent to π^* at P, so S_m would meet the relative interior of $\hat{K}$. This is impossible from the definition of m as the minimal cost. Thus π^* does separate S_m and $\hat{K}$.

Now suppose two distinct points P_1 and P_2 lie in $\hat{K} \cap S_m$. Hence the segment joining P_1 to P_2 lies in $\hat{K} \cap S_m$, and thus in the relative boundary of $\hat{K}$. Let $u_1(t)$ and $u_2(t)$ be extremal controls with responses $\hat{x}_1(t)$ and $\hat{x}_2(t)$ leading to P_1 and P_2, respectively. Note that $u_1(t)$ and $u_2(t)$ must be unequal on some set of positive duration on $t_0 \leq t \leq T$.

Consider the control $\tfrac{1}{2}[u_1(t) + u_2(t)]$ with response

$$\hat{x}(t) = (x^0(t), x(t)).$$

Here

$$x(T) = \tfrac{1}{2}[x_1(T) + x_2(T)]$$

but we show that

$$x^0(T) < \tfrac{1}{2}[x_1{}^0(T) + x_2{}^0(T)].$$

Compute

$$x^0(T) = \int_{t_0}^{T} \left\| \frac{x_1(s) + x_2(s)}{2} \right\|_W^2 + \left\| \frac{u_1(s) + u_2(s)}{2} \right\|_U^2 ds$$

$$= \int_{t_0}^{T} \{\tfrac{1}{4} \|x_1\|_W{}^2 + \tfrac{1}{2} x_1' W(s) x_2 + \tfrac{1}{4} \|x_2\|_W{}^2$$

$$+ \tfrac{1}{4} \|u_1\|_U{}^2 + \tfrac{1}{2} u_1' U(s) u_2 + \tfrac{1}{4} \|u_2\|_U{}^2\} ds.$$

Use the trivial inequalities

$$2x_1(s)' W(s) x_2(s) \leq \|x_1(s)\|_W{}^2 + \|x_2\|_W{}^2$$

and

$$2u_1(s)' U(s) u_2(s) < \|u_1(s)\|_U{}^2 + \|u_2(s)\|_U{}^2,$$

wherever $u_1(s) \neq u_2(s)$.
Then

$$x^0(T) < \tfrac{1}{2} \int_{t_0}^{T} \|x_1\|_W{}^2 + \|u_1\|_U{}^2 ds + \tfrac{1}{2} \int_{t_0}^{T} \|x_2\|_W{}^2 + \|u_2\|_U{}^2 ds$$

so

$$x^0(T) < \tfrac{1}{2}[x_1{}^0(T) + x_2{}^0(T)], \quad \text{as stated above.}$$

The half-line $x^0 > x^0(T)$, $x = x(T)$ lies in the relative interior of $\hat{K}$, which implies that the midpoint between P_1 and P_2 lies in the relative interior of $\hat{K}$. But $\tfrac{1}{2}[P_1 + P_2]$ lies on the relative boundary of $\hat{K}$. This contradiction proves that $\hat{K} \cap S_m$ consists of a single point P.

By Theorem 4 there exists a unique extremal control steering $(0, x_0)$ to the point $P \in \hat{K}$, and so this must be the optimal control $u^*(t)$. Therefore $P = (x^{0*}(T), x^*(T))$ must be attained by the optimal response $x^*(t)$.

The normal vector to S_m at $P = \hat{x}^*(T)$ is $\hat{\eta}^*(T) = (-\tfrac{1}{2}, \eta^*(T))$ where $\eta^*(T) = -\tfrac{1}{2} \operatorname{grad} g(P)$. By Theorem 3 the functions $x^*(t)$ and $\eta^*(t)$ satisfy

$$\dot{x} = A(t)x + B(t)\, U(t)^{-1}\, B(t)'\eta'$$

$$\dot{\eta} = x'\, W(t) - \eta\, A(t)$$

with boundary conditions

$$x^*(t_0) = x_0, \qquad \eta^*(T) = -\tfrac{1}{2} \operatorname{grad} g(x^*(T)).$$

Now let $x(t)$, $\eta(t)$ be any solution of this simultaneous differential system with the given boundary data. Then $\hat{x}(t) = (x^0(t), x(t))$ is the response determined by the extremal controller $u(t) = U(t)^{-1} B(t)' \eta(t)'$. Furthermore

$$\hat{\eta}(T)\,\hat{x}(T) = -\tfrac{1}{2}x^0(T) + \eta(T)\,x(T) > \hat{\eta}(T)\hat{\omega}$$

for all $\hat{\omega} \neq \hat{x}(T)$ in $\hat{K}$. Thus $\hat{\eta}(T) = (-\tfrac{1}{2}, \eta(T))$ is the exterior normal to a supporting hyperplane $\hat{\pi}$ to $\hat{K}$ at $\hat{x}(T)$. Also $\hat{\eta}(T)$ is the inward normal to the hypersurface S_c through $\hat{x}(T)$ since $\eta(T) = -\tfrac{1}{2}$ grad $g(x(T))$. Thus S_c is tangent to $\hat{K}$ at $\hat{x}(T)$, and $\hat{\pi}$ is the common supporting hyperplane. But then $S_c = S_m$ and $\hat{x}(T) = \hat{x}^*(T)$. By the uniqueness of the extremal controller steering $(0, x_0)$ to $\hat{x}^*(T)$ we find that $u(t) = u^*(t)$ almost everywhere, and hence $x(t) = x^*(t)$ on $t_0 \leq t \leq T$. Finally $\eta(t)$ is the unique solution of

$$\dot{\eta} = x^*(t)'\,W(t) - \eta\,A(t)$$

with

$$\eta(T) = -\tfrac{1}{2}\text{ grad } g(x^*(T)),$$

and so

$$\eta(t) = \eta^*(t) \quad \text{on} \quad t_0 \leq t \leq T. \qquad \text{Q.E.D.}$$

If we wish to steer a given initial point $x_0 \in R^n$ to a prescribed target point, it is natural to require that the process

$$(\mathfrak{L}) \qquad \dot{x} = A(t)x + B(t)u$$

be controllable. Here $\mathfrak{L}$ is controllable on $t_0 \leq t \leq T$ if: for each pair of points $x_0, x_1 \in R^n$, there exists a bounded measurable controller $u(t)$ steering $x(t_0) = x_0$ to $x(T) = x_1$. The controllable case is easier to analyse geometrically since here the set of attainability $\hat{K}$ has a nonempty interior in R^{n+1}, and hence the relative boundary of $\hat{K}$ in $L(\hat{K}) = R^{n+1}$ is the same as the total boundary.

THEOREM 6. *Consider the control process in* R^n

$$(\mathfrak{L}) \qquad \dot{x} = A(t)x + B(t)u$$

with set of attainability $\hat{K} \subset R^{n+1}$, *corresponding to cost functional* $C_0(u)$. *Then* $\mathfrak{L}$ *is controllable on* $t_0 \leq t \leq T$ *if and only if* $\hat{K}$ *has a nonempty interior in* R^{n+1}; *this is the case if and only if the matrix*

$$M(T) = \int_{t_0}^{T} \Phi(t)^{-1}\,B(t)\,B(t)'\,\Phi(t)^{-1\prime}\,dt$$

is nonsingular.

Proof. The orthogonal projection of $\hat{K}$ on $x^0 = 0$ is just the set of all endpoints in R^n

$$x(T) = \Phi(T)x_0 + \Phi(T)\int_{t_0}^{T} \Phi(t)^{-1}\,B(t)\,u(t)\,dt.$$

If $\mathcal{L}$ is controllable, the set of all $\{x(T)\}$ is all R^n, and so $\hat{K}$ has a nonempty interior. On the other hand, if $\hat{K}$ has a nonempty interior, then the lemma of Theorem 1 shows that $\{x(T)\} = R^n$. But this means that the set of all points of the form

$$x_1 = \int_{t_0}^{T} \Phi(T)\, \Phi(t)^{-1}\, B(t)\, u(t)\, dt$$

fills all R^n. Hence the responses initiating from any prescribed point in R^n terminate at all the points of R^n. [Here $u(t)$ varies over L_2—but, using approximations, we can take $u(t)$ bounded.] Thus $\mathcal{L}$ is controllable in this case.

Now consider the $n \times n$ matrix $M(T)$. Since

$$M(T)' = \int_{t_0}^{T} [\Phi(t)^{-1}\, B(t)\, B(t)'\, \Phi(t)^{-1\prime}]'\, dt = M(T),$$

and also

$$\zeta'\, M(T)\zeta = \int_{t_0}^{T} (B'\Phi^{-1\prime}\zeta)'(B'\Phi^{-1\prime}\zeta)\, dt \geq 0,$$

for each n-vector ζ, the matrix $M(T)$ is symmetric and positive semidefinite.

Assume that $M(T)$ is nonsingular and we show that $\mathcal{L}$ is controllable. Given points x_0 and x_1, define the control

$$u(t) = B(t)'\, \Phi(t)^{-1\prime}\xi,$$

where the constant vector ξ is defined by

$$\xi = M(T)^{-1}[\Phi(T)^{-1}x_1 - x_0].$$

In this case

$$x_1 = \Phi(T)x_0 + \Phi(T)\, M(T)\xi$$

or

$$x_1 = \Phi(T)x_0 + \Phi(T) \int_{t_0}^{T} \Phi(t)^{-1}\, B(t)\, u(t)\, dt,$$

as required.

On the other hand assume that $\mathcal{L}$ is controllable. If $M(T)$ is singular, there exists a constant n-vector $\zeta \neq 0$ such that

$$\zeta'\, M(T)\zeta = \int_{t_0}^{T} \|B(t)'\, \Phi(t)^{-1\prime}\zeta\|^2\, dt = 0.$$

But this requires that $B(t)'\, \Phi(t)^{-1\prime}\zeta \equiv 0$ on $t_0 \leq t \leq T$. Since $\mathcal{L}$ is controllable, the terminal point $\Phi(T)\zeta$ can be attained from the origin by a control $u(t)$, and

$$\zeta = \int_{t_0}^{T} \Phi(t)^{-1}\, B(t)\, u(t)\, dt.$$

Then compute

$$0 < \zeta'\zeta = \int_{t_0}^{T} \zeta' \Phi(t)^{-1} B(t)\, u(t)\, dt = \int_{t_0}^{T} (B(t)' \Phi(t)^{-1\prime}\zeta)' u(t)\, dt = 0.$$

This contradiction proves that $M(T)$ must be nonsingular. Q.E.D.

In particular, if $\Phi(t)^{-1} B(t) B(t)' \Phi(t)^{-1\prime}$ is nonsingular at just one instant t, then $M(T)$ is nonsingular and $\mathfrak{L}$ is controllable on $t_0 \leq t \leq T$.

3.3 ILLUSTRATIVE EXAMPLES AND SPECIAL PROBLEMS

In this section we construct a feedback synthesis for various optimal-control processes, based on the theory of the preceeding section. We first consider problems where no target is specified, then problems with a target set, and finally problems where the duration of control is unlimited.

The Cost Functional $C(u) = g(x(T)) + \int_{t_0}^{T} \|x(s)\|_W{}^2 + \|u(s)\|_U{}^2\, ds$

***Example* 1.** $C(u) = x(T)'G\, x(T) + x^0(T)$ where the constant symmetric matrix $G = G' \geq 0$, that is, $g(x) = x'Gx$ is a positive semidefinite quadratic form and hence a convex function. By Theorem 5 there exists a unique optimal controller $u^*(t)$ with response $x^*(t)$. These are determined by the unique solutions of

$$\dot{x} = A(t)x + B(t)\, U(t)^{-1}\, B(t)'\eta'$$

$$\dot{\eta} = x'\, W(t) - \eta\, A(t)$$

with the conditions $x(t_0) = x_0, \qquad \eta(T)' = -G\, x(T)$, where

$$u^*(t) = U(t)^{-1}\, B(t)'\, \eta^*(t)'.$$

The optimal response $\hat{x}^*(t) = (x^{0*}(t), x^*(t))$ leads to the unique point at which the quadric surface $S_m: x^0 + x'Gx = m$ is tangent to $\hat{K}$ (see Figure 3.1).

In this problem we can obtain a closed-form solution for the optimal controller by using a linear feedback control with a time-varying gain. The method is motivated by the analysis of Example 4 in the first section of Chapter 1.

We shall attempt to express the optimal control as

$$u^*(t) = E^*(t)\, x^*(t),$$

where $E^*(t)$ is a known matrix that does not depend on x_0. Define

$E^*(t) = U(t)^{-1} B(t)' E(t)$ where $E(t)$ is the solution of the nonlinear matrix differential equation

$$\dot{E}(t) = W(t) - A(t)'E - E\,A(t) - E\,B(t)\,U(t)^{-1}\,B(t)'E$$

with the initial data $E(T) = -G$. Since G is symmetric, and since the expression for $\dot{E}(t)$ is symmetric whenever $E(t)$ is, the solution $E(t)$ is a uniquely determined symmetric matrix.

We show that the solution $x(t)$ of

$$\dot{x} = A(t)x + B(t)[U(t)^{-1}\,B(t)'\,E(t)x], \qquad x(t_0) = x_0,$$

is just the optimal response $x^*(t)$ and thus the optimal controller is

$$u^*(t) = U(t)^{-1}\,B(t)'\,E(t)\,x^*(t).$$

Let $x(t)$ be the above solution and define $\eta(t) = x'\,E(t)$. Then a direct calculation shows that $x(t)$, $\eta(t)$ is a solution of the system

$$\dot{x} = A(t)x + B(t)\,U(t)^{-1}\,B(t)'\eta'$$

$$\dot{\eta} = x'\,W(t) - \eta\,A(t)$$

with data $x(t_0) = x_0$, $\eta(T)' = -G\,x(T)$. Thus $x(t) = x^*(t)$ and $\eta(t) = \eta^*(t)$ by the uniqueness property stated in Theorem 5.

Thus the feedback control

$$u^*(t) = E^*(t)x$$

automatically constructs the optimal response $x^*(t)$ for all initial states x_0. If the state of the process is abruptly perturbed by an external impulse, the feedback control will proceed to steer the response on an optimal response from the perturbed state. Of course, the matrix $E^*(t) = U(t)^{-1}\,B(t)'\,E(t)$ must be computed from the solution $E(t)$ of a nonlinear differential equation. This nonlinear differential equation is of Riccati type and can be solved in explicit elementary functions only in exceptional cases (see Exercises 1 and 2 below). However the Riccati equation can be solved by standard numerical methods to yield an effective gain matrix $E^*(t)$.

One delicate point remains—to prove that the solution $E(t)$ of the above nonlinear differential equation is defined on the entire interval $t_0 \le t \le T$. If this were not the case, the norm $|E(t)|$ would become unbounded as t decreases from the upper limit T. Then, for any prescribed α, there exists $\bar{t}_0$ and $\bar{x}_0$ such that

$$\bar{x}_0'\,E(\bar{t}_0)\bar{x}_0 > \alpha$$

with $|\bar{x}_0| = 1$ and $t_0 \le \bar{t}_0 < T$. But, since $E(t)$ is independent of x_0 or t_0, use the optimal response initiating at $\bar{x}_0$ on $\bar{t}_0 \le t \le T$ to write

$$\eta^*(\bar{t}_0;\,\bar{x}_0)\,x^*(\bar{t}_0;\,\bar{x}_0) = \bar{x}_0'\,E(\bar{t}_0)\bar{x}_0 > \alpha.$$

However a small change of x_0 to $x_0 + \delta x_0$ causes a small horizontal translation of the corresponding response; so $x^*(T, x_0 + \delta x_0)$ lies within a certain compact set contained beneath the hypersurface S_c, $c > m$. Therefore $|x^*(T, \bar{x}_0)|$ is uniformly bounded for $|\bar{x}_0| = 1$ and $t_0 \leq \bar{t}_0 < T$, and so the corresponding solutions $x^*(t, \bar{x}_0)$, $\eta^*(t, \bar{x}_0)$ of the above linear differential system are uniformly bounded. This contradicts the assertion

$$\bar{x}_0' \, E(\bar{t}_0)\bar{x}_0 > \alpha, \text{ for arbitary } \alpha,$$

and hence $|E(t)|$ is bounded and $E(t)$ exists on the entire interval $t_0 \leq t \leq T$.

***Example* 2.** $C(u) = e(T)'G\, e(T) + \int_{t_0}^{T} \|e(t)\|_W{}^2 + \|u(t)\|_U{}^2 \, dt$ where the error

$$e(t) = x(t) - \xi(t)$$

expresses the deviation of the response $x(t)$ from a desired ideal response $\xi(t)$ on $t_0 \leq t \leq T$. As earlier, we assume

$$G = G' \geq 0, \qquad W(t) = W(t)' \geq 0, \qquad U(t) = U(t)' > 0,$$

and $\xi(t)$ is a continuously differentiable vector function. We further generalize the basic linear control process $\mathfrak{L}$ by a known continuous disturbing force $v(t)$ so

$$\dot{x} = A(t)x + B(t)u + v(t).$$

We express this control process in terms of the error $e(t)$, instead of $x(t)$, to obtain

$$(\mathfrak{L}_+) \qquad \dot{e} = A(t)e + B(t)u + \omega(t), \qquad e(t_0) = e_0 = x_0 - \xi(t_0),$$

where we compute the known function

$$\omega(t) = A(t)\,\xi(t) - \dot{\xi}(t) + v(t).$$

Again define $\hat{e}(t) = (e^0(t), e(t))$ where

$$e^0(t) = \int_{t_0}^{t} \|e(s)\|_W{}^2 + \|u(s)\|_U{}^2 \, ds$$

$$e(t) = \Phi(t)e_0 + \Phi(t)\int_{t_0}^{t} \Phi(s)^{-1} B(s)\,u(s)\,ds + \int_{t_0}^{t} \Phi(t)\,\Phi(s)^{-1}\,\omega(s)\,ds,$$

and define the set of attainability $\hat{K}_+ = \{\hat{e}(T)\} = \{e^0(T), e(T)\}$. This $\hat{K}_+$ is just a translate of the set $\hat{K}$, which occurs for $\omega(t) \equiv 0$, by the constant horizontal vector $\left(0, \int_{t_0}^{T} \Phi(T)\Phi(s)^{-1}\,\omega(s)\,ds\right)$. Thus $\hat{K}_+$ is a closed convex set in R^{n+1}. The theory of the preceding section applies to the linear process $\mathfrak{L}_+$ with cost functional $C(u) = e(T)'G\,e(T) + \int_{t_0}^{T} \|e(s)\|_W{}^2 + \|u(s)\|_U{}^2 \, ds$.

In particular, there is a unique optimal controller

$$u^*(t) = U(t)^{-1} B(t)' \eta^*(t)'$$

with optimal response $e^*(t)$. In fact, $e^*(t)$ and $\eta^*(t)$ are the unique solution of the system

$$\dot{e} = A(t)e + B(t) U(t)^{-1} B(t)'\eta' + \omega(t)$$

$$\dot{\eta} = e' W(t) - \eta A(t)$$

with boundary conditions $e(t_0) = e_0 = x_0 - \xi(t_0)$ and $\eta(T)' = -Ge(T)$. The optimal control $u^*(t)$ is, of course, also the optimal control for the response $x^*(t) = e^*(t) + \xi(t)$.

We try to compute the optimal controller as a feedback control with time-varying gain functions,

$$u^*(t) = h^*(t) + E^*(t)e^*(t).$$

Here $h^*(t) = U(t)^{-1} B(t)' h(t)$ and $E^*(t) = U(t)^{-1} B(t)' E(t)$ where the functions $h(t)$ and $E(t)$ are defined by

$$\dot{E} = \omega(t) - A(t)'E - E A(t) - E B(t) U(t)^{-1} B(t)'E$$

with $E(T) = -G$, and

$$\dot{h} = -[E(t) B(t) U(t)^{-1} B(t)' + A(t)']h - E(t) \omega(t)$$

with $h(T) = 0$. Then $E(t)$ is a symmetric matrix on $t_0 \le t \le T$, as discussed in Example 1, and $h(t)$ is defined on $t_0 \le t \le T$ by the above linear differential equation. Note that $h(t)$ and $E(t)$, and hence $h^*(t)$ and $E^*(t)$, are independent of x_0.

It is an elementary direct calculation to show that the solution $e(t)$ of

$$\dot{e} = A(t)e + B(t)[h^*(t) + E^*(t)e] + \omega(t),$$

with $e(t_0) = e_0$, is just the optimal response $e^*(t)$. Define

$$\eta(t)' = h(t) + E(t)\, e(t)$$

and verify that the pair $(e(t), \eta(t))$ satisfy the differential system whose unique solution is $(e^*(t), \eta^*(t))$.

Thus the optimal controller has been synthesized as a feedback control by

$$u^*(t) = h^*(t) + E^*(t)e$$

or

$$u^*(t) = h^*(t) - E^*(t)\, \xi(t) + E^*(t)x.$$

In Figure 3.2 we indicate a block diagram of the control process $\mathfrak{L}_+$ with this feedback control.

Remarks. There is an interesting interpretation of the set $\hat{K}$ for the process of Example 1 (or for the translated set $\hat{K}_+$ for the process $\mathcal{L}_+$ of Example 2). Consider $\hat{K}$ in the $(n + 1)$-space of coordinates (x^0, x) so that the boundary of $\hat{K}$ defines a single valued function $\hat{K}(x)$ over the x-space, providing that $\mathcal{L}$ is controllable on $t_0 \leq t \leq T$. By definition, $\hat{K}(x)$ is the minimal cost $C_0(u) = x^0$ for steering from x_0 to a target point x.

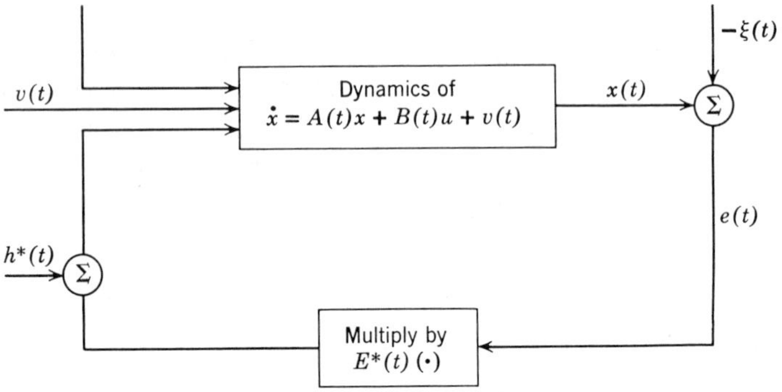

Figure 3.2 Feedback synthesis.

We next compute the minimal cost

$$V(x_0, t_0) = x^*(T)'G\,x^*(T) + \int_{t_0}^{T} \|x^*\|_W^{\,2} + \|u^*\|_U^{\,2}\,ds$$

for the control process $\mathcal{L}$ with initial data $x(t_0) = x_0$. Consider the derivative of $x'\,E(t)x$ along the optimal response $x^*(t)$ initiating at x_0, with optimal control $u^*(t) = U(t)^{-1}\,B(t)'\,E(t)\,x^*(t)$, to obtain

$$\begin{aligned}&\dot{x}'\,E(t)x + x'\,E(t)\dot{x} + x'\,\dot{E}(t)x\\ &\qquad = [Ax + BU^{-1}B'Ex]'\,Ex + x'E[Ax + BU^{-1}B'Ex] + x'\dot{E}x.\end{aligned}$$

Integrate and use the defining differential equation for $E(t)$ to obtain

$$x^*(T)'\,E(T)\,x^*(T) - x_0'\,E(t_0)x_0 = \int_{t_0}^{T} \|x^*(s)\|_W^2 + \|u^*\|_U^2\,ds.$$

When the total cost $C(u) = x(T)'G\,x(T) + x^0(T)$ is used, then the minimal value is $V(x_0, t_0) = -x_0'\,E(t_0)x_0$. This explicit elementary formula for the minimal cost in steering from x_0 to $x^*(T)$ justifies the dynamic programming of Example 4 in the first section of Chapter 1.

***Example* 3.** $C(u) = \zeta x(T) + x^0(T)$, where $\zeta \neq 0$ is a fixed row n-vector. By Theorem 5 there exists a unique optimal control $u^*(t)$ with response

$x^*(t)$. These are determined by the unique solutions of

$$\dot{x} = A(t)x + B(t)\,U(t)^{-1}\,B(t)'\eta'$$

$$\dot{\eta} = x'\,W(t) - \eta\,A(t)$$

with the conditions $x(t_0) = x_0$, $\eta(T) = -\frac{1}{2}\zeta$, where

$$u^*(t) = U(t)^{-1}\,B(t)'\,\eta^*(t)'.$$

We shall not compute the optimal controller $u^*(t)$ as a feedback control (see Exercise 9) but shall solve the two-point boundary value problem directly for an illustrative case with constant coefficients.

Let the linear process in R^2 be

$$(\mathfrak{L})\qquad \begin{aligned}\dot{x}^1 &= x^2\\ \dot{x}^2 &= u\end{aligned}$$

with $x^1(0) = x_0^{\,1}$, $x^2(0) = x_0^{\,2}$ prescribed. Let the cost functional be

$$C(u) = x^1(1) + x^2(1) + \int_0^1 \{(x^1(t))^2 + (u(t))^2\}\,dt,$$

for the scalar controller $u(t)$ on $0 \le t \le 1$. Then the system defining the optimal $x^*(t)$ and $\eta^*(t)$ is

$$\begin{aligned}\dot{x}^1 &= x^2\\ \dot{x}^2 &= \eta_2\\ \dot{\eta}_1 &= x^1\\ \dot{\eta}_2 &= -\eta_1\end{aligned}$$

with $x^1(0) = x_0^{\,1}$, $x^2(0) = x_0^{\,2}$, $\eta_1(1) = \eta_2(1) = -\frac{1}{2}$, and $u^*(t) = \eta_2^*(t)$. For each assignment of initial data $\eta_1(0) = \eta_{10}$, $\eta_2(0) = \eta_{20}$ a solution of the above linear system corresponding to the initial data $(x_0^{\,1}, x_0^{\,2}, \eta_{10}, \eta_{20})$ can be determined. In fact,

$$\eta_1(t) = \ddot{\phi}(t)x_0^{\,1} + \dot{\phi}(t)x_0^{\,2} + \dddot{\phi}(t)\eta_{10} + \phi(t)\eta_{20}$$

$$\eta_2(t) = -\dot{\phi}(t)x_0^{\,1} - \phi(t)x_0^{\,2} - \ddot{\phi}(t)\eta_{10} + \dddot{\phi}(t)\eta_{20}$$

where

$$\phi(t) = \frac{1}{\sqrt{2}}\left[\sin\frac{t}{\sqrt{2}}\cosh\frac{t}{\sqrt{2}} - \cos\frac{t}{\sqrt{2}}\sinh\frac{t}{\sqrt{2}}\right]$$

But (η_{10}, η_{20}) are restricted by the terminal condition

$$-\tfrac{1}{2} = \ddot{\phi}(1)x_0^{\,1} + \dot{\phi}(1)x_0^{\,2} + \dddot{\phi}(1)\eta_{10} + \phi(1)\eta_{20}$$

$$-\tfrac{1}{2} = -\dot{\phi}(1)x_0^{\,1} - \phi(1)x_0^{\,2} - \ddot{\phi}(1)\eta_{10} + \dddot{\phi}(1)\eta_{20}.$$

By these last two equations we define η_{10} and η_{20} as functions of (x_0^1, x_0^2). In this way the solution $(x^1(t), x^2(t), \eta_1(t), \eta_2(t))$ is completely determined by the control process $\mathcal{L}$, the cost function $C(u)$, and the initial state (x_0^1, x_0^2).

If we determine (η_{10}, η_{20}), or the corresponding quantities, at each instant t on $0 \leq t \leq 1$ as the state $(x^1(t), x^2(t))$ varies or is perturbed, then

$$u^*(t) = \eta_2(t, x_0^1, x_0^2)$$

can be interpreted as a feedback controller.

Restricted-endpoint Problems

There are linear control problems, with integral cost functionals, in which the process must be guided from a prescribed initial state to a prescribed target set. Here we consider the supplementary conditions that arise from the requirement that the terminal end of the response lie on the target. Again the process in R^n is

$$(\mathcal{L}) \qquad \dot{x} = A(t)x + B(t)u$$

with cost functional

$$C_0(u) = x^0(T) = \int_{t_0}^{T} \|x(s)\|_W^{\,2} + \|u(s)\|_U^{\,2}\, ds,$$

as in the theory of Section 2.

Let G be a fixed compact (nonempty), convex target set in R^n. Thus we must select the control $u(t) \subset R^m$ such that the response $x(t)$ steers $x(t_0) = x_0$ to some point $x(T) \in G$, and with $C_0(u)$ a minimum. For simplicity we shall assume that $\mathcal{L}$ is controllable on $t_0 \leq t \leq T$, so that the set of attainability $\hat{K} = \{x^0(T), x(T)\}$ is a closed convex set with interior in R^{n+1}. If $\mathcal{L}$ were not controllable a similar analysis would obtain in the linear manifold $L(\hat{K})$ spanned by $\hat{K}$ in R^{n+1}, provided G meets $L(\hat{K})$.

The set G lies in the x-space R^n and we define the cylindrical set $\hat{G} = G \times R^1$ in the (x^0, x) space R^{n+1}. Since $\mathcal{L}$ is controllable, $\hat{G}$ intersects $\hat{K}$ in a closed convex set. We wish to steer $(0, x_0)$ to $\hat{G} \cap \hat{K}$ so as to minimize $x^0(T)$. Clearly an optimal controller $u^*(t)$ exists.

The minimal value of x^0 in $\hat{G} \cap \hat{K}$ occurs at a common boundary point $\hat{x}^*(T) = (x^{0*}(T), x^*(T))$ of $\hat{G}$ and $\hat{K}$ [unless the optimal control minimizing $C_0(u)$ without any target steers x_0 to G—this special case is covered by Example 1 above, and will be henceforth discounted]. Thus the optimal control $u^*(t)$ steering $(0, x_0)$ to $\hat{x}^*(T)$ is given by

$$u^*(t) = U(t)^{-1} B(t)' \eta^*(t)'.$$

Here $\hat{\eta}(t) = (-\frac{1}{2}, \eta(t))$ is such that

$$\dot{\eta} = x' \, W(t) - \eta \, A(t)$$

and $\hat{\eta}(T)$ is normal to $\hat{K}$ at $\hat{x}^*(T)$.

Consider the horizontal cross section $P: x^0 = x^{0*}(T)$ in R^{n+1}. Then, considering G as the same as $\hat{G} \cap P$, we can show that G and $\hat{K} \cap P$ are separated by a common supporting $(n-1)$-plane π in P. In fact, $\eta(T)$ is normal to π and interior to G. By arguments similar to those of Theorem 5 one can prove that there exists a unique optimal controller $u^*(t)$; and $x^*(t)$, $\eta^*(t)$ are found as any solutions of

$$\dot{x} = A(t)x + B(t)\, U(t)^{-1}\, B(t)'\eta'$$
$$\dot{\eta} = x' \, W(t) - \eta \, A(t)$$

with $x(t_0) = x_0$, $x(T) \in \partial G$, and $\eta(T)$ an interior normal to G at $x(T)$.

Let the target G in R^n be defined by $\gamma(x) \leq 0$, where $\gamma(x)$ is a convex C^1 function with grad $\gamma \neq 0$ on ∂G. The boundary conditions then read $x(t_0) = x_0$, $\gamma(x(T)) = 0$, and $\eta(T) = -k \operatorname{grad} \gamma(x(T))$, for some $k > 0$. The last requirement is referred to as the transversality condition.

***Example* 1.** Consider the linear process in R^2

$$\dot{x}^1 = x^2, \qquad \dot{x}^2 = u$$

with the cost for the scalar control $u(t)$ on $0 \leq t \leq 1$

$$C(u) = \int_0^1 (u(t))^2 \, dt.$$

The initial point is $(x^1(0), x^2(0)) = (0, -3)$ and the target is the disk

$$G: (x^1)^2 + (x^2)^2 \leq 1 \quad \text{in} \quad R^2.$$

This system is controllable and therefore the unique optimal controller $u^*(t)$ exists. We shall compute

$$u^*(t) = \overset{*}{\eta}_2(t)$$

from the differential system

$$\dot{x}^1 = x^2, \qquad \dot{x}^2 = \eta_2, \qquad \dot{\eta}_1 = 0, \qquad \dot{\eta}_2 = -\eta_1$$

with the boundary conditions $x^1(0) = 0$, $x^2(0) = -3$, and the transversal conditions

$$\begin{pmatrix} \eta_1(1) \\ \eta_2(1) \end{pmatrix} = -k \begin{pmatrix} x^1(1) \\ x^2(1) \end{pmatrix} \quad \text{for} \quad x^1(1)^2 + x^2(1)^2 = 1, \qquad k > 0.$$

For each assignment of initial data $\eta_1(0) = \eta_{10}$, $\eta_2(0) = \eta_{20}$ we can obtain a unique solution of the given differential system,

$$\eta_1(t) = \eta_{10}, \qquad \eta_2(t) = -\eta_{10}t + \eta_{20}$$

$$x^1(t) = \frac{\eta_{10}t^3}{6} + \frac{\eta_{20}t^2}{2} - 3t, \qquad x^2(t) = -\frac{\eta_{10}t^2}{2} + \eta_{20}t - 3.$$

In order to satisfy the transversality conditions at $t = 1$ we have,

$$\eta_1(1) = \eta_{10} = -k\,x^1(1) = -k(-\tfrac{1}{6}\eta_{10} + \tfrac{1}{2}\eta_{20} - 3),$$

$$\eta_2(1) = -\eta_{10} + \eta_{20} = -k\,x^2(1) = -k(-\tfrac{1}{2}\eta_{10} + \eta_{20} - 3)$$

and

$$(\eta_{10})^2 + (-\eta_{10} + \eta_{20})^2 = k^2 \quad \text{for some} \quad k > 0.$$

The two linear conditions on (η_{10}, η_{20}) imply that

$$\eta_{10} = \frac{36[(k^2/2) + k]}{k^2 + 16k + 12}$$

$$\eta_{20} = \frac{12(k^2 + 6k)}{k^2 + 16k + 12}.$$

The final quadratic condition will be satisfied if

$$k^4 + 32k^3 - 80k^2 - 480k - 2448 = 0$$

has a positive root. But this quartic polynomial in k can be factored

$$(k - 6)(k^3 + 38k^2 + 148k + 408) = 0.$$

Thus $k = 6$, and there are no positive roots corresponding to the second factor. Hence $k = 6$, $\eta_{10} = 6$, $\eta_{20} = 6$, and the optimal control is

$$u^*(t) = -6t + 6 \quad \text{on} \quad 0 \le t \le 1.$$

***Example* 2.** Consider the autonomous, controllable process in R^n

$$(\mathfrak{L}) \qquad \dot{x} = Ax + Bu \quad \text{with} \quad C(u) = \int_0^T u(s)'U\,u(s)\,ds.$$

We wish to steer $x(0) = x_0$ to $x(T) = 0$ minimizing $C(u)$. We solve

$$\dot{x} = Ax + BU^{-1}B'\eta', \qquad \dot{\eta} = -\eta A$$

with $x(0) = x_0$, $x(T) = 0$. The optimal controller is $u^*(t) = U^{-1}B'\,\eta^*(t)'$. Here $\eta^*(t) = C'e^{-At}$ and the constant vector C is determined by the condition $x(T) = 0$ so

$$C = -\left[\int_0^T e^{-As}BU^{-1}B'e^{-A's}\,ds\right]^{-1} x_0.$$

If we consider $n = 2$ and $m = 1$ in the problem

$$\begin{pmatrix} \dot{x} \\ \dot{y} \end{pmatrix} = \begin{pmatrix} 0 & 1 \\ -1 & 0 \end{pmatrix} \begin{pmatrix} x \\ y \end{pmatrix} + \begin{pmatrix} 0 \\ 1 \end{pmatrix} u, \quad \text{with} \quad C(u) = \int_0^{2\pi} u(s)^2 \, ds,$$

we must compute

$$C = -\left[\int_0^{2\pi} \begin{pmatrix} \cos s & -\sin s \\ \sin s & \cos s \end{pmatrix} \begin{pmatrix} 0 \\ 1 \end{pmatrix} (0 \quad 1) \begin{pmatrix} \cos s & \sin s \\ -\sin s & \cos s \end{pmatrix} ds \right]^{-1} \begin{pmatrix} x_0 \\ y_0 \end{pmatrix}.$$

Then

$$C = \begin{pmatrix} -x_0/\pi \\ -y_0/\pi \end{pmatrix}$$

and

$$u^*(t) = \frac{x_0}{\pi} \sin t - \frac{y_0}{\pi} \cos t.$$

This example illustrates the regulation of an error $x(t)$ to zero in a finite duration with minimal energy. Explicit calculations for such an optimal controller $u^*(t)$, in terms of the initial data and various parameters, can be made for a wide variety of linear control processes.

Regulation on an Infinite Interval

When the basic interval $t_0 \leq t \leq T$ becomes infinite, that is $T = +\infty$, the above theory leads to the regulator problem: the problem of maintaining an optimally small total error for all time. We simplify the analysis by studying an autonomous linear process in R^n

$$(\mathfrak{L}) \qquad \dot{x} = Ax + Bu$$

where A and B are constant matrices. Furthermore, the cost functional for each control $u(t) \subset R^m$ on $0 \leq t < \infty$ is

$$C(u) = \int_0^{\infty} \|x(s)\|_W{}^2 + \|u(s)\|_U{}^2 \, ds$$

where $W = W' > 0$ and $U = U' > 0$ are also constant matrices.

An admissible controller $u(t)$ is measurable on $0 \leq t < \infty$ and is such that the cost functional is convergent to a finite value. In particular, $u(t)$ lies in the space of square integrable functions $L_2(0, \infty)$, and furthermore the corresponding response $x(t)$ also lies in $L_2(0, \infty)$. In fact, since both $\|x(t)\|_W{}^2$ and $\|u(t)\|_U{}^2$ contribute only a finite amount to the total cost on $0 \leq t < \infty$, it can be proved that $\lim_{t \to \infty} x(t) = 0$.

There might exist no admissible controllers, for instance, if $B = 0$, $A = I$, $x_0 \neq 0$. If $\mathfrak{L}$ is controllable, then any initial x_0 can be steered to the origin in a finite time, and thereafter held fixed with the zero control, so as to define an admissible control on $0 \leq t < \infty$.

The next theorem defines a feedback synthesis for the optimal controller of the regulator problem. However, first we present a lemma of Liapunov on negative definite matrices.

LEMMA. *For real $n \times n$ matrices consider the equation*

$$H'E' + EH = Q$$

where $Q = Q' > 0$. There exists a solution $E = E' < 0$ if and only if H is a stability matrix (that is, each eigenvalue of H has negative real part).

Proof. If H is a stability matrix, then define the required solution by the convergent integral

$$E = -\int_0^\infty e^{H't}Qe^{Ht}\,dt.$$

Clearly $E = E' < 0$ and we integrate by parts to obtain

$$H'E' = -\int_0^\infty H'e^{H't}Qe^{Ht}\,dt = -[e^{H't}Qe^{Ht}]_0^\infty + \int_0^\infty e^{H't}Qe^{Ht}H\,dt$$

or

$$H'E' = Q - EH, \qquad \text{as required.}$$

Conversely, assume $E = E' < 0$ is a matrix solution. Consider the linear differential system in R^n

$$\dot{x} = Hx.$$

Use the quadratic function $v(x) = -x'Ex$, which has ellipsoidal level surfaces enclosing the origin, and compute along a solution curve $x(t) \not\equiv 0$,

$$\frac{dv(x(t))}{dt} = -\dot{x}'Ex - x'E\dot{x} = x'[-H'E' - EH]x.$$

Then

$$\frac{dv}{dt} = -x'Qx < 0,$$

and $x(t) \to 0$. Thus H is a stability matrix. Q.E.D.

THEOREM 7. *Consider the autonomous controllable process in R^n*

$$(\mathfrak{L}) \qquad \dot{x} = Ax + Bu$$

with cost functional for each admissible controller $u(t) \subset R^m$ on $0 \leq t < \infty$

$$C(u) = \int_0^\infty \|x(t)\|_W^2 + \|u(t)\|_U^2\,dt, \qquad W = W' > 0, \qquad U = U' > 0.$$

Then there exists a unique symmetric negative definite matrix E satisfying

$$A'E + EA + EBU^{-1}B'E = W.$$

Then for each initial $x_0 \in R^n$ there exists a unique optimal controller $u^*(t)$, and this can be determined by a feedback synthesis

$$u^*(t) = U^{-1}B'E\,x^*(t).$$

Thus the optimal response $x^*(t)$ satisfies the asymptotically stable differential system

$$\dot{x} = (A + BU^{-1}B'E)x.$$

The minimal cost is $C(u^*) = -x_0'Ex_0$.

Proof. Suppose there exist solutions $x^*(t)$, $\eta^*(t)$ of

$$\dot{x} = Ax + BU^{-1}B'\eta'$$

$$\dot{\eta} = x'W - \eta A$$

with $x(0) = x_0$, $x(\infty) = 0$, $\eta(\infty) = 0$. Then we shall show the corresponding $u^*(t) = U^{-1}B'\,\eta^*(t)'$ is the unique optimal controller and $x^*(t)$ is the optimal response initiating at $x^*(0) = x_0$. Since $x^*(t)$, $\eta^*(t)$ are solutions of an autonomous linear system, and since they decay as $t \to \infty$, they must decay exponentially, and so $u^*(t)$ is an admissible controller.

Let $\omega(t)$ be the response to any admissible controller $u(t)$ on $0 \le t < \infty$. Define

$$\omega^0(t) = \int_0^t \|\omega(s)\|_W^2 + \|u(s)\|_U^2\,ds$$

and proceed just as in the proof of Theorem 3.

If $u(t)$ differs from $u^*(t)$ for some positive duration, then the argument of Theorem 3 shows that

$$-\tfrac{1}{2}x^0(\infty) + \eta(\infty)\,x(\infty) - \eta(0)x_0 > -\tfrac{1}{2}\omega^0(\infty) + \eta(\infty)\,\omega(\infty) - \eta(0)x_0,$$

and so

$$C_0(u^*) < C_0(u).$$

Hence $u^*(t)$ is the unique optimal controller.

We now construct the required solutions $x^*(t)$, $\eta^*(t)$ of the above differential system using the constant symmetric negative definite matrix E. Define $x^*(t)$ as the solution initiating at x_0 of the differential system

$$\dot{x} = (A + BU^{-1}B'E)x,$$

and define

$$\eta^*(t) = x^*(t)'E.$$

Then, using a solution E of

$$A'E + EA + EBU^{-1}B'E = W,$$

it is easy to verify that $x^*(t)$ and $\eta^*(t)$ are solutions of the required differential system. We prove that $x^*(\infty) = \eta^*(\infty) = 0$ by showing that $(A + BU^{-1}B'E)$ is a stability matrix.

The condition on E implies immediately that

$$(A + BU^{-1}B'E)'E + E(A + BU^{-1}B'E) = W + EBU^{-1}B'E.$$

Since $EBU^{-1}B'E = (U^{-1}B'E)'U(U^{-1}B'E) \geq 0$, as a symmetric matrix, the lemma implies that $(A + BU^{-1}B'E)$ is a stability matrix, as required.

To compute the optimal cost differentiate $x'Ex$ along the optimal response $x^*(t)$ to obtain

$$\frac{d}{dt}[x^*(t)'E\,x^*(t)] = \dot{x}'Ex + x'E\dot{x}$$
$$= (Ax + BU^{-1}B'Ex)'Ex + x'E(Ax + BU^{-1}B'Ex).$$

Use the algebraic condition on E and integrate to obtain

$$x^*(t)'E\,x^*(t) - x_0'Ex_0 = \int_0^t \|x^*(s)\|_W^2 + \|u^*(s)\|_U{}^2\,ds.$$

Thus $-x_0'Ex_0 = C(u^*)$.

The existence and uniqueness of the negative definite matrix E are asserted in the exercises below for the controllable case. Q.E.D.

***Example* 1.** Consider the scalar equation

$$\dot{x} = -x + u(t), \qquad x(0) = x_0$$

and the cost functional

$$C(u) = \int_0^\infty [x(t)^2 + u(t)^2]\,dt.$$

The determining equation for E is

$$(-1)E + E(-1) + E^2 = 1.$$

Choose $E = 1 - \sqrt{2} < 0$ and the optimal controller is

$$u^*(t) = (1 - \sqrt{2})x^*(t).$$

Furthermore $x^*(t) = x_0e^{-\sqrt{2}t}$.

EXERCISES

1. The dynamical equation for a rotor is

$$\dot{x} = u$$

where x is the angular momentum and $u(t)$ is the scalar controlling torque about the fixed axis of rotation. If $u(t)$ on $0 \leq t \leq 1$ is proportional to an electric current, the total energy expenditure is $\int_0^1 \alpha\, u(t)^2\, dt$, for a constant coefficient $\alpha > 0$. We wish to reduce the rate of rotation from an initial value of $x_0 > 0$.

(a) Use the cost functional $C(u) = x(1)^2 + \int_0^1 \alpha\, u(t)^2\, dt$ and synthesize the optimal controller as a feedback control. Compute the minimal cost.

(b) Use the cost functional $C(u) = x(1) + \int_0^1 \alpha\, u(t)^2\, dt$ and compute the minimal cost.

(c) Let the cost functional be $C(u) = \int_0^1 \alpha\, u(t)^2\, dt$ for controllers steering x_0 to rest. Compute the minimal cost.

2. The dynamical equation of an oscillator is

$$\ddot{x} + x = u$$

where $(x, \dot{x})$ is the state of the system and $u(t)$ on $0 \leq t \leq 2\pi$ is a controlling force. At $t = 0$ the initial state is $(0, 1)$ and we wish to advance the phase of the oscillation by $\pi/2$ to the desired response $\xi(t) = \cos t$. Let $C(u) = \int_0^{2\pi} [(x - \cos t)^2 + (\dot{x} + \sin t)^2 + u(t)^2]\, dt$. Write the quadratic differential equations for the components of the feedback matrix $E(t)$, as in the above examples. Replace $\dot{e}(t)$ by the difference $[e(t + h) - e(t)]/h$ for a small step $h > 0$ and design a rough flow chart for the numerical solution of the corresponding difference equations.

3. Synthesize the optimal control $u^*(t)$ for the regular problem

$$\ddot{x} + x = u \quad \text{with} \quad C(u) = \int_0^\infty [x(t)^2 + \dot{x}(t)^2 + u(t)^2]\, dt.$$

Find the minimal cost from the initial state $x(0) = 0$, $\dot{x}(0) = 1$.

4. Find the optimal control $u^*(t)$ on $0 \leq t \leq 1$ that steers the response from $(0, 0, -3)$ to the target $G\colon (x^1)^2 + (x^2)^2 + (x^3)^2 \leq 1$. The process is $\dddot{x} = u$, or the system in R^3

$$\dot{x}^1 = x^2, \qquad \dot{x}^2 = x^3, \qquad \dot{x}^3 = u \quad \text{with} \quad C(u) = \int_0^1 u(t)^2\, dt.$$

5. Consider the control process in R^n

$$(\mathfrak{L}) \qquad \dot{x} = A(t)x + B(t)u$$

with cost functional $C_0(u) = \int_{t_0}^{T} \|x(s)\|_W{}^2 + \|u(s)\|_U{}^2\, ds$ for each controller $u(t)$ on $t_0 \leq t \leq T$ in R^m. Let $\hat{K} = \{x^0(T), x(T)\}$ be the set of attainability in R^{n+1}. Show that each point $\hat{x} \in \hat{K}$ can be attained by using a continuous controller $u(t)$ on $t_0 \leq t \leq T$.

6. Consider the control process in R^n

$$(\mathfrak{L}) \qquad \dot{x} = A(t)x + B(t)u$$

with cost functional $\tilde{C}_0(u) = \left[\int_{t_0}^{T} \|x(s)\|_W{}^2 + \|u(s)\|_U{}^2\, ds\right]^{1/2}$ for each controller $u(t)$ on $t_0 \leq t \leq T$ in R^m. Let $\tilde{K} \subset R^{n+1}$ be the set of all points $\{\tilde{C}_0(u), x_u(T)\}$.

(a) Prove that there exists a unique optimal controller $u^*(t)$, namely the controller which minimizes $C_0(u) = [\tilde{C}_0(u)]^2$.

(b) Let $g(x)$ be a C^1 strictly convex function, that is,

$$g(\lambda x_1 + (1 - \lambda)x_2) < \lambda\, g(x_1) + (1 - \lambda)\, g(x_2) \quad \text{for}$$
$$0 < \lambda < 1, \qquad x_1 \neq x_2,$$

and assume that $g(x)$ is bounded below. Let the cost functional be $C(u) = g(x(T)) + \tilde{C}_0(u)$. Prove there exists a unique optimal controller, namely $u^*(t) = U(t)^{-1} B(t)' \eta^*(t)'$ with optimal response $x^*(t)$. Here $x^*(t)$, $\eta^*(t)$ are solutions of

$$\dot{x} = A(t)x + B(t)\, U(t)^{-1} B(t)'\eta'$$
$$\dot{\eta} = x'\, W(t) - \eta\, A(t)$$

with $x(t_0) = x_0$, $\eta(T) = -\sqrt{C_0(u^*)}\, \text{grad}\, g(x(T))$.

7. Consider the control process in R^n

$$(\mathfrak{L}) \qquad \dot{x} = A(t)x + B(t)u.$$

(a) Let the cost functional be $C(u) = \int_{t_0}^{T} \|x(s)\|_W^2\, ds$, for each control $u(t) \subset R^m$ on $t_0 \leq t \leq T$ satisfying the restraint $\int_{t_0}^{T} \|u(t)\|_U^2\, dt \leq 1$. Prove that there exists an optimal controller $u^*(t)$. If $W(t) > 0$, show that the optimal response $x^*(t)$ is unique. [Hint: use the weak compactness of the unit ball in $L_2(t_0, T)$.]

(b) Let the cost functional be $C(u) = \int_{t_0}^{T} \|u(s)\|_U^2\, ds$ for each controller $u(t) \subset R^m$ on $t_0 \leq t \leq T$ such that the response $x(t)$ satisfies the restraint $\int_{t_0}^{T} \|x(s)\|_W^2\, ds \leq 1$. Prove the existence of a unique optimal controller $u^*(t)$, provided there is at least one admissible controller. [Hint: let $u^{(k)} \to \bar{u}$ weakly so $C(u^{(k)}) \searrow \inf C(u)$, and note that $\lim C(u^{(k)}) \geq C(\bar{u})$. This last inequality follows from the observation that the generalized Fourier coefficients of $u^{(k)}$ approach those of $\bar{u}$.]

8. Let $E(t) = E(t)'$ on $t_0 \leq t \leq T$ be a solution of

$$\dot{E} = W(t) - A(t)'E - E\,A(t) - E\,B(t)\,U(t)^{-1}\,B(t)'E$$

with $E(T) = -G$, as in the first example of Section 3.3. Prove that

$$E(t) = [\psi_3(t) - \psi_4(t)G][\psi_1(t) - \psi_2(t)G]^{-1}$$

where

$$\Psi(t) = \begin{pmatrix} \psi_1 & \psi_2 \\ \psi_3 & \psi_4 \end{pmatrix}$$

is the fundamental matrix solution of

$$\dot{x} = A(t)x + B(t)\,U(t)^{-1}\,B(t)'\eta'$$

$$\dot{\eta} = x'\,W(t) - \eta\,A(t)$$

with $\Psi(T) = I$. The equality for $E(t)$ holds near $t = T$, in fact, as long as $[\psi_3 - \psi_4 G]$ and $[\psi_1 - \psi_2 G]$ are nonsingular.

9. Consider the control process in R^n

$$(\mathfrak{L}) \qquad \dot{x} = A(t)x + B(t)u$$

with cost functional $C(u) = \zeta\,x(T) + \int_{t_0}^{T} \|x(s)\|_W^2 + \|u(s)\|_U^2\, ds$ for a constant vector $\zeta \neq 0$. Verify the feedback synthesis for the optimal controller

$$u^*(t) = U(t)^{-1}\,B(t)'[h(t) + E(t)\,x^*(t)]$$

where

$$\dot{E} = W(t) - A(t)'E - E\,A(t) - E\,B(t)\,U(t)^{-1}\,B(t)'E$$

$$\dot{h} = -[E(t)\,B(t)\,U(t)^{-1}\,B(t)' + A(t)']h$$

with $E(T) = 0$, $h(T) = -\zeta'/2$.

10. Consider the autonomous process in R^n

$$(\mathfrak{L}) \qquad \dot{x} = Ax + Bu$$

with cost functional $C(u) = \int_0^{\infty} \|x(s)\|_W^2 + \|u(s)\|_U{}^2\, ds$, see Theorem 7.

Show that $V(x) = -x'Ex$ is a *Liapunov function* for the optimized system

$$\dot{x} = (A + BU^{-1}B'E)x.$$

That is, verify that

(a) $V(x) > 0$ for $x \neq 0$, and $V(0) = 0$,

(b) $\dot{V}(x(t)) < 0$ for $x \neq 0$.

11. Let $Q = Q' \geq 0$ and let A be a real $n \times n$ stability matrix. Show that

$$F_n = \frac{(-1)^{n-1}}{n!} \int_0^\infty t^n e^{A't} Q e^{At}\, dt, \qquad n = 0, 1, 2, \ldots$$

is the unique symmetric matrix solution of

$$(A')^{n+1}F + \binom{n+1}{1}(A')^n FA + \binom{n+1}{2}(A')^{n-1}FA^2$$
$$+ \cdots + \binom{n+1}{n}A'FA^n + FA^{n+1} = Q.$$

Here the binomial coefficients are, as usual, $\binom{n+1}{r} = \dfrac{(n+1)!}{(n+1-r)!\,r!}$.

[Hint: Verify $A'F_n + F_nA = F_{n-1}, F_{-1} = Q$. For uniqueness simplify the calculations by a coordinate transformation $\bar{A} = P^{-1}AP$, $\bar{F} = P'FP$, $\bar{Q} = P'QP$.]

12. Solve explicitly for $F = F'$ in $A'F + FA = Q$ where

$$A = \begin{pmatrix} 0 & 1 \\ -1 & -1 \end{pmatrix}, \qquad Q = \begin{pmatrix} 1 & 0 \\ 0 & 1 \end{pmatrix}.$$

13. Consider the autonomous process in R^n

$$(\mathcal{L}) \qquad \dot{x} = Ax + Bu$$

with $C(u) = \int_0^\infty \|x(s)\|_W^2 + \|u(s)\|_U{}^2\, ds$, $W = W' > 0$, $U = U' > 0$, as in Theorem 7. Let $u(t)$ on $0 \leq t \leq \infty$ be an admissible control, that is, $C(u) < \infty$. Prove $\lim_{t\to\infty} x(t) = 0$. [Hint: if $\overline{\lim}_{t\to\infty} |x(t)| > \epsilon > 0$, then for each $\tau > 0$ there exist infinitely many times $t_1 \to \infty$ such that $|x(t_1)| = \epsilon$ but $|x(t)| < \epsilon/2$ sometime on the interval $t_1 \leq t \leq t_1 + \tau$. Fix $\epsilon > 0$ and $\tau > 0$ small enough, then there exists $\delta > 0$ such that $\int_{t_1}^{t_1+\tau} |u(t)|\, dt > \delta$. Thus there exists $\xi > 0$ such that $\int_{t_1}^{t_1+\tau} \|u(t)\|_U{}^2\, dt > \xi > 0$ for infinitely many times $t_1 \to \infty$.]

14. Calculate a symmetric feedback matrix $E = \begin{pmatrix} e_1 & e \\ e & e_2 \end{pmatrix}$ for the control system

$$\ddot{x} - ax = u$$

with cost $C(u) = \int_0^\infty [\mathbf{x}'W\mathbf{x} + \gamma u^2]\,dt$, where

$$\mathbf{x} = \begin{pmatrix} x \\ \dot{x} \end{pmatrix}, \; W = \begin{pmatrix} w_1 & w \\ w & w_2 \end{pmatrix} > 0, \quad \text{and} \quad \gamma > 0.$$

[Hint: According to Theorem 7 we solve for E in

$$A'E + EA = W - \frac{1}{\gamma}\begin{pmatrix} e^2 & ee_2 \\ ee_2 & e_2^{\,2} \end{pmatrix} \quad \text{and} \quad A = \begin{pmatrix} 0 & 1 \\ a & 0 \end{pmatrix}.$$

Verify that $e = -\gamma a - \sqrt{\gamma^2 a^2 + \gamma w_1}$, $e_1 = w - \dfrac{1}{\gamma} ee_2 - ae_2$, and $e_2 = -\sqrt{\gamma w_2 - 2\gamma e}$ always define a real negative definite matrix E.]

15. Consider a controllable linear process in R^n

$$(\mathfrak{L}) \qquad \dot{x} = Ax + Bu$$

with cost $C(u) = \int_0^\infty [\|x\|_W{}^2 + \|u\|_U{}^2]\,dt$ for $W > 0$, $U > 0$, as in Theorem 7. Prove the existence and uniqueness of the negative definite matrix E satisfying

$$W - A'E - EA - EBU^{-1}B'E = 0$$

so that

$$u = U^{-1}B'Ex$$

synthesizes the optimal control with minimal cost $-x_0'Ex_0$. The following steps indicate the proof.

(a) In Example 1 of Section 3.3 above the symmetric matrix $D(t)$ on $0 \leq t \leq T$, satisfying

$$\dot{D} = W + A'D + DA - DBU^{-1}B'D, \qquad D(0) = 0,$$

defines the feedback gain $E(t) = -D(-t)$ for the corresponding optimal control process on the finite interval $0 \leq t \leq T$. The optimal cost for the initial state x_0 is $x_0'D(T)x_0$.

(b) If $0 < T_1 \leq T_2$, then $0 < D(T_1) \leq D(T_2)$ (in the sense, $x'\,D(T_1)x \leq x'\,D(T_2)x$ for all $x \in R^n$).

(c) Let $\bar{D}$ be a stabilizing matrix for

$$(\mathfrak{L}) \qquad \dot{x} = Ax + Bu \quad \text{with} \quad u = \bar{D}x,$$

that is $\bar{A} = A + BD$ is a stability matrix. Verify that the cost of the admissible controller

$$\bar{u}(t) = \bar{D}e^{\bar{A}t}x_0 \quad \text{with} \quad \bar{x}(t) = e^{\bar{A}t}x_0 \quad \text{on} \quad 0 \leq t < \infty$$

is

$$C(\bar{u}) = x_0'\theta x_0,$$

where

$$\theta = \int_0^\infty e^{\bar{A}'t}[W + \bar{D}'U\bar{D}]e^{\bar{A}t}\,dt.$$

Hence $D(T) \leq \theta$ for all $0 < T < \infty$.

(d) $\lim_{T\to\infty} D(T) = D_\infty$ exists and $D(T) \leq D_\infty \leq \theta$.

(e) D_∞ is a solution of $W + A'D + DA - DBU^{-1}B'D = 0$. Then $E = -D_\infty$ is the required negative definite matrix. The uniqueness of E follows from the formula $C(u^*) = -x_0'Ex_0$.

3.4 INTEGRAL CONVEX COST CRITERIA

We now treat the linear control process in R^n

$$(\mathfrak{L}) \qquad \dot{x} = A(t)x + B(t)u$$

with the integral cost functional

$$C(u) = g(x(T)) + \int_{t_0}^{T} f^0(t, x) + h^0(t, u)\,dt$$

where $A(t)$, $B(t)$, $g(x)$, $f^0(t, x)$, and $h^0(t, u)$ are continuous for all t in the fixed finite interval $t_0 \leq t \leq T$ and all $x \in R^n$ and $u \in R^m$. We also assume for the remainder of this section that $f^0(t, x)$ and $h^0(t, u)$ are convex functions for each fixed t; and furthermore

$$f^0(t, x) \geq 0 \quad \text{and} \quad h^0(t, u) \geq a\,|u|^p$$

for some constants $a > 0$ and $p > 1$. These positivity hypotheses, although they can be relaxed somewhat as in the exercises below, assure the existence of a (minimal cost) optimal controller among the class of all measurable controllers with finite cost.

We shall consider primarily the case $g(x) \equiv 0$. As before we shall study the geometry of the set of attainability $\hat{K} = \hat{K}(x_0, T) \subset R^{n+1}$ which consists of all endpoints $\hat{x}(T)$ for all responses $\hat{x}(t) = (x^0(t), x(t))$ of the system

$$\dot{x} = A(t)x + B(t)\,u(t)$$

$$\dot{x}^0 = f^0(t, x) + h^0(t, u(t)),$$

with the prescribed initial state $\hat{x}(t_0) = (0, x_0)$. Here $u(t)$ on $t_0 \leq t \leq T$ is any admissible control. Using

$$C_0(u) = \int_{t_0}^{T} f^0(t, x) + h^0(t, u)\, dt \geq a \int_{t_0}^{T} |u(t)|^p\, dt,$$

we note that every bounded measurable control $u(t)$ is admissible; also each admissible $u(t)$ lies in $L_p(t_0, T)$ and hence in $L_1(t_0, T)$. From the convexity of $f^0(t, x)$ and $h^0(t, u)$ it is easy to see that a convex combination of admissible controllers is also admissible.

The response $\hat{x}(T)$ can be computed from the customary variation of parameter formula

$$x(t) = \Phi(t)x_0 + \Phi(t) \int_{t_0}^{t} \Phi(s)^{-1} B(s)\, u(s)\, ds$$

and

$$x^0(T) = C_0(u).$$

In order to simplify the statements of our results we shall henceforth assume that $\mathfrak{L}$ is controllable on $t_0 \leq t \leq T$. From this it will follow that the projection of $\hat{K}(x_0, T)$ on the x-space is all of R^n—otherwise we should be forced to refer constantly to the geometry of $\hat{K}$ relative to the linear manifold it spans. Many of our results will be direct generalizations of the theorems of Section 3.2. However in the generality considered here we cannot obtain an elementary feedback synthesis for the optimal controller, as found for quadratic cost functionals; instead, we reduce the construction of the optimal controller to a two-point boundary-value problem that can be attacked by numerical methods.

LEMMA. *Consider the controllable process in* R^n

$$(\mathfrak{L}) \qquad \dot{x} = A(t)x + B(t)u$$

with cost functional $C_0(u)$ *and the set of attainability* $\hat{K} \subset R^{n+1}$. *Then the orthogonal projection of* $\hat{K}$ *on the hyperplane* $x^0 = 0$ *is all* R^n. *Further,* $\hat{K}$ *is the union of vertical rays, and also* $\hat{K}$ *lies above the hypersurface* $x^0 = a_1 |x|^p$ *for all large* $|x|$, *and some constant* $a_1 > 0$.

Proof. Since $\mathfrak{L}$ is controllable, the points $x(T)$ must fill all R^n as the control varies over the linear space of all bounded measurable functions. Since each admissible controller $u(t)$ in $L_1(t_0, T)$ can be approximated by a bounded controller, and since a convex combination of admissible controllers is also admissible, it follows that the projection of $\hat{K}$ is the entire space $x^0 = 0$.

Let $\bar{u}(t)$ steer $(0, x_0)$ to the point $\hat{y} = (y^0, y)$ in $\hat{K}$. We shall construct a control $u_\rho(t) = \bar{u}(t) + \rho\, u_\beta(t)$ such that

$$(1) \qquad \int_{t_0}^{T} \Phi(s)^{-1} B(s)\, u_\beta(s)\, ds = 0$$

and

(2) $$C_0(u_\rho) = y^0 + b$$

for any prescribed $b \geq 0$. This will show that $\hat{K}$ is the union of vertical rays. Let E be a closed set with positive duration in $t_0 \leq t \leq T$ such that $\bar{u}(t)$ is continuous and so bounded on E. Define $X_E(t) = 1$ for t in E and $X_E(t) = 0$ otherwise, and construct the partition of $[t_0, T]$ by $t_0 < t_1 < t_2 \cdots < t_{n+1} < T$ so that

$$\int_{t_k}^{T} X_E(t)\,dt = \frac{1}{k+1}\int_{t_0}^{T} X_E(t)\,dt \quad \text{for} \quad k = 1, \ldots, n+1.$$

Then, following the notation of the lemma before Theorem 1, we define

$$u_\beta^1(s) = [X(s, t_{n+1})\beta_1 + X(s, t_n)\beta_2 + \cdots + X(s, t_1)\beta_{n+1}]X_E(t)$$

and

$$u_\beta^j(s) = 0 \quad \text{for} \quad j = 2, 3, \ldots, m \quad \text{on} \quad t_0 \leq t \leq T.$$

Again choose $\beta = (\beta_1, \beta_2, \ldots, \beta_{n+1})$ to be nonzero and such that condition (1) holds with $u_\beta(t) \not\equiv 0$.

Since $u_\beta(t) = 0$ for t outside E, and $\bar{u}(t)$ is bounded on E, the controller $u_\rho(t)$ is admissible for each fixed $\rho \geq 0$. Note that $C_0(u_0) = C_0(\bar{u}) = y^0$ and $C_0(u_\rho) \geq a\int_{t_0}^{T} |u_\rho(t)|^p\,dt$ so that

$$\lim_{\rho\to\infty} C_0(u_\rho) = +\infty.$$

Since $C_0(u_\rho)$ varies continuously with ρ, there exists some $\rho \geq 0$ such that $C_0(u_\rho) = y^0 + b$, as required.

To obtain the estimate stated in the lemma note that, for each admissible controller $u(t)$,

$$|x(T)| \leq k_1 + l\int_{t_0}^{T} |u(t)|\,dt,$$

where $k_1 = |\Phi(T)x_0|$ and $l = \max\limits_{t_0 \leq t \leq T} |\Phi(T)\,\Phi(t)^{-1}\,B(t)|$. Since $|u|^p$ is a convex function of $|u|$, we have the immediate inequality

$$\left[\int_{t_0}^{T} |u(t)|\,dt\right]^p \leq \left[\int_{t_0}^{T} |u(t)|^p\,dt\right] |T - t_0|^{p-1}.$$

Thus

$$x^0(T) = C_0(u) \geq a\left[\int_{t_0}^{T} |u(t)|\,dt\right]^p \geq a\left[\frac{|x(T)| - k_1}{l}\right]^p |T - t_0|^{1-p}.$$

Hence, for all $|x(T)| \geq 2k_1$, we have

$$x^0(T) \geq \frac{a}{(2l)^p}\,|x(T)|^p\,|T - t_0|^{1-p}.$$ Q.E.D.

THEOREM 8. *Consider the controllable process in R^n*

$$(\mathfrak{L}) \qquad \dot{x} = A(t)x + B(t)u$$

with cost functional

$$C_0(u) = \int_{t_0}^{T} f^0(t, x) + h^0(t, u)\, dt.$$

Then the set of attainability $\hat{K} \subset R^{n+1}$ is closed and convex.

Proof. The proof that $\hat{K}$ is convex is the same as in Theorem 1. To prove that $\hat{K}$ is closed we shall use the inequality

$$\liminf_{k\to\infty} C_0(u_k) \geq C_0(\bar{u})$$

where $u_k(t)$ is a sequence of controllers converging weakly to $\bar{u}(t)$. The demonstration of this inequality, which is a general property of convex functions, is deferred to the end of the theorem.

Consider a sequence of points $\hat{x}_k(T) = (x_k^0(T), x_k(T))$, corresponding to controllers $u_k(t)$, and converging to $\hat{\bar{x}} = (\bar{x}^0, \bar{x})$ in R^{n+1}. Since the sequence $x_k^0(T) = C_0(u_k)$ is bounded, the functions $u_k(t)$ lie in some closed ball of the Banach space $L_p(t_0, T)$. Hence we can select a subsequence, still denoted $u_k(t)$, which converges weakly to $\bar{u}(t)$. Assuming the above inequality, we find that $\bar{u}(t)$ is an admissible controller with response $(\bar{x}^0(t), \bar{x}(t))$ and

$$\lim_{k\to\infty} x_k(t) = \bar{x}(t), \qquad \liminf_{k\to\infty} x_k^0(T) \geq \bar{x}^0(T).$$

Thus $\bar{x}(T) = \bar{x}$ and $x^0(T) \leq \bar{x}^0$. Since $\hat{K}$ is the union of vertical rays, we conclude that x lies in $\hat{K}$ and so $\hat{K}$ is closed.

We now prove the above stated inequality. Since $u_k(t)$ and $\bar{u}(t)$ lie in a given closed ball of $L_p(t_0, T)$, they satisfy a uniform bound in $L_1(t_0, T)$ and so $|x_k(t)|$ and $\bar{x}(t)$ satisfy some uniform bound. Thus

$$\lim_{k\to\infty} \int_{t_0}^{T} f^0(t, x_k(t))\, dt = \int_{t_0}^{T} f^0(t, \bar{x}(t))\, dt.$$

Hence we need only prove that

$$\liminf_{k\to\infty} \int_{t_0}^{T} h^0(t, u_k(t))\, dt \geq \int_{t_0}^{T} h^0(t, \bar{u}(t))\, dt.$$

Let C be a closed subset of $t_0 \leq t \leq T$ on which $\bar{u}(t)$ is continuous, and hence bounded; and let S be a compact ball in R^n that contains $\bar{u}(C)$ in its interior. For each fixed t consider the convex hypersurface in R^{m+1} defined by

$$u^0 = h^0(t, u).$$

For each $v \in S$ and $t_0 \leq t \leq T$ consider all the supporting hyperplanes to $u^0 = h^0(t, u)$ at the point $(v, h^0(t, v))$, say the hyperplanes

$$u^0 = \gamma(t)(u - v) + h^0(t, v).$$

The set H of all such hyperplanes is compact, where the metric in H is defined in terms of the $2m + 1$ coordinates (t, v, γ). Since no supporting hyperplane is vertical, there exists a uniform bound $|\gamma| \leq C_1$ in H.

Now prescribe $\epsilon > 0$ and define the control

$$v(t) = v_1 \mathrm{X}_{E_1}(t) + v_2 \mathrm{X}_{E_2}(t) + \cdots + v_q \mathrm{X}_{E_q}(t), \quad \text{for} \quad t \in C,$$

and $v(t) = \bar{u}(t)$ for t outside of C. Here $\mathrm{X}_{E_1}(t), \ldots, \mathrm{X}_{E_q}(t)$ are the characteristic functions of disjoint measurable sets $E_1, \ldots, E_q$, which partition C, and the vectors $v_1, \ldots, v_q$ are constants so chosen that

$$|\bar{u}(t) - v(t)| < \epsilon\, |T - t_0 + 1|^{-1} C_1^{-1},$$

$$v(t) \subset S$$

and

$$|h^0(t, \bar{u}(t)) - h^0(t, v(t))| < \epsilon\, |T - t_0 + 1|^{-1}.$$

For each $t \in E_1$ construct a supporting hyperplane to $u^0 = h^0(t, u)$ at the point $(v_1, h^0(t, v_1))$, say,

$$u^0 - h^0(t, v_1) = \gamma_1(t)(u - v_1).$$

We can require that $\gamma_1(t)$ be bounded and measurable on E_1 (see appendix to Chapter 2). Since the hypersurface $u^0 = h^0(t, u)$ lies above its supporting hyperplane, we obtain

$$h^0(t, u) - h^0(t, v_1) \geq \gamma_1(t)(u - v_1) \quad \text{for all} \quad u \in R^m.$$

For $t \in E_1$ we thus obtain

$$h^0(t, u_k(t)) - h^0(t, v_1) \geq \gamma_1(t)(u_k(t) - v_1) \quad \text{for} \quad k = 1, 2, 3, \ldots.$$

Using a similar inequality on each of $E_2, \ldots, E_q$ with

$$\gamma(t) = \gamma_1(t)\mathrm{X}_{E_1}(t) + \cdots + \gamma_q(t)\, \mathrm{X}_{E_q}(t),$$

we obtain

$$\int_C h^0(t, u_k(t))\, dt \geq \int_C h^0(t, v(t))\, dt + \int_C \gamma(t)\,(u_k(t) - v(t))\, dt.$$

For large $k \to \infty$, by weak convergence,

$$\left| \int_C \gamma(t)(u_k(t) - \bar{u}(t) + \bar{u}(t) - v(t))\, dt \right| < \epsilon + \epsilon$$

and also, by the previous estimate,

$$\int_C |h^0(t, v(t)) - h^0(t, \bar{u}(t))|\, dt < \epsilon.$$

Then

$$\int_C h^0(t, u_k(t))\, dt \geq \int_C h^0(t, \bar{u}(t))\, dt - 3\epsilon.$$

There exists a sequence of closed sets $C_1 \subset C_2 \subset C_3 \subset \cdots$ with $\lim_{l \to \infty} C_l = [t_0, T]$ and on each of which $\bar{u}(t)$ is continuous and we repeat the above argument for each C_l. Since $\int_{t_0}^T h^0(t, u_k(t))\, dt$ is a bounded sequence, and since $h^0(t, \bar{u}(t))\, X_{C_l}(t)$ is monotone in l, we find that the limit exists,

$$\lim_{l \to \infty} \int_{C_l} h^0(t, \bar{u}(t))\, dt = \int_{t_0}^T h^0(t, \bar{u}(t))\, dt.$$

Thus, given $\epsilon > 0$ there exists a C_l such that,

$$\int_{t_0}^T h^0(t, u_k(t))\, dt \geq \int_{C_l} h^0(t, u_k(t))\, dt \geq \int_{t_0}^T h^0(t, \bar{u}(t))\, dt - 4\epsilon$$

for all large k. Hence

$$\liminf_{k \to \infty} \int_{t_0}^T h^0(t, u_k(t))\, dt \geq \int_{t_0}^T h^0(t, \bar{u}(t))\, dt,$$

as required. Q.E.D.

COROLLARY. *Consider the controllable process in R^n*

$$(\mathcal{L}) \qquad \dot{x} = A(t)x + B(t)u$$

with cost functional

$$C(u) = g(x(T)) + C_0(u).$$

If either

(a) *$g(x) > b$, that is, $g(x)$ is bounded below on R^n, or*

(b) *$g(x)$ is convex on R^n,*

then there exists an optimal controller.

Proof. The proof follows closely that of Theorem 2 and uses the estimates in the lemma preceding Theorem 8. Q.E.D.

A controller $\bar{u}(t)$, for the controllable process $\mathcal{L}$ with cost $C_0(u)$, which steers $(0, x_0)$ to the boundary of $\hat{K}$ in R^{n+1}, is called an extremal controller and the corresponding response is also extremal. Of course, an optimal controller for $\mathcal{L}$ with cost $C(u) = g(x(T)) + C_0(u)$ also steers the response $\hat{x}(t) = (x^0(t), x(t))$ to the boundary of $\hat{K}$ and is hence extremal. As before, we characterize extremal controllers by the maximal principle.

In this part of the theory an adjoint response, corresponding to $u(t)$ and $x(t)$, is a row $(n + 1)$-vector $\hat{\eta}(t) = (\eta_0, \eta(t))$ satisfying the linear system

$$(\mathcal{A}) \qquad \dot{\eta}_0 = 0$$
$$\dot{\eta} = -\eta_0 \frac{\partial f^0}{\partial x}(t, x(t)) - \eta\, A(t).$$

In addition to the standing hypotheses of continuity listed at the beginning of Section 3.4, we henceforth demand that $\partial f^0/\partial x\ (t, x)$ be continuous in $t_0 \le t \le T$ and $x \in R^n$. The convexity of $f^0(t, x)$ then implies that

$$f^0(t, x) - f^0(t, \bar{x}) \ge \frac{\partial f^0}{\partial x}(t, \bar{x})\,(x - \bar{x}).$$

THEOREM 9. *Consider the controllable process in* R^n

$$(\mathfrak{L}) \qquad \dot{x} = A(t)x + B(t)u$$

with cost functional

$$C_0(u) = \int_{t_0}^{T} f^0(t, x) + h^0(t, u)\, dt.$$

A controller $\bar{u}(t)$ *with response* $\bar{x}(t)$ *is extremal if and only if there exists a vector* $\hat{\eta}(t) = (\eta_0, \eta(t))$ *satisfying*

$$(\mathcal{A}) \qquad \dot{\eta}_0 = 0, \qquad \eta_0 < 0$$
$$\dot{\eta} = -\eta_0 \frac{\partial f^0}{\partial x}(t, \bar{x}(t)) - \eta\, A(t)$$

and such that the maximal principle holds almost everywhere

$$\eta_0\, h^0(t, \bar{u}(t)) + \eta(t)\, B(t)\, \bar{u}(t) = \max_u\, [\eta_0\, h^0(t, u) + \eta(t)\, B(t)u].$$

Proof. Let $\bar{u}(t)$ with response $\hat{x}(t) = (\bar{x}^0(t), \bar{x}(t))$ and adjoint response $\hat{\eta}(t) = (\eta_0, \eta(t))$ satisfy

$$(\hat{\mathfrak{L}}) \qquad \dot{x} = A(t)x + B(t)\,\bar{u}(t)$$
$$\dot{x}^0 = f^0(t, x) + h^0(t, \bar{u}(t)), \qquad \hat{x}(t_0) = (0, x_0),$$

and also the system $\mathcal{A}$ with $\eta_0 < 0$, and the above maximal principle. We shall prove that

$$\hat{\eta}(T)\, \hat{x}(T) \ge \hat{\eta}(T)\, \hat{\omega}(T),$$

where $\hat{\omega}(T) = (\omega^0(t), \omega(t))$ is the response to an arbitrary admissible controller $u(t)$. From this inequality it follows that $\hat{x}(T)$ lies on the boundary of $\hat{K}$ and that $\hat{\eta}(T)$ is there an exterior normal to $\hat{K}$.

Use the differential equations $\hat{\mathfrak{L}}$ and $\mathcal{A}$ to evaluate

$$\frac{d}{dt}[\hat{\eta}(t)\,\hat{\omega}(t)] = \eta_0\dot{\omega}^0 + \dot{\eta}\omega + \eta\dot{\omega}$$

and compute

$$\hat{\eta}(T)\,\hat{\omega}(T) - \hat{\eta}(t_0)\hat{x}_0$$

$$= \int_{t_0}^{T} \eta_0\left[f^0(t, \omega) - \frac{\partial f^0}{\partial x}(t, \bar{x})\omega\right] + [\eta_0\, h^0(t, u) + \eta B u]\, dt.$$

Now specialize $u(t)$ to $\bar{u}(t)$ and $\hat{\omega}(t)$ to $\hat{x}(t)$ to obtain

$$\hat{\eta}(T)\,\hat{x}(T) - \hat{\eta}(t_0)\hat{x}_0$$

$$= \int_{t_0}^{T} \eta_0\left[f^0(t, \bar{x}) - \frac{\partial f^0}{\partial x}(t, \bar{x})\,\bar{x}\right] + [\eta_0\, h^0(t, \bar{u}) + \eta B \bar{u}]\, dt.$$

Use the maximal principle

$$\eta_0\, h^0(t, \bar{u}(t)) + \eta(t)\, B(t)\, \bar{u}(t) \geq \eta_0\, h^0(t, u(t)) + \eta(t)\, B(t) u(t)$$

and the convexity condition

$$f^0(t, \omega) - f^0(t, \bar{x}) \geq \frac{\partial f^0}{\partial x}(t, \bar{x})\,(\omega - \bar{x})$$

to deduce the required inequality $\hat{\eta}(T)\,\hat{x}(T) \geq \hat{\eta}(T)\,\hat{\omega}(T)$. Hence $\bar{u}(t)$ is an extremal controller.

Conversely, assume that $\bar{u}(t)$ is an extremal controller so that the corresponding response $\hat{x}(t) = (\bar{x}^0(t), \bar{x}(t))$ steers $(0, x_0)$ to $\hat{x}(T)$ on the boundary of $\hat{K}$. Let $\hat{\eta}(T) = (\eta_0, \eta(T))$ be an exterior normal to $\hat{K}$ at $\hat{x}(T)$. Clearly $\eta_0 < 0$ and we simplify the calculation by taking $\eta_0 = -1$. Let $\hat{\eta}(t)$ be defined as the solution of the adjoint system $\mathcal{A}$ with the given data $\hat{\eta}(T)$ at $t = T$. We must prove that

$$-h^0(t, \bar{u}(t)) + \eta(t)\, B(t)\, \bar{u}(t) = \max_u\, [-h^0(t, u) + \eta(t)\, B(t) u]$$

almost everywhere on $t_0 \leq t \leq T$.

The proof is complicated by the fact that the adjoint system $\mathcal{A}$ depends on the basic response $\bar{x}(t)$. The device introduced to overcome this difficulty is used later to prove the maximal principle for the most general nonlinear process. Essentially, this method consists in adding an impulse perturbation to $\bar{u}(t)$ over a short duration $t_1 \leq t \leq t_1 + \epsilon$, during which we suppose that $\bar{u}(t)$ fails to satisfy the maximal principle. The perturbed controller $u_\epsilon^*(t)$ yields an increment in the terms $\int_{t_0}^{T} [\eta_0\, h^0(t, u) + \eta(t)\, B(t) u]\, dt$

in the computation of $\hat{\eta}(T)\,\hat{\omega}(T)$, which contradicts the assertion that $\hat{x}(T)$ lies on the boundary of $\hat{K}$. We now offer the details of the proof.

Suppose that $\bar{u}(t)$ fails to satisfy the maximal principle for some positive duration on $t_0 \leq t \leq T$. Define a controller $\tilde{u}(t)$ by

$$-h^0(t, \tilde{u}(t)) + \eta(t)\, B(t)\, \tilde{u}(t) = \max_u \,[-h^0(t, u) + \eta(t)\, B(t)u].$$

It is clear that $\tilde{u}(t)$ is bounded, and $\tilde{u}(t)$ can be chosen measurable, as in the appendix to Chapter 2. Let C be a compact subset of positive duration in $t_0 < t < T$ whereon $\bar{u}(t)$ and $\tilde{u}(t)$ are continuous and where

$$-h^0(t, \bar{u}(t)) + \eta(t)\, B(t)\, \bar{u}(t) < -h^0(t, \tilde{u}(t)) + \eta(t)\, B(t)\, \tilde{u}(t) - \delta$$

for some constant $\delta > 0$. Pick a time $t_1 \in C$ for which the set $(t_1, t_1 + \epsilon) \cap C$ has a measure $\epsilon[1 + 0(\epsilon)]$ for all small $\epsilon > 0$ (here $\lim_{\epsilon \to 0} 0(\epsilon) = 0$). Define the perturbed control

$$u_\epsilon^*(t) = \begin{cases} \tilde{u}(t) & \text{on} \quad C \cap (t_1, t_1 + \epsilon) \\ \bar{u}(t) & \text{elsewhere on} \quad t_0 \leq t \leq T. \end{cases}$$

Then, for sufficiently small $\epsilon > 0$, the response $\hat{x}_\epsilon^*(t)$ to $u_\epsilon^*(t)$ uniformly approximates $\hat{x}(t)$. In fact, it is easy to show that

$$|\hat{x}_\epsilon^*(t) - \hat{x}(t)| < k\epsilon, \quad \text{for some} \quad k > 0,$$

on $t_0 \leq t \leq T$. Since $\partial f^0/\partial x\,(t, x)$ is continuous,

$$\left|f^0(t, x_\epsilon^*(t)) - f^0(t, \bar{x}(t)) - \frac{\partial f^0}{\partial x}(t, \bar{x}(t))(x_\epsilon^*(t) - \bar{x}(t))\right| < \epsilon\, O(\epsilon).$$

From earlier calculations for $\hat{\eta}(T)\,\hat{\omega}(T)$ we have

$$\begin{aligned} &\hat{\eta}(T)\,\hat{x}(T) - \hat{\eta}(T)\,\hat{x}_\epsilon^*(T) \\ &\qquad \leq \int_{t_0}^{T} \left[f^0(t, x_\epsilon^*(t)) - f^0(t, \bar{x}(t)) - \frac{\partial f^0}{\partial x}(t, \bar{x}(t))\,(x_\epsilon^*(t) - \bar{x}(t)) \right] dt \\ &\qquad\quad - \delta\epsilon[1 + 0(\epsilon)]. \end{aligned}$$

Then, for sufficiently small $\epsilon > 0$,

$$\hat{\eta}(T)\,\hat{x}_\epsilon^*(T) > \hat{\eta}(T)\,\hat{x}(T).$$

But this is impossible since $\hat{\eta}(T)$ is an exterior normal vector to $\hat{K}$ at the boundary point $\hat{x}(T)$. Therefore the extremal controller $\bar{u}(t)$ must satisfy the maximal principle with adjoint response $\hat{\eta}(t)$. Q.E.D.

COROLLARY. *Consider the controllable process in* R^n

$$(\mathfrak{L}) \qquad \dot{x} = A(t)x + B(t)u$$

with cost

$$C(u) = g(x(T)) + \int_{t_0}^{T} f^0(t, x) + h^0(t, u)\, dt$$

where $g(x)$ *is convex and* $h^0(t, u)$ *is strictly convex, that is, for* $0 < \lambda < 1$ *and* $u_1 \neq u_2$ *and each fixed* t,

$$h^0(t, \lambda u_1 + (1 - \lambda)u_2) < \lambda\, h^0(t, u_1) + (1 - \lambda)\, h^0(t, u_2).$$

Then any two extremal controllers steering $(0, x_0)$ *to the same boundary point of* $\hat{K}$ *must coincide almost everywhere. Furthermore, there exists a unique optimal controller.*

Proof. Let $\bar{u}(t)$ with response $\hat{x}(t) = (\bar{x}^0(t), \bar{x}(t))$ be extremal, and let $\hat{\eta}(t) = (-1, \eta(t))$ be the corresponding adjoint response, such that $\hat{\eta}(T)$ is an exterior normal to $\hat{K}$ at $\hat{x}(T)$. Then $\bar{u}(t)$ satisfies the maximal principle

$$-h^0(t, \bar{u}(t)) + \eta(t)\, B(t)\, \bar{u}(t) = \max_u \,[-h^0(t, u) + \eta(t)\, B(t)u] = m(t).$$

Now let $u(t)$ with response $\hat{\omega}(t)$ steer $(0, x_0)$ to the same endpoint $\hat{\omega}(T) = \hat{x}(T)$. If $u(t)$ fails to satisfy the maximal principle for the given $\eta(t)$, then

$$\int_{t_0}^{T} [-h^0(t, \bar{u}(t)) + \eta(t)\, B(t)\, \bar{u}(t)]\, dt > \int_{t_0}^{T} [-h^0(t, u(t)) + \eta(t)\, B(t)\, u(t)]\, dt.$$

Then the computation in Theorem 9 yields

$$\hat{\eta}(T)\,\hat{x}(T) > \hat{\eta}(T)\,\hat{\omega}(T) = \hat{\eta}(T)\,\hat{x}(T),$$

which is impossible. Thus we conclude that $\bar{u}(t)$ and $u(t)$ satisfy the same maximal principle almost everywhere on $t_0 \leq t \leq T$.

Next consider the controller $\frac{1}{2}[\bar{u}(t) + u(t)]$. The strict convexity of $h^0(t, u)$ shows that

$$-h^0(t, \tfrac{1}{2}\bar{u}(t) + \tfrac{1}{2}u(t)) + \eta(t)\, B(t)\tfrac{1}{2}[\bar{u}(t) + u(t)] > \tfrac{1}{2}m(t) + \tfrac{1}{2}m(t)$$

whenever $\bar{u}(t) \neq u(t)$. By the above argument we conclude that $\bar{u}(t) = u(t)$ almost everywhere.

Since $g(x)$ is convex there exists an optimal controller $u^*(t)$ that is extremal and that steers $(0, x_0)$ to the subset of the boundary of $\hat{K}$ where $x^0 + g(x)$ assumes its minimum. The argument of Theorem 5 shows that $x^0 + g(x)$ can assume its minimum at just one single point P of $\hat{K}$. Hence $u^*(t)$ is the unique extremal control which steers $(0, x_0)$ to the point P.

Q.E.D.

Remarks. Even if $h^0(t, u)$ is not strictly convex we can still specify a definite extremal controller $u^*(t, \eta(t))$ for each prescribed adjoint vector $\eta(t)$. Namely, for each fixed t and η select $u^*(t, \eta)$ by the maximum principle

$$-h^0(t, u^*) + \eta\, B(t)u^* = \max_u\, [-h^0(t, u) + \eta\, B(t)u].$$

If $h^0(t, u)$ is strictly convex for each t, then $u^*(t, \eta)$ is uniquely defined by the maximal principle. However, even if $h^0(t, u)$ is merely convex, we maintain a definite choice for $u^*(t, \eta)$ by selecting the point in R^m which has minimal coordinates among all those points satisfying the maximal principle. That is, choose $u^*(t, \eta) = (u^{*1}, u^{*2}, \ldots, u^{*m})$ so that u^{*1} is minimal among all possible solutions of the maximal principle, then choose u^{*2} as minimal among all solutions with the designated value of u^{*1}. Continue in this manner to define $u^*(t, \eta)$. If $\eta(t)$ is continuous, then $u^*(t) = u^*(t, \eta(t))$ is an admissible controller (see appendix to Chapter 2).

The next theorem shows how $u^*(t, \eta)$ can be interpreted as a determination of a feedback synthesis for an optimal controller.

THEOREM 10. *Consider the control process in* R^n

$$(\mathfrak{L}) \qquad \dot{x} = A(t)x + B(t)u$$

with cost functional

$$C(u) = g(x(T)) + \int_{t_0}^{T} f^0(t, x) + h^0(t, u)\, dt.$$

Assume $g(x) \in C^1$ *is convex in* R^n. *Then there exists a solution* $x^*(t), \eta^*(t)$ *of the system*

$$\dot{x} = A(t)x + B(t)\, u^*(t, \eta)$$

$$\dot{\eta} = \frac{\partial f^0}{\partial x}(t, x) - \eta\, A(t)$$

with $x(t_0) = x_0$, $\eta(T) = -\operatorname{grad} g(x(T))$. *Here* $u^*(t, \eta)$ *is defined by the maximal principle*

$$-h^0(t, u^*) + \eta\, B(t)u^* = \max_u\, [-h^0(t, u) + \eta\, B(t)u].$$

An optimal controller is $u^*(t) = u^*(t, \eta^*(t))$ *with the corresponding optimal response* $x^*(t)$.

If $h^0(t, u)$ *is strictly convex for each* t, *then the solution* $x^*(t), \eta^*(t)$ *is unique, and* $u^*(t)$ *is the unique optimal controller.*

Proof. Consider the hypersurfaces S_c: $x^0 + g(x) = C$ in R^{n+1}. Then, just as in Theorem 5, there exists a unique hypersurface S_m in this family such that S_m is tangent to $\hat{K}$, and m is the optimal cost. Let $\hat{\eta}^*(T) = (-1, \eta^*(T))$ be the normal to the tangent hyperplane to S_m at some point $P \in S_m \cap \hat{K}$. Let $u^*(t)$ be an extremal controller steering $(0, x_0)$ to $P = \hat{x}^*(T)$ by the response $\hat{x}^*(t) = (x^{0*}(t), x^*(t))$. Let $\hat{\eta}^*(t) = (-1, \eta^*(t))$ be defined as the solution of

$$\dot{\eta} = \frac{\partial f^0}{\partial x}(t, x^*(t)) - \eta\, A(t)$$

with $\eta^*(T) = -\text{grad}\, g(x^*(T))$.

By Theorem 9 we find that $u^*(t)$ satisfies the maximal principle with adjoint response $\eta^*(t)$, that is, $u^*(t) = u^*(t, \eta^*(t))$. Thus $x^*(t), \eta^*(t)$ is the required solution of the above nonlinear boundary-value problem.

If $h^0(t, u)$ is strictly convex for each fixed t, then $S_m \cap \hat{K}$ is just a single point P, according to the proof of Theorem 5. Then, by the corollary to Theorem 9, the optimal control $u^*(t)$ and its response $x^*(t)$ are unique. Also $\eta^*(t)$ is uniquely determined as the solution of a linear differential system with the data $\eta(T) = -\text{grad}\, g(x^*(T))$. Q.E.D.

Remarks on the Restricted Endpoint Problem. Consider the controllable process in R^n

$$(\mathfrak{L}) \qquad \dot{x} = A(t)x + B(t)u$$

with cost

$$C(u) = g(x(T)) + C_0(u),$$

as in Theorem 10. In this discussion the initial state x_0 is to be controlled with minimal cost to the prescribed target point $x_1 \in R^n$. We also shall assume that $h^0(t, u)$ is strictly convex for each fixed t on $t_0 \leq t \leq T$.

Under these conditions there exists a unique optimal controller $u^*(t)$ steering x_0 to x_1. In fact, consider the line l: $x = x_1$ in R^{n+1} so $l \cap \hat{K}$ is a closed vertical ray. Then $u^*(t)$ steers $(0, x_0)$ to $(x^{0*}(T), x_1)$ where $x^{0*}(T)$ is the lowest point on the ray $l \cap \hat{K}$, and $C(u^*) = g(x_1) + x^{0*}(T)$. Furthermore, if $x^*(t), \eta^*(t)$ is any solution of

$$\dot{x} = A(t)x + B(t)\, u^*(t, \eta)$$

$$\dot{\eta} = \frac{\partial f^0}{\partial x}(t, x) - \eta\, A(t)$$

with $x(t_0) = x_0, x(T) = x_1$, then $u^*(t) = u^*(t, \eta^*(t))$ and $x^*(t)$ is the optimal response corresponding to $u^*(t)$.

Again, under the same hypotheses that $\mathfrak{L}$ is controllable on $t_0 \leq t \leq T$ and that $h^0(t, u)$ is strictly convex for each t, consider the problem of steering x_0 to a fixed compact convex target set G: $\gamma(x) \leq 0$ in R^n. Here $\gamma(x)$ is a C^1 convex function and we assume that grad $\gamma(x) \neq 0$ on the boundary of G, a smooth convex hypersurface. As in the discussion of Section 3.3, we consider the cylinder $\hat{G} = G \times R^1$ in the (x^0, x) space R^{n+1}. Since $\mathfrak{L}$ is controllable, $\hat{G}$ intersects $\hat{K}$ in a closed convex set and hence there exists an optimal controller steering x_0 to G and minimizing $C(u)$.

For a further simplification of the problem let us require that $g(x) \equiv 0$ so $C(u) = C_0(u)$. Then the minimal value of x^0 in $\hat{G} \cap \hat{K}$ occurs at just one common boundary point $\hat{x}^*(T) = (x^{0*}(T), x^*(T))$, [unless the optimal controller minimizing $C_0(u)$ without reference to a target also steers x_0 to G—this special case, corresponding to $\eta(T) = 0$ in Theorem 10, will be disregarded since it occurs if and only if the subsequent method fails to have a solution]. Thus there is a unique optimal controller $u^*(t)$ steering x_0 to G and the corresponding response terminates at $\hat{x}^*(T)$. Just as in the corresponding discussion of Section 3.3, we can obtain $u^*(t)$ from any solution $x^*(t)$, $\eta^*(t)$ of the system

$$\dot{x} = A(t)x + B(t)\, u^*(t, \eta)$$

$$\dot{\eta} = \frac{\partial f^0}{\partial x}(t, x) - \eta\, A(t)$$

with $x(t_0) = x_0$, $\gamma(x(T)) = 0$ and the transversality condition

$$\eta(T) = -k \operatorname{grad} \gamma(x(T))$$

for some $k > 0$; $u^*(t) = u^*(t, \eta^*(t))$ and $x^*(t)$ is the optimal response.

***Example* 1.** A simple controlled dynamical system is modeled by the scalar equation

$$\dot{x} = x + u$$

and we employ the cost functional $C(u) = \frac{1}{4}\int_0^1 u(t)^4\, dt$ on the fixed interval $0 \le t \le 1$. The control problem consists in steering $x(t)$ from an initial state $x(0) = x_0$ to the target $x(1) = 0$, with minimal cost.

The maximal principle states

$$-\frac{u^{*4}}{4} + \eta u^* = \max_u \left[-\frac{u^4}{4} + \eta u\right]$$

or

$$u^* = \sqrt[3]{\eta}.$$

Thus we must solve

$$\dot{x} = x + \sqrt[3]{\eta}$$

$$\dot{\eta} = -\eta$$

with $x(0) = x_0$, $x(1) = 0$. Since $\eta = \eta_0 e^{-t}$ we have

$$x = e^t x_0 - \tfrac{3}{4}\eta_0^{1/3}[e^{-t/3} - e^t].$$

The boundary conditions yield

$$\eta_0^{1/3} = \frac{4x_0}{3}(e^{-4/3} - 1)^{-1}$$

and the optimal controller is

$$u^*(t) = \frac{4x_0}{3}(e^{-4/3} - 1)^{-1}e^{-t/3}$$

***Example* 2.** Consider the control process in R^n

$$(\mathfrak{L}) \qquad \dot{x} = A(t)x + B(t)u$$

with cost

$$C(u) = \|u\|_p = \left\{\int_{t_0}^{T} |u^1|^p + \cdots + |u^m|^p \, dt\right\}^{1/p},$$

for some given $1 < p < \infty$. We seek to minimize $C(u)$ while controlling the given initial state x_0 to the target point x_1 in R^n.

Consider the set $K(k) \subset R^n$ of attainability from x_0 in time $t_0 \le t \le T$, under the restraint $\|u\|_p \le k$. Then it is easy to prove that $K(k)$ is convex, compact, and continuously increasing with k since $K(k)$ consists of the horizontal slices of $\hat{K}(x_0, T) \subset R^{n+1}$ at the constant level $x^0 = k^p$.

The smallest k for which $K(k)$ meets the prescribed target point x_1 is the optimal cost. The optimal controller $u^*(t)$ is unique, as is shown in Theorem 10 and the subsequent remarks. In the case of scalar control $m = 1$, we solve

$$\dot{x} = A(t)x + B(t)\,u^*(t, \eta)$$

$$\dot{\eta} = -\eta\,A(t)$$

with $x(t_0) = x_0$, $x(T) = x_1$. Then $u^*(t) = u^*(t, \eta^*(t))$ where

$$u^*(t, \eta) = \begin{cases} \left|\dfrac{1}{p}\eta\,B(t)\right|^{1/p-1} & \text{if } \eta\,B(t) \ge 0 \\[2ex] -\left|\dfrac{1}{p}\eta\,B(t)\right|^{1/p-1} & \text{if } \eta\,B(t) < 0. \end{cases}$$

In the limiting case $p = \infty$ we define $\|u\|_\infty = \sup |u^i(t)|$ for $1 \le i \le m$, $t_0 \le t \le T$ (actually, $\lim_{p\to\infty} \|u\|_p = \sup |u^i(t)|$ where the supremum disregards values of $|u(t)|$ on any null set of times). For simplicity consider the autonomous system

$$(\mathfrak{L}) \qquad \dot{x} = Ax + Bu$$

with

$$C(u) = \|u\|_\infty \quad \text{on} \quad 0 \le t \le T;$$

and we assume that $\mathcal{L}$ is normal for the restraint cube $|u^i| \le 1$. By the results of Chapter 2 the set $K(k) \subset R^n$ of attainability from x_0 with $\|u\|_\infty \le k$ on $0 \le t \le T$, is convex, compact, and increases continuously with k. Since $\mathcal{L}$ is controllable, there exists a minimal k^* for which $K(k^*)$ first meets the target $x_1 = 0$. Thus there exists a unique $k^* > 0$ for which there is a solution of

$$\dot{x} = Ax + B \operatorname{sgn} (\eta B)'k$$

$$\dot{\eta} = -\eta A$$

with $x(0) = x_0$, $x(T) = 0$. The unique optimal controller is

$$u^*(t) = \operatorname{sgn} (\eta^*(t)B)'k^* \quad \text{on} \quad 0 \le t \le T,$$

since $u^*(t)$ is also the minimal time optimal controller for the restraint $\|u\|_\infty \le k^*$.

If we drop the condition that the time duration be fixed, and allow arbitrary finite intervals $t_0 \le t \le t_1 < \infty$, then there may not exist an optimal controller. For instance consider the scalar control process

$$\dot{x} = u$$

where we wish to steer $x(0) = 0$ to $x(t_1) = 1$ on some finite interval $0 \le t \le t_1$, so as to minimize the cost $C(u) = \|u\|_p$ for $1 < p \le \infty$. For each $\epsilon > 0$ take the control

$$u(t) = \frac{\epsilon}{t+1} \quad \text{on} \quad 0 \le t \le t_1.$$

Clearly $\lim_{\epsilon \to 0} \|u\|_p = 0$; but $\int_0^\infty \frac{\epsilon\, dt}{t+1}$ is divergent and so there exists an interval $0 \le t \le t_1(\epsilon)$ with $x(t_1(\epsilon)) = 1$ as required. But there exists no optimal controller on any $0 \le t \le t_1^*$ with $\|u^*(t)\|_p = 0$.

Remarks on Regulation Over an Infinite Interval. We next allow the time interval $t_0 \le t \le T$ to become infinite but retain all the other standing hypotheses listed at the beginning of Section 3.4. We combine all the results for this case in one rather long theorem.

THEOREM 11. *Consider the controllable autonomous process in* R^n

$$(\mathcal{L}) \qquad \dot{x} = Ax + Bu$$

with cost functional

$$C(u) = \int_0^\infty f^0(x) + h^0(u)\, dt,$$

where $f^0(x) \geq 0$ is convex, $f(x) = 0$ if and only if $x = 0$, $h^0(u) \geq a\,|u|^p$ is strictly convex and $h^0(0) = 0$. Then there exists an unique optimal controller $u^*(t)$ on $0 \leq t < \infty$ with response $x^*(t)$.

Assume that no eigenvalue of A has zero real part. Then a necessary and sufficient condition that an admissible controller $\bar{u}(t)$ with response $\bar{x}(t)$ on $0 \leq t < \infty$ be optimal is that the maximal principle obtains

$$\bar{\eta}_0\, h^0(\bar{u}(t)) + \bar{\eta}(t)B\,\bar{u}(t) = \max_u\,[\bar{\eta}_0\, h^0(u) + \bar{\eta}(t)Bu], \qquad \text{a.e.}$$

where $\hat{\eta}(t) = (\bar{\eta}_0, \bar{\eta}(t))$ satisfies the adjoint system

$$(\mathcal{A}) \qquad \dot{\eta}_0 = 0$$
$$\dot{\eta} = -\eta_0 \frac{\partial f^0}{\partial x}(\bar{x}(t)) - \eta A$$

with $\bar{\eta}_0 < 0$ and $\bar{\eta}(\infty) = 0$.

Proof. Since $\mathfrak{L}$ is controllable, the initial state x_0 can be steered to the origin at $t = 1$, and then kept fixed at the origin $x = 0$ by the null control $u = 0$. In this way there exists an admissible controller having finite cost M. We next construct the optimal controller $u^*(t)$ on $0 \leq t < \infty$ as the weak limit of the appropriate optimal controllers on finite time intervals.

Using the given initial state x_0 at $t = 0$ obtain the optimal controller $u_k^*(t)$ on each finite interval $0 \leq t \leq k$, for $k = 1, 2, 3, \ldots$, with the cost functional $C_k(u) = \int_0^k f^0(x) + h^0(u)\,dt$. Write this minimal cost $C_k(u_k^*) = m_k$ and note that $m_k \leq m_{k+1} \leq M$, since $u_{k+1}^*(t)$ on $0 \leq t \leq k$ cannot have a smaller cost than $u_k^*(t)$. Since $\int_0^\infty |u_k^*(t)|^p\,dt \leq M/a$ (we can define $u_k^*(t) \equiv 0$ for $t > k$), we can select a subsequence $u_{k_i}^*(t)$, which converges weakly to a limit $u^*(t)$ on each compact time interval. For each finite $T > 0$,

$$\int_0^T f^0(x^*(t)) + h^0(u^*(t))\,dt$$
$$\leq \liminf_{k_i \to \infty} \int_0^T f^0(x_{k_i}^*(t)) + h^0(u_{k_i}^*(t))\,dt \leq \lim_{k_i \to \infty} m_{k_i} \leq M.$$

Therefore $u^*(t)$ is an admissible controller with finite cost $C(u^*) = m \leq M$. We next show that $m = \lim_{k \to \infty} m_k$ and that $u^*(t)$ is the unique optimal controller on $0 \leq t < \infty$.

Since $u_{k_i}^*(t)$ converges weakly to $u^*(t)$ on a finite interval $0 \leq t \leq T$,

$$\int_0^T f^0(x^*(t)) + h^0(u^*(t))\,dt \leq \liminf_{k_i \to \infty} m_{k_i},$$

and so

$$m \leq \lim_{k \to \infty} m_k ,$$

But no admissible controller on $0 \leq t < \infty$ has a cost less than any m_k since this would contradict the optimality of $u_k^*(t)$ on $0 \leq t \leq k$. Thus $m = \lim_{k \to \infty} m_k$, and $u^*(t)$ on $0 \leq t < \infty$ is an optimal controller achieving the minimal cost m.

Let $\bar{u}^*(t)$ be another optimal controller, which differs from $u^*(t)$ on a positive duration in $0 \leq t < \infty$. Consider $\bar{\bar{u}}^*(t) = \frac{1}{2}[u^*(t) + \bar{u}^*(t)]$ on $0 \leq t < \infty$. Because of the strict convexity of $h^0(u)$ it is easy to show that

$$C(\bar{\bar{u}}^*) < \tfrac{1}{2}[C(u^*) + C(\bar{u}^*)] = m,$$

which is impossible. Thus $u^*(t)$ is the unique (almost everywhere) optimal controller.

We next show that $u^*(t)$ satisfies the maximal principle with $\hat{\eta}^*(t) = (\eta_0^*, \eta^*(t))$, the adjoint response that occurs as the limit of the appropriate adjoint responses for the controllers $u_{k_i}^*(t)$ on finite time intervals. For each k_i let $\hat{\eta}_{k_i}(t) = (\eta_{ok_i}, \eta_{k_i}(t))$ be an adjoint response to $u_{k_i}^*(t)$ on $0 \leq t \leq k_i$, where $\hat{\eta}_{k_i}(0)$ is a unit vector, $\eta_{ok_i} < 0$, and $\eta_{k_i}(k_i) = 0$, and

$$\eta_{0k_i}\, h^0(u_{k_i}^*(t)) + \eta_{k_i}(t)B\, u_{k_i}^*(t) = \max_u\, [\eta_{0k_i}\, h^0(u) + \eta_{k_i}(t)Bu].$$

Select a further subsequence, still called k_i, such that

$$\lim_{k_i \to \infty} \hat{\eta}_{k_i}(0) = \hat{\eta}^*(0)$$

and define $\hat{\eta}^*(t)$ as the adjoint response to $u^*(t)$ with this given initial unit vector.

Suppose $u^*(t)$ fails to satisfy the maximal principle with $\hat{\eta}^*(t)$ on $0 \leq t < \infty$. Then, for some finite $T > 0$ and $\delta > 0$, we have

$$\eta_0^*\, h^0(u^*(t)) + \eta^*(t)B\, u^*(t) + 2\delta < \eta_0^*\, h^0(\tilde{u}(t)) + \eta^*(t)B\, \tilde{u}(t),$$

for some controller $\tilde{u}(t)$ on a compact subset Δ of duration $\delta > 0$ in $0 \leq t \leq T$ [we can assume that both $u^*(t)$ and $\tilde{u}(t)$ are continuous and bounded on Δ]. Hence for sufficiently large k_i and for t in Δ we have

$$\eta_{0k_i}\, h^0(u^*(t)) + \eta_{k_i}(t)B\, u^*(t) + \delta < \eta_{ok_i}\, h^0(u_{k_i}^*(t)) + \eta_{k_i}(t)B\, u_{k_i}^*(t).$$

This obtains because $\hat{\eta}_{k_i}(t)$ uniformly approximates $\hat{\eta}^*(t)$ on $0 \leq t \leq T$ and $u_{k_i}^*(t)$ satisfies the maximal principle with $\hat{\eta}_{k_i}(t)$. Therefore, as in the calculations of Theorem 9,

$$\int_0^{k_i} f^0(x^*(t)) + h^0(u^*(t))\, dt - \delta^2 > \int_0^{k_i} f^0(x_{k_i}^*(t)) + h^0(u_{k_i}^*(t))\, dt = m_{k_i}.$$

But then $C(u^*) \geq \delta^2 + m$, which is impossible. Therefore $u^*(t)$ does satisfy the maximal principle with $\hat{\eta}^*(t)$ on $0 \leq t < \infty$.

We now show that $\eta_0^* < 0$ and $\hat{\eta}^*(\infty) = 0$. Clearly $\eta_0^* < 0$, for otherwise $\eta_0^* = 0$ and the maximal principle would involve a linear homogeneous condition which could not be fulfilled. We note that $\eta^*(t)$ satisfies the linear differential system

$$\dot{\eta} = -\eta_0^* \frac{\partial f^0}{\partial x}(x^*(t)) - \eta A.$$

Since $x^*(\infty) = 0$ (the response to any admissible controller must tend to the origin, see Exercise 13 of Section 3.3), and $f^0(x) = 0$ if and only if $x = 0$, we note that $\lim_{t\to\infty} (\partial f^0/\partial x)(x^*(t)) = 0$. In fact, a slight refinement of this argument shows that for each $\epsilon > 0$ there exists a time $T > 0$ such that $|(\partial f^0/\partial x)(x_{k_i}^*(t))| < \epsilon$ and $|(\partial f^0/\partial x)(x^*(t))| < \epsilon$ for all $t > T$ and all sufficiently large k_i. Also we recall that

$$\lim_{k_i\to\infty} \eta_{k_i}(t) = \eta^*(t)$$

uniformly on compact intervals, and furthermore $\eta_{k_i}(k_i) = 0$. Using the maximal principle we find that $u_{k_i}^*(t)$ converges to $u^*(t)$, $x_{k_i}^*(t)$ converges to $x^*(t)$, and hence $(\partial f^0/\partial x)(x_{k_i}^*(t))$ converges to $(\partial f^0/\partial x)(x^*(t))$, all uniformly on every compact time interval.

Now the variation of parameters formula yields

$$\eta^*(t) = \eta^*(0)e^{-At} + \int_0^t -\eta_0^* \frac{\partial f^0}{\partial x}(x^*(s))e^{-A(t-s)}\, ds.$$

If every eigenvalue of A has positive real part $> \lambda > 0$, then $|e^{-At}| \leq C_1 e^{-\lambda t}$ on $0 \leq t < \infty$, for a constant C_1. Using $(\partial f^0/\partial x)(x^*(t)) \to 0$, it is easy to prove that $\eta^*(\infty) = 0$ and we omit the details.

After a linear change of variables of η, we can suppose that

$$A = \begin{pmatrix} A_+ & 0 \\ 0 & A_- \end{pmatrix},$$

where every eigenvalue of A_+ has positive real part (and hence yields components of $\eta^*(t)$ that tend to zero as $t \to \infty$), and all eigenvalues of A_- have negative real parts. Using this partitioning of the components of η, we note that it is sufficient to prove that $\eta^*(\infty) = 0$ when $A = A_-$, which we henceforth assume.

Now

$$\eta_{k_i}(k_i) = \left[\eta_{k_i}(0) + \int_0^{k_i} -\eta_{0k_i} \frac{\partial f^0}{\partial x}(x_{k_i}^*(s))e^{As}\, ds\right] e^{-Ak_i} = 0,$$

and so $\eta_{k_i}(0) = \int_0^{k_i} \eta_{ok_i} \frac{\partial f^0}{\partial x}(x^*_{k_i}(s))e^{As}\,ds$. We next prove that

$$\eta^*(0) = \int_0^\infty \eta_0^* \frac{\partial f^0}{\partial x}(x^*(s))e^{As}\,ds.$$

Compute

$$\eta^*(0) = \lim_{k_i\to\infty} \eta_{k_i}(0) = \lim_{k_i\to\infty} \int_0^{k_i} \eta_{0k_i} \frac{\partial f^0}{\partial x}(x^*_{k_i}(s))e^{As}\,ds.$$

For each $\epsilon > 0$ there is a finite $T > 0$ such that $|(\partial f^0/\partial x)(x^*_{k_i}(t))| < \epsilon$ and $|(\partial f^0/\partial x)(x^*(t))| < \epsilon$ for $t > T$ and all large k_i. Then, for large k_i,

$$\left| \int_0^\infty \eta_0 \frac{\partial f^0}{\partial x}(x^*(s))e^{As}\,ds - \int_0^{k_i} \eta_{0k_i} \frac{\partial f^0}{\partial x}(x^*_{k_i}(s))e^{As}\,ds \right|$$

$$\leq \int_0^T \left| \eta_0^* \frac{\partial f^0}{\partial x}(x^*(s)) - \eta_{0k_i} \frac{\partial f^0}{\partial x}(x^*_{k_i}(s)) \right| |e^{As}|\,ds$$

$$+ \int_T^{k_i} \left| \eta_0^* \frac{\partial f^0}{\partial x}(x^*) - \eta_{0k_i} \frac{\partial f^0}{\partial x}(x^*_{k_i}) \right| |e^{As}|\,ds$$

$$+ \int_{k_i}^\infty \left| \eta_0^* \frac{\partial f^0}{\partial x}(x^*(s))e^{As} \right| ds \leq \epsilon + 3\epsilon \int_T^\infty |e^{As}|\,ds.$$

Therefore

$$\eta^*(0) = \int_0^\infty \eta_0^* \frac{\partial f^0}{\partial x}(x^*(s))e^{As}\,ds,$$

and so

$$\eta^*(t) = \int_t^\infty \eta_0^* \frac{\partial f^0}{\partial x}(x^*(s))e^{A(s-t)}\,ds.$$

Thus for $t > T$ we have

$$|\eta^*(t)| \leq \epsilon \int_t^\infty |e^{A(s-t)}|\,ds = \epsilon \int_0^\infty |e^{A\xi}|\,d\xi.$$

Therefore $\eta^*(\infty) = 0$, as required.

Finally we prove that an admissible controller $\bar{u}(t)$ with response $\bar{x}(t)$ and $(\bar{\eta}_0, \bar{\eta}(t))$ on $0 \leq t < \infty$, such that the maximal principle obtains and $\bar{\eta}_0 < 0$, $\bar{\eta}(\infty) = 0$, is the unique optimal controller. Let $u(t)$ be any admissible controller with response $\hat{\omega}(t) = (\omega^0(t), \omega(t))$. Note $\bar{x}(\infty) = \omega(\infty) = 0$ and use the calculations in Theorem 9 to show that

$$\bar{\eta}_0\, \bar{x}^0(T) + \bar{\eta}(T)\,\bar{x}(T) \geq \bar{\eta}_0\omega^0(T) + \bar{\eta}(T)\,\omega(T)$$

for each finite $T > 0$. Since each of these terms has a limit as $T \to \infty$, and $\bar{\eta}_0 < 0$, we find

$$C(\bar{u}) \leq C(u).$$

Thus $\bar{u}(t)$ is the required optimal controller. Q.E.D.

Example. Consider the regulator problem defined by the scalar control equation

$$\dot{x} = u$$

with the cost $C(u) = \int_0^\infty [x(s)^4/4 + u(s)^2/2]\, ds$. In this example $u^*(t, \eta) = \eta$, as determined by the maximal principle, and the differential system is

$$\dot{x} = \eta, \qquad \dot{\eta} = x^3,$$

If $x_0 = 0$ take $u^*(t) \equiv 0$. If $x_0 < 0$, then $x^*(t) = [1/x_0 - t/\sqrt{2}]^{-1}$, and $u^*(t) = \eta^*(t) = 2^{-1/2}[1/x_0 - t/\sqrt{2}]^{-2}$. If $x_0 > 0$, then $x^*(t) = [1/x_0 + t/\sqrt{2}]^{-1}$ and $u^*(t) = -2^{-1/2}[1/x_0 + t/\sqrt{2}]^{-2}$.

It is often important to compute u^* as a feedback control determined by the state x. To determine η as a function of x we integrate

$$\frac{d\eta}{dx} = \frac{x^3}{\eta} \quad \text{so} \quad \eta^2 = \frac{x^4}{2}.$$

Thus

$$u^*(x) = \begin{cases} \dfrac{-x^2}{\sqrt{2}} & \text{when} \quad x > 0 \\[2ex] \dfrac{x^2}{\sqrt{2}} & \text{when} \quad x < 0 \end{cases}$$

Remarks on Integral Restraints. Consider the linear process in R^n

$$(\mathfrak{L}) \qquad \dot{x} = A(t)x + B(t)u,$$

which we assume is controllable on each interval $t_0 \leq t \leq T < \infty$. We wish to steer the given initial state x_0 at t_0 to a target point $x_1 \neq x_0$ in a minimal time T. We restrict the controllers $u(t)$ on various intervals $t_0 \leq t \leq T$ by the integral restraint

$$C_0(u) = \int_{t_0}^{T} f^0(t, x) + h^0(t, u)\, dt \leq 1.$$

Here $h^0(t, u)$ is strictly convex for each $t \geq t_0$ and the other standard continuity and convexity conditions obtain. We assume that there exists some controller $u(t)$ steering $x(t_0) = x_0$ to $x(T) = x_1$ with $C_0(u) \leq 1$.

Let $u^{(k)}(t)$ on $t_0 \leq t \leq t^{(k)}$ be a sequence of such controllers with $t^{(k)}$ decreasing to the limit T^*. Extend each $u^{(k)}(t) \equiv 0$ on $t^{(k)} \leq t \leq T^* + 1$. From the Hölder inequality

$$\|u\|_1 \leq \|u\|_p \cdot |T - t_0|^{1-1/p} \leq \frac{1}{a} C_0(u)\, |T - t_0|^{1-1/p}$$

we can conclude that $T^* > t_0$ and that $\int_{t_0}^{T^*+1} |u^{(k)}(t)|^p$ is uniformly bounded.

Since $p > 1$ we can select a subsequence, still called $u^{(k)}(t)$, which converges weakly to $u^*(t)$ on $t_0 \leq t \leq T^* + 1$. Clearly $u^*(t) \equiv 0$ on $T^* < t \leq T^* + 1$. It is easy to compute, for the corresponding responses,

$$\lim_{k \to \infty} x^{(k)}(t) = x^*(t) \quad \text{on} \quad t_0 \leq t \leq T^* + 1$$

and also

$$x^*(T^*) = x_1 \quad \text{and} \quad C_0(u^*) \leq 1 \quad \text{on} \quad t_0 \leq t \leq T^*.$$

Therefore the optimal controller $u^*(t)$ exists.

Now consider the set of attainability $\hat{K}(x_0, T^*)$ in R^{n+1}. If $C_0(u^*) < 1$ then there exist $(n + 1)$ controllers $u_1(t), \ldots, u_{n+1}(t)$ on $t_0 \leq t \leq T^*$, which steer x_0 to the vertices of a simplex enclosing x_1, and $C_0(u_i) < 1$ for $i = 1, 2, \ldots, n + 1$. Take $\epsilon > 0$ such that $u_i(t)$ on $t_0 \leq t \leq T^* - \epsilon$ steers x_0 to the vertices of a simplex enclosing x_1, and still costs less than 1. By taking convex combinations of $u_1(t), \ldots, u_{n+1}(t)$ we can construct a controller steering x_0 to x_1 in time $T^* - \epsilon$ with cost less than 1. But this contradicts the optimality of T^*. Therefore $C_0(u^*) = 1$. Furthermore $u^*(t)$ minimizes the cost $C_0(u)$ among all controllers $u(t)$ on $t_0 \leq t \leq T^*$, which steer x_0 to x_1.

Thus the optimal time T^* is the minimal $T > t_0$ for which there exists a solution $x^*(t)$, $\eta^*(t)$ of

$$\dot{x} = A(t)x + B(t)\, u^*(t, \eta)$$

$$\dot{\eta} = \frac{\partial f^0}{\partial x}(t, x) - \eta\, A(t)$$

with $x(t_0) = x_0$, $x(T) = x_1$, and $C_0(u^*) = 1$. Here $u^*(t) = u^*(t, \eta^*(t))$ is the optimal controller. As an illustration of these ideas consider the following problem.

***Example* 1.** We seek to stop a trolley that moves along a smooth track according to the control process

$$\ddot{x} = u,$$

with the energy restraint $\int_0^T u(t)^2\, dt \leq 1$. We wish to steer the initial state $x_0 = 0, \dot{x}_0 = 3$ to the target $(0, 0)$ in minimal time $T^* > 0$. The appropriate differential system for consideration is

$$\dot{x}^1 = x^2, \qquad \dot{x}^2 = \tfrac{1}{2}\eta_2, \qquad \dot{\eta}_1 = 0, \qquad \dot{\eta}_2 = -\eta_1.$$

The solution with $x^1(0) = 0$, $x^2(0) = -3$, $\eta_1(0) = \eta_{10}$, $\eta_2(0) = \eta_{20}$ is

$$x^1(t) = \frac{1}{2}\left[-\frac{t^3}{6}\eta_{10} + \frac{t^2}{2}\eta_{20} - 3t\right], \qquad \eta_1(t) = \eta_{10}$$

$$x^2(t) = \frac{1}{2}\left[-\frac{t^2}{2}\eta_{10} + \eta_{20}t - 3\right], \qquad \eta_2(t) = -\eta_{10}t + \eta_{20}.$$

The condition $x^1(T) = x^2(T) = 0$ yields

$$\eta_{10} = \frac{18}{T^2}, \qquad \eta_{20} = \frac{12}{T} \quad \text{so} \quad u(t) = -\frac{9}{T^2}t + \frac{6}{T} \quad \text{on} \quad 0 \le t \le T.$$

The restraint $\int_0^T u(t)^2\,dt = 1$ determines the minimal optimal time of $T^* = 9$.

***Example* 2.** We consider a minimal time-optimal problem in which the constraint is expressed by prescribing the average energy $\alpha^2 > 0$ available for the control, that is,

$$\int_0^T u(t)^2\,dt \le \alpha^2 T.$$

The theory of linear control processes with such constraints is quite similar to the above discussion preceding Example 1. We illustrate the theory with the control process $\ddot{x} = u$ or

$$\dot{x}^1 = x^2, \qquad \dot{x}^2 = u$$

where we wish to steer $(x_0^{\,1}, x_0^{\,2}) = (0, -3)$ to $(0, 0)$ in a minimal $T^* > 0$, satisfying the average energy constraint. Thereby we seek the minimal $T > 0$ for which there is a solution of

$$\dot{x}^1 = x^2, \qquad \dot{x}^2 = \tfrac{1}{2}\eta_2, \qquad \dot{\eta}_1 = 0, \qquad \dot{\eta}_2 = -\eta_1$$

with $x^1(0) = 0$, $x^2(0) = -3$, $x^1(T) = 0$, $x^2(T) = 0$, and $\int_0^T u^2\,dt = \alpha^2 T$, where $u(t) = \frac{1}{2}\eta_2(t)$. The requirement

$$\int_0^T \left(-\frac{9}{T^2}t + \frac{6}{T}\right)^2 dt = \alpha^2 T$$

defines the optimal time $T^* = 3/\alpha$ and $u^*(t) = -9t/T^{*2} + 6/T^*$.

These two examples, and Exercise 3 below illustrate *free-time* control problems. A common technique is to reduce the unspecified free-time T to a fixed-time T^* for an equivalent optimal-control problem. In fact, one can often assume $T^* = 1$ after a change of scale, which must then be determined in solving the given problem.

EXERCISES

1. Use the scalar control process $\dot{x} = u$ and compute the optimal controller $u^*(t)$ for

   (a) $C(u) = x(1) + \frac{1}{4}\int_0^1 (x^4 + u^4)\,dt$ with $x(0) = -3^{1/4}[\exp 3^{-1/4}]^{-1}$;

   (b) $C(u) = \frac{1}{4}\int_0^\infty (x^4 + u^4)\,dt$ with $x(0) = 1$.

2. Consider the damped linear oscillator $\ddot{x} + \dot{x} + x = u$ and control from $x_0 = 0$, $\dot{x}_0 = 0$ to $(1, 1)$ on $0 \leq t \leq 2$. Compute the minimal cost $C(u) = \int_0^2 u(t)^2\,dt$.
3. Use the control process $\ddot{x} = u$ to steer from $x_0 = 1$, $\dot{x}_0 = 0$ to $(0, 0)$ on $0 \leq t \leq 1$ so as to minimize the cost $C(u) = \sup |u(t)|$. Show that the problem can be reduced to a minimal time $t^* = 1$ regulation of $(1/k, 0)$ to $(0, 0)$ under the restraint $|u| \leq 1$. Discuss the solution of the problem by means of the techniques of the switching locus.
4. Show that for any initial condition $(x^1(0), x^2(0))$ in the set $S = \{x^1, x^2 \mid x^1 = 0, |x^2| \leq 1\}$ there is a response for the process

$$\dot{x}^1 = x^2 + u$$

$$\dot{x}^2 = -x^1 - x^2 + u, \qquad |u| \leq 1,$$

which attains the minimum $C(u) = \int_0^T x^1(t)^2\,dt = 0$ for each $T > 0$. Find this optimal controller $u^*(t)$.
5. In the control process

$$\dot{x}^1 = x^2, \qquad \dot{x}^2 = -x^1 - x^2 + (u)^3$$

the control enters the system nonlinearly. Let the cost functional be $C(u) = \int_0^1 u^4\,dt$ and discuss the possibility of reducing the optimal-control problem to the form treated in Theorem 10.
6. Consider the scalar control process $\dot{x} = x + u$ with cost $C(u) = \int_0^1 |u(t)|\,dt$, with $x(0) = 0$. Show that the set of attainability $\hat{K}$, initiating at $x(0) = 0$, is not closed in R^2. [Hint: if $u(t)$ steers $x(0) = 0$ to $x(1) = e$, then $\int_0^1 e^{-t}\,u(t)\,dt = 1$ and so $\int_0^1 |u(t)| > 1$. But define $u_\epsilon(t) = (1 - e^{-\epsilon})^{-1}$ on $0 \leq t \leq \epsilon$ and $u_\epsilon(t) = 0$ for $\epsilon < t \leq 1$, and compute that $\lim_{\epsilon \to 0} C(u_\epsilon) = 1$. Thus there is no optimal controller steering $x(0) = 0$ to $x(1) = e$.]
7. The dynamical equations for a projectile in a planar linear central force field are

$$\ddot{r} = -r + r\dot{\theta}^2 + u$$

$$r\ddot{\theta} = -2\dot{r}\dot{\theta} \qquad + v,$$

where u and v are the radial and transversal components of the controlling force (per unit mass). We seek to transfer the projectile from the circular orbit $r = 1$, $\dot{\theta} = 1$ to the concentric orbit $r = 2$, $\dot{\theta} = 1$

while keeping $\dot{\theta}(t) \equiv 1$. If the cost of control is $C(u) = \int_0^1 (u^2 + v^2)\,dt$ on $0 \leq t \leq 1$, compute the optimal controller.

8. If $F(x)$ is in C^1 in R^n show that the following convexity hypotheses are equivalent:

(a) $F(\lambda x_1 + (1 - \lambda)x_2) \leq \lambda\, F(x_1) + (1 - \lambda)\, F(x_2),\ 0 \leq \lambda \leq 1$

(b) $F(x_1) - F(x_2) \geq \dfrac{\partial F}{\partial x}(x_2)(x_1 - x_2)$;

(c) the set $x^0 + F(x) \leq 0$ is convex in R^{n+1}.

9. Prove that $h^0(u) = |u^1|^p + |u^2|^p + \cdots + |u^m|^p$ in R^m is convex for $1 \leq p < \infty$ and strictly convex for $1 < p < \infty$.

10. Prove that the lower boundary of $\hat{K}$, as discussed in Theorem 8, is a continuous hypersurface over R^n.

11. Consider the nonlinear control process in R^n

$$\dot{x} = A(t)x + h(t, u), \qquad x(t_0) = x_0$$

with cost defined by $C_0(u) = x^0(T)$, where

$$\dot{x}^0 = f^0(t, x) + h^0(t, u) \quad \text{and} \quad x^0(t_0) = 0.$$

Let $A(t)$, $h(t, u)$, $f^0(t, x)$, $h^0(t, u)$, and $(\partial f^0/\partial x)(t, x)$ be continuous in all arguments. Assume that $f^0(t, x)$ is convex for each fixed t, and consider controllers $u(t)$ on the fixed interval $t_0 \leq t \leq T$.

Assume that $u^*(t)$ is an admissible control with response $x^*(t)$ satisfying the maximal principle

$$-h^0(t, u^*) + \eta(t)\, h(t, u^*) = \max_u\, [-h^0(t, u) + \eta(t)\, h(t, u)]$$

for almost all t, where $\eta(t)$ is a solution of

$$\dot{\eta} = \frac{\partial f^0}{\partial x}(t, x^*(t)) - \eta\, A(t),$$

and $\eta(T) = 0$. Prove that $u^*(t)$ is an optimal controller.

12. Consider the linear control process in R^n

$$(\mathfrak{L}) \qquad \dot{x} = A(t)x + B(t)u, \qquad x(t_0) = x_0$$

with cost functional $C_0(u) = x^0(T)$ where

$$\dot{x}^0 = f^0(t, x) + h^0(t, u), \qquad x^0(t_0) = 0,$$

as in Theorem 9. Assume that there exist admissible controllers $u(t)$ on

$t_0 \leq t \leq T$ with responses $x(t)$ lying within a given closed convex set $\Delta \subset R^n$, and we seek an optimal controller within these.

To solve this *bounded phase coordinate* problem let $F(x)$ be a continuous convex function in R^n with $F(x) = 0$ in Δ and $F(x) > 0$ outside Δ. Consider the modified cost

$$C_\lambda(u) = \int_{t_0}^{T} [f^0(t, x) + h^0(t, u) + \lambda F(x)]\, dt$$

for large constants $\lambda > 0$. Let $u_\lambda^*(t)$ be an optimal controller (without reference to Δ) for each $\lambda > 0$, and assume that

$$\lim_{k \to \infty} u_{\lambda_k}^*(t) = u^*(t),$$

for a subsequence converging in $L_1(t_0, T)$. Prove that $u^*(t)$ is an admissible controller, that is $C_0(u^*) < \infty$ and $x^*(t) \subset \Delta$. Prove that $u^*(t)$ is an optimal controller for the given bounded phase coordinate problem.

13. Consider the linear control problem

$$(\mathcal{L}) \qquad \dot{x} = A(t)x + B(t)u$$

with cost $C(u) = g(x(T)) + \int_{t_0}^{T} [\alpha(t) + \beta(t)x + f^0(t, x) + h^0(t, u)]\, dt$ where $\alpha(t)$ and $\beta(t)$ are continuous on $t_0 \leq t \leq T$ and the other standard continuity and convexity hypotheses obtain (except that one need not assume that $\mathcal{L}$ is controllable). State and prove the theorems analogous to Theorems 8, 9, and 10.

14. Consider the autonomous control process in R^n

$$(\mathcal{L}) \qquad \dot{x} = Ax + Bu$$

with initial state x_0 at $t_0 = 0$ for each square integrable controller $u(t)$ on $0 \leq t \leq T$. Obtain the estimate

$$\max_{0 \leq t \leq T, 1 \leq i \leq n} |x^i(t)| \leq M\left(\int_0^T [u(s)]^2\, ds\right)^{1/2}$$

Here

$$M = \max_i \left(\int_0^T w_i(T, s)^2\, ds\right)^{1/2}$$

is a constant where

$$(w_{ij}(t, s)) = e^{A(t-s)}B$$

and

$$w_i^2 = \sum_{j=1}^{m} w_{ij}^2, \qquad u^2 = \sum_{j=1}^{m} |u^j|^2$$

3.5 INTEGRAL CONVEX COST CRITERIA WITH BOUNDED CONTROLLERS

We now treat the linear control process in R^n

$$(\mathfrak{L}) \qquad \dot{x} = A(t)x + B(t)u$$

with the integral cost functional

$$C(u) = g(x(T)) + \int_{t_0}^{T} f^0(t, x) + h^0(t, u)\, dt.$$

We make the following standing hypotheses: $A(t)$ and $B(t)$ are real continuous matrices on the fixed finite interval $t_0 \leq t \leq T$; and $g(x), f^0(t, x)$, and $h^0(t, u)$ are continuous for all values of their arguments for $x \in R^n$ and $u \in R^m$. Also $f^0(t, x)$ and $h^0(t, u)$ are convex functions for each fixed value of $t_0 \leq t \leq T$. In addition to these hypotheses, which also held in Section 3.4, we further assume that each controller $u(t)$ on $t_0 \leq t \leq T$ lies in a given compact convex restraint set $\Omega \subset R^m$. This restraint $u(t) \subset \Omega$ enables us to eliminate any need for positivity or growth bounds on the functions $f^0(t, x)$ and $h^0(t, u)$.

For simplicity of exposition we shall also demand that $\{\mathfrak{L}, \Omega, x_0, t_0, T\}$ is a normal problem. Hence the set of attainability $K(T)$ in R^n is a strictly convex, compact set with nonempty interior (assuming that Ω contains more than a single point—see Theorem 3 of Chapter 2). Thus $\mathfrak{L}$ is controllable and, furthermore, each boundary point of $K(T)$ is reached by means of a unique extremal controller.

We follow the methods of Section 3.4. That is, consider the primary case where $g(x) \equiv 0$ and consider the set of attainability $\hat{K} \subset R^{n+1}$ of all endpoints $\hat{x}(T)$ of responses $\hat{x}(t) = (x^0(t), x(t))$ initiating at $\hat{x}(t_0) = (0, x_0)$. Here $\hat{x}(t)$ is the solution of the differential system

$$\dot{x} = A(t)x + B(t)\,u(t)$$

$$\dot{x}^0 = f^0(t, x) + h^0(t, u)$$

for any measurable controller $u(t) \subset \Omega$ on $t_0 \leq t \leq T$. Thus $x^0(T) = C_0(u)$ and $x(T)$ is determined by the variations of parameter formula

$$x(t) = \Phi(t)x_0 + \Phi(t) \int_{t_0}^{t} \Phi(s)^{-1} B(s)\, u(s)\, ds.$$

Because $u(t)$ lies in the compact restraint set Ω, the set of attainability $\hat{K}$ is bounded in R^{n+1}. The projection of $\hat{K}$ on the x-space R^n is just the convex domain $K(T)$, but the upper boundary of $\hat{K}$ might be very irregular. Since we seek the minimal cost controller, only the lower boundary of $\hat{K}$ is significant. We shall prove that this lower boundary is a convex hypersurface defined over $K(T)$.

Definition. Let $\hat{K} \subset R^{n+1}$ be the set of attainability for the control process

$$(\mathfrak{L}) \qquad \dot{x} = A(t)x + B(t)u$$

corresponding to the cost functional

$$C_0(u) = \int_{t_0}^{T} f^0(t, x) + h^0(t, u)\, dt,$$

and the convex compact restraint $\Omega \subset R^m$. Define $\hat{K}_v$ to be the vertical saturation of $\hat{K}$, that is, $\hat{K}_v$ consists of all points $(x^0, x) \in R^{n+1}$ for which there exists a point $(y^0, x) \in \hat{K}$ with $y^0 \leq x^0$. Then the lower boundary of $\hat{K}_v$ is clearly the lower boundary of $\hat{K}$ and a control, or corresponding response, leading to this lower boundary is called extremal.

THEOREM 12. *Consider the control process in* R^n

$$(\mathfrak{L}) \qquad \dot{x} = A(t)x + B(t)u$$

with cost functional

$$C_0(u) = \int_{t_0}^{T} f^0(t, x) + h^0(t, u)\, dt$$

and compact convex restraint $u(t) \subset \Omega \subset R^m$. *Let* $\hat{K} \subset R^{n+1}$ *be the set of attainability. Then its vertical saturation* $\hat{K}_v$ *is a closed convex set in* R^{n+1}. *Thus the lower boundary of* $\hat{K}_v$ *belongs to* $\hat{K}$ *and this consists of a convex hypersurface defined over* $K(T) \subset R^n$.

Proof. To show that $\hat{K}_v$ is closed consider a sequence of points $\hat{y}_k = (y_k^0, y_k)$ in $\hat{K}_v$ converging to $(\bar{y}^0, \bar{y})$ in R^{n+1}. Since $\hat{K}_v$ is the vertical saturation of $\hat{K}$, we can find a sequence of controllers $u^{(k)}(t)$ with responses $\hat{x}_k(t)$ such that $x_k(T) = y_k$ and $x_k^0(T) \leq y_k^0$. We can further suppose that (a subsequence) $u^{(k)}(t)$ converges weakly to an admissible controller $\bar{u}(t) \subset \Omega$, and the corresponding responses $x_k(t)$ converge to $\bar{x}(t)$, as in Chapter 2. From the inequality obtained in Theorem 8,

$$\bar{y}^0 \geq \liminf_{k\to\infty} x_k^0(T) \geq \bar{x}^0(T).$$

Hence the response $(\bar{x}^0(t), \bar{x}(t))$ for $\bar{u}(t)$ leads to the endpoint $(\bar{x}^0(T), \bar{y})$ in $\hat{K}$. Thus $(\bar{y}^0, \bar{y})$ lies in $\hat{K}_v$ and so $\hat{K}_v$ is closed in R^{n+1}.

In the above calculation suppose that $(\bar{y}^0, \bar{y})$ lies on the lower boundary of $\hat{K}$. Then $\bar{x}^0(T) = \bar{y}^0$ and $\bar{x}(T) = \bar{y}$ so that the control $\bar{u}(t)$ steers $(0, x_0)$ to $(\bar{y}^0, \bar{y})$. Hence the lower boundary of $\hat{K}$ belongs to $\hat{K}$.

The proof that the lower boundary of $\hat{K}$ is a convex hypersurface over $K(T)$, and hence that $\hat{K}_v$ is a convex set in R^{n+1}, follows the methods of Theorem 8 of this chapter. Q.E.D.

COROLLARY. *Consider the control process in* R^n

$$(\mathfrak{L}) \qquad \dot{x} = A(t)x + B(t)u$$

with cost functional

$$C(u) = g(x(T)) + \int_{t_0}^{T} f^0(t, x) + h^0(t, u)\, dt$$

and compact convex restraint $\Omega \subset R^m$. *Then there exists an optimal controller.*

Proof. We seek the minimum of the real continuous function $g(x) + x^0$ on the bounded set $\hat{K} \subset R^{n+1}$. Since $g(x) + x^0$ decreases monotonically in x^0 for each fixed x, the infimum of $g(x) + x^0$ is just the minimum of $g(x) + x^0$ on the lower boundary of $\hat{K}$. Using the inequality of the above theorem, we find that the required minimum is assumed. Q.E.D.

An optimal controller $u^*(t)$ for $\mathfrak{L}$, with cost $C(u) = g(x(T)) + C_0(u)$ and restraint Ω, must steer $(0, x_0)$ to the lower boundary of $\hat{K}$ and hence $u^*(t)$ must be extremal. As before, we shall characterize extremal controllers by the maximal principle. We henceforth assume that $(\partial f^0/\partial x)(t, x)$ is continuous and note that the convexity hypothesis implies

$$f^0(t, x) - f^0(t, \bar{x}) \geq \frac{\partial f^0}{\partial x}(t, \bar{x})(x - \bar{x}).$$

THEOREM 13. *Consider the normal control process in* R^n

$$(\mathfrak{L}) \qquad \dot{x} = A(t)x + B(t)u$$

with cost $C_0(u) = \int_{t_0}^{T} f^0(t, x) + h^0(t, u)\, dt$ *and compact convex restraint* $\Omega \subset R^m$. *A controller* $\bar{u}(t)$ *with response* $\bar{x}(t)$ *is extremal if and only if there exists a nonvanishing* $(n + 1)$*-row vector* $\hat{\eta}(t) = (\eta_0, \eta(t))$ *satisfying*

$$(\mathcal{A}) \qquad \dot{\eta}_0 = 0, \qquad \eta_0 \leq 0$$

$$\dot{\eta} = -\eta_0 \frac{\partial f^0}{\partial x}(t, \bar{x}(t)) - \eta\, A(t),$$

and satisfying the maximal principle

$$\eta_0\, h^0(t, \bar{u}(t)) + \eta(t)\, B(t)\, \bar{u}(t) = \max_{u \in \Omega} [\eta_0\, h^0(t, u) + \eta(t)\, B(t)u]$$

almost everywhere on $t_0 \leq t \leq T$.

Proof. Let $\bar{u}(t)$ with response $\hat{x}(t) = (\bar{x}^0(t), \bar{x}(t))$ and adjoint response $\hat{\eta}(t) = (\eta_0, \eta(t))$ satisfy the differential systems $\mathfrak{L}$ and $\mathcal{A}$ and the maximal principle. Then, just as in Theorem 9, we find that $\hat{\eta}(T)\,\hat{x}(T) \geq \hat{\eta}(T)\,\hat{\omega}(T)$ where $\hat{\omega}(t)$ is the response to any admissible controller $u(t)$. From this inequality it follows that $\hat{x}(T)$ lies on the lower boundary of $\hat{K}$ if $\eta_0 < 0$, and on the lateral boundary of $\hat{K}_v$ if $\eta_0 = 0$. But if $\eta_0 = 0$ the response $\bar{x}(t)$ of $\mathfrak{L}$ is extremal (in the sense of Chapter 2) and hence $\bar{x}(T)$ lies on the boundary of $K(T)$ in R^n. Moreover, since $\{\mathfrak{L}, \Omega, x_0, t_0, T\}$ is normal, $\bar{u}(t)$ is the only control steering x_0 to the boundary point $\bar{x}(T)$. Hence $\hat{x}(T) = (\bar{x}^0(T), \bar{x}(T))$ is the unique point of $\hat{K}$ above $\bar{x}(T)$. Thus $\hat{x}(T)$ lies on the lower boundary of $\hat{K}$ in all cases and so $\bar{u}(t)$ is extremal.

Conversely assume that $\bar{u}(t)$ is an extremal controller so that the corresponding response $\hat{x}(t) = (\bar{x}^0(t), \hat{x}(t))$ steers $(0, x_0)$ to $\hat{x}(T)$ on the lower boundary of $\hat{K}$. Let $\hat{\eta}(T) = (\eta_0, \eta(T))$ be an exterior normal to the convex set $\hat{K}_v$ at $\hat{x}(T)$. Clearly $\eta_0 \leq 0$ and $\eta_0 = 0$ just in case $\bar{x}(T)$ lies on the boundary of $K(T)$. Define $\hat{\eta}(t)$ as the solution of the adjoint system $\mathcal{A}$ with the given data $\hat{\eta}(T)$ at $t = T$. We must prove that, almost everywhere on $t_0 \leq t \leq T$,

$$\eta_0 h^0(t, \bar{u}(t)) + \eta(t)\, B(t)\, \bar{u}(t) = \max_{u \in \Omega} [\eta_0\, h^0(t, u) + \eta(t)\, B(t) u].$$

If $\eta_0 = 0$, then $\eta(T)$ is an exterior normal to $K(T)$ at $\bar{x}(T)$, and hence the maximal principle holds as in Chapter 2. If $\eta_0 < 0$, then the proof of the maximal principle is given in Theorem 9. Q.E.D.

COROLLARY. *Consider the normal control process in* R^n

$$(\mathfrak{L}) \qquad \dot{x} = A(t)x + B(t)u$$

with cost $C(u) = g(x(T)) + \int_{t_0}^{T} f^0(t, x) + h^0(t, u)\, dt$, *and the compact convex restraint* Ω *in* R^m. *Here* $g(x)$ *is convex and* $h^0(t, u)$ *is strictly convex for each* t. *Then any two extremal controllers steering* $(0, x_0)$ *to the same boundary point of* $\hat{K}$ *must coincide almost everywhere. Furthermore, there exists a unique optimal controller.*

Proof. Consider two extremal controllers $\bar{u}_1(t)$ and $\bar{u}_2(t)$ steering $(0, x_0)$ to the same point $\hat{x}(T)$ of the lower boundary of $\hat{K}$. If $(\eta_0 = 0, \eta(T))$ determines an exterior normal to $\hat{K}_v$ at $\hat{x}(T)$, the normality of $\{\mathfrak{L}, \Omega, x_0, t_0, T\}$ implies that $\bar{u}_1(t) = \bar{u}_2(t)$ almost everywhere. If $\eta_0 < 0$ for an exterior normal at $\hat{x}(T)$, then the proof of the corollary of Theorem 9 is valid so $\bar{u}_1(t) = \bar{u}_2(t)$.

The uniqueness of the optimal controller follows, as in Theorem 5, from the fact that $x^0 + g(x)$ assumes its minimum on $\hat{K}$ at just a single point. Q.E.D.

Just as in the discussion of Section 3.4 we can define the vector $u^*(t, \eta)$ by the maximum principle (for the case $\eta_0 = -1$)

$$-h^0(t, u^*) + \eta\, B(t)u^* = \max_{u \in \Omega} [-h^0(t, u) + \eta\, B(t)u].$$

If $\eta(t)$ is continuous, then $u^*(t) = u^*(t, \eta(t))$ is an admissible controller in Ω.

The next theorem shows how $u^*(t, \eta)$ can be interpreted as a determination of a feedback synthesis for an optimal controller.

THEOREM 14. *Consider the normal control process in* R^n

$$(\mathfrak{L}) \qquad \dot{x} = A(t)x + B(t)u$$

with cost functional

$$C(u) = g(x(T)) + \int_{t_0}^{T} f^0(t, x) + h^0(t, u)\, dt$$

and compact convex restraint $\Omega \subset R^m$. *Assume* $g(x) \in C^1$ *is convex in* R^n. *Then there exists a solution* $x^*(t)$, $\eta^*(t)$ *of the system*

$$\dot{x} = A(t)x + B(t)\, u^*(t, \eta)$$

$$\dot{\eta} = \frac{\partial f^0}{\partial x}(t, x) - \eta\, A(t)$$

with $x(t_0) = x_0$, $\eta(T) = -\text{grad}\, g(x(T))$. *Here* $u^*(t, \eta)$ *is defined by the maximal principle*

$$-h^0(t, u^*) + \eta\, B(t)u^* = \max_{u \in \Omega} [-h^0(t, u) + \eta\, B(t)u].$$

An optimal controller is $u^*(t) = u^*(t, \eta^*(t))$ *with the corresponding optimal response* $x^*(t)$.

If $h^0(t, u)$ *is strictly convex for each* t, *then the solution* $x^*(t)$, $\eta^*(t)$ *is unique, and* $u^*(t)$ *is the unique optimal controller.*

Proof. Among the family of hypersurfaces S_c: $x^0 + g(x) = c$ in R^{n+1} there is just one $c = m$ such that S_m is tangent to $\hat{K}_v$, and m is the optimal cost. Also S_m touches $\hat{K}_v$ at some point P, which lies on the lower boundary of $\hat{K}$. The tangent hyperplane to S_m is also a supporting hyperplane to $\hat{K}_v$ at P and hence we can let $\hat{\eta}^*(T) = (-1, \eta^*(T))$ be a normal to this hyperplane. Let $u^*(t)$ be an extremal controller steering $(0, x_0)$ to $P = \hat{x}^*(T)$ by the response $\hat{x}^*(t) = (x^{0*}(t), x^*(t))$. Let $\hat{\eta}^*(t) = (-1, \eta(t))$ be defined as a solution of

$$\dot{\eta} = \frac{\partial f^0}{\partial x}(t, x^*(t)) - \eta\, A(t)$$

with $\eta^*(T) = -\text{grad } g(x^*(T))$. By Theorem 13 we find that $u^*(t)$ satisfies the maximal principle with adjoint response $\hat{\eta}^*(t) = (-1, \eta^*(t))$, that is, $u^*(t) = u^*(t, \eta^*(t))$.

If $h^0(t, u)$ is strictly convex for each t, then S_m touches $\hat{K}$ at just the single point P. In this case $u^*(t)$, $x^*(t)$, and hence $\eta^*(t)$ are uniquely determined.

Q.E.D.

The following example shows how a minimal time-optimal problem, with integral constraints, can often be reduced to a minimal integral cost problem.

Example. Consider the normal control process in R^n

$$(\mathcal{L}) \qquad \dot{x} = A(t)x + B(t)u$$

with compact convex restraint set Ω: $|u^j| \leq 1$ for $j = 1, 2, \ldots, m$. We seek to steer the initial point x_0 at t_0 to $x_1 \neq x_0$ in a minimal time $T^* > t_0$. Also we impose the integral restraint

$$C_0(u) = (\|u\|_p)^p = \int_{t_0}^{T} |u^1|^p + \cdots + |u^m|^p \, dt \leq M,$$

for $1 \leq p < \infty$ and a given bound M.

For each $T > t_0$ consider the corresponding set of attainability $\hat{K}(T)$ consisting of endpoints $\hat{x}(T) = (x^0(T), x(T))$. Here $x(t)$ is the response of $\mathcal{L}$ to $u(t)$ on $t_0 \leq t \leq T$ and $x^0(T) = (\|u\|_p)^p$, where $u(t) \subset \Omega$ is the sole restraint. We must study the bounded set $\hat{K}(T) \cap [x^0 \leq M]$.

First we show that $\hat{K}(T)$ is compact, convex, and varies continuously with T. By the bang-bang principle every point in $K(T)$, the projection of $\hat{K}(T)$ on $x^0 = 0$, can be reached by a controller with $(\|u\|_p)^p = (mT)$. Thus the upper boundary of $\hat{K}(T)$ is the level hyperplane section $x^0 = (mT)$ above the convex domain $K(T)$ in R^n. The lower boundary of $\hat{K}(T)$ is a convex hypersurface over $K(T)$. These two boundaries meet over $\partial K(T)$.

Let $\bar{u}(t)$ on $t_0 \leq t \leq T$ steer $(0, x_0)$ to a point $(\bar{x}^0(T), \bar{x}(T))$ on the lower boundary of $\hat{K}(T)$. For each subinterval $t_0 \leq t \leq s$ let $v_s(t)$ be a bang-bang controller steering x_0 to $\bar{x}(s)$ at time $t = s$. Define, for each fixed s, the controller

$$u_s(t) = \begin{cases} v_s(t) & \text{on} \quad t_0 \leq t \leq s \\ \bar{u}(t) & \text{on} \quad s < t \leq T. \end{cases}$$

Then $u_s(t)$ steers x_0 to $\bar{x}(T)$ and $C_0(u_s)$ varies continuously with s from $C_0(u_0) = \bar{x}^0(T)$ to $C_0(v_T) = mT$. Therefore the set of attainability $\hat{K}(T)$ includes all points between its upper and lower boundaries. Therefore $\hat{K}(T)$ is a compact and convex. Since any controller $u(t)$ on $t_0 \leq t \leq T$ can be extended as $u(t) \equiv 0$ for $t > T$, it is easy to verify that $\hat{K}(T)$ varies continuously with T. The same holds for $\hat{K}(T) \cap [x^0 \leq M]$.

The minimal time T^* is the smallest $T > t_0$ for which $\hat{K}(T) \cap [x^0 \leq M]$ meets the vertical line $x = x_1$ in R^{n+1}. Thus, if there exists an admissible controller steering x_0 to x_1 in R^n, then there exists a minimal time-optimal controller $u^*(t)$ on $t_0 \leq t \leq T^*$. Moreover $u^*(t)$ can also be characterized as the controller on the fixed interval $t_0 \leq t \leq T^*$, which minimizes the cost $C_0(u) = (\|u\|_p)^p$, among all measurable controllers in Ω steering x_0 to x_1. If $p > 1$, then $u^*(t)$ is unique.

Thus T^* is characterized as the smallest $T > t_0$ for which there exists a solution $x^*(t)$, $\eta^*(t)$ satisfying

$$\dot{x} = A(t)x + B(t)\,u^*(t)$$

$$\dot{\eta} = -\eta\, A(t)$$

with $x(t_0) = x_0$, $x(t_1) = x_1$, and $C_0(u^*) \leq M$. Here the optimal control $u^*(t)$ maximizes

$$\eta_0[|u^1|^p + \cdots + |u^m|^p] + \eta(t)\, B(t)u$$

for all $u \subset \Omega$ and some constant $\eta_0 \leq 0$.

There are two cases of interest. If $M \geq mT^*$, then the integral restraint is superfluous and the problem reduces to a time-optimal control problem (take $\eta_0 = 0$) such as was treated in Chapter 2. If $M < mT^*$, then we can take $\eta_0 = -1$ and use $u^*(t) = u^*(t, \eta^*(t))$ in the maximal principle. In this case $C_0(u^*) = M$.

EXERCISES

1. Consider the scalar control process $\dot{x} = u$ with restraints $|u(t)| \leq 1$ and $\int_0^T u(t)^2\, dt \leq M$. Steer $x_0 = 0$ to $x_1 = 1$ in minimal time T^*. Compute the optimal controller $u^*(t)$ for each prescribed bound $M > 0$.
2. Consider the scalar control process $\dot{x} = u$ with the integral restraint $\int_0^1 u(t)^2\, dt \leq 4$ on the fixed interval $0 \leq t \leq 1$. We seek to steer $x_0 = 0$ to $x_1 = 1$ with a miminal cost $C(u) = \|u\|_\infty = \sup |u(t)|$. Find the optimal controller. [Hint: Find the smallest $k > 0$ for which there exists a control $u(t)$ on $0 \leq t \leq 1$, satisfying the two restraints $\int_0^1 u^2\, dt \leq 4$ and $|u(t)| \leq k$, steering x_0 to x_1].
3. Consider an autonomous process in R^n, normal for the restraint cube $|u^i| \leq 1$ in R^m,

   $$(\mathfrak{L}) \qquad \dot{x} = Ax + Bu$$

   with cost $C(u) = \int_0^\infty f^0(x) + h^0(u)\, dt$.

Assume that $f^0(x) \geq 0$ is convex and $f(x) = 0$ if and only if $x = 0$, $h^0(u) \geq 0$ is strictly convex, and $h^0(0) = 0$. Assume also that A is a stability matrix and prove that there exists a unique optimal controller $u^*(t)$ on $0 \leq t < \infty$. Prove also that an admissible controller $\bar{u}(t)$ on $0 \leq t < \infty$ is optimal if and only if it satisfies the maximal principle with some adjoint response $(\bar{\eta}_0, \bar{\eta}(t))$ such that $\bar{\eta}_0 < 0$ and $\bar{\eta}(\infty) = 0$. (See Theorem 11 of Section 3.4 above.)

CHAPTER 4

The Maximal Principle and the Existence of Optimal Controllers for Nonlinear Processes

In this chapter we consider the basic geometric properties of the set of attainability and prove the maximal principle—a response endpoint belongs to the boundary of the set of attainability only if a maximal condition obtains. The general results on the existence of optimal controllers with bounded restraints are given in the second section and the existence theorems for unbounded optimal controllers are treated in the third.

4.1 GEOMETRY OF THE SET OF ATTAINABILITY

Let us consider a nonlinear process defined by a differential system in R^n.

$$(\mathcal{S}) \qquad \dot{x} = f(x, t, u),$$

where f is in C^1 in R^{n+1+m}. The admissible controllers $u(t)$ on a specified finite interval $t_0 \leq t \leq T$ will constitute a certain family $\mathcal{F}$ of measurable m-vector functions. The initial point x_0 lies in a given compact initial set X_0 in R^n and we assume that each response $x(t, x_0, t_0) = x(t)$ for $u(t) \in \mathcal{F}$ exists on the interval $t_0 \leq t \leq T$.

For example, suppose for each $u(t) \in \mathcal{F}$ there is a bound

$$|x(t, x_0, t_0)| < b \quad \text{(where defined on } t_0 \leq t \leq T)$$

and also

$$|f(x, t, u(t))| + \left|\frac{\partial f}{\partial x}(x, t, u(t))\right| \leq m(t)$$

on $t_0 \leq t \leq T$, $|x| \leq b$, with $\int_{t_0}^{T} m(t)\, dt < \infty$. Then the unique (absolutely continuous) response $x(t, x_0, t_0)$ initiating at x_0 when $t = t_0$ is defined on the whole interval $t_0 \leq t \leq T$. In this case we say that the control $u(t)$ *admits a bound* for the response. If the bound b and the integrable function $m(t)$ can be chosen independently of the controller $u(t) \in \mathcal{F}$, then the problem $\{\mathcal{S}, x_0, \mathcal{F}, t_0, T\}$ has a uniform bound.

***Example* 1.** Consider the nonlinear process in R^n

$$(\mathcal{S}) \qquad \dot{x} = f(x, t, u)$$

with $f \in C^1$ in R^{n+1+m}. Let the family $\mathcal{F}$ of controllers be all measurable functions $u(t)$ on $t_0 \leq t \leq T$ with the restraint $u(t) \subset \Omega \subset R^m$ for a compact set Ω. Assume there exist positive constants A and B so

$$|f(x, t, u)| \leq A\,|x| + B$$

for $t_0 \leq t \leq T$, $u \in \Omega$, and $|x| \leq (B/A + |x_0|)e^{A(T-t_0)} - B/A$. Then every control $u(t) \in \mathcal{F}$ is admissible with a response $x(t)$ on $t_0 \leq t \leq T$. Moreover, the problem has a uniform bound. In fact, we compute

$$|x(t)| \leq |x_0| + \int_{t_0}^{t} |f(x(s), s, u(s))|\,ds \leq |x_0| + \int_{t_0}^{t} (A\,|x(s)| + B)\,ds$$

so

$$|x(t)| + \frac{B}{A} \leq \left(|x_0| + \frac{B}{A}\right) + \int_{t_0}^{t} A\left(|x(s)| + \frac{B}{A}\right) ds.$$

Then a standard inequality yields

$$|x(t)| + \frac{B}{A} \leq \left(|x_0| + \frac{B}{A}\right)e^{A(t-t_0)} \leq \left(|x_0| + \frac{B}{A}\right)e^{A(T-t_0)},$$

so the required bound on $|x(t)|$ is maintained. Thus $|f(x, t, u(t))|$ and $|\partial f/\partial x(x, t, u(t))|$ are bounded uniformly for all controllers $u(t) \in \mathcal{F}$.

Definition. Consider the nonlinear process in R^n

$$(\mathcal{S}) \qquad \dot{x} = f(x, t, u) \quad \text{in} \quad C^1 \quad \text{in} \quad R^{n+1+m}$$

with initial state x_0 at time t_0. Assume that the family $\mathcal{F}$ of admissible controllers $u(t)$ on $t_0 \leq t \leq T$ in R^m consists of certain measurable functions, for each of which the response $x(t) = x(t, x_0, t_0)$ is defined on $t_0 \leq t \leq T$. The set of attainability $K(x_0, t) = K(t)$, for each time on $t_0 \leq t \leq T$, consists of all the response endpoints $x(t)$ corresponding to all the controllers in $\mathcal{F}$.

In the control processes of Chapter 3 there was a bound on the response for each admissible control, yet there was no uniform bound and the set of attainability was unbounded. The next theorem analyses the behavior of $K(t)$ when the process has a uniform bound.

THEOREM 1. *Consider the nonlinear process in R^n*

$$(\mathcal{S}) \qquad \dot{x} = f(x, t, u) \quad \text{in} \quad C^1 \quad \text{in} \quad R^{n+1+m}$$

with initial state x_0 at time t_0 and admissible control family $\mathcal{F}$ on $t_0 \leq t \leq T$. Assume the process $\{\mathcal{S}, x_0, \mathcal{F}, t_0, T\}$ has a uniform bound. Then $\overline{K(t)}$ is a compact, continuously varying set in R^n for $t_0 \leq t \leq T$.

Proof. From the uniform bound each response satisfies

$$|x(t)| \leq |x_0| + \int_{t_0}^{T} m(t)\, dt,$$

where $m(t)$ is an integrable function so that

$$|f(x(t), t, u(t))| + \left| \frac{\partial f}{\partial x}(x(t), t, u(t)) \right| \leq m(t).$$

Thus the set of attainability $K(t)$ lies in some bounded region of R^n and $\overline{K(t)}$ is compact.

To show that $\overline{K(t)}$ varies continuously pick a point $P_1 \in \overline{K(t_1)}$ and $\epsilon > 0$. Then there is a response $x(t)$ such that $|x(t_1) - P_1| < \epsilon/2$ and

$$|x(t) - x(t_1)| \leq \int_{t_1}^{t} m(s)\, ds < \frac{\epsilon}{2}$$

for $|t - t_1| < \delta(\epsilon)$. Thus each point $P_1 \in \overline{K(t_1)}$ lies within an ϵ-distance of some point $x(t)$ in $K(t)$ for all $|t - t_1| < \delta(\epsilon)$. But similarly, each point of $K(t)$ lies within an ϵ-distance of $K(t_1)$, provided $\delta(\epsilon) > 0$ is sufficiently small. Thus

$$\text{dist}\,[\overline{K(t_1)}, \overline{K(t)}] < \epsilon \quad \text{whenever} \quad |t - t_1| < \delta(\epsilon),$$

and hence the function $t \to \overline{K(t)}$ is a continuous map of the real interval $t_0 \leq t \leq T$ into the metric space of nonempty compact subsets of R^n.

Q.E.D.

Remark. If the initial set X_0 is compact, let $K(X_0, t) = \bigcup_{x_0 \in X_0} K(x_0, t)$. Let us assume that the same uniform bound $m(t)$ holds for all controllers $u(t) \in \mathcal{F}$ and all initial points $x_0 \in X_0$. Then $\overline{K(X_0, t)}$ is clearly compact and varies continuously with time. Hence the sets

$$\bigcup_{t_0 \leq t \leq T} \overline{K(X_0, t)} \quad \text{in} \quad R^n$$

and

$$(t, \overline{K(X_0, t)}) \quad \text{in} \quad R^{n+1}$$

are also compact.

We shall now consider the control family $\mathcal{F}$ to be all measurable functions $u(t)$ on $t_0 \leq t \leq T$ with values $u(t) \subset \Omega$ for some compact restraint set in R^m. In this case, if $f(x, t, u) \in C^1$ in R^{n+1+m} and if there is a uniform bound $|x(t)| < b$ on all responses to $u(t) \in \mathcal{F}$, then there is a uniform bound on

$$|f(x(t), t, u(t))| + \left|\frac{\partial f}{\partial x}(x(t), t, u(t))\right|.$$

THEOREM 2. *Consider the nonlinear process in R^n*

$$(\mathcal{S}) \qquad \dot{x} = f(x, t, u) \quad \text{in} \quad C^1 \quad \text{in} \quad R^{n+1+m}$$

with initial state x_0 at time t_0. The admissible controllers are all measurable functions $u(t) \subset \Omega$ on $t_0 \leq t \leq T$, where Ω is a compact restraint set in R^m.

Assume

(a) $|x(t)| < b$, *there exists a uniform bound for all responses on $t_0 \leq t \leq T$.*
(b) $V(x, t) = \{f(x, t, u) \mid u \in \Omega\}$ *is convex for each fixed (x, t). The set $V(x, t)$ of velocity vectors at each (x, t) is hence compact and convex.*

Then the set of attainability $K(t)$ is compact and varies continuously with time on $t_0 \leq t \leq T$.

Proof. Under hypothesis (a) each response is defined on the interval $t_0 \leq t \leq T$ and $|f(x(t), t, u)| + |\partial f/\partial x(x(t), t, u)|$ is bounded. Thus $\overline{K(t)}$ is compact and varies continuously on $t_0 \leq t \leq T$. Under hypothesis (b), that the velocity set $V(x, t)$ is everywhere compact and convex, we shall prove that $K(t) = \overline{K(t)}$.

Consider responses $x_i(t)$ to controllers $u_i(t) \subset \Omega$ on $t_0 \leq t \leq T$. Then

$$x_i(t) = x_0 + \int_{t_0}^{t} f(x_i(s), s, u_i(s))\, ds$$

and

$$x_i(t) \in K(t) \quad \text{for} \quad i = 1, 2, 3, \ldots.$$

For each time t_1 on $t_0 \leq t \leq T$ we prove that any limit point $\bar{x}(t_1)$ of the sequence $\{x_i(t_1)\}$ lies in $K(t_1)$. Let us take a subsequence, still called $x_i(t)$ with $\lim_{i\to\infty} x_i(t_1) = \bar{x}(t_1)$.

Since $|f(x_i(t), t, u_i(t))| \leq m$, for some constant m, the sequence of indefinite integrals $\int_{t_0}^{t} f(x_i(s), s, u_i(s))\, ds$ forms a uniformly bounded and equicontinuous family of functions. By Ascoli's theorem a subsequence converges to some (Lipschitz) function. Simplifying the notation, we write

$$\lim_{i\to\infty}\int_{t_0}^{t} f(x_i(s), s, u_i(s))\, ds = \int_{t_0}^{t} \bar{\phi}(s)\, ds,$$

for an integrable function $\bar{\phi}(t)$. Thus

$$\lim_{i\to\infty}\int_{t_0}^{T} \chi_E(s) f(x_i(s), s, u_i(s))\, ds = \int_{t_0}^{T} \chi_E(s)\, \bar{\phi}(s)\, ds$$

where χ_E is the characteristic function of a subinterval E. But every measurable set can be approximated by a finite sum of disjoint open intervals, and so the above formula holds for an arbitrary measurable set E in $t_0 \le t \le T$. Hence, $f(x_i(t), t, u_i(t))$ converges weakly to $\bar{\phi}(t)$. If we define

$$\bar{x}(t) = x_0 + \int_{t_0}^{t} \bar{\phi}(s)\, ds,$$

then

$$\lim_{i\to\infty} x_i(t) = \bar{x}(t)$$

everywhere on $t_0 \le t \le T$. It remains to prove that $\bar{\phi}(t) = f(\bar{x}(t), t, \bar{u}(t))$ for some admissible controller $\bar{u}(t) \subset \Omega$.

We first show that $\bar{\phi}(t) \in V(\bar{x}(t), t)$ for almost all t. Suppose that $\bar{\phi}(t)$ lies outside the compact convex velocity set $V(\bar{x}(t), t)$ for some set W of positive measure on $t_0 \le t \le T$. Then at each $t \in W$ there is a hyperplane, even with rational unit normal, separating $\bar{\phi}(t)$ from $V(\bar{x}(t), t)$. Since the rationals form a denumerable set, there exists a constant unit row vector y such that

$$y\, \bar{\phi}(t) > \limsup_{i\to\infty} y\, f(\bar{x}(t), t, u_i(t)) = \limsup_{i\to\infty} y\, f(x_i(t), t, u_i(t))$$

for each t on a set W_1 of positive measure. Then

$$\int_{W_1} y\, \bar{\phi}(t)\, dt > \int_{W_1} \limsup_{i\to\infty} y\, f(x_i(t), t, u_i(t))\, dt$$

and, using Fatou's lemma on the Lebesgue integral,

$$\int_{W_1} y\, \bar{\phi}(t)\, dt > \limsup_{i\to\infty} \int_{W_1} y\, f(x_i(t), t, u_i(t))\, dt,$$

which contradicts the fact that $f(x_i(t), t, u_i(t))$ converges weakly to $\bar{\phi}(t)$. Thus $\bar{\phi}(t) \in V(\bar{x}(t), t)$ almost always.

We can redefine $\bar{\phi}(t)$ on a null set so that

$$\bar{\phi}(t) \in f(\bar{x}(t), t, \Omega) = V(\bar{x}(t), t)$$

for all t on the interval $t_0 \leq t \leq T$. Then, by Lemma 3A of Chapter 2, there exists a measurable m-vector $\bar{u}(t) \subset \Omega$ such that

$$\bar{\phi}(t) = f(\bar{x}(t), t, \bar{u}(t)) \quad \text{on} \quad t_0 \leq t \leq T.$$

Therefore the admissible controller $\bar{u}(t)$ produces the response $\bar{x}(t)$. In particular $\bar{x}(t_1) \in K(t_1)$. Hence $K(t_1)$ is compact and so $K(t) = \overline{K(t)}$ for all $t_0 \leq t \leq T$. Q.E.D.

The compactness (or at least closedness) of $K(t)$ is fundamental in the proof of a general existence theorem for optimal controllers. In Chapters 2 and 3 we have already noted the compactness of the set of attainability for various linear control processes. Later in this chapter we shall prove several existence theorems for optimal controllers for nonlinear processes under diverse conditions.

We next present several examples that illustrate the significance of the uniform bounds on responses and the compactness of the set of attainability. In some of these examples no optimal controller exists, that is, there do exist controllers $u \in \mathcal{F}$ with costs $C(u)$ decreasing towards a finite lower bound, but the greatest lower bound cannot be achieved by any admissible controller in the family $\mathcal{F}$.

***Example* 2.** Consider the system in R^4

$$\dot{x} = \sin 2\pi u, \qquad \dot{y} = \cos 2\pi u, \qquad \dot{z} = -1, \qquad \dot{w} = x^2 + y^2 + 1$$

with initial point $(0, 0, 1, 0)$ and control restraint $|u(t)| \leq 1$ on $0 \leq t \leq 1$. There exists a uniform bound for the responses

$$|x| + |y| + |z| + |w| \leq 1 + 1 + 1 + 3 = 6$$

and so $K(1)$ is bounded. We show that $K(1)$ is not closed. Choose controllers $u_l(t) = lt (\text{mod } 1)$ for $l = 1, 2, 3, \ldots$ so $\sin 2\pi u_l(t) = \sin 2\pi lt$ and $\cos 2\pi\, u_l(t) = \cos 2\pi lt$. For these controllers the corresponding responses are

$$x_l(t) = \frac{1 - \cos 2\pi lt}{2\pi l}, \qquad y_l(t) = \frac{\sin 2\pi lt}{2\pi l}, \qquad z_l(t) = 1 - t$$

and

$$w_l(t) = \int_0^t \left[\frac{1 - \cos 2\pi ls}{2\pi^2 l^2} + 1\right] ds.$$

The endpoints in $K(1)$ are $(0, 0, 0, (2\pi^2 l^2)^{-1} + 1)$. But for any admissible controller $u(t)$,

$$w(1) = \int_0^1 [x^2 + y^2 + 1]\, dt > 1.$$

Thus the point (0, 0, 0, 1) lies in $\overline{K(1)}$ but not in $K(1)$, and so $K(1)$ is not closed in R^4.

***Example* 3.** Consider the system in R^3

$$\dot{x} = \frac{\sin 2\pi u}{x^2 + y^2 + 1}, \qquad \dot{y} = \frac{\cos 2\pi u}{x^2 + y^2 + 1}, \qquad \dot{z} = \frac{-1}{x^2 + y^2 + 1}$$

with initial point (0, 0, 1) and control restraint $|u(t)| \leq 1$. We seek to steer the initial point (0, 0, 1) to the target (0, 0, 0) in minimal optimal time $t^* > 0$.

For each controller $u(t)$ and response $x(t)$, $y(t)$, $z(t)$ define a new independent variable τ by

$$\tau(t) = \int_0^t [x(s)^2 + y(s)^2 + 1]^{-1}\, ds.$$

Then define $u(\tau) = u(t(\tau))$ and compute

$$\frac{dx}{d\tau} = \sin 2\pi\, u(\tau), \qquad \frac{dy}{d\tau} = \cos 2\pi\, u(\tau), \qquad \frac{dz}{d\tau} = -1$$

and $dt/d\tau = x^2 + y^2 + 1$. Using the calculations of Example 2, we find that we can steer $x = y = t = 0$, $z = 1$ to the point $x = y = z = 0$ and $t > 1$. But the optimal time $t^* = 1$ cannot be attained, and so an optimal controller for this problem fails to exist.

***Example* 4.** Consider in R^2

$$\dot{x} = 1, \qquad \dot{y} = -xe^y u$$

with control restraint $0 \leq u(t) \leq 2$. The family $\mathcal{F}$ of all admissible controllers consists of all measurable $u(t)$ on $0 \leq t \leq 2$ that steer $(-1, 0)$ to $(1, 0)$. We seek to minimize the cost $C(u) = \int_0^2 (2 - y)\, dt = \int_{-1}^1 (2 - y)\, dx$.

For each response $x(t) = t - 1$, $y(t)$ we write $y(x) = y(t(x))$. Use $u(t) = 2$ to obtain $0 \leq y(x) \leq -\ln x^2$ for $x \neq 0$. But the curve $y = -\ln x^2$ is not bounded and so $K(2)$ is closed in R^2 but is not bounded.

The cost $C(u) > \int_{-1}^1 (2 + \ln x^2) = 0$. But the controllers $u_\epsilon(t) = 2 - \epsilon$ for small $\epsilon > 0$ yield a cost which approaches zero. Thus the optimal controller minimizing $C(u)$ fails to exist.

We next study the boundary of $K(X_0, t)$ and prove that an extremal controller $u(t)$, one that steers $x(t)$ to the boundary $\partial K(X_0, t)$, must satisfy the maximal principle. Since a point $x(t)$ lies in $\partial K(X_0, t)$ only if $x(t)$ lies in $\partial K(x_0, t)$, where $x(t_0) = x_0 \in X_0$, we henceforth take the initial set X_0 to consist of just the single point x_0 and write $K(t)$ for $K(x_0, t)$.

For convenience we first prove the maximal principle for an autonomous system in R^n

$$\text{(S)} \qquad \dot{x} = f(x, u)$$

with $f(x, u)$ and $(\partial f/\partial x)(x, u)$ continuous in R^{n+m}. The nonautonomous case will be discussed in the next chapter in the discussion of necessary conditions for an optimal controller. The admissible controls consist of all measurable functions $u(t)$ on the finite interval $0 \leq t \leq T$ with values in some restraint set $u(t) \subset \Omega \subset R^m$ where Ω is not necessarily compact. We assume that for each admissible controller there is an appropriate bound and hence a response $x(t)$, with $x(0) = x_0$, valid on $0 \leq t \leq T$. The maximal principle will be a direct generalization of the corresponding result for linear processes. For this purpose we develop a technique of linearizing the process S near a given response $x(t)$ by defining infinitesimal or tangent vectors and by using variational differential systems, as described below. For convenience the preliminaries to the maximal principle are presented in the following three parts involving displacements of tangent spaces, the tangent perturbation cone, and an approximation result.

Displacement of Tangent Spaces Along $\bar{x}(t)$

Let $\bar{u}(t)$ be an admissible control with response $\bar{x}(t)$ on $0 \leq t \leq T$. Along the flow defined by

$$\dot{x} = f(x, \bar{u}(t))$$

there is a transport or displacement of tangent vectors v along $\bar{x}(t)$, as determined by the variational differential system based on $\bar{x}(t)$,

$$\dot{v} = \frac{\partial f}{\partial x}(\bar{x}(t), \bar{u}(t))v.$$

That is, a (contravariant) tangent vector at a point $x_1 \in R^n$ is determined by a differentiable curve $x = \phi(\epsilon)$ for ϵ near 0, with $\phi(0) = x_1$; and $v_1 = \dot{\phi}(0)$ are the components of the tangent vector. (Actually, a tangent vector can be defined as a class of differentiable curves all of which have the same tangent components v_1 at x_1.) If $\phi(\epsilon)$ describes a tangent vector based at $x_1 = \bar{x}(t_1)$, we define the displaced curve

$$\bar{A}_{t_2 t_1}\phi(\epsilon) = x(t_2, \phi(\epsilon)),$$

where $x(t, z)$ is the solution of $\dot{x} = f(x, \bar{u}(t))$ with $x(t_1, z) = z$. Then we define the (parallel) displacement of $v_1 = \dot{\phi}(0)$ at x_1 to the vector v_2 at x_2 computed by

$$v_2 = A_{t_2 t_1} v_1 = \frac{d}{d\epsilon}[\bar{A}_{t_2 t_1}\phi(\epsilon)]_{\epsilon=0} = \frac{\partial x}{\partial z}(t_2, z)]_{z=x_1}\dot{\phi}(0).$$

Thus the real n-dimensional tangent space based at $x_1 = \bar{x}(t_1)$ is displaced onto the tangent space based at $x_2 = \bar{x}(t_2)$ by the linear transformation $A_{t_2 t_1}$, which is described by the matrix $(\partial x/\partial z)(t_2, x_1)$. But

$$\frac{d}{dt}\frac{\partial x}{\partial z}(t, x_1) = \frac{\partial f}{\partial x}(\bar{x}(t), \bar{u}(t))\frac{\partial x}{\partial z}(t, x_1)$$

and so $(\partial x/\partial z)(t, x_1)$ is the fundamental matrix solution of the variational differential system, with $(\partial x/\partial z)(t_1, x_1)$ equal to the identity matrix. Therefore the displaced vector $v(t) = A_{tt_1}\dot{\phi}(0)$, is the solution of the variational equation

$$(\mathcal{V}) \qquad \dot{v} = (\partial f/\partial x)(\bar{x}(t), \bar{u}(t))v$$

with $v(t_1) = \dot{\phi}(0)$. The linearity of $\mathcal{V}$ makes evident the linearity of the transformation A_{tt_1}. Clearly the matrix $(\partial x/\partial z)(t, \bar{x}(t_1))$ of A_{tt_1} is continuous in (t, t_1).

By the procedure of parallel displacement of tangent spaces along $\bar{x}(t)$ we define the displacement of an $(n-1)$-hyperplane π_t. (A hyperplane is the locus of zeros of a real linear functional on the tangent space.) Let $\eta(t_1)$ be the direction numbers of the normal to a hyperplane π_{t_1} at x_1 [that is, the real linear functional $\eta(t_1)v_1$ vanishes precisely for tangent vectors v_1 in π_{t_1}]. Define the solution $\eta(t)$ of the adjoint (row) system

$$(\mathcal{A}) \qquad \dot{\eta} = -\eta\frac{\partial f}{\partial x}(\bar{x}(t), \bar{u}(t))$$

with value $\eta(t_1)$ at $t = t_1$. Then $\eta(t)\, v(t) = 0$ for all $v(t)$ in π_t, since $\eta(t_1)\, v(t_1) = 0$ and

$$\frac{d}{dt}[\eta(t)\, v(t)] = \dot{\eta}v + \eta\dot{v} = -\eta\frac{\partial f}{\partial x}v + \eta\frac{\partial f}{\partial x}v = 0.$$

Thus each nontrivial (that is, nowhere vanishing) solution $\eta(t)$ of $\mathcal{A}$ defines a parallel displacement of hyperplanes π_t along $\bar{x}(t)$, and every such parallel field π_t arises in this way.

Elementary Perturbations and the Tangent Perturbation Cone

Let us perturb the basic controller $\bar{u}(t)$ by changing its value to some constant $u_1 \in \Omega$ near the time t_1. That is define

$$u_{\pi_1}(t, \epsilon) = \begin{cases} u_1 & \text{on} \quad t_1 - l_1\epsilon \le t \le t_1 \\ \bar{u}(t) & \text{elsewhere on} \quad 0 \le t \le T, \end{cases}$$

where the perturbation data are $\pi_1 = \{t_1, l_1, u_1\}$ for $0 < t_1 < T$, $l_1 \ge 0$, and $u_1 \in \Omega$. For suitably small $\epsilon \ge 0$ the perturbed function $u_{\pi_1}(t, \epsilon)$ is a well-defined admissible controller with response $x_{\pi_1}(t, \epsilon)$ initiating at $x_{\pi_1}(0, \epsilon) = x_0$. Moreover, it is easy to see that

$$\lim_{\epsilon \to 0} x_{\pi_1}(t, \epsilon) = \bar{x}(t), \quad \text{uniformly on} \quad 0 \leq t \leq T.$$

Furthermore $x_{\pi_1}(t, \epsilon)$ is a continuous function of $\{t_1, l_1, u_1, \epsilon, t\}$. We require t_1 to be a Lebesgue (or regular) point, that is,

$$\int_{t_1-\epsilon}^{t_1} |f(\bar{x}(t), \bar{u}(t)) - f(\bar{x}(t_1), \bar{u}(t_1))| \, dt = o(\epsilon)$$

so

$$\int_{t_1-\epsilon}^{t_1} f(\bar{x}(t), \bar{u}(t)) \, dt = f(\bar{x}(t_1), \bar{u}(t_1))\epsilon + o(\epsilon).$$

Such Lebesgue points t_1 are dense (in fact, almost all points) on $0 < t < T$, and we hereafter always choose such points for perturbations without further mention. Under these circumstances we define $u_{\pi_1}(t, \epsilon)$ to be an elementary perturbation of $\bar{u}(t)$ specified by the data $\pi_1 = \{t_1, l_1, u_1\}$ and $\epsilon \geq 0$.

Now let $u_{\pi_1}(t, \epsilon)$ be an elementary perturbation of $\bar{u}(t)$ at $\pi_1 = \{t_1, l_1, u_1\}$. Then the corresponding response $x_{\pi_1}(t, \epsilon)$ defines a tangent vector at t_1 by the curve $\phi(\epsilon) = x_{\pi_1}(t_1, \epsilon)$. Namely,

$$\dot{\phi}(0) = \lim_{\epsilon \to 0} \frac{1}{\epsilon} [x_{\pi_1}(t_1, \epsilon) - \bar{x}(t_1)] = [f(\bar{x}(t_1), u_1) - f(\bar{x}(t_1), \bar{u}(t_1))]l_1.$$

This follows from the estimate

$$x_{\pi_1}(t_1, \epsilon) = \bar{x}(t_1 - l_1\epsilon) + \int_{t_1 - l_1\epsilon}^{t_1} f(x_{\pi_1}(t, \epsilon), u_1) \, dt$$

or

$$x_{\pi_1}(t_1, \epsilon) = \bar{x}(t_1) - f(\bar{x}(t_1), \bar{u}(t_1))l_1\epsilon + f(\bar{x}(t_1), u_1)l_1\epsilon + o(\epsilon)$$

where

$$\lim_{\epsilon \to 0} \frac{o(\epsilon)}{\epsilon} = 0.$$

The tangent vector at $\bar{x}(t_1)$,

$$v_{\pi_1}(t_1) = [f(\bar{x}(t_1), u_1) - f(\bar{x}(t_1), \bar{u}(t_1))]l_1$$

is called the elementary perturbation vector for the data $\pi_1 = \{t_1, l_1, u_1\}$. Note that the data $\{t_1, \lambda l_1, u_1\}$ for $\lambda \geq 0$ yield a perturbation vector $\lambda v_{\pi_1}(t_1)$, and hence the elementary perturbation vectors fill a cone in the tangent space at $\bar{x}(t_1)$, that is, with each point of the cone the entire ray through that point also lies in the cone.

The parallel displacement of $v_{\pi_1}(t_1)$ to another time t is $v_{\pi_1}(t)$, the solution of the variational system $\mathcal{V}$ which agrees with $v_{\pi_1}(t_1)$ at $t = t_1$.

Definition. The tangent perturbation cone K_t, at any time $0 < t \leq T$, is the smallest closed convex cone in the tangent space at $\bar{x}(t)$ that contains all parallel displacements of all elementary perturbation vectors from all Lebesgue times t_1 on $0 < t_1 < t$.

Note that $A_{\bar{t}t}K_t \subset K_{\bar{t}}$ for $t < \bar{t}$, and

$$K_{\bar{t}} = \overline{\bigcup_{0<t<\bar{t}} A_{\bar{t}t}K_t}.$$

In particular, the final limit cone is

$$K_T = \overline{\bigcup_{0<t<T} A_{Tt}K_t}.$$

In order to appreciate the nature of the cone $K_{\bar{t}}$ let us consider a convex combination of elementary perturbation vectors in $K_{\bar{t}}$

$$v_\pi = \lambda_1 v_{\pi_1}(\bar{t}) + \lambda_2 v_{\pi_2}(\bar{t}) + \cdots + \lambda_s v_{\pi_s}(\bar{t})$$

with nonnegative λ_i such that $\sum_{i=1}^{s} \lambda_i = 1$. Here $\pi_i = \{t_i, l_i, u_i\}$ with $0 < t_i < \bar{t}$, $l_i \geq 0$, and $u_i \in \Omega$, and for simplicity we assume that all the t_i are distinct. Then define the data complex

$$\pi = \{t_1, \ldots, t_s, \lambda_1 l_1, \ldots, \lambda_s l_s, u_1, \ldots, u_s\},$$

and define the complex perturbation

$$u_\pi(t, \epsilon) = \begin{cases} u_i & \text{on} \quad t_i - \lambda_i l_i \epsilon \leq t \leq t_i \text{ for } i = 1, \ldots, s \\ \bar{u}(t) & \text{elsewhere on} \quad 0 \leq t \leq T. \end{cases}$$

Then, for small $\epsilon \geq 0$, the function $u_\pi(t, \epsilon)$ is an admissible control with response $x_\pi(t, \epsilon)$ on $0 \leq t \leq T$. Again $x_\pi(t, \epsilon)$ is a continuous function of the $4s$ arguments in π and (t, ϵ), which is an immediate consequence of the continuous dependence of the solutions of a differential system on the coefficients and on the initial data.

We shall now prove that the curve $\phi(\epsilon) = x_\pi(\bar{t}, \epsilon)$ has the tangent vector v_π at $\bar{x}(\bar{t})$. On the interval $0 \leq t \leq t_1 - \lambda_1 l_1 \epsilon$, we note $x_\pi(t, \epsilon) = \bar{x}(t)$ and then, as seen above,

$$x_\pi(t_1, \epsilon) = \bar{x}(t_1) + \epsilon \lambda_1 v_{\pi_1}(t_1) + o(\epsilon).$$

Thus $x_\pi(t_1, \epsilon)$ defines a curve, as $\epsilon \to 0$, which yields the tangent vector $\lambda_1 v_{\pi_1}(t_1)$ at $\bar{x}(t_1)$.

On $t_1 \leq t \leq t_2 - \lambda_2 l_2 \epsilon$ the controller $u_\pi(t, \epsilon) = \bar{u}(t)$ and so the curve $x_\pi(t_1, \epsilon)$ is displaced to a curve defining the tangent vector $\lambda_1 v_{\pi_1}(t_2 - \lambda_2 l_2 \epsilon)$ at $\bar{x}(t_2 - \lambda_2 l_2 \epsilon)$, that is,

$$x_\pi(t_2 - \lambda_2 l_2 \epsilon, \epsilon) = \bar{x}(t_2 - \lambda_2 l_2 \epsilon) + \epsilon \lambda_1 v_{\pi_1}(t_2) + o(\epsilon), \quad \text{or}$$

$$x_\pi(t_2 - \lambda_2 l_2 \epsilon, \epsilon) = \bar{x}(t_2) - f(\bar{x}(t_2), \bar{u}(t_2))\lambda_2 l_2 \epsilon + \epsilon \lambda_1 v_{\pi_1}(t_2) + o(\epsilon).$$

But $x_\pi(t_2, \epsilon) = x_\pi(t_2 - \lambda_2 l_2 \epsilon) + \int_{t_2-\lambda_2 l_2 \epsilon}^{t_2} f(x_\pi(t, \epsilon), u_2)\, dt$
or

$$x_\pi(t_2, \epsilon) = x_\pi(t_2 - \lambda_2 l_2 \epsilon) + f(x_\pi(t_2 - \lambda_2 l_2 \epsilon, \epsilon), u_2)\lambda_2 l_2 \epsilon + o(\epsilon).$$

Thus

$$x_\pi(t_2, \epsilon) = \bar{x}(t_2) - f(\bar{x}(t_2), \bar{u}(t_2))\lambda_2 l_2 \epsilon + \epsilon\lambda_1 v_{\pi_1}(t_2) + f(x_\pi(t_2 - \lambda_2 l_2 \epsilon, \epsilon), u_2)\lambda_2 l_2 \epsilon + o(\epsilon).$$

Therefore we compute at $t = t_2$,

$$x_\pi(t_2, \epsilon) = \bar{x}(t_2) + \epsilon\lambda_1 v_{\pi_1}(t_2) + \epsilon\lambda_2 v_{\pi_2}(t_2) + o(\epsilon).$$

Continuing in this way to $\bar{t} > t_3$, we obtain the basic perturbation formula

$$(*) \qquad x_\pi(\bar{t}, \epsilon) = \bar{x}(\bar{t}) + \epsilon\lambda_1 v_{\pi_1}(\bar{t}) + \cdots + \epsilon\lambda_s v_{\pi_s}(\bar{t}) + o(\epsilon),$$

and so $\lim_{\epsilon \to 0} \dfrac{x_\pi(\bar{t}, \epsilon) - \bar{x}(\bar{t})}{\epsilon} = v_\pi(\bar{t}) = \lambda_1 v_{\pi_1} + \cdots + \lambda_s v_{\pi_s}$.

It is important to notice that once the perturbation data complex $\{t_1, \ldots, t_s, l_1, \ldots, l_s, u_1, \ldots, u_s\}$ has been fixed,

$$\lim_{\epsilon \to 0} \frac{o(\epsilon)}{\epsilon} = 0,$$

uniformly on $0 \leq t \leq T$ and for all $0 \leq \lambda_i \leq 1$. This last observation follows from the trivial estimate

$$o(\lambda\epsilon) \quad \text{is uniformly} \quad o(\epsilon), \quad \text{for} \quad 0 \leq \lambda \leq 1,$$

and the fact that $x_\pi(t, \epsilon)$ lies uniformly near $\bar{x}(t)$ where the a priori bounds obtain for $f(x, \bar{u}(t))$ and $\partial f/\partial x(x, \bar{u}(t))$.

The fundamental perturbation formula (*) shows that any convex combination of elementary perturbations vectors (at distinct times) defines a point $\bar{x}(\bar{t}) + \epsilon v_\pi$, which lies in the set of attainability $K(\bar{t})$, within an error of $o(\epsilon)$. Thus the tangent perturbation cone $K_{\bar{t}}$, when enlarged to macroscopic size, serves as an accurate estimate for the set of attainability $K(\bar{t})$. Because of this relation we shall be able to state geometric properties concerning the boundary of $K(\bar{t})$ in terms of $K_{\bar{t}}$; namely, the maximal principle.

Definition. Let $v_1, \ldots, v_n$ be independent vectors in K_t, each arising as a convex combination of elementary perturbation vectors, with all perturbation times distinct (in forming each v_i and even among the different v_i). An elementary simplex cone $\mathfrak{C}$ consists of all convex combinations of the vectors $v_1, \ldots, v_n$.

Because of our requirement that all perturbation times are distinct (otherwise a more complicated limiting scheme is necessary), the fundamental perturbation formula (*) asserts the existence of a response

$$x(t, \epsilon, \lambda) = \bar{x}(t) + \epsilon(\lambda_1 v_1 + \cdots + \lambda_n v_n) + o(\epsilon),$$

corresponding to each vector $\lambda_1 v_1 + \cdots + \lambda_n v_n$ in $\mathfrak{C}$.

LEMMA 1. *Let v be a vector interior to K_t. Then there exists an elementary simplex cone $\mathfrak{C}$ which contains v in its interior.*

Proof. Since K_t is the closure of all convex combinations of elementary perturbation vectors and since v is interior to K_t (that is, v minus the origin lies interior to K_t minus the origin), there exist independent vectors $v_1, \ldots, v_n$, which span a cone in K_t containing v in its interior, and each of which is a convex combination of elementary perturbation vectors. We next modify $v_1, \ldots, v_n$ to obtain combinations of elementary perturbation vectors involving distinct perturbation times.

The elementary perturbation data $\pi_1 = \{t_1, l_1, u_1\}$ yield the perturbation vector

$$v_{\pi_1}(t_1) = [f(\bar{x}(t_1), u_1) - f(\bar{x}(t_1), \bar{u}(t_1))]l_1.$$

Since t_1 is a Lebesgue point, there are arbitrarily nearby Lebesgue times t_1', say $|t_1 - t_1'| < \xi$, with $|f(\bar{x}(t_1'), \bar{u}(t_1')) - f(\bar{x}(t_1), \bar{u}(t_1))| < \xi$, for any small $\xi > 0$. Thus the data $\pi_1' = \{t_1', l_1, u_1\}$ yield a vector v'_{π_1}. Since the linear transformation A_{tt_1} is continuous in t and $A_{t_1t_1}$ is the identity, we can require that $v'_{\pi_1}(t)$ closely approximates $v_{\pi_1}(t)$.

In this way we can modify all the elementary perturbations entering into $v_1, \ldots, v_n$ to obtain approximating vectors $v_1', \ldots, v_n'$ having all perturbation times distinct. Clearly $v_1', \ldots, v_n'$ span an elementary simplex cone $\mathfrak{C}$ that contains v in its interior. Q.E.D.

A Topological Approximation Result

We first present a topological substitute for the implicit function theorem, which can be used when differentiability hypotheses are difficult to verify.

SCHOLIUM. *Let $f(x)$ be a continuous map from a compact convex set B^n, having interior in R^n, into the space R^n. Let P be a point interior to B^n and assume*

$$\|f(x) - x\| < \|x - P\|$$

for each x in the boundary ∂B^n. Then the image $f(B^n)$ covers P.

Proof. We can assume that P is the origin of R^n since the hypotheses are unaffected by rigid motions of the Euclidean space R^n. Fix a topological map $x \to h(x)$ of B^n onto the unit ball $B_1{}^n$ centered at the origin, by a linear dilation or contraction along each ray from P.

For each point x in B^n assign the vector (directed line segment) $v(x)$ from x to $x + f(x)$. At the corresponding point $\bar{x} = h(x)$ in $B_1{}^n$ also assign the rigid translate of this vector $v(x)$. Then $v(h^{-1}(\bar{x})) = v(\bar{x})$ is a continuous vector field on the unit ball $B_1{}^n$. The hypothesis $\|f(x) - x\| < \|x\|$ asserts that $v(x)$ makes an acute angle with the outward ray Px, for each $x \in \partial B^n$. Thus $v(\bar{x})$ has an outward radial component at each point $x \in \partial B_1{}^n$. In this case $v(\bar{x})$ must vanish in $B_1{}^n$, or $v(x_0) = 0$ for some $x_0 \in B^n$. That is, $f(x_0) = 0$ and $P = 0$ is the image of x_0.

This last assertion, well known in the index theory of vector fields, also follows directly from the Brouwer fixed-point theorem. Consider the vector field $-v(\bar{x})$, which has a negative radial component, on and near the boundary sphere of $B_1{}^n$. Then for a sufficiently small positive number α, the vector field $-\alpha\, v(\bar{x})$ has the terminal end of the segment $\bar{x} \to \bar{x} - \alpha\, v(\bar{x})$ inside $B_1{}^n$. By the fixed-point theorem there exists a point $\bar{x}_0$ in $B_1{}^n$ where $\bar{x}_0 = \bar{x}_0 - \alpha\, v(\bar{x}_0)$ so $v(\bar{x}_0) = 0$. Then take $x_0 = h^{-1}(\bar{x}_0)$ in B^n as the required point at which $v(x_0) = f(x_0) = 0$.
Q.E.D.

As a special case we obtain the following result.

COROLLARY. *Let $f(x)$ be a continuous map from the ball $B_1{}^n$: $\|x\| \le 1$ into R^n, and assume*

$$\|f(x) - x\| \le 1 - \epsilon \quad \text{for all} \quad \|x\| = 1 \quad \text{and} \quad \epsilon > 0.$$

Then each point z in the ball $\|z\| < \epsilon$ is covered by the image of $B_1{}^n$.

The tangent perturbation cone K_t lies in the tangent space at $\bar{x}(t)$ and hence it consists of *infinitesimal* vectors. However, we can treat the tangent space at $\bar{x}(t)$ as the entire vector space R^n with origin at $\bar{x}(t)$. In this case K_t becomes a *macroscopic* cone in R^n with vertex at $\bar{x}(t)$ and we can use it as an estimate of $K(t)$, at least in a neighborhood of $\bar{x}(t)$.

LEMMA 2. *Let v be a nonzero vector interior to K_t. Then there exists an elementary simplex cone $\mathfrak{C}$ in K_t such that*

1. *$\mathfrak{C}$ contains v in its interior (as an infinitesimal cone);*
2. *$\mathfrak{C}$ lies interior to $K(t)$ (as a macroscopic cone—that is, a truncation of $\mathfrak{C}$ minus its vertex lies interior to $K(t)$ near $\bar{x}(t)$).*

Proof. By Lemma 1 there exists an elementary simplex cone $\mathfrak{C}_1$ in K_t that contains v in its interior (that is, v minus the origin lies interior to K_t). Let $v_1, \ldots, v_n$ be convex combinations of elementary perturbation vectors that generate $\mathfrak{C}_1$. For each convex combination $\lambda_1 v_1 + \cdots + \lambda_n v_n$ there is a response

$$x(t, \epsilon, \lambda) = \bar{x}(t) + \epsilon(\lambda_1 v_1 + \cdots + \lambda_n v_n) + o(\epsilon).$$

Let us consider $\mathfrak{C}_1$ in R^n as a macroscopic cone with vertex at $\bar{x}(t)$. We can choose the vectors $v_1, \ldots, v_n$ so that their endpoints all lie on a hyperplane through the endpoint of v and orthogonal to v. Then each point in $\mathfrak{C}_1$ can be uniquely prescribed by the barycentric coordinates $\lambda_1, \ldots, \lambda_n$ and the height $0 < l \leq \|v\|$ measured from $\bar{x}(t)$ along v.

Fix $\epsilon > 0$ so small that

$$x(t, \epsilon l, \lambda) = \bar{x}(t) + \epsilon\, l(\lambda_1 v_1 + \cdots + \lambda_n v_n) + o(\epsilon l)$$

lies in the half-space $l > 0$. Thus we define a map of $\mathfrak{C}_1$ (minus the vertex $\bar{x}(t)$) into the half-space $l > 0$ by

$$(\lambda_1 v_1 + \cdots + \lambda_n v_n) l \to x(t, \epsilon l, \lambda).$$

Let r be the horizontal vector (measured from the axis of v in hyperplanes orthogonal to v) and introduce the vector coordinate $\rho = r/l$ and the scalar $l > 0$. Then $\mathfrak{C}_1$, in the coordinates (ρ, l) is a prism P_1 and we write the given map

$$L = L(\rho, l) = l + o(l)$$

$$R = R(\rho, l) = \rho + o(l)/l = \rho + o(1),$$

for appropriately restricted coordinates $0 \leq \|\rho\| \leq a$, $0 < l \leq b$, and

$$\lim_{l \to 0} \frac{o(l)}{l} = 0 \quad \text{and} \quad \lim_{l \to 0} o(1) = 0$$

uniformly in ρ. Note that we can extend the definition of the map to the closed prism $\bar{P}_1$ by

$$L(\rho, 0) = 0, \qquad R(\rho, 0) = \rho.$$

Now further restrict the size of $\bar{P}_1$ by choosing $b < a$ so small that

$$\|R(\rho, l) - \rho\| < \frac{a}{4}$$

and

$$\|L(\rho, b) - b\| < \frac{b}{4}.$$

By the above topological scholium there is a cylindrical neighborhood U of the axis $\rho = 0$, $0 < l < b/4$, which is covered by the image of $\mathfrak{C}_1$. In the original coordinates (r, l) the set U defines a cone with vertex $\bar{x}(t)$. Within this conical neighborhood of the vector v we can find an elementary simplex cone $\mathfrak{C}$ about v such that the required two conclusions obtain.

Q.E.D.

With these preliminary results we now prove the maximal principle for nonlinear autonomous processes and arbitrary control restraints $u(t) \subset \Omega$, which need not be compact.

THEOREM 3. *Consider the process in* R^n

$$(\mathcal{S}) \qquad \dot{x} = f(x, u)$$

with $f(x, u)$ *and* $\partial f/\partial x(x, u)$ *continuous in* R^{n+m}. *Let* $\mathcal{F}$ *be the family of all measurable controllers* $u(t)$ *on* $0 \leq t \leq T$ *that satisfy the restraint* $u(t) \subset \Omega \subset R^m$ *and admit a bound for the response initiating at the point* x_0. *Let* $\bar{u}(t) \in \mathcal{F}$ *have a response* $\bar{x}(t)$ *with* $\bar{x}(T)$ *on the boundary of the set of attainability* $K(T)$. *Then there exists a nontrivial adjoint response* $\bar{\eta}(t)$ *of*

$$(\mathcal{A}) \qquad \dot{\eta} = -\eta \frac{\partial f}{\partial x}(\bar{x}(t), \bar{u}(t))$$

such that the maximal principle obtains, that is,

$$H(\bar{\eta}(t), \bar{x}(t), \bar{u}(t)) = M(\bar{\eta}(t), \bar{x}(t)) \quad \textit{almost everywhere.}$$

Further, if $\bar{u}(t)$ *is bounded,*

$$M(\bar{\eta}(t), \bar{x}(t))$$

is constant everywhere. Here the Hamiltonian function is

$$H(\eta, x, u) = \eta f(x, u) = \eta_1 f^1(x, u) + \cdots + \eta_n f^n(x,u)$$

and

$$M(\eta, x) = \max_{u \in \Omega} H(\eta, x, u) \quad (\textit{wherever it exists}).$$

Proof. Since $\bar{x}(T)$ lies on the boundary of $K(T)$, there is a sequence of points $\{P_n\}$ outside $K(T)$ such that $P_n \to \bar{x}(T)$ and the unit vectors along the segments $\bar{x}(T)$ to P_n approach a limit unit vector $w(T)$ at $\bar{x}(T)$.

Now $w(T)$ cannot be an interior vector in the perturbation tangent cone K_T for otherwise, by Lemma 2, there is a macroscopic conical neighborhood $\mathfrak{C}$ of $w(T)$ in $K(T)$. But this would contradict the assumption that the points P_n all lie outside $K(T)$.

Thus there exists a hyperplane $\pi(T)$ at $\bar{x}(T)$ such that $\pi(T)$ separates $w(T)$ from K_T. Let $\bar{\eta}(T)$ be the exterior unit normal to $\pi(T)$ at $\bar{x}(T)$ and then define $\bar{\eta}(t)$ as the corresponding solution of the linear differential system $\mathcal{A}$. Then

$$\bar{\eta}(T)\, v(T) = \bar{\eta}(t)\, v(t) \leq 0 \quad \text{for all} \quad t \leq T,$$

where $v(t)$ is any perturbation vector in K_t.

Suppose that the maximal principle fails, that is,

$$H(\bar{\eta}(t), \bar{x}(t), \bar{u}(t)) < H(\bar{\eta}(t), \bar{x}(t), u_1(t))$$

for $u_1(t) \in \Omega$ on some duration of positive measure on $0 \leq t \leq T$. Let t_1 on $0 < t_1 < T$ be a Lebesgue time for $f(\bar{x}(t), \bar{u}(t))$ when

$$\bar{\eta}(t_1) f(\bar{x}(t_1), \bar{u}(t_1)) < \bar{\eta}(t_1) f(\bar{x}(t_1), u_1)$$

for some point $u_1 \in \Omega$. Consider the elementary perturbation vector

$$v_{\pi_1}(t_1) = [f(\bar{x}(t_1), u_1) - f(\bar{x}(t_1), \bar{u}(t_1)]$$

for the data

$$\pi_1 = \{t_1, 1, u_1\}.$$

Then the denial of the maximal principle yields

$$\bar{\eta}(t_1)\, v_{\pi_1}(t_1) > 0,$$

which contradicts the assertion that

$$\bar{\eta}(t)\, v(t) \leq 0 \quad \text{for all} \quad t$$

and all $v(t) \in K_t$. Therefore

$$H(\bar{\eta}(t), \bar{x}(t), \bar{u}(t) = M(\bar{\eta}(t), \bar{x}(t))$$

almost everywhere on $0 \leq t \leq T$ (and the right member exists almost everywhere).

Finally we show that $M(\bar{\eta}(t), \bar{x}(t))$ is absolutely continuous and has a zero derivative everywhere on $0 \leq t \leq T$. Here we assume that $\bar{u}(t)$ is bounded, that is, $|\bar{u}(t)| \leq \beta$ on $0 \leq t \leq T$. Let

$$m(\eta, x) = \max_{|u| \leq \beta, u \in \Omega} H(\eta, x, u),$$

so

$$M(\eta, x) \geq m(\eta, x)$$

but

$$M(\bar{\eta}(t), \bar{x}(t)) = m(\bar{\eta}(t), \bar{x}(t)) \quad \text{almost everywhere.}$$

We first show that $m(\bar{\eta}(t), \bar{x}(t))$ is constant on $0 \leq t \leq T$. For (η, x) in a compact set Q in $R^n \times R^n \times R^m$ containing $(\bar{\eta}(t), \bar{x}(t))$ and all $|u| \leq \beta$ we obtain, for any two points (η, x, u) and (η', x', u),

$$|H(\eta, x, u) - H(\eta', x', u)| \leq kd$$

where

$$d = |\eta - \eta'| + |x - x'|,$$

and k is a Lipschitz constant majorizing $|f(x, u)|$ and $\left|\eta \dfrac{\partial f}{\partial x}(x, u)\right|$ in Q. Let u and u' in Ω, with $|u| \leq \beta$, $|u'| \leq \beta$, be selected so that

$$m(\eta, x) = H(\eta, x, u) \quad \text{and} \quad m(\eta', x') = H(\eta', x', u').$$

Then

$$H(\eta, x, u') \leq H(\eta, x, u) \quad \text{and} \quad H(\eta', x', u) \leq H(\eta', x', u').$$

Therefore in Q we compute

$$\begin{aligned} -kd &\le H(\eta, x, u') - H(\eta', x', u') \\ &\le H(\eta, x, u) - H(\eta', x', u') \\ &\le H(\eta, x, u) - H(\eta', x', u) \le kd \end{aligned}$$

and

$$|m(\eta, x) - m(\eta', x')| \le kd.$$

Hence $m(\eta, x)$ is Lipschitz continuous in Q and so $m(\bar{\eta}(t), \bar{x}(t))$ is absolutely continuous on $0 \le t \le T$.

Let $0 < \tau < T$ be a point at which $m(t) = m(\bar{\eta}(t), \bar{x}(t))$ and $\bar{x}(t)$ and $\bar{\eta}(t)$ all have derivatives. Then, for $t' > \tau$ we compute

$$m(t') \ge H(\bar{\eta}(t'), \bar{x}(t'), \bar{u}(\tau))$$

and

$$\begin{aligned} m(t') - m(\tau) \ge{}& H(\bar{\eta}(t'), \bar{x}(t'), \bar{u}(\tau)) - H(\bar{\eta}(t'), \bar{x}(\tau), \bar{u}(\tau)) \\ &+ H(\bar{\eta}(t'), \bar{x}(\tau), \bar{u}(\tau)) - H(\bar{\eta}(\tau), \bar{x}(\tau), \bar{u}(\tau)). \end{aligned}$$

Then

$$\lim_{t' \to \tau} \frac{m(t') - m(\tau)}{t' - \tau} = \frac{dm}{dt}\bigg]_{t=\tau} \ge \frac{\partial H}{\partial x^i}\frac{dx^i}{dt} + \frac{\partial H}{\partial \eta_i}\frac{d\eta_i}{dt}\bigg]_{t=\tau} = 0,$$

since

$$\frac{\partial H}{\partial x^i}\frac{dx^i}{dt} = \eta_j \frac{\partial f^j}{\partial x^i} f^i, \qquad \frac{\partial H}{\partial \eta_i}\frac{d\eta_i}{dt} = f^i\left(-\eta_j \frac{\partial f^j}{\partial x^i}\right).$$

Using $t' < \tau$ we compute $\dfrac{dm}{dt}\bigg]_{t=\tau} \le 0$ so

$$\frac{dm}{dt}(\bar{\eta}(t), \bar{x}(t)) = 0$$

almost everywhere. Since $m(\bar{\eta}(t), \bar{x}(t))$ is absolutely continuous with a zero derivative, it is a constant m everywhere on $0 \le t \le T$.

From the definition of $M(\eta, x)$ it is easy to verify that $M(\bar{\eta}(t), \bar{x}(t))$ is lower semicontinuous on $0 \le t \le T$, that is,

$$M(\bar{\eta}(t_1), \bar{x}(t_1)) \le M(\bar{\eta}(t), \bar{x}(t)) + \epsilon$$

for all t sufficiently near t_1 and a prescribed $\epsilon > 0$. (If $M(\bar{x}(t_1), \bar{\eta}(t_1)) = \infty$, a corresponding statement holds.) Thus

$$M(\bar{\eta}(t_1), \bar{x}(t_1)) \le m(\bar{\eta}(t), \bar{x}(t)) = m + \epsilon$$

for all $\epsilon > 0$, and so $M(\bar{\eta}(t), \bar{x}(t)) \le m$ everywhere on $0 \le t \le T$. Thus $M(\bar{\eta}(t), \bar{x}(t)) = m$ everywhere on $0 \le t \le T$. Q.E.D.

For linear processes a controller $u(t)$ satisfies the maximal principle if and only if $u(t)$ is extremal [the response $x(T) \in \partial K(T)$]. For nonlinear processes the maximal principle does not guarantee that $u(t)$ steers $x(t)$ to the boundary of the set of attainability, although it is easy to see that $x(T) \in \partial K(T)$ implies that $x(t) \in \partial K(t)$ for all $t \leq T$. The next two examples demonstrate this phenomenon for nonlinear control processes.

***Example* 5.** Consider in R^2

$$\dot{x} = yu - xv, \qquad \dot{y} = -xu - yv$$

with control restraints $|u(t)| \leq 1$ and $|v(t)| \leq 1$. In polar coordinates the differential system becomes

$$\dot{r} = -r\,v(t), \qquad \dot{\phi} = -u(t).$$

The initial point is $r_0 = 1$, $\phi_0 = 0$, and we study the duration $0 \leq t \leq \pi$. The control functions $u(t)$ and $v(t)$ enter independently in the angular and radial components. Thus it is easy to compute the set of attainability $K(\pi)$ as the annular ring $e^{-\pi} \leq r \leq e^{\pi}$, $0 \leq \phi < 2\pi$. Here the control process has a uniform bound, and also $K(\pi)$ is compact. However $K(\pi)$ is not convex, nor even simply connected.

Here the concept of the new frontier of $K(t)$, as developed for the linear systems of Chapter 2, is not significant. For example the point $(-1, 0)$ first appears in $K(\pi)$ as an interior point. Also $u(t) \equiv 1$, $v(t) \equiv 0$ satisfies the maximal principle on $0 \leq t \leq \pi$; yet the corresponding response does not lead to the boundary of $K(\pi)$.

***Example* 6.** Consider the control process in R^2, defined by a modification of Example 5,

$$\dot{r} = -r\,v(t)\,h(\phi)$$

$$\dot{\phi} = -u(t)\left[1 - \left(\frac{R-r}{2R}\right)^4 \left(\sin^2 \frac{1}{R-r}\right) h(\pi - \phi)\right]$$

where $h(\phi) = h(-\phi) \in C^\infty$ satisfies $0 \leq h(\phi) \leq 1$ with $h(\phi) = 0$ on $\pi/2 \leq \phi \leq \pi$ and $h(\phi) = 1$ for ϕ near 0. Also $R = \exp \int_0^{\pi/2} h(\phi)\,d\phi$. Restraints are $|u(t)| \leq 1$, $|v(t)| \leq 1$, and the initial point is $r_0 = 1$, $\phi_0 = 0$.

At $t = \pi/2$, the set $K(\pi/2)$ can meet the ray $\phi = \pi/2$ only if $u(t) \equiv -1$ so $\dot{\phi} = 1$. Then $\dot{r} = -r\,v(t)\,h(t)$ and so the segment $\phi = \pi/2$, $1/R \leq r \leq R$ is an edge of $K(\pi/2)$. Similarly the segment $\phi = -\pi/2$, $1/R \leq r \leq R$ in $K(\pi/2)$ can be attained only for $u(t) \equiv +1$. Thus $K(\pi/2)$ is a half of an annular ring, with radial width of $(e^{\pi/2} - e^{-\pi/2})$ at $\phi = 0$ tapering to a width of $(R - 1/R)$ at $\phi = \pm\pi/2$.

Now consider $K(\pi)$. Only some of the points on the rays $\phi = \pm\pi/2$ at time $t = \pi/2$ can reach the ray $\phi = \pi$ at time $t = \pi$. In the left half plane, the differential system is

$$\dot{r} = 0, \qquad \dot{\phi} = -u(t)\left[1 - \left(\frac{R - r}{2R}\right)^4 \left(\sin^2 \frac{1}{R - r}\right) h(\pi - \phi)\right].$$

Thus the intersection of $K(\pi)$ with the ray $\phi = \pi$ occurs only at radii satisfying $(R - r)^4 \sin^2 [1/(R - r)] = 0$; that is, at a countable set of points accumulating at $\phi = \pi$, $r = R$.

This analysis shows that $K(\pi)$ consists of an annular-like region, with radial width tapering to a minimum of $(R - 1/R)$ at $\phi = \pi$, and with an infinite number of disjoint open regions excised along the ray $\phi = \pi$. Thus $K(\pi)$ is infinitely connected and its boundary cannot be described by a finite number of simple, closed, continuous curves.

EXERCISES

1. Consider the control process in R^n

(S) $$\dot{x} = f(x, t, u) \quad \text{in} \quad C^1 \quad \text{in} \quad R^{n+1+m}$$

with compact restraint set $\Omega \subset R^m$. Assume, for constant k,

$$x'f(x, t, u) \leq k(|x|^2 + 1)$$

for all $x \in R^n$, all t in a compact interval $\mathfrak{J}$, and all $u \in \Omega$. Show each control $u(t) \subset \Omega$ on $\mathfrak{J}$ yields a response $x(t)$ for all $t \in \mathfrak{J}$. Also, for a fixed initial state x_0, the set $K(x_0, t)$ of attainable points is uniformly bounded.

2. Consider the control process in R^n

(S) $$\dot{x} = f(x, u) \quad \text{in} \quad C^1 \quad \text{in} \quad R^{n+m}$$

with initial state x_0 at time $t_0 = 0$ and compact restraint set $\Omega \subset R^m$. The admissible controllers are all measurable functions $u(t) \subset \Omega$ on $0 \leq t \leq t_1$, and assume that each control has a response $x(t)$ on all $0 \leq t \leq t_1$. Suppose $u^*(t)$ has a response $x^*(t)$ with $x^*(t_1)$ on the boundary of the set $K(t_1)$ of attainability. Then show that $x(t) \in \partial K(t)$ for all $0 \leq t \leq t_1$.

3. Consider the control process in R^n

(S) $$\dot{x} = f(x, t, u) \quad \text{in} \quad C^1 \quad \text{in} \quad R^{n+1+m}$$

with measurable controllers $u(t)$ on $0 \leq t \leq 1$ lying in the compact restraint set $\Omega \subset R^m$.

(a) Let $u^*(t)$ have a response $x^*(t)$ on $0 \leq t \leq 1$. Show there exists $\epsilon > 0$ such that: each control $u_\epsilon(t) \subset \Omega$ on $0 \leq t \leq 1$, with

$|u_\epsilon(t) - u^*(t)| < \epsilon$ on a set of measure $1 - \epsilon$, and initial state $|x_\epsilon(0) - x^*(0)| < \epsilon$ defines a response $x_\epsilon(t)$ on $0 \le t \le 1$. Moreover, $x_\epsilon(t) \to x^*(t)$ uniformly on $0 \le t \le 1$ as $\epsilon \to 0$.

(b) For each initial state $|x_0| \le \alpha$ and time $|t_0| \le \tau$, there exists a uniform duration $t_0 \le t \le t_0 + \zeta(\alpha, \tau)$ such that: the response $x(t, x_0, t_0)$ to any control $u(t) \subset \Omega$ on $t_0 \le t \le t_0 + \zeta$ is defined on all $t_0 \le t \le t_0 + \zeta$.

4. Consider the control process in R^n

$$(\mathcal{S}) \qquad \dot{x} = A(x, t) + B(x, t)u$$

with $A(x, t)$ and $B(x, t)$ in C^1 in R^{n+1} and controllers $u(t)$ on $0 \le t \le 1$ with $\|u\|_1 = \int_0^1 |u(t)|\, dt \le 1$.

(a) Let $u^*(t)$ have a response $x^*(t)$ on $0 \le t \le 1$. Show there exists $\epsilon > 0$ so that each control $u_\epsilon(t)$ on $0 \le t \le 1$, with $\|u_\epsilon - u^*\|_1 < \epsilon$ and initial state $|x_\epsilon(0) - x^*(0)| < \epsilon$, defines a response $x_\epsilon(t)$ on $0 \le t \le 1$. Moreover, $x_\epsilon(t) \to x^*(t)$ uniformly on $0 \le t \le 1$ as $\epsilon \to 0$.

(b) For each initial state $|x_0| \le \alpha$ and time $|t_0| \le \tau$, there exists a uniform duration $t_0 \le t \le t_0 + \zeta(\alpha, \tau)$ such that: the response $x(t, x_0, t_0)$ to any control $u(t)$ with $\int_{t_0}^{t_0+1} |u(t)|\, dt \le \zeta$ is defined on all $t_0 \le t \le t_0 + \zeta$.

4.2 EXISTENCE OF OPTIMAL CONTROL WITH MAGNITUDE RESTRAINTS

In this section we prove general existence theorems for optimal controllers for nonlinear processes where the restraint set is compact. We follow closely the methods of Theorem 2 above. In fact, the next existence theorem is an immediate corollary of Theorem 2 for the special case where the initial set X_0 is a fixed point and the initial time t_0 and restraint set Ω are also fixed. Later we extend our results to include relaxed controllers and impulse controllers.

THEOREM 4. *Consider the nonlinear process in* R^n

$$(\mathcal{S}) \qquad \dot{x} = f(x, t, u) \quad \text{in} \quad C^1 \quad \text{in} \quad R^{n+1+m}.$$

The data are as follows:

1. *The initial and target sets* $X_0(t)$ *and* $X_1(t)$ *are nonempty compact sets varying continuously in* R^n *for all* t *in the basic prescribed compact interval* $\tau_0 \le t \le \tau_1$.

2. *The control restraint set* $\Omega(x, t)$ *is a nonempty compact set varying continuously in* R^m *for* $(x, t) \in R^n \times [\tau_0, \tau_1]$.
3. *The state constraints are (possibly vacuous)* $h^1(x) \geq 0, \ldots, h^r(x) \geq 0$, *a finite or infinite family of constraints, where* $h^1, \ldots, h^r$ *are real continuous functions on* R^n.
4. *The family* $\mathfrak{F}$ *of admissible controllers consists of all measurable functions* $u(t)$ *on various time intervals* $t_0 \leq t \leq t_1$ *in* $[\tau_0, \tau_1]$ *such that each* $u(t)$ *has a response* $x(t)$ *on* $t_0 \leq t \leq t_1$ *steering* $x(t_0) \in X_0(t_0)$ *to* $x(t_1) \in X_1(t_1)$ *and* $u(t) \in \Omega(x(t), t)$, $h^1(x(t)) \geq 0, \ldots, h^r(x(t)) \geq 0$.
5. *The cost for each* $u \in \mathfrak{F}$ *is*

$$C(u) = g(x(t_1)) + \int_{t_0}^{t_1} f^0(x(t), t, u(t))\, dt + \max_{t_0 \leq t \leq t_1} \gamma(x(t))$$

where $f^0 \in C^1$ *in* R^{n+1+m}, *and* $g(x)$ *and* $\gamma(x)$ *are continuous in* R^n.

Assume

(a) *The family* $\mathfrak{F}$ *of admissible controllers is not empty.*
(b) *There exists a uniform bound*

$$|x(t)| \leq b \quad \text{on} \quad t_0 \leq t \leq t_1$$

for all responses $x(t)$ *to controllers* $u \in \mathfrak{F}$.
(c) *The extended velocity set*

$$\hat{V}(x, t) = \{f^0(x, t, u), f(x, t, u) | \; u \in \Omega(x, t)\}$$

is convex in R^{n+1} *for each fixed* (x, t).

Then there exists an optimal controller $u^*(t)$ *on* $t_0^* \leq t \leq t_1^*$ *in* $\mathfrak{F}$ *minimizing* $C(u)$.

Proof. Since $\Omega(x, t)$ lies within a bounded set in R^m when $|x| \leq b$ and $\tau_0 \leq t \leq \tau_1$, all $u(t) \in \mathfrak{F}$ and responses $x(t)$ are uniformly bounded. Thus there is a finite lower bound for the costs of admissible controllers. Select a sequence $u_k(t)$ on $t_0^k \leq t \leq t_1^k$ of controllers in $\mathfrak{F}$ with $C(u_k)$ decreasing monotonically to $\inf C(u)$ for $u \in \mathfrak{F}$, and let $x_k(t)$ be corresponding responses steering $X_0(t_0^k)$ to $X_1(t_1^k)$. Now select a subsequence u_k, without changing notation, so that $t_0^k \to t_0^*$, $t_1^k \to t_1^*$ and $x_k(t_0^k) \to x_0^* \in X_0(t_0^*)$ as $k \to \infty$. We must show that $u_k(t)$ leads to an admissible controller $u^*(t) \in \mathfrak{F}$, realizing the minimum cost. If $t_0^* = t_1^*$, then $\lim_{k \to \infty} C(u_k) = g(x_0^*) + \gamma(x_0^*)$ and $x_0^* \in X_0(t_0^*) \cap X_1(t_0^*)$ so that any $u^*(t_0^*) \in \Omega(x_0^*, t_0^*)$ realizes the minimum cost of $g(x_0^*) + \gamma(x_0^*)$. Henceforth we assume $t_0^* < t_1^*$.

As in the proof of Theorem 2 we select a subsequence of controls, still called $u_k(t)$, so that $\hat{f}(x_k(t), t, u_k(t))$ approaches an integrable $(n + 1)$-vector $\hat{\phi}(t) = (\phi^0(t), \phi(t))$ weakly on $t_0^* \leq t \leq t_1^*$. [Note: $\hat{f} = (f^0, f)$ and we here assume that $t_0^k \leq t_0^*$ and $t_1^k \geq t_1^*$ so that every $u_k(t)$ is defined on $t_0^* \leq t \leq t_1^*$; the other cases are easily treated later.] Let

$$\hat{x}^*(t) = \hat{x}_0^* + \int_{t_0^*}^{t} \hat{\phi}(s)\, ds \quad \text{on} \quad t_0^* \leq t \leq t_1^*$$

where

$$\hat{x}^*(t) = (x^{0*}(t), x^*(t)) \quad \text{and} \quad \hat{x}_0^* = (0, x_0^*).$$

Then

$$\lim_{k\to\infty} \hat{x}_k(t) = \hat{x}^*(t) \quad \text{everywhere on} \quad t_0^* \leq t \leq t_1^*$$

where we employ the notation

$$\hat{x}_k = (x_k^{\ 0}, x_k) \quad \text{and} \quad x_k^{\ 0}(t) = \int_{t_0^k}^{t} f^0(x_k(s), s, u_k(s))\, ds.$$

Since $\hat{x}_k(t)$ is an equicontinuous family of functions, $|\hat{x}_k(t_0^{\ k}) - \hat{x}_k(t_0^*)| \to 0$ and $|\hat{x}_k(t_1^{\ k}) - \hat{x}_k(t_1^*)| \to 0$ so that $x^*(t_0^*) = x_0^* \in X_0(t_0^*)$ and $x^*(t_1^*) \in X_1(t_1^*)$. By Ascoli's theorem we can assume that $\hat{x}_k(t) \to \hat{x}^*(t)$ uniformly on $t_0^* \leq t \leq t_1^*$ and so the constraints hold

$$h^1(x^*(t)) \geq 0, \ldots, h^r(x^*(t)) \geq 0 \quad \text{on} \quad t_0^* \leq t \leq t_1^*.$$

Moreover the stated convergence implies that

$$\lim_{k\to\infty} C(u_k) = g(x^*(t_1^*)) + \int_{t_0^*}^{t_1^*} \phi^0(s)\, ds + \max_{t_0^* \leq t \leq t_1^*} \gamma(x^*(t)).$$

Thus the demonstration is complete if we show that there exists $u^*(t) \in \mathcal{F}$ with response $x^*(t)$ such that $\hat{f}(x^*(t), t, u^*(t)) = \hat{\phi}(t)$.

In order to find $u^*(t)$ we first show that $\hat{\phi}(t) \in \hat{V}(x^*(t), t)$ for all times [after redefining $\hat{\phi}(t)$ on a null set, as in Theorem 2]. Suppose $\hat{\phi}(t)$ lies outside $\hat{V}(x^*(t), t)$ on some set of positive duration. Then there is a constant unit $(n + 1)$-row vector y such that

$$y\, \hat{\phi}(t) > \limsup_{k\to\infty} y \hat{f}(x^*(t), t, \bar{u}_k(t))$$

for all t in a positive duration W_1, and $\bar{u}_k(t)$ is a nearest point in $\Omega(x^*(t), t)$ to $u_k(t)$. But for each fixed $t \in W_1$,

$$\lim_{k\to\infty} x_k(t) = x^*(t)$$

and

$$\lim_{k\to\infty} |\bar{u}_k(t) - u_k(t)| = 0.$$

Thus

$$y\, \hat{\phi}(t) > \limsup_{k\to\infty} y \hat{f}(x_k(t), t, u_k(t)).$$

But this contradicts the weak convergence of $\hat{f}(x_k(t), t, u_k(t))$ to $\hat{\phi}(t)$. Therefore $\hat{\phi}(t) \in \hat{V}(x^*(t), t)$.

Now $\Omega(x^*(t), t)$ is a compact set in R^n, varying continuously with t. Also

$$\hat{\phi}(t) \in \hat{f}(x^*(t), t, \Omega(x^*(t), t)).$$

Then, by a slight extension of Lemma 3A of Chapter 2, which is left as an exercise, there exists a measurable function $u^*(t) \subset \Omega(x^*(t), t)$ such that

$$\hat{\phi}(t) = \hat{f}(x^*(t), t, u^*(t)).$$

Then $u^*(t)$ on $t_0^* \leq t \leq t_1^*$ is an admissible controller in $\mathscr{F}$ with the response $x^*(t)$ and with the minimum cost

$$C(u^*) = g(x^*(t_1^*)) + \int_{t_0^*}^{t_1^*} f^0(x^*(t), t, u^*(t))\, dt + \max_{t_0^* \leq t \leq t_1^*} \gamma(x^*(t)).$$

Finally let us remark that if $u_k(t)$ is not defined on all $t_0^* \leq t \leq t_1^*$, then we extend the domain of definition by assigning $u(t) \in \Omega(x, t)$ to obtain a bounded measurable function on the required interval. For suitably large k, the responses $x_k(t)$, with $x_k(t_0{}^k) \in X_0(t_0{}^k)$, will be defined on all $t_0^* \leq t \leq t_1^*$ and there constitute an equicontinuous family of functions. The rest of the argument then proceeds just as above. Q.E.D.

COROLLARY 1. *Let an initial time t_0^* on $\tau_0 \leq t_0^* < \tau_1$ be fixed, and let $\mathscr{F}_0 \subset \mathscr{F}$ consist of all the admissible controls $u(t)$ on various intervals $t_0^* \leq t \leq t_1$ in $[t_0^*, \tau_1]$. Assume* (1) *to* (5) *of the theorem and*

(a) *$\mathscr{F}_0$ is nonempty;*
(b) *$|x(t)| \leq b$ on $t_0^* \leq t \leq t_1$ for all $u(t) \in \mathscr{F}_0$;*
(c) *$\hat{V}(x, t)$ is convex in R^{n+1} for each (x, t).*

Then there exists an optimal control $u^*(t)$ on $t_0^* \leq t \leq t_1^*$ in $\mathscr{F}_0$ minimizing $C(u)$ among all $u \in \mathscr{F}_0$.

A similar existence theorem holds relative to $\mathscr{F}_{01} \subset \mathscr{F}_0$, the set of all admissible controllers on the fixed time interval $t_0^* \leq t \leq t_1^*$ in $[\tau_0, \tau_1]$.

COROLLARY 2. *Consider the control process in R^n*

$$(S) \qquad \dot{x} = A(x, t) + B(x, t)u$$

with cost

$$C(u) = g(x(t_1)) + \int_{t_0}^{t_1} [A^0(x(t), t) + B^0(x(t), t)\, u(t)]\, dt + \operatorname*{ess\,sup}_{t_0 \leq t \leq t_1} \gamma(x(t), u(t))$$

where the matrices A, B, A^0, B^0 are in C^1 in R^{n+1}, $g(x)$ and $\gamma(x, u)$ are continuous in R^{n+m}, and $\gamma(x, u)$ is a convex function of u for each fixed x. Assume that the restraint set $\Omega(x, t)$ is compact and convex for all (x, t). Then hypothesis (c) *is valid. If we also assume* (1) *to* (4) *and* (a), (b), *then the existence of an optimal control $u^*(t)$ on $t_0^* \leq t \leq t_1^*$ in $\mathscr{F}$ is assured.*

Proof. Let $u_k(t)$ on $t_0{}^k \leq t \leq t_1{}^k$ be a sequence of admissible controllers in $\mathcal{F}$ for which $C(u_k)$ decreases monotonically towards inf $C(u)$ for $u \in \mathcal{F}$, and let $x_k(t)$ be the corresponding responses steering $X_0(t_0{}^k)$ to $X_1(t_1{}^k)$. Now select a subsequence, still called $u_k(t)$, so that $t_0{}^k \to t_0^*$, $t_1{}^k \to t_1^*$, $x_k(t_0{}^k) \to x_0^* \in X_0(t_0^*)$, and $u_k(t)$ converges weakly to $u^*(t)$ on $t_0^* \leq t \leq t_1^*$. Again assume that $t_0{}^k \leq t_0^*$ and $t_1{}^k \geq t_1^*$, and use Ascoli's theorem to select a further subsequence of controllers so that the equicontinuous family of responses $x_k(t)$ converge,

$$\lim_{k\to\infty} \hat{x}_k(t) = \hat{x}^*(t) \quad \text{uniformly on} \quad t_0^* \leq t \leq t_1^*.$$

Here $\hat{x}_k(t) = (x_k{}^0(t), x_k(t))$, and $\hat{x}^*(t) = (x^{0*}(t), x^*(t))$ as in the theorem. By the stated convergence we note that

$$\lim_{k\to\infty} \hat{f}(x_k(t), t, u_k(t)) = \hat{f}(x^*(t), t, u^*(t))$$

weakly on $t_0^* \leq t \leq t_1^*$, where $\hat{f} = (A^0 + B^0u,\ A + Bu)$. Also, since $\hat{f}$ is linear in u, we compute

$$\hat{x}^*(t) = \lim_{k\to\infty} \left[\hat{x}_k(t_0{}^k) + \int_{t_0{}^k}^{t} \hat{f}(x_k(s), s, u_k(s))\, ds \right]$$

so

$$\hat{x}^*(t) = \hat{x}_0^* + \int_{t_0{}^*}^{t} f(x^*(s), s, u^*(s))\, ds.$$

Thus the function $\hat{\phi}(t)$ in the proof of Theorem 4 is just $\hat{f}(x^*(t), t, u^*(t))$ in this corollary.

We verify that $u^*(t)$ on $t_0^* \leq t \leq t_1^*$ is in $\mathcal{F}$ and that $\lim_{k\to\infty} C(u_k) = C(u^*)$. Just as in the proof of the theorem, $u^*(t)$ steers $x^*(t)$ from $x_0^* \in X_0(t_0^*)$ to $X_1(t_1^*)$ and the constraints are satisfied $h^1(x^*(t)) \geq 0, \ldots, h^r(x^*(t)) \geq 0$. Now suppose $u^*(t)$ lies outside $\Omega(x^*(t), t)$ for some positive time duration. Then there is a constant unit m-row vector y such that

$$y\, u^*(t) > \limsup_{k\to\infty} y\, \bar{u}_k(t)$$

for all t in a positive duration W_1, and $\bar{u}_k(t)$ is a nearest point in $\Omega(x^*(t), t)$ to $u_k(t)$. As in the proof of the theorem we deduce that

$$y\, u^*(t) > \limsup_{k\to\infty} y\, u_k(t)$$

for each t in W_1. But this contradicts the weak convergence of $u_k(t)$ to $u^*(t)$ on $t_0^* \leq t \leq t_1^*$. Therefore $u^*(t)$ on $t_0^* \leq t \leq t_1^*$ is an admissible controller in $\mathcal{F}$ with response $x^*(t)$. We now compute the cost $C(u^*)$.

Because

$$\lim_{k\to\infty} g(x_k(t_1^k)) + \int_{t_0^k}^{t_1^k} [A^0(x_k(t), t) + B^0(x_k(t), t)\, u_k(t)]\, dt$$
$$= g(x^*(t_1^*)) + \int_{t_0^*}^{t_1^*} [A^0(x^*(t), t) + B^0(x^*(t), t)\, u^*(t)]\, dt,$$

we need only verify that

$$\lim_{k\to\infty} \operatorname*{ess.\,sup}_{t_0^k\le t\le t_1^k} \gamma(x_k(t), u_k(t)) \ge \operatorname*{ess.\,sup}_{t_0^*\le t\le t_1^*} \gamma(x^*(t), u^*(t)).$$

In the contrary case there exists $\epsilon > 0$ and

$$\operatorname*{ess.\,sup}_{t_0^k\le t\le t_1^k} \gamma(x_k(t), u_k(t)) < \operatorname*{ess.\,sup}_{t_0^*\le t\le t_1^*} \gamma(x^*(t), u^*(t)) - \epsilon$$

for all large k. But then

$$\operatorname*{ess.\,sup}_{t_0^*\le t\le t_1^*} \gamma(x_k(t), u_k(t)) < \operatorname*{ess.\,sup}_{t_0^*\le t\le t_1^*} \gamma(x^*(t), u^*(t)) - \epsilon.$$

In fact, this implies that

$$\operatorname*{ess.\,sup}_{t_0^*\le t\le t_1^*} \gamma(x^*(t), u^k(t)) < \operatorname*{ess.\,sup}_{t_0^*\le t\le t_1^*} \gamma(x^*(t), u^*(t)) - \frac{\epsilon}{2}$$

for all large k. In this case there is a compact set W_2 of positive duration in $t_0^* \le t \le t_1^*$ whereon

$$\gamma(x^*(t), u_k(t)) < \gamma(x^*(t), u^*(t)) - \frac{\epsilon}{4}$$

for all large k. Now, by the argument on convex functions in Theorem 8 of Chapter 3 (wherein $h^0(t, u)$ plays the role of $\gamma(x^*(t), u)$) we find that

$$\liminf_{k\to\infty} \int_{W_2} \gamma(x^*(t), u_k(t))\, dt \ge \int_{W_2} \gamma(x^*(t), u^*(t))\, dt.$$

But this leads to the contradiction

$$\int_{W_2} \gamma(x^*(t), u^*(t))\, dt < \int_{W_2} \left[\gamma(x^*(t), u^*(t)) - \frac{\epsilon}{4}\right] dt.$$

Therefore we conclude that

$$\lim_{k\to\infty} C(u_k) \ge C(u^*)$$

and so $u^*(t)$ realizes the minimal cost.

Finally remark that if $u_k(t)$ is not defined on all $t_0^* \le t \le t_1^*$, then we extend its domain of definition as in the proof of Theorem 4. Proceed as above to define $u^*(t)$ on $t_0^* \le t \le t_1^*$ in $\mathcal{F}$ with response $x^*(t)$. Also in this case we must show that

$$\lim_{k\to\infty} \operatorname*{ess.\,sup}_{t_0{}^k \le t \le t_1{}^k} \gamma(x_k(t), u_k(t)) \ge \operatorname*{ess.\,sup}_{t_0{}^* \le t \le t_1{}^*} \gamma(x^*(t), u^*(t)).$$

In the contrary case, there exists $\eta > 0$ and $\epsilon > 0$

$$\operatorname*{ess.\,sup}_{t_0{}^k \le t \le t_1{}^k} \gamma(x_k(t), u_k(t)) < \operatorname*{ess.\,sup}_{t_0{}^*+\eta \le t \le t_1{}^*-\eta} \gamma(x^*(t), u^*(t)) - \epsilon$$

for all large k. But $u_k(t)$ is defined on $t_0^* + \eta \le t \le t_1^* - \eta$ and converges weakly to $u^*(t)$ on this interval. We then obtain a contradiction as above. Hence

$$\lim_{k\to\infty} C(u_k) \ge C(u^*)$$

in this case, and $u^*(t)$ on $t_0^* \le t \le t_1^*$ is the required optimal controller.

Q.E.D.

Remarks. The existence of an optimal controller for the problem of Corollary 2, with the initial time t_0^* or with the control duration $t_0^* \le t \le t_1^*$ specified as in Corollary 1, is easily proved. Note that the cost in Corollary 2 could be

$$C(u) = \operatorname*{ess.\,sup}_{t_0 \le t \le t_1} [\alpha \max_{1 \le i \le n} |x^i(t)| + \beta \max_{1 \le j \le m} |u^j(t)|]$$

for constants α and $\beta \ge 0$. For the nonlinear problem stated in Theorem 4, a permissible cost is the above $C(u)$ with $\beta = 0$. Thus we have proved the existence of optimal controllers for rather general *minimax problems*, that is, we minimize the maximum of $\gamma(x(t), u(t))$ over the control duration.

The importance of the convexity hypothesis in Theorem 4 is illustrated by the following example where the optimal control does not exist.

Example. Consider the control process in the plane

$$\dot{x} = -y^2 + u^2, \qquad \dot{y} = u$$

with restraint $|u(t)| \le 1$. We wish to steer $x(0) = y(0) = 0$ to the segment $X_1 = \{x = 1, |y| \le 1\}$ in minimal time $t^* > 0$. There is a uniform bound

$$|x(t)| + |y(t)| \le 12 \quad \text{on} \quad 0 \le t \le 2$$

for all measurable controllers $u(t)$ satisfying the restraint. Since $\dot{x}(t) \le 1$, we have a lower bound of $t^* \ge 1$. That is, for each controller $u(t)$ on $0 \le t \le t_1$ we compute

$$x(t_1) = \int_0^{t_1} [u(t)^2 - y(t)^2]\, dt = 1$$

only when $t_1 > 1$. To construct a minimizing sequence of controls, partition $0 \le t \le 2$ into segments of length $1/k$ and let $u_k(t)$ be $+1$ or -1 on alternate segments. Then the corresponding response satisfies

$$|y_k(t)| \leq \frac{1}{k} \quad \text{and} \quad \dot{x}_k(t) \geq 1 - \frac{1}{k^2} \quad \text{for} \quad k = 1, 2, 3, \ldots .$$

The response reaches the target X_1 at a time t_1^k on $1 < t_1^k \leq k^2/(k^2 - 1)$ and

$$\lim_{k\to\infty} t_1^k = 1.$$

Thus the minimal optimal time $t^* = 1$ is never achieved by any admissible controller. We note that the velocity set $\hat{V} = \{1, u, -y^2 + u^2\}$ is not convex in R^3 and so the basic existence theorem for optimal controllers is not applicable.

Intuitively, an "almost optimal" controller should hop rapidly back and forth between $u = +1$ and $u = -1$ so that $y(t) = \int_0^t u(s)\, ds$ is nearly zero and $x(t) = \int_0^t [u(s)^2 - y(s)^2]\, ds$ is nearly t. In each time interval $u(t)$ should spend half the time at $u = +1$ and half at $u = -1$; that is, $u(t) = +1$ with probability 1/2 and $u(t) = -1$ with probability 1/2 at each instant t. We next show that a relaxation of the concept of a controller to include probability distributions on Ω, at each instant t, enables us to prove a general existence theorem for optimal controllers even when the velocity set fails to be convex.

Definition. Consider a control process in R^n

$$\text{(S)} \qquad \dot{x} = f(x, t, u) \quad \text{in} \quad C^1 \quad \text{in} \quad R^{n+1+m}$$

with compact restraint set $\Omega(x, t) \subset R^m$ depending continuously on (x, t). A *relaxed controller* $\mu(t)$ on $t_0 \leq t \leq t_1$, with response $x(t)$, is an assignment of a probability measure on $\Omega(x(t), t)$ at each instant t. We consider only relaxed controllers of the form

$$\mu(t) = \alpha_1(t)\, \delta(u_1(t)) + \cdots + \alpha_{n+1}(t)\, \delta(u_{n+1}(t)),$$

where $\alpha_1(t) \geq 0, \ldots, \alpha_{n+1}(t) \geq 0$ are measurable functions and $\sum_{i=1}^{n+1} \alpha_i(t) \equiv 1$; $u_1(t), \ldots, u_{n+1}(t)$ are measurable functions in $\Omega(x(t), t)$, which are called a *chattering basis* for $\mu(t)$; and $\delta(u)$ is the δ-measure, which assigns a probability of 1 to each measurable subset of Ω containing u, and assigns zero to other sets. The response to $\mu(t)$ is defined by

$$x(t) = x_0 + \int_{t_0}^{t} \left[\int_{\Omega} f(x, t, u)\, d\mu \right] dt$$

or

$$x(t) = x_0 + \int_{t_0}^{t} f_\mu(x, t)\, dt$$

where $f_\mu(x, t) = \alpha_1(t) f(x, t, u_1(t)) + \cdots + \alpha_{n+1} f(x, t, u_{n+1}(t))$.

Let us note that a (classical) controller $u(t)$ can be interpreted as the relaxed controller $\delta(u(t))$, and hence a (classical) response is always a relaxed response. In order to interpret the response to classical and relaxed controllers we introduce the concept of a *multivalued differential equation*

$$\dot{x} \in U(x, t).$$

Here $U(x, t)$ is a nonempty set of tangent vectors at $x \in R^n$, for each time t on some interval $\tau_0 \leq t \leq \tau_1$. A solution $x(t)$ is an absolutely continuous curve, on a subinterval $t_0 \leq t \leq t_1$, whose tangent vector $\dot{x}(t)$ lies in the set $U(x(t), t)$ for almost all times.

LEMMA. *Consider the control process in* R^n

(S) $\qquad \dot{x} = f(x, t, u) \quad \text{in} \quad C^1 \quad \text{in} \quad R^{n+1+m}$

with compact restraint set $\Omega(x, t)$ *varying continuously with* (x, t) *in* $R^n \times [\tau_0, \tau_1]$.

Let the velocity set be $V(x, t) = f(x, t, \Omega(x, t))$. *Then a curve* $x(t)$ *on* $t_0 \leq t \leq t_1$ *is a classical response to* S *if and only if* $x(t)$ *is a solution of the multivalued differential equation*

$$\dot{x} \in V(x, t).$$

Let $H(V(x, t))$ *be the convex hull of* $V(x, t)$. *Then a curve* $x(t)$ *on* $t_0 \leq t \leq t_1$ *is a relaxed response to* S *if and only if* $x(t)$ *is a solution of*

$$\dot{x} \in H(V(x, t)).$$

If $V(x, t)$ *is convex at each* (x, t), *then every relaxed response* $x(t)$ *is a classical response. If* $V(x, t)$ *is not convex, but* $\Omega(x, t) = \Omega(t)$ *is independent of* x, *then every relaxed response* $x(t)$ *is the uniform limit of classical responses on* $t_0 \leq t \leq t_1$.

Proof. Let $u(t)$ on $t_0 \leq t \leq t_1$ be a classical controller with response $x(t)$. Then

$$\dot{x}(t) = f(x(t), t, u(t)) \in f(x(t), t, \Omega(x(t), t))$$

and so $x(t)$ is a solution of the multivalued differential equation

$$\dot{x}(t) \in V(x(t), t).$$

Conversely, let $x(t)$ on $t_0 \leq t \leq t_1$ satisfy the multivalued differential equation

$$\dot{x}(t) \in f(x(t), t, \Omega(x(t), t)).$$

We seek a measurable function $u(t) \subset \Omega(x(t), t)$ so $f(x(t), t, u(t)) = \dot{x}(t)$. But Lemma 3A of Chapter 2, with the modification of allowing Ω to depend continuously on t, guarantees the existence of the required controller $u(t)$, which has the response $x(t)$.

Next consider a relaxed controller

$$\mu(t) = \alpha_1(t)\,\delta(u_1(t)) + \cdots + \alpha_{n+1}(t)\,\delta(u_{n+1}(t))$$

with response $x(t)$ satisfying

$$\dot{x}(t) = \alpha_1(t) f(x, t, u_1(t)) + \cdots + \alpha_{n+1}(t) f(x, t, u_{n+1}(t)).$$

Then, at almost every instant t,

$$f(x(t), t, u_1(t)), f(x(t), t, u_2(t)), \ldots, f(x(t), t, u_{n+1}(t))$$

all lie in $V(x(t), t)$, and so

$$\dot{x}(t) \in H(V(x(t), t)).$$

Conversely let $x(t)$ on $t_0 \leq t \leq t_1$ be an absolutely continuous curve in R^n with

$$\dot{x}(t) \in H(V(x(t), t)).$$

Consider the continuous function

$$h(t, A) = \alpha_1 f(x(t), t, u_1) + \cdots + \alpha_{n+1} f(x(t), t, u_{n+1}),$$

where $A = (\alpha_1, \ldots, \alpha_{n+1}, u_1, u_2, \ldots, u_{n+1})$ varies over the subset of $R^{n+1+(n+1)m}$ defined by $\Sigma \times \Omega^{n+1}(t)$. Here Σ is the unit simplex in R^{n+1} and $\Omega^{n+1}(t) = \Omega(x(t), t) \times \cdots \times \Omega(x(t), t)$, with $n + 1$ factors.

At each time t and $A \in \Sigma \times \Omega^{n+1}(t)$, the point $h(t, A)$ lies in $H(V(x(t), t))$. In fact, since the convex hull of $V(x(t), t)$ is just the union of all simplices with vertices in $V(x(t), t)$, we note that

$$h(t, \Sigma \times \Omega^{n+1}(t)) = H(V(x(t), t).$$

Since $\dot{x}(t) \in h(t, \Sigma \times \Omega^{n+1}(t))$, Lemma 3A of Chapter 2 enables us to select a measurable

$$A(t) = (\alpha_1(t), \ldots, \alpha_{n+1}(t), u_1(t), \ldots, u_{n+1}(t)) \quad \text{on} \quad t_0 \leq t \leq t_1$$

so that

$$\dot{x}(t) = \alpha_1(t) f(x(t), t, u_1(t)) + \cdots + \alpha_{n+1}(t) f(x(t), t, u_{n+1}(t))$$

almost everywhere. That is, $x(t)$ is the response for the relaxed controller $\mu(t) = \alpha_1(t)\,\delta(u_1(t)) + \cdots + \alpha_{n+1}(t)\,\delta(u_{n+1}(t))$.

Suppose $V(x, t)$ is convex for all (x, t). Then $H(V(x, t)) = V(x, t)$. Hence a relaxed response $x(t)$ is an absolutely continuous curve in R^n with $\dot{x}(t) \in H(V(x(t), t) = V(x(t), t)$. Thus the relaxed response $x(t)$ is also a classical response.

Finally suppose that $V(x, t)$ is not necessarily convex. Let $x(t)$ on $t_0 \leq t \leq t_1$ be a relaxed response for a relaxed controller

$$\mu(t) = \alpha_1(t)\,\delta(u_1(t)) + \cdots + \alpha_{n+1}(t)\,\delta(u_{n+1}(t)).$$

We seek to approximate $x(t)$ by absolutely continuous curves $x_k(t)$ with $\dot{x}_k(t) \in V(x_k(t), t)$ for $k = 1, 2, 3, \ldots$. Now

$$x(t) = x_0 + \int_{t_0}^{t} [\alpha_1(s) f(x(s), s, u_1(s)) + \cdots + \alpha_{n+1}(s) f(x(s), s, u_{n+1}(s))]\, ds.$$

We can modify $\alpha(t) = (\alpha_1(t), \ldots, \alpha_{n+1}(t))$ on a small time duration so that $\alpha(t)$ is continuous and the response $x(t)$ is changed very slightly (in the uniform norm). We assume that such a modification has already been done and note that we still have $u_i(t) \in \Omega(t)$ for $i = 1, \ldots, n+1$ and $\dot{x}(t) \in H(V(x(t), t))$ as required.

Next define the vector

$$\tilde{\alpha}^{(k)}(t) = \begin{cases} (1, 0, 0, \ldots, 0, 0) & \text{on} \quad I_{k,1} \\ (0, 1, 0, \ldots, 0, 0) & \text{on} \quad I_{k,2} \\ \cdot \\ \cdot \\ \cdot \\ (0, 0, \ldots, 0, 1) & \text{on} \quad I_{k,n+1} \end{cases}$$

where $I_{k,j}$ is a finite sum of subintervals of $t_0 \leq t \leq t_1$ specified by $k \geq 1$ and $1 \leq j \leq n + 1$. To define $I_{k,j}$ divide $t_0 \leq t \leq t_1$ into k equal consecutive intervals by points $t_0 = t_{k0} < t_{k1} < t_{k2} < \cdots < t_{kk} = t_1$ and partition each of these k subintervals into $(n + 1)$ intervals whose lengths are proportional to $(\alpha_1(t_{kl}), \alpha_2(t_{kl}), \ldots, \alpha_{n+1}(t_{kl}))$ for $l = 0, 1, 2, \ldots, k - 1$. Then $I_{k,1}$ is the union of the first pieces in all the $t_{kl} < t < t_{k,l+1}$, and $I_{k,j}$ is similarly the union of the jth pieces in all of the $t_{kl} < t < t_{k,l+1}$. Then it is easy to verify that, for each interval I in $[t_0, t_1]$,

$$\lim_{k\to\infty} \int_I \tilde{\alpha}^{(k)}(t)\, dt = \int_I \alpha(t)\, dt,$$

and hence that

$$\lim_{k\to\infty} \tilde{\alpha}^{(k)}(t) = \alpha(t)$$

weakly on $t_0 \leq t \leq t_1$.

Now define the classical response $x_k(t)$ by

$$x_k(t) = x_0 + \int_{t_0}^{t} \sum_{i=1}^{n+1} \tilde{\alpha}_i^{(k)}(s) f(x_k(s), s, u_i(s))\, ds = x_0 + \int_{t_0}^{t} f(x_k(s), s, \tilde{u}(s))\, ds.$$

Note that $x_k(t)$ is the response (on all $t_0 \leq t \leq t_1$, as the estimate below shows) for the classical controller

$$\tilde{u}(t) = \begin{cases} u_1(t) & \text{on} \quad I_{k,1} \\ \cdot \\ \cdot \\ \cdot \\ u_{n+1}(t) & \text{on} \quad I_{k,n+1} \end{cases}$$

and $\tilde{u}(t) \in \Omega(t)$.

To show that $x_k(t)$ converges to $x(t)$ we estimate

$$|x(t) - x_k(t)| = \Bigg| \sum_{i=1}^{n+1} \int_{t_0}^{t} \alpha_i(s) f(x(s), s, u_i(s)) - \tilde{\alpha}_i^{(k)}(s) f(x(s), s, u_i(s))$$
$$+ \tilde{\alpha}_i^{(k)}(s) f(x(s), s, u_i(s)) - \alpha_i^{(k)}(s) f(x_k(s), s, u_i(s))\, ds \Bigg|$$

and

$$|x(t) - x_k(t)| \leq \epsilon_k + K \int_{t_0}^{t} |x(s) - x_k(s)|\, ds$$

where $\epsilon_k \to 0$ and K is bounded depending on the maximum of $|\partial f/\partial x|$. But then

$$|x(t) - x_k(t)| \leq \epsilon_k e^{K|t_1 - t_0|}$$

and

$$\lim_{k \to \infty} x_k(t) = x(t)$$

uniformly on $t_0 \leq t \leq t_1$. Q.E.D.

Remark. If $\Omega(x, t)$ depends on x, then the above proof merely constructs a sequence of absolutely continuous functions $x_k(t)$, which converge uniformly to the relaxed controller $x(t)$ on $t_0 \leq t \leq t_1$, but which are only approximate responses for controls $u_k(t)$, that is,

$$\operatorname{dist}\,(u_k(t), \Omega(x_k(t), t)) \leq \frac{1}{k}$$

and

$$\operatorname{dist}\,(\dot{x}_k(t), V(x_k(t), t)) \leq \frac{1}{k}.$$

However, in the case where $\Omega(x, t) = \Omega(t)$ we obtain the significant result that a classical optimal controller is also optimal among all relaxed controllers for the process (as in Theorem 4)

$$\text{(S)} \qquad \dot{x} = f(x, t, u)$$

with restraint $\Omega(t)$, and cost

$$C(u) = g(x(t_1)) + \int_{t_0}^{t_1} \left[\int_{\Omega} f^0(x(t), t, u)\, d\mu \right] dt + \max_{t_0 \leq t \leq t_1} \gamma(x(t)),$$

provided the target set $X_1(t) = R^n$. This follows easily from the expression

$$C(u) = g(x(t_1)) + x^0(t_1) + \max_{t_0 \leq t \leq t_1} \gamma(x(t))$$

where

$$x^0(t) = \int_{t_0}^{t} [\alpha_1(s) f^0(x(s), s, u_1(s)) + \cdots + \alpha_{n+1}(s) f^0(x(s), s\ u_{n+1}(s))]\, ds,$$

and the existence of classical responses $\hat{x}_k(t)$, which converge uniformly to the optimal relaxed response. If there was a target $X_1 \neq R^n$, say $X_1 = 0$ in R^n, then the classical responses $x_k(t)$ might fail to hit X_1 and so they would not be in the competition for the classical optimal minimal cost and hence no conclusion holds. However, even if $X_1 = 0$, we still can conclude that the classical optimal equals the relaxed optimal provided the process $\mathcal{S}$ is controllable near $x = 0$, $u = 0$, as is discussed in the next chapter.

The final result concerning relaxed controllers is a general existence theorem without the convexity hypothesis (c) of Theorem 4. However, with the inherent convexity properties of relaxed controllers, the following existence theorem is just an easy consequence of Theorem 4.

THEOREM 5. *Consider the nonlinear process in* R^n

$$(\mathcal{S}) \qquad \dot{x} = f(x, t, u) \quad \text{in} \quad C^1 \quad \text{in} \quad R^{n+1+m}.$$

The data are as follows:

1. *The initial and target sets $X_0(t)$ and $X_1(t)$ are nonempty compact sets varying continuously in R^n for all t in the basic prescribed compact interval $\tau_0 \leq t \leq \tau_1$.*
2. *The control restraint set $\Omega(x, t)$ is a nonempty compact set varying continuously in R^m for $(x, t) \in R^n \times [\tau_0, \tau_1]$.*
3. *The state constraints are (possibly vacuous, finite or infinite).*

$$h^1(x) \geq 0, \ldots, h^r(x) \geq 0$$

   where $h^1, \ldots, h^r$ are real continuous functions on R^1.

4. *The family $\mathcal{F}$ of admissible controllers consists of all relaxed controllers $\mu(t) = \alpha_1(t)\,\delta(u_1(t)) + \cdots + \alpha_{n+1}(t)\,\delta(u_{n+1}(t))$ on various time intervals $t_0 \leq t \leq t_1$ in $[\tau_0, \tau_1]$ such that each $\mu(t)$ has a response*

$$x(t) = x_0 + \int_{t_0}^{t} [\alpha_1(s) f(x(s), s, u_1(s)) + \cdots + \alpha_{n+1}(s) f(x(s), s, u_{n+1}(s))]\, ds$$

   on $t_0 \leq t \leq t_1$ steering $x(t_0) \in X_0(t_0)$ to $x(t_1) \in X_1(t_1)$, with some chattering basis

$$u_1(t) \in \Omega(x(t), t). \ldots, u_{n+1}(t) \in \Omega(x(t), t),$$

   and furthermore $h^1(x(t)) \geq 0, \ldots, h^r(x(t)) \geq 0$.

5. *The cost for each* $\mu(t) \in \mathcal{F}$ *is*

$$C(\mu) = g(x(t_1)) + \int_{t_0}^{t_1} [\alpha_1(s) f^0(x(s), s, u_1(s)) + \cdots + \alpha_{n+1}(s) f^0(x(s), s, u_{n+1}(s))] \, ds + \max_{t_0 \le t \le t_1} \gamma(x(t))$$

where $f^0 \in C^1$ *in* R^{n+1+m}, $g(x)$ *and* $\gamma(x)$ *are continuous in* R^n.

Assume

(a) *The set* $\mathcal{F}$ *of admissible relaxed controllers is not empty;*
(b) *There exists a uniform bound*

$$|x(t)| \le b \quad on \quad t_0 \le t \le t_1$$

for all responses $x(t)$ *to all relaxed controllers* $\mu(t) \in \mathcal{F}$.

Then there exists an optimal relaxed controller

$$\mu^*(t) = \alpha_1^*(t)\,\delta(u_1^*(t)) + \cdots + \alpha_{n+1}^*(t)\,\delta(u_{n+1}^*(t))$$

in $\mathcal{F}$ *minimizing* $C(\mu)$. *In this case the optimal controller chatters among the basis*

$$u_1^*(t), \ldots, u_{n+1}^*(t)$$

with probabilities $\alpha_1^*(t), \ldots, \alpha_{n+1}^*(t)$, *respectively, at each instant on* $t_0^* \le t \le t^*$.

Proof. Consider the control problem in R^n

$$(\mathcal{S}_r) \qquad \dot{x} = f_r(x, t, \tilde{u}) = \alpha_1 f(x, t, u_1) + \cdots + \alpha_{n+1} f(x, t, u_{n+1})$$

with classical controller

$$\tilde{u}(t) = (\alpha_1(t), \ldots, \alpha_{n+1}(t), u_1(t), \ldots, u_{n+1}(t))$$

in the compact restraint set $\Sigma \times \Omega^{n+1}(t)$. Here Σ is the unit simplex in R^{n+1} and $\Omega^{n+1}(t) = \Omega(x(t), t) \times \Omega(x(t), t) \times \cdots \times \Omega(x(t), t)$. The initial and target sets, the state constraints, and the cost

$$C_r(\tilde{u}) = g(x(t_1)) + \int_{t_0}^{t_1} f_r^0(x(s), s, \tilde{u}(s)) \, ds + \max_{t_0 \le t \le t_1} \gamma(x(t))$$

are just as above.

For each classical control $\tilde{u}(t)$ of $\mathcal{S}_r$ there corresponds a relaxed controller

$$\mu(t) = \alpha_1(t)\,\delta(u_1(t)) + \cdots + \alpha_{n+1}(t)\,\delta(u_{n+1}(t)) \quad \text{of} \quad \mathcal{S};$$

and, furthermore, every relaxed controller $\mu(t)$ of $\mathcal{S}$ appears in this manner. Moreover, the responses of $\mathcal{S}$ to $\mu(t)$ and of $\mathcal{S}_r$ to $\tilde{u}(t)$ are the same and these yield the same costs $C_r(\tilde{u}) = C(\mu)$.

However, for the problem $\mathcal{S}_r$, we note that the velocity set in R^{n+1}

$$\hat{V}_r(x, t) = \begin{cases} \alpha_1 f^0(x, t, \Omega(x, t)) + \cdots + \alpha_{n+1} f^0(x, t, \Omega(x, t)) \\ \alpha_1 f(x, t, \Omega(x, t)) + \cdots + \alpha_{n+1} f(x, t, \Omega(x, t)), \end{cases}$$

where $(\alpha_1, \ldots, \alpha_{n+1})$ varies over Σ, is necessarily convex for each (x, t). In fact, $\hat{V}_r(x, t) = H(\hat{V}(x, t))$ where $\hat{V}(x, t)$ is the velocity set for the original classical problem $\mathcal{S}$.

Therefore the classical control problem $\mathcal{S}_r$ satisfies all the hypotheses and conditions of Theorem 4 and hence a classical optimal controller $\tilde{u}^*(t) = (\alpha_1^*(t), \ldots, \alpha_{n+1}^*(t), u_1^*(t), \ldots, u_{n+1}^*(t))$ on $t_0^* \leq t \leq t_1^*$ exists minimizing $C_r(\tilde{u})$. But then

$$\mu^*(t) = \alpha_1^*(t)\, \delta(u_1^*(t)) + \cdots + \alpha_{n+1}^*(t)\, \delta(u_{n+1}^*(t))$$

is the required optimal relaxed controller for the given problem $\mathcal{S}$. Q.E.D.

COROLLARY. *Consider the control process in R^n*

$$(\mathcal{S}) \qquad \dot{x} = A(x, t) + B(x, t)u$$

with cost

$$C(u) = g(x(t_1)) + \int_{t_0}^{t_1} [A^0(x(t), t) + B^0(x(t), t)\, u(t)]\, dt + \max_{t_0 \leq t \leq t_1} \gamma(x(t))$$

where the matrices A, B, A^0, B^0 are in C^1 in R^{n+1}, $g(x)$ and $\gamma(x)$ are continuous in R^n. Assume that the compact restraint set $\Omega(x, t) \subset R^n$ depends continuously on (x, t) in $R^n \times [\tau_0, \tau_1]$. Then each relaxed controller in $\Omega(x(t), t)$

$$\mu(t) = \alpha_1(t)\, \delta(u_1(t)) + \cdots + \alpha_{n+1}(t)\, \delta(u_{n+1}(t)),$$

with response $x(t)$ of $\mathcal{S}$, determines a classical controller

$$\tilde{u}(t) = \alpha_1(t)\, u_1(t) + \cdots + \alpha_{n+1}(t)\, u_{n+1}(t)$$

in the relaxed restraint set $H(\Omega(x(t), t))$. Conversely, each classical controller $\tilde{u}(t) \subset H(\Omega(x(t), t)$ for $\mathcal{S}$ arises from some relaxed controller $\mu(t)$ in $\Omega(x(t), t)$ and, furthermore, $\tilde{u}(t)$ and $\mu(t)$ have the same cost.

Therefore an optimal relaxed controller $\mu^*(t)$ in $\Omega(x^*(t), t)$ determines an optimal classical controller $\tilde{u}^*(t)$ of $\mathcal{S}$ with the relaxed restraint $H(\Omega(x, t))$.

Proof. The correspondence $\mu(t) \rightarrow \tilde{u}(t)$ follows directly from the linearity of $\mathcal{S}$ and of the cost integrand as functions of u. The selection of some $\mu(t)$ which determines a given $\tilde{u}(t)$ is just as in the proof of the above theorem. The formulas for $C(\tilde{u})$ and $C(\mu)$ are identical and hence the corollary is proved. Q.E.D.

This corollary shows that, for linear processes, the relaxing of the restraint Ω to $H(\Omega)$ is equivalent to admitting relaxed controllers μ in Ω. Later we shall continue our study of measure controllers but first we show that such generalizations are not required for linear processes of the conventional type.

We obtain an existence theorem for optimal controllers, without convexity hypotheses and without any relaxation of the classical control restraints. The control enters the process in a nonlinear fashion but the basic dynamics of the process are linear. Thus we can use the underlying convexity that follows from Liapunov's results on the convexity of the range of a vector measure. We refer to the appendix of Chapter 2 and to some exercises below for these measure-theoretic ideas.

THEOREM 6. *Consider the process in* R^n,

$$(\mathcal{S}) \qquad \dot{x} = A(t)x + B(t, u),$$

where $A(t)$ *and* $B(t, u)$ *are continuous in* R^{1+m}. *The data are as follows:*

1. *The initial and target sets* $X_0(t)$ *and* $X_1(t)$ *are nonempty compact sets varying continuously in* R^n *for all* t *in the basic prescribed compact interval* $\tau_0 \leq t \leq \tau_1$.
2. *The control restraint* $\Omega(t)$ *is a nonempty compact set varying continuously in* R^m *for* $\tau_0 \leq t \leq \tau_1$.
3. *The integral constraints are (possibly vacuous or finite)*

$$\int_{t_0}^{t_1} h^1(t, u(t))\, dt \geq 0, \ldots, \int_{t_0}^{t_1} h^r(t, u(t))\, dt \geq 0,$$

   where $h^1, \ldots, h^r$ *are real continuous functions in* R^{1+m}.
4. *The family* $\mathcal{F}$ *of admissible controllers consists of all measurable functions* $u(t)$ *on various time intervals* $t_0 \leq t \leq t_1$ *in* $[\tau_0, \tau_1]$ *such that each* $u(t)$ *has a response* $x(t)$ *on* $t_0 \leq t \leq t_1$ *steering* $x(t_0) \in X_0(t_0)$ *to* $x(t_1) \in X_1(t_1)$ *and the restraint* $u(t) \subset \Omega(t)$ *on* $t_0 \leq t \leq t_1$ *as well as the integral constraints of* (3) *are fulfilled.*
5. *The cost of each* $u \in \mathcal{F}$ *is*

$$C(u) = g(x(t_1)) + \int_{t_0}^{t_1} A^0(t)\, x(t) + B^0(t, u(t))\, dt$$

   where $g(x)$, $A^0(t)$, $B^0(t, u)$ *are continuous in all* (x, t, u),

Assume that the set $\mathcal{F}$ *of admissible controllers is not empty. Then there exists an optimal controller* $u^*(t)$ *on* $t_0^* \leq t \leq t_1^*$ *in* $\mathcal{F}$ *minimizing* $C(u)$.

Proof. Let us augment the system $\mathcal{S}$ to obtain

$$(\hat{\mathcal{S}}) \qquad \dot{x}^0 = A^0(t)x + v^0(t)$$

$$\dot{x} = A(t)x + v(t)$$

$$\dot{x}^\alpha = v^\alpha(t) \qquad \alpha = 1, 2, \dots, r$$

and consider the family $\hat{\mathcal{F}}$ of all measurable controllers

$$\hat{v}(t) = (v^0(t), v(t), v^\alpha(t))$$

on $\tau_0 \le t_0 \le t \le t_1 \le \tau_1$ steering a response $\hat{x}(t) = (x^0(t), x(t), x^\alpha(t))$ in R^{1+n+r} from

$$\hat{X}_0(t_0) = (0, X_0(t_0), 0) \quad \text{to} \quad \hat{X}_1(t_1) = (x^0, X_1(t_1), x^\alpha(t_1))$$

with $x^\alpha(t_1) \ge 0$, and $\hat{v}(t)$ satisfying the relaxed restraint $\hat{v}(t) \subset H(\hat{\Omega}(t))$, where

$$\hat{\Omega}(t) = B^0(t, \Omega(t)) \times B(t, \Omega(t)) \times h^1(t, \Omega(t)) \times \cdots \times h^r(t, \Omega(t)).$$

Note that $\hat{\mathcal{F}}$ is not empty, since

$$v^0(t) = B^0(t, u(t))$$

$$v(t) = B(t, u(t))$$

$$v^\alpha(t) = h^\alpha(t, u(t)),$$

with $u(t) \in \mathcal{F}$, specifies an admissible augmented control. Also the basic linearity of $\hat{\mathcal{S}}$ in $\hat{x}$ establishes a uniform bound $|\hat{x}(t)| \le b$ for all responses. Since $\hat{\mathcal{S}}$ is linear in the control $\hat{v}$, and since $H(\hat{\Omega}(t))$ is a continuously varying convex restraint set in R^{1+m+r}, Theorem 4 applies directly to show that there exists an optimal $\hat{v}^*(t)$ on $t_0^* \le t \le t_1^*$ minimizing $g(x(t_1)) + x^0(t_1)$.

Let the optimal response be $\hat{x}^*(t) = (x^{0*}(t), x^*(t), x^{\alpha*}(t))$ on $t_0^* \le t \le t_1^*$. Then the set of attainability $K_{H(\hat{\Omega})}(t)$ in R^{1+n+r}, consisting of responses initiating at $\hat{x}^*(t_0^*)$, meets the target set $\hat{X}_1(t)$ (which is compact when $|x^0| \le b$, $|x^\alpha| \le b$) at the instant $t = t_1^*$ so as to minimize the real function $g(x) + x^0$ on the intersection. But Theorem 1A in the appendix of Chapter 2 asserts that $K_{\hat{\Omega}}(t) \equiv K_{H(\hat{\Omega})}(t)$ (the slight extension to the present case where $\hat{\Omega}(t)$ depends on t is left to an exercise below). That is, there exists a controller $\bar{v}^*(t) \subset \hat{\Omega}(t)$ on $t_0^* \le t \le t_1^*$ that equally well realizes the minimal cost.

But now, Lemma 3A of the same appendix—also generalized to the case where $\Omega(t)$ depends on t—asserts the existence of an admissible controller

$u^*(t)$ on $t_0^* \leq t \leq t_1^*$ in $\mathcal{F}$ such that

$$\bar{v}^{0*}(t) = B^0(t, u^*(t))$$

$$\bar{v}^*(t) = B(t, u^*(t))$$

$$\bar{v}^{\alpha*}(t) = h^\alpha(t, u^*(t)).$$

Thus $u^*(t)$ realizes the same minimal cost $C(u^*) = g(x^*(t_1^*)) + x^{0*}(t_1^*)$. Since every admissible controller $u(t) \in \mathcal{F}$ determines some augmented controller $\hat{v}(t) \in \hat{\mathcal{F}}$, the cost $C(u) \geq C(u^*)$ and $u^*(t)$ is the required optimal controller for $\mathcal{S}$.   Q.E.D.

The usual remarks apply concerning the existence of optimal controllers, under the conditions of Theorem 6, with fixed initial time t_0^* or on a fixed interval $t_0^* \leq t \leq t_1^*$ in $[\tau_0, \tau_1]$.

Note that the essential idea of Theorem 6 is that, for linear processes, the relaxation of the restraint Ω to $H(\Omega)$, or the admission of relaxed controllers μ in Ω, does not improve the performance over the use of classical controllers $u(t) \subset \Omega$.

We next obtain an existence theorem for optimal controllers for nonlinear processes admitting impulse and other generalized controls. The response will be defined by an integral equation and can be discontinuous, with jumps corresponding to the *impulse controllers*. Thus some special care is needed in specifying the technical description of this control problem.

Let $u(t)$, on an interval $\mathfrak{I}$ in R^1, lie in R^m and define the total variation

$$\mathrm{var}\ u(t) = \sup \sum_{j=0}^{k} |u(t_{j+1}') - u(t_j')|$$

where $t_0' < t_1' < \cdots < t_k' < t_{k+1}'$ is an arbitrary finite set of points in $\mathfrak{I}$ and the supremum is computed over all such finite sequences in $\mathfrak{I}$. The function $u(t)$ is of bounded variation in $\mathfrak{I}$ in case $\mathrm{var}\ u(t) < \infty$, and this obtains if and only if each component of $u(t)$ is of bounded variation in $\mathfrak{I}$. If $\mathfrak{I}$ is compact and $u(t)$ is Lipschitz continuous on $\mathfrak{I}$, then clearly $\mathrm{var}\ u(t) < \infty$. But a function $u(t)$ of bounded variation can have a countable number of (first kind) discontinuities. We shall always normalize (by redefining on a countable point set) such functions $u(t)$ on an open interval $\mathfrak{I}$ to be continuous from the right.

If $u(t)$ is of bounded variation on an open interval $\mathfrak{I}$, then we define the signed measure Du as the (generalized) derivative of $u(t)$. That is, on each subinterval $t_j' < t \leq t_{j+1}'$

$$Du(t_j', t_{j+1}'] = u(t_{j+1}') - u(t_j')$$

and then the signed measure is defined on all Lebesgue sets of $\mathfrak{J}$ by the usual requirements of countable additivity. Every (vector) signed measure on $\mathfrak{J}$ arises from some function of bounded variation and two such functions yield the same measure just in case they differ by a constant. If $u(t)$ is also continuous, then the signed measure Du assigns a zero weight to each point of $\mathfrak{J}$, but if $u(t)$ has a jump at t', then

$$Du[t'] = u(t') - u(t'-) \equiv J(u(t')).$$

Thus $\int_{\mathfrak{J}} Du$ is the ordinary Riemann-Stieltjes integral. For instance, in case $n = 1$ and

$$u(t) = \begin{cases} 0 & \text{on} \quad -\infty < t < 0 \\ 1 & \text{on} \quad 0 \leq t < \infty, \end{cases}$$

then Du is a δ-function, or more precisely, Du is the measure assigning the weight $+1$ to each measurable set containing $t = 0$ and assigning the weight zero to sets which do not contain $t = 0$.

Let $u(t)$ be of bounded variation on an open interval $\mathfrak{J}$ in R^1. Then we can consider the restriction of the signed measure Du to any subset of $\mathfrak{J}$. In particular, the norm of Du on a compact interval $t_0 \leq t \leq t_1$ in $\mathfrak{J}$ is defined to be

$$\|Du\| = \int_{t_0}^{t_1} |Du| = |J(u(t_0))| + \operatorname*{var}_{t_0 \leq t \leq t_1} u(t).$$

Now consider a generalized or *impulse differential system* in R^n

$$Dx = f(x, t, u) + e(t)\, Du,$$

where $u(t)$ is of bounded variation on an open interval $\mathfrak{J}$ and $f(x, t, u)$ and $e(t)$ are of class C^1 everywhere. Then a solution $x(t)$ through x_0 at time $t_0 \in \mathfrak{J}$ is any function of bounded variation on an open neighborhood of $t = t_0$ (and thereon right continuous) satisfying the integral equation

$$x(t) = x_0 + \int_{t_0}^{t} f(x(s), s, u(s))\, ds + \int_{t_0}^{t} e(s)\, Du,$$

where the integral is in the usual Riemann-Stieltjes sense. Note that

$$x(t_0-) = x_0$$

and it is in this special sense the initial value is assumed. The existence and uniqueness theorems for the solution of this integral equation, and the continuous dependence of the solution $x(t)$ upon the initial conditions, can be proved by successive approximations.

A controller for the impulse process

$$Dx = f(x, t, u) + e(t)\, Du$$

on a compact interval $t_0 \leq t \leq t_1$ is a function of bounded variation $u(t)$ on some open neighborhood $\mathfrak{I}$ of $[t_0, t_1]$ and thereon defining a response $x(t)$ steering the initial state $x(t_0-) = x_0$ to some prescribed target $x(t_1)$. Note that the norm of the corresponding signed measure Du depends on the jump $J(u(t_0))$ as well as the total variation of $u(t)$ on $t_0 \leq t \leq t_1$. Also note that the degenerate interval $t_0 \leq t \leq t_0$, consisting of just the single instant $t = t_0$, is allowed, and such instantaneous jumps into the target must be permitted as is illustrated by the following example.

Example. Consider the impulse process in R^1

$$Dx = u + Du$$

with the scalar controller $u(t)$ of bounded variation on some neighborhood of an interval $0 \leq t \leq t_1$. We wish to steer $x_0 = -2$ to $x_1 = 0$ with the control restraints $|u(t)| \leq 1$, $\|Du\| \leq 1$ so as to minimize the cost $C(u) = \int_0^{t_1} |u(t)|\, dt$.

It is easy to see that the control

$$u^*(t) = \begin{cases} 0 & \text{on} \quad -\infty < t < 0 \\ 1 & \text{on} \quad 0 \leq t \leq t_1 = 1 \end{cases}$$

with the response

$$x^*(t) = \begin{cases} -2 & \text{on} \quad -\infty < t < 0 \\ -1 + t & \text{on} \quad 0 \leq t \leq 1 \end{cases}$$

steers x_0 to x_1 in minimal possible time $t_1 = 1$, since when $u(t) = +1$ the response has the maximal possible positive velocity, and we here combine this with the maximal positive jump that is allowed. Thus the control $u^*(t)$ also achieves the minimal cost

$$C(u^*) = \int_0^1 |u^*(t)|\, dt = 1$$

since $\dot{x}(t) = u$ on the intervals between jumps of $u(t)$ and we have augmented $x_0 = -2$ as much as possible by jumps. Therefore the optimal controller is achieved by superimposing $u \equiv 1$ with a δ-function impulse at $t = 0$.

Now let us modify the control problem by weakening the restraint to $|u(t)| \leq 2$, $\|Du\| \leq 2$. Then the optimal control is

$$u^+(t) = \begin{cases} 0 & \text{on} \quad -\infty < t < 0 \\ 2 & \text{on} \quad 0 \leq t \leq t_1 \end{cases}$$

so

$$x^+(t) = \begin{cases} -2 & \text{on} \quad -\infty < t < 0 \\ 0 & \text{on} \quad t = 0 \end{cases}$$

Thus the minimal time $t_1 = 0$ and the minimal cost $C(u^+) = 0$. If this impulse jump from $x_0 = -2$ into the target $x_1 = 0$ were not allowed, then no optimal controller would exist. Of course, the impulse controller $u^*(t)$ can be approximated by a smooth C^1 controller $\tilde{u}(t)$ and response $\tilde{x}(t) = \int_0^{t_1} [\tilde{u}(t) + \dot{\tilde{u}}(t)]\, dt$, approximating the minimal cost. As the function $\tilde{u}(t)$ becomes steeper on a shorter interval $0 \leq t \leq t_1$, and $\tilde{u}(t)$ approaches the δ-function $u^+(t)$, the cost approaches zero. However, the zero cost cannot be achieved by any smooth controller, and for the existence of an optimal controller we require an impulse input.

THEOREM 7. *Consider the impulse control process in R^n*

$$(\mathcal{S}) \qquad Dx = f(x, t, u) + e(t)\, Du$$

where $f(x, t, u)$ are in C^1 in R^{n+1+m}. The data are as follows:

1. *The initial and target sets $X_0(t)$ and $X_1(t)$ are nonempty, compact sets varying continuously in R^n on the basic compact time interval $\tau_0 \leq t \leq \tau_1$.*
2. *The control restraint $\Omega(x, t)$ is a nonempty compact set in R^m that is continuous in (x, t) in $R^n \times [\tau_0, \tau_1]$.*
3. *The response constraints (possibly vacuous, finite, or infinite)*

$$h^1(x, t, u) \geq 0, \ldots, h^r(x, t, u) \geq 0$$

   where each $h(x, t, u)$ is continuous in R^{n+1+m}.
4. *The family $\mathcal{F}$ of controllers consists of functions of bounded variation $u(t)$ on various intervals $t_0 \leq t \leq t_1$ for $\tau_0 \leq t_0 \leq t_1 \leq \tau_1$ [actually, $u(t)$ is of bounded variation on an open neighborhood of $t_0 \leq t \leq t_1$ and is thereon right continuous with response $x(t)$ and induced signed measure Du]. Furthermore $u(t)$ on $(t_0-) \leq t \leq t_1$ satisfies the restraints $u(t) \subset \Omega(x(t), t)$, $h^1(x(t), t, u(t)) \geq 0, \ldots, h^r(x(t), t, u(t)) \geq 0$, and $\|Du\| \leq E$ for a prescribed finite bound $E \geq 0$. Moreover the response $x(t)$ steers $x(t_0-) \in X_0(t_0)$ to $x(t_1) \in X_1(t_1)$.*
5. *The cost of each controller $u(t)$ on $t_0 \leq t \leq t_1$ in $\mathcal{F}$ is*

$$C(u) = g(x(t_1), u(t_1)) + \int_{t_0}^{t_1} f^0(x(t), t, u(t))\, dt$$
$$+ \int_{t_0-}^{t_1} g^0(t)\, Du + \gamma(\sup |x(t)|, \|Du\|)$$

   where g, f^0, g^0, γ are continuous for all real arguments, and γ is monotone nonincreasing in each of its two arguments.

Assume

(a) *The control family $\mathfrak{F}$ is not empty;*
(b) *There exists a uniform bound $|x(t)| \leq b$ on $(t_0-) \leq t \leq t_1$ for all responses to all controllers in $\mathfrak{F}$.*

Then there exists an optimal controller $u^*(t)$ on $t_0^* \leq t \leq t_1^*$ in $\mathfrak{F}$ minimizing $C(u)$.

Proof. Let $u_k(t)$ with responses $x_k(t)$, $k = 1, 2, 3, \ldots,$ be a sequence of controllers on $t_0^k \leq t \leq t_1^k$ such that

$$t_0^k \to t_0^*, \qquad t_1^k \to t_1^*,$$

and $C(u_k)$ tends monotonically to the infimum of all possible costs. For convenience, redefine

$$u_k(t) = u_k(t_0^k-) \quad \text{for} \quad t < t_0^k$$

and

$$u_k(t) = u_k(t_1^k) \quad \text{for} \quad t > t_1^k$$

and let $x_k(t)$ be the response to this controller on some neighborhood of $t_0^* - \epsilon \leq t \leq t_1^* + \epsilon$, $\epsilon > 0$. Fix $\epsilon > 0$ suitably small, and then for all large k the responses $x_k(t)$ are defined and uniformly bounded on $t_0^* - 2\epsilon < t < t_1^* + 2\epsilon$.

Since $|u_k(t)| + \text{var } u_k(t)$ is uniformly bounded on $t_0^* - \epsilon \leq t \leq t_1^* + \epsilon$, the Ascoli theorem asserts the existence of a subsequence (still called $u_k(t)$) that converges pointwise to a limit function $u^*(t)$ of bounded variation. We normalize $u^*(t)$ to be right continuous on $t_0^* - \epsilon \leq t \leq t_1^* + \epsilon$ (by changing its values at only a countable set of points that do not contain the endpoints) and let $x^*(t)$ be the corresponding solution of

$$x^*(t) = x^*(t_0^* - \epsilon) + \int_{t_0^*-\epsilon}^{t} f(x(s), s, u^*(s))\, ds + \int_{t_0^*-\epsilon}^{t} e(s)\, Du^*$$

where $x_k(t_0^* - \epsilon) \to x^*(t_0^* - \epsilon)$. We also infer that

$$\text{var } u^* \leq \liminf_{k\to\infty} \text{var } u_k \quad \text{on} \quad t_0^* - \epsilon \leq t \leq t_1^* + \epsilon$$

and hence $\| Du^* \| \leq E$ on $t_0^* \leq t \leq t_1^*$.

Since the functions $x_k(t)$ are of uniform bounded variation, a subsequence [still called $x_k(t)$] converges, and, using the Lebesgue convergence theorem and the Helly-Bray theorem, we obtain

$$\lim_{k\to\infty} x_k(t) = x^*(t) \quad \text{on} \quad t_0^* - \epsilon \leq t \leq t_1^* + \epsilon,$$

excluding the points of discontinuity of $u^*(t)$. Hence $x^*(t)$ is defined on all $t_0^* - \epsilon \leq t \leq t_1^* + \epsilon$. It is easy to verify that

$$\lim_{k \to \infty} x_k(t_0^k -) = x^*(t_0^* -)$$

and that $u^*(t)$ satisfies all the restraints for an admissible controller in $\mathcal{F}$.

The Helly-Bray theorem asserts

$$\lim_{k \to \infty} \int_{t_0^* - \epsilon}^{t_1^* + \epsilon} g^0(s)\, Du_k = \int_{t_0^* - \epsilon}^{t_1^* + \epsilon} g^0(s)\, Du^*.$$

Since

$$\sup |x^*(t)| \leq \liminf_{k \to \infty} (\sup |x_k(t)|),$$

and

$$\| Du^* \| \leq \liminf_{k \to \infty} \| Du_k \|,$$

we compute

$$C(u_k) \to C(u^*).$$

Hence $u^*(t)$ on $t_0^* \leq t \leq t_1^*$ is the required optimal controller. Q.E.D.

COROLLARY 1. *Consider the linear impulse process in* R^n

$$(\mathcal{L}) \qquad Dx = A(t)x + B(t)u + e(t)\, Du$$

where $A(t)$, $B(t)$, *and* $e(t)$ *are continuous in* R^1. *Assume the theorem hypotheses* (1) *to* (5) *and* (*a*). *Then the uniform bound required in* (*b*) *necessarily exists and the optimal controller* $u^*(t)$ *on* $t_0^* \leq t \leq t_1^*$ *exists.*

COROLLARY 2. *Consider the nonlinear process in* R^n

$$(\mathcal{S}) \qquad \dot{x} = f(x, t, u)$$

as in the theorem with $e(t) \equiv 0$. *Assume* (1) *to* (5), *including the restraint* $\| Du \| \leq E$, *the same cost* $C(u)$, *and the conditions* (a) *and* (b). *Then an optimal controller* $u^*(t)$ *on* $t_0^* \leq t \leq t_1^*$ *exists in* $\mathcal{F}$.

The usual remarks apply concerning the existence of an optimal controller among the subclass of $\mathcal{F}$ where the initial time, or the time interval, is fixed.

EXERCISES

1. Generalize the existence Theorem 4 to the process in R^n

$$(\mathcal{S}) \qquad \dot{x} = f(x, t, u)$$

where $f(x, t, u)$ is piecewise continuous in t on $\tau_0 \leq t \leq \tau_1$. That is, there exists a finite partition $\tau_0 = \sigma_0 < \sigma_1 < \sigma_2 < \cdots \sigma_l = \tau_1$ such that for each closed interval $\sigma_i \leq t \leq \sigma_{i+1}$, the functions $f(x, t, u)$ and

$(\partial f/\partial x)(x, t, u)$ are continuous in $R^n \times [\sigma_i, \sigma_{i+1}] \times R^m$. The other conditions are as in Theorem 4.

2. Generalize Corollary 2 of Theorem 4 to the process in R^n

$$(\mathcal{S}) \qquad \dot{x} = A(x, t) + B(x, t)u$$

with cost

$$C(u) = g(x(t_1)) + \int_{t_0}^{t_1} [A^0(x(t), t) + h^0(t, u(t))]\, dt + \operatorname*{ess.\,sup.}_{t_0 \le t \le t_1} \gamma(x(t), u(t))$$

where $h^0(t, u)$ is continuous in (t, u) and convex in u for each fixed t. Assume that the restraint set $\Omega(x, t)$ is compact and convex for all (x, t) and depends on these arguments continuously. All the other conditions are just as in the corollary.

3. Consider the control process in R^n

$$(\mathcal{S}) \qquad \dot{x} = f(x, t, u) \quad \text{in} \quad C^1 \quad \text{in} \quad R^{n+1+m}$$

with initial state x_0 at time t_0 and fixed target set G. The admissible controllers are absolutely continuous vectors $u(t)$ on intervals $t_0 \le t \le t_1$ with restraints $|u(t)| \le 1$, $|\dot{u}(t)| \le 1$. Show that a change of notation leads to a *bounded phase coordinate* problem with admissible measurable controllers.

4. The *problem of Bolza* in the calculus of variations concerns the minimum of an integral $C = \int_{t_1}^{t_2} f^0(z, t, \dot{z})\, dt$ over all absolutely continuous paths $z(t) \subset R^n$ joining two points z_0 and z_1, and satisfying a differential constraint

$$\dot{z} = w(z, t).$$

Introduce new notation to reduce this problem to a standard optimal control problem.

5. Generalize the results of the measure theoretic Lemmas 1A, 2A, 3A, and Theorem 1A of the appendix to Chapter 2 to the case where $\Omega(t)$ is a compact set varying continuously with time.

6. Consider the control process in R^n

$$(\mathcal{S}) \qquad \dot{x} = f(x, t, u) \quad \text{in} \quad C^1 \quad \text{in} \quad R^{n+1+m}$$

with a compact (nonconvex) restraint set $\Omega \subset R^m$. The admissible controllers are all Lipschitz continuous functions $u(t) \subset \Omega$ on $t_0 \le t \le t_1$ satisfying

$$|u(t') - u(t'')| \le k\,|t' - t''| \quad \text{for} \quad t_0 \le t' \le t'' \le t_1$$

and a preassigned constant k. State and prove an existence theorem for an optimal controller for this process.

7. Let $U(x, t)$ be a nonempty convex compact set in R^n that depends continuously on $(x, t) \in R^n \times R^1$. Let $x_0 \in R^n$ be an initial state at time t_0 and prove the existence of a solution of the multivalued differential equation

$$\dot{x} \in U(x, t)$$

with $x(t_0) = x_0$.

Let $t_1 > t_0$ be such that every solution of the multivalued differential equation initiating at x_0 at time t_0 exists on $t_0 \leq t \leq t_1$ (prove existence of t_1.) Then the set $K(x_0, t_1)$ of attainability is compact.

8. Consider the control process in R^n

$$(\mathcal{S}) \qquad \dot{x} = A(x, t) + B(x, t)u \quad \text{in} \quad C^1 \quad \text{in} \quad R^{n+1+m},$$

with initial state x_0 at time t_0. Let the restraint set Ω be closed and convex in R^m and consider admissible controllers $u(t)$ on $t_0 \leq t \leq t_1$ with $u(t) \subset \Omega$ and $\int_{t_0}^{t_1} |u(t)|^p \, dt \leq 1$ for a given $p > 1$. Assume that each admissible controller yields a response $x(t)$ that is uniformly bounded, $|x(t)| \leq b$ on $t_0 \leq t \leq t_1$. State and prove the existence theorem for an optimal controller $u^*(t)$ on $t_0 \leq t \leq t_1$ steering x_0 to a compact target X_1 and minimizing the cost

$$C(u) = \int_{t_0}^{t} [A^0(x, t) + B^0(x, t)u] \, dt.$$

[Hint: Let $u_k(t)$ be a sequence of admissible controls with $C(u_k)$ tending to the infimum of costs, and $u_k \to u^*$ weakly in L_p. Select a further subsequence so the responses $x_k(t)$ converge weakly to some function $x^*(t)$. Show that $\lim_{k\to\infty} x_k(t) = x^*(t)$ on $t_0 \leq t \leq t_1$. Hölder's inequality shows that $u^*(t)$ has the response $x^*(t)$ and cost $C(u^*) = \lim_{k\to\infty} C(u_k)$.]

9. Consider the impulse control system in R^n

$$(\mathcal{L}) \qquad Dx = A(t)x + B(t)\, Du$$

where $A(t)$ and $B(t)$ are continuous matrices on R^1. We seek to steer x_0 at time $\tau_0 = 0$ to x_1 at time $\tau_1 = 1$ by an m-controller $u(t) \in \mathcal{G}_p$ $(1 \leq p \leq \infty)$ with minimum norm

$$\|u\|_{v,p} = STV_p u = \sup \sum_{i=1}^{\nu} |u(t_i) - u(t_{i-1})|_p.$$

Here the strong total variation is computed from the supremum over all finite partitions

$$0 = t_0 < t_1 < \cdots < t_\nu = 1, \quad \text{and} \quad |u|_p = \left(\sum_{i=1}^{m} |u^i|^p\right)^{1/p}$$

for $1 \leq p < \infty$, $|u|_\infty = \max\limits_{1 \leq i \leq m} |u^i|$; hence this is finite if and only if $u(t)$ is of bounded variation on $0 \leq t \leq 1$. The Banach space $\mathcal{G}_p$ consists of all functions $u(t)$ on $0 \leq t \leq 1$, with finite strong total p-variation, and normalized by $u(0) = 0$ and $u(t)$ right continuous on $0 < t < 1$. Then each $u(t) \in \mathcal{G}_p$ defines a Stieltjes-Lebesgue measure Du on $0 \leq t \leq 1$ (as in Theorem 7), and $\mathcal{G}_p$ is the dual space of $\mathcal{S}_q$. $\Big(1/p + 1/q = 1$, and $\mathcal{S}_q$ consists of all continuous m-vectors $y(t)$ on $0 \leq t \leq 1$ with norm $\|y\|_{\infty,q} = \sup\limits_{0 \leq t \leq 1} |y(t)|_q$; the action of $\mathcal{G}_p$ on $\mathcal{S}_q$ computed by $\displaystyle\int_0^1 y(t) \cdot Du.\Big)$

If there is one admissible controller in $\mathcal{G}_p$ steering x_0 to x_1, prove there exists an optimal $u^*(t)$ with minimal norm in $\mathcal{G}_p$. [Hint: Use the weak * compactness of the closed unit ball in $\mathcal{G}_p$ and proceed as usual.]

10. Consider the impulse control process in R^n

$$(\mathcal{S}) \qquad Dx = f(x, t, u) + e(t)\, Du$$

where $f(x, t, u)$ and $e(t)$ are in C^1 in R^{n+1+m}. Let $u(t)$ be of bounded variation and right continuous on an open interval about $t = t_0$. Prove there exists a unique (local) response $x(t, x_0)$ with $x(t_0-) = x_0$. Show that $x(t, x_0)$ is continuous in x_0 for each fixed t.

11. Consider the control process in R^n

$$(\mathcal{S}) \qquad \dot{x} = A(x) + B(x)u$$

where $A(x)$, $B(x)$ are in C^1 in R^n and $A(0) = 0$. The initial state is $x_0 \neq 0$ at time $t_0 = 0$ and the target is the origin $x_1 = 0$. Assume that there exists a measurable controller $u(t)$ on $0 \leq t \leq T$ steering x_0 to x_1 and satisfying the m-cube restraint

$$u(t) \subset \Omega_{\bar{c}}: \max |u^i| \leq \bar{c},$$

for some constant $\bar{c} > 0$. Also assume that all responses $x(t)$ to controls $u(t) \subset \Omega_{\bar{c}}$ on $0 \leq t \leq T$ are uniformly bounded.

(a) On the fixed interval $0 \leq t \leq T$ show that there exists an optimal controller $u^*(t)$ steering x_0 to x_1 and minimizing the cost

$$C(u) = \operatorname*{ess.\,sup}_{0 \leq t \leq T,\ 1 \leq i \leq m} |u^i(t)|.$$

[Hint: Let $K_c(T)$ be the set of attainability from x_0, using controls in Ω_c. Show that $K_c(T)$ is a compact set that grows continuously and monotonically with c on $0 \leq c \leq \bar{c}$. Then let c^* be the smallest c for which $x_1 \in K_c(T)$, and $u^*(t)$ is the corresponding controller with $C(u^*) = c^*$.]

(b) Assume the process $\mathcal{S}$ is controllable at the origin, that is,

$$\text{rank}\,[B, AB, A^2B, \ldots, A^{n-1}B] = n,$$

where $B = B(0)$ and $A = (\partial A/\partial x)(0)$—see Chapter 6 for details. Now consider the set of all measurable controllers $u(t) \in \Omega_{c^*}$ on various time intervals $0 \leq t \leq t_1$, steering x_0 to x_1. Show that the above $u^*(t)$ on $0 \leq t \leq T$ is a minimal time-optimal controller for the restraint set Ω_{c^*}. [Hint: Suppose $\hat{u}(t)$ on $0 \leq t \leq T - \epsilon$, $\epsilon > 0$, steers x_0 to x_1 with restraint $\hat{u}(t) \subset \Omega_{c^*}$. Then $\hat{u}(t)$ can be uniformly approximated by some $\bar{u}(t)$ on $0 \leq t \leq T - \epsilon$ steering x_0 to $\bar{x}_1$ in a prescribed neighborhood of $x_1 = 0$, with restraint $\bar{u} \subset \Omega_{c^*-\delta}$, for some small $\delta > 0$. Then the controllability condition insures that some extension of $\bar{u}(t)$ to the interval $0 \leq t \leq T - \epsilon/2$, with the restraint $\Omega_{c^*-\delta}$, steers x_0 to $x_1 = 0$.]

12. In Problem 11 assume that $B(x)$ has rank n everywhere. Then show that $K_{c_1}(T)$ lies in the interior of $K_{c_2}(T)$ for $0 < c_1 < c_2 \leq \bar{c}$. [Hint: Use the maximal principle.]

13. Consider the control process in R^n

$$(\mathcal{S}) \qquad \dot{x} = f(x, u)$$

with cost

$$C(u) = \int_0^\infty f^0(x, u)\, dt$$

Here $f(x, u)$ and $f^0(x, u) \geq 0$ are in C^1 in R^{n+m} and $f(0, 0) = 0$. For each initial state x_0 at $t_0 = 0$, the admissible controllers $u(t)$ on $0 \leq t < \infty$ are those for which the cost is finite and $x(\infty) = 0$. Assume that there exist functions $v(x)$ and $u^*(x)$ in C^1 near the origin in R^n and such that:

(a) $v(x) > 0$ for $x \neq 0$, $v(0) = 0$;

(b) $(\partial v/\partial x)(x) f(x, u) + f^0(x, u) \geq 0$, and equality holds for $u = u^*(x)$.

Then, for each x_0 near the origin, the response $x^*(t)$ defined by

$$\dot{x} = f(x, u^*(x)), \qquad x(0) = x_0$$

is optimal and the controller

$$u(t) = u^*(x^*(t))$$

is optimal (provided $x^*(t)$ and $u^*(t)$ are defined on $0 \leq t < \infty$ with $C(u^*) < \infty$ and $x^*(\infty) = 0$).

4.3 EXISTENCE OF OPTIMAL CONTROL, WITHOUT MAGNITUDE RESTRAINTS

In this section we investigate three control problems where the magnitude of the optimal controller is not limited by any bound on the restraint set. The first problem is an immediate extension of the existence theorems of Chapter 3 to nonlinear processes. The last two problems study closed-loop feedback controllers for nonlinear processes.

Consider a nonlinear process in R^n

$$(\mathcal{S}) \qquad \dot{x} = A(x, t) + B(x, t)u$$

with cost integral

$$C(u) = \int_0^T [A^0(x, t) + B^0(u, t)]\, dt.$$

We shall assume that $A^0(x, t) \geq 0$ and $B^0(u, t) \geq a\,|u|^p$ for some real constants $a > 0$, $p > 1$. Then the admissible controllers are all m-vector functions $u(t)$ in class L_p on the given finite interval $0 \leq t \leq T$, for which the response $x(t)$, initiating at x_0, is defined on $0 \leq t \leq T$ and yields a finite cost $C(u)$. By Hölder's inequality

$$\int_0^T |u|\, dt \leq \left(\int_0^T |u|^p\, dt\right)^{1/p} T^{1/q}, \qquad \frac{1}{p} + \frac{1}{q} = 1,$$

each admissible controller $u(t)$ lies in L_1 on $0 \leq t \leq T$, that is,

$$\|u\|_1 = \int_0^T |u(t)|\, dt < \infty.$$

THEOREM 8. *Consider the process in* R^n

$$(\mathcal{S}) \qquad \dot{x} = A(x, t) + B(x, t)u$$

with cost

$$C(u) = \int_0^T [A^0(x, t) + B^0(u, t)]\, dt$$

with A, A^0, B, B^0, $\partial A/\partial x$, $\partial B/\partial x$ *continuous for all* $x \in R^n$, $u \in R^m$, *and* $t \in R^1$.

Assume

(a) $A^0(x, t) \geq 0$;
(b) $B^0(u, t) \geq a\,|u|^p$ *for constants* $a > 0$, $p > 1$;
(c) $B^0(u, t)$ *is convex in* u *for each fixed* t.

The admissible controllers are all $u(t)$ *in* L_p *on the given finite interval* $0 \leq t \leq T$ *for which the response* $x(t)$ *initiating at* x_0 *yields a finite cost.*

Assume further

(d) $|x(t)| \leq \beta(\|u\|_1)$ *where the bound* β *is monotonic increasing in* $\|u\|_1$.

Then there exists an optimal controller $u^*(t)$ *minimizing the cost.*

Proof. Note that each bounded measurable function $u(t)$ on $0 \leq t \leq T$ yields a response $x(t)$ bounded by $\beta(\|u\|_1)$ on $0 \leq t \leq T$ and hence $u(t)$ is an admissible controller. Since $C(u) \geq 0$ there is a nonnegative infimum m for all costs $C(u)$. Let $u^{(k)}(t)$ be a sequence of admissible controllers such that $C(u^{(k)})$ tends monotonically towards the limit m. Note that

$$C(u^{(k)}) \leq m + 1$$

and so

$$a \int_0^T |u^{(k)}|^p \, dt \leq m + 1$$

for large k. Thus we can select a subsequence (still denoted $u^{(k)}(t)$) so that

$$\int_0^T |u^{(k)}|^p \, dt \leq \frac{m+1}{a}$$

and $u^{(k)}(t)$ converges weakly to a limit $u^*(t)$ in $L_p(0, T)$. Also $\|u^{(k)}\|_1 \leq [(m+1)/a]^{1/p} T^{1/q}$, where $1/p + 1/q = 1$, and so all the responses are uniformly bounded

$$|x^{(k)}(t)| \leq \beta\left(\left(\frac{m+1}{a}\right)^{1/p} T^{1/q}\right).$$

We next find that the uniformly bounded family of functions $x^{(k)}(t)$ is equicontinuous on $0 \leq t \leq T$. For any two times $0 \leq t_1 < t_2 \leq T$ we compute

$$|x^{(k)}(t_1) - x^{(k)}(t_2)| \leq \int_{t_1}^{t_2} |A(x^{(k)}(s), s)| + |B(x^{(k)}(s), s)| \, |u^{(k)}(s)| \, ds.$$

Thus there exists a constant $c > 0$ (independent of k) for which

$$|x^{(k)}(t_1) - x^{(k)}(t_2)| \leq c\,|t_2 - t_1| + c\left(\int_{t_1}^{t_2} |u^{(k)}(s)|^p \, ds\right)^{1/p} |t_2 - t_1|^{1/q}$$

and

$$|x^{(k)}(t_1) - x^{(k)}(t_2)| \leq c\,|t_2 - t_1| + c\left(\frac{m+1}{a}\right)^{1/p} |t_2 - t_1|^{1/q}.$$

Therefore Ascoli's theorem asserts that (for a subsequence)

$$\lim_{k \to \infty} x^{(k)}(t) = \bar{x}(t)$$

at each time $0 \leq t \leq T$.

We next show that $\bar{x}(t)$ is the response to $u^*(t)$. Write

$$\bar{x}(t) = x_0 + \lim_{k\to\infty} \int_0^t [A(x^{(k)}(s), s) + B(x^{(k)}(s), s)u^{(k)}(s)]\, ds.$$

Use the known limits

$$\lim_{k\to\infty} \int_0^T |A(x^{(k)}(s), s) - A(\bar{x}(s), s)|\, ds = 0,$$

$$\lim_{k\to\infty} \int_0^t B(\bar{x}(s), s)[u^{(k)}(s) - u^*(s)]\, ds = 0.$$

Also

$$\lim_{k\to\infty} B(x^{(k)}(t), t) = B(\bar{x}(t), t)$$

uniformly outside of some set S of arbitrarily small duration. Since

$$\int_S |u^{(k)}|\, ds \le \left(\int_0^T |u^{(k)}|^p\, ds\right)^{1/p} |S|^{1/q},$$

we can prove that

$$\lim_{k\to\infty} \int_0^T |B(x^{(k)}(s), s) - B(\bar{x}(s), s)|\, |u^{(k)}(s)|\, ds = 0.$$

From these results it follows that

$$\bar{x}(t) = x_0 + \int_0^t [A(\bar{x}(s), s) + B(\bar{x}(s), s)\, u^*(s)]\, ds$$

and $\bar{x}(t)$ is the response to $u^*(t)$ on $0 \le t \le T$.

The convexity of $B^0(u, t)$ shows that

$$C(u^*) \le \lim_{k\to\infty} C(u^{(k)}) = m,$$

according to Theorem 8 of Chapter 3. Hence

$$C(u^*) = m$$

and $u^*(t)$ is the required optimal controller. Q.E.D.

In the next two problems we consider a closed-loop feedback controller $u = u(x)$ for a nonlinear process in R^n. We have already encountered closed-loop controllers in earlier syntheses of open-loop optimal controllers. However here we seek the closed-loop controller directly, and with slightly different notation than that of our earlier Theorem 7 in Chapter 3.

We first turn to the problem of stabilizing a nonlinear process in R^n

(S) $$\dot{x} = f(x, u) = f(x, u(x))$$

with the linear controller $u(x) = Fx$, where the feedback matrix F is chosen to optimize the decay rate of the solutions near the origin. We make this problem precise in terms of the concepts of *generalized characteristic exponents* and *critical damping*, as defined below.

Consider an autonomous differential system in C^1 in R^n

$$(\mathcal{S}_1) \qquad \dot{x} = g(x)$$

where $g(0) = 0$, and denote $\partial g/\partial x(0) = G$. If the constant matrix G is stable (that is, if all eigenvalues of G have negative real parts, max Re $\lambda[G] < 0$), then the nonlinear system is stable about the origin. (For each neighborhood N_1 of the origin there is a subneighborhood N_2 such that every solution initiating in N_2 remains forever in N_1.) If G has an eigenvalue with positive real part then the nonlinear system fails to be stable about the origin. These topics of stability are discussed in Chapter 6 and in treatises on ordinary differential equations.

In case G is stable it can further be shown that there exists a neighborhood N of the origin such that each solution $x(t) \not\equiv 0$ initiating in N remains forever in N and

$$\lim_{t\to+\infty} x(t) = 0.$$

A refinement of this limiting analysis proves that

$$\limsup_{t\to+\infty} \frac{\log |x(t)|}{t} = \mu.$$

The numbers μ that arise in this manner are independent of the choice of N and the particular norm in R^n; they are called the generalized characteristic exponents of $\mathcal{S}_1$. It is known that the generalized characteristic exponents of $\mathcal{S}_1$ are precisely the real parts of the eigenvalues of G.

Let G and $\hat{G}$ be stability matrices with corresponding eigenvalues $\{\lambda_1, \lambda_2, \ldots, \lambda_n\}$ and $\{\hat{\lambda}_1, \hat{\lambda}_2, \ldots, \hat{\lambda}_n\}$, which we order by their real parts

$$\text{Re } \lambda_1 \leq \text{Re } \lambda_2 \leq \cdots \leq \text{Re } \lambda_n < 0$$

and

$$\text{Re } \hat{\lambda}_1 \leq \text{Re } \hat{\lambda}_2 \leq \cdots \leq \text{Re } \hat{\lambda}_n < 0.$$

We say that $G \prec \hat{G}$, G is more stable than $\hat{G}$, if

$$\text{Re } \lambda_n < \text{Re } \hat{\lambda}_n,$$

or

$$\text{Re } \lambda_n = \text{Re } \hat{\lambda}_n \quad \text{and} \quad \text{Re } \lambda_{n-1} < \text{Re } \hat{\lambda}_{n-1},$$

or

$$\text{Re } \lambda_n = \text{Re } \hat{\lambda}_n, \ldots, \text{Re } \lambda_j = \text{Re } \hat{\lambda}_j, \quad \text{and} \quad \text{Re } \lambda_{j-1} < \text{Re } \hat{\lambda}_{j-1}$$

for some $1 < j < n$. We also write $G \preceq \hat{G}$ to indicate that either $G \prec \hat{G}$ or else G and $\hat{G}$ have eigenvalues with the same real parts.

We now define the concept of critical damping for a differential system in terms of the above ideas and notations.

Definition. Consider an autonomous process in R^n

$$(\mathcal{S}) \qquad \dot{x} = f(x, u)$$

with $f(x, u)$ in C^1 near the origin in R^{n+m} and

$$f(0, 0) = 0, \qquad f_x(0, 0) = A, \qquad f_u(0, 0) = B.$$

Let $\mathcal{F}$ be a subset of the space $\mathcal{M}_{mn}$ of all real $m \times n$ matrices. A matrix $F^* \in \mathcal{F}$ defines critical damping for $\mathcal{S}$ with feedbacks in $\mathcal{F}$ in case:

$$(A + BF^*) \quad \text{is stable, and}$$

$$(A + BF^*) \preceq (A + BF) \quad \text{for all} \quad F \in \mathcal{F}.$$

Then $u = F^*x$ defines an optimal controller realizing critical damping for $\mathcal{S}$ in $\mathcal{F}$.

Remark. In order that critical damping for $\mathcal{S}$ may exist it is necessary that

$$(\mathcal{S}) \qquad \dot{x} = f(x, u) = Ax + Bu + \cdots$$

be stabilized by a linear feedback $u = Fx$ with $F \in \mathcal{F}$, that is, $A + BF$ be a stability matrix. If (A, B) is controllable, and if $\mathcal{F} = \mathcal{M}_{mn}$, then the system can always be stabilized; but no critical damping exists. This follows from the analysis of controllable linear systems in Chapter 2, where it is shown that $(A + BF)$ can assume arbitrary real eigenvalues.

THEOREM 9. *Consider the autonomous process in* R^n

$$(\mathcal{S}) \qquad \dot{x} = f(x, u)$$

with $f(x, u)$ *in* C^1 *near the origin in* R^{n+m} *and*

$$f(0, 0) = 0, \qquad f_x(0, 0) = A, \qquad f_u(0, 0) = B.$$

Let $\mathcal{F} \subset \mathcal{M}_{mn}$ *be a set of real* $m \times n$ *matrices. Assume that* $(A + BF_0)$ *is stable for some* $F_0 \in \mathcal{F}$. *Under these conditions if either*

1. $\mathcal{F}$ *is compact*

or

2. $\liminf_{F \to \infty} \{\max \operatorname{Re} \lambda[A + BF]\} \geq 0$, *in the special sense that: for each real* $\epsilon > 0$ *there is a compact set* $\mathcal{F}_\epsilon \subset \mathcal{F}$ *outside of which every matrix* $A + BF$ *has an eigenvalue with a real part greater than* $-\epsilon$,

then an optimal controller $u = F^*x$ *exists realizing critical damping.*

Proof. First let $\mathcal{F}$ be a compact subset of $\mathcal{M}_{mn}$ (where $\mathcal{M}_{mn}$ is topologically identified with R^{mn}). Since the eigenvalues of a matrix F depend continuously on F, there exists a matrix $F_1 \in \mathcal{F}$ that minimizes $\max \operatorname{Re} \lambda[A + BF]$, and

$$\max \operatorname{Re} \lambda[A + BF_1] \leq \max \operatorname{Re} \lambda[A + BF_0] < 0.$$

Let $\mathcal{F}_1$ be the compact subset of $\mathcal{F}$ where

$$\max \operatorname{Re} \lambda[A + BF] = \max \operatorname{Re} \lambda[A + BF_1].$$

Let $F_2 \in \mathcal{F}_1$ be such that $\operatorname{Re} \lambda_{n-1}[A + BF]$ is minimized among all $F \in \mathcal{F}_1$. Let $\mathcal{F}_2 \subset \mathcal{F}_1$ be the compact set where

$$\operatorname{Re} \lambda_n[A + BF] = \operatorname{Re} \lambda_n[A + BF_1]$$

and

$$\operatorname{Re} \lambda_{n-1}[A + BF] = \operatorname{Re} \lambda_{n-1}[A + BF_2]$$

(in terms of the eigenvalues $\operatorname{Re} \lambda_1 \leq \operatorname{Re} \lambda_2 \leq \cdots \leq \operatorname{Re} \lambda_n$ of the matrices $A + BF$). Continue in this way to define the compact sets

$$\mathcal{F}_n \subset \mathcal{F}_{n-1} \subset \cdots \subset \mathcal{F}_2 \subset \mathcal{F}_1 \subset \mathcal{F}$$

so that any matrix $F^* \in \mathcal{F}_n$ achieves the required critical damping.

Now let $\mathcal{F}$ be a noncompact set in $\mathcal{M}_{mn}$ and assume hypothesis (2). Then

$$\max \operatorname{Re} \lambda[A + BF] > \tfrac{1}{2} \max \operatorname{Re} \lambda[A + BF_0]$$

for all $F \in \mathcal{F}$ which lie outside some compact set $\mathcal{F}_0$. Then the optimal controller $u = F^*x$, which critically damps $\mathcal{S}$ in $\mathcal{F}_0$, also yields the critical damping for $\mathcal{S}$ in $\mathcal{F}$. Q.E.D.

COROLLARY. *Consider the autonomous process in* R^n

$$(\mathcal{S}) \qquad \dot{x} = f(x, u) = Ax + Bu + \cdots$$

with $f(x, u)$ *in* C^1 *near the origin in* R^{n+m}. *Let* $u = F_0 x$ *stabilize* $\mathcal{S}$ *and let* $\mathfrak{F}$ *consist of all matrices* $\{cF_0\}$ *for all real numbers* c. *If there exist two eigenvalues* μ_1 *and* μ_2 *of* BF_0 *with*

$$(\operatorname{Re} \mu_1)(\operatorname{Re} \mu_2) < 0,$$

then there exists an optimal controller $F^* = c^* F_0$ *realizing critical damping for* $\mathcal{S}$ *in* $\mathfrak{F}$. *On the other hand, if all eigenvalues of* BF_0 *have real parts that are all of the same sign, then no critical damping exists.*

Proof. Assume $\operatorname{Re} \mu_1 > 0$ for definiteness. Then the eigenvalues of $A + cBF_0$ are just the eigenvalues of $(1/c)A + BF_0$ multiplied by $c \neq 0$. But if $c > 0$ is very large, there is an eigenvalue $\tilde{\mu}$ of $(1/c)A + BF_0$ with

Re $\tilde{\mu} > \frac{1}{2}$ Re μ_1. Hence there is an eigenvalue of $A + cBF_0$ with positive real part, and $A + cBF_0$ is not stable. Similarly for $c \to -\infty$ the matrix $A + cBF_0$ is not stable. Hence, for some number $\gamma > 0$, the matrix $A + cBF_0$ is not stable for $|c| > \gamma$. Therefore critical damping must exist and correspond to a value c^* on the compact interval $-\gamma \leq c \leq \gamma$.

On the other hand, if all the eigenvalues of BF_0 have negative (or positive) real parts, then large values of $c \to +\infty$ (or $c \to -\infty$) yield stability matrices $A + cBF_0$ with

$$\lim_{c \to +\infty} \max \operatorname{Re} \lambda[A + cBF_0] = -\infty.$$

Hence no critical damping occurs in these cases. Q.E.D.

Example. Consider the scalar process

$$x^{(n)} + a_1 x^{(n-1)} + \cdots + a_n x = u$$

with the feedback controllers

$$u = cx^{(n-1)}, \quad \text{for real numbers } c.$$

In R^n we have the linear system for the vector x

$$\dot{x} = Ax + Bu, \qquad u = Fx$$

where

$$A = \begin{pmatrix} 0 & 1 & 0 & & 0 \\ 0 & 0 & 1 & & 0 \\ \vdots & & \vdots & & \vdots \\ -a_n & -a_{n-1} & \cdots & & -a_1 \end{pmatrix}, \qquad B = \begin{pmatrix} 0 \\ 0 \\ \vdots \\ 1 \end{pmatrix}$$

$$F = (0\,0\,0 \cdots 0\; c).$$

We assume that the free process is stable (hence the closed-loop process is stable for sufficiently small $c_0 > 0$), and we show that critical damping exists for $c \in R^1$.

For $c > a_1$ the process is not stable (since a necessary condition for stability is that all coefficients of the characteristic polynomial are positive). As $c \to -\infty$, the sum $(c - a_1)$ of the eigenvalues of $A + BF$ approaches $-\infty$. Thus some eigenvalue must have a real part with a large magnitude. But the product of the eigenvalues of $A + BF$ is the fixed constant a_n. Thus

$$\liminf_{c \to -\infty} \{\max \operatorname{Re} \lambda[A + BF]\} \geq 0.$$

Hence critical damping exists.

In the special case where $n = 2$,

$$\ddot{x} + a_1\dot{x} + a_2 x = c\dot{x}, \qquad a_1 > 0, \qquad a_2 > 0,$$

the critical damping occurs when

$$(a_1 - c)^2 - 4a_2 = 0 \quad \text{and} \quad a_1 - c > 0.$$

This yields the optimal value

$$c^* = a_1 - 2\sqrt{a_2}.$$

Remarks. A somewhat analogous problem concerns a differential system in R^n

$$(\mathcal{S}) \qquad \dot{x} = f(x, u)$$

where u is a constant m-vector chosen from a parameter set $\Omega \subset R^m$. Assume that $f(x, u)$ lies in C^1 in R^{n+m} and that for each constant u there is a response $x(t)$ (defined on some prescribed interval) initiating from a given state x_0. Let $C(u)$ be a real continuous function of $u \in \Omega$. We seek to choose an optimal parameter $u^* \subset \Omega$ that minimizes the cost $C(u)$.

If Ω is compact, then an optimal u^* exists. If Ω is not compact but

$$\liminf_{u \to \infty} C(u) > C(u_0), \quad \text{for some} \quad u_0 \in \Omega,$$

[in the sense of Theorem 9] then an optimal controller u^* must exist with $C(u^*) \leq C(u_0)$.

We now turn to the construction of an optimal closed-loop nonlinear feedback, over an infinite time duration. Consider a nonlinear system in R^n

$$(\mathcal{S}) \qquad \dot{x} = f(x, u)$$

with cost functional

$$C(u) = \int_0^\infty G(x, u)\, dt.$$

In place of an open loop controller $u(t)$ in R^m, chosen for each initial state x_0, we seek an optimal closed-loop controller $u(x)$ that minimizes the cost

$$J(x_0, u) = \int_0^\infty G(x(t, x_0), u(x(t, x_0)))\, dt$$

along the responses $x(t, x_0)$ of the closed-loop system

$$\dot{x} = f(x, u(x)), \qquad x(0, x_0) = x_0$$

for all initial states x_0 near the origin in R^n.

We assume that $f(x, u)$, $G(x, u)$, and $u(x)$ are real analytic functions near $x = u = 0$ in R^{n+m}. Thus these functions each have absolutely convergent real power series about the point $x = u = 0$ in R^{n+m}, and hence these power series also define complex analytic functions when the arguments range over a neighborhood of the origin in the space of $(n + m)$ complex variables. We assume that the lowest-order terms of $f(x, u)$ are linear

$$f(x, u) = Ax + Bu + h(x, u)$$

and those of $G(x, u)$ are quadratic

$$G(x, u) = x'Wx + u'Uu + H(x, u),$$

where $h(x, u)$ and $H(x, u)$ are power series of higher-order terms, beginning with second- and third-order terms in (x, u) respectively. The real constant matrices (A, B) are controllable (or at least stabilizable), and the symmetric matrices $W > 0$ and $U > 0$ are real constant positive definite matrices. We study closed-loop analytic feedback controllers

$$u = u(x) = Fx + \mathcal{H}(x)$$

for real constant matrices F and higher-order terms $\mathcal{H}(x)$. We shall always select matrices F so $u(x)$ stabilizes

$$\dot{x} = f(x, u(x)) = Ax + BFx + h(x, u(x)) + B\mathcal{H}(x),$$

that is, we demand that $A + BF$ should be a stability matrix. In this case $x(t, x_0)$ and $u(x(t, x_0))$ decay exponentially towards zero if $|x_0|$ is suitably small. In fact, if the eigenvalues λ of $A + BF$ have negative real parts less than a constant $-\mu$,

$$\operatorname{Re} \lambda[A + BF] < -\mu < 0,$$

then

$$|x(t, x_0)| \leq c_1 e^{-\mu t} |x_0| \quad \text{for} \quad 0 \leq t < \infty$$

for a positive constant $c_1 > 0$. Moreover, this basic estimate also holds for solutions with complex initial values z_0, so

$$|x(t, z_0)| \leq c_1 e^{-\mu t} |z_0| \quad \text{for} \quad 0 \leq t < \infty,$$

and there is a positive constant c_2 so

$$|u(x(t, z_0))| \leq c_2 \, |x(t, z_0)| \leq c_1 c_2 e^{-\mu t} |z_0|,$$

provided $|z_0|$ is suitably near the origin. These basic estimates for $|x(t, z_0)|$ and $|u(x(t, z_0))|$ show that the cost integral is convergent to a finite value $J(z_0, u)$. We seek an optimal controller $u(x)$ that minimizes $J(x_0, u)$ for all x_0 in a neighborhood of the origin of R^n.

Thus our problem is a generalization of the synthesis construction of Chapter 3 where the truncated process in R^n

$$\dot{x} = Ax + Bu$$

$$C(u) = \int_0^\infty [x'Wx + u'Uu]\,dt$$

was optimized by the linear feedback controller

$$u = F^*x.$$

We recall that $F^* = U^{-1}B'E^*$ was defined by the unique negative definite solution E^* of the matrix quadratic equation

$$A'E + EA + EBU^{-1}B'E = W.$$

We study the uniqueness and the construction of the optimal feedback controller $u^*(x)$ for the nonlinear process $\mathcal{S}$ following a basic lemma that analyzes the analytic properties of the cost $J(x_0, u)$. We note that $J(x_0, u)$ is a real-valued function of the real vector x_0 (near the origin in R^n) and the analytic function u. Once the function $u = u_1(x)$ is specified, then $J(z_0, u_1)$ becomes a complex-valued function of the complex variable z_0.

LEMMA. *Consider the real analytic process in* R^n

$$(\mathcal{S}) \qquad \dot{x} = f(x, u) = Ax + Bu + h(x, u)$$

with cost integral

$$J(x_0, u) = \int_0^\infty G(x, u)\,dt = \int_0^\infty [x'Wx + u'Uu + H(x, u)]\,dt$$

for the analytic feedback controller

$$\hat{u}(x) = \hat{F}x + \hat{\mathcal{H}}(x)$$

and the initial state x_0. *Assume that* $A + B\hat{F}$ *is a stability matrix. Then*

1. *There exists a neighborhood* $\hat{N}_c$ *about the origin in complex n-space wherein the cost*

$$J(z_0, \hat{u}) = -z_0'\hat{E}z_0 + \hat{J}^{(3)}(z_0) + \cdots$$

is analytic in z_0. *Moreover*

$$\hat{E} = -\int_0^\infty e^{(A'+\hat{F}'B')t}(W + \hat{F}'U\hat{F})e^{(A+B\hat{F})t}\,dt$$

depends only on the truncated process (the data A, B, W, U, $\hat{F}$*), and* $J(x_0, \hat{u})$ *is a real power series.*

2. *In $\hat{N}_c$ the functional equation obtains*

$$\frac{\partial J(z,\hat{u})}{\partial z} f(z, \hat{u}(z)) + G(z, \hat{u}(z)) \equiv 0.$$

Proof. Since $A + B\hat{F}$ is a stability matrix $\operatorname{Re}\lambda[A + B\hat{F}] < -\mu < 0$, for some $\mu > 0$. Hence there is a neighborhood $\hat{N}_c$ of the origin in complex n-space wherein each solution $x(t, z_0)$ of

$$\dot{x} = f(x, u(x)) = (A + B\hat{F})x + \cdots,$$

initiating at $x_0 \in \hat{N}_c$, remains in $\hat{N}_c$ forever and satisfies the basic estimate

$$|x(t, z_0)| \leq c_1 e^{-\mu t} |z_0| \quad \text{on} \quad 0 \leq t < \infty.$$

Using this estimate for $|x(t, z_0)|$ and

$$|\hat{u}(z)| \leq c_2 |z| \quad \text{in} \quad \hat{N}_c,$$

we have

$$|G(z, \hat{u}(z))| \leq c_3 e^{-2\mu t} |z_0|$$

for positive constants $c_1, c_2, c_3, \ldots$, which are independent of z_0 in $\hat{N}_c$.

The functions $x(t, z_0)$ and $\hat{u}(x(t, z_0))$ are analytic in $z_0 \in \hat{N}_c$ for each fixed $t \geq 0$. Since the integral

$$J(z_0, \hat{u}) = \int_0^\infty G(x(t, z_0), \hat{u}(x(t, z_0)))\, dt$$

is uniformly convergent in $\hat{N}_c$, we conclude that $J(z_0, \hat{u})$ is analytic for $z_0 \in \hat{N}_c$.

To compute the power series of $J(x_0, \hat{u})$ we must obtain $x(t, x_0)$ as a power series in x_0 in $\hat{N} = \hat{N}_c \cap R^n$. It is easy to see that

$$x(t, x_0) = e^{(A+B\hat{F})t} x_0 + \text{higher-order terms}$$

and

$$\hat{u}(x(t, x_0)) = \hat{F} e^{(A+B\hat{F})t} x_0 + \text{higher-order terms.}$$

If termwise integration is valid for $G(x(t, x_0), \hat{u}(x(t, x_0)))$, then it is easy to compute

$$J(x_0, \hat{u}) = \int_0^\infty x_0' e^{(A'+\hat{F}'B')t} W e^{(A+B\hat{F})t} x_0\, dt$$
$$+ \int_0^\infty x_0' e^{(A'+\hat{F}'B')t} \hat{F}' U \hat{F} e^{(A+B\hat{F})t} x_0\, dt$$
$$+ \text{cubic and higher-order terms in } x_0.$$

In this case $J(x_0, \hat{u})$ has the required form.

In order to justify this result we must estimate the higher-order terms in $x(t, x_0)$. For this purpose write

$$x(t, x_0) = x_0 + \int_0^t \hat{f}(x(s, x_0))\, ds$$

and

$$x_L(t, x_0) = x_0 + \int_0^t \hat{f}_x(0)\, x_L(s, x_0)\, ds$$

where

$$\hat{f}(x) = f(x, \hat{u}(x)) \quad \text{and} \quad \hat{f}_x(0) = A + B\hat{F}.$$

Then the difference is

$$\Delta(t, x_0) = x(t, x_0) - x_L(t, x_0) = \int_0^t [\hat{f}(x(s, x_0)) - \hat{f}_x(0)\, x_L(s, x_0)]\, ds$$

so

$$\Delta = \int_0^t [\hat{f}_x(0)\, x(s, x_0) + \epsilon(s, x_0) - \hat{f}_x(0)\, x_L(s, x_0)]\, ds$$

and

$$\Delta = \int_0^t [\hat{f}_x(0)\Delta(s, x_0) + \epsilon(s, x_0)]\, ds$$

where

$$|\epsilon(t, x_0)| \le c_4\, |x(t, x_0)|^2 \le c_5 e^{-2\mu t}\, |x_0|^2.$$

Therefore

$$\Delta(t, x_0) = e^{(A+B\hat{F})t} \int_0^t e^{-(A+B\hat{F})s}\, \epsilon(s, x_0)\, ds$$

and

$$|\Delta(t, x_0)| \le c_6 e^{-\mu t}\, |x_0|^2 \quad \text{for} \quad x_0 \in \hat{N}, \qquad t \ge 0.$$

This yields the desired estimate

$$x(t, x_0) = e^{(A+B\hat{F})t} x_0 + \Delta(t, x_0).$$

Now we note

$$G(x, \hat{u}(x)) = x'Wx + \hat{u}'U\hat{u} + \gamma(x)$$

with $|\gamma(x)| \le c_7\, |x|^3$. Thus

$$G(x(t, x_0), \hat{u}(x(t, x_0))) = x'Wx + \hat{u}'U\hat{u} + \gamma(x(t, x_0))$$

and

$$\int_0^\infty |\gamma(x(t, x_0))|\, dt \le c_8 \int_0^\infty e^{-3\mu t}\, |x_0|^3\, dt \le c_9\, |x_0|^3$$

Hence the power series for $J(x_0, \hat{u})$ collects the linear and quadratic terms in x_0 from the expression

$$\int_0^\infty [e^{(A+B\hat{F})t} x_0 + \Delta]' W [e^{(A+B\hat{F})t} + \Delta]\, dt + \int_0^\infty \hat{u}(x)' U\, \hat{u}(x)\, dt.$$

But

$$\hat{u}(x(t, x_0)) = \hat{F}[e^{(A+B\hat{F})t}x_0 + \Delta] + \Delta_1(t, x_0)$$

with

$$|\Delta_1(t, x_0)| \leq c_{10}\,|x(t, x_0)|^2 \leq c_{11}e^{-2\mu t}\,|x_0|^2.$$

Thus the quadratic terms of $J(x_0, \hat{u})$ are just $\hat{J}^{(2)}(x_0) = -x_0'\hat{E}x_0$, hence,

$$J(x_0, \hat{u}) = -x_0'\hat{E}x_0 + \hat{J}^{(3)}(x_0) + \cdots$$

as required.

Finally we verify the functional equation for $J(x, \hat{u})$ in $\hat{N}$. (Note that the first argument in J is often written x or z for convenience.) The solution $x(t, x_0)$ reaches a point x_{t_1} at time $t = t_1$, and this new point can then be used as an initial point in $\hat{N}$. Thus

$$J(x_t, \hat{u}) = \int_0^\infty G(x(s + t, x_0), \hat{u}(x(s + t, x_0)))\, ds = \int_t^\infty G(x(s), \hat{u}(x(s)))\, ds.$$

Differentiate with respect to t to obtain

$$\frac{\partial J}{\partial x}(x_t, \hat{u})\, f(x_t, \hat{u}(x_t)) = -G(x(t, x_0), \hat{u}(x(t, x_0))).$$

Evaluate at $t = 0$ to obtain the result

$$\frac{\partial J}{\partial x}(x_0, \hat{u})\, f(x_0, \hat{u}(x_0)) + G(x_0, \hat{u}(x_0)) \equiv 0.$$

By analyticity this same functional equation holds with x_0 replaced by $z \in \hat{N}_c$. Q.E.D.

Definition. An analytic feedback controller

$$u^*(x) = F^*x + \mathcal{H}^*(x),$$

which stabilizes the analytic process in R^n

$$\text{(S)} \qquad \dot{x} = f(x, u) = Ax + Bu + h(x, u)$$

is optimal in case the cost is minimized,

$$J(x_0, u^*) \leq J(x_0, u_1)$$

for every such analytic controller $u_1(x)$. [The inequality holds in some neighborhood N_1 of the origin of R^n, depending on $u_1(x)$.]

We shall seek the optimal controller u^* as a solution of the functional equation

$$\frac{\partial J}{\partial x}(x, u^*)\frac{\partial f}{\partial u}(x, u^*(x)) + \frac{\partial G}{\partial u}(x, u^*(x)) = 0.$$

Later the construction of a solution $u^*(x)$ will be investigated, and the matrix F^* will be shown to have the value $F^* = U^{-1}B'E^*$, just as for the truncated process. In order to assert the uniqueness of the optimal $u^*(x)$, we shall agree to consider two analytic functions to be the same (or equivalent) in case they coincide on some neighborhood of the origin in R^n.

THEOREM 10. *Let $u^* = u^*(x) = F^*x + \mathcal{H}^*(x)$ be an analytic feedback controller stabilizing the analytic process in R^n*

$$(\mathcal{S}) \qquad \dot{x} = f(x, u) = Ax + Bu + h(x, u),$$

with finite cost

$$J(x_0, u^*) = \int_0^\infty G(x, u^*(x))\, dt = \int_0^\infty [x'Wx + u^{*\prime}Uu^* + H(x, u^*)]\, dt.$$

If u^* is a solution of the functional equation

$$\frac{\partial J}{\partial x}(x, u^*)\frac{\partial f}{\partial u}(x, u^*(x)) + \frac{\partial G}{\partial u}(x, u^*(x)) = 0$$

near the origin, then u^* is an optimal feedback controller near the origin of R^n.

Moreover u^* is unique in that:

1. *u^* is the unique analytic solution of the functional equation;*
2. *u^* is the unique analytic feedback that is optimal;*
3. *u^* synthesizes the unique open-loop optimal controller. That is, there exists $\epsilon > 0$ and a neighborhood N^* such that for each $x_0 \in N^*$ the response $x^*(t)$ satisfies*

$$\dot{x} = f(x, u^*(x)), \qquad x^*(0) = x_0, \qquad x^*(t) \subset N^*$$

and the corresponding controller

$$u^*(t) = u^*(x^*(t))$$

is the unique open-loop controller achieving the minimal cost

$$C(u) = \int_0^\infty G(x, u(t))\, dt$$

among the measurable controllers $u(t)$ on $0 \le t < \infty$ with $|u(t)| \le \epsilon$ and $x(t) \subset N^*$.

Proof. Consider the real-valued function of $u \in R^m$, for each fixed x near the origin in R^n,

$$Q(u) = \frac{\partial J}{\partial x}(x, u^*)\, f(x, u) + G(x, u).$$

Compute the symmetric quadratic form

$$\left.\frac{\partial^2 Q}{\partial u^i\,\partial u^j}\right|_{\substack{x=0\\u=0}} = U > 0.$$

Then there exists an $\epsilon_1 > 0$ such that for all $|x_1| < \epsilon_1$ and $|u_1| < \epsilon_1$, the graph of $Q(u)$ lies above the tangent hyperplane at $u = u_1$ (at least for all $|u| < \epsilon_1$).

The hypothesis asserts that

$$\left.\frac{\partial Q}{\partial u^i}\right|_{u=u^*(x)} = 0$$

and so

$$\frac{\partial J}{\partial x}(x, u^*) f(x, u^*(x)) + G(x, u^*(x)) < \frac{\partial J}{\partial x}(x, u^*) f(x, u_1) + G(x, u_1)$$

for all $u_1 \neq u^*(x)$, provided $|x| < \epsilon_1$, $|u^*(x)| < \epsilon_1$, and $|u_1| < \epsilon_1$.

Now let $u_1(x) \neq u^*(x)$ be an analytic feedback controller and let N_1^0 be a neighborhood of the origin in R^n such that

$$|x| < \epsilon_1, \qquad |u_1(x)| < \epsilon_1, \qquad |u^*(x)| < \epsilon_1 \quad \text{in} \quad N_1^0$$

and each response $x^*(t)$ or $x_1(t)$ to the corresponding feedback controller which initiates in some $N_1 \subset N_1^0$, remains forever in N_1^0. Take $u_1(x_0) \neq u^*(x_0)$ at some point $x_0 \in N_1$ and then the lemma asserts

$$0 < \int_0^\infty \left[\frac{\partial J}{\partial x}(x_1(t), u^*) f(x_1(t), u_1(x_1(t))) + G(x_1(t), u_1(x_1(t)))\right] dt.$$

This yields the desired result

$$0 < -J(x_0, u^*) + J(x_0, u_1).$$

Therefore

$$J(x_0, u^*) < J(x_0, u_1),$$

and u^* is the unique optimal analytic feedback controller. Hence u^* is the unique analytic solution of the functional equation appearing in the hypotheses.

Finally let $\hat{u}(t)$ be any measurable open-loop controller applied to an initial state $\hat{x}_0$. Choose a positive number $\epsilon < \epsilon_1$ and a neighborhood $N^* \subset N_1$ of the origin such that

$$|x^*(t)| \leq \epsilon \quad \text{and} \quad |u^*(x^*(t))| \leq \epsilon \quad \text{on} \quad 0 \leq t < \infty$$

for the optimal response $x^*(t)$ initiating at $\hat{x}_0$. Also let N^* be an invariant neighborhood for responses to the optimal controller $u^*(x)$. Now require that $|\hat{u}(t)| \leq \epsilon$ and that this open-loop controller yields a response $\hat{x}(t)$ in N^*.

Then, as above,

$$0 = \frac{\partial J}{\partial x}(x, u^*) f(x, u^*(x)) + G(x, u^*(x))$$

$$< \frac{\partial J}{\partial x}(x, u^*) f(x, \hat{u}(t)) + G(x, \hat{u}(t))$$

wherever $\hat{u}(t) \neq u^*(\hat{x})$. If $\hat{u}(t) \equiv u^*(\hat{x}(t))$ almost everywhere on $0 \leq t < \infty$, then the uniqueness theorem for differential equations asserts $\hat{x}(t) \equiv x^*(t)$, which implies that $\hat{u}(t) \equiv u^*(x^*(t))$. Now assume that $\hat{u}(t) \neq u^*(\hat{x}(t))$ on some positive duration. Then

$$0 < \int_0^\infty \left[\frac{\partial J}{\partial x}(\hat{x}(t), u^*) f(\hat{x}(t), \hat{u}(t)) + G(\hat{x}(t), \hat{u}(t)) \right] dt,$$

Since $\hat{u}(t)$ has a finite cost

$$C(\hat{u}) = \int_0^\infty G(\hat{x}(t), \hat{u}(t))\, dt,$$

it is easy to show that $\lim_{t \to \infty} \hat{x}(t) = 0$. Hence

$$0 < -J(x_0, u^*) + C(\hat{u})$$

and

$$C(u^*) = J(x_0, u^*) < C(\hat{u}).$$

Thus $u^*(x^*(t))$ is the unique optimal open-loop controller for $\hat{x}_0$, with the required constraints. Q.E.D.

Remarks. For the truncated process

$$\dot{x} = Ax + Bu, \qquad J(x_0, u) = \int_0^\infty [x'Wx + u'Uu]\, dt$$

the optimal feedback controller is

$$u^* = F^*x$$

where $F^* = U^{-1}B'E^*$, and E^* is the unique negative definite solution of

$$A'E + EA + EBU^{-1}B'E = W.$$

We note here that for the nonlinear process

$$\text{(S)} \qquad \dot{x} = Ax + Bu + h(x, u)$$

$$J(x_0, u) = \int_0^\infty [x'Wx + u'Uu + H(x, u)]\, dt,$$

the optimal feedback controller

$$u^* = F^*x + \mathcal{H}^*(x)$$

still has the same first order term F^*x as for the truncated process (assuming the existence of some analytic solution of the functional equation in the theorem).

To show this we seek an optimal controller

$$\hat{u} = \hat{F}x + \hat{\mathcal{H}}(x)$$

with cost

$$J(x_0, \hat{u}) = -x_0'\hat{E}x_0 + \hat{J}^{(3)}(x_0) + \cdots$$

where

$$\hat{E} = -\int_0^\infty e^{(A'+\hat{F}'B')t}(W + \hat{F}'U\hat{F})e^{(A+B\hat{F})t}\,dt,$$

and we demand that the functional equation hold

$$\frac{\partial J}{\partial x}(x, \hat{u})\frac{\partial f}{\partial u}(x, \hat{u}(x)) + \frac{\partial G}{\partial u}(x, \hat{u}(x)) = 0.$$

The linear terms of this functional equation yield

$$-2x'\hat{E}B + 2x'\hat{F}'U = 0$$

or

$$\hat{F} = U^{-1}B'\hat{E}.$$

But the integral expression for $\hat{E}$ means that $\hat{E}$ is the unique negative definite solution of the Liapunov equation

$$\hat{E}(A + B\hat{F}) + (A' + \hat{F}'B')\hat{E} = W + \hat{F}'U\hat{F}.$$

We then conclude that $\hat{E}$ must solve

$$A'E + EA + EBU^{-1}B'E = W.$$

Hence $\hat{E} = E^*$ and so $\hat{F} = F^*$ as required.

As a final calculation we show that for an arbitrary analytic feedback controller

$$u = F^*x + \mathcal{H}(x),$$

which begins with F^*x, the cubic terms $J^{(3)}(x)$ of the cost

$$J(x, u) = -x'E^*x + J^{(3)}(x) + \cdots$$

are completely specified by the data $\{A, B, W, U, F^*, h^{(2)}, H^{(3)}\}$. We note that

$$\frac{\partial J}{\partial x}(x, u)\,f(x, u(x)) + G(x, u(x)) = 0$$

and equate the cubic terms to zero,

$$-2x'E^*[Bu^{(2)}(x) + h^{(2)}(x, F^*x)] + \frac{\partial J^{(3)}}{\partial x}[Ax + BF^*x]$$
$$+ (F^*x)'Uu^{(2)} + u^{(2)\prime}UF^*x + H^{(3)}(x, F^*x) \equiv 0.$$

Since $F^* = U^{-1}B'E^*$, we compute

$$-2x'E^*B + 2x'F^{*\prime}U = 0,$$

and so

$$\frac{\partial J^{(3)}}{\partial x}[Ax + BF^*x] = 2x'E^*h^{(2)}(x, F^*x) - H^{(3)}(x, F^*x).$$

But there can be at most one solution $J^{(3)}(x)$ of this partial differential equation, since the difference $\Delta J(x)$ of any two solutions must be a constant along each integral curve of the asymptotically stable ordinary differential system

$$\dot{x} = (A + BF^*)x.$$

In this case $\Delta J(x)$ must have the constant value $\Delta J(0) = 0$ everywhere on a neighborhood of the origin in R^n. Therefore any two solutions for $J^{(3)}(x)$ can differ by at most an additive constant. But $J^{(3)}(0) = 0$ and so $J^{(3)}(x)$ is uniquely determined by the above partial differential equation, which involves only the data $\{A, B, W, U, F^*, h^{(2)}, H^{(3)}\}$, as required.

The existence of a solution $u^*(x)$ of the functional equation

$$\frac{\partial J}{\partial x}(x, u^*)\frac{\partial f}{\partial u}(x, u^*(x)) + \frac{\partial G}{\partial u}(x, u^*(x)) = 0,$$

and hence the existence of the unique optimal feedback controller $u^*(x)$, will be demonstrated in a sequence of steps in Exercise 4 below.

EXERCISES

1. Compute the real parts of the eigenvalues for the critical damping of the scalar process

$$\ddot{x} + x = u, \qquad u = c\dot{x}.$$

2. Consider the scalar process

$$\dddot{x} + \ddot{x} = c_1\ddot{x} + (c_1 + c_2)\dot{x} + c_2 x, \quad \text{real} \quad c_1, c_2.$$

Show that there does not exist critical damping even though

$$\min\{\max \operatorname{Re} \lambda\} = -1.$$

3. Consider the autonomous system in C^1 in R^n

$$\dot{x} = g(x) = Gx + \cdots$$

where G is a stability matrix. It is known that there always exists a C^1 homeomorphism near the origin of R^n

$$x \rightarrow y(x) = x + \cdots,$$

which carries the given nonlinear differential system onto the linear system

$$\dot{y} = Gy.$$

Use this result to prove that the generalized characteristic exponents of the nonlinear system are just the real parts of the eigenvalues of G. Show that this holds regardless of the (equivalent) norm used in R^n.

4. In this exercise we indicate the proof of the existence of an optimal feedback controller $u^*(x)$ in R^m for the real analytic process in R^n

$$(\mathcal{S}) \qquad \dot{x} = f(x, u) = Ax + Bu + h(x, u)$$

with cost

$$J(x_0, u) = \int_0^\infty G(x, u)\, dt = \int_0^\infty [x'Wx + u'Uu + H(x, u)]\, dt.$$

Here $W > 0$, $U > 0$, (A, B) is controllable and $h(x, u)$, $H(x, u)$ are analytic functions of higher order near $x = u = 0$. We shall construct $u^*(x)$ as an analytic solution of the functional equation

$$(\mathcal{F}) \qquad \frac{\partial J}{\partial x}(x, u)\frac{\partial f}{\partial u}(x, u(x)) + \frac{\partial G}{\partial u}(x, u(x)) = 0,$$

as in Theorem 10.

In the truncated problem, where $h \equiv 0$, $H \equiv 0$, we know that the optimal feedback controller is the linear function $u^* = F^*x$ where $F^* = U^{-1}B'E^*$ and E^* is the unique negative definite solution of the matrix equation

$$A'E + EA + EBU^{-1}B'E = W,$$

and the cost is

$$J(x_0, F^*x) = -x_0'E^*x_0$$

(see Section 3.2, Exercise 15). To solve the complete nonlinear problem define the Hamiltonian function $H(\eta, x, u)$, a real analytic function near the origin of R^{2n+m}, by

$$H(\eta, x, u) = -G(x, u) + \eta f(x, u),$$

where x is a column n-vector and η is a row n-vector, as usual.

(a) Consider the truncated problem and verify that the linear feedback $u^* = F^*x$, where $F^* = U^{-1}B'E^*$ and $E^* < 0$ satisfies the above quadratic matrix equation, is a solution of the functional equation $\mathcal{F}$.

(b) For each (η, x) near $(0, 0)$ we seek a value u^* to maximize the Hamiltonian function $H(\eta, x, u)$. Consider the following defining equation for the feedback control $u^*(\eta, x)$,

$$H_u(\eta, x, u) = 0 \quad \text{or} \quad -G_u(x, u^*) + \eta f_u(x, u^*) = 0.$$

Use the implicit-function theorem to prove the existence of a unique analytic $u^*(\eta, x)$, vanishing at $\eta = 0, x = 0$, and satisfying the condition $H_u(\eta, x, u) = 0$. Verify that

$$u^*(\eta, x) = \tfrac{1}{2}U^{-1}B'\eta' + \cdots,$$

where the higher-order terms are zero for the truncated problem.

(c) Define the feedback Hamiltonian

$$H^*(\eta, x) = H(\eta, x, u^*(\eta, x))$$

and the Hamiltonian system in R^{2n}

$$(\mathcal{H}^*) \qquad \begin{aligned} \dot{x}^i &= \frac{\partial H^*}{\partial \eta_i} \\ \dot{\eta}_i &= -\frac{\partial H^*}{\partial x^i}, \qquad i = 1, 2, \ldots, n. \end{aligned}$$

Verify that the Hamiltonian system has the form

$$(\mathcal{H}^*) \qquad \begin{aligned} \dot{x} &= f(x, u^*(\eta, x)) = Ax + \tfrac{1}{2}BU^{-1}B'\eta' + \cdots \\ \dot{\eta}' &= 2Wx - A'\eta' + \cdots \end{aligned}$$

and that the higher-order terms are zero for the truncated problem.

(d) Show that the matrix

$$\begin{pmatrix} A & \tfrac{1}{2}BU^{-1}B' \\ 2W & -A' \end{pmatrix}$$

has exactly n eigenvalues with negative real parts. This can be done by a change of variables

$$\begin{pmatrix} y \\ \zeta \end{pmatrix} = M \begin{pmatrix} x \\ \eta' \end{pmatrix}$$

so that the truncated Hamiltonian system becomes

$$\begin{aligned} \dot{y} &= (A + BF^*)y \\ \dot{\zeta} &= -(A + BF^*)'\zeta. \end{aligned}$$

Here

$$M = \begin{pmatrix} I - 2QE^* & Q \\ 2E^* & -I \end{pmatrix}, \qquad M^{-1} = \begin{pmatrix} I & Q \\ 2E^* & 2E^*Q - I \end{pmatrix}$$

and $Q \leq 0$ satisfies the Liapunov equation

$$(A + BF^*)Q + Q(A + BF^*)' = \tfrac{1}{2}BU^{-1}B'.$$

(e) For the truncated problem let $\zeta = 0$ or

$$\eta' = 2E^*x$$

be the unique n-dimensional stability manifold of the Hamiltonian system $\mathcal{H}^*$. Define the feedback controller

$$u^*(x) = u^*(2E^*x, x) = U^{-1}B'E^*x.$$

Note that this is the optimal feedback controller.

(f) It is known from the general theory of nonlinear ordinary differential equations that there exists a nonsingular analytic coordinate change, near the origin in R^{2n},

$$y = (I - 2QE^*)x + Q\eta' + \cdots$$
$$\zeta = 2E^*x - I\eta' + \cdots,$$

which transforms the nonlinear Hamiltonian system $\mathcal{H}^*$ to

$$\dot{y} = (A + BF^*)y + \cdots$$
$$\dot{\zeta} = -(A + BF^*)'\zeta + \cdots,$$

where $y = 0$ and $\zeta = 0$ are invariant manifolds. Furthermore, there exists a unique stability manifold for $\mathcal{H}^*$ in R^{2n}, and this is determined by an analytic function

$$\eta = \eta^*(x) = 2x'E^* + \cdots.$$

Define the controller

$$u^*(x) = u'(\eta^*(x), x) = U^{-1}B'E^*x + \cdots.$$

The remainder of the existence proof consists of a verification that this analytic function $u^*(x)$ satisfies the required functional equation $\mathcal{F}$.

(g) Show that the nonlinear Hamiltonian system $\mathcal{H}^*$ is defined on the invariant stability manifold by the differential system

$$(\mathcal{H}_s^*) \quad \begin{aligned} \dot{x} &= f(x, u^*(x)) = (A + BF^*)x + \cdots \\ \dot{\eta}' &= -[-G_x(x, u^*(x)) + \eta f_x(x, u^*(x))]'. \end{aligned}$$

Denote the solutions of $\mathcal{H}_s^*$ initiating at x_0, $\eta_0 = \eta^*(x_0)$ by $x(t, x_0)$, $\eta(t, x_0)$ and show that

$$|x(t, x_0)| \leq K_1 e^{-\lambda t}|x_0|$$
$$|\eta(t, x_0)| \leq K_1 e^{-\lambda t}|x_0|$$

for some positive constants K_1 and λ, with $(-\lambda)$ greater than any of the real parts of the eigenvalues of the stability matrix $(A + BF^*)$; and show that these estimates hold for all suitably small $|x_0| > 0$.

(h) Compute

$$\frac{\partial}{\partial x_0} J(x_0, u^*) = \int_0^\infty \frac{\partial}{\partial x_0} G(x(t, x_0), u^*(x(t, x_0)))\, dt = -\eta^*(x_0).$$

Use the invariance of the stability manifold, so

$$\eta(t) = \eta^*(x(t)),$$

the defining equation for $u^*(x, \eta)$, and the integration by parts

$$\int_0^\infty \dot{\eta}(t) \frac{\partial x}{\partial x_0}\, dt = -\eta(0) - \int_0^\infty \eta(t) \frac{\partial}{\partial t}\left(\frac{\partial x}{\partial x_0}\right) dt$$

to simplify the computation.

(i) From the equation

$$-G_u(x, u^*) + \eta f_u(x, u^*) = 0$$

show that

$$\frac{\partial J}{\partial x}(x, u^*) \frac{\partial f}{\partial u}(x, u^*(x)) + \frac{\partial G}{\partial u}(x, u^*(x)) = 0.$$

Thus $u^*(x)$ is the required solution of the functional equation $\mathcal{F}$, and so $u^*(x)$ is the unique optimal feedback control for the process $\mathcal{S}$.

CHAPTER 5

Necessary and Sufficient Conditions for Optimal Control

In the first section of this chapter the maximal principle and the transversality conditions are shown to be necessary conditions for an optimal controller. Various extensions of the maximal principle to processes with impulse controllers and to processes with smoothly varying controllers are then obtained.

In the second section some sufficient conditions for optimality are proved. These include global results using convexity hypotheses, and also local theorems in which nonlinear processes are approximated by linear ones. Under suitable hypotheses the methods of dynamic programming are shown to determine an optimal controller.

5.1 THE MAXIMAL PRINCIPLE WITH THE NECESSARY TRANSVERSALITY CONDITIONS

In this section we prove the maximal principle for the general case of nonlinear autonomous control systems with moving targets and finite or infinite time durations. The maximal principle, together with the transversality conditions, is a necessary criterion satisfied by an optimal controller. The proof of these results rests on the constructions introduced in Theorem 3 of Chapter 4, which already asserts much of the content of the maximal principle. We prove the main theorems first for autonomous systems, and then deduce the corresponding results for nonautonomous systems by introducing t as an additional spatial coordinate.

Hence we consider the autonomous control process

$$(\mathcal{S}) \qquad \dot{x} = f(x, u)$$

with $f(x, u)$ and $(\partial f/\partial x)(x, u)$ continuous in R^{n+m}. Let X_0 and $X_1 \subset R^n$ be given initial and target sets and let the nonempty control restraint set be $\Omega \subset R^m$. An admissible controller $u(t) \subset \Omega$ on some finite interval $0 \leq t \leq t_1$ is a bounded measurable function with a response $x(t, x_0)$, which transfers $x(0, x_0) = x_0 \in X_0$ to $x(t_1, x_0) = x_1 \in X_1$. The terminal time t_1, the initial point $x_0 \in X_0$, and the terminal point $x_1 \in X_1$ vary with the controller. The class of all admissible controllers is denoted by Δ.

For each $u(t)$ on $0 \leq t \leq t_1$ in Δ, with response $x(t)$, let us assign a cost

$$C(u) = \int_0^{t_1} f^0(x(t), u(t))\, dt$$

where $f^0(x, u)$ and $(\partial f^0/\partial x)(x, u)$ are continuous in R^{n+m}. An admissible controller $\bar{u}(t)$ in Δ is (minimal) optimal if

$$C(\bar{u}) \leq C(u) \quad \text{for all} \quad u \in \Delta.$$

We shall prove that an optimal controller $u^*(t)$ on $0 \leq t \leq t^*$ satisfies the maximal principle,

$$\hat{H}(\hat{\eta}^*(t), \hat{x}^*(t), u^*(t)) = \hat{M}(\hat{\eta}^*(t), \hat{x}^*(t)) \quad \text{almost everywhere,}$$

and

$$\hat{M}(\hat{\eta}^*(t), \hat{x}^*(t)) \equiv 0 \quad \text{and} \quad \eta_0^* \leq 0 \quad \text{everywhere.}$$

Here the augmented state

$$\hat{x}^*(t) = \begin{pmatrix} x^{0*}(t) \\ x^*(t) \end{pmatrix}$$

is the response to the augmented system

$$(\hat{\mathcal{S}}) \qquad \begin{aligned} \dot{x}^0 &= f^0(x, u) \\ \dot{x}^i &= f^i(x, u) \qquad i = 1, \ldots, n. \end{aligned}$$

Also $\hat{\eta}^*(t)$ is a nontrivial solution of the augmented adjoint system

$$(\hat{\mathcal{A}}) \qquad \begin{aligned} \dot{\eta}_0 &= 0 \\ \dot{\eta}_i &= -\sum_{j=0}^{n} \eta_j \frac{\partial f^j}{\partial x^i}(x^*(t), u^*(t)) \qquad i = 1, \ldots, n \end{aligned}$$

where the last n equations (with $f^0 \equiv 0$) form the adjoint system $\mathcal{A}$ of Section 4.1. The augmented Hamiltonian function is

$$\hat{H}(\hat{\eta}, \hat{x}, u) = \eta_0 f^0(x, u) + \eta_1 f^1(x, u) + \cdots + \eta_n f^n(x, u)$$

and

$$\hat{M}(\hat{\eta}, \hat{x}) = \max_{u \in \Omega} \hat{H}(\hat{\eta}, \hat{x}, u) \qquad \text{(when } \hat{M} \text{ exists).}$$

With these notations we now state and prove the basic maximal principle. We use the concepts and notations of Section 4.1.

In the proof we first consider the augmented tangent perturbation cone $\hat{K}_\tau$, at any Lebesgue time $0 < \tau < t^*$, to take the cost coordinate x^0 into account. Then we further enlarge $\hat{K}_\tau$ to $\hat{K}_\tau^{\pm}$ to introduce time variations, since the optimal $u^*(t)$ is minimal on $0 \leq t \leq t^* + \epsilon$. Finally we enlarge $\hat{K}_\tau^{\pm}$ to $\mathcal{K}_\tau$ to account for the variations of initial and final points in the proof of the transversality conditions. We define these various cones carefully later.

In each case we must define the limit cone and use the earlier topological scholium, or some generalization, to show that generic intersections or overlappings of the infinitesimal perturbation cones imply the intersections of the corresponding nonlinear approximating sets of attainability.

THEOREM 1. *Consider the control process in* R^n

$$(\mathcal{S}) \qquad \dot{x} = f(x, u)$$

with bounded measurable controllers $u(t)$ *on various intervals* $0 \leq t \leq t_1$ *in the restraint set* $\Omega \subset R^m$. *Let* Δ *be all admissible controllers that steer some initial point of* X_0 *to a final point in the target set* X_1. *For each* $u(t)$ *on* $0 \leq t \leq t_1$ *in* Δ *with response* $x(t)$ *let the cost functional be*

$$C(u) = \int_0^{t_1} f^0(x(t), u(t))\, dt.$$

If $u^*(t)$ *on* $0 \leq t \leq t^*$ *is minimal optimal in* Δ, *with augmented response* $\hat{x}^*(t) = (x^{0*}(t), x^*(t))$, *then there exists a nontrivial augmented adjoint response* $\hat{\eta}^*(t) = (\eta_0^*, \eta^*(t))$ *such that*

$$\hat{H}(\hat{\eta}^*(t), \hat{x}^*(t), u^*(t)) = \hat{M}(\hat{\eta}^*(t), \hat{x}^*(t)) \quad \textit{almost everywhere,}$$

and

$$\hat{M}(\hat{\eta}^*(t), \hat{x}^*(t)) \equiv 0 \quad \textit{and} \quad \eta_0^* \leq 0 \quad \textit{everywhere on} \quad 0 \leq t \leq t^*.$$

Also if X_0 *and* X_1 *(or just one of them) are manifolds with tangent spaces* T_0 *and* T_1 *at* $x^*(0)$ *and* $x^*(t^*)$, *then* $\hat{\eta}^*(t) = (\eta_0^*, \eta^*(t))$ *can be selected to satisfy the transversality conditions at both ends (or just one end)*

$$\eta^*(0) \quad \textit{orthogonal to} \quad T_0,$$
$$\eta^*(t^*) \quad \textit{orthogonal to} \quad T_1.$$

Proof. Let $u^*(t)$ on $0 \leq t \leq t^*$, with response $x^*(t)$ steering from $x^*(0) = x_0^* \in X_0$ to $x^*(t^*) = x_1^* \in X_1$, be optimal in Δ. Consider the augmented system in R^{n+1}

$$\dot{x}^0 = f^0(x^1, \ldots, x^n, u)$$
$$\dot{x}^i = f^i(x^1, \ldots, x^n, u) \qquad i = 1, 2, \ldots, n,$$

or

$$(\hat{S}) \qquad \dot{\hat{x}} = \hat{f}(x, u),$$

with corresponding response $\hat{x}^*(t) = (x^{0*}(t), x^*(t))$ where

$$x^{0*}(t) = \int_0^t f^0(x^*(s), u^*(s))\, ds.$$

Since each control $u(t)$ in Δ determines some augmented response $\hat{x}(t)$ leading from $(0, X_0)$ to $(0, X_1)$, we remark that $u^*(t)$ steers $(0, x_0^*)$ to the lowest possible point on the line $R^1 \times x_1^*$ in R^{n+1}; that is, $C(u^*) = x^{0*}(t^*)$ is the minimal cost for controllers in Δ.

Therefore $(x^{0*}(t^*), x^*(t^*))$ lies on the boundary of the set of attainability $\hat{K}(t^*)$ of $\hat{S}$ in R^{n+1}. Hence, by Theorem 3 of Chapter 4, the maximal principle holds for $\hat{S}$. That is, there exists a nontrivial adjoint response $\hat{\eta}^*(t) = (\eta_0^*, \eta^*(t))$ so

$$H(\hat{\eta}^*(t), \hat{x}^*(t), u^*(t)) = \hat{M}(\hat{\eta}^*(t), \hat{x}^*(t)) \quad \text{almost everywhere,}$$

and

$$\hat{M}(\hat{\eta}^*(t), \hat{x}^*(t)) = \hat{M}$$

is constant everywhere on $0 \le t \le t^*$.

We now prove that $\hat{M} = 0$ as a consequence of the minimizing property of $u^*(t)$. Define the time perturbation cone $\hat{K}_\tau{}^{\pm}$, at each Lebesgue point $0 < \tau < t^*$, as the smallest closed cone in the tangent space at $\hat{x}^*(\tau)$ containing the perturbation tangent cone $\hat{K}_\tau$ of $\hat{S}$ and the two vectors $v_+(\tau) = \hat{f}(x^*(\tau), u^*(\tau))$ and $v_-(\tau) = -\hat{f}(x^*(\tau), u^*(\tau))$. Consider in $\hat{K}_\tau{}^{\pm}$ the vector $v = v_\pi(\tau) + v_+(\tau)\,\delta t$, where π defines elementary perturbation data (at distinct times on $0 < t < \tau$, as in Section 4.1 preceding Lemma 1 of Chapter 4) corresponding to the perturbed controller $u_\pi(t, \epsilon)$ and δt is any real number. Then the response to $u_\pi(t, \epsilon)$ on $0 \le t \le \tau + \epsilon\,\delta t$, for small $\epsilon \ge 0$, is $\hat{x}_\pi(t, \epsilon)$ with endpoint

$$\hat{x}_\pi(\tau + \epsilon\,\delta t, \epsilon) = \hat{x}^*(\tau) + \epsilon v_\pi(\tau) + \epsilon v_+(\tau)\,\delta t + o(\epsilon).$$

This fundamental formula was proved in Section 4.1 (for $\delta t = 0$) and it shows that $\hat{K}_\tau{}^{\pm}$ lies in the time union of the augmented sets of attainability $\bigcup_{0<t<t^*} \hat{K}(t)$, to within errors of order $o(\epsilon)$.

The proof of Lemma 2 of Chapter 4 applies to $\hat{K}_\tau{}^{\pm}$ and asserts that any vector w interior to $\hat{K}_\tau{}^{\pm}$, must define a line segment from $\hat{x}^*(\tau)$, which lies interior to $\bigcup_{0<t<t^*} \hat{K}(t)$. In particular, the vector $w_\tau = (-1, 0)$ does not lie interior to $\hat{K}_\tau{}^{\pm}$, for otherwise there would exist a perturbed controller $\bar{u}_\pi(t, \epsilon)$ steering $(0, x_0)$ to some point $(\bar{x}^0, x^*(\tau + \epsilon\,\delta t))$ with partial cost $\bar{x}^0 < x^{0*}(\tau + \epsilon\,\delta t)$. Because $\hat{f}(x, u)$ does not depend on t or x^0, the controller $\bar{u}_\pi(t, \epsilon)$ could then be supplemented by $u^*(t - \epsilon\, \delta t)$ on $\tau + \epsilon\, \delta t < t \leq t^* + \epsilon\, \delta t$ to yield an admissible controller of total cost less than $C(u^*) = x^{0*}(t^*)$. Hence the cone $\hat{K}_\tau^{\pm}$ is separated from w_τ by a hyperplane with a normal vector $\hat{\eta}_\tau(\tau)$. Define $\hat{\eta}_\tau(t) = (\eta_{0\tau}, \eta_\tau(t))$ by the adjoint equations $\hat{\mathcal{A}}$ with value $\hat{\eta}_\tau(\tau)$ and $\eta_{0\tau} \leq 0$; $\hat{\eta}_\tau(t)$ then satisfies the maximal principle

$$\hat{H}(\hat{\eta}_\tau(t), \hat{x}^*(t), u^*(t)) = \hat{M}(\hat{\eta}_\tau(t), x^*(t))$$

almost everywhere on $0 \leq t \leq \tau$ and $\hat{M}(\hat{\eta}_\tau(t), \hat{x}^*(t)) \equiv \hat{M}_\tau$ is constant. Since $\hat{\eta}_\tau(\tau)\, v_\pm(\tau) \leq 0$, and $v_+(\tau) = -v_-(\tau)$, we conclude that $\hat{\eta}_\tau(\tau)\, v_\pm(\tau) = 0$ and so $\hat{M}_\tau = 0$.

We now construct the limit cone $\hat{K}_{t^*}^{\pm}$ in order to obtain an adjoint vector $\hat{\eta}^*(t)$ with $\hat{\eta}^*(t)\, v(t) \leq 0$ for all $v(t) \in \hat{K}_t^{\pm}$ on $0 \leq t \leq t^*$. In order to show that the parallel displacement of tangent vectors along $\hat{x}^*(t)$, by the matrices $\hat{A}_{t_2 t_1}$, as in preliminaries of Section 4.1, carries $\hat{K}_{t_1}^{\pm}$ into $\hat{K}_{t_2}^{\pm}$ for $0 < t_1 < t_2 < t^*$, we need only prove that $\hat{A}_{t_2 t_1} v_+(t_1) \in \hat{K}_{t_2}^{\pm}$. If $\hat{A}_{t_2 t_1} v_+(t_1)$ does not belong to $\hat{K}_{t_2}^{\pm}$, then there is a hyperplane that separates them; that is, there exists a vector $\hat{\xi}$ with

$$\hat{\xi}\hat{A}_{t_2 t_1} v_+(t_1) > 0 \quad \text{and} \quad \hat{\xi}\hat{K}_{t_2}^{\pm} \leq 0,$$

that is, $\hat{\xi}\hat{v} \leq 0$ for all vectors $\hat{v} \in \hat{K}_{t_2}^{\pm}$. Let $\hat{\xi}(t)$ be the corresponding adjoint solution so $\hat{\xi}(t)\, v_+(t_1) > 0$. But the maximal principle holds for $\hat{\xi}(t)$ on $0 \leq t \leq t_2$, and hence

$$\max_{u \in \Omega} \hat{\xi}(t) \hat{f}(x^*(t), u^*(t)) = \hat{M}(\hat{\xi}(t), \hat{x}^*(t)) \equiv 0.$$

Thus $\hat{\xi}(t_1)\, v_+(t_1) = 0$, which is a contradiction. Therefore

$$\hat{A}_{t_2 t_1} \hat{K}_{t_1}^{\pm} \subset \hat{K}_{t_2}^{\pm} \quad \text{for} \quad 0 < t_1 < t_2 < t^*,$$

and we define the limit cone (at t^* or even at any t)

$$\hat{K}_{t^*}^{\pm} = \bigcup_{0<\tau<t^*} \hat{A}_{t^*\tau} \hat{K}_\tau^{\pm}.$$

Now let $w_{t^*} = (-1, 0)$ be a vector in the tangent space at $\hat{x}^*(t^*)$. Note that the parallel displacement of w_{t^*} is just $w_\tau = (-1, 0)$, since $\partial\hat{f}/\partial x^0 \equiv 0$. If w_{t^*} lies interior to the limit cone $\hat{K}_{t^*}^{\pm}$, then there exists a polyhedral cone in $\hat{K}_\tau^{\pm}$, which contains w_τ in its interior, for some Lebesgue point $\tau < t^*$ (see Lemma 2 above). But we have already proved that w_τ cannot lie interior to $\hat{K}_\tau^{\pm}$ and hence w_{t^*} can be separated from $\hat{K}_{t^*}^{\pm}$ by some vector $\hat{\eta}^* = (\eta_0^*, \eta^*)$ with $\eta_0^* \leq 0$ and $\hat{\eta}^*\hat{K}_{t^*}^{\pm} \leq 0$. Define the required adjoint vector $\hat{\eta}^*(t)$ as the solution of $\hat{\mathcal{A}}$ with any such data $\hat{\eta}^*(t^*) = \hat{\eta}^*$. Then, as above, the maximal principle holds for $\hat{\eta}^*(t)$

$$\hat{H} = \hat{M} \quad \text{almost everywhere on} \quad 0 \leq t \leq t^*$$

and

$$\hat{M} \equiv 0 \quad \text{everywhere on} \quad 0 \leq t \leq t^*.$$

Finally we must select $\hat{\eta}^*$ so that $\hat{\eta}^*(t)$ satisfies the transversality conditions. We discuss the case where X_0 and X_1 have tangent spaces T_0 and T_1 at x_0^* and x_1^*, respectively. The case for just one end transversality is an easy modification of the argument.

Let T_0 be the tangent space to X_0 at x_0^* in R^n and let $\hat{T}_0$ be the linear space at $(0, x_0^*)$ spanned by vectors of the form $(0, T_0)$. Similarly let $\hat{T}_1$ be all vectors $(0, T_1)$ at $(x^{0*}(t^*), x_1^*)$. Let $\mathcal{K}_t$ be the smallest closed cone in the tangent space at $\hat{x}^*(t)$ generated by $\hat{A}_{t0}\hat{T}_0$ and $\hat{K}_t^{\pm}$.

Let $\mathbf{T}_1$ be the cone in the tangent space at $\hat{x}^*(t^*)$ generated by $\hat{T}_1$ and the downward vector $w_{t^*} = (-1, 0)$. Let us suppose, as will be shown later, that $\mathcal{K}_t^*$ and $\mathbf{T}_1$ are separated by a hyperplane π. In this case take a normal vector $\hat{\eta}^* = (\eta_0^*, \eta^*)$ at $\hat{x}^*(t^*)$ with $\eta_0^* \leq 0$ and

$$\hat{\eta}^*\mathcal{K}_{t^*} \leq 0, \qquad \hat{\eta}^*\mathbf{T}_1 \geq 0.$$

The corresponding solution $\hat{\eta}^*(t)$ of $\hat{\mathcal{A}}$ then satisfies the maximal principle and we next show that $\hat{\eta}^*(t)$ satisfies the transversality conditions.

The linear space $\hat{T}_1$, which lies in $\mathbf{T}_1$, must lie in the hyperplane π. Thus the vector $(0, \eta^*)$ satisfies $\eta^*T_1 = 0$, the required transversality condition at x_1^*. Also the parallelly displaced linear space $\hat{A}_{t^*0}\hat{T}_0 \subset \mathcal{K}_{t^*}$, and hence this is separated from $\hat{\eta}^*$ by π. Thus $\hat{\eta}^*\hat{A}_{t^*0}(0, \alpha) = 0$ for each vector $\alpha \in T_0$. Therefore $\hat{\eta}^*(0)(0, \alpha) = 0$ or $\eta^*(0)\alpha = 0$. Thus $\eta^*(0)T_0 = 0$, the required transversality condition at x_0^*.

The entire proof will be completed when we justify the separation of $\mathcal{K}_{t^*}$ and $\mathbf{T}_1$. Suppose $\mathcal{K}_{t^*}$ and $\mathbf{T}_1$ cannot be separated. Then together they span all the tangent space at $\hat{x}^*(t^*)$ and, moreover, there is a vector common to both their relative interiors. We thus suppose that $\dim \mathcal{K}_{t^*} = r$ and $\dim \mathbf{T}_1 = n - r$. If $\dim \mathbf{T}_1 > n - r$, select an appropriate closed subcone of dimension $n - r$ which meets $\mathcal{K}_{t^*}$ generically, that is, meets $\mathcal{K}_{t^*}$ at a common relative interior vector.

Since $\mathcal{K}_{t^*}$ and $\mathbf{T}_1$ meet generically, so do $\hat{A}_{t^*\tau}\mathcal{K}_\tau$ and $\mathbf{T}_1$, for some Lebesgue time τ on $0 < \tau < t^*$. Then $\mathcal{K}_\tau$ and $\mathbf{T}_\tau = \hat{A}_{t^*\tau}\mathbf{T}_1$ meet generically in the tangent space at $\hat{x}^*(\tau)$.

Let π be an elementary perturbation data complex on $0 < t < \tau$, and let $u_\pi(t, \epsilon)$ be the corresponding perturbed controller for small $\epsilon \geq 0$. Let $\hat{x}_{\pi,\alpha}(t, \epsilon)$ be the corresponding response initiating at the point $(0, \epsilon\alpha)$ on $(0, X_0) \subset R^{n+1}$. (Here $\hat{\alpha} = (0, \alpha)$, a vector in $\hat{T}_0$, and we coordinatize X_0 near x_0^* by the orthogonal projection of T_0 on X_0 in R^n.) Then the customary basic perturbation formula is

$$\hat{x}_{\pi,\alpha}(\tau + \epsilon\,\delta t, \epsilon) = \hat{x}^*(\tau) + \epsilon[v_\pi(\tau) + v_+(\tau)\,\delta t + \hat{A}_{\tau 0}\hat{\alpha}] + o(\epsilon).$$

Thus this set $\mathcal{K}(\tau)$ of points attainable from X_0 agrees with $\mathcal{K}_\tau$ to within errors of order $o(\epsilon)$.

The manifold $\mathcal{X}_1$, defined as all points (x^0, X_1) in R^{n+1} with $x^0 < x^{0*}(t^*)$, can be parallelly displaced by the solutions of

$$\dot{\hat{x}} = \hat{f}(x, u^*(t))$$

to define a submanifold $\mathcal{X}(\tau)$ with tangent half-space $\mathbf{T}_\tau$ (on the limiting upper edge). But $\mathcal{X}(\tau)$ cannot meet $\mathcal{K}(\tau)$, for in such a case a perturbed controller $u_\pi(t, \epsilon)$ on $0 \le t \le \tau + \epsilon\,\delta t$, acting on some initial point in X_0, yields a point in $\mathcal{X}(\tau)$, and then using $u^*(t - \epsilon\,\delta t)$ to extend the control, we obtain a target point in $\mathcal{X}_1$, that is, we obtain a point (x^0, x_1) with $x^0 < x^{0*}(t^*)$ and $x_1 \in X_1$. This is impossible because of the minimal cost of $u^*(t)$.

Thus $\mathcal{K}(\tau)$ agrees with $\mathcal{K}_\tau$ and $\mathcal{X}(\tau)$ agrees with $\mathbf{T}_\tau$ with errors of order $o(\epsilon)$. If $\mathcal{K}_\tau$ and $\mathbf{T}_\tau$ meet generically, then $\mathcal{K}(\tau)$ and $\mathcal{X}(\tau)$ must intersect. This last conclusion follows from a topological approximation result, which generalizes the earlier scholium and is proved in the exercises below.

Therefore $\mathcal{K}_{t^*}$ and $\mathbf{T}_1$ do not intersect generically, and so they can be separated by a hyperplane. As noted above the transversality conditions can be satisfied once this separation is realized. Q.E.D.

An important special case of optimal-control theory occurs when the cost functional is $C(u) = t_1$ for a controller $u(t)$ on $0 \le t \le t_1$ in Δ. This is the problem of minimal time-optimal control, and the corresponding maximal principle is obtained from Theorem 3 of Chapter 4 by setting $f^0(x, u) \equiv 1$. We state our result in terms of the Hamiltonian

$$H(\eta, x, u) = \eta_1 f^1(x, u) + \cdots + \eta_n f^n(x, u)$$

and

$$M(\eta, x) = \max_{u \in \Omega} H(\eta, x, u).$$

COROLLARY 1. *Consider the control process in R^n*

$$(\mathcal{S}) \qquad \dot{x} = f(x, u)$$

with bounded measurable controllers $u(t) \subset \Omega$ on various intervals $0 \le t \le t_1$ in Δ steering points of X_0 to X_1, as above. Let the cost be the time duration of control

$$C(u) = \int_0^{t_1} 1\, dt = t_1.$$

If $u^*(t)$ on $0 \le t \le t^*$ is minimal time-optimal in Δ with response $x^*(t)$, then there exists a nontrivial adjoint response $\eta^*(t)$ of $\mathcal{A}$ such that

$$H(\eta^*(t), x^*(t), u^*(t)) = M(\eta^*(t), u^*(t)) \quad \textit{almost everywhere,}$$

and

$$M(\eta^*(t), u^*(t)) \geq 0$$

is constant everywhere.

Also, if X_0 and X_1 are manifolds with tangent spaces T_0 and T_1 at $x^*(0)$ and $x^*(t^*)$, respectively, then $\eta^*(t)$ can be selected to satisfy the transversality conditions

$$\eta^*(0) \perp T_0 \quad \text{and} \quad \eta^*(t^*) \perp T_1.$$

Proof. Let $\hat{\eta}^*(t) = (\eta_0^*, \eta^*(t))$ be the adjoint response determined in the theorem. Then

$$\eta_0^* + \eta^*(t) f(x^*(t), u^*(t)) = \max_{u \in \Omega} [\eta_0^* + \eta^*(t) f(x^*(t), u)]$$

or

$$H(\eta^*(t), x^*(t), u^*(t)) = M(\eta^*(t), x^*(t)) \quad \text{almost everywhere.}$$

Also $\hat{M}(\hat{\eta}^*(t), \hat{x}^*(t)) \equiv 0$ or $M(\eta^*(t), x^*(t)) = -\eta_0^* \geq 0$ everywhere. Here

$$\hat{H}(\hat{\eta}, \hat{x}, u) = \eta_0 + H(\eta, x, u), \qquad \hat{M}(\hat{\eta}, \hat{x}) = \eta_0 + M(\eta, x)$$

as usual.

If $\eta^*(t)$ vanished at some point on $0 \leq t \leq t^*$, then it vanishes identically because it is a solution of the homogeneous linear differential system $\mathcal{A}$. But in such a case the condition $\hat{M}(\hat{\eta}^*(t), \hat{x}^*(t)) \equiv 0$ implies that $\eta_0 = 0$ so $\hat{\eta}^*(t) \equiv 0$, which is impossible. Therefore $\eta^*(t)$ is a nonvanishing solution of $\mathcal{A}$.

The transversality conditions on $\eta^*(t)$ have already been obtained in Theorem 1. Q.E.D.

Remark 1. If the target set X_1 is all R^n then the control problem is known as the *free-endpoint problem*. Of course, an optimal controller $u^*(t)$ on $0 \leq t \leq t^*$, with response $\hat{x}^*(t)$ and adjoint response $\hat{\eta}^*(t) = (\eta_0^*, \eta^*(t))$, satisfies the maximal principle and transversality conditions stated in Theorem 1. The final transversality condition requires $\eta^*(t^*) = 0$.

Remark 2. Consider the control process in R^n

$$(\mathcal{S}) \qquad \dot{x} = f(x, u)$$

with bounded measurable controllers $u(t)$ on a fixed finite time duration $0 \leq t \leq T$, satisfying the restraint Ω. Let Δ_T be all such controllers that steer some initial point in X_0 to a final point in X_1, with the cost functional

$$C(u) = \int_0^T f^0(x(t), u(t))\, dt,$$

as in Theorem 1.

Let $u^*(t)$ on $0 \leq t \leq T$ be an optimal controller for this *fixed-time problem*, and let $x^*(t)$ be the corresponding response. Then there exists a nontrivial adjoint response $\hat{\eta}^*(t) = (\eta_0^*, \eta^*(t))$ on $0 \le t \le T$ such that

$$\hat{H}(\hat{\eta}^*(t), \hat{x}^*(t), u^*(t)) = \hat{M}(\hat{\eta}^*(t), \hat{x}^*(t)) \quad \text{almost everywhere,}$$

and

$$\hat{M}(\hat{\eta}^*(t), \hat{x}^*(t))$$

is constant and $\eta_0^* \le 0$. Also, if X_0 and X_1 are manifolds with corresponding tangent spaces T_0 and T_1 in R^n, then the transversality conditions obtain

$$\eta^*(0) \perp T_0 \quad \text{and} \quad \eta^*(T) \perp T_1.$$

The proofs of these assertions are precisely those of Theorem 1 except that no time perturbations are allowed, that is, $\hat{K}_t^{\pm}$ is replaced by $\hat{K}_t$. Hence we are unable to maintain the vanishing of $\hat{M}(\hat{\eta}^*(t), \hat{x}^*(t))$.

If the fixed-time duration becomes infinite we obtain an interesting result, which we state for the fixed-endpoint problem. We fix an initial point $x_0 \in R^n$ and consider measurable controllers $u(t)$ on $0 \le t < \infty$, bounded on each compact time interval, satisfying the restraint Ω, and each of which defines a response $x(t)$ on $0 \le t < \infty$, with $\lim_{t \to \infty} x(t) = x_1$ in R^n. The set Δ_∞ of all admissible controllers consists of all such controllers for which the cost is convergent

$$C(u) = \int_0^\infty f^0(x(t), u(t))\, dt < \infty.$$

The augmented response $\hat{x}(t) = (x^0(t), x(t))$, where $\dot{x}^0 = f^0(x(t), u(t))$, $x^0(0) = 0$, is defined as usual.

COROLLARY 2. *Consider the control process in* R^n

$$\text{(S)} \qquad \dot{x} = f(x, u).$$

The measurable controllers $u(t) \subset \Omega$ *on* $0 \le t < \infty$, *with responses* $x(t)$ *steering* x_0 *to* x_1, *having finite cost*

$$C(u) = \int_0^\infty f^0(x(t), u(t))\, dt,$$

belong to the admissible class Δ_∞, *as above.*

Let $u^*(t)$, *with augmented response* $\hat{x}^*(t)$, *be optimal in* Δ_∞. *Then there exists a nontrivial augmented adjoint response* $\hat{\eta}^*(t) = (\eta_0^*, \eta^*(t))$ *such that*

$$\hat{H}(\hat{\eta}^*(t), \hat{x}^*(t), u^*(t)) = \hat{M}(\hat{\eta}^*(t), \hat{x}^*(t)) \quad \textit{almost everywhere on} \quad 0 \le t < \infty,$$

and

$$\hat{M}(\hat{\eta}^*(t), \hat{x}^*(t)) \equiv 0 \quad \textit{everywhere on} \quad 0 \le t < \infty,$$

and $\eta_0^* \le 0$.

Proof. For each finite interval $0 \leq t \leq T$ consider the class Δ_T of bounded measurable controllers in Ω that steer x_0 to $x^*(T)$. Then $u^*(t)$ on $0 \leq t \leq T$ is optimal in Δ_T, for otherwise any less cost controller in Δ_T could be supplemented by $u^*(t)$ on $T < t < \infty$ to contradict the optimality of $u^*(t)$ in Δ_∞.

Then let $\hat{\eta}^*(T) = (\eta^*_{0T}, \eta^*_T(t))$ be an adjoint response to

$$(\hat{\mathcal{A}}) \qquad \dot{\hat{\eta}} = -\hat{\eta}\,\frac{\partial \hat{f}}{\partial \hat{x}}(x^*(t), u^*(t))$$

such that

$$\hat{H}(\hat{\eta}^*_T(t), x^*(t), u^*(t)) = \hat{M}(\hat{\eta}^*_T(t), \hat{x}^*(t)) \quad \text{almost everywhere}$$

and

$$\hat{M}(\hat{\eta}^*_T(t), \hat{x}^*(t)) \equiv 0 \quad \text{everywhere on} \quad 0 \leq t < \infty.$$

Also $\eta^*_{0T} \leq 0$ and we can choose $\hat{\eta}^*_T(0)$ to be a unit vector.

Now let $T = 1, 2, 3, \ldots, r, \ldots$, and select a subsequence of the unit vectors $\hat{\eta}^*_T(0)$ to converge to some unit vector $\hat{\eta}^*(0)$ with $\hat{\eta}^*_0 \leq 0$. Let $\hat{\eta}^*(t) = (\eta^*_0, \eta^*(t))$ be the correspondingly determined solution of $\hat{\mathcal{A}}$ on $0 \leq t < \infty$. Note that

$$\lim_{T\to\infty} \hat{\eta}^*_T(t) = \hat{\eta}^*(t) \quad \text{on} \quad 0 \leq t < \infty$$

where the convergence is uniform on compact time intervals. We prove that $\hat{\eta}^*(t)$ satisfies the required maximal principle with $u^*(t)$ and $x^*(t)$ on $0 \leq t < \infty$.

Suppose

$$\hat{H}(\hat{\eta}^*(t), \hat{x}^*(t), u^*(t)) < \hat{M}(\hat{\eta}^*(t), \hat{x}^*(t))$$

on some set of positive duration on $0 \leq t < \infty$. Then there is a (Lebesgue) time t_1 when

$$\hat{H}(\hat{\eta}^*(t_1), \hat{x}^*(t_1), u^*(t_1)) < H(\hat{\eta}^*(t_1), \hat{x}^*(t_1), u_1) - \delta$$

for some point $u_1 \in \Omega$ and $\delta > 0$. Then, for sufficiently large $T > t_1$, we obtain

$$\hat{\eta}^*_T(t_1)\hat{f}(x^*(t_1), u^*(t_1)) < \hat{\eta}^*_T(t_1)\hat{f}(x^*(t_1), u_1) - \frac{\delta}{2}.$$

But the proof of Theorem 1 shows that this inequality is impossible, and so

$$\hat{H}(\hat{\eta}^*(t), \hat{x}^*(t), u^*(t)) = \hat{M}(\hat{\eta}^*(t), \hat{x}^*(t))$$

almost everywhere on $0 \leq t < \infty$.

The proof of Theorem 1 shows that

$$\hat{M}(\eta^*(t), \hat{x}^*(t)) \quad \text{is constant on} \quad 0 \leq t < \infty.$$

It is then easy to show that

$$\hat{M}(\hat{\eta}^*(t), \hat{x}^*(t)) \equiv 0 \quad \text{on} \quad 0 \leq t < \infty. \qquad \text{Q.E.D.}$$

We finally turn to the most general nonlinear nonautonomous control process. The maximal principle for such processes will be obtained as an immediate consequence of Theorem 3 of Chapter 4 by introducing the time as a new coordinate $x^{n+1} = t$.

In the following analysis we assume:

1. $(\mathcal{S})$ $\dot{x} = f(x, t, u)$ is a control process in R^n with $f \in C^1$ in R^{n+1+m}.
2. The initial and target sets X_0 and X_1 are nonempty in R^n.
3. The admissible controllers Δ are bounded measurable functions $u(t)$ on various finite time intervals $t_0 \leq t \leq t_1$, satisfying some restraint $u(t) \subset \Omega \subset R^m$, and each steering some point $x_0 \in X_0$ to some point $x_1 \in X_1$.
4. The cost of a controller $u(t)$ on $t_0 \leq t \leq t_1$ in Δ with response $x(t)$ is

$$C(u) = \int_{t_0}^{t_1} f^0(x(t), t, u(t))\, dt$$

where $f^0 \in C^1$ in R^{n+1+m}.

The time-augmented response to $u(t)$ is

$$\tilde{x}(t) = (x^0(t), x(t), x^{n+1}(t)),$$

which is the solution of

$$(\tilde{\mathcal{S}}) \qquad \dot{\tilde{x}} = \tilde{f}(\tilde{x}, u)$$

or

$$\begin{aligned} \dot{x}^0 &= f^0(x, x^{n+1}, u) \\ \dot{x} &= f(x, x^{n+1}, u) \\ \dot{x}^{n+1} &= 1 \end{aligned}$$

with $\tilde{x}(t_0) = (0, x_0, t_0)$. The time-augmented adjoint system, based on $u(t)$ and $\tilde{x}(t)$, is

$$(\tilde{\mathcal{A}}) \qquad \dot{\tilde{\eta}} = -\tilde{\eta} \frac{\partial \tilde{f}}{\partial \tilde{x}} (\tilde{x}(t), u(t))$$

or

$$\begin{aligned} \dot{\eta}_0 &= 0 \\ \dot{\eta}_j &= -\sum_{i=0}^{n} \eta_i \frac{\partial f^i}{\partial x^j} (x(t), t, u(t)) \qquad j = 1, \ldots, n. \\ \dot{\eta}_{n+1} &= -\sum_{i=0}^{n} \eta_i \frac{\partial f^i}{\partial t} (x(t), t, u(t)). \end{aligned}$$

The time-augmented Hamiltonian function is

$$\tilde{H}(\tilde{\eta}, \tilde{x}, u) = \eta_0 f^0(x, x^{n+1}, u) + \cdots + \eta_n f^n(x, x^{n+1}, u) + \eta_{n+1}$$

and

$$\tilde{M}(\tilde{\eta}, \tilde{x}) = \max_{u \in \Omega} \tilde{H}(\tilde{\eta}, \tilde{x}, u).$$

We also write

$$\tilde{x} = (\hat{x}, x^{n+1}), \qquad \tilde{\eta} = (\hat{\eta}, \eta_{n+1})$$

and

$$\tilde{H}(\tilde{\eta}, \tilde{x}, u) = \hat{H}(\hat{\eta}, \hat{x}, t, u) + \eta_{n+1}$$
$$\tilde{M}(\tilde{\eta}, \tilde{x}) = \hat{M}(\hat{\eta}, \hat{x}, t) + \eta_{n+1}.$$

THEOREM 2. *Consider the control process in* R^n

$$(\mathcal{S}) \qquad \dot{x} = f(x, t, u).$$

Let Δ *be all bounded measurable controllers* $u(t) \subset \Omega \subset R^m$ *on various finite time intervals* $t_0 \leq t \leq t_1$, *steering points of* X_0 *to* X_1, *as above, with cost*

$$C(u) = \int_{t_0}^{t_1} f^0(x(t), t, u(t))\, dt.$$

If $u^*(t)$ *on* $t_0^* \leq t \leq t_1^*$ *with time-augmented response* $\tilde{x}^*(t)$ *is optimal in* Δ, *then there exists a nontrivial time-augmented adjoint response* $\tilde{\eta}^*(t)$ *of* $\tilde{\mathcal{A}}$ *such that*

$$\tilde{H}(\tilde{\eta}^*(t), \tilde{x}^*(t), u^*(t)) = \tilde{M}(\tilde{\eta}^*(t), \tilde{x}^*(t)) \quad \textit{almost everywhere}$$

and

$$\tilde{M}(\tilde{\eta}^*(t), \tilde{x}^*(t)) \equiv 0, \qquad \eta_0^* \leq 0 \quad \textit{everywhere on} \quad t_0^* \leq t \leq t_1^*.$$

These conclusions can also be written

$$\hat{H}(\hat{\eta}^*(t), \hat{x}^*(t), t, u^*(t)) = \hat{M}(\hat{\eta}^*(t), \hat{x}^*(t), t) \quad \textit{almost everywhere}$$

and

$$\hat{M}(\hat{\eta}^*(t), \hat{x}^*(t), t) \equiv \int_{t_0^*}^{t} \sum_{i=0}^{n} \eta_i^*(s) \frac{\partial f^i}{\partial t} (x^*(s), s, u^*(s))\, ds.$$

The transversality conditions yield

$$\eta_{n+1}^*(t_0^*) = \eta_{n+1}^*(t_1^*) = 0,$$

so

$$\hat{M}(\hat{\eta}^*(t_1^*), \hat{x}^*(t_1^*), t_1^*) = 0.$$

If X_0 *and* X_1 *(or just one of them) are manifolds in* R^n *with tangent spaces* T_0 *and* T_1 *at* x_0^* *and* x_1^*, *respectively, then* $\tilde{\eta}^*(t)$ *can be selected to satisfy the further transversality conditions (or at just one end)*

$$\eta^*(t_0^*) \perp T_0 \quad \text{and} \quad \eta^*(t_1^*) \perp T_1.$$

Proof. In the space R^{n+1} of (x, x^{n+1}) the control problem

$$\dot{x} = f(x, x^{n+1}, u)$$
$$\dot{x}^{n+1} = 1$$

with cost

$$C(u) = \int_{t_0}^{t_1} f^0(x(t), x^{n+1}(t), u(t))\, dt$$

is an autonomous process as treated by Theorem 3 of Chapter 4. The initial and target sets are the cylinders $X_0 \times R^1$ and $X_1 \times R^1$. Since $\dot{x}^{n+1} = 1$, each controller $u(t)$ on $0 \leq t \leq t_1 - t_0$ of this autonomous problem, which steers (x_0, t_0) to (x_1, t_1), can be defined instead on the time interval $t_0 \leq t \leq t_1$; hence $u(t)$ is a controller in Δ. Thus the optimal $u^*(t)$ on $t_0^* \leq t \leq t_1^*$ in Δ is also the optimal controller for this autonomous problem in R^{n+1}. From Theorem 3 of Chapter 4 we obtain the necessary conditions

$$\tilde{H}(\tilde{\eta}^*(t), \tilde{x}^*(t), u^*(t)) = \tilde{M}(\tilde{\eta}^*(t), \tilde{x}^*(t)) \quad \text{almost everywhere}$$

and

$$\tilde{M}(\tilde{\eta}^*(t), \tilde{x}^*(t)) \equiv 0, \qquad \eta_0^* \leq 0 \quad \text{everywhere on} \quad t_0^* \leq t \leq t_1^*.$$

The assertions

$$\hat{H} = \hat{M} \quad \text{and} \quad \hat{M} = \int_{t_0^*}^{t} \sum_{i=0}^{n} \eta_i^*(s) \frac{\partial f^i}{\partial t}(x^*(s), s, u^*(s))\, ds$$

follow directly from the definitions preceding this theorem and the calculation

$$\eta_{n+1}^*(t) = -\int_{t_0^*}^{t} \sum_{i=0}^{n} \eta_i^*(s) \frac{\partial f^i}{\partial t}(x^*(s), s, u^*(s))\, ds + \eta_{n+1}^*(t_0^*).$$

The vanishing of $\eta_{n+1}^*(t_0^*)$ follows from the transversality conditions.

In fact, the transversality conditions assert that $\tilde{\eta}^*(t)$ can be chosen so that $(\eta^*(t_0^*), \eta_{n+1}^*(t_0^*))$ is orthogonal to the line $x_0^* \times R^1$ and $(\eta^*(t_1^*), \eta_{n+1}^*(t_1^*))$ is orthogonal to the line $x_1^* \times R^1$. This means that

$$\eta_{n+1}^*(t_0^*) = \eta_{n+1}^*(t_1^*) = 0, \quad \text{as required.}$$

Then, using an alternate computation for $\hat{M} = -\eta_{n+1}^*$,

$$\hat{M}(\hat{\eta}^*(t), \hat{x}^*(t), t) = \int_{t_1^*}^{t} \sum_{i=0}^{n} \eta_i^*(s) \frac{\partial f^i}{\partial t}(x^*(s), s, u^*(s))\, ds$$

and

$$\hat{M}(\hat{\eta}^*(t_0^*), \hat{x}^*(t_0^*), t_0^*) = \hat{M}(\hat{\eta}^*(t_1^*), \hat{x}^*(t_1^*), t_1^*) = 0.$$

If X_0 and X_1 are manifolds in R^n, then the cylinders $X_0 \times R^1$ and

$X_1 \times R^1$ are orthogonal to $(\eta^*(t_0^*), \eta_{n+1}^*(t_0^*))$ and $(\eta^*(t_1^*), \eta_{n+1}^*(t_1^*))$, respectively. Thus, if q_0 is a vector tangent to X_0 at x_0^* in R^n, then $\eta^*(t_0^*)q_0 = 0$, and $\eta^*(t_0^*)$ is orthogonal to X_0. A similar result holds for X_1. Q.E.D.

Remark 1. Consider the nonautonomous control problem of Theorem 2, but fix the initial time $t_0 = t_0^*$ while allowing various final times $t_1 > t_0$ for controllers steering X_0 to X_1. Then the corresponding set of admissible controllers is Δ_{t_0}. Let $u^*(t)$ on $t_0 \le t \le t_1^*$ with time-augmented response $\tilde{x}^*(t)$ be optimal in Δ_{t_0}.

Then the maximal principle holds

$$\tilde{H}(\tilde{\eta}^*(t), \tilde{x}^*(t), u^*(t)) = \tilde{M}(\tilde{\eta}^*(t), \tilde{x}^*(t)) \quad \text{almost everywhere}$$

and

$$\tilde{M}(\tilde{\eta}^*(t), \tilde{x}^*(t)) \equiv 0, \qquad \eta_0^* \le 0 \quad \text{everywhere on} \quad t_0 \le t \le t_1^*.$$

This again implies that

$$\hat{H} = \hat{M} \quad \text{and} \quad \hat{M} = \int_{t_1^*}^{t} \sum_{i=0}^{n} \eta_i^*(s) \frac{\partial f^i}{\partial t} (x^*(s), s, u^*(s))\, ds$$

so $\hat{M}(\hat{\eta}^*(t_1^*), \hat{x}^*(t_1^*), t_1^*) = 0$, since $\eta_{n+1}^*(t_1^*) = 0$. However, we cannot assert that $\eta_{n+1}^*(t_0)$ vanishes.

If X_0 and X_1 are manifolds in R^n, then it follows that

$$\eta^*(t_0) \perp X_0 \quad \text{and} \quad \eta^*(t_1^*) \perp X_1$$

as before.

Remark 2. We now consider the nonautonomous control problem of Theorem 2 with the extra complication of varying initial and target sets $X_0(t)$ and $X_1(t)$. Let us suppose that $X_0(t)$ and $X_1(t)$ are differentiable manifolds in the R^{n+1} of (x, x^{n+1}). Let $u^*(t)$ on $t_0^* \le t \le t_1^*$ with time-augmented response $\tilde{x}^*(t)$ be optimal in Δ. Then the maximal principle holds as before,

$$\tilde{H} = \tilde{M} \quad \text{and} \quad \tilde{M} \equiv 0 \quad \text{with} \quad \eta_0 \le 0.$$

Again this implies

$$\hat{H} = \hat{M} \quad \text{and} \quad \hat{M} = \int_{t_1^*}^{t} \sum_{i=0}^{n} \eta_i^*(s) \frac{\partial f^i}{\partial t} (x^*(s), s, u^*(s))\, ds - \eta_{n+1}^*(t_1^*).$$

The transversality conditions assert that

$$(\eta^*(t_0^*), \eta_{n+1}^*(t_0^*)) \perp X_0(t_0^*) \quad \text{at} \quad (x_0^*, t_0^*) \quad \text{in} \quad R^{n+1}$$

and

$$(\eta^*(t_1^*), \eta_{n+1}^*(t_1^*)) \perp X_1(t_1^*) \quad \text{at} \quad (x_1^*, t_1^*) \quad \text{in} \quad R^{n+1}.$$

If $t_0 = t_0^*$ is fixed and the admissible controllers are correspondingly restricted to Δ_{t_0}, wherein $u^*(t)$ on $t_0 \le t \le t_1^*$ is optimal, then only the transversal condition at t_1^* is fulfilled (but also $\eta^*(t_0^*) \perp X_0$ in R^n).

In particular, let X_0 be the point x_0 and $X_1(t)$ be the curve $(x_1(t), t)$ in R^{n+1} and let $u^*(t)$ on $t_0 \leq t \leq t_1^*$ be optimal in Δ_{t_0}.

Then the transversality conditions at t_0 are vacuous, but at t_1^* we obtain

$$\eta^*(t_1^*)q_1 + \eta_{n+1}^*(t_1^*) = 0$$

where $q_1 = \dot{x}_1(t_1^*)$ is the velocity of the target point. In this case

$$\hat{M}(\hat{\eta}^*(t_1^*), \hat{x}^*(t_1^*), t_1^*) = \eta^*(t_1^*)q_1.$$

COROLLARY 1. *Consider the control process in* R^n

$$(S) \qquad \dot{x} = f(x, t, u).$$

Let Δ *be all bounded measurable controllers* $u(t) \subset \Omega \subset R^m$ *on various finite time intervals* $t_0 \leq t \leq t_1$, *steering points of* X_0 *to* X_1, *as above, with cost*

$$C(u) = \int_{t_0}^{t_1} 1\, dt = t_1 - t_0.$$

If $u^*(t)$ *on* $t_0^* \leq t \leq t_1^*$ *with response* $x^*(t)$ *is optimal in* Δ, *then there exists a nontrivial adjoint response* $\eta^*(t)$ *of* $\mathcal{A}$ *such that*

$$H(\eta^*(t), x^*(t), t, u^*(t)) = M(\eta^*(t), x^*(t), t) \quad \text{almost everywhere}$$

and

$$M(\eta^*(t), x^*(t), t) - \int_{t_1^*}^{t} \sum_{i=1}^{n} \eta_i^*(s) \frac{\partial f^i}{\partial t}(x^*(s), s, u^*(s))\, ds$$

is a nonnegative constant everywhere on $t_0 \leq t \leq t_1^*$, *so*

$$M(\eta^*(t_1^*), x^*(t_1^*), t_t^*) \geq 0.$$

If X_0 *and* X_1 *are manifolds in* R^n *with tangent spaces* T_0 *and* T_1 *at* x_0^* *and* x_1^* *respectively, then we can choose* $\eta^*(t)$ *so*

$$\eta^*(t_0^*) \perp T_0 \quad \text{and} \quad \eta^*(t_t^*) \perp T_1.$$

Proof. Here

$$H(\eta, x, t, u) = \sum_{i=1}^{n} \eta_i f^i(x, t, u) = \hat{H}(\hat{\eta}, x, t, u) - \eta_0$$

and

$$M(\eta, x, t) = \max_{u \in \Omega} H(\eta, x, t, u) = \hat{M}(\hat{\eta}, x, t) - \eta_0.$$

Also note that the nontrivial adjoint response $\tilde{\eta}^*(t) = (\eta_0^*, \eta^*(t), \eta_{n+1}^*(t))$ of $\tilde{\mathcal{A}}$ defines the solution $\eta^*(t)$ of

$$(\mathcal{A}) \qquad \dot{\eta}_j = -\sum_{i=1}^{n} \eta_i \frac{\partial f^i}{\partial x^j}(x^*(t), t, u^*(t)), \qquad j = 1, \ldots, n.$$

Because

$$\dot{\eta}_{n+1} = -\sum_{i=1}^{n} \eta_i \frac{\partial f^i}{\partial t}(x^*(t), t, u^*(t)),$$

the vanishing of $\eta^*(t)$ implies that $\eta_0 \leq 0$ and η_{n+1} are constants. But then the transversality conditions imply that $\eta_{n+1}(t) \equiv 0$, and from $\tilde{M} \equiv 0$ we obtain $\eta_0 = 0$ so $\tilde{\eta}(t) \equiv 0$, which is impossible. Thus $\eta^*(t)$ is a nontrivial solution of $\mathcal{A}$.

The results of Theorem 2 assert that

$$H(\eta^*(t), x^*(t), t, u^*(t)) = M(\eta^*(t), x^*(t), t) \quad \text{almost everywhere,}$$

and

$$M(\eta^*(t), x^*(t), t) = \int_{t_1^*}^{t} \sum_{i=1}^{n} \overset{*}{\eta_i}(s) \frac{\partial f^i}{\partial t}(x^*(s), s, u^*(s))\, ds - \overset{*}{\eta_0},$$

so

$$M(\eta^*(t_1), x^*(t_1^*), t_1^*) = -\overset{*}{\eta_0} \geq 0, \quad \text{as required.}$$

If X_0 and X_1 are manifolds in R^n, then the transversality conditions of Theorem 2 yield the required orthogonality conditions. Q.E.D.

Remark 1. If the initial time $t_0 = t_0^*$ is fixed and the correspondingly restricted controllers Δ_{t_0} are used, then the conclusions of the corollary hold as stated.

Remark 2. Consider the minimal time-optimal control problem of the corollary, but with varying initial and target sets $X_0(t)$ and $X_1(t)$. Then

$$H(\eta^*(t), x^*(t), t, u^*(t)) = M(\eta^*(t), x^*(t), t) \quad \text{almost everywhere,}$$

and

$$M(\eta^*(t), x^*(t), t) = \int_{t_1^*}^{t} \sum_{i=1}^{n} \overset{*}{\eta_i}(s) \frac{\partial f^i}{\partial t}(x^*(s), s, u^*(s))\, ds - \overset{*}{\eta}_{n+1}(t_1) - \eta_0.$$

Again if $X_0(t)$ is arbitrary but $X_1(t)$ or $(x_1(t), t)$ is a curve in R^{n+1}, then the transversality conditions yield

$$\eta^*(t_1^*)q_1 + \overset{*}{\eta}_{n+1}(t_1^*) = 0$$

where $q_1 = \dot{x}_1(t_1^*)$ is the velocity of the target point. This computation for

$$\overset{*}{\eta}_{n+1}(t_1^*) = -\eta^*(t_1^*)q_1$$

holds even when the initial time t_0 is fixed and the restricted controllers Δ_{t_0} are used.

Remark 3. Consider the control process in R^n

$$(\mathcal{S}) \qquad \dot{x} = f(x, t, u)$$

with bounded measurable controllers $u(t)$ on a fixed finite time duration $0 \leq t \leq T$, satisfying the restraint $\Omega \subset R^m$. Let Δ_T be all such controllers that some initial point in X_0 to some final point in X_1, with cost functional

$$C(u) = \int_0^T f^0(x(t), t, u(t))\, dt,$$

as in Theorem 2.

Let $u^*(t)$ on $0 \leq t \leq T$ be an optimal controller for this fixed-time problem, and let $\tilde{x}^*(t)$ be the corresponding time-augmented response. Then there exists a nontrivial time-augmented adjoint response $\tilde{\eta}^*(t)$ of $\tilde{\mathcal{A}}$ on $0 \leq t \leq T$ such that

$$\tilde{H}(\tilde{\eta}^*(t), \tilde{x}^*(t), u^*(t)) = \tilde{M}(\tilde{\eta}^*(t), \tilde{x}^*(t)) \quad \text{almost everywhere}$$

Considering the initial and final targets as sets $(X_0, 0)$ and (X_1, T) in R^{n+1}, we can ignore the time restrictions and conclude that

$$\tilde{M}(\tilde{\eta}^*(t), \tilde{x}^*(t)) \equiv 0 \quad \text{with} \quad \eta_0^* \leq 0 \quad \text{on} \quad 0 \leq t \leq T.$$

If X_0 and X_1 are manifolds in R^n, we choose $\tilde{\eta}^*(t) = (\eta_0^*, \eta^*(t), \eta_{n+1}^*(t))$ so $\eta^*(t_0^*) \perp X_0$ and $\eta^*(t_1^*) \perp X_1$. But $\eta_{n+1}^*(0)$ and $\eta_{n+1}^*(T)$ might fail to vanish. Nevertheless $\hat{\eta}^*(t) = (\eta_0^*, \eta^*(t))$ is nonvanishing [since $\tilde{M} \equiv 0$ and $\tilde{\eta}^*(t)$ is nonvanishing] and is a nontrivial solution of

$$(\hat{\mathcal{A}}) \quad \dot{\eta}_0 = 0, \qquad \dot{\eta}_j = -\sum_{i=0}^{n} \eta_i \frac{\partial f^i}{\partial x^j}(x^*(t), t, u^*(t)), \qquad j = 1, \ldots, n.$$

The necessary conditions satisfied by $u^*(t)$ in this fixed-time problem are then just those stated in Theorem 2 [with the expression for $\tilde{M}$ diminished by $\eta_{n+1}^*(0)$, which may not vanish]. If $X_1 = R^n$, then $\eta^*(T) = (0, 0, 0, \ldots, 0)$ and we can choose $\hat{\eta}^*(T) = (-1, 0, 0, \ldots, 0)$.

We next turn to a version of the maximal principle that applies to linear control processes with impulse controllers. The concepts and notations are those introduced in Section 4.2 (see Theorem 7 of Chapter 4).

Consider the linear impulse control process in R^n

$$(\mathfrak{L}) \qquad Dx = A(t)x + B(t)\, Du$$

or

$$x(t) = \Phi(t)x_0 + \int_0^t \Phi(t)\, \Phi(s)^{-1}\, B(s)\, Du(s),$$

with initial state $x(0-) = x_0$, on a fixed compact time interval $0 \leq t \leq T$. The controllers $u(t)$ are right continuous, bounded variation, m-vector functions, each on some open interval about $0 \leq t \leq T$. Each such controller defines a (Lebesgue-Stieltjes) signed measure Du on $0 \leq t \leq T$, and $x(t)$ is the response initiating at $x(0-) = x_0$. The coefficients $A(t)$ and

$B(t)$ are continuous matrices, and $\Phi(t)$ is the fundamental matrix solution of the homogeneous differential system with $\Phi(0) = I$. We seek a controller $u(t)$ (or corresponding signed measure Du) that steers $x(0-) = x_0$ to a prescribed target point $x(T) = x_1$ with minimal cost

$$\|Du\| = \sum_{j=1}^{m} \int_0^T |Du^j| = \sum_{j=1}^{m} |u^j(0) - u^j(0-)| + \underset{0 \le t \le T}{\text{var}}\ u^j.$$

The linear space $\mathcal{M}$ of all such signed measures, with the above norm, is a Banach space (a complete normed vector space). Moreover the space $\mathcal{M}$ can be identified with the dual space (space of all continuous linear functionals) of the Banach space $\mathfrak{C}$. Here $\mathfrak{C}$ is the space of all real continuous m-vectors $y(t)$ on $0 \le t \le T$ with norm

$$\|y(t)\|_\infty = \max_{0 \le t \le T,\ 1 \le j \le m} |y^j(t)|,$$

and the operation of $\mathcal{M}$ on $\mathfrak{C}$ is computed by

$$Du(y(t)) = \int_0^T y(t)\, Du.$$

Let Du_k be a sequence of measures in $\mathcal{M}$ with uniform bound $\|Du_k\| \le \beta$. Then for each $y(t) \in \mathfrak{C}$ there exists a subsequence Du_{ki} that converges weakly to some limit $D\bar{u}$ with $\|D\bar{u}\| \le \beta$, that is,

$$\lim_{k_i \to \infty} \int_0^T y(t)\, Du_{k_i} = \int_0^T y(t)\, D\bar{u}.$$

If $m = 1$ this result is known as the Helly-Bray theorem.

If we restrict attention to absolutely continuous controllers $u(t)$, then $Du = \dot{u}(t)\, dt$ and the impulse-control process reduces to an ordinary differential system with absolutely continuous response

$$x(t) = \Phi(t)x_0 + \int_0^t \Phi(t)\, \Phi(s)^{-1}\, B(s)\, \dot{u}(s)\, ds$$

and cost

$$\|\dot{u}(t)\, dt\| = \int_0^T |\dot{u}(t)|\, dt.$$

However, within this class of controllers, an optimal may not exist and so we are led to the more general impulse controllers.

We shall assume that $\mathfrak{L}$ is controllable on $0 \le t \le T$, that is, the rows of $\Phi(s)^{-1}\, B(s)$ are linearly independent in $\mathfrak{C}$. This is the case if and only if the set of attainability $K(T)$ has a nonempty interior. If A and B are constant, then the customary controllability condition

$$\text{rank}\,[B, AB, \ldots, A^{n-1}B] = n$$

is necessary and sufficient that the rows of $\Phi(s)^{-1} B(s)$ are independent in $\mathfrak{C}$.

LEMMA. *Consider the controllable impulse process in* R^n

$$(\mathfrak{L}) \qquad Dx = A(t)x + B(t)\,Du$$

with $x(0-) = x_0$ *and restraint* $\|Du\| \leq \alpha$ *on* $0 \leq t \leq T$. *Then the set* $K(T, \alpha)$ *of attainability from* x_0 *is a compact convex set which varies continuously with* $\alpha \geq 0$. *Also* $K(T, \alpha_1)$ *lies in the interior of* $K(T, \alpha_2)$ *whenever* $0 \leq \alpha_1 < \alpha_2$.

Proof. The compactness of $K(T, \alpha)$ follows from the Helly-Bray theorem and the convexity follows from an elementary computation on variations. Since a controller $Du_2(t)$ with norm $\|Du_2\| \leq \alpha_2$ can be approximated by a controller Du_1 with $\|Du_1\| \leq \alpha_1$ (merely define $Du_1 = Du_2$ except on a neighborhood of $t = T$ where we take $Du_1 = 0$), we note that $K(T, \alpha_1) \subset K(T, \alpha_2)$ and $K(T, \alpha)$ depends continuously on $\alpha \geq 0$. In fact, $K(T, \alpha) - \Phi(T)x_0 = \alpha[K(T, 1) - \Phi(T)x_0]$.

Finally let $\alpha_1 < \alpha_2$ and take a controller $u_1(t)$, with $\|Du_1\| \leq \alpha_1$, leading to a point $x_1 \in K(T, \alpha_1)$. The controllability of $\mathfrak{L}$ assures the existence of $n + 1$ smooth controllers $w_1(t), w_2(t), \ldots, w_{n+1}(t)$ with

$$\|Dw_k(t)\| < \alpha_2 - \alpha_1,$$

such that $u_1(t) + w_k(t)$ lead to the vertices of a simplex centered about x_1 for $k = 1, 2, \ldots, n + 1$. Thus x_1 lies in the interior of $K(T, \alpha_2)$, as required. Q.E.D.

This lemma yields an existence guarantee for an optimal controller and also a formula for the optimal cost. This formula is computed from the maximization of a linear function on a spheroid hypersurface $H \subset R^n$. The set H is defined by all row n-vectors η such that

$$\|\eta\, \Phi(T)\, \Phi(t)^{-1}\, B(t)\|_\infty = 1.$$

Because of the controllability of $\mathfrak{L}$, each unit vector or direction in R^n specifies a nonzero vector $\eta \in H$.

COROLLARY. *There exists an optimal controller* $u*(t)$ *with minimal cost* $\|Du^*\| = \alpha^*$. *Also*

$$\alpha^* = \max_{\eta \in H} \eta[x_1 - \Phi(T)x_0].$$

Proof. Since $K(T, \alpha)$ is compact and grows continuously with $\alpha \geq 0$ from a single point $\Phi(T)x_0$ to reach any point x_1, there exists a minimal value $\alpha = \alpha^*$ for which $x_1 \in K(T, \alpha)$. In fact x_1 lies on the boundary of $K(T, \alpha^*)$.

Let η^* be an exterior normal to the convex set $K(T, \alpha^*)$ at x_1, and take η^* normalized to lie in H. Then

$$\eta^* x_1 = \max_{x \in K(T,\alpha^*)} \eta^* x = \max_{\|Du\| \le \alpha^*} \eta^* \left[\Phi(T)x_0 + \int_0^T \Phi(T)\,\Phi(t)^{-1}\,B(t)\,Du \right],$$

and so

$$\eta^*[x_1 - \Phi(T)x_0] = \max_{\|Du\| \le \alpha^*} \int_0^T \eta^*\,\Phi(T)\,\Phi(t)^{-1}\,B(t)\,Du.$$

Since $\mathcal{M}$ is the dual space of $\mathfrak{C}$, this maximum is just

$$\alpha^* \|\eta^*\,\Phi(T)\,\Phi(t)^{-1}\,B(t)\|_\infty = \alpha^*.$$

Hence

$$\alpha^* = \eta^*[x_1 - \Phi(T)x_0],$$

and we conclude the rule: take any point x_1 on $\partial K(T, \alpha^*)$ and any external normal $\eta^* \in H$, then the value α^* is computed as above.

Now choose any vector $\eta \in H$ based at x_1. The hyperplane $\tilde{\pi}$ normal to η is then a supporting hyperplane at some point $\tilde{x}$ on the boundary of some $K(T, \tilde{\alpha})$ for $0 \le \tilde{\alpha} \le \alpha^*$. If η is an exterior normal

$$\eta[\tilde{x} - \Phi(T)x_0] = \tilde{\alpha}$$

and so

$$\eta[x_1 - \Phi(T)x_0] = \tilde{\alpha} \le \alpha^*.$$

If η is an interior normal, then $-\eta$ is an exterior normal and so

$$\eta[x_1 - \Phi(T)x_0] = -\tilde{\alpha} \le 0.$$

Thus in all cases

$$\eta[x_1 - \Phi(T)x_0] \le \alpha^* = \eta^*[x_1 - \Phi(T)x_0],$$

and the formula for α^* is verified. Q.E.D.

We conclude that a controller $u^*(t)$ steering x_0 to x_1 is optimal if and only if

$$\|Du^*\| = \max_{\eta \in H} \eta[x_1 - \Phi(T)x_0].$$

This formula will serve as a basis in the computation of optimal controllers as a sum of δ-functions. But first we shall show that every point in $K(T, \alpha)$ can be attained by an impulse controller, which is a linear combination of (at most) $n + 1$ scalar δ-functions, and which has norm α.

It is sufficient to prove that every point in $K(T, 1)$ is attained by a convex combination of $n + 1$ scalar δ-functions. By a scalar δ-function $\pm\Delta_k(t - t')$ we mean the signed measure corresponding to the step function

$$u^j(t) = \delta_k{}^j \cdot \begin{cases} 0 & \text{for} \quad t < t' \\ \pm 1 & \text{for} \quad t \ge t' \end{cases}$$

where $0 \le t' \le T$ and δ_k^j is the usual Kronecker symbol. In other words $\Delta_k(t - t')$ has a weight of $+1$ in the kth component at instant $t = t'$, and no weight otherwise.

Now let $g(t)$ be a controller with $\|Dg\| \le 1$ and response endpoint $x_g(T) \in K(T, 1)$. Give $\epsilon > 0$ and choose a finite time partition

$$(0-) = t_0 < t_1 < \cdots < t_\nu = T$$

such that

$$\left| \int_0^T \Phi(T)\,\Phi(t)^{-1}\,B(t)\,Dg(t) - \sum_{\sigma=0}^{\nu-1} \Phi(T)\,\Phi(t_\sigma)^{-1}\,B(t_\sigma)[g(t_{\sigma+1}) - g(t_\sigma)] \right| < \epsilon.$$

Consider the linear combination of scalar δ-functions

$$\sum_{j=1}^{m} [g^j(t_{\sigma+1}) - g^j(t_\sigma)]\,\Delta_j(t - t_\sigma),$$

which has a vector weight of $[g(t_{\sigma+1}) - g(t_\sigma)]$ at the instant $t = t_\sigma$, for each $\sigma = 0, 1, 2, \ldots, \nu$. Now take the sum of all these impulses

$$\sum_{\sigma=0}^{\nu} \sum_{j=1}^{m} [g^j(t_{\sigma+1}) - g^j(t_\sigma)]\,\Delta_j(t - t_\sigma) = Dg_\Delta(t).$$

Then Dg_Δ is a linear combination of scalar δ-functions, and

$$\|Dg_\Delta\| = \sum_{\sigma=0}^{\nu} |g(t_{\sigma+1}) - g(t_\sigma)| \le \|Dg\| \le 1.$$

Also, with an appropriate choice of sign in $\pm\Delta_j(t - t_\sigma)$, and with the possible addition of a positive multiple of $\Delta_1(t) - \Delta_1(t) \equiv 0$, we can consider Dg_Δ as a convex combination of scalar δ-functions. Also the response endpoint $x_\Delta(T)$ approximates $x_g(T)$, that is

$$|x_g(T) - x_\Delta(T)| < \epsilon.$$

Let $\mathfrak{D}$ be the set of all points in $K(T, 1)$ that are attained by scalar δ-function controllers. We have shown that the convex hull $H(\mathfrak{D})$ is dense in $K(T, 1)$. Thus every point in the interior of $K(T, 1)$ lies in $H(\mathfrak{D})$. Using standard results on convex combinations of points in R^n, we note that every point in the interior of $K(T, 1)$ lies in an n-simplex with vertices in $\mathfrak{D}$. Thus each point interior to $K(T, 1)$ is attainable by a convex combination of $n + 1$ scalar δ-function controllers. By taking appropriate limits it is easy to show that every boundary point of $K(T, 1)$ is attainable by a convex combination of (at most) $n + 1$ scalar δ-function controllers.

THEOREM 3. *Consider the controllable impulse process in* R^n

$$(\mathfrak{L}) \qquad Dx = A(t)x + B(t)\,Du$$

with initial and target states $x(0-) = x_0$ and $x(T) = x_1$, and controllers $u(t)$ of bounded variation with cost $\|Du\|$ on the compact time interval $0 \leq t \leq T$.

Let $\eta^* \in H$ satisfy

$$\eta^*[x_1 - \Phi(T)x_0] = \max_{\eta \in H} \eta[x_1 - \Phi(T)x_0],$$

and define the closed time sets

$$\Gamma_j = \{t \mid (\eta^* \Phi(T) \Phi(t)^{-1} B(t))^j = \pm 1\}, \qquad j = 1, \ldots, m.$$

Assume that a controller $u^*(t)$, with

$$Du^*(t) = c_1 \Delta_{k_1}(t - t_1) + \cdots + c_n \Delta_{k_n}(t - t_n)$$

as a linear combination of n scalar δ-functions, satisfies the conditions;

(a). *$u^*(t)$ steers x_0 to x_1, that is,*

$$(x_1 - \Phi(T)x_0)^j = c_1 \delta_{k_1}{}^j \Phi(T) \Phi(t_1)^{-1} B(t_1) + \cdots + c_n \delta_{k_n}{}^j \Phi(T)\Phi(t_n)^{-1} B(t_n).$$

(b). *All the impulses for the jth component of $u^*(t)$ lie in the set Γ_j, and the signs of the real coefficients $c_1, \ldots, c_n$ are such that*

$$\int_0^T \eta^* \Phi(T) \Phi(t)^{-1} B(t)\, Du^* = |c_1| + \cdots + |c_n|.$$

Then $u^*(t)$ is an optimal controller.

On the other hand there always exists an optimal controller with such a representation as a combination of n scalar δ-functions (with the prescribed η^* and Γ_j).

Proof. If $u^*(t)$ satisfies hypotheses (a) and (b), then $u^*(t)$ steers x_0 to x_1 and it achieves the minimal cost $\|Du^*\| = \alpha^* = \eta^*[x_1 - \Phi(T)x_0]$. This last assertion follows from

$$|c_1| + \cdots + |c_n| = \int_0^T \eta^* \Phi(T) \Phi(t)^{-1} B(t)\, Du^*.$$

Therefore $u^*(t)$ is an optimal controller.

Now let η^* and the sets D_j be chosen as above. There exists an optimal controller steering x_0 to x_1 with the minimal cost α^*. By the argument preceding the theorem, we can find an optimal $\hat{u}(t)$ that is a linear combination of $n + 1$ scalar δ-functions, in fact, $\hat{u}(t)$ is a convex combination of $n + 1$ modified δ-functions each of which has an impulse of $\pm\alpha^*$ in just one component (that is, an impulse controller of the form $\pm\alpha^* \Delta_k(t - t'))$. Since x_1 lies on the boundary of $K(T, \alpha^*)$, it cannot lie in the interior of the n-simplex whose vertices correspond to the modified

δ-function controllers. Therefore x_1 lies on a face of this simplex, or the simplex has no interior but falls on a hyperplane in R^n; in either case x_1 can be attained by a convex combination of only n modified δ-functions (and possibly the zero controller).

Therefore $\hat{u}(t)$ is an optimal controller with $D\hat{u}$ written as a linear combination of n scalar δ-functions. The impulses of the jth component of $D\hat{u}$ must occur in the set Γ_j, for otherwise we compute that

$$\alpha^* = \int_0^T \eta^* \Phi(T)\,\Phi(t)^{-1}\,B(t)\,Du < \|D\hat{u}\| = \alpha^*,$$

which is impossible. Thus $\hat{u}(t)$ must be an optimal controller of the form prescribed in the theorem. Q.E.D.

We next compute an impulse optimal controller to illustrate the significance of the above theorem.

Example. Consider the impulse-control process

$$D^2x = Du,$$

or the phase-plane system

$$Dx^1 = x^2, \qquad Dx^2 = Du$$

where we wish to steer (0, 0) to (1, -1) in the duration $0 \le t \le 1$ with minimal cost $\|Du\|$. Here $\Phi(t) = \begin{pmatrix} 1 & t \\ 0 & 1 \end{pmatrix}$ and $B = \begin{pmatrix} 0 \\ 1 \end{pmatrix}$ so the condition H is

$$\max_{0 \le t \le 1} |(1 - t)\eta_1 + \eta_2| = 1.$$

For each angular coordinate θ, we seek to compute $c(\theta)$ $(\cos\theta, \sin\theta)$ in H, [here $c(\theta) > 0$]. This yields

$$c(\theta) = \min\left\{\frac{1}{|\cos\theta + \sin\theta|}, \frac{1}{|\sin\theta|}\right\}.$$

Then the value of θ defining η^* is specified by maximizing

$$(\eta_1, \eta_2)\begin{pmatrix} 1 \\ -1 \end{pmatrix} = \begin{cases} \dfrac{\cos\theta - \sin\theta}{|\cos\theta + \sin\theta|} \\[2ex] \dfrac{\cos\theta - \sin\theta}{|\sin\theta|} \end{cases}, \quad \text{whichever has the larger denominator.}$$

A careful study of this trigonometric function yields a unique value θ^*, where $\cos\theta^* = 2/\sqrt{5}$, $\sin\theta^* = -1/\sqrt{5}$, so that $\eta^* = (2, -1)$ is uniquely specified.

Then the set Γ where $1 - 2t = \pm 1$ is just

$$\Gamma = \{t = 0, t = 1\}.$$

Therefore the optimal controller must be

$$Du^* = c_1\,\delta(t) + c_2\,\delta(t-1).$$

The unknown coefficients c_1 and c_2, which yield a cost

$$\|Du^*\| = \eta^* x_1 = \alpha^* = 3$$

and steer $(0, 0)$ to $(1, -1)$, are computed to be

$$c_1 = 1, \qquad c_2 = -2.$$

Thus

$$Du^* = \delta(t) - 2\delta(t-1).$$

This is the unique optimal controller (because any optimal controller must have all its weight in the set Γ, and hence has the prescribed form of the combination of two scalar δ-function controllers).

We note that the optimal response first jumps from the origin to $(0, 1)$, then moves freely to $(1, 1)$, and then jumps to the target $(1, -1)$.

As a final study in the maximal principle we investigate linear control processes with smooth controllers having a limited rate of variation. In this way the inertia of the control mechanism is realistically considered and instantaneous switching is eliminated. The analysis leads to a bounded phase coordinate problem, but of such a special nature that the maximal principle is immediately applicable. In order to assure that certain controllability hypotheses are explicitly verifiable, we restrict attention to autonomous linear processes. More general bounded phase coordinate problems are also discussed.

Consider a linear autonomous process in R^n

$$(\mathfrak{L}) \qquad \dot{x} = Ax + Bu,$$

where we wish to steer the initial state x_0 to the target state x_1 by some optimal controller $u^*(t)$ in R^m defined for some minimal time duration $0 \leq t \leq t^*$. The admissible controllers are absolutely continuous functions $u(t)$ on various finite time durations $0 \leq t \leq t_1$ satisfying the restraints

1. $u(t) \subset \Omega$, where Ω is a given closed convex set containing the origin of R^m
2. $v(t) = \dot{u}(t)$ are measurable and $|v^j(t)| \leq 1$ for all $j = 1, \ldots, m$ almost everywhere
3. $u(0) = u(t_1) = 0$.

We can introduce the $(n + m)$-dimensional state

$$z(t) = \begin{pmatrix} x^1(t) \\ \cdot \\ \cdot \\ \cdot \\ x^n(t) \\ u^1(t) \\ \cdot \\ \cdot \\ \cdot \\ u^m(t) \end{pmatrix}$$

and write the extended control process

$$(\tilde{\mathfrak{L}}) \qquad \dot{z} = \begin{pmatrix} A & B \\ 0 & 0 \end{pmatrix} z + \begin{pmatrix} 0 \\ I \end{pmatrix} v.$$

For the process $\tilde{\mathfrak{L}}$ we take the initial state $z_0 = \begin{pmatrix} x_0 \\ 0 \end{pmatrix}$ and the target $z_1 = \begin{pmatrix} x_1 \\ 0 \end{pmatrix}$ and use measurable controllers $v(t)$ on $0 \leq t \leq t_1$ such that

$$|v^j(t)| \leq 1 \quad \text{for all} \quad j = 1, \ldots, m$$

and the response $z(t)$ satisfies the phase constraint $z(t) \subset \begin{pmatrix} R^n \\ \Omega \end{pmatrix}$. An optimal controller $u^*(t)$ of $\mathfrak{L}$ then defines an optimal controller $v^*(t) = \dot{u}^*(t)$ of $\tilde{\mathfrak{L}}$, and vice versa. So it is clear that an optimal controller $u^*(t)$ on $0 \leq t \leq t^*$ exists for $\mathfrak{L}$, provided there is some admissible controller steering x_0 to x_1 (see Theorem 4 in Chapter 4).

We shall prove that an optimal controller $u^*(t)$ on $0 \leq t \leq t^*$ for $\mathfrak{L}$ satisfies at least one of the conditions

$$u^*(t) \in \partial\Omega$$

or

$$|\dot{u}^{*j}(t)| = 1 \quad \text{for all} \quad j = 1, \ldots, m$$

at almost every instant. This property of the optimal controller is sometimes referred to as *pang-bang* behavior.

If $\mathfrak{L}$ is controllable, and if Ω contains $u = 0$ in its interior, then every initial x_0 which lies sufficiently near the origin in R^n can be steered to $x_1 = 0$ by an admissible controller. If, in addition, A is stable, then every initial state $x_0 \in R^n$ can be steered to the origin by an admissible controller, and hence an optimal controller exists in all these cases (see Exercise 8 below). In order to simplify the presentation of Theorem 4 we shall assume that $\mathcal{L}$ is controllable, that is,

$$\text{rank}\,[B, AB, A^2B, \ldots, A^{n-1}B] = n.$$

LEMMA. *Consider the autonomous linear process in R^n*

$$(\mathcal{L}) \qquad \dot{x} = Ax + Bu$$

with initial state x_0 at $t = 0$ and absolutely continuous controllers $u(t)$ on $0 \le t \le t_1$ satisfying the restraints $u(t) \subset \Omega$, and $|\dot{u}^j(t)| \le 1$, and $u(0) = u(t_1) = 0$, as above. Then the set $K(t_1)$ of attainable endpoints $x(t_1)$ is a compact convex set in R^n.

Moreover, $\bar{u}(t)$ on $0 \le t \le t_1$ is extremal, that is, the response $\bar{x}(t)$ terminates at $\bar{x}(t_1) \in \partial\, K(t_1)$ if and only if the maximal principle holds:

$$\int_0^{t_1} \bar{\eta}(s)B\,\bar{u}(s)\,ds = \max \int_0^{t_1} \bar{\eta}(s)\,B\,\,u(s)\,ds.$$

Here the maximum is taken over all admissible controllers $u(t)$ on $0 \le t \le t_1$ satisfying the restraints, and $\bar{\eta}(t)$ is some nontrivial adjoint response of

$$(\mathcal{A}) \qquad \dot{\eta} = -\eta A.$$

Proof. The proof that $K(t_1)$ is convex follows from the customary techniques presented in the theory of linear processes in Chapter 2. The compactness of $K(t_1)$ can be proved by selecting a subsequence of controllers $u(t)$, with corresponding derivatives $\dot{u}(t) = v(t)$ converging weakly to some limit $\hat{v}(t)$. It is easy to verify that $\hat{u}(t) = \int_0^t \hat{v}(s)\,ds$ is an admissible controller that steers x_0 to a desired point of $\overline{K(t_1)}$. The details are left as exercises.

The proof of the maximal principle is just as in Chapter 2. As usual, $\bar{\eta}(t_1)$ is an outward unit vector to the convex set $K(t_1)$ at the boundary point $\bar{x}(t_1)$. Q.E.D.

We now assume that $\mathcal{L}$ is normal for the standard m-cube, that is,

$$\det |Bv, ABv, A^2Bv, \ldots, A^{n-1}Bv| \ne 0,$$

for each of the m-vectors v along the coordinate axes in R^m (for all edges v of the standard m-cube). We say that $\mathcal{L}$ is *cube-normal* and refer to discussions of normality in Chapter 2 where it is shown that this hypothesis implies the controllability of $\mathcal{L}$. It is also easy to show that for a cube-normal process $\mathcal{L}$, each component of $\eta_0 e^{-At}B$, for every $\eta_0 \ne 0$, vanishes only at a discrete set of time instants.

THEOREM 4. *Consider the cube-normal autonomous process in R^n*

$$(\mathfrak{L}) \qquad \dot{x} = Ax + Bu$$

with initial state x_0 at $t = 0$ and target state x_0. The controllers are absolutely continuous functions $u(t)$ on various finite durations $0 \leq t \leq t_1$ steering the response $x(t)$ from $x(0) = x_0$ to $x(t_1) = x_1$ and satisfying the restraints:

1. *$u(t) \subset \Omega$, a closed convex set containing the origin of R^m;*
2. *$|\dot{u}^j(t)| \leq 1$ for all $j = 1, \ldots, m$ almost always;*
3. *$u(0) = u(t_1) = 0$.*

If $u^*(t)$ on $0 \leq t \leq t^*$ is a minimal time-optimal controller, then either

$$u^*(t) \in \partial\Omega, \quad \text{or} \quad |\dot{u}^{j*}(t)| = 1 \quad \text{for all} \quad j = 1, \ldots, m$$

at almost every instant.

Proof. Suppose, to the contrary, that there is a time interval $\mathfrak{J}$ on $0 \leq t \leq t^*$, wherein $u^*(t)$ lies in the interior of Ω and, say, $|\dot{u}^{1*}(t)| < 1$ on a subset of positive duration. Let $\eta^*(t) = \eta_0{}^*e^{-At}$ be the nontrivial adjoint response, so that

$$\int_0^{t^*} \eta_0^* e^{-At}B\, u^*(t)\, dt = \max \int_0^{t^*} \eta_0^* e^{-At}B\, u(t)\, dt,$$

as in the lemma.

Since $\mathfrak{L}$ is cube-normal the first component $\zeta(t)$ of $\eta_0{}^*e^{-At}B$ is nonvanishing on $\mathfrak{J}$ (or on a subinterval still called $\mathfrak{J}$). We shall define a new admissible controller $\hat{u}(t)$ on $0 \leq t \leq t^*$ such that

$$\hat{u}^j(t) = u^{j*}(t) \quad \text{for} \quad j = 2, 3, \ldots, m \quad \text{on} \quad 0 \leq t \leq t^*,$$

$$\hat{u}^1(t) = u^{1*}(t) \quad \text{outside} \quad \mathfrak{J},$$

and

$$\int_{\mathfrak{J}} \zeta(t)\, u^{1*}(t)\, dt < \int_{\mathfrak{J}} \zeta(t)\, \hat{u}^1(t)\, dt.$$

Thus the construction of $\hat{u}(t)$ will contradict the maximal principle holding for $u^*(t)$ and the theorem will follow.

It is sufficient to examine the case where $\zeta(t) > 0$ on $\mathfrak{J}$: $t_0 \leq t \leq t_1$, and the initial point t_0 of $\mathfrak{J}$ is a point of positive metric density for the set where $|\dot{u}^{1*}(t)| < 1 - \epsilon$, for some fixed $\epsilon > 0$. Then the graph of $u^{1*}(t)$ versus $t \in \mathfrak{J}$ lies within a sector bounded by lines with slopes $\pm(1 - \epsilon/2)$, at least if $\mathfrak{J}$ is sufficiently short. Now define $\hat{u}^1(t) = u^{1*}(t_0) + t - t_0$ for a short duration and then define $\hat{u}^1(t)$ by the linear function with slope -1 until its graph meets that of $u^{1*}(t)$. In this way we construct the continuous and piecewise linear real function $\hat{u}^1(t)$ over a subinterval $\mathfrak{J}_1 \subset \mathfrak{J}$ such that $\hat{u}^1(t) \geq u^{1*}(t)$ with equality only at the endpoints of $\mathfrak{J}_1$. Thus

$$\int_{\mathfrak{J}_1} \zeta(t)(\hat{u}^1(t) - u^{1*}(t))\, dt > 0.$$

We define $\hat{u}^1(t) = u^{1*}(t)$ outside of $\mathfrak{J}_1$ and then $\hat{u}(t)$ is an admissible controller that has the desired property, namely,

$$\int_0^{t^*} \eta_0^* e^{-At} B\, u^*(t)\, dt < \int_0^{t^*} \eta_0^* e^{-At} B\, \hat{u}(t)\, dt.$$

This contradiction proves the theorem. Q.E.D.

It is not easy to decide which intervals of $0 \leq t \leq t^*$ are *pang* where $|\dot{u}^j{}^*(t)| = 1$, and which intervals are *bang*, where $u^*(t) \in \partial\Omega$. The synthesis of the optimal controller by numerical and graphical schemes is still to be developed and perfected.

We conclude this section with a discussion of a general bounded phase coordinate process.

Let us consider a control process in R^n

$$(\mathcal{S}) \qquad \dot{x} = f(x, u)$$

where the initial state x_0 is to be steered to the target state x_1 by some measurable controller $u(t) \subset \Omega$ on some $0 \leq t \leq t_1$. The restraint set Ω is compact in R^m and the cost integral is

$$C(u) = \int_0^{t_1} f^0(x(t), u(t))\, dt$$

with $f(x, u)$ and $f^0(x, u)$ in class C^1 in $R^n \times \Omega$. Under these circumstances an optimal controller $u^*(t)$ with response $x^*(t)$ on $0 \leq t \leq t^*$ satisfies the maximal principle almost everywhere,

$$\hat{H}(\hat{\eta}^*(t), x^*(t), u^*(t)) = \max_{u \in \Omega} \hat{H}(\hat{\eta}^*(t), x^*(t), u) \equiv 0.$$

Here the Hamiltonian function is

$$\hat{H}(\hat{\eta}, x, u) = \sum_{\alpha=0}^{n} \eta_\alpha f^\alpha(x, u),$$

and $\hat{\eta}^*(t) = (\eta_0^*, \eta^*(t))$ is a nontrivial solution of

$$\dot{\eta}_0 = 0, \qquad \eta_0 \leq 0$$

$$\dot{\eta}_i = -\sum_{\alpha=0}^{n} \eta_\alpha \frac{\partial f^\alpha}{\partial x^i}(x^*(t), u^*(t)), \qquad 1 \leq i \leq n.$$

Now let us further assume that the state variable x is constrained to a closed subset $\Lambda \subset R^n$, with x_0 and x_1 in Λ. The optimal control problem $\{\mathcal{S}, C, x_0, x_1, \Omega, \Lambda\}$ is called a bounded phase coordinate process. The original optimal response $x^*(t)$ might not lie in Λ in which case we seek a new optimal controller for this constrained process.

We recall that the existence problem of optimal controllers for bounded phase coordinate processes has been settled in Chapter 4 (Theorem 4 and its obvious consequences). Here we discuss the corresponding modifications of the maximal principle that arise in bounded phase coordinate processes. The strongest results apply only in the cases where $\mathcal{S}$ is linear (or perhaps convex in u) and we present these analyses later. In this section we present a general study of the maximal principle for bounded phase coordinate processes.

Let us assume that $u^*(t)$ on $0 \leq t \leq t^*$ is an optimal controller, with response $x^*(t) \subset \Lambda$ joining the states x_0 and x_1, which lie interior to Λ. Assume that the interval $0 \leq t \leq t^*$ has a finite subdivision into subintervals $0 = t_0 < t_1 \leq t_2 < t_3 \leq t_4 < \cdots < t_{2k+1} = t^*$ such that:

$$x^*(t) \subset \text{interior } \Lambda \quad \text{on} \quad t_i < t < t_{i+1}, \quad \text{even} \quad i$$
$$x^*(t) \subset \text{boundary } \Lambda \quad \text{on} \quad t_i \leq t \leq t_{i+1}, \quad \text{odd} \quad i.$$

Then on each interior segment (even i) the response $x^*(t)$ joins $x^*(t_i)$ to $x^*(t_{i+1})$ without phase constraint (except at the endpoints, which are not relevant in the usual maximal principle); and here the usual maximal principle applies as described above. We shall then discuss only the modified maximal principle that refers to the boundary segments (odd i). Of course this still leaves the unresolved problem of determining the endpoints of the time subdivision, even assuming that the optimal response has this simple type of intersection with the boundary of Λ.

Consider the optimal response $x^*(t)$ on a boundary segment, say $t_1 \leq t \leq t_2$. Assume that the boundary of Λ in R^n is a smooth hypersurface (a C^∞ submanifold of dimension $n - 1$ in R^n) and let $(\xi^1, \xi^2, \ldots, \xi^{n-1})$ be local coordinates on $\partial\Lambda$, valid in an open set $\mathcal{O}$ of $\partial\Lambda$, which contains the curve $x^*(t) = \xi^{*1}(t), \ldots, \xi^{*n-1}(t))$ on $t_1 \leq t \leq t_2$. The differential system $\mathcal{S}$ can now be written $\dot{\xi} = \check{f}(\xi, u)$ in $\mathcal{O}$, at least for controllers $u(t)$, which produce a response lying in $\mathcal{O}$. We next parametrize the control variables u which describe motions in $\mathcal{O}$ by points of a compact set $W \subset R^s$; that is, let $u(\xi, w)$ be a C^1 function from $\mathcal{O} \times W$ into Ω such that $\check{f}(\xi, u(\xi, w))$ is tangent to $\partial\Lambda$. We are thus led to the optimal-control problem for

$$(\check{\mathcal{S}}) \qquad \dot{\xi} = \check{f}(\xi, u(\xi, w)) = \phi(\xi, w)$$

with measurable controllers $w(t) \subset W$ on $t_1 \leq t \leq t_2$ steering $\xi^*(t_1)$ to $\xi^*(t_2)$ in $\mathcal{O}$. The cost integral is interpreted similarly by

$$\check{C}(w) = \int_{t_1}^{t_2} \check{f}^0(\xi(t), w(t))\, dt,$$

and we assume that $u^*(t)$ lies in the compact set $u(\xi^*(t), W)$ for each $t_1 \leq t \leq t_2$. Then Lemma 3A of the appendix to Chapter 2 asserts that there is a controller $w^*(t) \subset W$ for which $u(\xi^*(t), w^*(t)) = u^*(t) \subset \Omega$. Moreover $w^*(t)$ is an optimal controller, since every $w(t) \subset W$ [with response $\xi(t) \subset \mathcal{O}$] yields some corresponding $u(t) \subset \Omega$ with the same corresponding response in $\mathcal{O}$. Therefore $w^*(t)$ satisfies the maximal principle for the process $\check{\mathcal{S}}$:

$$\sum_{\alpha=0}^{n-1} \xi_\alpha^*(t) \check{f}^\alpha(\xi^*(t), u(\xi^*(t), w^*(t))) = \max_{w\in W} \sum_{\alpha=0}^{n-1} \xi_\alpha^*(t)\, \check{f}^\alpha(\xi^*(t), u(\xi^*(t), w)) \equiv 0$$

where $\zeta_\alpha^*(t)$ is a nontrivial solution of

$$\dot{\zeta}_0 = 0 \qquad (\zeta_0 \leq 0)$$

$$\dot{\zeta}_i = -\sum_{\alpha=0}^{n-1} \zeta_\alpha \left[\frac{\partial \check{f}^\alpha}{\partial \xi^i} + \frac{\partial \check{f}^\alpha}{\partial u^j}\frac{\partial u^j}{\partial \xi^i}\right], \qquad \begin{matrix} 1 \leq i \leq n-1 \\ 1 \leq j \leq m \end{matrix}.$$

In this way we obtain a maximal principle satisfied by the optimal controller $u^*(t)$ on each interior interval and by the corresponding $w^*(t)$ on each boundary interval of the subdivision of $0 \leq t \leq t^*$. More detailed and useable conditions will be formulated later for linear processes.

EXERCISES

1. State and prove the maximal principle, Theorem 1, for the cost functional

$$C(u) = g(x(t_1)) + \int_0^{t_1} f^0(x(t), u(t))\, dt$$

where $g \in C^2$ and $f^0 \in C^1$.

$$\text{[Hint: } C(u) = \int_0^{t_1} [f^0 + (\partial g/\partial x) f]\, dt + g(x_0)]$$

2. Let the admissible controllers form a family $\mathcal{M}$ of bounded measurable controllers $u(t)$ on various finite intervals $0 \leq t \leq t_1$ within the restraint set Ω, satisfying the requirements:

(a) $\mathcal{M}$ contains all constant functions

(b) $\mathcal{M}$ contains the "splice" of any two controls in $\mathcal{M}$, that is, if $u_1(t)$ and $u_2(t)$ on $0 \leq t \leq t_1$ are in $\mathcal{M}$, then

$$u(t) = \begin{cases} u_1(t) & \text{on} \quad 0 \leq t \leq \tau \\ u_2(t) & \text{on} \quad \tau < t \leq t_1 \end{cases}$$

also lie in $\mathcal{M}$.

Prove that an optimal controller within the family $\mathcal{M}$ must satisfy the maximal principle, Theorem 1. Verify that the set of all piecewise continuous controllers forms an admissible family $\mathcal{M}$.

3. Let the restraint set $\Omega(t)$ be the closure of a bounded open set in R^m, and let $\Omega(t)$ vary continuously with the time t. State and prove the maximal principle, Theorem 1, for the corresponding control problem.

4. In the unit cube C: $\max_{1 \le i \le n} |x^i| \le 1$ consider the generically intersecting linear spaces

$$\pi_1\colon x^{k+1} = 0, \ldots, x^n = 0$$

$$\pi_2\colon x^1 = 0, \ldots, x^k = 0.$$

Let h_1 be a continuous map of the k-plane π_1 into R^n, and let h_2 be a continuous map of the $(n-k)$-plane π_2 into R^n such that

$$|h_1(x_1) - x_1| < \tfrac{1}{4} \quad \text{and} \quad |h_2(x_2) - x_2| < \tfrac{1}{4}$$

when $x_1 = (x^1, x^2, \ldots, x^k, 0, \ldots, 0)$ and $x_2 = (0, \ldots, 0, x^{k+1}, \ldots, x^n)$ lie in C. Prove that $h_1(\pi_1)$ meets $h_2(\pi_2)$ in C and use this topological result to complete the final part of the transversality proof of Theorem 1. [Hint: At each point $x = (x_1, x_2)$ in C attach the vector $v(x) = x_1 + (h_1(x_1) - x_1) - x_2 - (h_2(x_2) - x_2)$; so $v(x) = 0$ where $h_1(x_1) = h_2(x_2)$, which yields a point on $h_1(\pi_1) \cap h_2(\pi_2)$. But the vector field $v(x)$ on ∂C can be deformed into the vector field $v_0(x)$ corresponding to the case where $h = (h_1, h_2)$ is the identity. Hence the index of $v(x)$ on ∂C is that of $v_0(x)$ and can be computed directly to be nonzero. Since $v(x)$ is a continuous vector field in C and the index of $v(x)$ on ∂C is nonzero, there exists a point $\bar{x}$ in C where $v(\bar{x}) = 0$.] For another proof see the remarks on the references of Appendix B.

5. Consider the impulse control process in R^2 described by

$$D^2 x = Du \quad \text{on} \quad 0 \le t \le 1.$$

Compute the optimal controller steering the origin $(x = 0, \dot{x} = 0)$ to $(x = 1, \dot{x} = 1)$ with minimal cost $\|Du\|$.

6. Consider the controllable linear process in R^n

$$(\mathfrak{L}) \qquad \dot{x} = A(t)x + B(t)u$$

with essentially bounded measurable controllers $u(t)$ on $0 \le t \le T$ in R^m with

$$\operatorname*{ess\,sup}_{0 \le t \le T,\, 1 \le j \le m} |u^j(t)| \le \alpha.$$

The initial state x_0 is steered to endpoints filling the set $K(T, \alpha)$ of attainability. Prove the analogue of the lemma preceding Theorem 3, that $K(T, \alpha_1)$ lies interior to $K(T, \alpha_2)$ whenever $0 \leq \alpha_1 < \alpha_2$, and that $K(T, \alpha)$ is a compact convex set varying continuously with $\alpha > 0$.

7. For an m-vector function $u(t)$ of bounded variation on $0 \leq t \leq T$ define the STV_p norm (strong total variation norm of order $1 < p < \infty$) by

$$\| Du \|_p = \sup \sum_{\alpha=0}^{k} \|u(t'_{\alpha+1}) - u(t'_\alpha)\|_p$$

where $t'_0 = 0-$, $t'_1 = 0 < t'_2 < \cdots < t'_k$ are an arbitrary finite set of points in $0 \leq t \leq T$, and the supremum is computed over all such finite sequences of times. The p-norm of a single vector is denoted by $\|u\|_p = [\sum_{j=1}^{m} |u^j|^p]^{1/p}$. The linear space of all signed measures Du, with the STV_p norm, is then the dual space of $\mathfrak{C}_q$. Here $\mathfrak{C}_q$ is the space of all real continuous m-vectors $y(t)$ on $0 \leq t \leq T$ with the norm $\|y(t)\|_q = \max_{0 \leq t \leq T} \|y(t)\|_q$, where $1/p + 1/q = 1$.

Use these data to state and prove the analogue of Theorem 3 where the cost of a controller is $\| Du \|_p$.

8. Consider the autonomous linear process in R^n

$$(\mathfrak{L}) \qquad \dot{x} = Ax + Bu$$

with $|u^j(t)| \leq 1$ and $|\dot{u}^j(t)| \leq 1$ for absolutely continuous controllers $u(t)$ on $0 \leq t \leq 1$, as in Theorem 4. Assume that $\mathfrak{L}$ is cube-normal and prove that $\mathfrak{L}$ is controllable and, furthermore, that there exists a neighborhood N of $x_1 = 0$ in R^n such that: each point $x_0 \in N$ can be steered to $x_1 = 0$ by a C^1 controller $u(t)$ satisfying the restraints and with $\dot{u}(t) \equiv 0$ for t near 0 and for t near 1.

9. Consider the autonomous linear process in R^1

$$\dot{x} = -x + u, \quad \text{with} \quad |u| \leq 1 \quad \text{and} \quad |\dot{u}(t)| \leq 1$$

for scalar absolutely continuous controllers having terminal values $u(0) = u(t_1) = 0$. Compute the optimal controller $u^*(t)$ steering $x_0 = 10$ to $x_1 = 0$ in minimal time $t^* > 0$.

10. Consider the control process in C^1 in R^n

$$\dot{x} = A(x) + B(x)u$$

with cost functional

$$C(u) = \int_0^T [A^0(x) + B^0(x)u]\, dt$$

for measurable controllers $u(t) \subset \Omega \subset R^m$ on $0 \leq t \leq T$. Assume that

Ω is a compact convex body and that

$$\hat{B}(x) = \begin{pmatrix} B(x) \\ B^0(x) \end{pmatrix}$$

has rank $(n + 1)$ everywhere. Show that an optimal controller $u^*(t)$ lies on the boundary $\partial\Omega$ almost always.

11. Consider the control process in C^1 in R^2

$$\dot{x} = A(x) + Bu$$

with constant $2 \times m$ matrix B, and measurable controllers $u(t)$ in a convex polyhedron $\Omega \subset R^m$. Assume the vectors

$$Bw \quad \text{and} \quad \frac{\partial A}{\partial x}(x)\, Bw$$

are independent for every $x \in R^2$, and for each vector $w \neq 0$ along an edge of Ω (or along Ω itself if Ω is a segment). Then a time-optimal control $u^*(t)$ on $0 \leq t \leq t^*$ lies at the vertices of Ω almost always. [Hint: If $u^*(t)$ fails to lie in the vertices of Ω, then $\eta^*(t)Bw = 0$ for a positive duration, where $\eta^*(t)$ is a nontrivial C^1 adjoint solution and $w \neq 0$ lies along an edge of Ω. Then $\dot{\eta}^*(t)Bw = -\eta^*(\partial A/\partial x)Bw = 0$ which, together with $\eta^* Bw = 0$, forces $\eta^* = 0$.]

12. Consider the control process in C^1 in R^n described by

$$x^{(n)} - f(x, \dot{x}, \ldots, x^{(n-1)}) = u$$

with measurable controllers satisfying $|u(t)| \leq 1$. Then a time optimal controller $u^*(t)$ on $0 \leq t \leq t^*$ satisfies $|u^*(t)| = 1$ and $u^*(t)$ has only a countable number of discontinuities (after redefining $u^*(t)$ on a null set of times). Hence $u^*(t)$ is a *generalized bang-bang* controller. [Hint: $\eta_n^*(t)\, u^*(t) = |\eta_n^*(t)|$ for all $0 \leq t \leq t^*$, where $\eta^*(t)$ is the nonvanishing C^1 adjoint response. If $\eta_n^*(t)$ had a noncountable number of zeros, then the set of such zeros contains a perfect set Σ on the time axis. By examining the special form of the adjoint differential system, it is easy to show that $\eta^*(t) = 0$ on Σ, which is impossible.]

5.2 SUFFICIENCY CONDITIONS FOR AN OPTIMAL CONTROLLER

In Chapter 2 the maximal principle was shown to be necessary and sufficient for the optimality of a controller for certain linear processes. We shall here prove a similar result for processes in which the control is effected through a convex function.

We shall consider a control process in R^n

(S) $$\dot{x} = A(t)x + h(u, t)$$

with initial state $x(t_0) = x_0$ and closed convex target set $G \subset R^n$ (possibly $G = R^n$). The cost of a controller $u(t)$ on $t_0 \leq t \leq T$ is

$$C(u) = \int_{t_0}^{T} [f^0(x(t), t) + h^0(u(t), t)]\, dt$$

or

$$C(u) = x^0(T)$$

where $x^0(t)$ is defined by the scalar differential equation

$$\dot{x}^0 = f^0(x, t) + h^0(u, t) \quad \text{and} \quad x^0(t_0) = 0.$$

The coefficients f^0, $\partial f^0/\partial x$, h^0, A, and h are assumed continuous in all (x, t, u) in R^{n+1+m}.

The admissible controllers $u(t)$ are all bounded measurable m-vector functions on the fixed finite duration $t_0 \leq t \leq T$, steering $x(t_0) = x_0$ to some point in G, and lying in some nonempty restraint set $\Omega \subset R^m$. The basic linearity of $\mathcal{S}$ assures the existence of the response $\hat{x}(t) = (x^0(t), x(t))$ for the entire duration $t_0 \leq t \leq T$.

We further assume $f^0(x, t)$ is convex in x for each fixed t in $t_0 \leq t \leq T$, that is,

$$\frac{\partial f^0}{\partial x}(x, t)(\omega - x) \leq f^0(\omega, t) - f^0(x, t)$$

for all endpoints ω and x in R^n.

THEOREM 5. *Consider the control process in* R^n

$$(\mathcal{S}) \qquad \dot{x} = A(t)x + h(u, t),$$

with initial state x_0 *and closed convex target* $G \subset R^n$. *The cost* $C(u) = x^0(T)$, *of an admissible controller* $u(t)$ *on* $t_0 \leq t \leq T$ *lying in the restraint set* $\Omega \subset R^m$, *is defined by*

$$\dot{x}^0 = f^0(x, t) + h^0(u, t), \qquad x^0(t_0) = 0.$$

The coefficients f^0, $\partial f^0/\partial x$, h^0, A, h *are everywhere continuous, and also* $f^0(x, t)$ *is convex in* x *for each fixed* t *in the finite interval* $t_0 \leq t \leq T$.

Assume that $u^*(t)$ *is a controller with response* $\hat{x}^*(t) = (x^{0*}(t), x^*(t))$ *satisfying the maximal principle:*

$$-h^0(u^*(t), t) + \eta(t)\, h(u^*(t), t) = \max_{u \in \Omega} [-h^0(u, t) + \eta(t)\, h(u, t)]$$

for almost all t. *Here* $\eta(t)$ *is any nontrivial solution of*

$$\dot{\eta} = \frac{\partial f^0}{\partial x}(x^*(t), t) - \eta\, A(t)$$

satisfying the transversality condition, namely

> $\eta(T)$ *is an inward normal of G at the boundary point* $x^*(T)$. *(If* $G = R^n$, *then* $\eta(T) = 0$; *if* $G = x_1$ *is a single point, the condition is vacuous.)*

Then $u^*(t)$ *is an optimal controller achieving the minimal cost*

$$C(u^*) = x^{0*}(T).$$

Proof. Let $u^*(t)$, $\hat{x}^*(t)$, and $\eta(t)$ satisfy the maximal principle and transversality conditions, and let $u(t)$ be any admissible controller with response $\hat{x}(t) = (x^0(t), x(t))$ on $t_0 \leq t \leq T$. We shall first prove the basic inequality

$$-x^{0*}(T) + \eta(T)\,x^*(T) \geq -x^0(T) + \eta(T)\,x(T).$$

Compute the derivative

$$\frac{d}{dt}[-x^0(t) + \eta(t)\,x(t)] = -\dot{x}^0(t) + \dot{\eta}(t)\,x(t) + \eta(t)\,\dot{x}(t).$$

Use the differential systems for $\dot{x}^0$ and $\dot{x}(t)$ and integrate over the basic interval $t_0 \leq t \leq T$ to obtain

$$-x^0(T) + \eta(T)\,x(T) - \eta(t_0)x_0$$
$$= \int_{t_0}^{T}\left[\frac{\partial f^0(x^*, t)}{\partial x}\,x - f^0(x, t) - h^0(u, t) + \eta\,h(u, t)\right] dt.$$

Next specialize this formula to the control $u^*(t)$ with response $\hat{x}^*(t)$ and subtract these inequalities to obtain

$$[-x^{0*}(T) + \eta(T)\,x^*(T)] - [-x^0(T) + \eta(T)\,x(T)]$$
$$= \int_{t_0}^{T}\Big\{[-h^0(u^*, t) + \eta\,h(u^*, t)] - [-h^0(u, t) + \eta\,h(u, t)]$$
$$+ f^0(x, t) - f^0(x^*, t) + \frac{\partial f^0(x^*, t)}{\partial x}(x^* - x)\Big\}\,dt.$$

But the integrand is almost everywhere positive because of the assumptions of the maximal principle for $u^*(t)$ and the convexity of $f^0(x, t)$. Thus the basic inequality is proved.

If $G = R^n$ the transversality condition asserts that $\eta(T) = 0$, and hence

$$-x^{0*}(T) \geq -x^0(T),$$

or $C(u^*) \leq C(u)$, for every admissible controller $u(t)$. Hence $u^*(t)$ is optimal in this case.

Next let G be a closed convex set in R^n and let π be a supporting plane to G at $x^*(T)$ with inward normal $\eta(T)$ (which could be zero). Then

$$x^0(T) - x^{0*}(T) \geq \eta(T)(x(T) - x^*(T)).$$

But $x(T)$ lies in G, hence $x(T)$ lies on the inward side of π and $\eta(T)(x(T) - x^*(T)) \geq 0$. Thus

$$x^{0*}(T) \leq x^0(T)$$

and $u^*(t)$ is an optimal controller. Q.E.D.

COROLLARY. *Consider the control process in* R^n

$$(\mathcal{S}) \qquad \dot{x} = A(t)x + h(u, t)$$

with initial state x_0, *target* $G = R^n$, *and cost*

$$C(u) = g(x(T)) + x^0(T),$$

where $g(x)$ *is a differentiable convex function and* $x^0(T)$ *is specified in the theorem. Let* $u^*(t)$ *on* $t_0 \leq t \leq T$ *satisfy the maximal principle, as in the theorem, and the transversality condition*

$$\eta(T) = -\operatorname{grad} g(x^*(T)).$$

Then $u^*(t)$ *is an optimal controller.*

Proof. The basic inequality relating $u^*(t)$ and its response $x^*(t)$ to any other admissible control and response still obtains,

$$-x^{0*}(T) + \eta(T)\, x^*(T) \geq -x^0(T) + \eta(T)\, x(T).$$

Using the transversality condition, we conclude

$$-x^{0*}(T) - \frac{\partial g}{\partial x}(x^*(T))\, x^*(T) \geq -x^0(T) - \frac{\partial g}{\partial x}(x^*(T))\, x(T)$$

and

$$-x^{0*} - g(x^*) + g(x^*) - \frac{\partial g}{\partial x}(x^*)x^* \geq -x^0 - g(x) + g(x) - \frac{\partial g}{\partial x}(x^*)x.$$

But the convexity of g assures us that

$$\frac{\partial g}{\partial x}(x^*)(x - x^*) \leq g(x) - g(x^*).$$

Therefore

$$x^{0*} + g(x^*) \leq x^0 + g(x),$$

and $u^*(t)$ is an optimal controller. Q.E.D.

The next result yields a sufficient condition for an optimal controller in a linear process with bounded phase coordinates.

Consider the linear process in R^n

$$(\mathcal{L}) \qquad \dot{x} = A(t)x + B(t)u$$

with $A(t)$ and $B(t)$ real continuous matrices on R^1, with initial state $x(t_0) = x_0$ and compact target set G both lying in the interior of a closed convex phase constraint set $\Lambda \subset R^n$. The admissible controllers are all measurable m-vectors $u(t)$ on various finite durations $t_0 \leq t \leq t_1$, with $u(t)$ lying in a given compact convex restraint set $\Omega \subset R^m$, and with the corresponding responses $x(t)$ in Λ. We seek a minimal time-optimal controller $u^*(t)$ on $t_0 \leq t \leq t^*$ steering x_0 to G within Λ.

If there exists an admissible controller steering x_0 to G within Λ, then there exists a minimal time-optimal controller $u^*(t)$ on $t_0 \leq t \leq t^*$. This assertion follows from the general existence theorems proved in Chapter 4. In fact, let $K_\Lambda(t_1)$ be the set of attainability, consisting of all endpoints $x(t_1)$ of responses initiating at $x(t_0) = x_0$, as produced by admissible controllers $u(t)$ on $t_0 \leq t \leq t_1$. Then it is easy to show that $K_\Lambda(t_1)$ is a compact convex set; and if a point P_1 lies in the interior of $K_\Lambda(t_1)$, then P_1 also lies interior to $K_\Lambda(t)$ for all t sufficiently near to t_1. From this observation we deduce that the optimal response $x^*(t)$ must lead to a point $x^*(t^*)$ that lies on the boundaries of both $K_\Lambda(t^*)$ and G in R^n.

We say that a control $u(t)$ on $t_0 \leq t \leq t_1$ has a response $x(t)$, which meets the boundary of Λ in intervals in case there exists a finite time partition

$$t_0 = \tau_0 \leq \tau_1 \leq \tau_2 \leq \cdots \leq \tau_r = t_1$$

such that during the even-numbered closed intervals $x(t)$ lies on the boundary of Λ, that is,

$$x(t) \in \partial\Lambda \quad \text{for} \quad \tau_{k-1} \leq t \leq \tau_k \quad \text{for even} \quad k;$$

and $x(t)$ lies interior to Λ during the odd-numbered open intervals, that is,

$$x(t) \in \text{interior } \Lambda \quad \text{for} \quad \tau_{k-1} < t < \tau_k \quad \text{for} \quad k \quad \text{odd}.$$

Of course, if $x(t)$ lies always in the interior of Λ, then it satisfies these conditions with a vacuous partitioning of $[t_0, t_1]$.

Definition. A controller $u(t)$ on $t_0 \leq t \leq t_1$, having a response $x(t)$ meeting the boundary of Λ in intervals, is called maximal in case:

1. The maximal principle obtains almost always

$$\eta(t)\, B(t)\, u(t) = \max_{u \in \Omega} \eta(t)\, B(t) u.$$

2. Here $\eta(t)$ is some solution of

$$\dot{\eta} = -\eta\, A(t) \quad \text{on each} \quad \tau_{k-1} < t < \tau_k \quad \text{for} \quad k \quad \text{odd}$$

and

$$\dot{\eta} = -\eta\, A(t) + \zeta(t)\, \theta(x(t)) \quad \text{on each} \quad \tau_{k-1} \leq t \leq \tau_k \quad \text{for} \quad k \quad \text{even},$$

for some integrable function $\zeta(t) \geq 0$, and for $\theta(x)$ an outward normal unit vector from Λ at x, depending piecewise continuously on $x \in \partial\Lambda$.

3. $\eta(t)$ is continuous on $t_0 \leq t \leq t_1$ except possibly at the junction points $\tau_1, \tau_2, \ldots, \tau_{r-1}$ where $x(t) \in \partial\Lambda$ and here the jump conditions hold

$$\eta(\tau_k + 0) - \eta(\tau_k - 0) = \nu_k \, \theta(x(\tau_k))$$

for some constants $\nu_k \geq 0$.

4. $\eta(t_1) \neq 0$.

LEMMA. *Consider the linear process in* R^n

$$(\mathfrak{L}) \qquad \dot{x} = A(t)x + B(t)u$$

with initial state $x(t_0) = x_0$ *interior to the closed convex constraint set* $\Lambda \subset R^n$, *and with controllers in the compact convex restraint set* $\Omega \subset R^m$. *Let* $u^*(t)$ *on* $t_0 \leq t \leq t^*$, *with response* $x^*(t)$ *meeting the boundary of* Λ *in intervals, be a maximal controller with the adjoint response* $\eta^*(t)$. *Then*

$$\eta^*(t^*) \, x^*(t^*) \geq \eta^*(t^*)x$$

for all points x *in* $K_\Lambda(t^*)$. *Thus* $\eta^*(t^*)$ *is an outward unit normal to* $K_\Lambda(t^*)$ *at the boundary point* $x^*(t^*)$.

Proof. We use the time partition $t_0 = \tau_0 \leq \tau_1 \leq \cdots \leq \tau_r = t^*$ and let $x(t)$ be any admissible response to find

$$\begin{aligned}
&\eta^*(t^*)[x^*(t^*) - x(t^*)] - \eta^*(t_0)[x^*(t_0) - x(t_0)] \\
&\qquad = \eta^*(\tau_r)[x^*(\tau_r) - x(\tau_r)] - \eta^*(\tau_{r-1} + 0)[x^*(\tau_{r-1}) - x(\tau_{r-1})] \\
&\qquad + \eta^*(\tau_{r-1} + 0)[x^*(\tau_{r-1}) - x(\tau_{r-1})] - \eta^*(\tau_{r-1} - 0)[x^*(\tau_{r-1}) - x(\tau_{r-1})] \\
&\qquad + \cdots + \eta^*(\tau_1 - 0)[x^*(\tau_1) - x(\tau_1)] - \eta^*(\tau_0)[x^*(\tau_0) - x(\tau_0)].
\end{aligned}$$

Use the differential equation for $\eta(t)$ in computing the differences that correspond to the ends of a partition interval, that is,

$$\begin{aligned}
&\int_{\tau_{k-1}}^{\tau_k} \frac{d}{dt} \eta^*(t)[x^*(t) - x(t)] \, dt \\
&\qquad = \eta^*(\tau_k - 0)[x^*(\tau_k) - x(\tau_k)] - \eta^*(\tau_{k-1} + 0)[x^*(\tau_{k-1}) - x(\tau_{k-1})],
\end{aligned}$$

and use the jump conditions in computing the differences that correspond to the junction points $\tau_1, \ldots, \tau_{r-1}$. A standard computation, together with the maximal principle, then yields the result

$$\eta^*(t^*)[x^*(t^*) - x(t^*)] - \eta^*(t_0)[x^*(t_0) - x(t_0)] \geq 0.$$

Since $x^*(t_0) = x(t_0) = x_0$, and since $x(t^*)$ represents an arbitrary point in $K_\Lambda(t^*)$, we conclude

$$\eta^*(t^*) \, x^*(t^*) \geq \eta^*(t^*)x \quad \text{for all} \quad x \in K_\Lambda(t^*).$$

Therefore $x^*(t^*)$ lies on the boundary of $K_\Lambda(t^*)$ in R^n and the vector $\eta^*(t^*)$ is an outward normal to some supporting hyperplane to $K_\Lambda(t^*)$ at $x^*(t^*)$.

Q.E.D.

In the following theorem we require a "penetrating hypothesis" similar to that of Theorem 19 in Chapter 2. This asserts that each point $\bar{x}$ in the target G, at arbitrary time $\bar{t}$, can be steered along $\bar{x}(t)$ so that $\bar{x}(t) \in [\text{interior } K_\Lambda(\bar{x}, t) \cup \text{interior } G]$ for all $\bar{t} < t < \infty$, by an admissible controller $\bar{u}(t)$ [that is, $\bar{u}(t)$ is admissible on every finite subinterval of $t > \bar{t}$]. If the velocity set $\{A(t)x + B(t)u \mid u \in \Omega\}$, for $x \in \partial G$, always contains vectors directed into the interior of G, this hypothesis is satisfied. However if G is merely a single point, say $\bar{x} = 0$, the hypothesis can also be satisfied if $\mathfrak{L}$ is controllable on each time interval and $u = 0$ lies in the interior of Ω.

THEOREM 6. *Consider the linear process in* R^n

$$(\mathfrak{L}) \qquad \dot{x} = A(t)x + B(t)u$$

with initial state $x(t_0) = x_0$ *and compact convex target set* G *interior to the closed convex constraint set* $\Lambda \subset R^n$, *and with measurable controllers in the compact convex restraint set* $\Omega \subset R^m$. *Let* $u^*(t)$ *on* $t_0 \le t \le t^*$ *be an admissible controller with response* $x^*(t)$ *steering* x_0 *to* $x^*(t^*) \in \partial G$ *while meeting the boundary of* Λ *in intervals.*

Assume

(a) $u^*(t)$ *is a maximal controller with adjoint response* $\eta^*(t)$;
(b) $\eta^*(t)$ *satisfies the transversality condition that is,* $\eta^*(t^*)$ *is an inward normal to* G *at the boundary point* $x^*(t^*)$;
(c) *the penetrating hypothesis obtains, that is, for each point* $\bar{x}$ *in* G *and instant* $\bar{t}$ *there exists an admissible controller on* $\bar{t} < t < \infty$ *steering* $\bar{x}$ *forever after into the interior of* $K_\Lambda(\bar{x}, t)$ *or into the interior of* G.

Then $u^*(t)$ *is an optimal controller with the minimal time* t^*.

Proof. Since $u^*(t)$ is a maximal controller, $x^*(t^*)$ lies on the boundary of $K_\Lambda(t^*)$, as well as on the boundary of G. Moreover $\eta^*(t^*)$ is normal to a common supporting hyperplane to $K_\Lambda(t^*)$ and G at $x^*(t^*)$, by the transversality condition.

If $K_\Lambda(t_1)$ had met G for some time $t' < t^*$, then the penetrating hypothesis would guarantee that the interior of $K_\Lambda(t^*)$ meets G, or else that $K_\Lambda(t^*)$ meets the interior of G. In neither of these cases could $K_\Lambda(t^*)$ and G be separated by a common supporting hyperplane. Thus $t = t^*$ is the first instant at which $K_\Lambda(t)$ meets G. Hence $u^*(t)$ on $t_0 \le t \le t^*$ is an optimal controller.

Q.E.D.

We now turn to sufficiency conditions related to the principles of dynamic programming. Consider a control process in R^n

$$(\mathcal{S}) \qquad \dot{x} = f(x, t, u)$$

where the admissible controllers $u(t)$ are all bounded measurable functions on a fixed finite time interval $t_0 \leq t \leq T$, with values in some restraint set $\Omega \subset R^m$, and steering the initial state x_0 to a target set $G \subset R^n$. The cost is

$$C(u) = g(x(T)) + \int_{t_0}^{T} f^0(x(t), t, u(t))\, dt,$$

where $g, f,$ and f^0 are in class C^1 in all arguments.

Consider the Hamiltonian function

$$H(\eta, x, t, u) = -f^0(x, t, u) + \eta f(x, t, u).$$

(In Section 5.1 this function is designated by $\hat{H}$ with $\eta_0 = -1$.) We shall seek a feedback control $u^0(\eta, x, t)$ that maximizes $H(\eta, x, t, u)$ for each fixed $(\eta, x, t) \in R^{n+n+1}$, and u varying over Ω.

Definition. The control process in R^n

$$(\mathcal{S}) \qquad \dot{x} = f(x, t, u),$$

with restraint set $\Omega \subset R^m$ and Hamiltonian

$$H(\eta, x, t, u) = -f^0(x, t, u) + \eta f(x, t, u),$$

has a *feedback control* $u^0(\eta, x, t)$ in case

$$H^0(\eta, x, t) \equiv \max_{u \in \Omega} H(\eta, x, t, u) \equiv H(\eta, x, t, u^0(\eta, x, t)).$$

If $\eta(x, t)$ is then specified, $\bar{u}(x, t) = u^0(\eta(x, t), x, t)$ is then called a *control law*. Given a control law $\bar{u}(x, t)$ in C^1, a response $\bar{x}(t)$ satisfying $\dot{x} = f(x, t, \bar{u}(x, t)), \bar{x}(t_0) = x_0$, and a control $\bar{u}(t) = \bar{u}(\bar{x}(t), t)$ are determined.

In this development we consider a control process $\mathcal{S}$ with a feedback control $u^0(\eta, x, t)$ in C^1 in R^{n+n+1}. Processes with this property, where $\Omega = R^m$ and $G = R^n$, were analyzed in Chapters 3 and 4.

In order to motivate our discussion let us suppose that each state $x_0 \in R^n$ with initial time t_0 can be steered by an optimal controller on $t_0 \leq t \leq T$ (with a fixed T and various $t_0 < T$) to the target G. Let the minimal cost be $V(x_0, t_0)$ and suppose $V(x, t)$ is in class C^2 for $x \in R^n$ and $t \leq T$. Then the method of dynamic programming, as illustrated in Chapter 1, asserts that

$$V(x_0, t_0) = \min \left[\int_{t_0}^{t_0+\delta} f^0(x(t), t, u(t))\, dt + V(x(t_0 + \delta), t_0 + \delta) \right].$$

Here the minimum is considered over all admissible controllers $u(t)$ on $t_0 \leq t \leq T$ with response $x(t)$ steering x_0 to $x(T) \in G$. If we expand the above functions in terms of the small parameter $\delta > 0$, ignore discontinuities and higher-order terms in δ, then

$$V(x_0, t_0) = \min_{u \in \Omega} \{f^0(x_0, t_0, u)\delta + V(x_0, t_0) + \delta \cdot [V_x(x_0, t_0)f + V_t]\}.$$

This yields the functional equation for $V(x, t)$

$$-V_t(x, t) = \min_{u \in \Omega} [f^0(x, t, u) + V_x(x, t)f(x, t, u)].$$

If we write $S(x, t) = -V(x, t)$, then

$$S_t = -\max_{u \in \Omega} [-f^0(x, t, u) + S_x f(x, t, u)]$$

or

$$S_t = -H^0(S_x, x, t).$$

Thus the negative cost $S(x, t)$ satisfies the partial differential equation

$$\frac{\partial S}{\partial t} + H^0\left(\frac{\partial S}{\partial x}, x, t\right) = 0$$

with boundary data $S(x, T) = -g(x)$ for $x \in G$. This partial differential equation, referred to as the *Hamilton-Jacobi equation* in classical analytical dynamics, is the main result of the dynamic programming approach to our optimization problem.

THEOREM 7. *Consider the control process in* R^n

$$(\mathcal{S}) \qquad \dot{x} = f(x, t, u)$$

with initial state x_0 *and target* $G \subset R^n$. *Admissible controllers are all bounded measurable functions* $u(t)$ *on* $t_0 \leq t \leq T$, *with values in the restraint set* $\Omega \subset R^m$, *steering the response* $x(t)$ *from* $x(t_0) = x_0$ *to* $x(T) \in G$. *The cost is*

$$C(u) = g(x(T)) + \int_{t_0}^{T} f^0(x(t), t, u(t))\, dt$$

where g, f, f^0 *are in* C^1 *in all arguments. Assume that there exists a feedback control* $u^0(\eta, x, t)$ *in* C^1 *in* R^{n+n+1} *so that*

$$H^0(\eta, x, t) = H(\eta, x, t, u^0(\eta, x, t)).$$

(a) *Let* $S(x, t) \in C^2$ *for* $x \in R^n$, $t \leq T$ *be a solution of the Hamilton-Jacobi equation*

$$S_t + H^0(S_x, x, t) = 0, \quad \textit{with} \quad S(x, T) = -g(x) \quad \textit{for} \quad x \in G.$$

Assume that the control law

$$\bar{u}(x, t) = u^0(S_x(x, t), x, t)$$

determines a response $\bar{x}(t)$ *steering* (x_0, t_0) *to* (G, T). *Then* $\bar{u}(t) = \bar{u}(\bar{x}(t), t)$ *is an optimal controller, provided it lies in* Ω, *with optimal response* $\bar{x}(t)$ *and with cost*

$$C(\bar{u}(t)) = -S(x_0, t_0).$$

(b) *On the other hand, assume there exists an optimal controller for each initial state* $x_0 \in R^n$, *and arbitrary initial time* $t_0 < T$ *(here* T *is fixed), leading to the target set* G *with minimal cost* $V(x_0, t_0) \in C^2$. *Then* $S(x, t) = -V(x, t)$ *satisfies*

$$S_t + H^0(S_x, x, t) = 0, \quad \textit{with} \quad S(x, T) = -g(x) \quad \textit{for} \quad x \in G.$$

Proof. Let $S(x, t) \in C^2$ in $x \in R^n$, $t \leq T$ be a solution of the Hamilton-Jacobi equation with $S(x, T) = -g(x)$ for $x \in G$, and assume that the control law $\bar{u}(x, t) = u^0(S_x, x, t)$ determines a response $\bar{x}(t)$ steering (x_0, t_0) to (G, T) and a corresponding controller $\bar{u}(t) = \bar{u}(\bar{x}(t), t)$ in Ω. Then the cost of $\bar{u}(t)$ is

$$C(\bar{u}) = g(\bar{x}(T)) + \int_{t_0}^{T} f^0(\bar{x}(t), t, \bar{u}(t))\, dt$$

so

$$C(\bar{u}) = \int_{t_0}^{T} \left[f^0(\bar{x}(t), t, \bar{u}(\bar{x}(t), t)) - \frac{\partial S}{\partial t} - \frac{\partial S}{\partial x} f(\bar{x}, t, \bar{u}) \right] dt - S(x_0, t_0).$$

Thus

$$C(\bar{u}) = \int_{t_0}^{T} -[S_t + H^0(S_x, x, t)]\, dt - S(x_0, t_0)$$

and

$$C(\bar{u}) = -S(x_0, t_0), \quad \text{as required.}$$

Now let $u(t)$ be any admissible controller with response $x(t)$ steering x_0 to G. The cost is then

$$C(u) = \int_{t_0}^{T} [f^0(x(t), t, u(t)) - S_x f(x, t, u) - S_t]\, dt - S(x_0, t_0).$$

Note that

$$H^0(S_x(x(t), t), x(t), t) \geq -f^0(x(t), t, u(t)) + S_x(x(t), t) f(x(t), t, u(t))$$

and

$$C(u) \geq \int_{t_0}^{T} [-H^0(S_x, x, t) - S_t]\, dt - S(x_0, t_0) = -S(x_0, t_0).$$

Thus $\bar{u}(t)$ is an optimal controller, as required.

Next let $V(x, t) = -S(x, t)$ be the minimal optimal cost in steering (x, t) to (G, T), as in hypothesis (b). Suppose there is a point x_0 in R^n and time $t_0 < T$, for which

$$S_t(x_0, t_0) + H^0(S_x, x_0, t_0) < 0,$$

and this inequality holds in an open neighborhood N of (x_0, t_0) in R^{n+1}. Let $x^*(t)$ be an optimal response from (x_0, t_0) to (G, T), corresponding to an optimal controller $u^*(t)$. Then

$$f^0(x^*(t), t, u^*(t)) - S_x f - S_t \geq -[S_t + H^0(S_x, x^*(t), t)] > \epsilon > 0,$$

for t near t_0, say on $t_0 \leq t \leq t_0 + \delta < T$, and some constant $\epsilon > 0$. For this duration

$$S_t + S_x f < f^0(x^*(t), t, u^*(t)) - \epsilon$$

and

$$S(x^*(t_0 + \delta), t_0 + \delta) - S(x_0, t_0) < \int_{t_0}^{t_0+\delta} f^0(x^*(t), t, u^*(t))\, dt - \epsilon\delta.$$

This means that

$$V(x_0, t_0) < \int_{t_0}^{t_0+\delta} f^0(x^*(t), t, u^*(t))\, dt + V(x^*(t_0 + \delta), t_0 + \delta) - \epsilon\delta.$$

But we know that

$$V(x_0, t_0) = \int_{t_0}^{t_0+\delta} f^0(x^*(t), t, u^*(t))\, dt + V(x^*(t_0 + \delta), t_0 + \delta),$$

and this contradiction proves that

$$S_t + H^0(S_x, x, t) \geq 0$$

in the considered domain.

If, in some neighborhood N of (x_0, t_0), we have

$$S_t(x, t) + H^0(S_x(x, t), x, t) > 0,$$

then we compute

$$S_t + S_x f(x^*(t), t, u^*(t)) > f^0(x^*(t), t, u^*(t)) + \epsilon/2.$$

Then, integrating over the short duration $t_0 \leq t \leq t_0 + \delta$, we obtain

$$S(x^*(t_0 + \delta), t_0 + \delta) - S(x_0, t_0) > \int_{t_0}^{t_0+\delta} f^0(x^*(t), t, u^*(t)) + \epsilon\delta/2,$$

or

$$V(x_0, t_0) > \int_{t_0}^{t_0+\delta} f^0(x^*(t), t, u^*(t)) + V(x^*(t_0 + \delta), t_0 + \delta) + \epsilon\delta/2.$$

This contradiction proves that

$$S_t + H^0(S_x, x, t) = 0,$$

as required, and it is evident that

$$S(x, T) = -g(x) \quad \text{for} \quad x \in G. \qquad \text{Q.E.D.}$$

We remark that the existence of an appropriate solution $S(x, t)$ of the Hamilton-Jacobi equation, in a region W of the (x, t)-space, is sufficient for the construction of a controller that is optimal among all those with responses in W.

In order to relate the maximal principle of Section 5.1 to the maximized Hamiltonian function $H^0(\eta, x, t)$ we shall simplify the control problem.

COROLLARY. *Consider the autonomous process in* R^n

$$(\mathcal{S}) \qquad \dot{x} = f(x, u)$$

with initial state x_0 *and target* $G = R^n$. *The admissible controllers are all bounded measurable functions* $u(t)$ *on the fixed finite interval* $0 \le t \le T$ *with values in the restraint set* $\Omega \subset R^m$, *steering the response* $x(t)$ *on* $0 \le t \le T$. *The cost is*

$$C(u) = \int_0^T f^0(x(t), u(t))\, dt$$

where f^0 *and* f *are everywhere in* C^1.

Assume

(a) *There exists a feedback control* $u^0(\eta, x)$ *in* C^1 *in* R^{n+n}, *which yields the unique point* u^0 *in* Ω *where*

$$H^0(\eta, x) = \max_{u \in \Omega} [-f^0 + \eta f] = H(\eta, x, u^0(\eta, x))\,;$$

(b) Ω *is either an open set or else the closure of an open set with* C^1 *smooth boundary in* R^m.

Then an optimal controller $u^*(t)$ *with response* $x^*(t)$ *is necessarily associated with an adjoint response* $\eta^*(t)$ *satisfying the Hamiltonian system*

$$\dot{x}^i = \frac{\partial H^0}{\partial \eta_i}, \qquad \dot{\eta}_i = -\frac{\partial H^0}{\partial x^i} \qquad i = 1, \ldots, n$$

with boundary conditions $x^*(0) = x_0$, $\eta^*(T) = 0$; *and the maximal principle*

$$H^0(\eta^*(t), x^*(t)) = H(\eta^*(t), x^*(t), u^*(t))$$

holds almost everywhere on $0 \le t \le T$.

Proof. By Theorem 1 of Section 5.1 above, the optimal controller $u^*(t)$ has responses $x^*(t)$ and $\eta^*(t)$ satisfying

$$\dot{x}^i = \frac{\partial H}{\partial \eta_i}(\eta, x, u^*(t)), \qquad \dot{\eta}_i = -\frac{\partial H}{\partial x^i}(\eta, x, u^*(t)) \qquad i = 1, \ldots, n.$$

[Note that the function H is denoted in Section 5.1 by $\hat{H}$ and that the transversality conditions allow us to assume $\eta_0^* \equiv -1$ and $\eta^*(T) = 0$.] The maximal principle also obtains

$$H(\eta^*(t), x^*(t), u^*(t)) = \max_{u \in \Omega} H(\eta^*(t), x^*(t), u) = H^0(\eta^*(t), x^*(t)),$$

almost everywhere, so $u^*(t) = u^0(\eta^*(t), x^*(t))$.

We must show that $(x^*(t), \eta^*(t))$ is also a solution of the Hamiltonian differential system specified by the Hamiltonian function $H^0(\eta, x)$. For this we shall prove that

$$\frac{\partial H}{\partial \eta}(\eta^*(t), x^*(t), u^*(t)) = \frac{\partial H^0}{\partial \eta}(\eta^*(t), x^*(t))$$

and

$$\frac{\partial H}{\partial x}(\eta^*(t), x^*(t), u^*(t)) = \frac{\partial H^0}{\partial x}(\eta^*(t), x^*(t))$$

almost everywhere on $0 \leq t \leq T$.

Compute the derivatives

$$\frac{\partial H^0}{\partial \eta}(\eta, x) = \frac{\partial H}{\partial \eta}(\eta, x, u^0) + \frac{\partial H}{\partial u}(\eta, x, u^0)\frac{\partial u^0}{\partial \eta}(\eta, x)$$

and

$$\frac{\partial H^0}{\partial x} = \frac{\partial H}{\partial x} + \frac{\partial H}{\partial u}\frac{\partial u^0}{\partial x}.$$

If $u^*(t)$ at some time lies interior to Ω, then $(\partial H/\partial u) = 0$ at that point $(\eta^*(t), x^*(t), u^*(t)) = (\eta^*, x^*, u^0(\eta^*, x^*))$ because u^0 maximizes $H(\eta^*, x^*, u)$. Hence at those times when $u^*(t)$ lies interior to Ω, the required Hamiltonian differential equations are verified. Thus, if Ω is open in R^m, the corollary is proved.

Now fix a point $(\eta^*(t_1), x^*(t_1)) = (\eta_1, x_1)$, discarding the null set of times for which $u^*(t)$ fails to equal $u^0(\eta^*(t), x^*(t))$. If $u^0(\eta_1, x_1)$ lies interior to Ω, then $\partial H/\partial u = 0$, as remarked above. Furthermore, if (η_1, x_1) is the limit of points (η, x) in R^{n+n}, at which $u^0(\eta, x)$ lies interior to Ω, then continuity considerations assure that $(\partial H/\partial u)(\eta_1, x_1, u^0(\eta_1, x_1)) = 0$.

Thus the only remaining case concerns the situation when $u^0(\eta, x)$ lies on the boundary of Ω for all (η, x) in a neighborhood N of (η_1, x_1). In this case, the vectors $(\partial u^0/\partial \eta)(\eta_1, x_1)$ and $(\partial u^0/\partial x)(\eta_1, x_1)$ are tangent to the boundary of Ω. But $H(\eta^*(t), x^*(t), u)$ is maximized at the boundary point $u^0(\eta^*(t), x^*(t))$ for each t near t_1, which means that the u-gradient of $H(\eta^*(t), x^*(t), u)$ is a normal vector to the boundary of Ω. Hence

$$\frac{\partial H}{\partial u}(\eta_1, x_1, u^0(\eta_1, x_1))\frac{\partial u^0}{\partial \eta}(\eta_1, x_1) = 0$$

and

$$\frac{\partial H}{\partial u}(\eta_1, x_1, u^0(\eta_1, x_1))\frac{\partial u^0}{\partial x}(\eta_1, x_1) = 0.$$

Therefore the Hamiltonian differential system is satisfied almost always.

Q.E.D.

Under the hypotheses of the corollary, the search for an optimal controller is reduced to the solution of a two-point boundary-value problem for a nonlinear Hamiltonian system specified by the Hamiltonian function $H^0(\eta, x)$. If the process $\mathcal{S}$ has, in addition, the convexity hypotheses of Chapter 3, and $\Omega = R^m$ and G is a fixed compact convex target set in R^n, then the corresponding corollary is valid. Other cases are also of interest, whenever $\eta_0 \neq 0$ in the appropriate maximal principle.

Of course this approach to the optimal-control problem requires the existence of a smooth feedback control $u^0(\eta, x)$. The next theorem describes a class of control problems for which such a feedback control can be determined.

For simplicity we consider only the minimal time-optimal problem in steering a given initial state x_0 to a target set G in R^n. Hence the Hamiltonian is $H(\eta, x, u) = -1 + \eta f(x, u)$. The compact restraint set Ω is also in R^n and is *diffeomorphic* (for each fixed $x \in R^n$) with the velocity set $V = \{f(x, u) \mid u \in \Omega\}$. This means that the map

$$\Omega \to V: \qquad u \to f(x, u)$$

is 1-to-1 onto V and is in class C^1 with a nonvanishing Jacobian determinant (moreover, the *diffeomorphism* is extendible to an open neighborhood of Ω in R^n).

In the exercises it is shown that a given linear control process can be suitably approximated by a process for which the hypotheses of the following theorem are valid. Thus, after a suitable approximation, such linear processes admit a smooth feedback control $u^0(\eta, x)$. Hence, they can be solved by a two-point boundary-value problem of the Hamiltonian system specified by $H^0(\eta, x)$, or treated by the corresponding Hamilton-Jacobi theory of dynamic programming. More general nonlinear systems also can be suitably approximated, provided V is always convex, and then treated by the methods of dynamic programming.

THEOREM 8. *Consider the control process in R^n*

$$(\mathcal{S}) \qquad \dot{x} = f(x, u) \quad \text{in } C^1 \quad \text{in } R^{n+n},$$

with compact restraint set $\Omega \subset R^n$. For each $x \in R^n$ let the velocity set be

$$V(x) = \{f(x, u) \mid u \in \Omega\}.$$

Assume that for each $x \in R^n$

(a) *there is a diffeomorphism of Ω onto V*

$$\Omega \to V: \qquad u \to f(x, u)$$

(b) *$V(x)$ is a strictly convex body in R^n with a smooth C^2 boundary manifold ∂V having positive Gaussian curvature.*

Then there exists a smooth feedback control $u^0(\eta, x)$ in C^1 for $\eta \neq 0$, $x \in R^n$ describing the unique point in Ω where

$$H^0(\eta, x) = \max_{u \in \Omega} [-1 + \eta f(x, u)].$$

Moreover $u^0(\eta, x)$ always lies on the boundary $\partial\Omega$.

Proof. We seek $u^0 = u^0(\eta, x)$ in Ω, which maximizes the function $\eta f(x, u)$. Fix $\bar{\eta} \neq 0$, $\bar{x} \in R^n$, and then $V(\bar{x})$ is a strictly convex set whereon $\bar{\eta} f$ assumes its maximum at the unique point $\bar{f}$, which lies on $\partial V(\bar{x})$, where the outward normal is in the direction of $\bar{\eta}$. This point $\bar{f}$ corresponds to a unique point $u^0(\bar{\eta}, \bar{x})$ on the boundary of Ω. Hence $\eta f(x, u)$ assumes its maximum at the unique point $u^0(\eta, x) \in \partial\Omega$. Since $V(x)$ is a strictly convex set in R^n, which varies continuously with x, the map $(\bar{\eta}, \bar{x}) \to \bar{f}$ is continuous, and so the map $(\bar{\eta}, \bar{x}) \to u^0(\bar{\eta}, \bar{x})$ is also continuous. We next show that $u^0(\eta, x)$ is in class C^1.

Consider the function

$$Q(\eta, x, u) = \nu(f(x, u)) - \frac{\eta}{|\eta|},$$

where $\nu(f(x, u))$ is the unit outward normal to $\partial V(x)$ at the boundary point $f(x, u)$, for $x \in R^n$ and $u \in \partial\Omega$. Note that the n-vector function $Q(\eta, x, u)$ is of class C^1 for $\eta \neq 0$, $x \in R^n$, and $u \in \partial\Omega$. [Since $\partial V(x)$ is a C^2 hypersurface, the map $f \to \nu(f)$ of ∂V into the unit sphere S^{n-1} is of class C^1.]

The unique solution of $Q = 0$ is $u = u^0(\eta, x)$ and we use the implicit function to prove that this is of class C^1. Compute

$$\det \left| \frac{\partial Q}{\partial u} (\bar{\eta}, \bar{x}, u^0(\bar{\eta}, \bar{x})) \right| = \det \left| \frac{\partial \nu}{\partial f} \right| \cdot \det \left| \frac{\partial f}{\partial u} \right|$$

[using local coordinates on $\partial\Omega$, $\partial V(x)$, and S^{n-1}]. But $\nu(f)$ describes the varying outward unit normal to $V(\bar{x})$ at an arbitrary boundary point f; hence det $|\partial v/\partial f|$ is the nonzero Gaussian curvature of $V(\bar{x})$. Also $\partial f/\partial u$ is the nonsingular Jacobian matrix of the diffeomorphism of $\partial\Omega$ onto $\partial V(x)$, and so det $|\partial f/\partial u| \neq 0$. Therefore det $|\partial Q/\partial u| \neq 0$, and the implicit function theorem asserts that $u^0(\eta, x)$ is in class C^1 for $\eta \neq 0$, $x \in R^n$, as a map into $\partial\Omega$ and hence as a map into R^n. Q.E.D.

The final sufficiency theorem for optimality will involve the second variation of the cost functional and will yield a local rather than global optimal controller.

Consider the autonomous control process in R^n

$$\text{(S)} \qquad \dot{x} = f(x, u)$$

with initial state $x(0) = x_0$ and cost functional

$$C(u) = \int_0^T f^0(x, u)\, dt,$$

with smooth coefficients $f(x, u)$ and $f^0(x, u)$. The admissible controllers $u(t)$ are each bounded and measurable on the fixed finite interval $0 \leq t \leq T$, with the restraint $u(t) \subset \Omega \subset R^m$, and produce a response $x(t)$ on $0 \leq t \leq T$. We seek a sufficiency theorem to characterize an optimal controller $u^*(t)$ for this free-endpoint problem.

The maximal principle

$$H(\eta^*(t), x^*(t), u^*(t)) = \max_{u \in \Omega} H(\eta^*(t), x^*(t), u)$$

is a necessary condition for the optimality of $u^*(t)$, where the responses $x^*(t)$ and $\eta^*(t)$ satisfy

$$\dot{x} = \frac{\partial H}{\partial \eta}(\eta, x, u^*(t))$$

$$\dot{\eta} = -\frac{\partial H}{\partial x}(\eta, x, u^*(t))$$

with $x(0) = x_0$, $\eta(T) = 0$. The Hamiltonian function is here

$$H(\eta, x, u) = -f^0(x, u) + \eta f(x, u).$$

As we have seen in Theorem 5, the maximal principle, together with some convexity hypotheses on $f^0(x, u)$ and $f(x, u)$, yields a sufficiency condition for an optimal controller $u^*(t)$. We shall next replace these global convexity hypotheses by local convexity conditions asserted in terms of certain second derivatives of f and f^0, and then we shall seek a sufficient condition for a locally optimal controller.

Definition. A controller $u^*(t)$ of $\mathcal{S}$ is locally optimal in case there exists an $\epsilon > 0$ such that: for every admissible controller $u(t)$ with

$$|u^*(t) - u(t)| \leq \epsilon \quad \text{on} \quad 0 \leq t \leq T,$$

the cost is $C(u) \geq C(u^*)$.

Since we wish to impose local convexity conditions, it is reasonable to assume that the candidate for optimality $u^*(t)$ lies everywhere in the interior of the restraint set Ω. In this case the maximal principle asserts that $H(\eta^*(t), x^*(t), u)$ is maximized at $u = u^*(t)$ and so the gradient vanishes at this point,

$$\frac{\partial H}{\partial u}(\eta^*(t), x^*(t), u^*(t)) = 0.$$

This is the form in which the maximal principle usually enters into the classical calculus of variations in which Ω is usually taken to be an open set in R^m. In this classical study the local convexity condition—primarily emphasizing the convex nature of $f^0(x, u)$—is the *Weierstrass condition*

$$(v - u)\frac{\partial H}{\partial u}(\eta, x, u) \geq H(\eta, x, v) - H(\eta, x, u)$$

whenever (η, x, u) are near any of the values $(\eta^*(t), x^*(t), u^*(t))$. This Weierstrass condition is traditionally expressed in terms of the E-function,

$$E(\eta, x, u, v) \geq 0$$

where

$$E(\eta, x, u, v) = H(\eta, x, u) - H(\eta, x, v) + (v - u)\frac{\partial H}{\partial u}(\eta, x, u).$$

We shall not use this Weierstrass condition, but shall replace it with a more easily verified convexity condition that makes the proof of our next theorem elementary. In addition we shall assume the customary *Legendre condition* on the positivity of the second variation of the cost integral.

In order to motivate all the assumptions let us carry out some preliminary calculations in which we vary the given controller from $u^*(t)$ to $u(t) = u^*(t) + \epsilon\,\theta(t)$, for some small $\epsilon > 0$ and an arbitrary measurable control variation $\theta(t)$ with $|\theta(t)| \leq 1$ on $0 \leq t \leq T$. The response $x(t)$ is then defined and it is easy to compute $|x^*(t) - x(t)| \equiv |\Delta\, x(t)| \leq k_1\epsilon$, where k_1 is a constant depending only on the given data $\{\mathcal{S}, C, u^*, x^*, \eta^*\}$. Also

$$\Delta x(t) = x(t) - x^*(t) = \int_0^t \left[\frac{\partial f}{\partial x}(x^*, u^*)(x(s) - x^*(s)) + \frac{\partial f}{\partial u}(x^*, u^*)\epsilon\,\theta(s) \right.$$
$$\left. + \overline{\frac{\partial^2 f}{\partial x^2}}\Delta x^2 + 2\,\overline{\frac{\partial^2 f}{\partial x\,\partial u}}(\Delta x)(\epsilon\theta) + \overline{\frac{\partial^2 f}{\partial u^2}}(\epsilon\theta)^2\right] ds$$

where the bar notation indicates that the second derivatives are evaluated at some point near $(x^*(s), u^*(s))$. If we define $\psi(t)$ by

$$\dot{\psi} = \frac{\partial f}{\partial x}(x^*(t), u^*(t))\psi + \frac{\partial f}{\partial u}(x^*(t), u^*(t))\,\theta(t)$$

with $\psi(0) = 0$, then

$$\Delta x(t) - \epsilon\,\psi(t) = \int_0^t \frac{\partial f}{\partial x}(x^*, u^*)[\Delta x - \epsilon\,\psi(s)]\,ds$$
$$+ \int_0^t \left[\overline{\frac{\partial^2 f}{\partial x^2}}\Delta x^2 + 2\overline{\frac{\partial^2 f}{\partial x\,\partial u}}(\Delta x)(\epsilon\theta) + \overline{\frac{\partial^2 f}{\partial u^2}}(\epsilon\theta)^2\right] ds,$$

hence

$$\Delta x(t) = \epsilon\,\psi(t) + k_2(t)\epsilon^2,$$

where the function $|k_2(t)| \leq k_2$, a constant depending only on the given data $\{\mathcal{S}, C, \theta^*, x^*, \eta^*\}$.

Now compute the variation in the cost integral due to the control variation $\theta(t)$. Here

$$\Delta C = C(u^* + \epsilon\theta) - C(u^*) = \int_0^T \left[\frac{\partial f^0}{\partial x}(x^*, u^*)\,\Delta x(s) + \frac{\partial f^0}{\partial u}(x^*, u^*)\epsilon\,\theta(s)\right.$$
$$\left. + \overline{\frac{\partial^2 f^0}{\partial x^2}}\Delta x^2 + 2\overline{\frac{\partial^2 f^0}{\partial x\,\partial u}}(\Delta x)(\epsilon\theta) + \overline{\frac{\partial^2 f^0}{\partial u^2}}(\epsilon\theta)^2\right] ds.$$

The first variation of $C(u^*)$ is (ΔC up to terms of order ϵ)

$$\delta C = \epsilon \int_0^T \left[\frac{\partial f^0}{\partial x}(x^*, u^*)\,\psi(s) + \frac{\partial f^0}{\partial u}(x^*, u^*)\,\theta(s)\right] ds.$$

Use $\partial f^0/\partial x = \dot{\eta}^* + \eta^*\,(\partial f/\partial x)(x^*, u^*)$, and integrate by parts to compute

$$\delta C = \epsilon \int_0^T -\frac{\partial H}{\partial u}(\eta^*, x^*, u^*)\,\theta(t)\,dt = 0.$$

The second variation is

$$\delta^2 C = \epsilon^2 \int_0^T \left[\frac{\partial^2 f^0}{\partial x^2}\Delta x^2 + 2\frac{\partial^2 f^0}{\partial x\,\partial u}(\Delta x)(\epsilon\theta) + \frac{\partial^2 f^0}{\partial u^2}(\epsilon\theta)^2\right] ds,$$

which we shall assume to be positive, as guaranteed by the positivity of the symmetric matrix

$$\begin{pmatrix} f_{xx}{}^0 & f_{xu}{}^0 \\ f_{ux}{}^0 & f_{uu}{}^0 \end{pmatrix}$$

evaluated on the locus $(x^*(s), u^*(s))$. The extra complication in the following sufficiency theorem stems from the second-order terms $\Delta x - \epsilon\psi$ arising in the simplifying of the expression for the first variation δC.

THEOREM 9. *Consider the autonomous process in* R^n

$$(\mathcal{S}) \qquad \dot{x} = f(x, u)$$

with initial state $x(0) = x_0$ *and cost*

$$C(u) = \int_0^T f^0(x, u)\, dt,$$

where f, f^0 *are in* C^2 *in* R^{n+m}. *The admissible controllers are each bounded measurable functions* $u(t)$ *on the fixed finite interval* $0 \le t \le T$, *with response* $x(t)$, *and satisfying the restraint* $u(t) \subset \Omega \subset R^m$.

Let $u^*(t)$ *be a controller interior to* Ω, *and assume:*

1. $$\frac{\partial H}{\partial u}(\eta^*(t), x^*(t), u^*(t)) = 0$$

almost always, where

$$H(\eta, x, u) = -f^0(x, u) + \eta f(x, u)$$

and (η^*, x^*) *satisfy*

$$\dot{x} = \frac{\partial H}{\partial \eta}(\eta, x, u^*(t)), \qquad \dot{\eta} = -\frac{\partial H}{\partial x}(\eta, x, u^*(t)),$$

$$x(0) = x_0, \qquad \eta(T) = 0.$$

2. $$f_{xx}{}^0 p^2 + 2 f_{xu}{}^0 pq + f_{uu}{}^0 q^2 \ge c(p^2 + q^2)$$

for arbitrary real constant n and m-vectors p and q, and for a fixed constant $c > 0$, *where these second partial derivatives are evaluated at almost every point* $(x^*(t), u^*(t))$. *This implies that the symmetric matrix is positive definite*

$$\begin{pmatrix} f_{xx}{}^0 & f_{xu}{}^0 \\ f_{ux}{}^0 & f_{uu}{}^0 \end{pmatrix} > 0.$$

3. *Either of the following two conditions holds along* $(x^*(t), u^*(t))$

(α) $f_{xx} = f_{xu} = f_{uu} = 0$

(β) $f_x{}^0 = 0$.

Then $u^*(t)$ *is a locally optimal controller.*

Proof. Vary the controller $u^*(t)$ to $u(t) = u^*(t) + \epsilon\,\theta(t)$, where $|\theta(t)| \le 1$ on $0 \le t \le T$ and $\epsilon > 0$ is suitably small (we specify ϵ later in terms of the given data $\{\mathcal{S}, C, u^*, x^*, \eta^*\}$). The corresponding response is $x(t)$ and

$$\Delta x(t) = \epsilon\, \psi(t) + k_2(t)\epsilon^2, \quad \text{with} \quad |k_2(t)| \leq k_2$$

as above, where

$$\dot{\psi} = \frac{\partial f}{\partial x}(x^*(t), u^*(t))\psi + \frac{\partial f}{\partial u}(x^*(t), u^*(t))\,\theta(t)$$

and $\psi(0) = 0$.

The change in cost is then

$$\Delta C = C(u^* + \epsilon\theta) - C(u^*) = \int_0^T \frac{\partial f^0}{\partial x}(x^*, u^*)[\Delta x(t) - \epsilon\, \psi(t)]\, dt$$

$$+ \int_0^T \left[\overline{\frac{\partial^2 f^0}{\partial x^2}}\Delta x^2 + 2\,\overline{\frac{\partial^2 f^0}{\partial x\, \partial u}}(\Delta x)(\epsilon\theta) + \overline{\frac{\partial^2 f^0}{\partial u^2}}\epsilon^2\theta^2\right] dt.$$

Here we have used the preliminary calculations preceding the theorem and have noted the vanishing of $(\partial H/\partial u)(\eta^*(t), x^*(t), u^*(t)) = 0$. Note that hypothesis (2) asserts that

$$\overline{\frac{\partial^2 f^0}{\partial x^2}}(\Delta x)^2 + \overline{\frac{\partial^2 f^0}{\partial x\, \partial u}}(\Delta x)(\epsilon\theta) + \overline{\frac{\partial^2 f^0}{\partial u^2}}\epsilon^2\theta^2 \geq \frac{c}{2}(|\Delta x|^2 + |\epsilon\theta|^2)$$

provided $\epsilon > 0$ is sufficiently small (the restriction of ϵ depends only on the given data $\{\mathcal{S}, C, u^*, x^*, \eta^*\}$).

Consider the hypothesis

$$(\alpha) \qquad f_{xx} = f_{xu} = f_{uu} = 0 \quad \text{on} \quad (x^*(t), u^*(t)).$$

The preliminary calculation for $\Delta x(t) - \epsilon\, \psi(t)$ preceding the theorem yields

$$\left|\frac{\partial f^0}{\partial x}(x^*, u^*)\right| \cdot |\Delta x(t) - \epsilon\, \psi(t)| \leq \frac{c\epsilon^2}{2(1+T)}\int_0^T |\theta(t)|^2\, dt.$$

In this case

$$\Delta C \geq \frac{c}{2}\epsilon^2 \int_0^T |\theta(t)|^2\, dt - \frac{cT\epsilon^2}{2(1+T)}\int_0^T |\theta(t)|^2\, dt > 0,$$

so $\Delta C > 0$ whenever $\theta(t) \not\equiv 0$, and $u^*(t)$ is locally optimal.

Consider next the hypothesis

$$(\beta) \qquad f_x^0 = 0 \quad \text{on} \quad (x^*(t), u^*(t)).$$

In this case

$$\Delta C = \int_0^T [\bar{f}_{xx}{}^0\, \Delta x^2 + \bar{f}_{xu}{}^0(\Delta x)(\epsilon\theta) + \bar{f}_{uu}{}^0\epsilon^2\theta^2]\, dt \geq \frac{c}{2}\epsilon^2 \int_0^T |\theta(t)|^2\, dt$$

so

$$\Delta C > 0 \quad \text{whenever} \quad \theta(t) \not\equiv 0. \qquad \text{Q.E.D.}$$

Remark. In the case of hypothesis (α), we have assumed that the variations from $u^*(t)$ are related by $u(t) - u^*(t) = \epsilon\theta$, $\Delta x(t) = \epsilon\psi$,

$$\dot{\psi} = f_x(x^*(t), u^*(t))\psi + f_u(x^*(t), u^*(t))\theta$$

$$C(\theta) = \int_0^T [f_{xx}{}^0\psi^2 + 2f_{xu}\psi\theta + f_{uu}\theta^2]\, dt,$$

when all terms beyond ϵ^2 are neglected. Thus we have reduced the investigation of a local optimal to the study of a linear process with a quadratic cost functional, just as was studied in Chapter 3. Our result here is analogous to that of Chapter 3, namely, $u^*(t)$ is the unique controller to achieve the minimal cost $C(u^*)$, in some appropriately restricted neighborhood $|u(t) - u^*(t)| \le \epsilon$

EXERCISES

1. Combine the methods of Theorems 5 and 6 to prove a sufficiency condition for optimal controls of a convex system with bounded phase coordinates.
2. Consider the control process in R^n

   $$(\mathcal{S}) \qquad \dot{x} = f(x, u) \quad \text{in} \quad C^1 \quad \text{in} \quad R^{n+m}$$

   with measurable controllers in a compact set Ω in R^m. The initial state is x_0 and the compact target is G in R^n. We seek a minimal-time controller steering x_0 to G. Assume there exists a real function $T(x)$ in C^1 in R^n such that
   (a) $T(x) \ge 0$ in R^n with $T(x)$ vanishing if and only if $x \in G$,
   (b) $\max_{u \in \Omega} [-\operatorname{grad} T(x)] f(x, u) = 1$ in $R^n - G$.
   Let $u^*(t)$ on $0 \le t \le t^*$ be a controller with response $x^*(t)$ and

   $$-\operatorname{grad} T(x^*(t)) f(x^*(t), u^*(t)) = 1.$$

   Prove that $u^*(t)$ is an optimal controller and that $t^* = T(x_0)$. We further remark that the loci $T(x) =$ constant are *isochronal hypersurfaces*. With suitable relaxation of the differentiability requirements on $T(x)$, very general time-optimal processes can be treated by these methods.
3. Consider the control process in R^n

   $$(\mathcal{S}) \qquad \dot{x} = f(x, u)$$

   with measurable controllers $u(t)$ in a closed region Ω with a smooth boundary in R^m, as in the corollary to Theorem 7. Assume the existence of a smooth feedback control $u^0(\eta, x)$ realizing the unique point $u^0 \in \Omega$ where

$$\max_{u\in\Omega} [-1 + \eta f(x, u)] = H^0(\eta, x).$$

Let $u^*(t)$ on $0 \leq t \leq t^*$ be a minimal time-optimal controller steering $x^*(t)$ from the given initial state x_0 to the target x_1, and let $\eta^*(t)$ be a corresponding nonvanishing adjoint response satisfying the maximal principle

$$-1 + \eta^*(t) f(x^*(t), u^*(t)) = H^0(\eta^*(t), x^*(t))$$

almost everywhere. Then prove that $(x^*(t), \eta^*(t))$ is a solution of the Hamiltonian system

$$\dot{x} = \frac{\partial H^0}{\partial \eta}, \qquad \dot{\eta} = -\frac{\partial H^0}{\partial x} \quad \text{and} \quad x(0) = x_0, \qquad x(t^*) = x_1.$$

4. Consider the control process in R^n

$$(\mathcal{S}) \qquad \dot{x} = f(x, t, u) \quad \text{in} \quad C^1 \quad \text{in} \quad R^{n+1+m}$$

with initial and target states x_0 and x_1, and bounded measurable controllers $u(t)$ on a fixed finite interval $t_0 \leq t \leq T$, lying in a restraint set $\Omega \subset R^m$. Two cost functionals,

$$C_1(u) = \int_{t_0}^{T} f_1^0(x, t, u)\, dt \quad \text{and} \quad C_2(u) = \int_{t_0}^{T} f_2^0(x, t, u)\, dt,$$

with f_1^0 and f_2^0 in C^1 in all variables, are defined to be equivalent if

$$f_2^0(x, t, u) = f_1^0(x, t, u) - \frac{\partial S}{\partial t} - \frac{\partial S}{\partial x} f(x, t, u)$$

for some real function $S(x, t)$ in C^1 in R^{n+1}. Prove that an optimal controller for $\mathcal{S}$ with cost C_1 is also optimal for $\mathcal{S}$ with cost C_2.

5. Consider the control process in R^n

$$(\mathcal{S}) \qquad \dot{x} = f(x, t, u)$$

with initial state x_0 and target $G \subset R^n$, and bounded measurable controllers $u(t)$, on a fixed finite time duration $t_0 \leq t \leq T$, lying in a restraint set $\Omega \subset R^m$. Let the cost be determined by

$$C(u) = \int_{t_0}^{T} f^0(x, t, u)\, dt,$$

where f and f^0 are in class C^1 in all arguments. Assume that there is a function $\bar{u}(x, t)$ in C^1 in R^{n+1} such that

(a) $\bar{u}(x, t)$ lies in Ω for all $(x, t) \in R^{n+1}$;

(b) $f^0(x, t, \bar{u}(x, t)) \equiv 0$;
(c) $f^0(x, t, u) > 0$ if $u \neq \bar{u}(x, t)$ for $u \in \Omega$ and all (x, t)).
Let $x^*(t)$ be the solution of

$$\dot{x} = f(x, t, \bar{u}(x, t)), \qquad x(t_0) = x_0,$$

and assume $x^*(T) \in G$. Prove $u^*(t) = \bar{u}(x^*(t), t)$ is an optimal controller and $x^*(t)$ is the corresponding response.

6. Consider control processes in R^n of the form

$$(\mathcal{S}) \qquad \dot{x} = f(x, t, u) \quad \text{in} \quad C^1 \quad \text{in} \quad R^{n+1+m}$$

with initial state x_0 at $t = t_0$ and compact target set $G \subset R^n$. The controllers are measurable functions $u(t)$, on various finite intervals $t_0 \leq t \leq t_1$, lying in a compact restraint set $\Omega \subset R^m$. Assume the usual conditions that guarantee the existence of a minimal time optimal controller:
(a) $xf(x, t, u) \leq$ Const. $(|x|^2 + 1)$ for all relevant (x, t, u);
(b) $V(x, t) = \{f(x, t, u) \mid u \in \Omega\}$ is always convex;
(c) there exists at least one admissible controller which steers x_0 to G.
Define two problems $P = \{f, x_0, t_0, G, \Omega, V\}$ and $\hat{P} = \{\hat{f}, x_0, t_0, G, \hat{\Omega}, \hat{V}\}$ to be equivalent in case $V(x, t) = \hat{V}(x, t)$ for all (x, t). Then an optimal response $x^*(t)$ for P is also an optimal response for $\hat{P}$ and hence the minimal optimal time t^* (but not the optimal controllers) must be the same for the two problems.

7. Consider control problems of the type described in Problem 6 above.
(a) Suppose that $\hat{V}(x,t) \supset V(x,t)$ for all (x, t). Show that every response $x(t)$ realizable by the control data P is also a response realizable by the data $\hat{P}$. Hence $\hat{t}^* \leq t^*$.
(b) Next assume that for each $\epsilon > 0$, there is a control problem $P(\epsilon) = \{f_\epsilon, x_0, t_0, G, \Omega_\epsilon, V_\epsilon\}$ with $V_\epsilon \supset V$ and dist $(V_\epsilon, V) \leq \epsilon$. Show that for some subsequence $\epsilon(k) \to 0$ the optimal responses $x^*_{\epsilon(k)}(t)$ converge uniformly to $x^*(t)$, an optimal response for $P(0) = P$. In this case $t^*_{\epsilon(k)} \to t^*$.
(c) Consider the problem P in R^2

$$\dot{x}^1 = x^2, \qquad \dot{x}^2 = -x^1 + u$$

with given initial point $(x_0{}^1, x_0{}^2)$ at $t_0 = 0$ and target $(0, 0)$. The control restraint is $\Omega\colon -1 \leq u(t) \leq 1$. For each $\epsilon > 0$ consider $P(\epsilon)$, defined by

$$\dot{x}^1 = x^2 + u^1, \qquad \dot{x}^2 = -x^1 + u^2$$

with restraint $(u^1)^2 + \epsilon^2(u^2)^2 \leq \epsilon^2$ in R^2. Show that $P(\epsilon)$ converges to $P(0) = P$ in the sense indicated above. Show that a smooth feedback control $u^0(\eta, x, \epsilon)$ exists for $P(\epsilon)$, maximizing the Hamiltonian

$$\max_{u\in\Omega(\epsilon)} [-1 + f(x, u, \epsilon)] = H^0(\eta, x, \epsilon)$$

in the sense of Theorem 8. Compute $u^0(\eta, x, \epsilon)$ and $H^0(\eta, x, \epsilon)$.

8. Consider the autonomous linear process in R^n

$$(\mathfrak{L}) \qquad \dot{x} = Ax + Bu$$

with $C(u) = \int_0^T f^0(x, u)\, dt$, as in Theorem 9. Assume that $u^*(t)$ on $0 \leq t \leq T$ is a controller in the interior of the restraint set $\Omega \subset R^m$, and state a sufficient condition that $u^*(t)$ is a locally optimal controller.

CHAPTER 6

Control System Properties: Controllability, Observability, and Stability

In the first section the concepts of controllability and observability are examined for general nonlinear control processes. Extensions of the results of Chapter 2 are obtained, usually in only a local sense. In Section 6.2 we consider various concepts of stability important in the qualitative theory of control systems and in the study of the domain of controllability.

6.1 CONTROLLABILITY AND OBSERVABILITY FOR NONLINEAR PROCESSES

The basic qualitative concepts of (complete) controllability and observability, which were developed in Chapter 2 for linear processes, will here be extended to general nonlinear processes. For the nonlinear processes we usually shall be able to obtain only local criteria and results, rather than the global theory that holds for linear systems. We shall begin by considering the problem of regulating an initial state, near the origin, to the precise origin, first by a smooth controller and then by a bang-bang controller.

Definition. Consider the control process in R^n

$$(\mathcal{S}) \qquad \dot{x} = f(x, u) \quad \text{in} \quad C^1 \quad \text{in} \quad R^n \times \Omega,$$

where Ω is a restraint set in R^m. The domain $\mathcal{C}$ of null controllability is defined as the set of all initial points $x_0 \in R^n$, each of which can be steered to $x_1 = 0$ by some bounded measurable controller $u(t) \subset \Omega$ on some finite time duration. If $\mathcal{C}$ contains an open neighborhood of $x_1 = 0$, then $\mathcal{S}$ is said to be locally controllable (near the origin).

Remark. It is clear that $\mathcal{C}$ is connected since each point in $\mathcal{C}$ is joined to the origin by a continuous solution curve that lies entirely in $\mathcal{C}$. Also $\mathcal{C}$ is open in R^n if and only if $\mathcal{C}$ contains a neighborhood of $x_1 = 0$. This follows from the continuous dependence of the solutions of a differential equation upon the initial conditions. It is clear that the local analysis near $x_1 = 0$ requires that $f(x, u)$ be defined only for x near zero, but we leave such deductions to the reader.

For a linear system in R^n

$$(\mathcal{L}) \qquad \dot{x} = Ax + Bu$$

with $u = 0$ lying interior to $\Omega \subset R^m$, the domain $\mathcal{C}$ of null controllability is open if and only if $\mathcal{L}$ is controllable in the algebraic sense

$$\operatorname{rank}\,[B, AB, A^2B, \ldots, A^{n-1}B] = n.$$

The following example shows that this algebraic controllability condition is not necessary for the geometrical property that $\mathcal{C}$ is open, when nonlinear phenomena are involved.

Example. Consider the nonlinear system in R^2

$$\dot{x} = -x + u$$

$$\dot{y} = -y - x^3,$$

with the restraint Ω: $-1 \leq u \leq 1$. The linear approximation to this system, near the origin, is

$$\begin{pmatrix} \dot{x} \\ \dot{y} \end{pmatrix} = A\begin{pmatrix} x \\ y \end{pmatrix} + Bu$$

where $A = \begin{pmatrix} -1 & 0 \\ 0 & -1 \end{pmatrix}$ and $B = \begin{pmatrix} 1 \\ 0 \end{pmatrix}$. Note that the linear approximation is nondegenerate in that A is nonsingular, but the algebraic controllability criterion fails since

$$\operatorname{rank}\,[B, AB] = \operatorname{rank}\begin{pmatrix} 1 & -1 \\ 0 & 0 \end{pmatrix} = 1 < 2.$$

Nevertheless, the full nonlinear system has an open domain $\mathcal{C}$ of null controllability, that is, every initial state (x_0, y_0) near $(0, 0)$ can be steered to the origin in a finite time.

To prove that $\mathcal{C}$ is open we first examine the two curves Γ_+ and Γ_-, which lead directly into the origin, in finite time, corresponding to $u(t) \equiv -1$ and $u(t) \equiv +1$. These two curves $\Gamma_\pm$ closely approximate the positive and negative rays of the x-axis, in the vicinity of the origin. Now the radial coordinate along any free solution [where $u(t) \equiv 0$] satisfies

$$r\dot{r} = x\dot{x} + y\dot{y} = -(x^2 + y^2) - yx^3 < 0$$

and so there is a disk D wherein each solution approaches the origin monotonically. We show that each initial point in D can be steered in D to meet Γ_+ or Γ_- and hence steered to the origin.

Follow the free solution initiating at (x_0, y_0) in D until $(x(t), y(t))$ reaches (x_1, y_1) very near $(0, 0)$. Suppose (x_1, y_1) lies above $\Gamma_- \cup \Gamma_+$. Then apply the control $u(t) = +1$ for a short time to reach a point (x_2, y_2) above $\Gamma_- \cup \Gamma_+$ (unless this curve $\Gamma_\pm$ is intersected) with $x_2 > 0$. Now apply the control $u(t) = x_2$ so $\dot{x} = 0$ and $\dot{y} = -y - (x_2)^3$. Then the response travels down the line $x = x_2$ until it meets the curve Γ_+. The case where (x_1, y_1) lies below $\Gamma_- \cup \Gamma_+$ is similar. In every case (x_0, y_0) can be steered to the origin in finite time and so $D \subset \mathcal{C}$ and $\mathcal{C}$ contains a neighborhood of the origin and is an open set in R^2.

THEOREM 1. *Consider the control process in* R^n

$$(\mathcal{S}) \qquad \dot{x} = f(x, u) \quad in \quad C^1 \quad in \quad R^{n+m}$$

with $u = 0$ *interior to the restraint set* $\Omega \subset R^m$.

Assume

(a) $f(0, 0) = 0$;
(b) *rank* $[B, AB, A^2B, \ldots, A^{n-1}B] = n$
where $A = (\partial f/\partial x)(0, 0)$ *and* $B = (\partial f/\partial u)(0, 0)$.

Then the domain $\mathcal{C}$ *of null controllability is open in* R^n.

Proof. We consider the differential system $\mathcal{S}$ with the time reversed,

$$(\mathcal{S}_-) \qquad \dot{x} = -f(x, u),$$

and we prove that the endpoints $x(1)$, of responses of $\mathcal{S}_-$ initiating at $x(0) = 0$, cover an open neighborhood N about $x = 0$. Then, by reversing the time sense again on each of the appropriate controllers, we note that each point in N can be steered to the origin along solutions of $\mathcal{S}$. Then $\mathcal{C} \supset N$ is open, as required.

Because $f(0, 0) = 0$, each response $x(t)$ of $\mathcal{S}_-$ with $x(0) = 0$ is defined on $0 \le t \le 1$ provided we take controllers with $|u(t)| < \epsilon$ suitably small; we henceforth impose this restriction on all controllers. Next note that

$$(\mathcal{L}_-) \qquad \dot{x} = -Ax - Bu$$

is controllable and so there exist controllers in R^m, $u_1(t), u_2(t), \ldots, u_n(t)$ on $0 \le t \le 1$ steering the origin by solutions of $\mathcal{L}_-$ to independent points on the positive coordinate axes of R^n. In fact, we can take each of these controllers to be infinitely differentiable (in class C^∞) and so small that

$$u(t, \xi_1, \xi_2, \ldots, \xi_n) = \xi_1 u_1(t) + \xi_2 u_2(t) + \cdots + \xi_n u_n(t)$$

satisfies the restriction

$$|u(t, \xi)| < \epsilon \quad \text{for} \quad 0 \leq t \leq 1 \quad \text{and} \quad \max_{1 \leq i \leq n} |\xi_i| \leq 1.$$

Now let $x(t, \xi_1, \ldots, \xi_n) = x(t, \xi)$ be the response of $\mathcal{S}_-$, initiating at $x(0, \xi) = 0$, for each of the controllers $u(t, \xi)$. Consider the differentiable map from a neighborhood of $\xi = 0$ into R^n,

$$\xi \to x(1, \xi).$$

We show that these images $x(1, \xi)$ cover an open neighborhood N of the origin in R^n. Note that $x(t, 0) = 0$ and so $x(1, 0) = 0$. We shall show that the matrix

$$Z(t) = \frac{\partial x}{\partial \xi}(t, \xi)\bigg|_{\xi=0}$$

is nonsingular at $t = 1$, and then the conclusion follows from the implicit function theorem.

Because $u(t, \xi)$ yields the response $x(t, \xi)$ with $x(0, \xi) = 0$ we have

$$\frac{\partial}{\partial t} x(t, \xi) = -f(x(t, \xi), u(t, \xi))$$

and

$$\frac{\partial}{\partial t}\frac{\partial x}{\partial \xi} = -f_x(x(t, \xi), u(t, \xi))\frac{\partial x}{\partial \xi} - f_u(x(t, \xi), u(t, \xi))\frac{\partial u}{\partial \xi}.$$

Use $x(t, 0) = 0$, $u(t, 0) = 0$, and write

$$\dot{Z}(t) = -AZ - B[u_1, u_2, \ldots, u_n],$$

where the columns of the last matrix are

$$u_1 = \begin{pmatrix} u_1^{\,1} \\ \cdot \\ \cdot \\ \cdot \\ u_1^{\,m} \end{pmatrix}, \ldots, u_n = \begin{pmatrix} u_n^{\,1} \\ \cdot \\ \cdot \\ \cdot \\ u_n^{\,m} \end{pmatrix}.$$

Let the columns of Z_1 be $z_1, \ldots, z_n$ so

$$\dot{z}_j(t) = -Az_j - Bu_j, \qquad z_j(0) = 0.$$

But then $z_1(1), z_2(1), \ldots, z_n(1)$ are independent vectors, since they denote points on the positive coordinate axes of R^n. Thus $Z(1)$ is nonsingular and the implicit function theorem asserts that the points $x(1, \xi)$ cover an open set N when ξ varies near zero. Hence $\mathcal{C}$ contains N and $\mathcal{C}$ is an open set in R^n. Q.E.D.

In the theorem we could employ controllers $u(t)$ of class C^∞, which vanish near $t = 0$ and $t = 1$ and still steer points of N to $x = 0$ along solutions of $\mathcal{S}$. This enables us to use smooth controllers for the global regulation of nonlinear processes, as indicated in the next example.

Example. Consider the regulation of a nonlinear oscillator

$$\ddot{x} + f(x, \dot{x})\dot{x} + g(x) = u$$

or the phase plane system

$$\dot{x} = y, \qquad \dot{y} = -g(x) - f(x, y)y + u$$

with damping coefficient $f(x, y)$ and restoring force $g(x)$ in C^1, and control restraint Ω: $|u| \leq 1$. We assume

$$f(x, y) > 0$$

$$xg(x) > 0 \quad \text{for} \quad x \neq 0,$$

and

$$\lim_{|x| \to \infty} G(x) = \infty \quad \text{where} \quad G(x) = \int_0^x g(s)\, ds.$$

We seek to steer initial states (x_0, y_0) to the origin in R^2. For the free system, where $u(t) \equiv 0$, define the mechanical energy

$$E(x, y) = \frac{y^2}{2} + G(x)$$

and along each solution we find

$$\dot{E} = y\dot{y} + g(x)\dot{x} = -y^2 f(x, y) \leq 0.$$

Since each locus $E(x, y) = E_0 > 0$ is a simple closed curve encircling the origin, and these loci are ordered by inclusion according to the constant value E_0, the free motion $(x(t), y(t))$ initiating at any (x_0, y_0) must remain bounded as $t \to +\infty$ and penetrate ever inwards across these energy loci. If

$$\lim_{t \to \infty} E(x(t), y(t)) = E_\infty > 0,$$

then some point (x_∞, y_∞) on $E(x, y) = E_\infty$ must be approached arbitrarily closely by $(x(t), y(t))$ as $t \to +\infty$. But the solution beginning at (x_∞, y_∞) must enter the interior of the curve $E = E_\infty$ (since $\dot{E} < 0$ unless $y = 0$, and no free solution remains always on the x-axis except the unique critical point $x = y = 0$). By continuity $(x(t), y(t))$ also enters the interior of $E = E_\infty$, which contradicts the construction of $E_\infty > 0$. Therefore $E_\infty = 0$ and every free solution must approach the origin asymptotically.

Now the linear approximation to the nonlinear control process at the origin is

$$\begin{pmatrix} \dot{x} \\ \dot{y} \end{pmatrix} = \begin{pmatrix} 0 & 1 \\ -\dfrac{dg}{dx}(0) & -f(0,0) \end{pmatrix} \begin{pmatrix} x \\ y \end{pmatrix} + \begin{pmatrix} 0 \\ 1 \end{pmatrix} u,$$

which satisfies the controllability condition of the above theorem. Thus we conclude that the domain of null controllability for the nonlinear oscillator is $\mathcal{C} = R^2$. In fact, each initial state in R^2 can be controlled to the origin in finite time by a C^∞ controller $u(t)$ satisfying any previously assigned bound $|u(t)| < \epsilon$.

COROLLARY. *Consider the scalar process*

$$x^{(n)} - \not{f}(x, \dot{x}, \ddot{x}, \ldots, x^{(n-1)}, u) = 0$$

or the corresponding system $\mathcal{S}$ *in the phase space* R^n, *where* $\not{f} \in C^1$ *in* R^{n+1} *and the restraint* $\Omega \subset R^1$ *is* $|u| \leq 1$.

Assume

(a) $\not{f}(0, 0, \ldots, 0, 0) = 0$;
(b) $(\partial \not{f}/\partial u)(0, 0, \ldots, 0, 0) \neq 0$.

Then the domain $\mathcal{C}$ *of null controllability is open in* R^n.

Proof. The system in R^n is

$$(\mathcal{S}) \qquad \begin{aligned} \dot{x}^1 &= x^2 \\ \dot{x}^2 &= x^3 \\ &\vdots \\ \dot{x}^{n-1} &= x^n \\ \dot{x}^n &= \not{f}(x^1, x^2, \ldots, x^n, u). \end{aligned}$$

The linear approximation near the origin is described by the matrices

$$A = \begin{pmatrix} 0 & 1 & 0 & 0 & \cdots & 0 \\ 0 & 0 & 1 & 0 & \cdots & 0 \\ \cdot & \cdot & \cdot & \cdot & \cdot & \cdot \\ \cdot & \cdot & \cdot & \cdot & \cdot & \cdot \\ \cdot & \cdot & \cdot & \cdot & \cdot & \cdot \\ \not{f}_1^{(0)} & \not{f}_2^{(0)} & & & \cdots & \not{f}_n^{(0)} \end{pmatrix}, \qquad B = \begin{pmatrix} 0 \\ 0 \\ \cdot \\ \cdot \\ \cdot \\ \not{f}_u^{(0)} \end{pmatrix}.$$

The controllability of this linear system has already been demonstrated in

Chapter 2, and it can easily be verified by direct matrix computations. Therefore the hypotheses of the above theorem are met and $\mathcal{C}$ is open in R^n.

Q.E.D.

We next turn to the problem of bang-bang controllability. The bang-bang principle states that any response of a controlled system that can be achieved by an arbitrary controller varying over the total control domain can equally well be achieved by a controller that is restricted to the extreme values of the control domain. The term bang-bang refers to the abrupt switching of the controller from one of these extreme values to another. In engineering design it is often simpler to construct a control device with only a finite number of positions (say, the vertices of a polyhedron) rather than a continuum of possible positions (say, all the points of a solid polyhedron), and hence the bang-bang principle is of great practical importance, when it is applicable.

Definition. Consider a control process in R^n

$$(\mathcal{S}) \qquad \dot{x} = f(x, u) \quad \text{in} \quad C^1 \quad \text{in} \quad R^{n+m}$$

with restraint set $\Omega \subset R^m$, and the initial state $x_0 \in R^n$ at $t_0 = 0$. For each subset $Z \subset \Omega$ consider the set $K_Z(t_1)$ of attainability consisting of all points $\{x(t_1)\}$ attained by responses to bounded measurable controllers $u(t) \subset Z \subset \Omega$ on $0 \leq t \leq t_1$. We say that Z has the bang-bang property if

$$K_Z(t_1) \equiv K_\Omega(t_1) \quad \text{for all} \quad t_1 \geq 0.$$

The main result concerning the bang-bang control of linear systems is the following theorem, which is proved as Theorem 2.4 and the Problem 14 immediately following in Chapter 2.

THEOREM 2. *Consider the linear autonomous process in* R^n

$$(\mathcal{L}) \qquad \dot{x} = Ax + Bu$$

with initial state x_0 *at* $t_0 = 0$ *and compact restraint set* $\Omega \subset R^m$. *If a compact subset* $Z \subset \Omega$ *has the same convex hull*

$$H(Z) = H(\Omega),$$

then Z *has the bang-bang property*

$$K_Z(t_1) \equiv K_\Omega(t_1) \quad \textit{for all} \quad t_1 \geq 0.$$

On the other hand, if $\mathcal{L}$ *is controllable,* $H(\Omega)$ *has a nonempty interior in* R^m, *and* rank $B = m$, *then a compact subset* $Z \subset \Omega$, *having the bang-bang property*

$$K_Z(t_1) \equiv K_\Omega(t_1) \quad \textit{for all} \quad t_1 \geq 0,$$

necessarily has the same convex hull

$$H(Z) = H(\Omega).$$

COROLLARY. *Consider the linear autonomous process in* R^n

$$(\mathcal{L}) \qquad \dot{x} = Ax + Bu$$

with compact restraint set $\Omega \subset R^m$. *If*

1. $H(\Omega)$ *contains* $u = 0$ *in its interior,*
2. $\mathcal{L}$ *is controllable,*

then the domain $\mathcal{C}$ *of null controllability is open in* R^n.

Proof. Since $H(\Omega) = H(H(\Omega))$, the origin is attainable from an initial x_0 using controllers in $H(\Omega)$ just in case it is attainable from x_0 using controllers in Ω. But there is an open neighborhood N of $x_1 = 0$ of points which can be steered to the origin using controllers in $H(\Omega)$. Thus $\mathcal{C}$ contains N and $\mathcal{C}$ is open in R^n. Q.E.D.

We shall not discuss the bang-bang principle for nonlinear processes in terms of the global structure K_Ω, but we shall concentrate on the local problem of steering points to the origin of R^n. Thus we seek those subsets $Z \subset \Omega$ for which the domain $\mathcal{C}_Z$ of null controllability, with controllers in Z, is an open set in R^n. If $\mathcal{C}_Z$ is open then we say that Z has the *null bang-bang* property.

The following example shows that some care is needed in generalizing the above corollary to nonlinear processes, even when the analysis is restricted to a local vicinity of the origin.

Example. Consider the scalar process in R^1

$$\dot{x} = u + u^2, \quad \text{with restraint } \Omega\colon |u| = 2.$$

The convex hull $H(\Omega)$ is the segment $-2 \leq u \leq 2$ joining the two points which comprise Ω, and certainly $H(\Omega)$ contains $u = 0$ in its interior. The linear approximation to this process near $x = 0$, $u = 0$ is

$$\dot{x} = u,$$

which is easily seen to satisfy the customary controllability criterion. Yet this nonlinear process has the domain of null controllability $\mathcal{C}$

$$x \leq 0,$$

which fails to contain a neighborhood of the origin. This is true because $u + u^2 > 0$ for $|u| = 2$.

In order to prove an interesting result on the bang-bang control of nonlinear processes we must first generalize some of the theorems of A. Lyapunov on the convexity of the range of a vector measure (which are discussed in the appendix to Chapter 2).

Consider a compact time interval $\mathfrak{J}: 0 \leq t \leq T$ and consider the σ-algebra $\mathfrak{B}$ of all Lebesgue measurable subsets of $\mathfrak{J}$ (modulo null sets—according to the concepts and notations of Lemma 4A in the appendix to Chapter 2). Let μ be the usual Lebesgue measure on $\mathfrak{B}$ so $\{\mathfrak{J}, \mathfrak{B}, \mu\}$ is a measure σ-algebra. On $\mathfrak{B}$ we specify the customary metric by the distance formula for the distance ρ

$$\rho(E, F) = \mu(E \cup F) - \mu(E \cap F),$$

for sets E, F in $\mathfrak{B}$. As we noted in Chapter 2, there exists in $\mathfrak{B}$ (or in any nonatomic σ-subalgebra $\mathfrak{A} \subset \mathfrak{B}$ with the induced measure and metric) a topological image of a line segment $0 \leq \alpha \leq 1$ by sets $D_\alpha \in \mathfrak{B}$ with the monotone properties: $\mu(D_\alpha) = \alpha\mu(\mathfrak{J})$ and $D_{\alpha_1} \subset D_{\alpha_2}$ if and only if $\alpha_1 \leq \alpha_2$.

Let $\{\mathfrak{J}, \mathfrak{A}, \mu\}$ be a measure of σ-algebra with the above metric. A k-partition of $\mathfrak{J}$ is an ordered collection of k sets $A_1, \ldots, A_k$ in $\mathfrak{A}$ with $A_1 \cup A_2 \cup \cdots \cup A_k = \mathfrak{J}$ and $A_i \cap A_j$ empty for $i \neq j$. The collection $\mathfrak{P}_k$ of all k-partitions of $\mathfrak{J}$ is a subset of the k-fold product of $\mathfrak{B}$ with itself, and so we define the corresponding topology on $\mathfrak{P}_k$ which it inherits from this k-fold product. Now let S be any topological space, and we define a continuous family of k-partitions of $\mathfrak{J}$ to be a continuous map of S into $\mathfrak{P}_k$.

LEMMA. *Let $h_1(t), \ldots, h_k(t)$ be integrable n-vector functions on the finite real interval $\mathfrak{J} : 0 \leq t \leq T$. Let S be the $(k - 1)$-simplex with barycentric coordinates*

$$\alpha = (\alpha_1, \ldots, \alpha_k), \qquad \alpha_i \geq 0, \qquad \sum_{i=1}^{k} \alpha_i = 1.$$

Then there exists a continuous family of k-partitions of $\mathfrak{J}$ in $\mathfrak{B}$

$$\alpha \to \{A_1(\alpha), \ldots, A_k(\alpha)\}$$

such that the integrable function

$$h(t, \alpha) = \begin{Bmatrix} h_1(t) & \text{for} & t \in A_1(\alpha) \\ & \cdot & \\ & \cdot & \\ & \cdot & \\ h_k(t) & \text{for} & t \in A_k(\alpha) \end{Bmatrix}$$

satisfies the convexity condition

$$\int_0^T h(t, \alpha)\, dt = \alpha_1 \int_0^T h_1(t)\, dt + \cdots + \alpha_k \int_0^T h_k(t)\, dt$$

Proof. There exists a nonatomic σ-subalgebra $\mathcal{A} \subset \mathcal{B}$ such that the kn-vector $h^* = (h_1, \ldots, h_k)$ satisfies the identity of the earlier Lemma 4A,

$$\int_D h^*\, dt = \frac{\mu(D)}{\mu(\mathfrak{J})} \int_{\mathfrak{J}} h^*\, dt$$

for every set $D \in \mathcal{A}$.

Now let D_β be a topological image of the segment $0 \leq \beta \leq 1$ into $\mathcal{A}$ such that $\mu(D_\beta) = \beta\mu(\mathfrak{J})$, and $D_{\beta_1} \subset D_{\beta_2}$ if and only if $\beta_1 \leq \beta_2$. For each point $\alpha = (\alpha_1, \ldots, \alpha_k)$ of S we define the k-partition of $\mathfrak{J}$ in $\mathcal{A}$ by

$$\begin{aligned}
A_1(\alpha) &= D_{\alpha_1}, && \text{so } \mu(A_1) = \alpha_1\mu(\mathfrak{J}),\\
A_2(\alpha) &= D_{\alpha_1+\alpha_2} - D_{\alpha_1}, && \text{so } \mu(A_2) = (\alpha_1 + \alpha_2)\mu(\mathfrak{J}) - \alpha_1\mu(\mathfrak{J}) = \alpha_2\mu(\mathfrak{J}),\\
A_3(\alpha) &= D_{\alpha_1+\alpha_2+\alpha_3} - D_{\alpha_1+\alpha_2}, && \text{so } \mu(A_3) = \alpha_3\mu(\mathfrak{J}),\\
&\vdots\\
A_k(\alpha) &= D_1 - D_{1-\alpha_k}, && \text{so } \mu(A_k) = \alpha_k\mu(\mathfrak{J}).
\end{aligned}$$

Then it is easy to verify that

$$\alpha \to \{A_1(\alpha), \ldots, A_k(\alpha)\}$$

is a continuous family of k-partitions of $\mathfrak{J}$ in $\mathcal{A}$.

Since each $A_i(\alpha) \in \mathcal{A}$, we integrate the components of h^* to obtain

$$\int_{A_i(\alpha)} h_i\, dt = \alpha_i \int_{\mathfrak{J}} h_i\, dt, \qquad i = 1, \ldots, k.$$

Thus

$$\int_0^T h(t, \alpha)\, dt = \alpha_1 \int_0^T h_1(t)\, dt + \cdots + \alpha_k \int_0^T h_k(t)\, dt,$$

as required. Q.E.D.

THEOREM 3. *Consider the control process in R^n*

$$(\mathcal{S}) \qquad \dot{x} = f(x, u) \quad \text{in} \quad C^1 \quad \text{in} \quad R^{n+m}.$$

Assume

(a) $f(0, 0) = 0$
(b) *rank* $[B, AB, \ldots, A^{n-1}B] = n$

where $A = f_x(0, 0)$, $B = f_u(0, 0)$.

Let π be a fixed convex polytope containing the origin in its interior in R^m. There then exists an $\epsilon > 0$ such that, for the restraint set Ω consisting of the finite set of vertices of the radially similar polytope $\epsilon\pi$, the domain $\mathcal{C}$ of null controllability is open in R^n.

Proof. We shall write the proof for the case where π is an m-simplex with vertices $\hat{u}_1, \hat{u}_2, \ldots, \hat{u}_{m+1}$, and the same method of proof holds for other such convex polytopes. All time durations of control are $0 \leq t \leq 1$ and we first restrict the magnitudes of controllers $u(t)$ so that the corresponding responses initiating at $x = 0$ will be defined on $0 \leq t \leq 1$, and their endpoints will cover an open ball about $x = 0$. Then, upon reversing the time sense as in Theorem 1, we show that $\mathcal{C}$ is open.

There is a bound $\epsilon_0 > 0$ such that for each controller $u(t)$ satisfying

$$|u(t)| \leq \epsilon < \epsilon_0 \quad \text{on} \quad 0 \leq t \leq 1,$$

the response $x(t)$ of

$$(\mathcal{S}) \qquad \dot{x} = f(x, u(t)), \qquad x(0) = 0$$

and the response $x_L(t)$ of the linear system

$$(\mathcal{L}) \qquad \dot{x} = Ax + B\,u(t), \qquad x_L(0) = 0$$

are defined on $0 \leq t \leq 1$ and there satisfy a corresponding bound

$$|x(t)| + |x_L(t)| \leq c(\epsilon) < 1,$$

where $\lim_{\epsilon \to 0} c(\epsilon) = 0$.

Consider the restraint set $\bar{\Omega}$, the vertices $\bar{u}_1, \ldots, \bar{u}_{m+1}$ of a polytope similar to π but with diameter less than ϵ_0. For the linear process $\mathcal{L}$ the set of attainability $\bar{K}_L(1)$, for solutions $x_L(t)$ initiating at the origin and for controllers in $\bar{\Omega}$, is a convex set containing $x = 0$ in its interior in R^n. Let $\bar{u}_1(t), \ldots, \bar{u}_{n+1}(t)$ on $0 \leq t \leq 1$ be such controllers whose linear responses $\bar{x}_{L,1}(t), \ldots, \bar{x}_{L,n+1}(t)$ lead to the vertices of an n-simplex $\bar{S}$ centered at the origin. Denote the inscribed and circumscribed radii of $\bar{S}$ by $c_1 > 0$ and $c_2 > 0$, respectively.

Take barycentric coordinates $\alpha = (\alpha_1, \ldots, \alpha_{n+1})$ in $\bar{S}$ and use the above lemma to obtain a continuous family of $(n + 1)$-partitions of the time interval $\mathcal{J} = [0 \leq t \leq 1]$ for the functions

$$h_i(t) = e^{-At}B\,\bar{u}_i(t), \qquad i = 1, \ldots, n + 1$$

so that

$$h(t, \alpha) = h_i(t) \quad \text{for} \quad t \in A_i(\alpha), \qquad i = 1, \ldots, n + 1$$

satisfies the convexity condition

$$\int_0^1 h(t, \alpha)\, dt = \alpha_1 \int_0^1 h_1(t)\, dt + \cdots + \alpha_{n+1} \int_0^1 h_{n+1}(t)\, dt.$$

But this means that the control family

$$\bar{u}(t, \alpha) = \bar{u}_i(t) \quad \text{for} \quad t \in A_i(\alpha), \qquad i = 1, \ldots, n+1,$$

determines a linear response $\bar{x}_L(t, \alpha)$ with

$$\bar{x}_L(1, \alpha) = \alpha_1 \bar{x}_{L,1}(1) + \cdots + \alpha_{n+1}\bar{x}_{L,n+1}(1).$$

Therefore the map of $\bar{S}$ into R^n defined by the linear responses

$$\alpha \to \bar{x}_L(1, \alpha)$$

is the identity map on $\bar{S}$.

Now repeat this entire construction for the restraint set $\Omega = \epsilon\bar{\Omega}$, for a suitably small $\epsilon > 0$. We use the control family $u(t, \alpha) = \epsilon\, \bar{u}(t, \alpha)$ to obtain the linear responses $x_L(t, \alpha) = \epsilon\, \bar{x}_L(t, \alpha)$. Then, if α are the barycentric coordinates of the simplex $S = \epsilon\bar{S}$ we find that

$$\alpha \to x_L(1, \alpha)$$

is the identity map of S onto itself. We shall compare this map of S to the corresponding map of S defined by the nonlinear responses

$$\alpha \to x(1, \alpha).$$

Here $x(t, \alpha)$ is the response of $\mathcal{S}$, with $x(0, \alpha) = 0$, for the controller $u(t, \alpha)$; just as $x_L(t, \alpha)$ is the response of the linear system $\mathcal{L}$ approximating $\mathcal{S}$. Clearly the map $\alpha \to x(1, \alpha)$ is continuous on S. We shall show that the two maps of S into R^2 defined by $x_L(1, \alpha)$ and $x(1, \alpha)$ are in close agreement on the boundary of S (provided only that $\epsilon > 0$ is suitably small). The topological scholium of Chapter 4 (which is based on the Brouwer fixed-point theorem) then yields the result that the set $K(1) = \{x(1, \alpha) \mid \alpha \in S\}$ covers a neighborhood of the origin in R^n. Thus the main technical difficulty remaining in the proof is to obtain an estimate for $|x(1, \alpha) - x_L(1, \alpha)|$ when α describes the boundary of S.

For the required estimates we restrict $\epsilon > 0$ so

$$|u(t, \alpha)| < \epsilon < \epsilon_0$$

and

$$|x(t, \alpha)| + |x_L(t, \alpha)| < c(\epsilon), \qquad |x_L(t, \alpha)| < c_3\epsilon$$

and all these functions on $0 \leq t \leq 1$ lie in a region where

$$|f(x_L(t, \alpha), u(t, \alpha)) - Ax_L(t, \alpha) - B\, u(t, \alpha)| \leq c_4\, |x_L(t, \alpha)| + c_4\, |u(t, \alpha)|$$

and

$$\left| \frac{\partial f}{\partial x}(x, u) \right| \leq |A| + 1,$$

where c_3 is constant and c_4 is determined (explicitly, below) in terms of the constants $|A|$, $|B|$, c_1, c_2, and c_3. We now compute

$$|x(t, \alpha) - x_L(t, \alpha)| \leq \int_0^t |f(x(s, \alpha), u(s, \alpha)) - Ax_L(s, \alpha) - B\, u(s, \alpha)|\, ds$$

$$\leq \int_0^t |f(x(s, \alpha), u(s, \alpha)) - f(x_L(s, \alpha), u(s, \alpha))|\, ds$$

$$+ \int_0^t |f(x_L(s, \alpha), u(s, \alpha)) - Ax_L(s, \alpha) - B\, u(s, \alpha)|\, ds,$$

so

$$|x(t, \alpha) - x_L(t, \alpha)| \leq (|A| + 1)\int_0^t |x(s, \alpha) - x_L(s, \alpha)|\, ds + c_4[c_3 + 1]\epsilon t.$$

But standard analysis of this integral inequality implies

$$|x(1, \alpha) - x_L(1, \alpha)| \leq e^{|A|+1} c_4[c_3 + 1]\epsilon.$$

Now we choose $\epsilon > 0$ so small that

$$c_4 = \frac{c_1}{2(c_3 + 1)}\, e^{-|A|-1},$$

and then

$$|x(1, \alpha) - x_L(1, \alpha)| \leq \frac{c_1}{2}\, \epsilon.$$

Thus the Euclidean norm of $x(1, \alpha) - x_L(1, \alpha)$ is less than $c_1\epsilon/2$ for $\alpha \in S$, and we recall that the inscribed radius of S is $c_1\epsilon$.

By the topological scholium proved in Chapter 4, we conclude that the image of S, under the nonlinear response map $\alpha \to x(1, \alpha)$, covers an open ball N about the origin in R^n. Finally, we could apply the entire construction and argument developed for $\mathcal{S}$ to the time-reversed system

$$(\mathcal{S}_-) \qquad \dot{x} = -f(x, u)$$

and obtain the analogous open ball N_- about the origin of R^n. By reversing the time sense again we find that every initial point $x_0 \in N_-$ can be steered to the origin by the process $\mathcal{S}$ with some controller $u(t)$ on $0 \leq t \leq 1$ taking values only at the vertices of the polytope $\epsilon\pi$, for sufficiently small $\epsilon > 0$. Hence the required domain $\mathcal{C}$ of null controllability of $\mathcal{S}$ is open in R^n. Q.E.D.

Let us remark that the set of vertices of the polytope $\epsilon\pi$ has the null bang-bang property whenever $\epsilon > 0$ is sufficiently small. Thus, if π is the cube $|u^i| \leq 1$ for $i = 1, \ldots, m$, there exists an $\epsilon_1 > 0$ such that: for each $\epsilon < \epsilon_1$ there is a full neighborhood of $x = 0$ consisting of points which can be steered to the origin by controllers $u(t)$ on $0 \leq t \leq 1$ with

$$|u^i(t)| = \epsilon \qquad i = 1, \ldots, m.$$

Very recent work has extended the techniques of bang-bang-control theory to the study of controllers that are piecewise constant with only a finite number of switches. Also there are some results on closed-loop bang-bang control. However these recent studies apply primarily to linear systems and no comparable treatment is currently available for the general nonlinear theory; and so we shall not present these researches here.

Let us now turn to the concept of an observed process, which was developed for linear systems in Chapter 2. Consider the real autonomous control process in R^n

$$(\mathcal{S}) \qquad \dot{x} = f(x, u)$$

with input or control vector $u \in R^m$ and response or state vector $x \in R^n$. In many physical situations the state x is not known directly, but only some function $h(x)$ of the state is observed or measured as the output of the system. In such a case we augment the description of the process by supplying an observation or output equation

$$\omega = h(x).$$

The total complex of ideas

$$(\mathcal{S}) \qquad \dot{x} = f(x, u) \quad \text{and} \quad \omega = h(x)$$

is called an observed process with input $u \in R^m$, output $\omega \in R^r$, and state $x \in R^n$.

Definition. The observed process in R^n

$$(\mathcal{S}) \qquad \dot{x} = f(x, u) \quad \text{and} \quad \omega = h(x)$$

is (completely) observable in case: for each bounded measurable input $u(t)$ on some interval $0 \leq t \leq t_1$ and for any two responses $x(t)$ and $\bar{x}(t)$ with distinct initial states, the outputs $h(x(t))$ and $h(\bar{x}(t))$ are distinct; that is, for each control $u(t)$ we find

$$h(x(t)) \equiv h(\bar{x}(t)) \quad \text{implies} \quad x(t) \equiv \bar{x}(t).$$

The significance of an observable process $\mathcal{S}$ is that here the state $x(t)$ is uniquely determined from a knowledge of the input $u(t)$ and the output $\omega(t) = h(x(t))$, without any measurement of the state at the initial instant or at any later time. Thus the experimental data collected by input-output observations lead to a complete analysis of the internal structure and dynamics of the process $\mathcal{S}$ involving the state $x(t)$ and its transient behavior under various controls.

In Chapter 2 we considered a linear observed process in R^n, with input $u \in R^m$ and output $\omega \in R^r$,

$$(\mathcal{L}) \qquad \dot{x} = Ax + Bu \quad \text{and} \quad \omega = Hx$$

with constant matrices A, B, and H. The process was defined to be (completely) observable if, for the null input $u(t) \equiv 0$, the null output $Hx(t) \equiv 0$ implies that $x(t) \equiv 0$. However, because of the linearity of $\mathcal{L}$, this definition is entirely equivalent to the current definition of an observable process. Theorem 13 of Chapter 2 asserts that

$$(\mathcal{L}) \qquad \dot{x} = Ax + Bu \quad \text{and} \quad \omega = Hx$$

is observable if and only if

$$\text{rank}\,[H', A'H', A'^2H', \ldots, A'^{n-1}H'] = n$$

Since we here wish to investigate the concept of observability for a nonlinear process $\mathcal{S}$ we shall restrict the analysis to a neighborhood of the origin, for both input $u(t)$ and output $\omega(t)$. To facilitate the discussion of this problem we introduce the concept of local observability near the origin.

Definition. Consider the observed process in R^n

$$(\mathcal{S}) \qquad \dot{x} = f(x, u) \quad \text{and} \quad \omega = h(x)$$

with $f(x, u)$ and $h(x)$ in C^1 in a neighborhood of $x = u = 0$, and $f(0, 0) = 0$, $h(0) = 0$. The process $\mathcal{S}$ is locally observable (near the origin) in case: there exists an $\epsilon > 0$ such that for each measurable input $u(t)$ on $0 \le t \le 1$ in R^m with $|u(t)| < \epsilon$—and for any two distinct responses $x(t) \not\equiv \bar{x}(t)$ with $|x(t)| < \epsilon$, $|\bar{x}(t)| < \epsilon$—we find that the outputs are distinct,

$$h(x(t)) \not\equiv h(\bar{x}(t)) \quad \text{on} \quad 0 \le t \le 1.$$

Note that $|u(t)| < \epsilon$ and $|x(0)| < \epsilon$ implies that $|x(t)|$ and $|h(x(t))|$ are suitably small on the fixed duration $0 \le t \le 1$. Thus the process $\mathcal{S}$ is locally observable just in case there exists $\epsilon > 0$ such that

$$|u(t)| < \epsilon, \qquad |x(0)| < \epsilon, \qquad |\bar{x}(0)| < \epsilon,$$

and

$$h(x(t)) \equiv h(\bar{x}(t))$$

implies that

$$x(0) = \bar{x}(0).$$

THEOREM 4. *Consider the observed process in* R^n

$$(\mathcal{S}) \qquad \dot{x} = f(x, u) \quad \textit{and} \quad \omega = h(x) \quad \textit{in} \quad C^1 \quad \textit{near} \quad x = u = 0,$$

with inputs $u(t)$ *on* $0 \le t \le 1$ *in* R^m *and outputs* $h(x(t))$ *in* R^r.

Assume

(a) $f(0, 0) = 0$ *and* $h(0) = 0$;
(b) rank $[H', A'H', A'^2H', \ldots, A'^{n-1}H'] = n$
where $A = f_x(0, 0), H = h_x(0)$.

Then $\mathcal{S}$ *is locally observable near the origin.*

Proof. For each measurable input $u(t)$ on $0 \leq t \leq 1$ and each initial state x_0 there corresponds a response $x(t)$ and an output $\omega(t) = h(x(t))$, at least when the given data are suitably small. We indicate this correspondence by

$$x_0, u((t) \rightarrow \Omega(x_0, u) = \omega.$$

We seek to show that a given input-output pair uniquely specifies the state by determining the value of x_0. In other words,

$$\Omega(x_0, u) = \omega$$

has at most one small solution $x_0 \in R^n$ when the functions u and ω are appropriately specified. Thus the conclusion will follow immediately from the implicit function theorem, when this theorem is suitably phrased for function spaces.

Let L_∞ be the Banach space of all essentially bounded measurable functions $u(t)$ on $0 \leq t \leq 1$ with the norm

$$\|u(t)\|_\infty = \underset{0 \leq t \leq 1}{\text{ess. sup.}} [|u^1(t)| + |u^2(t)| + \cdots + |u^m(t)|],$$

and let $C_r[0, 1]$ be the Banach space of all continuous functions $\omega(t)$ on $0 \leq t \leq 1$ with the norm

$$\|\omega(t)\| = \max_{0 \leq t \leq 1} [|\omega^1(t)| + \cdots + |\omega^r(t)|].$$

Then $\Omega(x_0, u)$ is a function defined on a neighborhood U of the origin in $R^n \times L_\infty$ with values in the space $C_r[0, 1]$. We shall show that $\Omega(x_0, u)$ is in class C^1 (in the sense of the Frechet derivative—see the exercises below for these concepts of differential calculus in Banach spaces), and that the partial derivative $(\partial\Omega/\partial x_0)(0, 0)$ has rank n. From this deduction the local observability of $\mathcal{S}$ will follow.

Let $(\bar{x}_0, \bar{u})$ be an element of U and consider the nearby point $x_0 = \bar{x}_0 + \Delta x_0$, $u = \bar{u} + \Delta u$ and we compute the corresponding responses $\bar{x}(t)$ and $x(t) = \bar{x}(t) + \Delta x(t)$. Here

$$\Delta x(t) = \int_0^t [f(x(s), u(s)) - f(\bar{x}(s), \bar{u}(s))]\, ds$$

or

$$\Delta x(t) = \int_0^t \left\{ \left[\frac{\partial f}{\partial x}(\bar{x}(s), \bar{u}(s)) + \epsilon(s) \right] \Delta x(s) + \left[\frac{\partial f}{\partial u} + \epsilon(s) \right] \Delta x(s) \right\} ds$$

where $|\epsilon(t)| \to 0$ as $(|\Delta x_0| + \|\Delta u\|_\infty) \to 0$. Thus, $\Delta x(t)$ satisfies a linear differential system so

$$\Delta x(t) = \Phi(t)\,\Delta x_0 + \int_0^t \Phi(t)\,\Phi(s)^{-1} \frac{\partial f}{\partial u}(\bar{x}(s), \bar{u}(s))\,\Delta u(s)\,ds$$

$$+\, o(|\Delta x_0| + \|\Delta u\|_\infty),$$

where $\Phi(t)$ is the matrix solution of $\dot{\Phi} = (\partial f/\partial x)(\bar{x}(t), \bar{u}(t))\Phi$, $\Phi(0) = I$. Hence the map into the space of continuous n-vectors,

$$U \to C_n[0, 1]: \qquad x_0, u(t) \to x(t)$$

is of class C^1 and the (total) derivative at $(\bar{x}_0, \bar{u})$ is the bounded linear transformation

$$R^n \times L_\infty \to C_n[0, 1]: \qquad \Delta x_0, \Delta u \to \Phi(t)\,\Delta x_0 + \int_0^t \Phi(t)\,\Phi(s)^{-1} \frac{\partial f}{\partial u}\,\Delta u(s)\,ds$$

In particular, this derivative at (0, 0) is just

$$\Delta x_0, \Delta u \to e^{At}\Delta x_0 + \int_0^t e^{A(t-s)}B\,\Delta u(s)\,ds.$$

We must compose this map $(x_0, u(t)) \to x(t)$ with the map

$$C_n[0, 1] \to C_r[0, 1]: \qquad x(t) \to h(x(t)),$$

which is easily seen to be of class C^1 near the origin with the derivative at $x = 0$ of

$$\Delta x \to H\,\Delta x.$$

Hence the composite map $(x_0, u) \to \Omega(x_0, u)$ is also of class C^1 near the origin and the derivative, which is even uniformly continuous in some neighborhood of the origin $x_0 = 0$, $u = 0$, can be evaluated at (0, 0) as the linear transformation

$$\Delta x_0, \Delta u \to He^{At}\Delta x_0 + H\int_0^t e^{A(t-s)}B\,\Delta u(s)\,ds.$$

In particular

$$\frac{\partial \Omega}{\partial x_0}(0, 0): \qquad R^n \to C_r[0, 1]: \qquad \Delta x_0 \to He^{At}\Delta x_0.$$

The rank of $(\partial\Omega/\partial x_0)(0, 0)$ is precisely the number of linearly independent columns of He^{At}, considering each column as a vector in $C_r[0, 1]$. We shall prove that the number of independent columns of He^{At} is n under the hypothesis that rank $[H', A'H', \ldots, A'^{n-1}H'] = n$.

Suppose the rank of $(\partial\Omega/\partial x_0)(0, 0)$ is less than n. Then there exists a constant n-vector $\Delta x_0 \neq 0$ such that

$$He^{At}\Delta x_0 \equiv 0 \quad \text{on} \quad 0 \leq t \leq 1.$$

Set $t = 0$, after repeated differentiations, to obtain

$$H\,\Delta x_0 = 0, \quad HA\,\Delta x_0 = 0, \ldots, HA^{n-1}\,\Delta x_0 = 0,$$

or

$$\Delta x_0'\,H' = 0, \quad \Delta x_0'\,A'H' = 0, \ldots, \Delta x_0'\,A'^{n-1}H' = 0.$$

So the n-rows of the controllability matrix $[H', A'H', \ldots, A'^{n-1}H']$ are linearly dependent, which contradicts the hypothesis of the theorem. Therefore $(\partial\Omega/\partial x_0)(0, 0)$ has rank n and this linear transformation is a nonsingular map of R^n onto an n-dimensional subspace of $C_r[0, 1]$.

In this case the implicit function theorem (see exercises below) then asserts that there exists an $\epsilon > 0$ such that $\|u\|_\infty < \epsilon$, $|x_0| < \epsilon$, $|\bar{x}_0| < \epsilon$, and

$$\Omega(x_0, u) = \Omega(\bar{x}_0, u),$$

implies that $x_0 = \bar{x}_0$. But this is just the required conclusion that $\mathcal{S}$ is locally observable. Q.E.D.

This theorem asserts that the plant dynamics (as interpreted by the state vector $x(t)$) of the locally observable system $\mathcal{S}$ can be identified or recognized from a complete knowledge of the input-output relation. In certain cases it may be difficult to observe the output r-vector $\omega(t)$ at all times $0 \leq t \leq 1$ or we can make just n-observations at instants $0 < t_1 < t_2 < \cdots < t_n < 1$ to obtain the sample output data $\omega(t_1), \omega(t_2), \ldots, \omega(t_n)$. In this way, for each n-observation program $\mathsf{P} = \{t_1, t_2, \ldots, t_n\}$ we obtain a sample output $\mathsf{P}\omega = \begin{pmatrix} \omega(t_1) \\ \cdot \\ \cdot \\ \cdot \\ \omega(t_n) \end{pmatrix}$, which we interpret as an rn-column vector, or a point in the vector space R^{rn}. Thus the n-observation program $0 < t_1 < t_2 < \cdots < t_n < 1$ specifies a linear operator P on $C_r[0, 1]$ into R^{rn}, namely, $\omega(t) \to \mathsf{P}\omega$.

If there exists an $\epsilon > 0$ such that

$$\|u\|_\infty < \epsilon, |x_0| < \epsilon, |\bar{x}_0| < \epsilon, \quad \text{and} \quad \mathsf{P}\Omega(x_0, u) = \mathsf{P}\Omega(\bar{x}_0, u)$$

$$\text{implies } x_0 = \bar{x}_0,$$

then the process $\mathcal{S}$ is called *locally n-observable* for the given program P.

We note that each such program $\bar{\mathsf{P}} = \{\bar{t}_1, \ldots, \bar{t}_n\}$ is a point in the open region $\mathcal{T}$ in the first quadrant of R^n, as specified by the inequalities $0 < t_1 < t_2 < \cdots < t_n < 1$.

We shall show that for a *generic* or *non-exceptional* program P in $\mathcal{T}$ the process $\mathcal{S}$ is locally n-observable. The exceptional programs will fill a closed and nowhere dense set in $\mathcal{T}$ (or in the unit n-cube $|t_1| \leq 1, \ldots, |t_n| \leq 1$); that is, the nonexceptional programs are open and dense in $\mathcal{T}$. In particular, every program P in $\mathcal{T}$ can be approximated by non-exceptional programs.

COROLLARY. *Consider the observed process in* R^n

$$(\mathcal{S}) \qquad \dot{x} = f(x, u) \quad \text{and} \quad \omega = h(x) \quad \text{in} \quad C^1 \quad \text{near} \quad x = u = 0,$$

with inputs $u(t)$ *on* $0 \leq t \leq 1$ *in* R^m *and outputs* $h(x(t))$ *in* R^r.

Assume

(a) $f(0, 0) = 0$ *and* $h(0) = 0$.
(b) *rank* $[H', A'H', \ldots, A'^{n-1}H'] = n$

where $A = f_x(0, 0)$ *and* $H = h_x(0)$. *Then for every n-observation program*

$$\mathsf{P}: 0 < t_1 < t_2 < \cdots < t_n < 1,$$

with the exception of a nowhere dense set, the process $\mathcal{S}$ *is locally n-observable.*

Proof. Note that each P in $\mathcal{T}$ is a continuous linear map (and thus is in class C^1)

$$\mathsf{P}: C_r[0, 1] \to R^{rn}: \qquad \omega(t) \to \begin{pmatrix} \omega(t_1) \\ \cdot \\ \cdot \\ \cdot \\ \omega(t_n) \end{pmatrix}.$$

Since R^{rn} is a Banach space of finite dimension rn, we can apply the implicit function theorem to the map (in class C^1, uniformly near the origin)

$$R^n \times L_\infty \to R^{rn}: \qquad x_0, u \to \mathsf{P}\Omega(x_0, u),$$

as in the theorem.

We need only verify that the partial derivative matrix $\partial(\mathsf{P}\Omega)/\partial x_0(0, 0)$ has rank n as a linear map of R^n into R^{rn}. But this partial derivative is expressed by the n-tuple of $r \times n$ matrices $He^{At_1}, He^{At_2}, \ldots, He^{At_n}$ as indicated by the formula

$$\Delta x_0 \to \begin{pmatrix} He^{At_1} \\ \cdot \\ \cdot \\ \cdot \\ He^{At_n} \end{pmatrix} \Delta x_0.$$

Hence we need only show that there are n linearly independent rows among these matrices. Thus the corollary is proved if we can verify that

$$[e^{A't_1}H', \ldots, e^{A't_n}H']$$

has n independent columns, or n independent rows. Hence the programs P for which $\mathcal{S}$ is locally n-observable fill an open subset of $\mathcal{T}$, since the condition of linear independence is preserved under small perturbations of vectors.

Let $D(t_1, t_2, \ldots, t_n)$ be the sum of the squares of all $n \times n$ subdeterminates of $[e^{A't_1}H', \ldots, e^{A't_n}H']$. Thus this matrix has rank n just in case $D(t_1, \ldots, t_n) \neq 0$. Now consider $D(t_1, \ldots, t_n)$, for various programs P in $\mathcal{T}$, as a real analytic function of the n real variables $(t_1, \ldots, t_n)$. Either $D(t_1, \ldots, t_n)$ is nonvanishing everywhere, except possibly on a closed subset of $\mathcal{T}$ without interior, or else $D(t_1, \ldots, t_n) \equiv 0$ on $\mathcal{T}$. In the first alternative the corollary is proved; we show that the second alternative is impossible.

Hence we need only show that the function D is somewhere positive on $\mathcal{T}$. Take the n-times $t_1 = t$, $t_2 = 2t, \ldots, t_n = nt$ for some small value $t > 0$. We shall prove that

$$[e^{A't}H', \ldots, e^{A'nt}H'] =$$

$$[H', A'H', \ldots, A'^{n-1}H'] \begin{bmatrix} I & \cdots & I \\ t & \cdots & nt \\ \vdots & & \vdots \\ \left\{\dfrac{t^{n-1}}{(n-1)!} + 0(t^n)\right\} & \cdots & \left\{\dfrac{(nt)^{n-1}}{(n-1)!} + 0(t^n)\right\} \end{bmatrix}$$

has rank n for some small $t > 0$ and so $D \neq 0$. In the van der Mond determinant

$$\begin{vmatrix} I & \cdots & I \\ t & \cdots & nt \\ \vdots & & \vdots \\ \dfrac{t^{n-1}}{(n-1)!} & \cdots & \dfrac{(nt)^{n-1}}{(n-1)!} \end{vmatrix} = V \cdot t^{(n-1)nr/2}$$

each entry is understood to be a scalar $r \times r$ matrix, and $V \neq 0$ so this entire $nr \times nr$ matrix is nonsingular for every $t > 0$. But

$$\begin{vmatrix} I & \cdots & I \\ t & \cdots & nt \\ \cdot & & \\ \cdot & & \\ \cdot & & \\ \left\{\dfrac{t^{n-1}}{(n-1)!} + 0(t^n)\right\} & \cdots & \left\{\dfrac{(nt)^{n-1}}{(n-1)!} + 0(t^n)\right\} \end{vmatrix}$$

$$= V \cdot t^{(n-1)nr/2} + 0(t^{[(n-1)nr/2]+r})$$

and so this matrix is nonsingular for all small $t > 0$.

Since $[H', A'H', \ldots, A'^{n-1}H']$ has rank n, its product with a nonsingular matrix of rank nr still has rank n. Thus

$$\text{rank } [e^{A't}H', \ldots, e^{A'nt}H'] = n$$

for all small $t > 0$, and the corollary is proved. Q.E.D.

Let us consider a physical process with unknown internal dynamics; even the dimension of the state space may be uncertain. However, from some basic theoretical viewpoint, we might expect the process to be described by some category of mathematical systems, for instance, ordinary differential systems—perhaps linear and autonomous. Then we conduct a series of experiments on the process by introducing various inputs $u(t)$ and observing the resulting outputs $\omega(t)$, which depend on the indirectly measured state $x(t)$ of the process. From this experimentally determined input-output relation we seek to analyze the internal structure of the dynamics of the state space of the given physical process.

In order to demonstrate the mathematical aspects of this input-output analysis of a process, we first present the fundamental theory for linear autonomous processes, and then turn to nonlinear processes where the analysis is local in nature and technically more difficult.

Definition. Consider a linear autonomous observed process in R^n

$$(\mathfrak{L}) \qquad \dot{x} = Ax + Bu \quad \text{and} \quad \omega = Hx.$$

Each input $u(t)$ is a piecewise continuous m-vector vanishing off some compact subinterval of $t \leq 0$, and the corresponding response $x(t)$ with $x(-\infty) = 0$ determines a continuous r-vector output $\omega(t) = Hx(t)$. The set of all such inputs forms the real linear input space $\mathfrak{J}$, and the outputs for $0 \leq t \leq 1$ belong to $C_r[0, 1]$. The linear transformation

$$\mathbf{T}: u(t) \rightarrow \omega(t): \qquad \mathfrak{J} \rightarrow C_r[0, 1]$$

is the input-output relation or transfer-operator of the process $\mathfrak{L}$.

We shall next show that any two linear autonomous observed processes

$$(\bar{\mathfrak{L}}) \qquad \dot{\bar{x}} = \bar{A}\bar{x} + \bar{B}u \quad \text{and} \quad \omega = \bar{H}\bar{x} \quad \text{in} \quad R^{\bar{n}}$$

and

$$(\hat{\mathfrak{L}}) \qquad \dot{\hat{x}} = \hat{A}\hat{x} + \hat{B}u \quad \text{and} \quad \omega = \hat{H}\hat{x} \quad \text{in} \quad R^{\hat{n}},$$

which both realize the same input-output relation **T** most efficiently, are linearly equivalent. This means that $\bar{n} = \hat{n}$ and there exists a nonsingular linear transformation

$$\hat{x} = P\bar{x} \quad \text{carrying} \quad \bar{\mathfrak{L}} \quad \text{to} \quad \hat{\mathfrak{L}},$$

that is,

$$P\bar{A}P^{-1} = \hat{A}, \qquad P\bar{B} = \hat{B}, \qquad \bar{H}P^{-1} = \hat{H}.$$

In such a case the process $\bar{\mathfrak{L}}$ is the same as $\hat{\mathfrak{L}}$ when the coordinates in the state space $R^{\hat{n}}$ are transformed linearly. If $\mathbf{T} \equiv 0$, then this input-output relation is realized by the degenerate "zero-dimensional process" and we ignore this case.

THEOREM 5. *Consider the linear autonomous observed process in* R^n

$$(\mathfrak{L}) \qquad \dot{x} = Ax + Bu \quad \text{and} \quad \omega = Hx,$$

with m-vector inputs $u(t) \in \mathfrak{I}$ *and r-vector outputs* $\omega(t)$ *specifying elements of the linear space* $C_r[0, 1]$. *Let the input-output relation be*

$$\mathbf{T}: u(t) \to \omega(t): \qquad \mathfrak{I} \to C_r[0, 1].$$

Then there exists a linear autonomous observed process, in some state space $R^{\hat{n}}$,

$$(\hat{\mathfrak{L}}) \qquad \dot{\hat{x}} = \hat{A}\hat{x} + \hat{B}u \quad \text{and} \quad \omega = \hat{H}\hat{x}$$

such that

1. $\hat{\mathfrak{L}}$ *is (completely) controllable and observable;*
2. $\hat{\mathfrak{L}}$ *has the same input-output relation* **T**.

Moreover, $\hat{\mathfrak{L}}$ *is the unique process, up to linear equivalence, with these properties.*

Proof. The existence of $\hat{\mathfrak{L}}$ having the required properties has been demonstrated in Chapter 2 and we merely sketch these ideas here. First let $C \subset R^n$ be the linear subspace of all points of R^n to which responses of $\mathfrak{L}$ initiating at the origin can be steered by a continuous controller on a finite time duration. Then C is an invariant subspace (any response meeting

C lies entirely in C for all times) and we consider the restriction of $\mathcal{L}$ to C,

$$(\mathcal{L}_c) \qquad \dot{x}_c = A_c x_c + B_c u \quad \text{and} \quad \omega = H_c x_c.$$

Here new coordinates $\begin{pmatrix} x_c \\ x_u \end{pmatrix}$ have been introduced in R^n so that the subspace C is specified by the equations $x_u = 0$. Hence the state vectors in C are of the form $\begin{pmatrix} x_c \\ 0 \end{pmatrix}$, which we write merely as x_c, and the coefficient matrices are first transformed to the new coordinates in R^n and are then truncated appropriately to describe the observed process $\mathcal{L}_c$ on the space C. Note that $\mathcal{L}_c$ is controllable on C and has the prescribed input-output relation $\mathbf{T}$.

Now consider the free system obtained from $\mathcal{L}_c$ with $u(t) \equiv 0$. Let $C_0 \subset C$ consist of those initial states for which $\omega(t) \equiv 0$ on $t \geq 0$. We shall define a projection of $\mathcal{L}_c$ onto the factor space C/C_0, the vector space arising from the identification of any two states of C whose difference lies in C_0. Let us partition the coordinates of C into $\begin{pmatrix} x_a \\ x_b \end{pmatrix}$, where $x_b = 0$ describes C_0. Then, after an appropriate linear coordinate change in C, the controllable process $\mathcal{L}_c$ can be written

$$\dot{x}_a = A_{aa} x_a + A_{ab} x_b + B_a u$$

$$\dot{x}_b = A_{bb} x_b + B_b u$$

and

$$\omega = H_b x_b.$$

Note that two initial states whose difference lies in C_0 yield responses, under an arbitrarily chosen but fixed controller $u(t)$ on $0 \leq t \leq 1$, whose difference is always in C_0. This means that the cosets of C_0, or the points of the factor space C/C_0, are each steered as a whole by an arbitrary controller. Hence we need only study the coordinate x_b to observe this control of the points of C/C_0. Thus we consider the projected system

$$(\hat{\mathcal{L}}) \qquad \dot{x}_b = A_{bb} x_b + B_b u \quad \text{and} \quad \omega = H_b x_b$$

in the space $R^{\hat{n}} = C/C_0$. Since the zero coset $x_b = 0$ can be steered to an arbitrary coset of C/C_0, the process $\hat{\mathcal{L}}$ is controllable in $R^{\hat{n}}$. Because of the identifications introduced into the construction of C/C_0, the process $\hat{\mathcal{L}}$ is observable. Since ω depends only on x_b in the process $\mathcal{L}_c$, the input-output relation of $\hat{\mathcal{L}}$ is just $\mathbf{T}$. Therefore $\hat{\mathcal{L}}$ is a controllable observable process with the given input-output relation $\mathbf{T}$, and the existence assertion of the theorem is proved.

We next prove the uniqueness of $\hat{\mathfrak{L}}$ in $R^{\hat{n}}$ up to linear equivalence. Let another observable controllable process be

$$(\bar{\mathfrak{L}}) \qquad \dot{\bar{x}} = \bar{A}\bar{x} + \bar{B}u \quad \text{and} \quad \omega = \bar{H}\bar{x}$$

in the state space $R^{\bar{n}}$, with input-output relation **T**.

Both $\hat{\mathfrak{L}}$ and $\bar{\mathfrak{L}}$ use the same input space $\mathfrak{J}$ and have the input-output relation **T**. Let $\mathfrak{J}_0$ consist of those inputs of $\mathfrak{J}$ for which the output $\omega(t) \equiv 0$ on $0 \leq t \leq 1$. Since $\hat{\mathfrak{L}}$ and $\bar{\mathfrak{L}}$ are both controllable and observable, $\mathfrak{J}/\mathfrak{J}_0$ is a linear space isomorphic with both $R^{\hat{n}}$ and $R^{\bar{n}}$. Hence $\hat{n} = \bar{n}$ and we use the above isomorphism to define a nonsingular linear transformation $\hat{x} = P\bar{x}$ of $R^{\bar{n}}$ onto $R^{\hat{n}}$.

We must now verify that the transformation $\hat{x} = P\bar{x}$ carries each response of $\bar{\mathfrak{L}}$ onto a response of $\hat{\mathfrak{L}}$ for an arbitrarily chosen but fixed controller $\tilde{u}(t)$ on $0 \leq t \leq \tilde{t}$. But the action of $\tilde{u}(t)$ on an initial state of $\bar{\mathfrak{L}}$ or $\hat{\mathfrak{L}}$ can be computed from the action of $\tilde{u}(t)$ on $\mathfrak{J}/\mathfrak{J}_0$ under concatenation. That is, the action of $\tilde{u}(t)$ on a point $[u_0(t)] \in \mathfrak{J}/\mathfrak{J}_0$, where $u_0(t)$ on $-t_0 \leq t \leq 0$ belongs to $\mathfrak{J}$, is just the point in $\mathfrak{J}/\mathfrak{J}_0$ defined by

$$\begin{aligned} &u_0(t + \tilde{t}) \quad \text{on} \quad -t_0 - \tilde{t} \leq t < -\tilde{t} \\ &\tilde{u}(t + \tilde{t}) \quad \text{on} \quad -\tilde{t} \leq t \leq 0. \end{aligned}$$

Since this action of $\tilde{u}(t)$ on $\mathfrak{J}/\mathfrak{J}_0$ corresponds to the action of $\tilde{u}(t)$ on the state spaces $R^{\hat{n}}$ and $R^{\bar{n}}$ under the processes $\hat{\mathfrak{L}}$ and $\bar{\mathfrak{L}}$, we conclude that $\hat{\mathfrak{L}}$ and $\bar{\mathfrak{L}}$ are identical after the isomorphism $\hat{x} = P\bar{x}$ is used to map $R^{\bar{n}}$ onto $R^{\hat{n}}$. Q.E.D.

Thus the given input-output relation **T** can be realized by a unique controllable observable linear process $\hat{\mathfrak{L}}$ in $R^{\hat{n}}$. The dimension $\hat{n}$ is clearly the smallest dimension for a state space in which **T** can be realized by a linear autonomous process, and so the realization by $\hat{\mathfrak{L}}$ is also the most efficient realization of **T**.

The input-output relation **T** is a type of transfer function, and Theorem 5 has somewhat the same significance as has Theorem 14 of Chapter 2. We next extend these concepts and methods to nonlinear observed processes.

Definition. Consider an observed process in R^n

$$(\mathcal{S}) \qquad \dot{x} = f(x, u) \quad \text{and} \quad \omega = h(x) \quad \text{in} \quad C^1 \quad \text{near} \quad x = u = 0,$$
$$\text{and} \quad f(0, 0) = 0, \qquad h(0) = 0.$$

Each input $u(t)$ is a piecewise continuous m-vector, vanishing outside

$-1 \leq t \leq 0$ and satisfying a bound $|u(t)| \leq \delta$, and the corresponding response $x(t)$ with $x(-1) = 0$ determines a continuous r-vector output $\omega(t) = h(x(t))$ on $0 \leq t \leq 1$ (at least for suitably small $\delta > 0$). The convex input set $\mathfrak{J}_\delta$, with the uniform topology, is thus mapped continuously into Banach space $C_r[0, 1]$ by the input-output relation or transfer-operator

$$\mathbf{T}: u(t) \to \omega(t): \qquad \mathfrak{J}_\delta \to C_r[0, 1].$$

We shall prove the uniqueness of the locally controllable and observable process $\mathcal{S}$ realizing the input-output relation $\mathbf{T}$. The uniqueness will be specified up to local topological equivalence.

Definition. Observed processes

$$(\mathcal{S}) \qquad \dot{x} = f(x, u) \quad \text{and} \quad \omega = h(x) \quad \text{in} \quad R^n, \quad \text{in} \quad C^1 \quad \text{near}$$
$$x = u = 0 \quad \text{with} \quad f(0, 0) = 0, \qquad h(0) = 0$$

and

$$(\hat{\mathcal{S}}) \qquad \dot{\hat{x}} = \hat{f}(\hat{x}, u) \quad \text{and} \quad \omega = \hat{h}(\hat{x}) \quad \text{in} \quad R^{\hat{n}} \quad \text{in} \quad C^1 \quad \text{near}$$
$$\hat{x} = u = 0 \quad \text{with} \quad \hat{f}(0, 0) = 0, \qquad \hat{h}(0) = 0$$

are locally topologically equivalent in case: there exists a topological map Ψ of a neighborhood N of the origin of R^n onto a neighborhood $\hat{N}$ of the origin of $R^{\hat{n}}$, with $\Psi(0) = 0$, and with Ψ carrying $\mathcal{S}$ onto $\hat{\mathcal{S}}$. That is, there exists an $\epsilon > 0$ such that for each initial state $x_0 \in N$ with $|x_0| < \epsilon$ and each piecewise continuous controller $u(t)$ on $0 \leq t \leq \epsilon$ where $|u(t)| < \epsilon$, the responses $x(t)$ and $\hat{x}(t)$ of $\mathcal{S}$ and $\hat{\mathcal{S}}$ initiating at x_0 and $\hat{x}_0 = \Psi(x_0)$ correspond:

$$\Psi(x(t)) = \hat{x}(t) \quad \text{on} \quad 0 \leq t \leq \epsilon.$$

Moreover

$$h(x) \equiv \hat{h}(\Psi(x)).$$

THEOREM 6. *Consider the observed processes*

$$(\mathcal{S}) \quad \dot{x} = f(x, u) \quad \textit{and} \quad \omega = h(x) \quad \textit{in} \quad R^n, \quad \textit{in} \quad C^1 \quad \textit{near} \quad x = u = 0$$
$$\textit{with} \quad f(0, 0) = 0, \qquad h(0) = 0$$

and

$$(\hat{\mathcal{S}}) \quad \dot{\hat{x}} = \hat{f}(\hat{x}, u) \quad \textit{and} \quad \omega = \hat{h}(\hat{x}) \quad \textit{in} \quad R^{\hat{n}} \quad \textit{in} \quad C^1 \quad \textit{near} \quad \hat{x} = u = 0$$
$$\textit{with} \quad \hat{f}(0, 0) = 0, \qquad \hat{h}(0) = 0.$$

Assume that $\mathcal{S}$ *and* $\hat{\mathcal{S}}$ *are (algebraically) locally controllable and observable and have the same input-output relation*

$$\mathbf{T}: u(t) \to \omega(t): \qquad \mathfrak{J}_\delta \to C_r[0, 1],$$

as defined above. Then $\mathcal{S}$ *and* $\hat{\mathcal{S}}$ *are locally topologically equivalent.*

Proof. Let $\mathfrak{J}_L$ consist of those inputs in $\mathfrak{J}_\delta$ that satisfy the Lipschitz condition

$$|u(t) - u(s)| \leq |t - s|,$$

so $u(t)$ is continuous on $-\infty < t \leq 0$. Then $\mathfrak{J}_L$, with the uniform norm, is a compact metric space. The map of $\mathfrak{J}_L$ into R^n, as defined by the responses of $\mathcal{S}$ with $x(-1) = 0$,

$$u(t) \to x(0) = x_0: \qquad \mathfrak{J}_L \to R^n$$

is a uniformly continuous map onto a compact subset of R^n.

Now consider the space $[\mathfrak{J}_L]$ of all equivalence classes in $\mathfrak{J}_L$, where two controllers of $\mathfrak{J}$ are equivalent in case they produce the same output. Then $[\mathfrak{J}_L]$, with the decomposition or identification topology, is a compact space. Consider the map of $[\mathfrak{J}_L]$ into R^n;

$$\psi: u(t) \to x(0) = x_0: \qquad [\mathfrak{J}_L] \to R^n,$$

where $u(t)$ represents an element of $[\mathfrak{J}_L]$ and $x(t)$ is its response with $x(-1) = 0$, under the process $\mathcal{S}$. We note that ψ is well defined and one-to-one because $\mathcal{S}$ is locally observable (perhaps with $\delta > 0$ further reduced, if necessary) and the image of ψ covers an open neighborhood about the origin, since $\mathcal{S}$ is locally controllable (see Theorem 1 and subsequent remarks). Therefore ψ is a topological map of $[\mathfrak{J}_L]$ onto some subset of R^n (a one-to-one continuous map of a compact space onto a Hausdorff space is a topological map).

We define an analogous map

$$\hat{\psi}: u(t) \to \hat{x}(0) = \hat{x}_0: \qquad [\mathfrak{J}_L] \to R^{\hat{n}}$$

using the responses of $\hat{\mathcal{S}}$. Therefore $\Psi = \hat{\psi}\psi^{-1}$ is a topological map of some ball neighborhood N about the origin in R^n onto some neighborhood $\hat{N}$ of the origin in $R^{\hat{n}}$. Also $\Psi(0) = 0$ since both origins $x_0 = \hat{x}_0 = 0$ correspond to the zero control $u(t) \equiv 0$. Thus we conclude that $n = \hat{n}$ and we must now show that $\mathcal{S}$ and $\hat{\mathcal{S}}$ correspond under the mapping Ψ.

Choose a positive number

$$\epsilon < \min \left\{ \frac{1}{2}, \frac{\delta}{2}, \frac{\text{radius } N}{2} \right\}$$

such that

$$|x_0| < \epsilon \quad \text{and} \quad |u(t)| < \epsilon \quad \text{on} \quad 0 \leq t \leq 1$$

yields a response $x(t)$ of $\mathcal{S}$, with $x(0) = x_0$, which lies in N for all $0 \leq t \leq 1$. We now examine the action or response of $\mathcal{S}$, initiating at x_0, to an arbitrary piecewise continuous controller $\tilde{u}(t)$ satisfying the bound $|\tilde{u}(t)| < \epsilon$ on $0 \leq t \leq \epsilon$.

Let us define an action of $\tilde{u}(t)$ on a subset $U \subset [\mathfrak{J}_L]$. This will be done in such a way that the response in $[\mathfrak{J}_L]$ will correspond to $x(t)$ under ψ; and to the $\hat{\mathcal{S}}$ response $\hat{x}(t)$, with $\hat{x}(0) = \Psi(x_0)$, under $\hat{\psi}$. In this case

$$\Psi(x(t)) = \hat{x}(t) \quad \text{on} \quad 0 \leq t \leq \epsilon$$

and the topological equivalence of $\mathcal{S}$ and $\hat{\mathcal{S}}$ will be demonstrated.

Let $\epsilon > 0$ be further restricted so that the neighborhood $|x_0| < \epsilon$ in R^n corresponds to an open set $U \subset [\mathfrak{J}_L]$, in which each equivalence class can be represented by a controller $u_0(t)$, which also vanishes on $-1 \leq t \leq -\frac{1}{2}$. Since $\mathcal{S}$ is locally controllable this representation of elements in U is possible. We now define the action of $\tilde{u}(t)$ for $0 \leq t \leq \epsilon$ on an element of U represented by $u_0(t)$, which vanishes on $-1 \leq t \leq -\frac{1}{2}$. This action of $\tilde{u}(t)$ yields the element of $[\mathfrak{J}_L]$ represented by the concatenation

$$[u_0 \# \tilde{u}] \equiv \begin{cases} u_0(t + \epsilon) & \text{on} \quad -\epsilon - \frac{1}{2} \leq t < -\epsilon, \\ \tilde{u}(t + \epsilon) & \text{on} \quad -\epsilon \leq t \leq 0, \end{cases}$$

with the control vanishing elsewhere on R^1. It is easy to see that the response $x(t)$ of $\mathcal{S}$ with $x(0) = x_0 = \psi(u_0)$ reaches the point $x(\epsilon)$ under the control $\tilde{u}(t)$, and that $\psi[u_0 \# \tilde{u}] = x(\epsilon)$ (which lies in N if $\epsilon > 0$ is small enough). This holds because $\mathcal{S}$ is autonomous rather than time dependent. If we wish to compute the response in $[\mathfrak{J}_L]$, with initial state $[u_0] \in U$, to the controller $\tilde{u}(t)$ we obtain the final state $[u_0 \# \tilde{u}] \in [\mathfrak{J}_L]$. The response in $[\mathfrak{J}_L]$ at any intermediate instant t_1 on $0 \leq t_1 \leq \epsilon$, to the control $\tilde{u}(t)$ can be obtained by using t_1 in place of ϵ in the above computation.

In this way we define an action of $\tilde{u}(t)$ on states $[u_0] \in U \subset [\mathfrak{J}_L]$, and this action is mapped by ψ onto the corresponding responses of $\mathcal{S}$ to $\tilde{u}(t)$. Since the action of $\tilde{u}(t)$ on $U \subset [\mathfrak{J}_L]$ is defined without reference to $\mathcal{S}$ or $\hat{\mathcal{S}}$, but depends only on the input-output relation **T**, we conclude that Ψ carries $\mathcal{S}$ onto $\hat{\mathcal{S}}$ and that these two observed processes are locally topologically equivalent. Q.E.D.

EXERCISES

1. Consider the process in R^2

$$\dot{x} = y, \qquad \dot{y} = x^2 + y^2 + u^2 \quad \text{with} \quad u \in R^1.$$

Show that the domain $\mathcal{C}$ of null controllability lies in the fourth quadrant and so $\mathcal{C}$ does not contain an open neighborhood of the origin.

2. Consider the process in R^2

$$\dot{x} = y^3, \qquad \dot{y} = -x + u \quad \text{with} \quad |u(t)| \leq 1.$$

Show that $\mathcal{C} = R^2$ even though the algebraic conditions for controllability fail.

3. Consider the process in R^2 defined by

$$\ddot{x} + f(x, \dot{x}) = u \quad \text{with} \quad |u(t)| \leq 1,$$

that is,

$$\dot{x} = y, \qquad \dot{y} = -f(x, y) + u.$$

Assume $f(x, y)$ in C^1 in R^2 satisfies
(a) $xf(x, 0) > 0$ for $x \neq 0$
(b) $(\partial f/\partial y)(x, y) \geq 0$.
Then $\mathcal{C} = R^2$. [Hint: Compute the derivative of

$$V = y^2/2 + \int_0^x f(x, 0)\, dx$$

along the free motion.]

4. Consider the controlled Van der Pol equation

$$\ddot{x} + (x^2 - 1)\dot{x} + x = u$$

or the equivalent system in R^2

$$\dot{x} = y - \left(\frac{x^3}{3} - x\right), \qquad \dot{y} = -x + u.$$

(a) For a fixed control bound $|u(t)| \leq k$, show that there is a uniform bound for all responses initiating at (x_0, y_0) for a finite time duration. Thus there is a time-optimal controller steering $(x_0, y_0) \in \mathcal{C}$ to $(0, 0)$. [Hint: $r\dot{r} = x\dot{x} + y\dot{y} = -(x^4/3 - x^2) + yu$, so we have $r\dot{r} \leq 1 + kr$, where $r^2 = x^2 + y^2$.]

(b) Show that if $k > 0$ is sufficiently small, then $\mathcal{C}$ contains only a small neighborhood of the origin; but if $k > 0$ is sufficiently large, then $\mathcal{C} = R^2$. [Hint: It is known that the free van der Pol oscillator has a unique limit cycle towards which spiral all solutions other than the critical point. Thus assume x_0, $y_0 = 0$ lies near the limit cycle and take $u = x(t)$ until $x(t) \to \pm\sqrt{3}$; then take $u = (x^2 - 1)\dot{x} - \dot{x}$.]

5. Caratheodory originated the concept of controllability in his theory of thermodynamics. To illustrate this application in a simple case consider an ideal gas sample with pressure p, volume V, temperature T satisfying $pV = RT$, for a positive constant R. As the time t varies the history of the gas is described by the gas law and the first law of thermodynamics.

$$\frac{dQ}{dt} = p\frac{dV}{dt} + c_v\frac{dT}{dt}$$

where dQ/dt is the rate of influx of heat energy and c_v is the constant specific heat. Take $u^1(t) = \dot{p}(t)$ and $u^2(t) = \dot{Q}(t)$ as the control functions, and compute the system describing the history of the gas to be

$$\dot{p} = u^1, \qquad \dot{V} = \frac{1}{(1 + R/c_v)}\left[\frac{R}{c_v p}u^2(t) - \frac{V}{p}u^1(t)\right].$$

If the control is adiabatic, $u^2(t) \equiv 0$, show that there exist states between which the system cannot be steered.

6. Consider the nonlinear process in R^n

$$(\mathcal{S}) \qquad \dot{x} = f(x, u) \quad \text{in} \quad C^1 \quad \text{near} \quad x = u = 0.$$

Assume $f(x, u) = Ax + Bu + o(x, u)$ and

$$\text{rank}\,[B, AB, \ldots, A^{n-1}B] = n.$$

Prove that there exists a constant matrix D such that the control law $u = Dx$ stabilizes the process $\mathcal{S}$ near the origin, that is, $\dot{x} = f(x, Dx)$ is asymptotically stable about the origin.

7. The control process in R^n

$$\dot{x} = A(t, x) + B(t, x)u \quad \text{in} \quad C^1$$

is locally controllable along a solution $x = \phi(t)$ corresponding to the bounded measurable controller $v(t)$ on $t_0 \leq t \leq t_1$ in case: for each $\epsilon > 0$ there exists $\delta > 0$ such that each point x_1 with $|x_1 - \phi(t_1)| < \delta$ can be attained from x_0 by a measurable controller $u(t)$ satisfying $|u(t) - v(t)| < \epsilon$ on $t_0 \leq t \leq t_1$. Prove that the given process is locally controllable along $x = \phi(t)$ if the variational equation

$$\dot{y} = [A_x(t, \phi(t)) + B_x(t, \phi(t))\,v(t)]y + B(t, \phi(t))u$$

is completely controllable on $t_0 \leq t \leq t_1$.

8. Consider the process in R^n

$$\dot{x} = A(x) + Bu \quad \text{in} \quad C^1$$

with $|A(x)| + |\partial A/\partial x| < k$ (a constant) in R^n, and B a constant $n \times m$ matrix with BB' nonsingular. Prove that each initial point $x_0 \in R^n$ can be steered to each target point $x_1 \in R^n$ by a continuous controller $u(t)$ on $0 \leq t \leq 1$. [Hint: Let $u(t) = B'\xi$ for constant $\xi \in R^n$ with response $x = \phi^\xi(t)$. Seek ξ as a fixed point of the map

$$\xi \to (BB')^{-1}\left[x_1 - x_0 - \int_0^1 A(\phi^\xi(t))\,dt\right].]$$

9. Show that the input-output relation **T** of the scalar observed process

$$\dot{x} = u, \qquad \omega = x^2$$

cannot be realized by any locally observable and controllable process.

The next three problems indicate the development of differential calculus in Banach spaces and lead up to the implicit function theorem, which is required for the theory of local observability of nonlinear processes.

10. Let E and F be Banach spaces and let f be a function from an open neighborhood U about the origin of E into F. We say that f is differentiable at a point $u_0 \in U$ if there exists a continuous (bounded) linear transformation T of E into F such that

$$f(u_0 + \Delta u) = f(u_0) + T\,\Delta u + o(|\Delta u|),$$

for all small $\Delta u \in E$. We write $f'(u_0) = T$ and note that $f'(u_0)$ belongs to the Banach space $L(E, F)$ of all continuous linear maps of E into F. If $f'(u)$ exists at each $u \in U$ and

$$u \to f'(u): \qquad U \to L(E, F)$$

is continuous, then we say that $f \in C^1$ in U.

(a) Show that the composition of maps of class C^1 is again of class C^1 and that the *chain rule* yields the derivative of the composition.

(b) If $f(u) = Tu$ is a continuous linear transformation of E into F, prove that $f \in C^1$ and $f'(u)\,\Delta u = T\,\Delta u$. If T is a one-to-one map of E onto F, then we remark that the *closed-graph theorem* asserts that T^{-1} is continuous from F onto E.

(c) If $f \in C^1$ the theorem of the mean holds

$$f(u_0 + \Delta u) - f(u_0) = \int_0^1 f'(u_0 + t\,\Delta u)\,\Delta u\,dt,$$

for the segment $u_0 + t\,\Delta u$, $0 \le t \le 1$, in U. [The Riemann integral is defined as usual for continuous functions, and the two members of the equality yield the same value under every linear functional of $L(F, R^1)$].

(d) If $f \in C^1$ and the segment $u_0 + t\,\Delta u$, $0 \le t \le 1$, lies in U, then the theorem of the mean yields

$$|f(u_0 + \Delta u) - f(u_0)| \le |\Delta u| \sup f'(\xi),$$

for ξ on the segment. Also

$$|f(u_0 + \Delta u) - f(u_0) - f'(\bar{u})\,\Delta u| \le |\Delta u| \sup |f'(\xi) - f'(\bar{u})|$$

where $\bar{u}$ is fixed on the segment and the supremum is taken for all ξ on the segment.

11. Let $f: U \rightarrow F$ be a C^1 map from an open neighborhood U of the origin of a Banach space E into a Banach space F. Assume that
(a) $U \rightarrow L(E, F): u \rightarrow f'(u)$ is uniformly continuous on U, and
(b) $f'(0)$ is a one-to-one continuous linear transformation of E onto a closed subspace of F.

Then prove that there exists a subneighborhood $V \subset U$ whereon $f(u_1) = f(u_2)$ implies that $u_1 = u_2$. [Hint: The closed-graph theorem asserts that $f'(0)^{-1}$ is continuous from $f'(0)E$ onto E. Thus there exists a constant $c > 0$ such that $|f'(0)\,\Delta u| \geq c\,|\Delta u|$ for all Δu in E, and so $|f'(u)\,\Delta u| \geq (c/2)\,|\Delta u|$ for all u in some neighborhood $V \subset U$ about the origin. Now the theorem of the mean yields

$$|f(u_2) - f(u_1) - f'(\bar{u})(u_2 - u_1)| \leq |u_2 - u_1| \sup |f'(\xi) - f'(\bar{u})|$$

fo $\bar{u}$ and ξ on the segment joining u_1 to u_2. By the uniform continuity of $f'(u)$ we can restrict V so that $|f'(\xi) - f'(u)| \leq c/4$. Then

$$|f(u_2) - f(u_1)| \geq (c/4)\,|u_2 - u_1|$$

from which the result follows.]

12. Apply the results on C^1 functions from one Banach space into another to the proof of Theorem 4. Here

$$R^n \times L_\infty \rightarrow C_r[0, 1]: \qquad x_0, u \rightarrow \Omega(x_0, u)$$

has a uniformly continuous derivative in some neighborhood U of the origin, and at $x_0 = 0$, $u = 0$ the derivative is

$$\Delta x_0, \Delta u \rightarrow He^{At}\Delta x_0 + H\int_0^t e^{A(t-s)}B\,\Delta u(s)\,ds.$$

Show that the map (near the origin)

$$R^n \times L_\infty \rightarrow C_r[0, 1] \times L_\infty: \qquad x_0, u \rightarrow \Omega(x_0, u), u$$

satisfies the hypotheses of the preceding exercise. Then there is a neighborhood $V \subset U$ whereon

$$\Omega(x_0, u) = \Omega(\bar{x}_0, u) \text{ implies } x_0 = \bar{x}_0.$$

6.2 GLOBAL STABILITY OF NONLINEAR PROCESSES

In the preceding section we studied the local controllability of the nonlinear process in R^n

$$(\mathcal{S}) \qquad \dot{x} = f(x, u).$$

The domain $\mathcal{C}$ of null controllability consisted of all initial states $x_0 \in R^n$

that could be steered to the origin $x_1 = 0$ by admissible controllers in a finite time duration. We found that, under an algebraic controllability hypothesis on the linear approximation to $\mathcal{S}$, the domain $\mathcal{C}$ contained an open neighborhood of the origin. In this section we describe processes for which $\mathcal{C} = R^n$, generally $\mathcal{S}$ will be globally asymptotically stable and also locally controllable to the origin.

In Chapter 2 the linear process in R^n

$$(\mathcal{L}) \qquad \dot{x} = Ax + Bu$$

with Ω containing $u = 0$ in its interior in R^m, was shown to have $\mathcal{C} = R^n$ if

1. A is stable, that is, each eigenvalue λ of A has a negative real part, and
2. rank $[B, AB, \ldots, A^{n-1}B] = n$.

Our results here will extend these concepts and methods to nonlinear processes.

We first recall certain standard terminologies describing stability of autonomous differential systems.

Definition. The differential system

$$\dot{x} = f(x) \quad \text{in} \quad C^1 \quad \text{in} \quad R^n$$

is stable about the origin in case: for each $\epsilon > 0$ there exists a $\delta > 0$ such that $|x_0| < \delta$ implies that the solution $x(t)$ initiating at $x(0) = x_0$ satisfies $|x(t)| < \epsilon$ on $0 \leq t < \infty$.

Clearly a necessary condition for stability is that $f(0) = 0$, the origin is a critical, or rest, or equilibrium point. If $A = (\partial f/\partial x)(0)$ has all eigenvalues with negative real parts, then it is easy to prove that the differential system is stable, even asymptotically stable about the origin.

Definition. The differential system

$$\dot{x} = f(x) \quad \text{in} \quad C^1 \quad \text{in} \quad R^n$$

is asymptotically stable about the origin in case: for each $\epsilon > 0$ there exists $\delta > 0$ such that $|x_0| < \delta$ implies that the solution $x(t)$ initiating at $x(0) = x_0$ satisfies $|x(t)| < \epsilon$ on $0 \leq t < \infty$, and $\lim_{t \to \infty} x(t) = 0$. If, in addition, every solution of $\mathcal{S}$ in R^n is defined on $0 \leq t < \infty$ and tends towards the origin as $t \to \infty$, then $\mathcal{S}$ is called globally asymptotically stable.

Note that the requirements of stability about the origin, asymptotic stability, and global asymptotic stability are progressively stronger in that each includes the preceding. The next two theorems yield global asymptotic stability for differential systems arising in nonlinear control processes.

The first result is an immediate extension of the famous stability criterion of Liapunov, and the second result is a stability criterion depending on the negative nature of the eigenvalues of a certain matrix. We remark that these theorems on global asymptotic stability, together with earlier results on local controllability, will yield the desired property that $\mathcal{C} = R^n$.

THEOREM 7. *Consider the control process in* R^n

$$(\mathcal{S}) \qquad \dot{x} = f(x, u) \quad \text{in} \quad C^1 \quad \text{in} \quad R^{n+m}$$

with control restraint set $\Omega \subset R^m$. *Assume there exists a scalar function* $V(x)$ *and an m-vector function* $U(x)$ *in* C^1 *in* R^n *such that*

(a) $V(x) \geq 0$, *and* $V(x) = 0$ *if and only if* $x = 0$;
(b) $\lim_{|x| \to \infty} V(x) = +\infty$;
(c) $U(x) \subset \Omega$;
(d) $(\partial V/\partial x^i)\, f^i(x, U(x)) < 0$ *for* $x \neq 0$.

Then the differential system

$$(\mathcal{S}_U) \qquad \dot{x} = f(x, U(x))$$

is globally asymptotically stable about the origin. Hence, for each initial $x_0 \in R^n$ *the response* $x(t)$ *of* $\mathcal{S}_U$ *tends to* $x_1 = 0$ *as* $t \to +\infty$, *and the controller*

$$u(t) = U(x(t)) \subset \Omega \quad \text{on} \quad 0 \leq t < \infty$$

yields the same response $x(t)$ *of* $\mathcal{S}$ *steering* x_0 *towards the origin.*

Proof. First note that the level loci

$$V(x) = c > 0$$

are compact subsets of R^n and locally they are each smooth hypersurfaces since grad $V(x) \neq 0$ thereon. Thus each such level locus is a compact smooth submanifold (a hypersurface) that separates the space R^n into two regions, and the origin lies in the interior region where $V(x) < c$. [A gradient curve of $V(x)$ leads from the exterior region to a neighborhood of the origin with only one intersection with the locus $V(x) = c$.]

For each initial point $x_0 \in R^n$ consider the solution curve $x(t)$ of $\mathcal{S}_U$ initiating at x_0. Along $x(t)$ we compute the rate of change of V

$$\frac{dV}{dt} = \frac{\partial V}{\partial x^i}(x(t))\, f^i(x(t), U(x(t))) \leq 0$$

and so $x(t)$ remains within the level hypersurface $V(x) = V(x_0)$. Thus $x(t)$ is defined on $0 \leq t < \infty$. Moreover if $x_0 = 0$ then $V(x(t)) \equiv 0$ so $x(t) \equiv 0$, and hence $f(0, U(0)) = 0$ and the origin is a critical point of the differential system $\mathcal{S}_U$.

If $x_0 \neq 0$, then $x(t)$ tends inwards across the level hypersurfaces $V(x) = c > 0$ but never reaches the origin. Let

$$\lim_{t \to \infty} V(x(t)) = V_\infty \geq 0.$$

If $V_\infty > 0$, then some point x_∞ on the compact hypersurface $V(x) = V_\infty$ must be approached arbitrarily closely by $x(t)$. However the solution of $\mathcal{S}_U$ commencing at x_∞ penetrates into the interior of $V(x) = V_\infty$, and continuity considerations show that $x(t)$ must also penetrate this interior. Thus we conclude that $V_\infty = 0$ and

$$\lim_{t \to \infty} x(t) = 0.$$

We have proved that $\mathcal{S}_U$ is globally asymptotically stable about the origin. It is evident that the C^1 controller

$$u(t) = U(x(t)) \quad \text{on} \quad 0 \leq t < \infty$$

yields the response $x(t)$ for the control process $\mathcal{S}$. Q.E.D.

In the special case where $f(x, u)$ does not depend on u, and where no condition is imposed on $U(x)$ (which is thereby omitted from consideration), the above theorem is known as the stability criterion of Liapunov. The function $V(x)$ is called a *Liapunov function* for the differential system.

COROLLARY. *Consider the control process in* R^n

$$(\mathcal{S}) \qquad \dot{x} = f(x, u) \quad \textit{in} \quad C^1 \quad \textit{in} \quad R^{n+m}$$

with restraint set $\Omega \subset R^m$.

Assume the existence of functions $U(x)$ *and* $V(x)$ *in* C^1 *in* R^n, *satisfying hypotheses* (*a*), (*b*), (*c*), *and* (*d*) *of the theorem. Assume further that*

(e) $f(0, 0) = 0$;
(f) $u = 0$ *lies interior to* Ω;
(g) *rank* $[B, AB, \ldots, A^{n-1}B] = n$, *where* $A = f_x(0, 0)$, $B = f_u(0, 0)$.

Then the domain of null controllability for $\mathcal{S}$ *is* $\mathcal{C} = R^n$.

Proof. The proof follows immediately from the theorem, and the earlier results on local controllability. Q.E.D.

THEOREM 8. *Consider the control process in* R^n

$$(\mathcal{S}) \qquad \dot{x} = f(x, u) \quad \textit{in} \quad C^1 \quad \textit{in} \quad R^{n+m}$$

with control restraint set $\Omega \subset R^m$.

Assume

(a) $f(0, 0) = 0$;
(b) *there exists an m-vector function* $U(x)$ *in* C^1 *in* R^n *with* $U(x) \subset \Omega$ *and* $U(0) = 0$;
(c) *every eigenvalue* $\lambda(x)$ *of* $J(x) + J'(x)$ *satisfies* $\lambda(x) \leq -\epsilon < 0$ *for all* $x \in R^n$ *and some constant* $\epsilon > 0$, *where*

$$J(x) = \frac{\partial f}{\partial x}(x, U(x)) + \frac{\partial f}{\partial u}(x, U(x))\frac{\partial U}{\partial x}$$

and $J'(x)$ *is the transpose matrix of* $J(x)$.

Then the differential system

$$(\mathcal{S}_U) \qquad \dot{x} = f(x, U(x))$$

is globally asymptotically stable about the origin. Hence, for each $x_0 \in R^n$ *the response* $x(t)$ *of* $\mathcal{S}_U$ *tends towards* $x_1 = 0$ *as* $t \to \infty$, *and the controller*

$$u(t) = U(x(t)) \subset \Omega \quad \text{on} \quad 0 \leq t < \infty$$

yields the same response $x(t)$ *of* $\mathcal{S}$ *steering* x_0 *towards the origin.*

Proof. The origin $x_1 = 0$ is a critical point for the differential system $\mathcal{S}_U$ and we shall find that this is the only critical point.

Let $\bar{x}$ be a critical point of $\mathcal{S}_U$ in R^n so $F(\bar{x}) = 0$, where $F(x) = f(x, U(x))$, and $J(\bar{x}) = \partial F/\partial x(\bar{x})$ is the coefficient matrix of the linear approximation to $\mathcal{S}_U$ near this critical point. If $J(\bar{x})$ is singular then there exists a nonzero constant vector $w \in R^n$ for which $J(\bar{x})w = 0$. This implies that

$$w'Jw = 0 \quad \text{and} \quad w'J'w = 0$$

so

$$w'(J + J')w = 0,$$

which contradicts the hypothesis that

$$w'(J + J')w \leq -\epsilon w'w < 0.$$

Therefore $J(\bar{x})$ is nonsingular and hence the map

$$x \to F(x) \quad \text{of} \quad R^n \quad \text{into} \quad R^n$$

has an isolated zero at $\bar{x}$. Thus the critical points of $\mathcal{S}_U$ are each isolated.

Let $x(t)$ be the solution of $\mathcal{S}_U$ initiating at the arbitrary initial state x_0. Define the velocity vector along this solution $v(t) = \dot{x}(t)$. Then

$$\dot{v} = J(x(t))v$$

and so

$$v'\dot{v} = v'Jv, \qquad \dot{v}'v = v'J'v.$$

Write $\|v\|^2 = v'v$ and then compute

$$\frac{d\,\|v\|^2}{dt} = v'(J + J')v \le -\epsilon\,\|v\|^2.$$

Therefore the speed of the solution $x(t)$ satisfies

$$\|v(t)\| \le \|v(0)\|\, e^{-\epsilon t/2},$$

and

$$\|x(t)\| \le \|x_0\| + \int_0^t \|v(s)\|\, ds \le \|x_0\| + \|v(0)\|\,(2/\epsilon).$$

Hence $x(t)$ remains within a compact ball in R^n for $0 \le t \le \infty$ and $x(t)$ must approach a critical point of $\mathcal{S}_U$ as $t \to \infty$, since the speed $\|v(t)\| \to 0$.

Since the speed $\|f(x, U(x))\|$ of $\mathcal{S}_U$ vanishes only at the isolated critical points, and since $\|v(t)\|$ is monotone decreasing, each critical point of $\mathcal{S}_U$ is locally asymptotically stable. But every solution $x(t)$ of $\mathcal{S}_U$ must tend to some critical point as $t \to \infty$. It is easy to see that the set of all initial states $x_0 \in R^n$, for which the motion $x(t)$ of $\mathcal{S}_U$ tends towards a given critical point $\bar{x}$, is an open set $\mathcal{O}(\bar{x}) \subset R^n$. Since the space R^n is connected it cannot be decomposed into two disjoint nonempty open subsets. Therefore there is just one critical point $x_1 = 0$ of $\mathcal{S}_U$ and so the origin is globally asymptotically stable.

For each initial point x_0 the motion $x(t)$ of $\mathcal{S}_U$ can also be recovered as the response of the control process $\mathcal{S}$ to the controller $u(t) = U(x(t))$.

Q.E.D.

COROLLARY. *Consider the control process in* R^n

($\mathcal{S}$) $$\dot{x} = f(x, u) \quad \text{in} \quad C^1 \quad \text{in} \quad R^{n+m}$$

with restraint set $\Omega \subset R^m$.

Assume hypotheses (a), (b), and (c) of the theorem and also

(d) $u = 0$ *lies interior to* Ω *in* R^m

(e) *rank* $[B, AB, \ldots, A^{n-1}B] = n$ *where* $A = f_x(0, 0)$, $B = f_u(0, 0)$.

Then the domain of null controllability for $\mathcal{S}$ *is* $\mathcal{C} = R^n$.

Proof. The corollary follows immediately from the theorem and the earlier results on local controllability. Q.E.D.

In Chapter 2 we studied the problem of stabilizing a linear autonomous process in R^n

($\mathcal{L}$) $$\dot{x} = Ax + bu$$

with the observed linear feedback data

$$\sigma = cx$$

and the control law

$$u = \sigma = cx.$$

Here A is a real $n \times n$ matrix, b is a real column vector, c is a real row vector, and σ is a real scalar. In Theorem 9 of Chapter 2 we showed that if (A, b) is (completely) controllable, then the vector c can be selected so that $A + bc$ is a stability matrix; that is, if

$$\det |b, Ab, A^2b, \ldots, A^{n-1}b| \neq 0,$$

then we can choose c so that all eigenvalues of $A + bc$ have negative real parts and

$$\dot{x} = (A + bc)x$$

is asymptotically stable about the origin.

In this section we shall analyze the stability of the above process, but with the nonlinear control laws

$$u = u(\sigma) \quad \text{where} \quad \sigma u(\sigma) > 0 \quad \text{for} \quad \sigma \neq 0.$$

This study is referred to as the *Luré problem* for direct control.

Definition. Consider the control process in R^n

$$(\mathfrak{I}) \qquad \dot{x} = Ax + b\,u(\sigma), \qquad \sigma = cx$$

for a real $n \times n$ matrix A and vectors b and c. Assume the control law is such that $-u(\sigma)$ yields a negative feedback; that is,

$$u(\sigma) \quad \text{is in} \quad C^1 \quad \text{on} \quad -\infty < \sigma < \infty,$$
$$\sigma u(\sigma) > 0 \quad \text{for} \quad \sigma \neq 0.$$

These conditions specify the control class $\mathfrak{A}$, and the process $\mathfrak{I}$ is absolutely stable in case the differential system

$$\dot{x} = Ax + b\,u(cx)$$

is globally asymptotically stable for every $u(\sigma) \in \mathfrak{A}$. If A and b are prescribed, then the absolute stability domain $\mathfrak{C} \subset R^n$ consists of all those vectors c for which the process $\mathfrak{I}$ is absolutely stable.

Remarks. If one eigenvalue of A has a positive real part, then the domain $\mathfrak{C}$ is empty since the class $\mathfrak{A}$ contains the linear control laws $u = \epsilon\sigma$ for arbitrarily small $\epsilon > 0$. Therefore we shall consider the problem of Luré only where A is a stability matrix (although a more refined analysis could permit certain pure imaginary eigenvalues of A), and here $\mathfrak{C}$ always contains $c = 0$. The domain $\mathfrak{C}$ consists of the union of half-rays from the origin. To note this convexity property of $\mathfrak{C}$ consider the feedback

$u(\lambda cx)$ with $\lambda > 0$ and $c \in \mathfrak{C}$. Define $u_\lambda(\sigma) = u(\lambda\sigma)$ and u_λ lies in $\mathfrak{A}$ whenever u lies in $\mathfrak{A}$. Hence the process $\mathfrak{L}$ is globally asymptotically stable with the feedback $u_\lambda(cx) = u(\lambda cx)$ and so $\lambda c \in \mathfrak{C}$, which is the desired result. The geometry of $\mathfrak{C}$ is not easily ascertained and the problem of Luré consists in determining subregions in $\mathfrak{C}$ for given data (A, b).

The usual method of proving absolute stability for the process $\mathfrak{L}$ utilizes a Liapunov function (see Theorem 7) of the form

$$V(x, \sigma) = x'Bx + \int_0^\sigma u(s)\, ds,$$

where x' is the transpose of the column vector x. The positive definite matrix $B > 0$ is sometimes computed from the relation

$$A'B + B'A = -C$$

where $C > 0$. We recall that in Theorem 7 of Chapter 3 it was shown that for each such $C > 0$ there exists one and only one $B > 0$, provided A is a stability matrix. In fact,

$$B = \int_0^\infty e^{A't} C e^{At}\, dt.$$

THEOREM 9. *Consider the process in* R^n

$$(\mathfrak{L}) \qquad \dot{x} = Ax + b\, u(\sigma), \qquad \sigma = cx$$

with control laws $u(\sigma)$ *in class* $\mathfrak{A}$, *where* A *is a real stability matrix and* b *and* c *are real vectors. Assume that there exists a positive definite matrix* $B > 0$ *such that* $\{A, b, c\}$ *satisfy, for some* $\alpha > 0$,

$$-cb > (Bb + \tfrac{1}{2}A'c' + \alpha c\tfrac{1}{2})'\, C^{-1}(Bb + \tfrac{1}{2}A'c' + \alpha c\tfrac{1}{2})$$

where

$$A'B + B'A = -C, \text{ and } C > 0$$

Then $c \in \mathfrak{C}$ *so* $\mathfrak{L}$ *is absolutely stable.*

Proof. Fix a control law $u(\sigma) \in \mathfrak{A}$. For each response $x(t)$ of $\mathfrak{L}$ we then define $\sigma(t) = cx(t)$ to obtain a solution of

$$\dot{x} = Ax + b\, u(\sigma)$$
$$\dot{\sigma} = cAx + cbu(\sigma)$$

in R^{n+1}. Define the real function in R^{n+1}

$$V(x, \sigma) = x'Bx + \int_0^\sigma u(s)\, ds,$$

which is positive except at $x = \sigma = 0$. We now compute the derivative of $V(x, \sigma)$ along a solution $x(t)$, $\sigma(t)$ to obtain

$$\dot{V} = x'\, B(Ax + bu) + (x'A' + ub')Bx + u \cdot (cAx + cbu)$$

or, noting that $x' A' c' = cAx$, and adding and subtracting $\alpha\sigma u(\sigma)$,

$$-\dot{V} = \{x'Cx - 2x'(Bb + \tfrac{1}{2}A' c' + \alpha c\tfrac{1}{2})u - cbu^2\} + \alpha\sigma u(\sigma).$$

In order that $V(x, \sigma)$ serve as a Liapunov function we require that $-\dot{V} > 0$, except at $x = \sigma = 0$. Notice that $-\dot{V}$ depends on a quadratic form in the $(n + 1)$ variables (x, u), with the matrix

$$\begin{pmatrix} C & d \\ d' & -cb \end{pmatrix}$$

where $d = -(Bb + \frac{1}{2}A'c' + \alpha c\frac{1}{2})$. This matrix is positive definite in case all the principal minor determinants are positive. The hypothesis $C > 0$ guarantees the positivity of all the principal minor determinants except for the full determinant requirement

$$\begin{vmatrix} C & d \\ d' & -cb \end{vmatrix} > 0.$$

We now prove that this final positivity demand also follows from the hypotheses of the theorem.

Since $C > 0$ we need only establish the positivity of the determinant of

$$\begin{pmatrix} C^{-1} & 0 \\ 0 & 1 \end{pmatrix}\begin{pmatrix} C & d \\ d' & -cb \end{pmatrix} = \begin{pmatrix} I & C^{-1}d \\ d' & -cb \end{pmatrix}.$$

But this last determinant is easily computed to be

$$-cb - d'C^{-1}d = -cb - (Bb + \tfrac{1}{2}A'c' + \alpha c\tfrac{1}{2})'\, C^{-1}(Bb + \tfrac{1}{2}A'c' + \alpha c\tfrac{1}{2}),$$

which is asserted to be positive in the hypotheses.

Finally we prove the global asymptotic stability of Л. For any initial state $x_0 \in R^n$ the response $x(t)$ of

$$\dot{x} = Ax + b\,u(cx)$$

remains in the region of R^n where

$$x'Bx \leq V(x_0, cx_0),$$

and hence $x(t)$ is defined on $0 \leq t < \infty$, and also the origin is a stable critical point. Since $\dot{V}(x(t), \sigma(t)) < 0$ along a response $x(t) \neq 0$, $\sigma(t) = c\,x(t) \neq 0$, we conclude that

$$\lim_{t \to \infty} x(t) = 0.$$

Thus the process is globally asymptotically stable about the origin and Л is absolutely stable. Q.E.D.

Example. The problem of Luré is often phrased in terms of an indirect control involving the derivative of an actuating signal. We here state the indirect control problem of Luré and show how it can be reduced to an equivalent direct control problem of the type analyzed in Theorem 9.

Consider the indirect control problem in R^{n+1}

$$(\text{Л}_1) \qquad \dot{x} = Ax + bv$$
$$\dot{v} = u(\sigma), \qquad \sigma = cx - \rho v$$

where $u(\sigma)$ belongs to class $\mathfrak{A}$, that is, $\sigma\, u(\sigma) > 0$ for $\sigma \neq 0$. We seek values of the real constant $n \times n$ matrix A, n-vectors b and c, and scalar ρ, so that Л_1 is globally asymptotically stable for every suitable $u(\sigma)$. Note that the indirect control problem Л_1 is a direct control problem in R^{n+1}

$$\dot{w} = \hat{A}w + \hat{b}\, u(\sigma), \qquad \sigma = \hat{c}w$$

where

$$w = \begin{pmatrix} x \\ v \end{pmatrix}, \qquad \hat{A} = \begin{pmatrix} A & b \\ 0 & 0 \end{pmatrix}, \qquad \hat{b} = \begin{pmatrix} 0 \\ \cdot \\ \cdot \\ \cdot \\ 0 \\ 1 \end{pmatrix}, \qquad \hat{c} = (c, -\rho).$$

Here the matrix $\hat{A}$ is singular and this leads to certain additional restrictions on the solution of the Luré indirect control problem.

For linear control laws $u(\sigma) = \epsilon\sigma$, for some small $\epsilon > 0$, the process Л_1 becomes

$$\dot{x} = Ax + bv$$
$$\dot{v} = \epsilon cx - \epsilon\rho v.$$

To insure the asymptotic stability of this system in R^{n+1} for all small $\epsilon > 0$ we assume that A is a stability matrix. Also take $\rho > 0$. Since the trace of this coefficient matrix is $(Tr A - \epsilon\rho)$, which is the sum of all its eigenvalues, the condition $\rho > 0$ insures that n eigenvalues are near those of A and the last eigenvalue is negative. We also shall require that $x = 0$, $v = 0$ is the unique critical point of Л_1, that is

$$Ax + bv = 0 \quad \text{and} \quad cx - \rho v = 0$$

have a unique solution. This obtains if

$$\begin{vmatrix} A & b \\ c & -\rho \end{vmatrix} \neq 0,$$

or if

$$\begin{vmatrix} A^{-1} & 0 \\ 0 & 1 \end{vmatrix} \begin{vmatrix} A & b \\ c & -\rho \end{vmatrix} = \begin{vmatrix} I & A^{-1}b \\ c & -\rho \end{vmatrix} \neq 0.$$

But this last determinant is easily computed (by induction),

$$-\rho - cA^{-1}b \neq 0.$$

Thus we consider the process Л_1 under the *standing hypotheses*

$$A \quad \text{is a stability matrix}$$

$$\rho > 0, \qquad \rho \neq -cA^{-1}b.$$

We introduce new coordinates in R^{n+1} by

$$y = Ax + bv, \qquad s = cx - \rho v$$

to obtain the direct control process

$$(\text{Л}_2) \qquad \dot{y} = Ay + b\,u(\sigma)$$

$$\dot{s} = cy - \rho\,u(\sigma), \qquad \sigma = s.$$

Clearly Л_2 is globally asymptotically stable just in case Л_1 is also. The coefficient matrices of Л_2 are

$$\tilde{A} = \begin{pmatrix} A & 0 \\ c & 0 \end{pmatrix}, \qquad \tilde{b} = \begin{pmatrix} b \\ -\rho \end{pmatrix}, \qquad \tilde{c} = (0, 1),$$

and hypotheses of Theorem 9 are not fulfilled, but we proceed by an analogous calculation.

Assertion. Assume there exists a symmetric positive definite matrix $B > 0$ such that

$$\rho > (Bb + \tfrac{1}{2}c')'\, C^{-1}(Bb + \tfrac{1}{2}c')$$

where

$$A'B + B'A = -C, \quad \text{and} \quad C > 0.$$

Then Л_1 is globally asymptotically stable for each control law

$$u(\sigma) \quad \text{in} \quad \mathfrak{A} \quad \text{and} \quad \int_0^{\pm\infty} u(\sigma)\, d\sigma = \infty.$$

We indicate the proof of this assertion, which is the analogue of Theorem 9 for indirect control problems. Define the Liapunov function for Л_2 in R^{n+1}

$$V(y, s) = y'By + \int_0^s u(s)\, ds > 0$$

for $(y, s) \neq (0, 0)$. Next compute the derivative of $V(y, s)$ along the solutions of $\mathfrak{J}\mathfrak{I}_2$ to obtain

$$-\dot{V} = y'Cy + \rho\, u(s)^2 - 2(Bb + \tfrac{1}{2}c')'y\, u(s).$$

Thus $-\dot{V} > 0$ at $(y, u) \neq (0, 0)$ provided

$$\begin{pmatrix} C & Bb + \frac{1}{2}c' \\ (Bb + \frac{1}{2}c')' & \rho \end{pmatrix} > 0,$$

which is guaranteed by the hypothesis

$$\rho > (Bb + \tfrac{1}{2}c')'\, C^{-1}(Bb + \tfrac{1}{2}c').$$

Since

$$\lim_{|y|+|s|\to\infty} V(y, s) = \infty,$$

each locus $V(y, s) \leq V(y_0, s_0)$ is compact in R^{n+1} and the proof of Theorem 9 demonstrates that $\mathfrak{J}\mathfrak{I}_2$ is globally asymptotically stable, as required.

We next attack the direct control problem of Luré by techniques suggested by the transform-function methods of linear analysis. Indeed, consider the transfer-function from input u to output σ in the autonomous linear system

$$\dot{x} = Ax + bu, \qquad \sigma = cx;$$

namely the rational complex function

$$c\, A(z)^{-1}b \quad \text{where} \quad A(z) = zI - A.$$

If we permit the linear control laws

$$u(\sigma) = \epsilon\sigma \quad \text{for} \quad \epsilon > 0,$$

then the absolute stability criterion, for the scalar case that we discuss first for simplicity, requires that $(A + \epsilon bc)$ be a stability matrix for each $\epsilon > 0$. This means that

$$|zI - A - \epsilon bc| = |A(z)| \cdot |I - \epsilon c\, A(z)^{-1}b|$$

has no zeros in the closed right-half plane $\operatorname{Re} z \geq 0$. Since A is a stability matrix, the absolute stability criterion for the linear control laws $u(\sigma) = \epsilon\sigma$ becomes:

$$\{-c\, A(z)^{-1}b\} \quad \text{nonnegative for} \quad \operatorname{Re} z \geq 0.$$

It is this condition, as appropriately modified for the nonlinear Luré problem, which occurs in the next theorem as the stability criterion of Popov. We shall restrict our attention to the case where (A, b) is a controllable pair, although this hyothesis can be eliminated with some care.

We first recall some special properties of (completely) controllable and observable linear processes. Consider the linear autonomous process in R^n

$$(\mathcal{L}) \qquad \dot{x} = Ax + bu, \qquad \sigma = cx,$$

with the above notations. If (A, b) is controllable, then there exists a real nonsingular coordinate transformation $x = P\bar{x}$ in the state space R^n so that the process $\mathcal{L}$ has the description

$$(\bar{\mathcal{L}}) \qquad \dot{\bar{x}} = \bar{A}\bar{x} + \bar{b}u, \qquad \sigma = \bar{c}\bar{x}$$

where

$$\bar{A} = P^{-1}AP = \begin{pmatrix} 0 & 1 & 0 & \cdots & 0 \\ 0 & 0 & 1 & & \\ 0 & 0 & & & \\ \cdot & & & & \\ \cdot & & & & \\ \cdot & & & & \\ -a_n & -a_{n-1} & & & -a_1 \end{pmatrix}$$

and

$$\bar{b} = P^{-1}b = \begin{pmatrix} 0 \\ 0 \\ \cdot \\ \cdot \\ \cdot \\ 1 \end{pmatrix}, \qquad \bar{c} = cP.$$

Using these coordinates it is easy to compute the transfer function from u to σ,

$$\bar{c}\,\bar{A}(z)^{-1}\bar{b} = \frac{\bar{c}(z)}{|A(z)|}.$$

Here $|A(z)|$ is the determinant of $A(z)$, or of $\bar{A}(z) = zI - \bar{A}$, and the polynomial

$$\bar{c}(z) = \bar{c}_n z^{n-1} + \bar{c}_{n-1}z^{n-2} + \cdots + \bar{c}_2 z + \bar{c}_1$$

is formed directly from the components of $\bar{c} = (\bar{c}_1, \bar{c}_2, \ldots, \bar{c}_n)$. This important calculation is facilitated by noting that

$$\bar{A}(z)\begin{pmatrix}1\\ z\\ \cdot\\ \cdot\\ \cdot\\ z^{n-2}\\ z^{n-1}\end{pmatrix} = |A(z)|\begin{pmatrix}0\\ 0\\ \cdot\\ \cdot\\ \cdot\\ 0\\ 1\end{pmatrix}$$

so

$$\bar{A}(z)^{-1}\bar{b} = \begin{pmatrix}1\\ z\\ \cdot\\ \cdot\\ \cdot\\ z^{n-1}\end{pmatrix}|A(z)|^{-1}.$$

Thus, given any real polynomial $\bar{c}(z)$ of degree $\leq n - 1$, there is a unique vector c satisfying the relation

$$\bar{c}\,\bar{A}(z)^{-1}\bar{b} = \frac{\bar{c}(z)}{|A(z)|}.$$

Now revert to the original coordinates x in R^n and let $\mu(z)$ be any real polynomial of degree $\leq n - 1$. Consider the equation with the vector unknown p

$$p\,A(z)^{-1}b = \frac{\mu(z)}{|A(z)|}.$$

But this equation for p can be written

$$(pP)(P^{-1}A(z)^{-1}P)(P^{-1}b) = \frac{\mu(z)}{|A(z)|}$$

or

$$(pP)\,\bar{A}(z)^{-1}\bar{b} = \frac{\mu(z)}{|A(z)|},$$

which has a unique solution for (pP), and so p is uniquely determined.

The process $\bar{\mathfrak{L}}$—and hence the process $\mathfrak{L}$ or the pair (A, c)—is (completely) observable if and only if

$$\det|c', A'c', \ldots, A'^{n-1}c'| \neq 0.$$

By Lemma 2 of Theorem 14 in Chapter 2 the process $\bar{\mathfrak{L}}$ is observable if and only if the complex polynomial

$$\bar{c}(z) = \bar{c}_n z^{n-1} + \bar{c}_{n-1} z^{n-2} + \cdots + \bar{c}_2 z + \bar{c}_1$$

and the characteristic polynomial

$$|A(z)| = |\bar{A}(z)| = \det |zI - A| = z^n + a_1 z^{n-1} + \cdots + a_n$$

are relatively prime (have no common zeros). Thus $(\bar{A}, \bar{c})$ is an observable pair in $\bar{\mathfrak{L}}$ if and only if the transfer-function

$$\bar{c}\,\bar{A}(z)^{-1}\bar{b} = \frac{\bar{c}(z)}{|A(z)|}$$

is a rational expression in which the numerator and denominator have no common roots.

Now revert to the original coordinates x in R^n and consider the process $\mathfrak{L}$. The transfer function is unchanged

$$\bar{c}\,\bar{A}(z)^{-1}\bar{b} = (cP)(P^{-1}\,A(z)^{-1}P)(P^{-1}b) = c\,A(z)^{-1}b,$$

and so (A, c) is observable if and only if

$$c\,A(z)^{-1}b = \frac{\bar{c}(z)}{|A(z)|}$$

is such that $\bar{c}(z)$ and $|A(z)|$ are relatively prime.

With these preliminary notations established we now return to the Luré problem of stabilizing systems with nonlinear control laws. We first prove a complicated algebraic lemma concerning matrices.

LEMMA. *Let A be a real $n \times n$ stability matrix, b and c real n-vectors, and $\tau \geq 0$ a scalar. Assume (A, b) is controllable and that the rational complex function*

$$T(z) = \tau - 2c\,A(z)^{-1}b \not\equiv 0$$

satisfies $\operatorname{Re} T(i\omega) \geq 0$ *for* $-\infty < \omega < \infty$. *Then there exists two $n \times n$ real symmetric matrices*

$$B > 0 \quad \textit{and} \quad D \geq 0$$

and there exists a real n-vector q such that

1. $A'B + B'A = -qq' - D;$
2. $Bb + c' + \sqrt{\tau}\,q = 0;$
3. (A, q') *is observable.*

Proof. Note that b and q are column vectors but that c is a row vector to conform to earlier notations. We begin by defining the real polynomial

$$\eta(z) = |A(z)|\,|A(-z)|\,\{\tau - c\,A(z)^{-1}b - b'\,A'(-z)^{-1}c'\}.$$

Note that the degree of $\eta(z)$ is $2n$ since the highest-degree polynomial arising in $|A(z)|\,A(z)^{-1} = \text{adj}\,A(z)$ is $n-1$, and that the leading term of $\eta(z)$ is $(-1)^n\tau z^{2n}$. [If $\tau = 0$, then $\eta(z)$ has degree $< 2n$; but $\eta(z) \not\equiv 0$ according to an exercise below, and the proof remains valid.] The proof proceeds by a detailed analysis of the factors of $\eta(z)$ into real prime linear and quadratic polynomials.

Since $A(z)' = A'(z)$ and $|A(z)|\,A(z)^{-1}$ is a real polynomial matrix, $\eta(z)$ is a polynomial with real coefficients. Since $\eta(z) = \eta(-z)$, it is an even polynomial. Also on the imaginary axis

$$\text{Re}\,\eta(i\omega) = |A(i\omega)|\,|A(-i\omega)|\,\{\tau - 2\text{Re}\,cA(i\omega)^{-1}b\} \geq 0$$

for $-\infty < \omega < \infty$, as follows from the hypotheses. Thus the zeros of $\eta(z)$ are symmetrically distributed about both the real and imaginary axes, and the zeros of $\eta(i\omega)$ have even multiplicity (since $\text{Re}\,\eta(i\omega)$ does not change sign on the imaginary axis). Therefore

$$\eta(z) = \theta(z)\,\theta(-z)$$

where $\theta(z)$ is a real polynomial with no zeros in $\text{Re}\,z > 0$. Finally define $\theta_1(z)$ and $\theta_2(z)$ by

$$\theta(z) = \theta_1(z)\,\theta_2(z)$$

where $\theta_1(z)$ and $\theta_2(z)$ are real polynomials with zeros only in the half-plane $z < 0$ and on the pure imaginary axis, respectively. Also take the leading coefficient of $\theta_2(z)$ to be $+1$. This factorization

$$\eta(z) = \theta_1(z)\,\theta_2(z)\,\theta_1(-z)\,\theta_2(-z)$$

follows directly from the known distribution of the zeros of $\eta(z)$.

Since $\theta_1(i\omega)$ is a polynomial in ω, which does not vanish,

$$\theta_1(i\omega)\,\theta_1(-i\omega) > \epsilon_0 > 0 \quad \text{for} \quad -\infty < \omega < \infty.$$

Let α be a real positive number such that

$$\alpha^2 \leq \epsilon_0 \quad \text{and} \quad \alpha^2 \neq \theta_1(\lambda_i)\,\theta_1(-\lambda_i)$$

where $\lambda_1, \lambda_2, \ldots, \lambda_n$ are the eigenvalues of A. If $\theta_1(z)$ is constant, take $\alpha = 0$.

Using the controllability of (A, b), define the real vector g by

$$g'\,A(z)^{-1}b = \frac{\alpha\,\theta_2(z)}{|A(z)|}.$$

This is possible since the degree of $\theta_2(z)$ is less than n unless $\theta_1(z)$ is constant and $\alpha = 0$. The required symmetric matrix D is defined $D = gg'$.

Define the real even polynomial

$$\Gamma(z) = \theta_2(z)\,\theta_2(-z)[\theta_1(z)\,\theta_1(-z) - \alpha^2].$$

Since

$$\theta_2(i\omega)\,\theta_2(-i\omega) \geq 0$$

and

$$\theta_1(i\omega)\,\theta_1(-i\omega) > \epsilon_0 \geq \alpha^2,$$

we find that $\Gamma(i\omega) \geq 0$ for all real ω. Also $\Gamma(z)$ is relatively prime to $|A(z)| \cdot |A(-z)|$ since the zeros of $\Gamma(z)$ are not equal to $\pm\lambda_1$, or $\pm\lambda_2, \ldots,$ or $\pm\lambda_n$.

Because $\Gamma(z)$ is a real even polynomial and Re $\Gamma(i\omega) \geq 0$, there exists a real polynomial $\nu(z)$ with zeros only in Re $z \leq 0$ and

$$\Gamma(z) = \nu(z)\,\nu(-z).$$

Since the leading term of $\Gamma(z)$ is the same as that of $\eta(z)$, namely $(-1)^n \tau z^{2n}$, we can take $\nu(z)$ with leading term $\sqrt{\tau}\, z^n$.

Now divide $\nu(z)$ by $|A(z)|$. The quotient is $\sqrt{\tau}$ with a remainder called $-\mu(z)$,

$$\frac{\nu(z)}{|A(z)|} = -\frac{\mu(z)}{|A(z)|} + \sqrt{\tau}.$$

Here $\mu(z)$ is a real polynomial of degree less than n. Clearly $\mu(z) \not\equiv 0$ for in such a case

$$\nu(z) = \sqrt{\tau}\,|A(z)| \quad \text{and} \quad \Gamma(z) = \tau\,|A(z)| \cdot |A(-z)|,$$

which contradicts the fact that $\Gamma(z)$ is relatively prime to $|A(z)| \cdot |A(-z)|$. [If $\tau = 0$, then $\nu(z) = -\mu(z) \not\equiv 0$].

Define the required vector q by

$$q'\,A(z)^{-1}b = \frac{-\mu(z)}{|A(z)|}.$$

Since $\mu(z)$ and $|A(z)|$ are relatively prime, it follows from the theory of observability that (A, q') is completely observable. Define the positive definite matrix $B > 0$ by

$$A'B + B'A = -qq' - D$$

or

$$B = \int_0^\infty e^{A't}\{qq' + D\}e^{At}\,dt.$$

The rest of the proof will establish the final assertion

$$Bb + c' + \sqrt{\tau}\,q = 0.$$

Recall that

$$\nu(z)\,\nu(-z) = \Gamma(z) = \eta(z) - \alpha^2\,\theta_2(z)\,\theta_2(-z).$$

Dividing by $|A(z)| \cdot |A(-z)|$ we obtain

$$\left[\frac{-\mu(z)}{|A(z)|} + \sqrt{\tau}\right]\left[\frac{-\mu(-z)}{|A(-z)|} + \sqrt{\tau}\right] = \{\tau - c\,A(z)^{-1}b - b'\,A'(-z)^{-1}c'\} - b'\,A'(z)^{-1}gg'\,A(-z)^{-1}b.$$

Replace $-\mu(z)/|A(z)|$ by $q'\,A(z)^{-1}b$, evaluate on the imaginary axis $z = i\omega$, and simplify to obtain

$$c\{A(i\omega)^{-1} + A(-i\omega)^{-1}\}b + \sqrt{\tau}\,q'\{A(i\omega)^{-1} + A(-i\omega)^{-1}\}b = b'\,A'(i\omega)^{-1}\{-qq' - gg'\}\,A(-i\omega)^{-1}b.$$

Now use

$$A'B + B'A = -qq' - gg'$$

and the formulas

$$A = zI - A(z), \qquad A(z)^{-1}A = AA(z)^{-1} = z\,A(z)^{-1} - I.$$

Then compute

$$(c + \sqrt{\tau}\,q')\{A(i\omega)^{-1} + A(-i\omega)^{-1}\}b = -b'\,B\{A(i\omega)^{-1} + A(-i\omega)^{-1}\}b.$$

Thus

$$\operatorname{Re}\,(b'B + c + \sqrt{\tau}\,q')\,A(i\omega)^{-1} = 0.$$

But $A^{-1}(0) = -A^{-1}$ is real and nonsingular, so

$$b'B + c + \sqrt{\tau}\,q' = 0.$$

Therefore

$$Bb + c' + \sqrt{\tau}\,q = 0,$$

as required. Q.E.D.

We insert a brief corollary to display another form for the hypothesis of the above lemma.

COROLLARY. *Let A be a real $n \times n$ stability matrix and b and c real n-vectors. If*

$$T(i\omega) \geq 0 \quad \textit{for} \quad -\infty < \omega < \infty,$$

then

$$\operatorname{Re} T(z) \geq 0 \quad \textit{for} \quad \operatorname{Re} z \geq 0.$$

Proof. We note that $T(z) = \tau - 2c\,A(z)^{-1}b$ is a complex analytic function without any singularities in $\operatorname{Re} z \geq 0$, and so $\operatorname{Re} T(z)$ is a harmonic function in the right half-plane. But

$$\lim_{|z|\to\infty} T(z) = \tau \geq 0 \quad \text{for} \quad \operatorname{Re} z \geq 0,$$

and so Re $T(z)$ is nonnegative at infinity in the right half-plane. Also

$$\text{Re } T(i\omega) \geq 0 \quad \text{for} \quad -\infty < \omega < \infty.$$

If Re $T(z_0) < 0$ for some point z_0 in the right half-plane, then there would be a local maximum of the harmonic function $-$Re $T(z)$ somewhere in the region Re $z > 0$. This construction contradicts the principle of the maximum for harmonic functions. Therefore

$$\text{Re } T(z) \geq 0 \quad \text{for} \quad \text{Re } z \geq 0,$$

as required. Q.E.D.

The corollary shows that the hypothesis of the Lemma concerning $T(z)$ could be phrased

$$\text{Re } T(z) \geq 0 \quad \text{for} \quad \text{Re } z \geq 0,$$

rather than for $z = i\omega$, without any loss of generality; and the condition $T(z) \not\equiv 0$ is equivalent to $\tau + |c| > 0$ (see exercises). In this second formulation $T(z)$ is said to be a positive real function, and this class of functions has a classical theory within the theory of complex analytic functions. Also with this formulation the condition bears a close resemblance to the stability criterion for linear control laws, which we examined as motivation for the statement of the lemma and the following theorem.

THEOREM 10. *Consider the process in* R^n

$$(\text{Л}) \qquad \dot{x} = Ax + b\, u(\sigma), \qquad \sigma = cx$$

with control laws $u(\sigma)$ *in class* $\mathfrak{A}$, *where* A *is a real stability matrix,* b *and* c *are real vectors, and* (A, b) *is controllable. Assume that there exist non-negative real numbers* α *and* β *with* $\alpha + \beta > 0$ *such that the rational complex function*

$$T(z) = -(\alpha + \beta z)c\, A(z)^{-1}b$$

satisfies

$$\text{Re } T(i\omega) \geq 0 \quad \textit{for} \quad -\infty < \omega < \infty.$$

Then the origin is a stable critical point of Л *for each* $u(\sigma)$ *in* $\mathfrak{A}$. *If, in addition,* $\alpha > 0$ *then* Л *is globally asymptotically stable, and hence is absolutely stable.*

Proof. Use $zI = A(z) + A$ to write

$$T(z) = -\beta cb - 2\left(\frac{\beta cA + \alpha c}{2}\right) A(z)^{-1}b.$$

Since $\lim_{|z| \leftarrow \infty} A(z)^{-1} = 0$ and Re $T(i\omega) \geq 0$, we find that $-\beta cb \geq 0$. Note that $T(z) \not\equiv 0$, for otherwise $c\, A(z)^{-1}b \equiv 0$, which implies (see exercise below) that $c = 0$. But we exclude the case $c = 0$, since here the theorem is trivially true. Thus the lemma is applicable with

$$\tau = -\beta cb \quad \text{and} \quad k = \frac{\beta cA + \alpha c}{2} \quad \text{(replacing } c\text{).}$$

By the above lemma there exist symmetric matrices

$$B > 0 \quad \text{and} \quad D \geq 0 \quad \text{and vector } q$$

such that

$$A'B + B'A = -qq' - D, \qquad Bb + \left(\frac{\beta cA + \alpha c}{2}\right)' + \sqrt{\tau}\, q = 0.$$

Define the real function in R^{n+1}

$$V(x,\sigma) = x'Bx + \beta \int_0^\sigma u(s)\, ds \geq 0,$$

which is positive for $x \neq 0$, and

$$\lim_{|x| \to \infty} V(x, \sigma) = \infty.$$

Along any solution curve $x(t)$, $\sigma(t) = c\,x(t)$ of the system in R^{n+1}, for a fixed $u(\sigma) \in \mathfrak{A}$,

$$\dot{x} = Ax + b\,u(\sigma)$$
$$\dot{\sigma} = cAx + cb\,u(\sigma),$$

we compute the derivative of V to be

$$\dot{V} = x'\,B(Ax + b\,u(\sigma)) + (x'A' + u(\sigma)b')Bx + \beta\,u(\sigma)\cdot(cAx + cb\,u(\sigma)).$$

This yields

$$-\dot{V} = -x'(BA + A'B)x - (2b'B + \beta cA)x\,u(\sigma) - \beta cb\,u(\sigma)^2,$$

or

$$-\dot{V} = x'Dx + x'qq'x + 2\left(-Bb - \frac{\beta A'c' + \alpha c'}{2}\right)'x\,u(\sigma) + \alpha cx\,u(\sigma) + \tau\,u(\sigma)^2.$$

Then we compute

$$-\dot{V} = x'Dx + (\sqrt{\tau}\,u(\sigma) + q'x)^2 + \alpha\sigma\,u(\sigma) \geq 0.$$

Let x_0, $\sigma_0 = cx_0$ be an initial state in R^{n+1} and then $x(t)$ satisfies

$$x'Bx \leq V(x_0, cx_0)$$

and so $x(t)$ is defined on $0 \leq t < \infty$ and also the origin is stable.

Assume $\alpha > 0$. If $x(t)$ fails to approach the origin, it must come arbitrarily near to some point $\bar{x} \in R^n$ where

$$\lim_{t \to \infty} V(x(t), cx(t)) = V(\bar{x}, c\bar{x}) > 0.$$

But the solution $(\bar{x}(t), \bar{\sigma}(t))$ initiating at $(\bar{x}, c\bar{x})$ satisfies

$$V(\bar{x}(t), c\bar{x}(t)) < V(\bar{x}, c\bar{x}) \quad \text{for some} \quad t > 0,$$

and by continuity $V(x(t), cx(t)) < V(\bar{x}, c\bar{x})$, unless $\sigma(t) \equiv 0$. But in this exceptional case $\bar{x}(t)$ satisfies $\dot{x} = Ax$ or $\bar{x}(t) = e^{At}\bar{x}$, and $-\dot{V} = x'Dx + (q'x)^2 = 0$. But this implies that $q'e^{At}\bar{x} \equiv 0$. The observability of (A, q') then forces $\bar{x} = 0$, which is a contradiction. Therefore

$$\lim_{t\to\infty} x(t) = 0,$$

and the process Л is absolutely stable. Q.E.D.

The next theorem extends the Luré problem to nonlinear processes with nonlinear control laws $u(\sigma)$ in $\mathfrak{U}$. We prove that the algebraic conditions of Theorem 10 guarantee asymptotic stability about the origin for every $u(\sigma) \in \mathfrak{U}$. Of course, the asymptotic stability refers only to a neighborhood of the origin for this nonlinear process.

THEOREM 11. *Consider the process in R^n*

$$(\Sigma) \qquad \dot{x} = f(x, u) = Ax + bu + o(x, u) \quad \text{in} \quad C^1$$

with feedback data

$$\sigma = \sigma(x) = cx + o(x) \quad \text{in} \quad C^1$$

and with control laws $u(\sigma)$ in the class $\mathfrak{U}$. Assume A is a real stability matrix, b and c are real vectors, with (A, b) controllable. Assume there exist real numbers

$$\alpha > 0, \qquad \beta > 0 \quad \text{so} \quad \tau = -\beta cb > 0$$

and the rational function

$$T(z) = -(\alpha + \beta z)c\, A(z)^{-1}b$$

satisfies

$$\operatorname{Re} T(i\omega) \geq 0 \quad \text{on} \quad -\infty < \omega < \infty.$$

Then the process Σ is asymptotically stable about the origin for each $u(\sigma) \in \mathfrak{U}$.

Proof. The notation $o(x, u)$ indicates a term that has magnitude less than $\epsilon(|x| + |u|)$, for a prescribed $\epsilon > 0$, when $|x|$ and $|u|$ are suitably small. In other words, $A = (\partial f/\partial x)(0, 0)$, $b = (\partial f/\partial u)(0, 0)$, $c = (\partial \sigma/\partial x)(0)$. We use the matrices $B > 0$, $D \geq 0$, and vector $q \neq 0$ guaranteed by the hypotheses, as in Theorem 10.

Fix a control law $u(\sigma) \in \mathfrak{U}$ and define the real function in R^{n+1}

$$V(x, \sigma) = x'Bx + \beta \int_0^{\sigma} u(s)\, ds > 0,$$

except at $x = \sigma = 0$. Take an initial state $x_0 \neq 0$ in R^n and let $x(t)$ be the corresponding solution of

$$\dot{x} = f(x, u(\sigma(x))).$$

Define $\sigma(t) = \sigma(x(t))$. Compute the derivative of $V(t) = V(x(t), \sigma(x(t)))$ to obtain

$$-\dot{V} = x'Dx + (\sqrt{\tau}\, u + q'x)^2 + \alpha(\sigma - o(x))u + uo(x, u) + xo(x, u).$$

Thus there exists a number $a > 0$ for which

$$-\dot{V} > a(|u|^2 + |x|^2) - uo(x) - uo(x, u) - xo(x, u).$$

Therefore there exists a $\delta > 0$ such that for $|x| < \delta$ and $|u| < \delta$ we have

$$-\dot{V} > \frac{a}{2}(|x|^2 + |u|^2) > 0.$$

In fact, there exists a positive $\delta_1 < \delta$ such that $|x| < \delta_1$ implies $|u(\sigma(x))| < \delta$, and so $\dot{V}$ is negative along solutions in this domain.

Take x_0 so near the origin that the ellipsoid in R^n

$$x'Bx \leq V(x_0, \sigma(x_0))$$

lies entirely within the set $|x| < \delta_1$. Then $x(t)$ remains forever in this ellipsoid and the origin is a stable critical point. Also $V(t) = V(x(t), \sigma(x(t)))$ is monotonic decreasing towards zero. Therefore

$$\lim_{t \to \infty} x(t) = 0,$$

and the process Σ is asymptotically stable about the origin. Q.E.D.

As a final topic in stability theory we consider the correctness of the optimal-control problem itself. Is the optimal control problem well posed in the sense that small changes in the problem data produce only small changes in the optimal cost?

Let us consider a control problem in R^n

$$P = \{\mathcal{S}, C, X_0, X_1, \Omega\},$$

for the process

$$(\mathcal{S}) \qquad \dot{x} = A(x) + B(x)u$$

with cost

$$C(u) = \int_0^{t_1} [A^0(x) + B^0(x)u]\, dt,$$

where $A(x)$, $B(x)$, $A^0(x)$, $B^0(x)$ are in C^1 in R^n. The controllers are all measurable functions $u(t)$ on various subintervals $0 \leq t \leq t_1$ of a given finite interval $0 \leq t \leq T$, and $u(t) \subset \Omega$ for a compact convex restraint set $\Omega \subset R^m$. We seek to steer some point of the compact initial set X_0 to a point in the compact target set X_1 in R^n, with optimal minimal cost $C(u)$.

Another such control problem

$$\hat{P} = \{\hat{\mathcal{S}}, \hat{C}, \hat{X}_0, \hat{X}_1, \hat{\Omega}\},$$

with the same type of data, will be said to lie within a distance $\delta > 0$ of P in case:

$$|A(x) - \hat{A}(x)| + |B(x) - \hat{B}(x)| + |A^0(x) - \hat{A}^0(x)| + |B^0(x) - \hat{B}^0(x)| < \delta$$

whenever $|x| < 1/\delta$, and

$$\text{dist}\,(X_0, \hat{X}_0) + \text{dist}\,(X_1, \hat{X}_1) + \text{dist}\,(\Omega, \hat{\Omega}) < \delta$$

(using the Hausdorff distance between nonempty compact subsets of R^n and R^m). The distance between P and $\hat{P}$ is the infimum of all $\delta > 0$ for which these conditions hold, and thus we treat the set $\mathcal{T}$ of all such control problems as a metric space.

THEOREM 12. *Consider the control problem P in the space $\mathcal{T}$ specified in R^n by the process*

$$(\mathcal{S}) \qquad \dot{x} = A(x) + B(x)u$$

with cost

$$C(u) = \int_0^{t_1} [A^0(x) + B^0(x)u]\, dt$$

for $A(x)$, $A^0(x)$, $B(x)$, $B^0(x)$ in C^1 in R^n. The measurable controllers $u(t)$ on various intervals $0 \leq t \leq t_1 \leq T$ satisfy $u(t) \subset \Omega$ for a compact convex restraint set $\Omega \subset R^m$. The initial point $x_0 \in R^n$ is to be steered optimally to the target origin $x_1 = 0$.

Assume

(a) *there exists a uniform bound $|x(t)| \leq \beta$ for all responses of $\mathcal{S}$ initiating at $x(0) = x_0$ and controlled by any $u(t) \subset \Omega$ on $0 \leq t \leq T + 1$;*
(b) *there exists an admissible response of $\mathcal{S}$ steering x_0 to $x_1 = 0$ on $0 \leq t \leq t_1$ with cost less than that of any controller $u(t) \subset \Omega$ on $0 \leq t \leq \tau$ with $T \leq \tau \leq T + 1$;*
(c) *rank $[B, AB, A^2B, \ldots, A^{n-1}B] = n$ where $A(0) = 0$, $A = (\partial A/\partial x)(0)$, $B = B(0)$;*
(d) *Ω contains $u = 0$ in its interior.*

Then there exists a neighborhood $\mathcal{N}$ of P in $\mathcal{T}$ such that each control problem

$\hat{P} \in \mathcal{N}$ has an optimal controller $\hat{u}^*(t)$ on $0 \leq t \leq \hat{t}^* \leq T$ and the optimal cost $\hat{C}(\hat{u}^*)$ approaches $C(u^*)$ when $\hat{P}$ approaches P.

Proof. The first two hypotheses assure the existence of an optimal controller $u^*(t) \subset \Omega$ on $0 \leq t \leq t^*$ for the problem P. Moreover $C(u^*) < \inf C(u)$ for all controllers of P with $u(t) \subset \Omega$ on $0 \leq t \leq \tau$ with $T \leq \tau \leq T + 1$.

Now let $\hat{P}$ lie in some suitably small neighborhood $\mathcal{N}$ of P in the metric space $\mathcal{T}$. Routine estimates show that there is a uniform bound on all responses of $\hat{\mathcal{S}}$ for controllers $\hat{u}(t) \subset \hat{\Omega}$ on $0 \leq t \leq T + 1$. Thus, if $\hat{X}_0$ can be steered to $\hat{X}_1$ by a controller in $\hat{\Omega}$ on $0 \leq t \leq \hat{t} \leq T$, then $\hat{P}$ has an optimal controller $\hat{u}^*(t)$ on $0 \leq t \leq \hat{t}^* \leq T$.

Let $u^*(t)$ on $0 \leq t \leq t^*$ steer x_0 optimally to $x_1 = 0$ along a response of $\mathcal{S}$. Let $\hat{u}_1(t)$ be the nearest point in $\hat{\Omega}$ to $u^*(t)$ and let $\hat{x}_0 \in \hat{X}_0$ be very near to x_0. Then $\hat{u}_1(t)$ steers $\hat{x}_0$ to some point $\hat{x}_1$, very near the origin, along a response of $\hat{\mathcal{S}}$. By hypotheses (c) and (d) x_1 can be steered to any point in a neighborhood of the origin by small controllers in $\Omega \cap \hat{\Omega}$, along responses of $\mathcal{S}$. By the approximation techniques used in local controllability theory we find that $\hat{x}_1$ can be steered to meet $\hat{X}_1$, along a response of $\hat{\mathcal{S}}$, in a short time duration. Therefore, for $\mathcal{N}$ sufficiently small there exists an optimal controller $\hat{u}^*(t)$ on $0 \leq t \leq \hat{t}^* < T$ for each problem $\hat{P} \in \mathcal{N}$.

By the above construction we find that $\hat{C}(\hat{u}^*) < C(u^*) + \epsilon$ for a prescribed $\epsilon > 0$, provided $\mathcal{N}_1 \subset \mathcal{N}$ is suitably small. Now suppose that $\hat{C}(\hat{u}^*) < C(u^*) - \epsilon$ for some $\hat{P} \in \mathcal{N}_1$, no matter how small the neighborhood $\mathcal{N}_1$ is chosen about P. Then, using the controller $\tilde{u}(t) \subset \Omega$ nearest to $\hat{u}^*(t)$, the point x_0 can be steered near to $x_1 = 0$ in the time $0 \leq t \leq \hat{t}^* \leq T$ and thence steered just to $x_1 = 0$ by some controller $\tilde{\tilde{u}}$ on $0 \leq t \leq T + 1$ with cost $C(\tilde{\tilde{u}}) < C(u^*) - \epsilon/2$. But this contradicts hypothesis (b), and so $C(\hat{u}^*) \geq C(u^*) - \epsilon$ whenever $\mathcal{N}_1$ is sufficiently small. Therefore $\hat{C}(\hat{u}^*)$ varies continuously with $\hat{P}$ at the point $P \in \mathcal{T}$. Q.E.D.

COROLLARY 1. *Let $P^{(k)}$ be a sequence of control problems in $\mathcal{N} \subset \mathcal{T}$ with*

$$\lim_{k \to \infty} P^{(k)} = P.$$

Then there exists a subsequence $P^{(k_i)}$, for which $t^*_{(k_i)} \to t^*$ and

$$\lim_{i \to \infty} u^*_{(k_i)}(t) = u^*(t) \quad \text{weakly}$$

and

$$\lim_{i \to \infty} x^*_{(k_i)}(t) = x^*(t) \quad \text{uniformly,}$$

on each compact subinterval of $0 \leq t < t^*$. Here $u^*(t) \subset \Omega$ on $0 \leq t \leq t^*$ is an optimal controller of P with response $x^*(t)$.

Proof. Since all $t^*_{(k)}$ lie on the compact interval, a subsequence $t^*_{(k_i)}$ converges to some instant $t^* \leq T$. Since all $u^*_{(k_i)}(t) \subset \Omega_{(k_i)}$ and $\Omega_{(k_i)} \to \Omega$, the $u^*_{(k_i)}(t)$ are uniformly bounded on a compact duration $0 \leq t \leq \bar{t} < t^*$ and a subsequence converges weakly thereon. Note that all $u^*_{(k_i)}(t)$ are defined on $0 \leq t \leq \bar{t}$, for large k_i, and that we continue to write the subsequence as $u^*_{(k_i)}(t)$ with the weak limit $u^*(t)$. Clearly $u^*(t) \subset \Omega$ (almost always) since, if $u^*(t)$ lay for a positive duration on the external side of one of the countable supporting planes that determine Ω, this would contradict the assertion that $u^*_{(k_i)}(t) \subset \Omega_{(k_i)} \to \Omega$.

The corresponding responses $x^*_{(k_i)}(t)$ are uniformly bounded with uniformly bounded derivatives, and so a subsequence (still denoted by k_i) converges uniformly on each subinterval $0 \leq t \leq \bar{t} < t^*$ to some limit function $x^*(t)$. Using the methods of the basic existence theorems of Chapter 4, we find that $u^*(t)$ on $0 \leq t \leq t^*$ is an admissible controller for $\mathcal{S}$ steering x_0 to $x_1 = 0$ along the response $x^*(t)$. Also $C_{(k_i)}(u^*_{(k_i)}) \to C(u^*)$, and hence $u^*(t)$ is an optimal controller for $\mathcal{S}$. Q.E.D.

COROLLARY 2. *Consider the control problem P in the space $\mathcal{T}$ specified in R^n by the process*

$$(\mathcal{S}) \qquad \dot{x} = A(x) + B(x)u$$

with cost

$$C(u) = t_1.$$

Under the hypotheses of the theorem, there exists a neighborhood $\mathcal{N}$ of P in $\mathcal{T}$ such that each control problem $\hat{P} \in \mathcal{N}$ has an optimal minimal cost $\hat{t}^* < T$, and $\hat{t}^*$ approaches t^* when $\hat{P}$ approaches P.

EXERCISES

1. Consider the attitude control of a rotating rigid body

$$I_1\dot{\omega}_1 = (I_2 - I_3)\omega_2\omega_3 + u_1(t)$$

$$I_2\dot{\omega}_2 = (I_3 - I_1)\omega_3\omega_1 + u_2(t)$$

$$I_3\dot{\omega}_3 = (I_1 - I_2)\omega_1\omega_2 + u_3(t)$$

where I_1, I_2, I_3 are the principal moments of inertia and $\omega_1, \omega_2, \omega_3$ are the corresponding components of the angular velocity. With the control restraint $|u_i(t)| \leq 1$, construct a Liapunov function to prove that any given initial state can be steered towards the origin.

2. Consider the autonomous differential system

$$(\mathcal{S}) \qquad \dot{x} = f(x) \quad \text{in} \quad C^1 \quad \text{in} \quad R^n.$$

Assume that $V(x)$ in C^1 in R^n satisfies

(a) $V(x) \geq 0$ in R^n and $V(x) = 0$ only at $x = 0$;

(b) $\lim_{|x| \to \infty} V(x) = \infty$;

(c) $(\partial V/\partial x^i) f^i(x) \leq \theta(x, V)$ in C^1 in R^{n+1}

where every solution of the scalar differential equation $\dot{V} = \theta(x, V)$ tends to zero as $t \to +\infty$.

Prove $\mathcal{S}$ is globally asymptotically stable about the origin.

3. Consider the differential system

$$(\mathcal{S}) \qquad \dot{x} = f(x) \quad \text{in} \quad C^1 \quad \text{in} \quad R^n.$$

Let D be a compact set in R^n, which is future-invariant (every solution initiating in D remains always in D). Let $V(x)$ be a real C^1 function in D with $(\text{grad } V) f(x) \leq 0$. Let E be the maximal invariant set in D whereon $(\text{grad } V) f(x) = 0$. Prove every solution of $\mathcal{S}$ in D approaches E as $t \to +\infty$.

4. Consider the control process in R^n

$$(\mathcal{S}) \qquad \dot{x} = f(x, u) \quad \text{in} \quad C^1 \quad \text{in} \quad R^{n+m}$$

for u in a compact set $\Omega \subset R^m$. Show that the ball of radius $\rho > 0$,

$$D(\rho) = \{x \mid x'x \leq \rho^2\},$$

is future invariant for all controllers if

$$\max_{\substack{u \in \Omega \\ x'x = \rho^2}} x' f(x, u) \leq 0.$$

5. Show that the substitutions

$$\bar{A} = P^{-1}AP, \qquad \bar{B} = P'BP, \qquad \bar{C} = P'CP, \qquad \bar{b} = P^{-1}b, \qquad \bar{c} = cP,$$

where P is an orthogonal matrix, preserve the stability criterion of Theorem 9.

6. In theorem 9 show that the hypothesis corresponding to $\alpha = 0$ is vacuous since

$$-\dot{V} = -(2Bx + uc')'(Ax + bu)$$

is not positive definite in (x, u).

But assume

$$-cb = (Bb + \tfrac{1}{2}A'c')' \, C^{-1} \, (Bb + \tfrac{1}{2}A'c')$$

and assume that the system of equations

$$Ax + bu(cx) = 0$$

has only the solution $x = 0$, for all $u \in \mathcal{U}$. Then show that r is absolutely stable.

7. Consider the linear autonomous open-loop process

$$\dot{x} = Ax + Bu$$

where x and u are in R^n. Now consider a closed-loop feedback yielding the process

$$\dot{x} = Ax + B(u - x).$$

Let the transfer matrix of the open loop be

$$T(z) = (z - A)^{-1}B$$

and the transfer matrix of the closed loop be

$$T_c(z) = [z - (A - B)]^{-1}B.$$

Prove the *Nyquist formula*

$$T_c(z) = [I + T(z)]^{-1}\, T(z).$$

Show that the closed loop is stable [that is, $(A - B)$ is a stability matrix] if $T_c(z)$ has n poles in the left half-plane.

8. Let A be a real stability $n \times n$ matrix, b and c real vectors with (A, b) controllable, and

$$c\, A(z)^{-1}b + c\, A(-z)^{-1}b \equiv 0.$$

Prove, as needed in the lemma of Theorem 10, that $c = 0$. [Hint: For large $|z|$ power-series analysis yields

$$cAb = 0,\ cA^3b = 0, \ldots,\ cA^{2n-1}b = 0.$$

Thus, if (A^2, b) is controllable, it follows that $c = 0$. But (A, b) is controllable, which is equivalent to the nonvanishing of certain components of $\hat{b}$ when $\hat{A}$ is reduced to complex Jordan canonical form (see Exercise 7 of Section 2.3) and from this Jordan form it is easy to prove that (A^2, b) is controllable.]

9. Consider the control process in R^n

$$(\mathcal{S}) \qquad \dot{x} = A(x) + B(x)u$$

with cost

$$C(u) = \int_0^T [A^0(x) + B^0(x)u]\, dt$$

for measurable controllers $u(t)$ on the fixed finite duration $0 \leq t \leq T$ lying in the compact convex restraint set $\Omega \subset R^m$. Assume $A(x)$, $B(x)$, $A^0(x)$, $B^0(x)$ are in C^1 in R^n and the given initial state x_0 lies in a closed constraint set $\Delta \subset R^n$. We seek an optimal controller $u^*(t)$ on $0 \leq t \leq T$ with response $x^*(t) \subset \Delta$. We solve this bounded phase coordinate problem by the method of *penalty functions* (see Exercise 12 in Section 3.4).

Let $F(x)$ be a real continuous function that vanishes on Δ but is positive outside Δ. For each real $\lambda > 0$ let $u_\lambda^*(t) \subset \Omega$ on $0 \leq t \leq T$ be an optimal controller for $\mathcal{S}$ with cost

$$C_\lambda(u) = \int_0^T [A^0(x) + B^0(x)u + \lambda\, F(x)]\, dt,$$

without regard to the phase constraint Δ. Assume

(a) there is a uniform bound $|x(t)| \leq \beta$ for all responses to $\mathcal{S}$ for controllers in Ω;

(b) $|C_\lambda(u_\lambda^*)|$ are uniformly bounded for all real $\lambda > 0$.

Show that there is a subsequence $\lambda_i \to \infty$ such that

$$u_{\lambda_i}^*(t) \to u^*(t) \quad \text{weakly,}$$

$$x_{\lambda_i}^*(t) \to x^*(t) \quad \text{uniformly,}$$

and

$$C_{\lambda_i}(u_{\lambda_i}^*) \to C(u^*).$$

Here $u^*(t)$ is an optimal controller for the bounded phase coordinate problem with the response $x^*(t) \subset \Delta$ (which is thereby proved to exist).

10. Consider a linear autonomous process in R^n

$$(\mathcal{L}) \qquad \dot{x} = Ax + Bu$$

with measurable controllers $u(t)$ on various finite time durations $0 \leq t \leq t_1$ in a compact convex polytope $\Omega \subset R^m$, containing $u = 0$ in its interior. We seek to steer the initial state x_0 to $x_1 = 0$ in minimal time. Assume there exists an optimal controller $u^*(t)$ on $0 \leq t \leq t^*$ and that $\mathcal{L}$ is normal so that $u^*(t)$ is unique.

A controller $\hat{u}(t) \subset \Omega$ on $0 \leq t \leq \hat{t}$ is called δ-suboptimal in case $\hat{t} \leq t^* + \delta$ and $|\hat{x}(\hat{t})| \leq \delta$. Show that for a given $\delta > 0$ there exists $\epsilon > 0$ such that: if there exist absolutely continuous functions $x(t)$, $\eta(t)$ satisfying

(a) $|\dot{x}(t) - A\,x(t) - B\,\hat{u}(t)| < \epsilon$,

(b) $|x(0) - x_0| < \epsilon$ and $|x(\hat{t})| < \epsilon$,

(c) $|\dot{\eta}(t) + \eta(t)A| < \epsilon$,

(d) $|\eta(t)B\,\hat{u}(t) - \max_{u\in\Omega} \eta(t)Bu| < \epsilon$ almost always,

(e) $\eta(t)$ is nonvanishing,

(f) $\eta(t)Bu$ assumes its maximum on an entire edge of Ω for only a finite set of times,

then $\hat{u}(t)$ is δ-suboptimal, that is,

$$\hat{t} \le t^* + \delta \quad \text{and} \quad |\hat{x}(\hat{t})| < \delta.$$

11. Consider the differential system in R^n

$$\dot{x} = f(x) + w(t)$$

with continuous disturbances $w(t)$ on $0 \le t < \infty$. The system is stable (at the origin) for permanent disturbances in case: for each $\epsilon > 0$, there exists $\delta > 0$ such that

$$|x_0| < \delta \quad \text{and} \quad |w(t)| < \delta \quad \text{on} \quad 0 \le t < \infty$$

implies that

$$|x(t)| < \epsilon \quad \text{on} \quad 0 \le t < \infty.$$

Now consider the control process in R^n

$$(\mathrm{S}) \qquad \dot{x} = f(x, u) \quad \text{in} \quad C^1 \quad \text{in} \quad R^{n+m}$$

with $f(0, 0) = 0$ and rank $[B, AB, \ldots, A^{n-1}B] = n$ where $A = f_x(0, 0)$, $B = f_u(0, 0)$. Show that there exists a linear feedback control

$$u = Dx$$

such that

$$\dot{x} = f(x, Dx) + w(t)$$

is stable for permanent disturbances.

CHAPTER 7

Synthesis of Optimal Controllers for Some Basic Nonlinear Control Processes

In this chapter applications of the previously developed general theory of optimal control will be made to a number of control problems of technology and science. The synthesis is in terms of specifying the feedback controller, when this can be done, or by specifying a routine numerical problem from which the control function, perhaps as a feedback controller, can be determined. More extensive discussions of computational methods for handling the two-point boundary-value problem of optimal control can be found in Appendix A and in the references.

In Appendix A we give a brief discussion of techniques available for determining the optimal controller by the use of a computer machine, that is, by solving the remaining two-point boundary-value problem that arises in application of the maximal principle. This gives us an open-loop (see Chapter 1) control function for the given initial condition. One technique for obtaining a feedback controller synthesis from knowledge of open-loop controllers is to measure the current control process state and then compute very rapidly for the open-loop control function. The first portion of this function is then used during a short time interval, after which a new measurement of the process state is made and a new open-loop control function is computed for this new measurement. The procedure is then repeated. In this way external disturbances and other unknowns are taken into account in much the same way as is done by a feedback controller. If no disturbances or other unknowns are encountered, the recomputed control function should agree with the appropriate portion of the previously computed controller. This is essentially the principle of optimality [Bellman] in the theory of dynamic programming, a feedback principle.

Principle of Optimality

An optimal-control policy has the property that, whatever the initial state and initial control policy, the remaining control policy (that is, the policy after a short lapse of time) must constitute an optimal policy with regard to the state that results from use of the initial policy during the short lapse of time.

There are control-decision problems that, when directly stated, would not satisfy this principle. For example, when considering the problem of minimal time control to the origin, with a constraint on the average energy used, after a short time duration along the optimal path the average may no longer be available. This shortcoming is easily removed by adding another coordinate to the problem. We now turn to problems of directly constructing the feedback controller synthesis.

Many of the peculiarities of nonlinear control problems appear in systems of only second order, and there are many systems of technology and science whose evolution can be summarized as a solution of an ordinary second-order differential equation. The synthesis of optimal feedback controllers for nonlinear second-order processes with one degree of freedom is considered in the Section 7.1 without reference to specific control problems.

Section 7.2 contains a classical example from the field of rocketry, the problem of propelling a sounding rocket to a prescribed altitude above the earth with the least amount of fuel.

Section 7.3 is an application having to do with the angular velocity control of a space vehicle and the Section 7.4 is an application to the problem of optimal guidance between planetary orbits.

Further applications can be found in available books in the fields of economics, chemical engineering, operations research, and so on. Of course the theory has application in all problems concerned with decision making in influencing some dynamical process, say a manager and his plant or a governor and his state. Generally the ingredient lacking in the application to such problems is an adequate mathematical model for the controlled process. With the efficient computational machines now available more and more applications will be made to such problems once an understanding of the basic dynamics is achieved.

The journals of all of the technical societies (for example the AIAA, IEEE, ASME, AIChE, ORS, SIAM *Journal on CONTROL*) and the publications, such as *Automation and Remote Control*, contain many examples of the application of our theory.

7.1 SYNTHESIS OF TIME-OPTIMAL FEEDBACK CONTROLLERS FOR SECOND-ORDER NONLINEAR SYSTEMS WITH ONE DEGREE OF FREEDOM

Consider the single-degree-of-freedom differential equation

$$\ddot{x} + f(x, \dot{x}) = u$$

or the equivalent system

$$(S) \qquad \begin{aligned} \dot{x} &= y \\ \dot{y} &= -f(x, y) + u. \end{aligned}$$

Assume that the function $f(x, y)$ is in the class C^1 in the real number plane R^2, and that the control variable u is restricted to the compact interval Ω: $-1 \leq u \leq 1$.

Definition. For given initial data (x_0, y_0) at time $t = 0$ define Δ as the class of all measurable controllers on various finite time interval $0 \leq t \leq t_1$ with $u(t)$ lying in Ω such that the response $(x(t), y(t))$ is defined on $0 \leq t \leq t_1$ with $x(0) = x_0$, $y(0) = y_0$, and that it first reaches the origin $(0, 0)$ at $t = t_1$. The response $(x(t), y(t))$ is an absolutely continuous solution of

$$\begin{aligned} \dot{x} &= y \\ \dot{y} &= -f(x, y) + u(t) \end{aligned}$$

with $x(0) = x_0$, $y(0) = y_0$.

A controller $u(t)$ on $[0, t_1]$ in Δ is called (minimal time) optimal if for each $\hat{u}(t)$ in Δ on $[0, \hat{t}_1]$ we find $t_1 \leq \hat{t}_1$. By the maximal principle [Theorem 2 of Chapter 5] it is known that an optimal controller is necessarily a maximal controller. That is,

1. there exists an adjoint response, a nowhere-vanishing, absolutely continuous vector $\boldsymbol{\eta}(t) = (\eta_1(t), \eta_2(t))$ on $[0, t_1]$ such that $\mathbf{x}(t) = (x(t), y(t))$, $\boldsymbol{\eta}(t)$, and $u(t)$ satisfy the Hamiltonian system

$$\dot{x} = \frac{\partial H}{\partial \eta_1} \qquad \dot{\eta}_1 = -\frac{\partial H}{\partial x}$$

$$\dot{y} = \frac{\partial H}{\partial \eta_2} \qquad \dot{\eta}_2 = -\frac{\partial H}{\partial y}.$$

2. $H(\eta_1(t), \eta_2(t), x(t), y(t), u(t)) = M(\eta_1(t), \eta_2(t), x(t), y(t))$ for almost all t on $[0, t_1]$. Here

$$H(\eta_1, \eta_2, x, y, u) = \eta_1 y - \eta_2 f(x, y) + \eta_2 u$$

and

$$M(\eta_1, \eta_2, x, y) = \max_{-1 \leq u \leq 1} H(\eta_1, \eta_2, x, y, u).$$

3. $0 \leq M(\eta_1(0), \eta_2(0), x(0), y(0)) = M(\eta_1(t), \eta_2(t), x(t), y(t))$ for all t on $[0, t_1]$.

We want to construct a function $\Psi(x, y)$ in the differential system

$$\dot{x} = y$$
$$\dot{y} = -f(x, y) + \Psi(x, y)$$

so that each of its responses $(x(t), y(t)) \to (0, 0)$ in the minimal time interval. Generally we find controllers $u(t)$ for the system ($\mathcal{S}$) which steer (x_0, y_0) to $(0, 0)$ time optimally and then indicate how the controller $u(t)$, so found for each initial condition (x_0, y_0), can be used to construct the *feedback controller synthesis*, which we call Ψ (x, y).

We consider first the time-optimal problem, as indicated above, finding controllers $u(t) \subset \Omega$ steering (x_0, y_0) to the target $(0, 0)$ along solutions $(x(t), y(t))$ of $\mathcal{S}$ in the minimum time interval. The required feedback controller synthesis $\Psi(x, y)$ is then accomplished by following backwards in time along such solution curves and marking where $u(t)$ changes magnitude in the (x, y)-plane. A switching boundary can usually be found since maximal controllers assume only the values $+1$ and -1 and there is a certain smoothness associated with the changes in magnitude of the controller as viewed in the (x, y)-plane. Two examples indicate the method of construction of $\Psi(x, y)$ in detail. We now show that an optimal controller assumes essentially only the two values $+1$, and -1.

Maximal Controllers are Relay Controllers

Consider the control problem for the system

$$(\mathcal{S}) \qquad \begin{aligned} \dot{x} &= y, \\ \dot{y} &= -f(x, y) + u, \qquad f(x, y) \in C^1, \qquad \Omega\colon -1 \leq u \leq 1 \end{aligned}$$

with measurable controllers $u(t) \in \Delta$ steering an initial point (x_0, y_0) to the origin, as discussed above. The response of $\mathcal{S}$ corresponding to a maximal controller is called a maximal response.

Definition. A controller $u(t)$ on $0 \leq t \leq t_1$ in Δ is called a relay controller in case there exists a finite number of switching times

$$0 = \tau_0 < \tau_1 < \cdots < \tau_k = t_1$$

such that on each open interval $\tau_{i-1} < t < \tau_i$, $i = 1, \ldots, k$, $u(t)$ is constant and equals $+1$ or -1, and $u(t)$ switches values on successive intervals.

Using the maximal principle we obtain the following immediate result.

COROLLARY. *Let $u(t)$ on $0 \le t \le t_1$ in Δ be a maximal controller for the system $\mathcal{S}$, as defined above. Then $u(t)$ is almost everywhere equal to the relay control sgn $\eta_2(t)$. Furthermore the adjoint response $\eta_2(t)$ on $0 \le t \le t_1$ has only a finite number of zeros and each of these zeros is simple.*

Proof. Since $H(\boldsymbol{\eta}(t), \mathbf{x}(t), u(t)) = M(\boldsymbol{\eta}(t), \mathbf{x}(t))$ for almost all t on $0 \le t \le t_1$, we find that $u(t) = \operatorname{sgn} \eta_2(t)$ for almost all t. In fact, we can redefine $u(t)$ on a null set without changing the response, so that

$$u(t) = \operatorname{sgn} \eta_2(t) = \begin{cases} -1 & \text{if} \quad \eta_2(t) < 0 \\ 0 & \text{if} \quad \eta_2(t) = 0 \\ +1 & \text{if} \quad \eta_2(t) > 0. \end{cases}$$

Now the adjoint response $\boldsymbol{\eta}(t)$ satisfies

$$\dot{\eta}_1 = \eta_2 \frac{\partial f}{\partial x}, \qquad \dot{\eta}_2 = -\eta_1 + \eta_2 \frac{\partial f}{\partial y},$$

and so the nonvanishing vector solution $\boldsymbol{\eta}(t)$ is in class C^1 on $0 \le t \le t_1$ and satisfies the above linear differential system everywhere on that interval. If $\eta_2(t)$ had an infinite set of zeros on $0 \le t \le t_1$, then at an accumulation point $\bar{t}$ we find $\eta_2(\bar{t}) = 0$, $\dot{\eta}_2(\bar{t}) = 0$ so that $\eta_1(\bar{t}) = 0$, which is impossible. Hence $\eta_2(t)$ has only a finite number of zeros on $0 \le t \le t_1$. At each such zero $t = \tau$ we have $\eta_1(\tau) \ne 0$ so that $\dot{\eta}_2(\tau) \ne 0$ and $t = \tau$ is a simple zero of $\eta_2(t)$. Q.E.D.

Remark. Henceforth we shall always consider a maximal control $u(t)$ to be modified on a null set so as to be equal everywhere to the corresponding relay control,

$$u(t) = \operatorname{sgn} \eta_2(t).$$

On the closed interval between switching times we note that the response $\mathbf{x}(t)$, to the maximal control $u(t)$, is of class C^1.

The next theorem relates the switching times for a maximal response to the geometry of the phase plane. This result, which states that the zeros of $y(t)$ and $\eta_2(t)$ are interlaced, is basic in establishing the properties of the switching locus W needed for the construction of the feedback controller.

THEOREM 1. *Let $u(t)$ on $0 \le t \le t_1$ in Δ be a maximal controller for the system $\mathcal{S}$, as defined above. Let the corresponding response be $\mathbf{x}(t) = (x(t), y(t))$ with adjoint response $\boldsymbol{\eta}(t) = (\eta_1(t), \eta_2(t))$ on $0 \le t \le t_1$.*

Let ξ_1, ξ_2 be times on $0 \le \xi_1 < \xi_2 \le t_1$. Then four assertions hold:

1. *If $\eta_2(\xi_1) = \eta_2(\xi_2) = 0$ and if $y(\xi_1) = 0$, then $y(\xi_2) = 0$.*
2. *If $\eta_2(\xi_1) = \eta_2(\xi_2) = 0$ and if $y(\xi_1) \ne 0$, then $y(\xi_2) \ne 0$, but there is a zero of $y(t)$ on the open interval $\xi_1 < t < \xi_2$.*
3. *If $y(\xi_1) = y(\xi_2) = 0$, $y(t) \ne 0$ on $\xi_1 < t < \xi_2$, and if $\eta_2(\xi_1) = 0$, then $\eta_2(\xi_2) = 0$.*
4. *If $y(\xi_1) = y(\xi_2) = 0$, $y(t) \ne 0$ on $\xi_1 < t < \xi_2$, and if $\eta_2(\xi_1) \ne 0$, then $\eta_2(\xi_2) \ne 0$ but there is a zero of $\eta_2(t)$ on the open interval $\xi_1 < t < \xi_2$.*

Thus, provided the zeros of $y(t)$ are isolated, they either coincide with the zeros of $\eta_2(t)$ or else no zero of $y(t)$ is a zero of $\eta_2(t)$, but these two sets of zeros are interlaced.

Proof. Assume $\eta_2(\xi_1) = \eta_2(\xi_2) = 0$. Now

$$M(\boldsymbol{\eta}(0), \mathbf{x}(0)) = M(\boldsymbol{\eta}(t), \mathbf{x}(t)) = \eta_1 y - \eta_2 f(x, y) + |\eta_2| \ge 0$$

for all t on $0 \le t \le t_1$. Then $\eta_1(\xi_1)\,\eta_1(\xi_2) < 0$ and

$$\eta_1(\xi_1)\,y(\xi_1) = \eta_1(\xi_2)\,y(\xi_2) \ge 0.$$

Thus $y(\xi_1) = 0$ if and only if $y(\xi_2) = 0$. If $y(\xi_1) \ne 0$, then $y(\xi_1)\,y(\xi_2) < 0$, and so there is a zero of $y(t)$ on $\xi_1 < t < \xi_2$. Therefore assertions (1) and (2) are proved.

Next assume that $y(\xi_1) = y(\xi_2) = 0$ and $y(t) \ne 0$ on $\xi_1 < t < \xi_2$. Suppose $\eta_2(\xi_1) = 0$. Then we have already shown that $\eta_2(t)$ does not vanish on $\xi_1 < t < \xi_2$. Thus on the closed interval $\xi_1 \le t \le \xi_2$ we have $y(t) \in C^2$ where the derivatives at the endpoints are evaluated from the interior of the interval. On this closed interval we verify

$$\frac{d}{dt}[y\eta_1 + \dot{y}\eta_2] = 0.$$

Therefore

$$\dot{y}(\xi_1)\,\eta_2(\xi_1) = \dot{y}(\xi_2)\,\eta_2(\xi_2).$$

Now $y(t)^2 + \dot{y}(t)^2 \ne 0$ on $\xi_1 < t < \xi_2$, for otherwise the uniqueness property for the differential system $\mathcal{S}$ requires that $y(t) \equiv 0$ on the interval. Since $\eta_2(\xi_1) = 0$ we find that $\eta_2(\xi_2) = 0$. Thus assertion (3) is proved.

Finally assume $y(\xi_1) = y(\xi_2) = 0$, $y(t) \ne 0$ on $\xi_1 < t < \xi_2$, and $\eta_2(\xi_1) \ne 0$. By the above argument $\eta_2(\xi_2) \ne 0$. But if $\eta_2(t)$ vanishes nowhere on $\xi_1 \le t \le \xi_2$ then $\dot{y}(\xi_1)\,\dot{y}(\xi_2) > 0$, which is impossible since ξ_1 and ξ_2 are consecutive zeros of $y(t)$. Q.E.D.

Remark. In the specific cases analysed later in this section the zeros of $y(t)$ are isolated since there a switching time, for a maximal control, never occurs when the response is at a critical point of $\mathcal{S}_+$ or $\mathcal{S}_-$. Here $\mathcal{S}_+$ and $\mathcal{S}_-$ are the differential system $\mathcal{S}$ with controller $u \equiv +1$ and $u \equiv -1$ respectively.

COROLLARY. *Let $u(t)$ on $0 \le t \le t_1$ in Δ be a maximal controller for the system $\mathcal{S}$, as defined above. Let the corresponding response be $\mathbf{x}(t) = (x(t), y(t))$ with adjoint response $\boldsymbol{\eta}(t) = (\eta_1(t), \eta_2(t))$ on $0 \le t \le t_1$. At a time $t = \xi$ on $0 \le \xi \le t_1$ let $\eta_2(\xi) = 0$.*

$$\textit{If } y(\xi) > 0, \textit{ then } \dot{\eta}_2(\xi) < 0.$$
$$\textit{If } y(\xi) < 0, \textit{ then } \dot{\eta}_2(\xi) > 0.$$

Proof. Now $M(\boldsymbol{\eta}(t), \mathbf{x}(t)) = \eta_1 y - \eta_2 f(x, y) + |\eta_2| \ge 0$. If $\eta_2(\xi) = 0$, $y(\xi) > 0$, then $\eta_1(\xi) > 0$, and so $\dot{\eta}_2(\xi) = -\eta_1(\xi) < 0$. The other case is similar. Q.E.D.

Remark. This corollary establishes that changes in the relay maximal controllers are from $+1$ to -1 in $y > 0$, going forward in time along a maximal solution curve, and from -1 to $+1$ in $y \le 0$.

Domains of Controllability and the Existence of Optimal Controllers

Definition. Consider the differential system

$$(\mathcal{S}) \qquad \dot{x} = y, \qquad \dot{y} = -f(x, y) + u, \qquad f(0, 0) = 0, \qquad f(x, y) \in C^1,$$

where the control variable satisfies $-1 \le u \le 1$, as above. The set $\mathcal{C}$ of all points $(x_0, y_0) \in R^2$, for which there exists a measurable controller $-1 \le u(t) \le 1$ on a finite interval $0 \le t \le t_1$, steering (x_0, y_0) to the origin $(0, 0)$, is called the domain of (null) controllability.

Remark. We agree that $(0, 0)$ is in $\mathcal{C}$ and take the maximal controller to be $u(t) = 0$ for the time $t = 0$.

THEOREM 2. *For the system*

$$(\mathcal{S}) \qquad \dot{x} = y, \qquad \dot{y} = -f(x, y) + u, \qquad f(0, 0) = 0, \qquad f(x, y) \in C^1$$

with $-1 \le u \le 1$, the domain $\mathcal{C}$ of controllability is an open connected subset of R^2.

Proof. Compute the real constant matrices

$$A = \begin{pmatrix} 0 & 1 \\ -\dfrac{\partial f}{\partial x}(0, 0) & -\dfrac{\partial f}{\partial y}(0, 0) \end{pmatrix} \quad \text{and} \quad B = \begin{pmatrix} 0 \\ 1 \end{pmatrix}$$

as in Theorem 1 of Chapter 6. The criterion that establishes the desired conclusion is that B and AB are linearly independent. Thus the theorem follows. Q.E.D.

THEOREM 3. *Consider the system*

$$(\mathcal{S}) \qquad \dot{x} = y, \qquad \dot{y} = -f(x, y) + u, \qquad f(x, y) \in C^1$$

and $u(t)$ *measurable on* $-1 \leq u \leq 1$, *as above. Assume that* $f(x, y)$ *is an attractive force with nonnegative friction, that is,* $xf(x, 0) > 0$ *for* $x \neq 0$ *and* $(\partial f/\partial y)(x, y) \geq 0$ *in* R^2. *Then the domain of controllability is* $\mathcal{C} = R^2$. *Furthermore each point in* R^2 *can be steered to the origin by an optimal control* $u(t) \in \Delta$.

Proof. There exists a neighborhood N of the origin which lies in $\mathcal{C}$. Take a point P in $y > 0$ (the case $y < 0$ is similar, and $y = 0$ leads immediately to one of these cases) and we show that P can be steered along a solution S of $\mathcal{S}$ by some controller $u(t)$ until S meets N.

First follow the solution S_0 of

$$(\mathcal{S}_0) \qquad \dot{x} = y, \qquad \dot{y} = -f(x, y)$$

initiating at P until S_0 enters the first quadrant. Then apply the control $u(t) \equiv -1$ until P meets the positive x-axis at a point P_1. If the solution of $\mathcal{S}$ so generated meets N during these processes, the solution can then be steered to the origin (see Figure 7.1).

Next follow the solution of $\mathcal{S}_0$ through P_1 until the third quadrant is reached and then apply the controller $u(t) \equiv +1$ until the solution meets the negative x-axis at a point P_2. Let $S_0{}^1$ be the solution of $\mathcal{S}_0$ through P_2. Follow $S_0{}^1$ backwards in time to the first intersection point $P_1{}^1$ with the positive x-axis. If $S_0{}^1$ followed forward from P_2 fails to meet the positive x-axis, then we obtain a region bounded by $S_0{}^1$ and a vertical segment through $P_1{}^1$ that violates the area condition of Bendixson, since $df/dy \geq 0$. Thus $S_0{}^1$ must be a periodic solution of $\mathcal{S}_0$ or must spiral inwards, crossing the positive and negative x-axis alternately, towards N or towards a limit cycle of $\mathcal{S}_0$.

But as $S_0{}^1$ approaches the limit cycle of $\mathcal{S}_0$, an appropriate slight impulse from $u(t)$, and thereafter a return to a solution of $\mathcal{S}_0$, makes a solution S of $\mathcal{S}$, which penetrates the interior of the given limit cycle of $\mathcal{S}_0$. Now the limit cycles of $\mathcal{S}_0$ are linearly ordered by inclusion. Let Σ be the infimum of all limit cycles of $\mathcal{S}_0$ that can be approached by such a solution S of $\mathcal{S}$. But clearly Σ can also be penetrated by using an appropriate small impulse for $u(t)$. Thus the solution S of $\mathcal{S}$, through P, can be forced by an appropriate controller $u(t)$ to enter N and then to be steered to the origin.

To prove the existence of an optimal control in Δ for each initial point $P \in R^2$, we must show that solutions of $\mathcal{S}$, which initiate at P and which reach

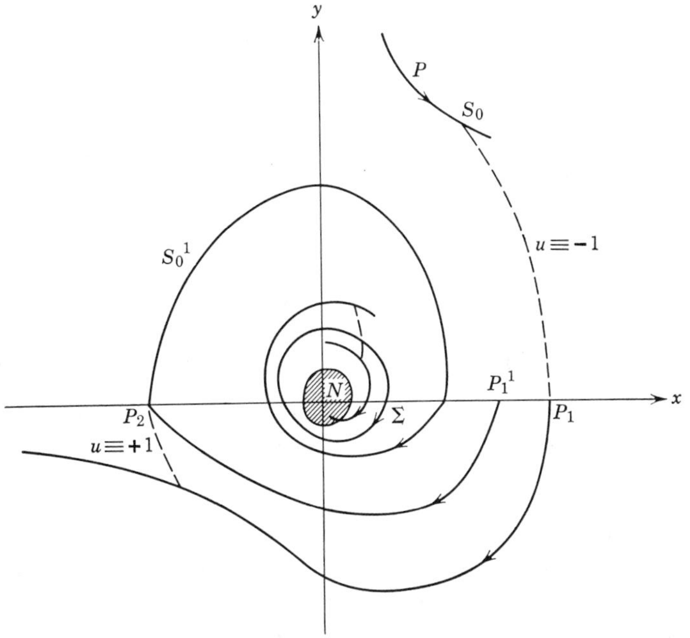

Figure 7.1 Construction of domain of controllability in the plane with attractive force and nonnegative friction.

the origin before the prescribed time $t_1 + 1$, cannot travel far from the origin, the boundedness condition of Theorem 4 of Chapter 4.

Let $g(x) = f(x, 0)$ and $G(x) = \int_0^x g(s)\, ds \geq 0$ and define

$$E(x, y) = \frac{y^2}{2} + G(x).$$

Then

$$\dot{E} = y[g(x) - f(x, y)] + u(t)y \leq u(t)y.$$

Thus

$$\dot{E} \leq |y| \leq \frac{y^2}{2} + 1 \leq E + 1,$$

and so

$$E(t) + 1 \leq [E(x_0, y_0) + 1)]e^{t_1+1}$$

along any solution initiating at $P = (x_0, y_0)$ and defined for a time subinterval of $0 \leq t \leq t_1 + 1$. Therefore

$$\frac{y^2}{2} \leq [E(x_0, y_0) + 1]e^{t_1+1}$$

and $|y(t)|$, and therefore $|x(t)|$, is bounded for all such solutions of $\mathcal{S}$. Q.E.D.

For the study of differential systems with a repulsive force we shall assume the following conditions, which could be relaxed slightly in the manner suggested by Theorem 3:

$$(\mathcal{S}_0) \qquad \dot{x} = y,$$

$$\dot{y} = -f(x, y), \qquad f(0, 0) = 0, \qquad f(x, y) \in C^1,$$

with $(\partial f/\partial x)(x, 0) < -\epsilon < 0$, $\partial f/\partial y \geq 0$, for some $\epsilon > 0$. This system has the origin as the unique critical point. Near the origin the solution-curve family of $\mathcal{S}_0$ is topologically equivalent [Hartman] to the solution-curve family of the linear system

$$(\mathcal{L}) \qquad \dot{x} = y,$$

$$\dot{y} = x - y.$$

Thus there are four solution curves of $\mathcal{S}_0$ which approach the origin as $t \to +\infty$ or as $t \to -\infty$. Call these curves I, II, III, IV, corresponding to the quadrants in which they lie. A study of the geometry of $\mathcal{S}_0$, shows that I and III are each defined and single valued over a half x-axis. Also II and IV are single valued over an interval of the x-axis and on these $|y| \to \infty$ as $|x|$ increases.

In fact, we find that the critical point and the curves I, II, III, IV are the only separatrices of $\mathcal{S}_0$ and the remainder of the solution-curve family of $\mathcal{S}_0$ consists of four parallel canonical regions [see Markus]. Thus $\mathcal{S}_0$ is globally homeomorphic to $\mathcal{L}$ in R^2. The separatrices II and IV are called the principal separatrices of $\mathcal{S}_0$ (see Figure 7.2).

THEOREM 4. *Consider the system*

$$(\mathcal{S}) \qquad \dot{x} = y,$$

$$\dot{y} = -f(x, y) + u, \qquad f(0, 0) = 0, \qquad f(x, y) \in C^1,$$

and $u(t)$ measurable on $-1 \leq u \leq 1$, as above. Assume $f(x, y)$ is a repulsive force with nonnegative friction, that is

$$\frac{\partial f}{\partial x}(x, 0) < -\epsilon < 0, \qquad \frac{\partial f}{\partial y}(x, y) \geq 0 \quad in \quad R^2$$

for some $\epsilon > 0$. Then each of $\mathcal{S}_+$ and $\mathcal{S}_-$ (see below for systems $\mathcal{S}_+$ and $\mathcal{S}_-$) is homeomorphic with the linear system $\mathcal{L}$ and has corresponding separatrices I_+, II_+, III_+, IV_+ and I_-, II_-, III_-, IV_-, as indicated above. The domain $\mathcal{C}$ of controllability of $\mathcal{S}$ is precisely the open topological band $\mathcal{B}$ bounded

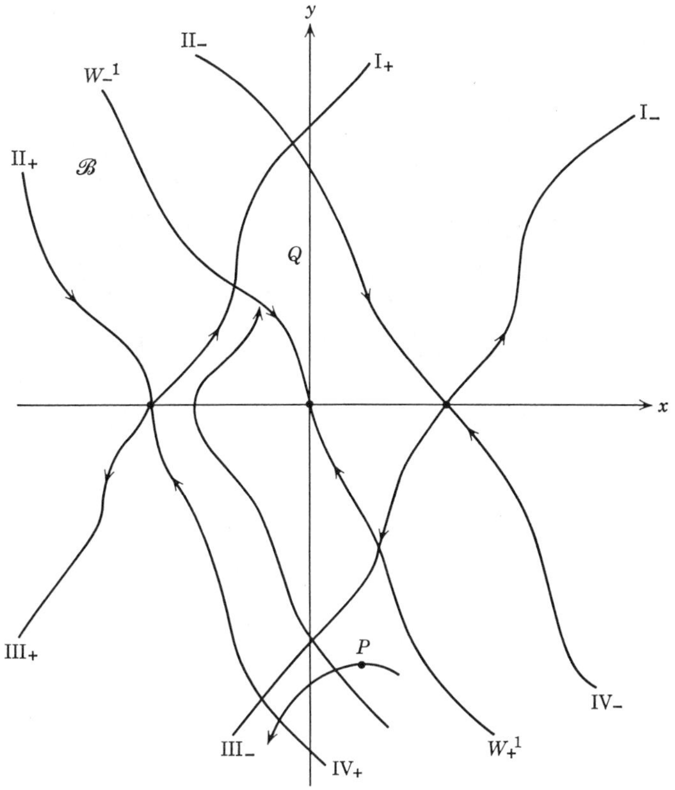

Figure 7.2 Domain of controllability with repulsive force and nonnegative friction.

by the principal separatrices II_+, IV_+ *and* II_-, IV_-, *and the two critical points of* $\mathcal{S}_+$ *and* $\mathcal{S}_-$. *Furthermore, each point in* $\mathcal{B}$ *can be steered to the origin by an optimal control* $u(t) \in \Delta$.

Proof. Each of

$$(\mathcal{S}_+) \qquad \dot{x} = y$$
$$\dot{y} = -f(x, y) + 1$$

and

$$(\mathcal{S}_-) \qquad \dot{x} = y,$$
$$\dot{y} = -f(x, y) - 1$$

has just one critical point and four other separatrices $\mathrm{I}_\pm$, $\mathrm{II}_\pm$, $\mathrm{III}_\pm$, $\mathrm{IV}_\pm$, and is homeomorphic of $\mathcal{L}$. Then, the principal separatrices II_+, IV_+, and II_-, IV_-, together with the two critical points of $\mathcal{S}_+$ and $\mathcal{S}_-$, bound an open band $\mathcal{B}$ that is homeomorphic to a plane strip between two parallel lines. Moreover the domain of controllability is $\mathcal{C} = \mathcal{B}$.

The separatrices I_+ and III_- bound a relatively closed curvilinear quadrilateral Q in $\mathcal{B}$ (see Figure 7.2). An initial point in Q, which is steered to the origin by a controller $u(t) \in \Delta$, can never leave Q in its sojourn to the origin. Furthermore an initial point P in $\mathcal{B}$, which is steered to the origin by a controller $u(t) \in \Delta$, can never leave the bounded set consisting of the union of Q and a subset of $\mathcal{B}$ bounded by (1) the solution of $\mathcal{S}_+$ and $\mathcal{S}_-$ through P, for $t \geq 0$, (2) boundary separatrices of $\mathcal{B}$, and (3) the edge of Q nearest P. Thus, in any case, P and its future trajectory lie in a bounded set of R^2. This a priori bound assures the existence of an optimal controller in Δ for each initial point P in $\mathcal{B} = \mathcal{C}$. Q.E.D.

The Switching Locus

Consider the optimal-control problem for

$$(\mathcal{S}) \qquad \dot{x} = y,$$
$$\dot{y} = -f(x, y) + u, \qquad f(0, 0) = 0, \qquad f(x, y) \in C^1,$$

and $u(t)$ measurable on $-1 \leq u \leq 1$, as above. If $u(t)$ is a maximal control in Δ, then $u(t) = \operatorname{sgn} \eta_2(t)$, where the adjoint response $\boldsymbol{\eta}(t) = (\eta_1(t), \eta_2(t))$ satisfies

$$\dot{\eta}_1 = \eta_2 \frac{\partial f}{\partial x},$$
$$\dot{\eta}_2 = -\eta_1 + \eta_2 \frac{\partial f}{\partial y}.$$

Definition. Consider the set of all maximal relay controllers in Δ, for the system $\mathcal{S}$, steering points of the domain $\mathcal{C}$ of controllability to the origin, as above. The switching locus W is the set of all points in $\mathcal{C}$ at which the corresponding responses $x(t)$ fail to have derivatives, that is, W consists of the points at which the maximal responses switch from the solution family of $\mathcal{S}_+$ to $\mathcal{S}_-$ or vice versa.

For definiteness we include the origin in W. We next describe a construction for W. Let W_+^1 be the solution (or segment of a solution) of $\mathcal{S}_+$ through the origin and lying in the fourth quadrant $x \geq 0, y \leq 0$. Let W_-^1 be the solution (or segment of a solution) of $\mathcal{S}_-$ through the origin and lying in the second quadrant $x \leq 0, y \geq 0$.

Now reflect W_+^1 through a conjugate interval along solutions of $\mathcal{S}_-$. That is, from each point on W_+^1 follow backwards in time along the corresponding solution of $\mathcal{S}_-$ for a time duration equal to the interval between zeros of $\eta_2(t)$. This means that we use

$$\dot{\eta}_1 = \eta_2 \frac{\partial f}{\partial x}(x(t), y(t)),$$

$$\dot{\eta}_2 = -\eta_1 + \eta_2 \frac{\partial f}{\partial y}(x(t), y(t))$$

with $\mathbf{x}(t) = (x(t), y(t))$ the appropriate solution of $\mathcal{S}_-$ and $\eta_1(0) = -1$, $\eta_2(0) = 0$. [Note that the adjoint linear system is homogeneous and that we start the response $\mathbf{x}(t)$ to the autonomous system $\mathcal{S}_-$ at $t = 0$.] We obtain the first zero of $\eta_2(t)$ in $t < 0$, which determines the time interval through which we follow the solution $\mathbf{x}(t)$ of $\mathcal{S}_-$, which initiates on W_+^1. Denote the endpoints of such solutions of $\mathcal{S}_-$ as the reflection W_-^2 of W_+^1.

Now reflect W_-^1 through a conjugate interval along solutions of $\mathcal{S}_+$. That is, from each point on W_-^1 follow backwards in time along the corresponding solution of $\mathcal{S}_+$ for a time duration equal to the interval between zeros of $\eta_2(t)$. This means that we use

$$\dot{\eta}_1 = \eta_2 \frac{\partial f}{\partial x}(x(t), y(t)),$$

$$\dot{\eta}_2 = -\eta_1 + \eta_2 \frac{\partial f}{\partial y}(x(t), y(t)),$$

with $\mathbf{x}(t) = (x(t), y(t))$ the appropriate solution of $\mathcal{S}_+$ initiating on W_-^1 and $\eta_1(0) = 1$, $\eta_2(0) = 0$, to obtain the first zero of $\eta_2(t)$ in $t < 0$. Denote the reflection of W_-^1 along S_+ by W_+^2.

If $W_+^1, W_+^2, \ldots, W_+^k$ and $W_-^1, W_-^2, \ldots, W_-^k$ have been defined, let W_+^{k+1} be the reflection of W_-^k by the solutions of $\mathcal{S}_+$ and let W_1^{k+1} be the reflection of W_+^k by solutions of $\mathcal{S}_-$ through conjugate time intervals as indicated above. Of course it might happen that $W_+^k \cup W_-^k$ is empty for every k greater than some positive integer.

THEOREM 5. *Consider the system*

$$(\mathcal{S}) \qquad \begin{aligned} \dot{x} &= y, \\ \dot{y} &= -f(x, y) + u, \qquad f(0, 0) = 0, \qquad f(x, y) \in C^1, \end{aligned}$$

with measurable controllers $u(t)$ in $-1 \leq u \leq 1$. The switching locus W is precisely the union of the sets $W_+^k \cup W_-^k$ for $k = 1, 2, 3, \ldots$, as described above.

Proof. Using Theorem 1 we note that for each point P on W_+^1 in $y < 0$ we can select an adjoint response $\boldsymbol{\eta}(t) = (\eta_1(t), \eta_2(t))$ with $\eta_2(t)$

vanishing when $\mathbf{x}(t) = P$ and nowhere else on W_+^1 in $y < 0$. Also if P is an endpoint of W_+^1 where $y = 0$ we can select $\boldsymbol{\eta}(t)$ so that $\eta_2(t)$ vanishes at times corresponding to $\mathbf{x}(t) = P$ and $\mathbf{x}(t) = 0$. Then each point of W_+^1, including the endpoints, can occur as a switching point for a response to a maximal relay controller in Δ. Thus $W_+^1 \subset W$ and similarly $W_-^1 \subset W$.

Since every maximal relay controller which steers a point of $\mathcal{C}$ to the origin must have a response which enters the origin along either W_+^1 or W_-^1, and since the response must follow alternately the solution curve families of $\mathcal{S}_+$ and $\mathcal{S}_-$, or vice versa, with switches occurring at the zeros of $\eta_2(t)$, we find that W is the union of all W_+^k and W_-^k for $k = 1, 2, 3, \ldots$.

Q.E.D.

Synthesis of an Optimal Controller for the Case of an Attractive Force

We consider here the synthesis of the optimal controller for a system with one degree of freedom subjected to an attractive force. Thus consider the system

$$(\mathcal{S}) \qquad \begin{aligned} \dot{x} &= y, \\ \dot{y} &= -f(x, y) + u, \qquad f(0, 0) = 0, \qquad f(x, y) \in C^1, \end{aligned}$$

with measurable controllers $u(t)$ on $-1 \leq u \leq 1$, and

$$\frac{\partial f}{\partial x}(x, 0) > \epsilon > 0, \qquad \frac{\partial f}{\partial y}(x, y) \geq 0,$$

for some $\epsilon > 0$. These hypotheses, somewhat stronger than those of Theorem 3, assure that each of the maximal systems $\mathcal{S}_+$ and $\mathcal{S}_-$ has exactly one critical point; and these hypotheses generally simplify the exposition of the problem. For example, every solution curve of $\mathcal{S}_\pm$ other than the unique critical point, is either a periodic solution or else spirals inwards towards a limit cycle or approaches the critical point as $t \to +\infty$. If, further $\partial f/\partial y > 0$, then the area condition of Bendixson states that there are no limit cycles.

THEOREM 6. *Consider the system*

$$(\mathcal{S}) \qquad \begin{aligned} \dot{x} &= y, \\ \dot{y} &= -f(x, y) + u, \qquad f(0, 0) = 0, \qquad f(x, y) \in C^2, \end{aligned}$$

with $(\partial f/\partial x)(x, 0) > \epsilon > 0$, *for some* $\epsilon > 0$, *and* $\partial f/\partial y \geq 0$ *in* R^2. *Consider maximal controllers* $u(t) \in \Delta$ *steering points of* $\mathcal{C} = R^2$ *to the origin and construct the switching locus* W *as the union of the sets* $W_+^k \cup W_-^k$, *as described in Theorem 5. Then* W *contains a homomorph of a line, a piecewise*

C^1 curve that separates the plane. Also W lies in $y \leq 0$ for $x \geq 0$ and in $y \geq 0$ for $x \leq 0$.

Proof. Now W is exactly the union of all $W_+^k \cup W_-^k$ for $k = 1, 2, 3, \ldots$. Certainly W_+^1 and W_-^1 are curves in C^2. Because the conjugate time interval used in the reflection process is a C^1 function of the initial conditions, we find that W_+^2 and W_-^2 are C^1 curves, and similarly W_+^k and W_-^k for $k = 1, 2, 3, \ldots$ are C^1 curves, as functions of the initial abscissas on W_+^1 and W_-^1.

We next show that W contains a line homeomorph that separates the plane. For simplicity of exposition we first consider the conservative case where $\partial f/\partial y \equiv 0$ in R^2. Here each solution curve of $\mathcal{S}_\pm$ other than the critical point is a periodic solution containing the critical point in its interior. Also it is easy to see that each such periodic solution is a convex closed curve that is symmetric about the x-axis. For in $y > 0$ we compute

$$\frac{\dot{y}}{\dot{x}} = y' = \frac{dy}{dx} = \frac{-f(x) + u}{y}$$

and

$$\frac{d^2y}{dx^2} = \frac{-yf'(x) - [f(x) - u]^2/y}{y^2} < 0.$$

Similarly in $y < 0$ we compute $y'' > 0$, and so the convexity and also the symmetry are easily verified.

Let S_+^1 be the solution of $\mathcal{S}_+$ through the origin so that W_+^1 is an arc of S_+^1. Call R_1 the right x-intercept of W_+^1. Let S_-^1 be the solution of $\mathcal{S}_-$ through the origin with subarc W_-^1 and left x-intercept L_1. Let S_-^2 be the solution of $\mathcal{S}_-$ through R_1 and S_+^2 the solution of $\mathcal{S}_+$ through L_1 (see Figure 7.3).

Let S_+^k and S_-^k, $k = 1, 2, \ldots, 1$, be defined as above as solutions of $\mathcal{S}_+$ and $\mathcal{S}_-$, respectively, having right and left x-intercepts R_k and L_k. Then S_-^{k+1} is the solution of $\mathcal{S}_-$ through R_k and S_+^{k+1} is the solution of $\mathcal{S}_+$ through L_k. It is easy to see that S_+^{k+1} contains S_+^k and that S_-^{k+1} contains S_-^k. Also

$$0 < R_1 < R_2 < \cdots < R_k \to +\infty$$

and

$$0 > L_1 > L_2 > \cdots > L_k \to -\infty.$$

Now we shall describe W for each segment $[0, R_1]$, $[R_1, R_2], \ldots$, $[R_k, R_{k+1}], \ldots$, so as to form a continuous curve for $x \geq 0$, lying in $y < 0$ except at the vertices $0, R_1, R_2, \ldots, R_k, \ldots$. Also we shall define W for $x \leq 0$ for each $[0, L_1], \ldots, [L_k, L_{k+1}], \ldots$ in $y > 0$ except at the vertices $0, L_1, L_2, \ldots, L_k, \ldots$.

On $[0, R_1]$ take W to be W_+^1, the arc of S_+^1 in $y \leq 0$. By Theorem 1, W_+^2 is a C^1 curve joining R_1 to R_2 and lying between S_+^1 and S_+^2 in

$y \leq 0$. Similarly W_-^2 is a C^1 curve joining L_1 to L_2 and lying between S_-^1 and S_-^2 in $y \geq 0$. By induction we find that W_+^k is a C^1 curve joining R_{k-1} to R_k and lying between S_+^{k-1} and S_+^k in $y \leq 0$; and W_-^k is a C^1 curve joining L_{k-1} to L_k and lying between S_-^{k-1} and S_-^k in $y \geq 0$, for each $k = 2, 3, 4, \ldots$. Thus $W = \bigcup_{k=1}^{\infty}(W_+^k \cup W_-^k)$ is a countable union of C^1 curves and W separates the plane in two components.

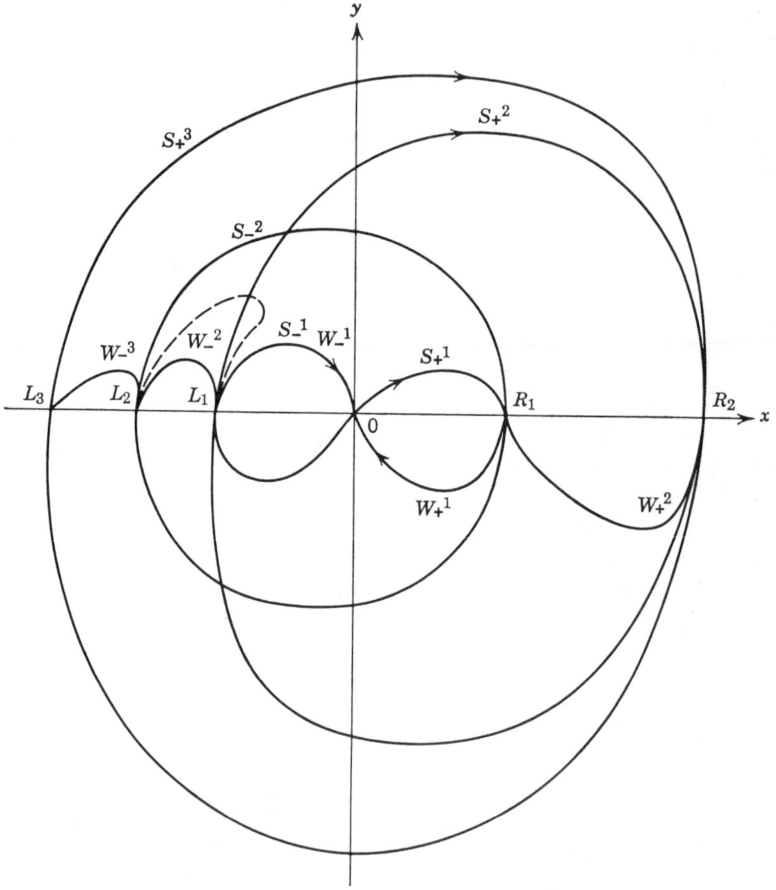

Figure 7.3 Switching locus construction for nonlinear attractive force.

Now consider the case with $\partial f/\partial y \not\equiv 0$. The analysis is the same as for the conservative case except that here S_+^1 or S_-^1, or one of the successive reflections $\hat{W}_\pm^k$, might have an endpoint at $\pm\infty$. That is, $W_\pm^k$ might not reintersect the x-axis, but necessarily approach infinity. In such a case the set of nonempty $W_{\pm}{}^{k}$ may be finite (see Figure 7.4) but the conclusions of the theorem hold just as in the conservative case. Q.E.D.

Special Hypothesis on the Geometry of the Switching Locus

We are here forced to assume that the switching locus $W = W(x)$ is a single-valued curve for $-\infty < x < +\infty$. This certainly holds if $W_+{}^1$ and $W_-{}^1$ extend to infinity without reintersecting the x-axis, as in the repulsive

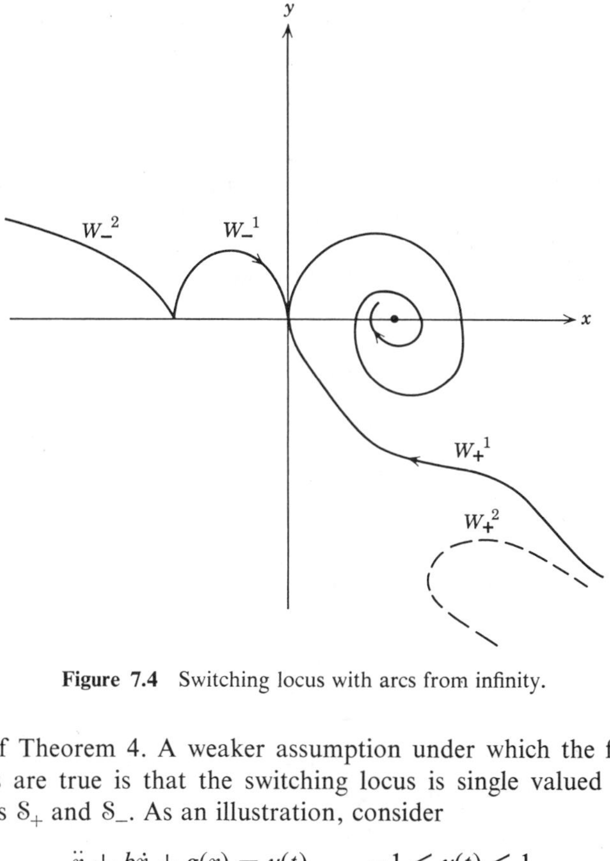

Figure 7.4 Switching locus with arcs from infinity.

system of Theorem 4. A weaker assumption under which the following theorems are true is that the switching locus is single valued over the two flows $\mathcal{S}_+$ and $\mathcal{S}_-$. As an illustration, consider

$$\ddot{x} + b\dot{x} + g(x) = u(t), \qquad -1 \le u(t) \le 1,$$

where $b > 0$ is constant and $g(x) \in C^1$ with $g(0) = 0$, $g'(x) > 0$, and $|g(x)| > 1$ for all large $|x|$. We shall further assume that the derivative $g'(x) \le b^2/4$ for all $-\infty < x < +\infty$. As an example of such a weak spring with linear friction take

$$\ddot{x} + 2\dot{x} + (x - \tfrac{1}{2} \text{Arc tan } x) = u(t), \qquad -1 \le u(t) \le 1,$$

where

$$-\frac{\pi}{2} < \text{Arc tan } x < \frac{\pi}{2}.$$

If W_+^1 did re-intersect the x-axis, then the variational equation based on $x(t) = W_+^1$ has a solution $v^1(t) = \dot{x}(t)$, which vanishes twice. But this variational equation

$$\ddot{v} + b\dot{v} + g'(x(t))v = 0$$

is easily seen to be disconjugate by letting $z = e^{bt/2}v$ and computing

$$\ddot{z} + \left(g'(x(t)) - \frac{b^2}{4}\right)z = 0,$$

that is, a nontrivial solution $z(t)$ has at most one zero. Therefore W_+^1 does not reintersect the positive x-axis, and an easy estimate for its slope shows that W_+^1 is defined for all $x > 0$. Similarly W_-^1 is single valued over the negative x-axis.

A maximal solution traced back from $W(x) = W_-^1 \cup W_+^1$ has no further switches. This follows from the disconjugacy of the adjoint equation

$$\ddot{\eta}_2 - b\dot{\eta}_2 + g'(x(t))\eta_2 = 0.$$

Therefore the switching locus is exactly

$$W(x) = W_-^1 \cup W_+^1$$

and the special hypothesis is valid in this case. The hypothesis also holds in the case of isochronous periodic solutions for $S_\pm$, say $f(x, y) = x$. For here consider the variational equation along S_+ or S_-.

$$\dot{v}^1 = v^2,$$

$$\dot{v}^2 = -\frac{\partial f}{\partial x}v^1 - \frac{\partial f}{\partial y}v^2$$

where $\mathbf{v}(t) = (v^1(t), v^2(t))$ represents a vector and its parallel transport along the flow of $S_\pm$. But

$$v^1\eta_1 + v^2\eta_2 = \text{constant}$$

and so

$$v^1(0)\,\eta_1(0) = v^1(t_1)\,\eta_1(t_1)$$

for successive switching times $t = 0$ and $t = t_1$. Now

$$\eta_1(0)\eta_1(t_1) < 0$$

and so

$$v^1(0)v^1(t_1) \leq 0.$$

Therefore, taking $\mathbf{v}(0)$ as the tangent vector along $W_{\pm}^{k}$, $\mathbf{v}(t_1)$ is the tangent vector along $W_{\pm}^{k+1}$, and thus $W_{\pm}^{k+1}$ is single valued over the x-axis if $W_{\pm}^{k}$ is single valued over the x-axis. By induction we can then show that $W = W(x)$ is single valued on $-\infty < x < +\infty$.

COROLLARY. *Consider the system* $\mathcal{S}$ *of Theorem* 6. *Assume that* $W = W(x)$ *is a single-valued curve on* $-\infty < x + \infty$. *Then for each point* $P \in R^2$ *there is one and only one maximal relay controller in* Δ *which steers* P *to the origin.*

Proof. We must only show that no solution curve S of $\mathcal{S}_-$ (or of $\mathcal{S}_+$) intersects W_+^{k} (or W_-^{k}) on the interior of the arc of S, which is used in producing the reflection of W_+^{k} (or W_-^{k}). Since W_+^{k} lies in $y < 0$ except at its endpoints, the solution S of $\mathcal{S}_-$ initiating at a point P of W_+^{k} cannot re-intersect W_+^{k} in $y < 0$ at an abscissa x lying to the left of P. Moreover, since there is just one point of W_+^{k} on each solution of $\mathcal{S}_+$ in the appropriate annular ring, and since the solutions of $\mathcal{S}_-$ have a negative tangent vector $(-y, f + 1)$ as contrasted to $(-y, f - 1)$ for solutions of $\mathcal{S}_+$, we find that S cannot re-intersect W_+^{k} in $y < 0$.

A similar argument holds for solutions of $\mathcal{S}_+$ reintersecting W_-^{k}.

Q.E.D.

THEOREM 7. *Consider the system*

$$(\mathcal{S}) \qquad \dot{x} = y,$$
$$\dot{y} = -f(x, y) + u, \qquad f(0, 0) = 0, \qquad f(x, y) \in C^1,$$

with $(\partial f/\partial x)(x, 0) > \epsilon > 0$, *for some* $\epsilon > 0$, *and* $\partial f/\partial y \geq 0$ *in* R^2. *Assume the switching locus* $W = W(x)$ *is single valued on* $-\infty < x < +\infty$. *Then for each point* P *in* R^2 *there is exactly one optimal controller* $u(t)$ *on* $-1 \leq u \leq 1$, *in the admissible class* Δ, *which steers* P *to the origin. Define the synthesizer*

$$\Psi(x, y) = \begin{cases} 1 & \text{for} \quad y < W(x), \\ 0 & \text{for} \quad y = W(x), \\ -1 & \text{for} \quad y > W(x). \end{cases}$$

Then the optimal response for P *is the unique solution of*

$$\ddot{x} + f(x, \dot{x}) = \Psi(x, \dot{x})$$

initiating at P *at* $t = 0$.

Proof. By the general theory of the existence of optimal, and thereby maximal relay controllers, as developed above, it is known that each point $P \in R^2$ lies on at least one response to a maximal relay controller steering P to the origin. But the construction of Theorem 6, together with the corollary, shows that P lies on a unique response to such a maximal relay controller and hence this controller must be the unique optimal controller for P in Δ.

The properties of the synthesizer $\Psi(x, y)$ follow from the construction in Theorem 6. Q.E.D.

Synthesis of an Optimal Controller for the Case of a Repulsive Force

Consider the control problem for

$$(\mathcal{S}) \qquad \begin{aligned} \dot{x} &= y, \\ \dot{y} &= -f(x, y) + u, \qquad f(0, 0) = 0, \qquad f(x, y) \in C^1 \end{aligned}$$

and $u(t)$ measurable on $-1 \leq u \leq 1$. Assume $(\partial f/\partial x)(x, y) < -\epsilon < 0$, $(\partial f/\partial y)(x, y) \geq 0$ in R^2, for some $\epsilon > 0$. These hypotheses, somewhat stronger than those of Theorem 4, assure the existence of a band $\mathcal{B}$, between the principal separatrices of $\mathcal{S}_+$ and $\mathcal{S}_-$, which is the domain $\mathcal{C}$ of controllability. For each point $P \in \mathcal{B}$ there exists an optimal control $u(t)$, in the admissible class Δ, which steers P to the origin.

The switching locus W of $\mathcal{S}$ is defined as in Theorem 5. To construct W let W_+^1 be the solution of $\mathcal{S}_+$ through the origin and lying in the fourth quadrant. By a comparison of the slopes of the solutions of $\mathcal{S}_+$ and $\mathcal{S}_-$, we find that W_+^1 lies always interior to the band $\mathcal{B}$ in $y < 0$ (except for the origin) and $y \to -\infty$ as x increases on W_+^1. Let W_-^1 be the solution of $\mathcal{S}_-$ through the origin and let it lie in the second quadrant. Clearly W_-^1 lies interior to $\mathcal{B}$ and $y \to +\infty$ as $(-x)$ increases on W_-^1. The reflections $W_\pm^k$ are defined as in Theorem 5, but we shall show that these are all empty for $k = 2, 3, 4, \ldots$ and that $W = W_+^1 \cup W_-^1$ (see Figure 7.2).

THEOREM 8. *Consider the system*

$$(\mathcal{S}) \qquad \begin{aligned} \dot{x} &= y, \\ \dot{y} &= -f(x, y) + u, \qquad f(0, 0) = 0, \qquad f(x, y) \in C^1 \end{aligned}$$

with $(\partial f/\partial x)(x, y) < -\epsilon < 0$, $(\partial f/\partial y)(x, y) \geq 0$ *in* R^2, *for some* $\epsilon > 0$. *Consider maximal relay controllers* $u(t)$ *on* $-1 \leq u \leq 1$ *in* Δ *and construct the switching locus* W. *Then* $W = W(x)$ *is a continuous single-valued curve over a segment of the* x*-axis, and* W *separates* $\mathcal{B}$ *in two regions. Moreover*

$$W(x) = \begin{cases} W_+^1(x) & \text{for} \quad x \geq 0, \\ W_-^1(x) & \text{for} \quad x \leq 0 \end{cases}$$

so $W(x) \in C^1$ *except at* $x = 0$.

Proof. We first prove that $W_{\pm}^2$ are empty. This follows if a nontrivial solution $\boldsymbol{\eta}(t) = (\eta_1(t), \eta_2(t))$ of

$$\dot{\eta}_1 = \eta_2 \frac{\partial f}{\partial x}(x(t), y(t)),$$

$$\dot{\eta}_2 = -\eta_1 + \eta_2 \frac{\partial f}{\partial y}(x(t), y(t))$$

has at most one zero for $\eta_2(t)$, where $\mathbf{x}(t) = (x(t), y(t))$ is a solution of $\mathcal{S}_-$ or of $\mathcal{S}_+$. Let $t_1 < t_2$ be successive zeros of $\eta_2(t)$ and say $\eta_2(t) > 0$ on $t_1 < t < t_2$. Then $\eta_1(t_1) < 0$ and $\eta_1(t_2) > 0$. Since $\partial f/\partial x < -\epsilon < 0$, $\dot{\eta}_1(t) < 0$ on $t_1 < t < t_2$. This is impossible. The case where $\eta_2(t) < 0$ on $t_1 < t < t_2$ is similarly impossible.

It is clear that $W_+^1 \cup W_-^1$ belongs to W by Theorem 5. Thus

$$W = W(x) = \begin{cases} W_+^1(x) & \text{for} \quad x \geq 0, \\ W_-^1(x) & \text{for} \quad x \leq 0. \end{cases}$$

It is clear that W is a line homeomorph which separates the topological band $\mathcal{B}$ in two components. Q.E.D.

COROLLARY. *Consider the system $\mathcal{S}$ of Theorem 8. Then for each point $P \in \mathcal{B}$ there is one and only one maximal relay controller in Δ that steers P to the origin.*

Proof. No solution of $\mathcal{S}_-$ can intersect W_+^1 in two distinct points, as follows from a comparison of the tangent vectors of the solutions of $\mathcal{S}_+$ and $\mathcal{S}_-$. A similar property holds for W_-^1. The construction of W then shows that each maximal relay controller in Δ, steering $P \in \mathcal{B}$ to the origin, produces a response that intersects W and then never leaves W but follows it into the origin. Q.E.D.

THEOREM 9. *Consider the system*

$$(\mathcal{S}) \qquad \begin{aligned} \dot{x} &= y, \\ \dot{y} &= -f(x, y) + u, \qquad f(0, 0) = 0, \qquad f(x, y) \in C^1, \end{aligned}$$

with $(\partial f/\partial x)(x, y) < -\epsilon < 0$, for some $\epsilon > 0$ and $(\partial f/\partial y)(x, y) \geq 0$ in R^2. Consider the domain $\mathcal{C} = \mathcal{B}$ of controllability for measurable controllers $u(t)$ on $-1 \leq u \leq 1$ in Δ, and construct the switching locus $W = W(x)$.

Then for each point $P \in \mathcal{B}$ there exists exactly one optimal controller $u(t)$ in Δ that steers P to the origin. Define the synthesizer

$$\Psi(x, y) = \begin{cases} 1 & \textit{for} \quad y < W(x), \\ 0 & \textit{for} \quad y = W(x), \\ -1 & \textit{for} \quad y > W(x). \end{cases}$$

Then the optimal response for P is the unique solution of

$$\ddot{x} + f(x, \dot{x}) = \Psi(x, \dot{x})$$

initiating at P at $t = 0$.

Proof. The general theory of optimal and maximal control, as developed above, shows that each point $P \in \mathcal{B}$ lies on at least one response to maximal control steering P to the origin. But the construction of Theorem 8, together with the corollary, shows that P lies on a unique response to such a maximal relay controller, and hence this control must be the unique optimal controller for P in Δ. The properties of $\Psi(x, y)$ follow from the construction of Theorem 8. Q.E.D.

Examples of Construction of the Feedback Controller

***Example* 1.** (Hard spring) Consider *Duffing's equation* with control

$$\ddot{x} + x + 2x^3 = u(t),$$

or the equivalent system

$$(\mathcal{S}_d) \qquad \begin{aligned} \dot{x} &= y \\ \dot{y} &= -x - 2x^3 + u, \end{aligned}$$

with $-1 \leq u \leq 1$. This corresponds to a mechanical system constrained by a nonlinear hard spring with some external force $u(t)$.

For the time-optimal problem the optimal controller will be restrained by $u(t) = \pm 1$ on various intervals of time. Thus consider

$$(\mathcal{S}_\pm) \qquad \begin{aligned} \dot{x} &= y \\ \dot{y} &= -x - 2x^3 \pm 1. \end{aligned}$$

Clearly every point in the phase space can be connected to the origin by picking paths only from this $+, -$ family, assuring us of the existence of the optimal controller. Also, consider the adjoint system (as in the previous analysis)

$$(\mathcal{A}) \qquad \begin{aligned} \dot{\eta}_1 &= -\frac{\partial H}{\partial x} = \eta_2(1 + 6x^2(t)) \\ \dot{\eta}_2 &= -\frac{\partial H}{\partial y} = -\eta_1, \end{aligned}$$

from which we find the response $\eta_2(t)$, which determines the interval of time that the solutions of $\mathcal{S}_+$ and $\mathcal{S}_-$ are followed to obtain maximal responses.

We shall construct W, the switching locus or boundary, as in the previous theory. Here W_+^1 is the solution of $\mathcal{S}_+$ through the origin and lying in the fourth quadrant (see Figure 7.5) and W_-^1 is the solution of $\mathcal{S}_-$ through the origin and lying in the second quadrant.

We now reflect W_-^1 along solutions of $\mathcal{S}_+$. From the points 1, 2, . . . , 9 of W_-^1 follow backwards in time along the corresponding solutions of $\mathcal{S}_+$ for a time duration equal to the interval between zeros of $\eta_2(t)$, that is, from, say, point 2 with $\eta_1(0) = 1$, $\eta_2(0) = 0$ follow S_+ and consider the response of the adjoint system $\mathcal{A}$ until $\eta_2(t)$ is again zero. This determines a point $2a$ on W_+^2. We repeat this for each of the points 3, . . . , 9 and obtain the corresponding points $3a, \ldots, 9a$ of W_+^2. Since we know that W_+^2 is a C^1 curve we can fill in the values between $2a, 3a, \ldots, 8a$ by interpolation. For this reason we need consider only a finite number of maximal responses in generating W.

Now that W_+^2 has been found, W_-^2 is the reflection through the origin see (Figure 7.5). From W_+^2 we find W_-^3 by reflecting W_+^2 along a conjugate interval as defined above. The remaining segments of W are built up in exactly the same manner.

Thus we have found the feedback controller $u = \Psi(x, y)$ for the Duffing equation ($\mathcal{S}_d$). It is very easy to store this function and therefore the synthesis is complete. $\Psi(x, y) = +1$ below the switching boundary W and $\Psi(x, y) = -1$ above the switching boundary W.

There is some question as to the accuracy with which we have determined W in Figure 7.5 (it was constructed by use of a digital-analog computer), but if it is nearly correct, then one sees that the switching locus is indeed a single-valued function over both flows $\mathcal{S}_+$ and $\mathcal{S}_-$. The numerical method for constructing the feedback controller using a digital computer is discussed in detail by H. L. Burmeister. Compare his Figure 3 with our Figure 7.5.

For Duffing's equation with control

$$\dot{x} = y$$

$$\dot{y} = -x - x^3 + u$$

$-1 \leq u \leq 1$, we have also constructed in Figure 7.6 the isochronal curves for the steering to the origin. The isochronal function is discussed just before Theorem 21 of Chapter 2 for the time-optimal linear problem, and in Exercise 2 of Section 5.2 for nonlinear systems. The interior and each of these curves correspond to the attainable set, going backwards in time from the origin during the time interval $[0, T]$, where $T > 0$ is the value of the isochrone. Here we see that the attainable set is not convex, but appears to have a smooth boundary.

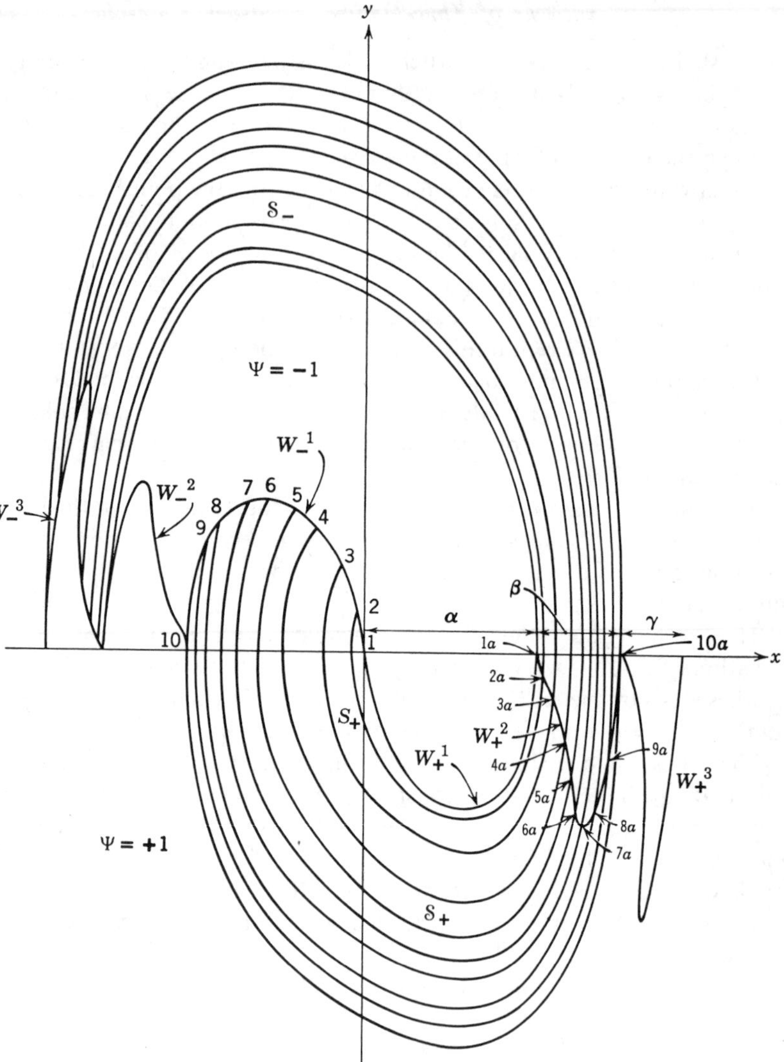

Figure 7.5 Feedback synthesis for the Duffing equation.

***Example* 2.** (Pendulum) Consider the equations of motion of the forced pendulum of Figure 7.7, with motion constrained to the plane and mass m attached to the pivot by means of a rigid (massless) member:

$$I\ddot{\theta} + b\dot{\theta} + mgl \sin \theta = M(t).$$

Here 1. $I = ml^2$ is the moment of inertia;
2. $b \geq 0$ is the damping coefficient;

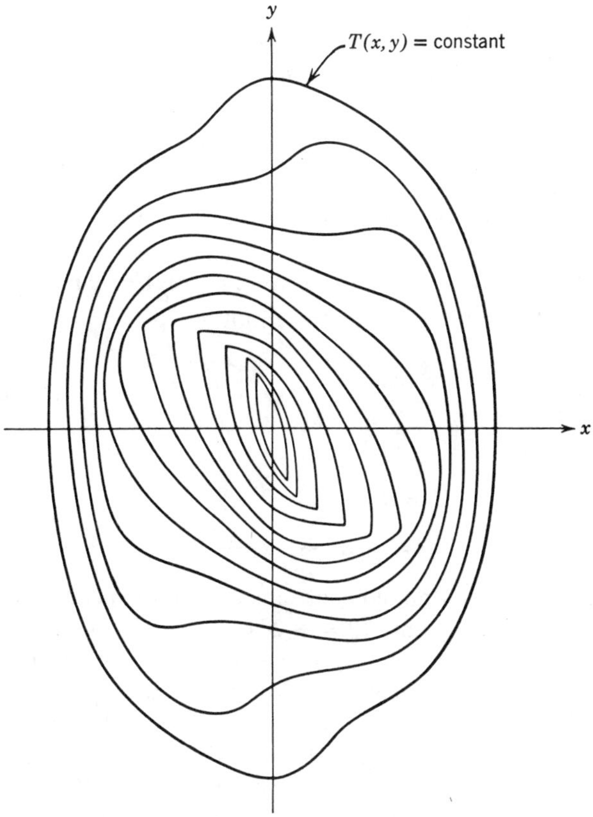

Figure 7.6 Isochrones for $\ddot{x} + \dot{x} + x^3 = u$, $|u| \leq 1$.

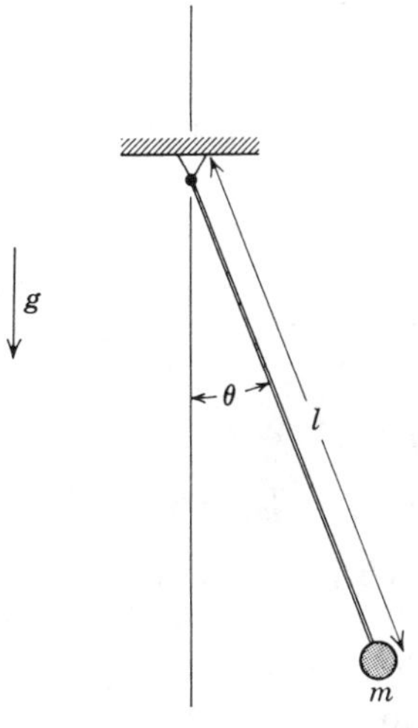

Figure 7.7 Controlled pendulum parameters.

3. $M(t)$ is the external controlling torque;
4. g is the gravitational constant;
5. l is the length of the pendulum support member;
6. m is the mass located at the end of the support member.

To put the equation in a standard form we make the change of time scale $\tau = t\sqrt{mgl/I}$ obtaining the equation of motion

$$\ddot{\theta} + \alpha\dot{\theta} + \sin\theta = \beta(t)$$

with $\alpha = b/\sqrt{Imgl} \geq 0$, $\beta(t) = M(t)/mgl$.

The equivalent system is

$$\dot{\theta} = z$$

$$\dot{z} = -\sin\theta - \alpha z + \beta(t).$$

The problem of time-optimal control is to stop the pendulum at one of the stable equilibrium points ($\theta = \pm 2\pi k$, $z = 0$) for $k = 0, 1, 2, \ldots$ in minimal time duration by choice of the controller torque $\beta(t)$, which is subject to the restraint $|\beta(t)| \leq B$, $B > 0$. Thus the target G is the set of points $G = \{(\theta, z) \mid \theta = \pm 2\pi k, z = 0, k = 0, 1, 2, \ldots\}$.

It is first shown that the target can be reached by application of some admissible controller from each initial point of the (θ, z)-plane. This can always be accomplished by application of the following control rule:

1. Use $\beta(t) = -B \operatorname{sgn} z(t)$ until $z = 0$ in $2\pi k \leq \theta \leq 2(k+1)\pi$ for some $k = 0, 1, 2, \ldots$.

2. Then take $\beta(t) = -\beta_0 z(t)$ where $\beta_0 > 0$ is so small that $B \geq |\beta(t)|$ (after displacing the motion from an unstable equilibrium if necessary).

3. In the vicinity of $\theta = 2k\pi$, $z = 0$ we apply the controllability test (Chapter 6) for

$$\dot{\theta} = z$$

$$\dot{z} = -\theta - \alpha z + \beta(t)$$

with

$$\left(\begin{pmatrix}0\\1\end{pmatrix} \begin{pmatrix}0 & 1\\ -1 & -\alpha\end{pmatrix}\begin{pmatrix}0\\1\end{pmatrix}\right)$$

nonsingular for all $\alpha \geq 0$. This ensures that the equilibrium point can be reached exactly, during a finite time duration, from each point of some neighborhood of the equilibrium point.

Next we estimate the velocity and find a uniform bound so that only a finite number of points of the target need to be considered in the problem of optimal control, that is, the target can be considered to be a compact set. Clearly

$$|\dot{z}| \leq 1 + \alpha\,|z| + B,$$

$$|z(t)| \leq ce^{\gamma t}, \quad \text{for some} \quad c > 0, \qquad \gamma > 0,$$

and

$$|\theta(t)| \leq c_1 e^{\gamma t} \quad \text{for some} \quad c_1 > 0, \qquad 0 \leq t < \infty.$$

Thus the solutions are defined for each time duration with a uniform bound. Also points $\theta = 2\pi k$, $z = 0$ for large $|k|$ can be achieved only for very long time durations and are therefore not candidates for the end points of time-optimal trajectories.

By standard theorems of Chapter 4 an optimal controller $\beta^*(t)$ exists. Moreover by the previous results of this section the optimal controller is a relay controller assuming only the two values $+B$ and $-B$ on various intervals of time. The results of the section on the geometry of the switching locus are applicable and we now turn to construction of the optimal switching locus. For convenience we study first the case when $\alpha = 0$ and later show that the results are even simpler when $\alpha > 0$.

With $\alpha = 0$ every target point can be reached, that is, one can make the swinging motions of the pendulum larger and larger, even if B is very small, until it is possible to go over the unstable equilibrium point and from there to any other equilibrium point. We first consider steering optimally to only one equilibrium point, which we take to be the origin.

There are two cases of particular interest: (1) $B > 2/\pi$, which corresponds to the situation in which the switching locus (for the one target point) intersects the θ-axis only at the origin; (2) $B \leq 2/\pi$, where the switching locus intersects the θ-axis at more than one point. The case when $B > 2/\pi$ is dealt with first.

Consider the Hamiltonian system of equations (as in Chapter 5)

$$\dot{\eta}_1 = -\frac{\partial H}{\partial \theta} = (\cos\theta)\eta_2$$

$$\dot{\eta}_2 = -\frac{\partial H}{\partial z} = \alpha\eta_2 - \eta_1$$

$$\dot{\theta} = \frac{\partial H}{\partial \eta_1} = z$$

$$\dot{z} = \frac{\partial H}{\partial \eta_2} = -\sin\theta - \alpha z + \beta(t)$$

where $H = H(\eta_1, \eta_2, \theta, z, \beta) = \eta_1 z + \eta_2[-\sin\theta - \alpha z + \beta]$.

The optimal controller is a relay maximal controller of the form

$$\beta(t) = \operatorname{sgn} \eta_2(t).$$

The question of whether the switching locus reintersects the θ-axis is the same as whether, along the time backwards equation starting at the origin in the (θ, z)-plane with $\eta_2 = 0$, and $\eta_1 = 1$, the response $\eta_2(t)$ has more than one real zero. By previous results we know that the zeros of $\eta_2(t)$

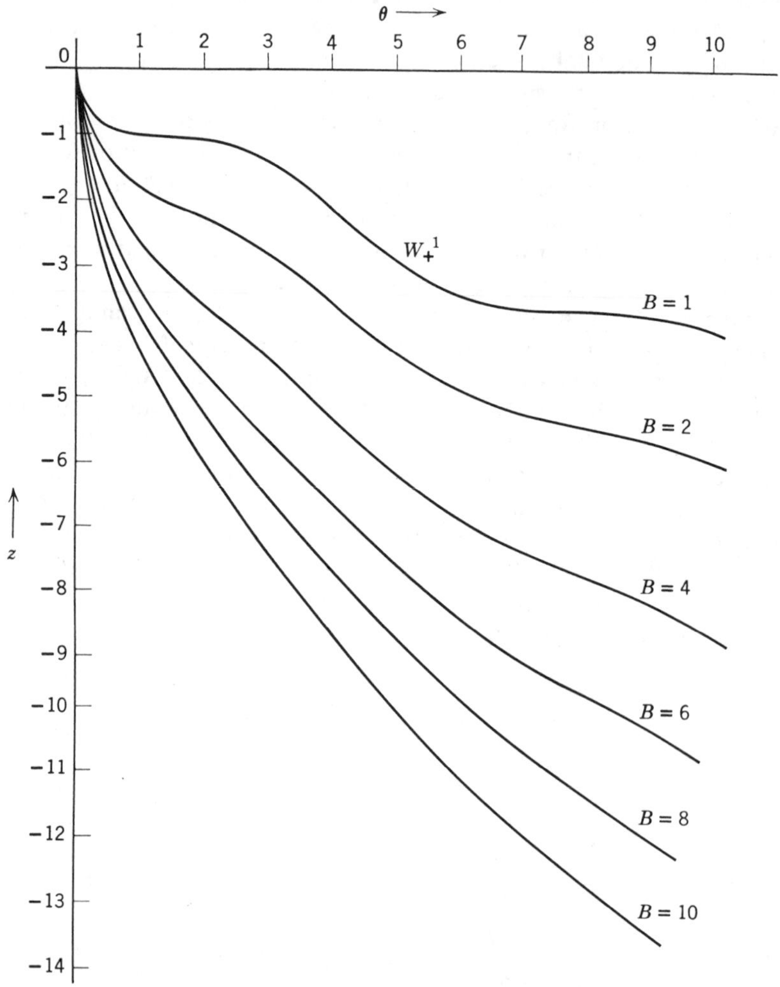

Figure 7.8 Switching locus for pendulum without damping.

and $z(t)$ are interlaced along such a solution curve. For the particular initial condition $(0, 0, \pm 1, 0) = (\theta(0), z(0), \eta_1(0), \eta_2(0))$ the zeros of $\eta_2(t)$ are identical to those of $z(t)$.

The time-backward equations to consider are

$$\begin{aligned}\dot{\theta} &= -z \\ \dot{z} &= \sin\theta - B \operatorname{sgn} \eta_2 \\ \dot{\eta}_1 &= -\eta_2 \cos\theta \\ \dot{\eta}_2 &= \eta_1\end{aligned}$$

with initial data $(0, 0, 1, 0)$, (or $(0, 0, -1, 0)$). Immediately $\eta_2(t)$ becomes positive (negative) and cannot have another zero until $z(t)$ does. Consider the phase-plane curve of the first two coordinates when $\eta_2(t) > 0$, that is, the curve through the origin satisfying the equation

$$\frac{dz}{d\theta} = \frac{\sin\theta - B}{-z}$$

and lying in the fourth quadrant. Integration yields

$$z = -\sqrt{2(\cos\theta + B\theta - 1)},$$

which is plotted in Figure 7.8 for several values of B and clearly does not reintersect the θ-axis if $B > 2/\pi$, and is a portion of the switching locus, namely W_+^1.

To determine whether there are pieces of the switching curve sticking back from infinity we have to carry out the steps outlined for construction of the switching locus, except in special cases. Namely, from a point of the curve W_+^1 with $(\eta_1(0) = -1,\ \eta_2(0) = 0)$, we follow forward in time along the solution of the backward equations for a time duration until $\eta_2(t)$ is again zero and mark this point in the (θ, z)-plane as belonging to the switching curve. As in Example 1, for the case $\alpha = 0$, $B = 1$, ten points along W_+^1 were used and the construction of the switching locus gave us Figure 7.9 with pieces of the switching curve sticking back from infinity.

By symmetry one constructs the switching locus in the second quadrant. Also, we can translate the origin to each of the target points, if it is the end point, and the switching locus is similarly translated.

However, we have yet to specify the procedure if all points of the target are considered simultaneously. Let us show that we need only consider the switching locus to the left or the switching locus to the right of the initial

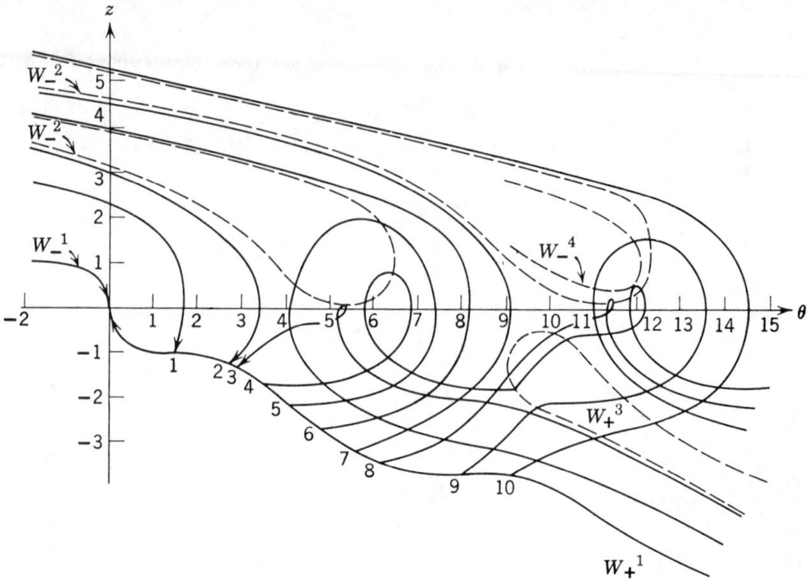

Figure 7.9 Construction of switching locus for pendulum with large control restraints.

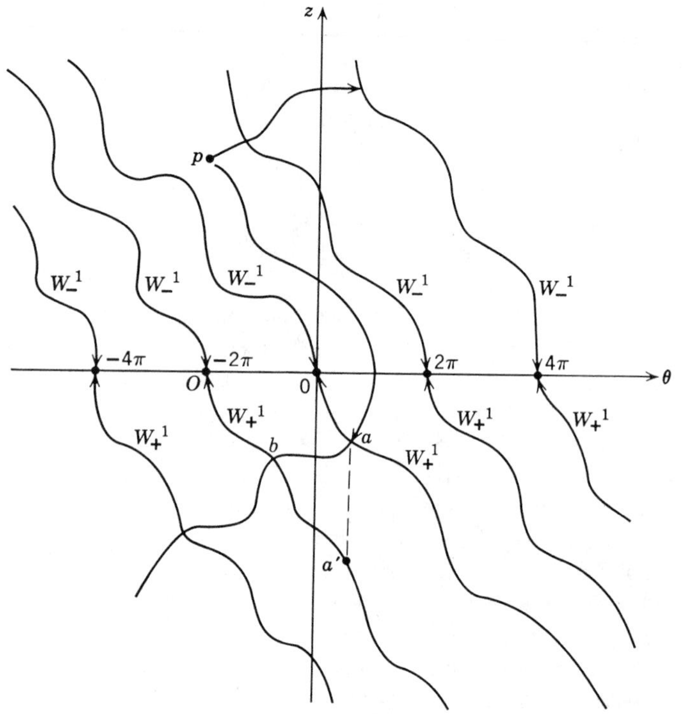

Figure 7.10 Construction of feedback controller for pendulum with global switching decisions.

point in deciding which point to take as target, if there are no parts of the switching locus sticking back from infinity. The situation is indicated in Figure 7.10 for all switching curves.

Consider the point p of Figure 7.10. If $\beta \equiv +B$ is selected initially there can be no switching until the point a or the point b etc. are reached. Obviously switching at a and then going into the origin is better than switching at b and going into the origin. This is true because it takes less

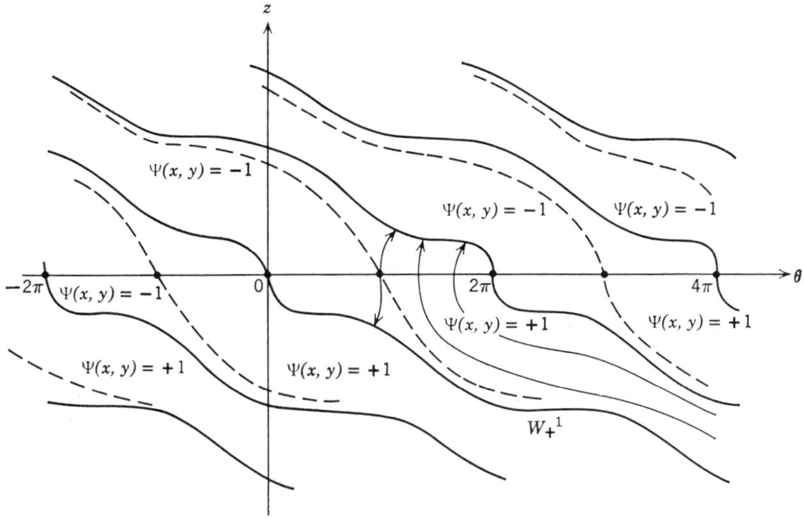

Figure 7.11 Feedback controller synthesis for pendulum.

time to go from a' to 0 along the path through a' with $\beta \equiv +B$ than along the path a to b to 0_1, and therefore the path from a to 0 is a shorter time path than the path a to b to 0_1. Similarly if β is initially selected to be $-B$ the switching occurs at the first switch curve crossed. The complete construction of the optimal feedback controller synthesis is as indicated in Figure 7.11, where the dashed curve is found by considering the above two possibilities and by assigning the equal time paths to, say, the lowest angular coordinate.

If $\alpha > 0$ then the slope of the switch curves is steeper and the construction technique is identical to that for $\alpha = 0$. In fact for $\alpha > 2$ it is easy to show that the switching locus has no parts sticking back from infinity. We need only show that the equation

$$\ddot{\eta}_2 + \alpha\dot{\eta}_2 + \eta_2(\cos\theta(t)) = 0$$

has solutions with at most one real zero. We make the change of variable $\eta_2 = e^{-(\alpha/2)t}v$ with v zero if and only if $\eta_2 = 0$. The equation for v is

$$\ddot{v} + v\left(-\frac{\alpha^2}{4} + \cos\theta(t)\right) = 0,$$

which, for $\alpha > 2$, has solutions with at most one real zero.

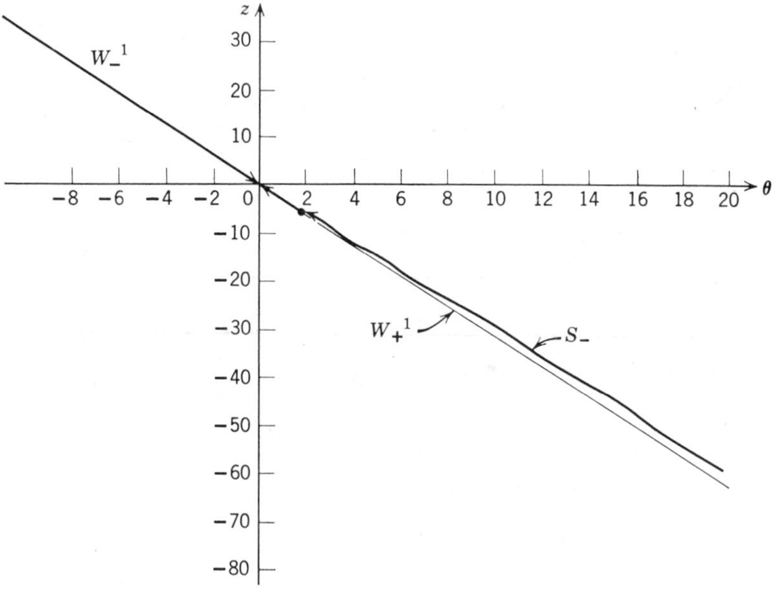

Figure 7.12 Switching locus for pendulum with $B = 1$, $\alpha = 3$.

The switching locus and feedback controller synthesis for $\alpha = 3$, $B = 1$ are shown in Figure 7.12.

Finally, consider the case when $B \leq 2/\pi$ with $\alpha \geq 0$. With $B = 0.1$ and $\alpha = 0$ the switching locus for steering to the origin is constructed as in the previous example and as shown in Figure 7.13. Again, if $\alpha > 0$ the same techniques can be used in finding the switching locus.

It is of general interest to remark that the phase or state space for the pendulum, when all stable equilibrium points are considered equivalent, is a cylinder. The optimal-control problem is thus resolved in a phase space that is topologically different from the plane, and it is this topological complexity that causes the confusion and profusion of switching loci.

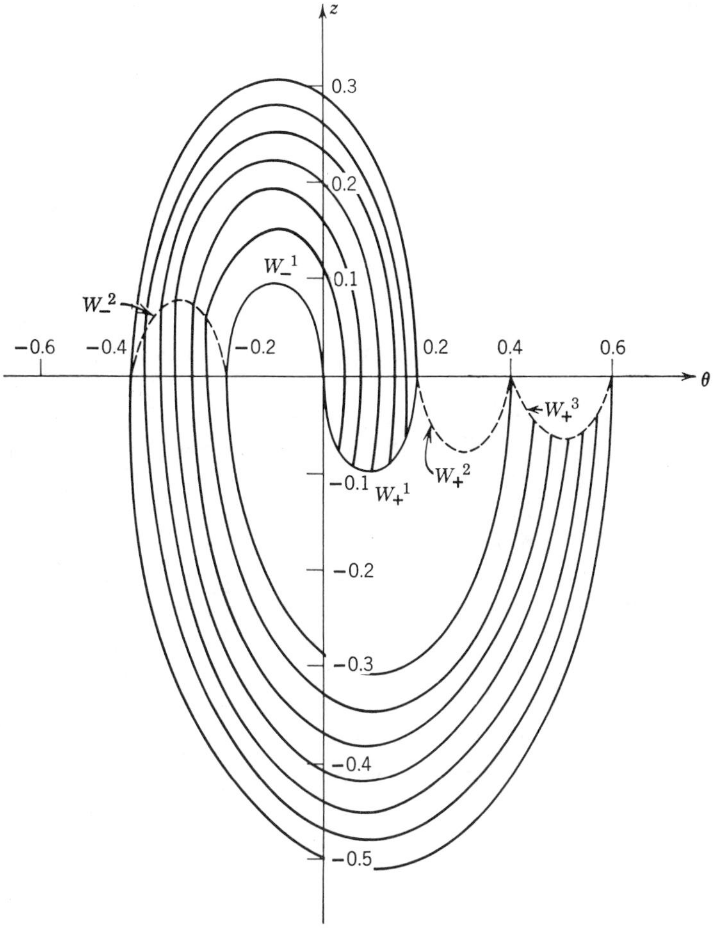

Figure 7.13 Switching locus for pendulum with $B = 0.1$, $\alpha = 0$.

EXERCISES

1. Discuss the controllability of the second-order system

$$\dot{x} = y^3$$

$$\dot{y} = u, \qquad |u| \leq 1$$

at the origin.

2. In Figure 7.5 we have labeled the distances between the points where the switching locus is fastened to the x-axis, for x increasing, by α, β, γ. Obviously, $\alpha \geq \beta \geq \gamma$ just by visual inspection of the diagram for the

Duffing equation in Example 1. Prove that this is in general true as x increases (for this example) and that this distance converges to zero as $x \to \infty$. [Hint: compare with the series $\sum 1/n^2$.]

3. Construct the switching locus for the Duffing equation with control and damping:

$$\dot{x} = y$$

$$\dot{y} = -x - x^3 - y + u, \qquad |u| \leq 1,$$

as in the example of Figure 7.5.

7.2 OPTIMAL STEERING FOR SOUNDING ROCKETS

In 1919 Goddard considered the problem of sending a rocket-propelled vehicle to a prescribed altitude above the earth using the least amount of fuel. He established that this problem was not solved by the standard methods of the calculus of variations.

A variant of this problem is to maximize the altitude when the fuel resource is prescribed. We shall consider this later problem when the rocket-motor thrust is bounded in amplitude and the mathematical model is the standard one for a class of drag forces. The case when there is no bound on the thrust force has been treated [Ewing], and reduced to the bounded case by assuming a certain uniform Lipschitz condition. In practice there is always a limit on the available thrust.

Consideration will be given to the system of equations

$$(\mathcal{S}_0) \qquad \frac{dh}{dt} = v, \qquad \frac{dv}{dt} = \frac{T - D(v, h)}{m} - c_1, \qquad \frac{dm}{dt} = -\frac{T}{c_2}$$

where h is the vehicle altitude above the earth, v is its vertical velocity, and m is its mass. Here c_1 and c_2 are positive constants. The control T is the rocket-thrust force, which can be varied between the limits $0 \leq T \leq T_1$, and $D(v, h)$ is the drag force defined for $h \geq 0$, and has the properties:

1. $D(0, h) = 0$.
2. $|D(v, h_1)| > |D(v, h_2)|$ if $h_2 > h_1$.
3. $-D(-v, h) = D(v, h) \geq 0$ if $v \geq 0$.

4. $D(v, h) \to 0$ as $h \to \infty$ all v.
5. $D(v, h) \in C^1$ everywhere.
6. $v\, D(v, h) \geq 0$.
7. $|D(v_1, h)| \geq |D(v_2, h)|$ if $|v_1| \geq |v_2|$.

A model drag force is $v\,|v|\,e^{-\alpha h}$ with $\alpha =$ constant > 0. The mass m is initially m_0 and can decrease to $m_1 > 0$ by the expenditure of fuel. More general drag forces can be handled in somewhat the same manner as we handle this one (see for example [Ewing]). Also, other classes of these problems have been recently considered (see [Lawden]).

It is assumed that there is a time $\bar{t}$, $0 < \bar{t} < \infty$, beyond which the experiment has no meaning, and therefore we consider only problems of maximizing the altitude for times smaller or equal to $\bar{t}$.

The problem is to find a (Lebesgue) measurable steering function $T(t)$ on $[0, t_1]$ satisfying the restraint $0 \leq T \leq T_1$, and the time $t_1 \leq \bar{t}$ such that the corresponding response $(h(t), v(t), m(t))$, defined on $[0, t_1]$ with initial data $(0, 0, m_0)$ and satisfying the phase constraints $m_0 \geq m \geq m_1 > 0$, $h \geq 0$, attains the maximum altitude with respect to all such admissible steering functions and corresponding responses.

We shall be concerned first with the problem of existence of the optimal steering policy, since the necessary conditions for optimal control, Chapter 5, can then be applied. Sufficiency results are also included, and the synthesis is indicated for one assumed drag force.

Existence

The closure of the attainable set (see Chapter 4) will be the basic device used to establish the existence of the optimal steering function. We present a proof of existence that is longer than necessary because some of the results will be needed later in discussing the form of the optimal controller (see Exercise 3). The constraint on $m(t)$ will be handled by exploiting its monotone decreasing characteristic. The constraint on h will be handled by initially allowing h to assume negative values, proving the existence of an optimal steering policy in that case, and then showing that there is always an optimum with $h \geq 0$.

It is assumed that $D(v, h)$ is defined with the above properties for $h \leq 0$ as in the model. The definition of $m(t)$ is extended outside of the interval, $m_1 \leq m \leq m_0$, by defining the function

$$F(M) = \frac{1}{M + m_1} \quad \text{for} \quad M \geq 0$$

$$= \frac{m_1 - M}{{m_1}^2} \quad \text{for} \quad M \leq 0$$

and replacing the system S_0 by the system

$$(S_1) \qquad \begin{aligned} \frac{dh}{dt} &= v \\ \frac{dv}{dt} &= F(M)[T - D(v, h)] - c_1 \\ \frac{dM}{dt} &= -\frac{T}{c_2}. \end{aligned}$$

This system coincides with the system S_0 for $0 \leq M \leq m_0 - m_1$, that is, for $m_1 \leq m \leq m_0$, and in this interval $M(t) = m(t) - m_1$.

To this system we add another coordinate x to take account of the independent time variable. That is, consider the augmented system

$$(\hat{S}) \qquad \begin{aligned} \frac{dx}{dt} &= 1 \\ \frac{dh}{dt} &= v \\ \frac{dv}{dt} &= F(M)[T - D(v, h)] - c_1 \\ \frac{dM}{dt} &= -T/c_2 \end{aligned}$$

with initial data $x(0) = 0$, $v(0) = h(0) = 0$, $M(0) = m_0 - m_1$.

For the present, disregard the above restraints on h and M and consider the attainable set (see Chapter 4) $\hat{R}(\bar{t})$ for the system $\hat{S}$ in the space of variables (x, h, v, M). This set is compact in R^{3+1} if each response of $\hat{S}$ is bounded on $[0, \bar{t}]$.

Corresponding to each admissible steering function $T(t)$, that is, measurable controller $T(t)$ with range in the interval $[0, T_1]$, there exists a unique, bounded, absolutely continuous response $(x(t), h(t), v(t), M(t))$ of $\hat{S}$ with $(x(0), h(0), v(0), M(0)) = (0, 0, 0, m_0 - m_1)$ defined on the interval $[0, \bar{t}]$ and satisfying the differential equation $\hat{S}$ almost everywhere on $[0, \bar{t}]$.

The boundedness of the response is proved as follows. For each admissible steering function $T(t)$ on $[0, \bar{t}]$ the function $M(t)$ is bounded and therefore so is $F(M(t))$.

Consider

$$\frac{d\,|v/2|^2}{dt} = v\frac{dv}{dt} = v\,F(M)[T(t) - D(v, h)] - vc_1.$$

Therefore

$$\frac{d\,|v/2|^2}{dt} \leq |v|\,|F(M)\,T(t) - c_1|$$

or

$$\frac{d\,|v|}{dt} \leq |F(M)T - c_1|,$$

imposing a bound on $v(t)$ on the finite interval $[0, \bar{t}]$ and, therefore, a bound on $h(t)$ on $[0, \bar{t}]$.

The restraint on M is now imposed. Consider the intersection between the closed set

$$H = \{x, h, v, M \mid 0 \leq x \leq \bar{t}, v \text{ and } h \text{ in } R^1, 0 \leq M \leq m_0 - m_1\}$$

and $\hat{R}(\bar{t})$. That is, consider the compact set

$$\hat{K}(\bar{t}) = H \cap \hat{R}(\bar{t}).$$

Since $M(t)$ is a nonincreasing function on $[0, \bar{t}]$ it is obvious that $\hat{K}(\bar{t})$ is the attainable set when the restraint is imposed.

The linear function $\mathfrak{L}(x, h, v, M) \equiv h$ assumes both its maximum and its minimum value on the compact set $\hat{K}(\bar{t})$. *This establishes the existence of an optimal steering policy if $h(t)$ is allowed to take on negative values.*

We now wish to show that along an optimal solution the velocity and altitude are positive. A lemma is first established.

LEMMA 1. *Consider the system $\mathcal{S}_1$, as above, with initial point $(h_0, v_0 \geq 0, m_0 - m_1)$ at time t_0. Let $T(t)$ be an admissible steering function with corresponding response $(h(t), v(t), M(t))$ on $[t_0, t_1]$ with $v(t) \geq 0$ on $[t_0, t_1]$. Then for, each initial point $(\bar{h}_0, \bar{v}_0, \bar{M}_0)$ with $(\bar{h}_0 \geq h_0,\ \bar{v}_0 \geq v_0,\ \bar{M}_0 = m_0 - m_1)$ there exists an admissible steering function $\bar{T}(t)$ on $[t_0, t_1]$ with corresponding response $(\bar{h}(t), \bar{v}(t), \bar{M}(t))$ passing through the initial point $(\bar{h}_0, \bar{v}_0, \bar{M}_0)$ at time t_0 and having $\bar{h}(t) \geq h(t)$, $\bar{v}(t) \geq v(t)$ on $[t_0, t_1]$ and $\bar{M}(t_1) = M(t_1)$. Moreover, the strict inequality $\bar{h}(t) > h(t)$ can hold if $\bar{h}_0 > h_0$. (That is, it is always possible to stay above a response that has a positive velocity if one is initially above it.)*

Proof. Choose $\bar{T}(t) = T(t)$ on $[t_0, t_1]$. Then $\bar{M}(t) = M(t)$ and $F(\bar{M}(t)) = F(M(t))$ everywhere on $[t_0, t_1]$. Since $D(v, h) \geq D(v, \bar{h})$ for $v \geq 0$, and $\bar{h} \geq h$, $\bar{v}(t)$ cannot get smaller than $v(t)$. This keeps $\bar{h}(t)$ at least as large as $h(t)$. Since $D(v, h) > D(v, \bar{h})$, if $\bar{h} > h$, the strict inequality also follows.

For physical reasons we now assume that

$$T_1 \geq m_0 c_1, \qquad m_1 < m_0.$$

This enables us to show that even if the physical restraint on h is not used in the problem, the maximum altitude $h^* > 0$, and h^* can be achieved along a response in which the velocity is positive. However, if $T_1 < m_0 c_1$ we would consider the hard restraint on h and burn fuel until the mass m had decreased to the point where $T_1 = mc_1 = (M + m_1 c_1)$, if possible, and until this point is reached every steering policy that uses up fuel is optimal. When this point is reached the problem is as above and we then remove the phase restraint. It therefore suffices to consider only the case when $T_1 \geq m_0 c_1$, with no restraint on h. Q.E.D.

THEOREM 10. *Consider the system* $\mathcal{S}_1$, *as above, with initial point* $h_0 = v_0 = 0$, $M_0 = m_0 - m_1$. *Let* $T^*(t)$ *be an optimal steering function with response* $(h^*(t), v^*(t), M^*(t))$ *on the optimal interval* $[0, t^*]$. *Then* $h^*(t) \geq 0$, $v^*(t) \geq 0$, *on* $[0, t^*]$, *and* $T^*(t)$ *is an optimal controller satisfying all constraints.*

Proof. If $m_1 = m_0$ no fuel can be used and the optimal altitude $h^* = 0$. This occurs, in the restrained-altitude problem, when $t^* = 0$.

If $m_1 < m_0$ then $h^* > 0$, and this occurs at $t^* > 0$. Since $h^*(t^*) > 0$, $v(t^*) \geq 0$ for the absolutely continuous response $v^*(t)$ on $[0, t^*]$.

Suppose $v^*(t) < 0$ on some interval of positive duration in $[0, t^*]$, and let t_2 and t_3 be the endpoints of the last interval of positive duration in $[0, t^*]$ on which $v^*(t) < 0$, that is, $v^*(t) < 0$, $t \in (t_2, t_3)$, with $v^*(t_2) = v^*(t_3) = 0$, and $v^*(t) \geq 0$, $t \in [t_3, t^*]$. Clearly $M^*(t_2) \geq M^*(t_3) \geq 0$.

Because $v^*(t) < 0$ on (t_2, t_3), $h^*(t_3) < h^*(t_2)$. Let $M^*(t_2) - M^*(t_3) = \beta$. From the point $(h^*(t_2), v^*(t_2) = 0, M^*(t_2))$ pick the steering function $T(t) = T_1$ for a time interval $[t_2, \bar{t}_3]$ of length such that $(T_1/c_2)(\bar{t}_3 - t_2) = \beta$. Clearly $\bar{t}_3 \leq t_3$, $v(\bar{t}_3) > 0$, $h(\bar{t}_3) \geq h^*(t_2) > h^*(t_3)$, and $M(\bar{t}_3) = M(t_3)$. On the interval $[\bar{t}_3, t^* - (t_3 - \bar{t}_3)]$ pick $T(t)$ as in the proof of the lemma, with the appropriate shift in time scale. Because $v^*(t) \geq 0$ on $[t_3, t^*]$, $v(t) \geq 0$ on $[\bar{t}_3, t_1]$, $(t_1 = t^* - (t_3 - \bar{t}_3))$, and as in Lemma 1 we find $h(t_1) > h^*(t^*)$, $t_1 \leq t^*$, contradicting the optimality of $T^*(t)$. Q.E.D.

Necessary and Sufficient Conditions for Optimal Steering

We now formulate the problem in the standard form required for the necessary conditions of Chapter 5. Various properties of the optimal steering policy are then developed.

Consider the system

$$(\bar{\mathcal{S}}) \qquad \begin{aligned} \frac{dh}{dt} &= v \\ \frac{dv}{dt} &= \frac{T - D_0(v)e^{-\alpha h}}{m} - c_1 \\ \frac{dm}{dt} &= -\frac{T}{c_2}, \qquad 0 \leq T \leq T_1 \end{aligned}$$

where the drag force is now assumed to have the special form $D_0(v)e^{-\alpha h}$ still satisfying conditions (1) to (7) of the previous section, with $D_0(v) \in C^2$. It is further assumed that the drag belongs to the positive class:

8. $$v\left[\frac{\partial^2 D_0}{\partial v^2} + \frac{1}{c_2}\frac{\partial D_0}{\partial v}\right] + \frac{D_0}{c_2} > 0 \quad \text{for} \quad v > 0,$$

α is a positive constant, as are c_1 and c_2. The initial data is $h(0) = v(0) = 0$, $m(0) = m_0$, and we take the constraints $h \geq 0$, $v \geq 0$, $0 < m_1 \leq m \leq m_0$. We also assume that $T_1 > m_0 c_1$ (see discussion before Theorem 10).

The Hamiltonian function for the maximal principle (Chapter 5) is

$$H(\eta_1, \eta_2, \eta_3, h, v, m, T) = \eta_1 v + \eta_2\left[\frac{T - D_0(v)e^{-\alpha h}}{m} - c_1\right] + \eta_3\left[-\frac{T}{c_2}\right],$$

and the adjoint differential system is

$$(\bar{\mathcal{A}}) \qquad \begin{aligned} \dot{\eta}_1 &= -\eta_2 \frac{\alpha}{m} D_0(v)e^{-\alpha h} \\ \dot{\eta}_2 &= -\eta_1 + \eta_2 \frac{\partial D_0(v)}{\partial v}\frac{e^{-\alpha h}}{m} \\ \dot{\eta}_3 &= \eta_2 \frac{1}{m^2}[T - D_0(v)e^{-\alpha h}]. \end{aligned}$$

Let t_1 be the time at which the maximal altitude is attained, then we have the boundary conditions (Chapter 5): $\eta_1(t_1) \geq 0$, $\eta_2(t_1) = 0$, $\eta_3(t_1) \geq 0$, $m(t_1) \geq m_1$, $v(t_1) \geq 0$, $h(t_1) \geq 0$.

Define the switching function

$$W = \left[\eta_2 \frac{1}{m} - \eta_3 \frac{1}{c_2}\right].$$

From the maximal principle it follows that if $T(t)$ is a maximal steering function on $[0, t_1]$, then for almost all t on $[0, t_1]$, $(t_1 \leq \bar{t})$,

$$\begin{aligned} T(t) &= T_1 \quad \text{if} \quad W(t) > 0 \\ T(t) &= 0 \quad \text{if} \quad W(t) < 0 \\ 0 \leq T(t) &\leq T_1 \quad \text{if} \quad W(t) = 0. \end{aligned}$$

For such a maximal controller, if $W(t) < 0$ on some interval $[\xi_1, \xi_2]$ of positive duration in $[0, t_1]$ we call that interval a *coasting subarc*, if $W(t) > 0$ on $[\xi_1, \xi_2]$ we call it a *full-thrust subarc*, and if $W(t) \equiv 0$ on $[\xi_1, \xi_2]$ we call it a *variable-thrust subarc*. Note that $W(t)$ is the sum of absolutely continuous functions. Moreover $H \equiv 0$ along the maximal response $(h(t), v(t), m(t))$, $t \in [0, t_1]$, and the nonvanishing adjoint response is absolutely continuous on $[0, t_1]$.

Define, for convenience, the function

$$E = \left[v\frac{\partial D_0}{\partial v} + \frac{D_0 v}{c_2} - D_0\right]e^{-\alpha h} - mc_1$$

$$= \lambda(v)e^{-\alpha h} - mc_1,$$

with

$$\lambda(v) = \left[v\frac{\partial D_0}{\partial v} + \frac{D_0 v}{c_2} - D_0\right].$$

Note, that condition (8) on the drag force then becomes

$$\frac{d\lambda}{dv} > 0 \quad for \quad v > 0,$$

and further that

$$\lambda(0) = 0 \quad \text{so} \quad \lambda(v) > 0 \quad \text{for} \quad v > 0.$$

THEOREM 11. *Consider the system* $\bar{\mathbb{S}}$, *as above, satisfying conditions* (1) *to* (8) *of the previous subsection, with initial data* $h(0) = v(0) = 0$, $m(0) = m_0$. *Let* $T(t)$ *on* $[0, t_1]$ *be an optimal steering function with response* $(h(t), v(t), m(t))$ *satisfying constraints* $h(t) \geq 0$, $v(t) \geq 0$, $m_0 \geq m(t) \geq m_1$ *on* $[0, t_1]$. *Then the maximal controller* $T(t)$ *begins with a full-thrust subarc, that is, there exists* $\epsilon > 0$ *such that* $W(t) > 0$ *for* $t \in (0, \epsilon)$.

Proof. In order for the velocity and altitude to remain positive (and they can do this with $T_1 > m_0 c_1$) for $t \in [0, t_1]$ it is necessary that the optimal steering begin with $T(t) > 0$ on some compact set T_{ϵ_1} of positive measure in $[0, \epsilon_1]$, for a small $\epsilon_1 > 0$.

Suppose $W(t) \equiv 0$ on $[0, \epsilon_2]$ for some $\epsilon_2 > 0$. Then $dW/dt \equiv 0$ on $[0, \epsilon_2]$ (evaluating the derivative from the interior of the interval), that is.

$$-\frac{1}{m}\eta_1 + \eta_2\frac{1}{m^2}\left[\frac{\partial D_0}{\partial v}e^{-\alpha h} + \frac{D_0(v)e^{-\alpha h}}{c_2}\right] = 0$$

on $[0, \epsilon_2]$. From the conditions that $H \equiv 0$, $W(0) = 0$, and $\dot{W}(0) = 0$ one obtains $\eta_1(0) = \eta_2(0) = \eta_3(0) = 0$. This contradicts the nonvanishing of the adjoint response. Hence if $W(0) = 0$, $\dot{W}(0) \neq 0$. Obviously in this case $\dot{W}(0) > 0$ yields the positivity conditions on (v, h) [($\dot{W}(t)$ is also absolutely continuous on $[0, t_1]$]. Thus if $W(0) = 0$, $\dot{W}(0) > 0$, and the existence of $\epsilon > 0$ is assured.

The case when $W(0) < 0$ leads to a nonpositive condition and the case when $W(0) > 0$ ensures the existence of the duration $(0, \epsilon)$. Q.E.D.

Let $\hat{t}$ be such that

$$\frac{T_1}{c_2}\hat{t} = m_0 - m_1.$$

If $\bar{t} \leq \hat{t}$ the optimal steering is obviously to choose $T(t) \equiv T_1$ on $[0, t_1]$ and the optimal time $t_1 = \bar{t}$.

We now consider the case when $\bar{t} > \hat{t}$. The following lemma will be needed in establishing further properties of the optimal steering policy.

LEMMA 2. *Consider any response $(\eta_1(t), \eta_2(t), \eta_3(t))$ of the adjoint equation system $\bar{\mathcal{A}}$, as given above, corresponding to an admissible controller $T(t)$ on $[0, t_1]$ with response $(h(t), v(t), m(t))$ of the system $\bar{\mathrm{S}}$ having $h(t) \geq 0$, $v(t) \geq 0$ on $[0, t_1]$, and satisfying the boundary conditions $\eta_1(t_1) = k_1 > 0$, $\eta_2(t_1) = 0$, $\eta_3(t_1) \geq 0$. Then $\eta_2(t) > 0$ on $[0, t_1)$.*

Proof. Suppose $\eta_2(t_2) = 0$, $t_2 < t_1$. Consider first the case when $\eta_1(t_2) > 0$. But then $\dot{\eta}_2(t_2) = -\eta_1(t_2) < 0$, and $\dot{\eta}_1(t_2) = 0$. This indicates that $\eta_1(t)$ must become negative on $[t_2, t_1)$ and $\eta_2(t)$ is negative at the same point. Thus we need only consider the case when $\eta_1(t_2) < 0$, with $\eta_2(t_2) \leq 0$.

When $\eta_2(t) \geq 0$ we have $\dot{\eta}_1(t) \leq 0$. Since $\eta_1(t_1) > 0$ it is necessary that $\eta_1(t)$ become positive before $\eta_2(t)$ if $\eta_1(t)$ is to increase to this end value. But then $\dot{\eta}_2(t) < 0$ and η_2 cannot achieve the end value $\eta_2(t_1) = 0$.

Thus if the required solution of $\bar{\mathcal{A}}$ exists, then $\eta_2(t) > 0$ on $[0, t_1)$.

Q.E.D.

We now indicate a number of properties that obtain along an optimal trajectory. Later these will be used to construct an optimal steering policy for an example.

THEOREM 12. *Consider the above system $\bar{\mathrm{S}}$ satisfying conditions* (1) *to* (8), *with adjoint response satisfying the boundary conditions of Lemma 2. Let $T(t)$ on $[0, t_1]$ be an optimal steering function and $(h(t), v(t), m(t))$ the corresponding response satisfying the boundary condition $v(0) = h(0) = 0$, $m(0) = m_0$, and the phase constraints $h(t) \geq 0$, $v(t) \geq 0$, $m(t) \geq m_1$ on $[0, t_1]$. Then switching from a full-thrust subarc to a coasting subarc is only possible at an altitude where the fuel is exhausted $(m(t) = m_1)$; that is, $W(t) > 0$ for $t \in [\xi_1, t_2)$ and $W(t) < 0$ for $t \in (t_2, t_1)$, $\xi_1 \leq t_2 \leq t_1$, can only occur if $m(t_2) = m_1$.*

Proof. At the switching point from the full-thrust subarc to coasting the absolutely continuous switching function $W(t)$ must vanish. Since

$\eta_2(t) > 0$ except at the final time, by Lemma 2, it follows that

$$\frac{dW}{dt} = \frac{\eta_2 E}{m^2 v} \leq 0$$

at the switching point, that is $E \leq 0$. Along the coasting subarc $\dot{m} = 0$, $\dot{v} < 0$, and

$$\frac{dE}{dt} = \dot{\lambda}(v)e^{-\alpha h}\dot{v} - \alpha\, \lambda(v)e^{-\alpha h}\dot{h}$$

$$= [\dot{\lambda}(v)\dot{v} - \alpha\, \lambda(v)v]e^{-\alpha h} < 0.$$

Thus $\dfrac{dW}{dt} < 0$ beyond the switch point, so $W(t)$ can never become positive. We now use the principle which indicates that, if time permits along the optimal trajectory, all of the fuel is used. Q.E.D.

COROLLARY. *Switching from a variable-thrust subarc to a coasting subarc can take place only if the fuel is exhausted.*

The last result of this section is a decomposition of the optimal program into at most three subarcs when the maximal thrust is large enough.

THEOREM 13. *Consider the system* $\bar{S}$ *of Theorem* 12. *If*

$$T_1 \geq (\alpha c_2^{\,2}/c_1)\, \lambda(c_2 \log (m_0/m_1)) \log (m_0/m_1),$$

then the optimal steering policy is composed of a full-thrust subarc, $W > 0$, *followed by a variable-thrust subarc* $W \equiv 0$, *and then a coasting subarc; or a full-thrust subarc followed by a coasting subarc; or just a full-thrust subarc when* $\bar{t} \leq \hat{t}$. (*The thrust can have its maximal value on a variable thrust arc, as is noted below.*)

Proof. Because of the previous results we need only verify that there is no variable-thrust subarc followed by a full-thrust subarc. First we establish a bound on the attainable velocity. From

$$\dot{v} < -c_2 \frac{\dot{m}}{m}$$

and the required boundary conditions along an optimum it follows that

$$v_{\max} < c_2 \log \left(\frac{m_0}{m_1}\right) = \hat{v}$$

At a switching at t_s from a variable-thrust subarc to a full-thrust subarc $E(t_s) = 0$. Along the full-thrust subarc $\dot{v} > 0$ and

$$\frac{dE}{dt} = e^{-\alpha h}\left[\dot{\lambda}(v)\dot{v} - \alpha\, \lambda(v)v + e^{\alpha h}\frac{T_1}{c_2}c_1\right] > 0$$

if $T_1 \geq (\alpha c_2/c_1)\, \lambda(\hat{v})\hat{v}$.

But then $dW/dt > 0$ for $t \geq t_s$, and since $W(t_s) = 0$, $W(t)$ cannot decrease to the required boundary condition

$$W(t_1) = \left[\frac{\eta_2(t_1)}{m(t_1)} - \eta_3(t_1)\right] \leq 0.$$

Thus the switching from a variable-thrust subarc to a full-thrust subarc cannot take place. Q.E.D.

Remarks. Better estimates for the condition on T_1 of Theorem 13 are easily found [Munick].

During the variable-thrust subarc $W(t) \equiv 0$ so the optimal controller must have a form such that $E(t) \equiv 0$. This requires

$$T(t) = \frac{\dot{\lambda}(v)[D_0(v)e^{-\alpha h} + mc_1] + \alpha mv\,\lambda(v)}{\dot{\lambda}(v) + (c_1/c_2)me^{\alpha h}}.$$

Further details and other cases for this problem can be found in [Munick]

Numerical Example for Goddard Problem

In Figures 7.14 and 7.15 we have constructed the response for four different thrust programs, labeled 1, 2, 3 and 4, for the drag force $D_0(v)e^{-\alpha h}$ with $D_0(v) = v^2$, and $\alpha = 1$. Notice the improvement in the maximal altitude achieved by changing from the program of "full thrust till the fuel is burned" to the program 4, which is near the above optimal program (see Exercise 4).

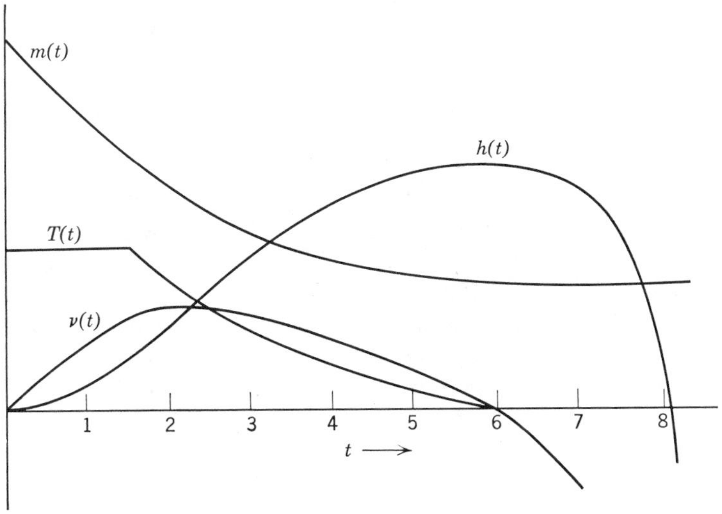

Figure 7.14 Altitude-velocity responses for Goddard problem with drag force $D(v, h) = v^2e^{-h}$.

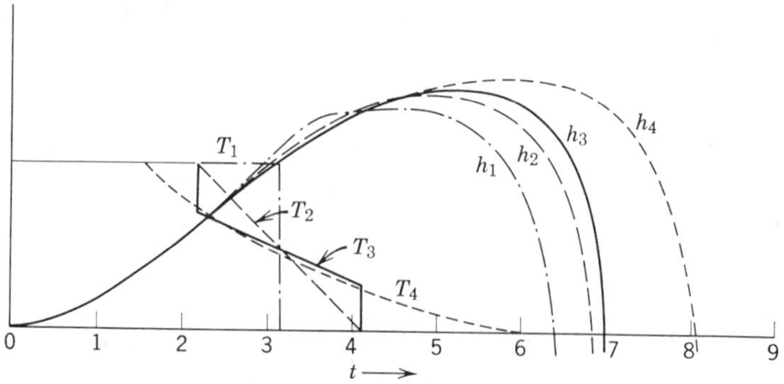

Figure 7.15 Comparison of altitudes for Goddard problem with thrust program 1-4.

EXERCISES

1. Show that for a constant-density atmosphere, $D(v, h) = D_1(v)$, the conclusion of Theorem 13 is valid even without the requirement on T_1.
2. Calculate the optimal thrust program for a typical sounding rocket when the drag force increases sharply near a velocity equal to that of the velocity of sound.
3. Prove existence of the optimal steering for the equation system $\mathcal{S}_0$ under the assumptions that $D(v, h)$ is piecewise differentiable and $v \ D(v, h) \geq 0$. (See Theorem 4 of Chapter 4.)
4. Complete the sequence of thrust programs by constructing the optimal response of the numerical example for the Goddard problem as considered in the text above.

7.3 ANGULAR VELOCITY CONTROL FOR A RIGID BODY

Consider a rigid body in an inertial reference frame and there free from all external noninertial forces. Let ω_x, ω_y, ω_z be the components of angular velocity along the body-fixed x, y, z rectangular-coordinate system located at the center of gravity, and let I_x, I_y, I_z denote the principal moments of inertia (positive real numbers). Assume that we are able to generate torques on the body by couples obtained from pairs of steering jets. Call these torques u_x, u_y, and u_z depending on the axis about which they act. The so-called Euler equations of motion are

$$(Q) \qquad \begin{aligned} I_x\dot{\omega}_x &= \omega_y\omega_z[I_y - I_z] + b_x u_x \\ I_y\dot{\omega}_y &= \omega_z\omega_x[I_z - I_x] + b_y u_y \\ I_z\dot{\omega}_z &= \omega_x\omega_y[I_x - I_y] + b_z u_z \end{aligned}$$

where b_x, b_y, b_z are positive constants.

It is natural to assume that the thrust available from the steering jets is limited. We consider two cases:

1. For each axis there is a pair of steering jets, creating an independent couple for each axis with $|u_i| \leq 1$, $i = x, y, z$.
2. Only one pair of steering jets is available, but they can assume any angular position with respect to the rigid body, imposing the restraint $\|u\|^2 = u_x^2 + u_y^2 + u_z^2 \leq 1$.

The problem of control is to stop the spinning vehicle in an optimal fashion, in particular to find a feedback controller synthesis that provides the optimal control for each initial angular velocity. The optimality will be in terms of minimal time, or minimal fuel or energy during a prescribed time interval. In the next section we consider the problem of existence of optimal controllers.

Existence of Optimal Controllers

Consider the above equation system Q with initial angular velocity $(\omega_{x0}, \omega_{y0}, \omega_{z0})$ at time $t = 0$. We show first that in either case (1) or (2) there exists an admissible controller steering $(\omega_x(t), \omega_y(t), \omega_z(t))$, a response of Q with initial data $(\omega_{x0}, \omega_{y0}, \omega_{z0})$, to $(0, 0, 0)$ in a finite time duration. Define $J = \frac{1}{2}(I_x\omega_x^2 + I_y\omega_y^2 + I_z\omega_z^2)$ and choose the feedback controller synthesis

$$u_x = -\frac{1}{2}\frac{\alpha I_x\omega_x}{b_x J^{1/2}}$$

$$u_y = -\frac{1}{2}\frac{\alpha I_y\omega_y}{b_y J^{1/2}}$$

$$u_z = -\frac{1}{2}\frac{\alpha I_z\omega_z}{b_z J^{1/2}}$$

where

$$\alpha = \min\left\{\frac{b_x}{I_x^{1/2}}, \frac{b_y}{I_y^{1/2}}, \frac{b_z}{I_z^{1/2}}\right\}.$$

Clearly $(u_x^2 + u_y^2 + u_z^2) \leq 1$ and $|u_x| \leq 1$, $|u_y| \leq 1$, $|u_z| \leq 1$. We calculate dJ/dt along the solution curves of Q with the above feedback controller to obtain

$$\frac{dJ}{dt} = \frac{-\frac{1}{2}\alpha}{J^{1/2}}[I_x\omega_x^2 + I_y\omega_y^2 + I_z\omega_z^2] = -\alpha J^{1/2}.$$

Let $W = J^{1/2}$. Then $\dot{W} = -\frac{1}{2}\alpha$ (for $W \geq 0$). Thus $W(t)$ goes to zero in finite time for each given initial condition, and therefore so does $J = W^2$. Because of the positive nature of J, $J = 0$ implies $\omega_x = \omega_y = \omega_z = 0$.

It is now shown that each response of Q under either of the restraints 1 or 2 is uniformly bounded on each finite time interval $[0, \tau]$. Consider the function J as above. Then

$$\frac{dJ}{dt} = \omega_x b_x u_x + \omega_y b_y u_y + \omega_z b_z u_z.$$

Obviously

$$\left|\frac{dJ}{dt}\right| \leq \gamma J^{1/2}$$

for some constant $0 \leq \gamma < \infty$, if we take into account the restraint 1 which contains 2 as a subset. So

$$0 \leq J(t) \leq (\gamma t + c)^2$$

for some constant $0 \leq c < \infty$. Thus each response starting at $(\omega_{x0}, \omega_{y0}, \omega_{z0})$ is uniformly bounded on each finite time duration $0 \leq t \leq \tau$.

When the time-optimal problem is considered, obviously the convexity and continuity conditions of our existence theorem 4 of Chapter 4 are satisfied. Further, we have just shown that there is always an admissible controller steering each initial state to the origin in finite time duration and that all responses are uniformly bounded on each finite time interval. These are valid even in the extended-phase space when considering the time-cost coordinate. Thus the existence of the time optimal controller has been established with either of the restraints 1 or 2.

The minimal-energy problem† we consider is the transfer of the system Q from the initial point $(\omega_{x0}, \omega_{y0}, \omega_{z0})$ to $(0, 0, 0)$ during the fixed time interval $[0, T]$, minimizing the cost coordinate

$$x_E{}^0(T) = \tfrac{1}{2}\int_0^T \|u(t)\|^2 \, dt.$$

The minimal-fuel problem is the same except the cost coordinate is

$$x_F{}^0(T) = \int_0^T \|u(t)\| \, dt.$$

If we add the cost coordinate to the basic system the convexity hypothesis of our basic existence theorem 4 of Chapter 4 is violated. However, both of the cost coordinates and either case (1) or (2) satisfy the existence theorem of Exercise 2.2 of Chapter 4. We have already established the existence of the uniform bound on the basic response $(\omega_x(t), \omega_y(t), \omega_z(t))$ on each finite time interval, and it is easy to see that the cost coordinate is also bounded if the two restraints 1 or 2 are considered. Thus the optimal fuel

† The concepts of fuel and energy used here are mathematical conventions and may bear little relation to the actual energy and fuel.

and energy controllers exist if the time T is at least as large as the time required for transfer from $(\omega_{x0}, \omega_{y0}, \omega_{z0})$ to $(0, 0, 0)$ in the time optimal problem.

We will hereafter consider only the time-optimal problem to illustrate the application of the previous theory. The minimal-fuel and energy problems can be treated in a similar manner and we prefer to leave the actual execution of the synthesis to the exercises. The introduction of the fuel and energy problems serves to call attention to the existence theorem of Exercise 2.2 of Chapter 4.

Synthesis of Time Optimal Controllers

Consider the above equation system Q with the controller restraint $u_x^2 + u_y^2 + u_z^2 \leq 1$, and assume that $b_1 = b_2 = b_3$, that is, the control has equal influence in each direction.

We now show directly that the *solution to the time-optimal problem with the restraint* 2 is to pick the feedback controller synthesis

$$u_x^* = -\frac{I_x\omega_x}{M^{1/2}}$$

$$u_y^* = -\frac{I_y\omega_y}{M^{1/2}}$$

$$u_z^* = -\frac{I_z\omega_z}{M^{1/2}}$$

where

$$M = I_x^2\omega_x^2 + I_y^2\omega_y^2 + I_z^2\omega_z^2.$$

Clearly $u_x^{*2} + u_y^{*2} + u_z^{*2} \leq 1$.

It is easily shown, using the Schwarz inequality, that

$$-1 \leq \frac{dM^{1/2}}{dt} \leq 1$$

along any path and control pair of system Q with the above restraint 2 on the controller.

Moreover, if the controller u^* is used

$$\frac{dM^{*1/2}}{dt} = -1$$

and u^* steers to the origin in finite time by an analysis similar to that of the previous section. Comparison of the above differential inequalities indicates the time optimality of the above feedback controller synthesis.

Thus the optimal controller is at its maximal amplitude limit 1 and is directed oppositely to the angular-momentum vector $(I_x\omega_x, I_y\omega_y, I_z\omega_z)$.

More general problems, called norm-invariant problems, can be treated in a manner similar to the above time optimal problem with restraint 2 (see Exercise 1).

We now turn to a method for constructing the feedback controller synthesis for the time-optimal problem under the restraint 1, that is when $|u_x| \leq 1$, $|u_y| \leq 1$, $|u_z| \leq 1$. To expedite the discussion it is assumed that the rigid body has a certain symmetry, namely, $I_x = I_y$.

For this case, where the first two principal moments of inertia are equal, the Euler equations are

$$(Q_s)\qquad \begin{aligned} \dot{\omega}_x &= \alpha\omega_y\omega_z + \beta_x u_x \\ \dot{\omega}_y &= -\alpha\omega_x\omega_z + \beta_y u_y \\ \dot{\omega}_z &= \beta_z u_z, \end{aligned}$$

where $\alpha = (I_y - I_z)/I_x$ and $\beta_x = b_x/I_x, \beta_y = b_y/I_y, \beta_z = b_z/I_z$. As in Chapter 4, define the Hamiltonian

$$H = \eta_1[\alpha\omega_y\omega_z + \beta_x u_x] + \eta_2[-\alpha\omega_x\omega_z + \beta_y u_y] + \eta_3\beta_z u_z$$

where (η_1, η_2, η_3) satisfy the differential system

$$(\mathcal{A})\qquad \begin{aligned} \dot{\eta}_1 &= -\frac{\partial H}{\partial \omega_x} = \alpha\omega_z\eta_2 \\ \dot{\eta}_2 &= -\frac{\partial H}{\partial \omega_y} = -\alpha\omega_z\eta_1 \\ \dot{\eta}_3 &= -\frac{\partial H}{\partial \omega_z} = -\alpha\omega_y\eta_1 + \alpha\omega_x\eta_2. \end{aligned}$$

A first integral for this system is $\eta_1^2 + \eta_2^2 = c$, and because η_1, η_2, η_3 enter H linearly, it is assumed that $|\eta_1(0)| + |\eta_2(0)| + |\eta_3(0)| = 1$, which gives a solution that has the same zeros.

The maximal principle states that if $(u_x(t), u_y(t), u_z(t))$ is optimal, it must maximize H. Therefore, for known $\omega_i(t), \eta_i(t)$

$$\begin{aligned} u_i(t) &= 1 \quad \text{if} \quad \eta_i(t) > 0 \\ &= -1 \quad \text{if} \quad \eta_i(t) < 0 \qquad i = x, y, z, \quad \text{or} \quad 1, 2, 3. \end{aligned}$$

Thus $\eta_1(t)$, $\eta_2(t)$ uniquely specify $u_x(t)$, $u_y(t)$, $u_z(t)$, except when $\eta_1(t)$ $\eta_2(t)$, or $\eta_3(t)$ is zero. Thus it appears that the zeros of $\eta_1(t)$, $\eta_2(t)$, $\eta_3(t)$ play a special role in specifying maximal controllers. If the total time for which $\eta_1(t)$, $\eta_2(t)$, $\eta_3(t)$ are zero has measure zero, then the control is specified almost everywhere.

Equations $\mathcal{A}$ are now considered. It is supposed that $\eta_1(0) = \eta_2(0) = 0$; then $\dot{\eta}_1 = 0$, $\dot{\eta}_2 = 0$, $\dot{\eta}_3 = 0$ implies that $\eta_1(t) = 0$, $\eta_2(t) = 0$, $\eta_3(t) = 1$ or -1. Thus the only control that is specified is $u_z(t)$. However, this is no real problem. Consideration is given the following method, which will be used in constructing optimal trajectories that lead to the origin. Namely, a start is made at the origin [that is, with $\omega_x(0) = \omega_y(0) = \omega_z(0) = 0$]; any $\eta_1(0)$, $\eta_2(0)$, $\eta_3(0)$ is chosen with $|\eta_1(0)| + |\eta_2(0)| + |\eta_3(0)| = 1$; and the response of equations Q_s and $\mathcal{A}$ to maximal control on $-\infty < \tau \leq t \leq 0$ is considered. As $t \to \tau$, the response $\omega_x(t)$, $\omega_y(t)$, $\omega_z(t)$ traces out a certain set of points called a maximal trajectory. As $\eta_1(0)$, $\eta_2(0)$, $\eta_3(0)$ range over the sphere $|\eta_1(0)| + |\eta_2(0)| + |\eta_3(0)| = 1$, all maximal trajectories that pass through the origin are obtained. These maximal trajectories contain the optimal trajectories. Therefore, if there are only a few maximal trajectories that connect a point to the origin, an optimum can be determined. Along this optimal trajectory the control is known and can be specified as a function of the variables $(\omega_x, \omega_y, \omega_z)$, which is the desired control for all points that happen to lie on this trajectory. If a dense set of nonintersecting optimal trajectories is found, the optimal control is known almost everywhere. This is the information required for the storage of the control feedback function. This is the backing-out method used in finding the feedback controllers in Section 7.1.

Return now to the specifying of maximal controllers when $\eta_1(t) = 0$, $\eta_2(t) = 0$, $\eta_3(t) = 1$ or -1. Consideration is given the case when $\eta_3(t) = 1$. Starting at the point $\omega_x(0) = \omega_y(0) = \omega_z(0) = 0$, with $u_z(t) = 1$, and $|u_x(t)| \leq 1$, $|u_y(t)| \leq 1$ on the interval $t_1 \leq t \leq 0$, a certain cone K of points is attained. It is easy to show that this cone increases continuously as $t_1 \to \tau$. Obviously if t^* is the first t_1 for which a point $(\omega_x, \omega_y, \omega_z)$ is in this cone, this will specify the smallest time interval for controlling from this point to the origin. But it is noted that $u_z(t)$ must equal $+1$, and $u_x(t)$, $u_y(t)$ are any allowable controls that bring (ω_x, ω_y) to $(0, 0)$ in the interval $[t^*, 0]$.

This supports the conjecture that maximal controllers are optimal controllers. It is not difficult to show that the point $(\omega_x, \omega_y, \omega_z)$ lies in the base corner of some smaller cone that is achieved by keeping $u_x(t) = u_y(t) = 0$ on an interval $[t_a, 0]$ and then using any allowable controller on $[t^*, t_a]$. By restricting the analysis to the (ω_x, ω_y)-plane, it can be shown that if $u_x(t) = u_y(t) = 0$ on $[t_a, 0]$, then for $t \in [t^*, t_a]$, $u_x(t)$ and $u_y(t)$ can be relay controls and can be uniquely specified by $\eta_1(t)$, $\eta_2(t)$ with

$$|\eta_1(t_a)| + |\eta_2(t_a)| > 0.$$

The explanation of this nonuniqueness is that if the initial value of ω_z is so large that it takes longer to control it than to control the other two variables, it can be considered by itself and the other two variables controlled by any allowable control that brings them to zero in the same (or smaller) time interval.

The other cases where the η_i's do not specify the control uniquely are now considered. Suppose that $\eta_1(t)$ has a collection of zeros; then $\dot{\eta}_1(t) \equiv 0 \equiv \alpha\omega_z\eta_2$. This implies that either $\eta_2(t) \equiv 0$ or $\omega_z \equiv 0$, or both. If $\omega_z \equiv 0$, then $\dot{\omega}_z \equiv 0 = \beta_z u_z$ implies $\eta_3 \equiv 0$, implies $\dot{\eta}_3 \equiv 0$, implies $\eta_1\omega_y = \omega_x\eta_2$, implies $\eta_1\beta_x \operatorname{sgn} \eta_2 \equiv \eta_2\beta_y \operatorname{sgn} \eta_1$, which is easily shown to be impossible. Thus $\eta_2(t) \equiv 0$ implies $\dot{\eta}_2(t) \equiv 0$, implies $\dot{\eta}_3(t) \equiv 0$, which is the case considered in the foregoing, since $|\eta_3| = 1$. A similar argument holds for any collection of zeros of $\eta_2(t)$. Lastly, it is assumed that $\eta_3(t) \equiv 0$ implies $\dot{\eta}_3(t) \equiv 0$, implies $\eta_1\omega_y \equiv \omega_x\eta_2$, which is impossible. Thus the only case in which the control is not uniquely specified by

$$[\eta_1(t), \eta_2(t), \eta_3(t)]$$

is when $\eta_1(0) = \eta_2(0) = 0$, that is, in the cones as described previously. Therefore, optimal control can be relay control except along the ω_z axis, and it is indicated how optimal control can be determined experimentally.

We will not carry this analysis any further. It has been indicated how optimal trajectories can be found, and once this information is available one can store it by, for example, training a logical network (see [Smith] for a description of such a training and storage procedure). The only question of construction is whether two maximal (or optimal) trajectories pass through the same point. Because of the symmetry of our example it is easy to see that there are initial points through which two optimal trajectories pass. The training method must take account of this.

EXERCISES

1. Consider the nonlinear differential system (in vector notation)

$$(\mathcal{S}) \qquad \dot{x} = f(x, t) + u(t) \quad \text{in} \quad R^n$$

with controller restraint in R^m

$$\Omega\colon \|u\| \leq k \quad \text{for} \quad \text{some } k > 0.$$

Here $\|u\|^2 = \sum_{i=1}^{m} (u^i)^2$.

Assume the above system $\mathcal{S}$ has the property that all solutions of the homogeneous equation

$$\dot{x} = f(x, t)$$

lie on a sphere in R^n; that is,

$$\|x(t)\| = \|x(0)\| \quad \text{all} \quad t \geq 0.$$

Consider the minimal-time problem to the origin and show that the time-optimal feedback controller synthesis is

$$u^*(t) = \frac{-k\, x(t)}{\|x(t)\|}$$

where $x(t)$ is the (present) system state.

2. Consider the Euler equations of motion Q for the rigid body in free space. Suppose $I_x \neq I_y \neq I_z \neq I_x$ and that the angular velocity $\omega_x(t)$ can be measured on $[0, t_1]$, $(t_1 > 0)$ by a gyroscopic instrument.

(a) Show that if $u_i(t) \equiv 0$, $i = x, y, z$ on $[0, t_1]$, it is impossible to calculate $\omega_y(0)$, $\omega_z(0)$ from the knowledge of Q, $u(t)$, $\omega_x(t)$ on $[0, t_1]$.

(b) Show that if $u_i = 0$ on $[0, t_1]$ and $u_i(t) = k_i$ on $[t_1, t_2]$ for any $k_i \neq 0$, $0 < t_1 < t_2$, $i = x, y, z$, it is possible to determine both $\omega_y(0)$ and $\omega_z(0)$ from the knowledge of Q and $(\omega_x(t), u(t))$ on $[0, t_2]$.

7.4 OPTIMAL GUIDANCE BETWEEN PLANETARY ORBITS

We consider the problem of guiding a rocket ship through interplanetary space with a minimal fuel consumption. The fundamental problem of this type requires the guidance of the ship from an initial phase (position and velocity) in the solar system to some assigned target phase. This is the general rendezvous problem, with or without a prescribed flight duration. Such a general problem has not been solved in any effective manner, and we shall simplify this space-guidance problem so that an elementary but significant solution can be computed.

We seek to transfer the ship from one elliptical *Kepler orbit* to another, with minimal specific impulse. No rendezvous or time restriction is imposed on the process. We consider only orbits that all lie in one fixed plane that contains the sun at the origin of (x, y) coordinates. We consider only ellipses that have one focus fixed at the sun and the other focus lying on a given line, the x-axis. Thus each such ellipse is completely specified by two real parameters (all ellipses are traced in the same sense of rotation)

$$f = \text{abscissa of the second focus}, \qquad -\infty < f < \infty$$

$$l = \text{length of major axis}, \qquad l > 0.$$

The usual parameters of the ellipse are

$$c = \frac{|f|}{2}, \qquad a = \frac{l}{2}, \qquad e = \frac{c}{a}.$$

Circular orbits correspond to $f = 0$ (or eccentricity $e = 0$), and they have diameters measured by l.

We admit only impulsive thrusts that are fired tangential along the elliptical orbit at the exact instant of perihelion or aphelion. Each such impulse control will modify an elliptical orbit of the given type to a new elliptical orbit of the type considered (provided the solar escape velocity is not attained). Thus an admissible controller will consist of a finite sequence of impulsive thrusts, tangential to the orbit at alternating transits across the positive and negative x-axes.

The cost of each admissible controller will be defined as the sum of all specific impulses, that is, the sum of all velocity changes when each change is counted as a positive number.

Let us represent each Kepler orbit, specified by the data (f, l), by a point in an upper half-plane with cartesian coordinates f, $l > 0$. Then each controller can be represented as a finite sequence of points in this plane, or a polygon joining the initial to the terminal orbit coordinates.

If an orbit has coordinates $f > 0$, $l > 0$, then a thrust (backwards or forwards) at aphelion (at $x > 0$ transit) preserves the aphelion distance $(l + f)/2$. A similar inspection of all possible cases leads to the conclusion that thrusts where $x > 0$ preserve $(l + f)/2$, but thrusts where $x < 0$ preserve $(l - f)/2$. Therefore the polygon representing any admissible controller must consist of alternate segments with slope -1 and slope $+1$ in the (f, l)-plane.

We next turn to the computation of the cost of each such polygonal control path. From celestial mechanics we describe the Kepler orbits as the solutions of *Newton's equations*

$$\ddot{x} = \frac{-kx}{r^3}, \qquad \ddot{y} = \frac{-ky}{r^3}$$

where k is the solar gravitational constant. Then the polar coordinate (r, θ) presentation of the orbit is known to be

$$r = \frac{h^2/k}{1 + e\cos\theta}$$

where

$$h^2 = k\,a(1 - e^2) \quad \text{and} \quad r^2\dot{\theta} = h.$$

The perihelion velocity is then $\dfrac{h^2/k}{1 + e}\, v_{\text{per.}} = h$

$$v_{\text{per.}} = \sqrt{\frac{2k}{l}}\sqrt{\frac{l + |f|}{l - |f|}},$$

and the aphelion velocity is

$$v_{\text{aph.}} = \sqrt{\frac{2k}{l}}\sqrt{\frac{l - |f|}{l + |f|}}.$$

Consider a thrust at $x > 0$. If $f > 0$ this corresponds to aphelion and the specific impulse is

$$\Delta v = \Delta\sqrt{\frac{2k}{l}}\sqrt{\frac{l-f}{l+f}}.$$

However if $f < 0$ at perihelion, the specific impulse is

$$\Delta v = \Delta\sqrt{\frac{2k}{l}}\sqrt{\frac{l-f}{l+f}},$$

which is the same in the two cases. Therefore a thrust at $x > 0$ corresponds

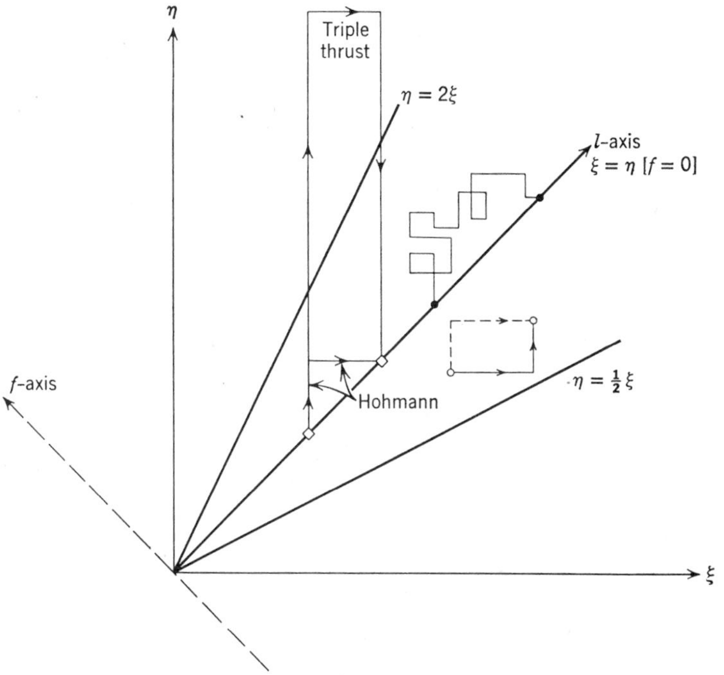

Figure 7.16 Impulse control between elliptical orbits.

to a polygon segment with slope -1 and with cost computed by the magnitude of the change of $\sqrt{2k/l}\sqrt{(l-f)/(l+f)}$ between the ends of the segment. A similar analysis for a thrust at $x < 0$ yields a polygon segment with slope $+1$ and cost computed by the magnitude of the change of $\sqrt{2k/l}\sqrt{(l+f)/(l-f)}$ between the ends of the segment.

As an example of our theory consider the transfer between two given elliptical orbits (f_0, l_0) and (f_1, l_1) by two impulses. This controller is represented by a polygon with two sides, and there are two possibilities [see Figure 7.16, in which the (f, l)-coordinates have been rotated through 45°].

In the case where $f_0 = f_1 = 0$ so the initial and final orbits are circular, the two possibilities yield the same cost (by symmetry) and such a controller is called a Hohmann orbit transfer. We compute the cost of the Hohmann orbit transfer between circular orbits of diameters l_0 and $l_1 > l_0$. The single intermediate Hohmann orbit is an ellipse with $f = (l_1 - l_0)/2$ and $l = (l_1 + l_0)/2$, which is tangent to the initial and final circles.

The first segment, at $x < 0$, has cost

$$\Delta v = \sqrt{\frac{4k}{l_0 + l_1}}\sqrt{\frac{l_1}{l_0}} - \sqrt{\frac{2k}{l_0}} \cdot 1$$

and the second segment, at $x > 0$, has cost

$$\Delta v = \sqrt{\frac{2k}{l_1}} \cdot 1 - \sqrt{\frac{4k}{l_0 + l_1}}\sqrt{\frac{l_0}{l_1}}.$$

If we set the diameter ratio $l_1/l_0 = \rho > 1$, and choose the time units so that the initial orbital speed is unity

$$v_0^2 = \frac{2k}{l_0} = 1,$$

then the total Hohmann cost is

$$C(\rho) = \sqrt{\frac{2\rho}{1 + \rho}} - 1 + \sqrt{\frac{1}{\rho}} - \sqrt{\frac{2}{\rho(\rho + 1)}}.$$

Of course, $C(1) = 0$, and we easily compute $C(\frac{3}{2}) = 0.18$, $C(2) = 0.29$. The orbital range $1 < \rho < 2$ is of particular importance in the Hohmann transfer theory. For within this range we shall find that the Hohmann double-thrust transfer is the optimal controller between two circular orbits. For further reference we note the cost derivative

$$\frac{dC}{d\rho} = \sqrt{\frac{1}{2\rho}}\frac{1}{(1 + \rho)^{3/2}} - \frac{1}{2\rho\sqrt{\rho}} + \frac{(2\rho + 1)}{\sqrt{2}[\rho(\rho + 1)]^{3/2}},$$

and in $1 < \rho < 2$ the first and last terms dominate the second term so that $dC/d\rho > 0$.

A further computation for $\rho = 12$ yields $C(12) = 0.534$. We compare this with the cost of a triple-thrust program, first a very large thrust on $x < 0$, then a small thrust on $x > 0$, and finally a large thrust on $x < 0$. This controller takes the rocket ship out of the first circular orbit into a very elongated ellipse, the next orbit is a slightly modified ellipse tangent to the final circle, and the last thrust slows the ship to fit into the final circular orbit. A maneuver of this type can be selected to give an arbitrarily close approximation to a parabolic escape from the first circle, an adjustment of zero cost at infinity, and a parabolic return to the target circle. Using the fact that the parabolic escape velocity is always $\sqrt{2}$ times the circular orbital velocity we can compute the cost of this limiting triple-thrust controller to be

$$C_{\infty}(12) = [\sqrt{2} - 1] + 0 + \left[\sqrt{2}\sqrt{\frac{1}{12}} - \sqrt{\frac{1}{12}}\right] = 0.533\,.$$

For an arbitrary value of $\rho > 1$ this triple-thrust transfer out near infinity costs $C_{\infty}(\rho) = 0.414\,[1 + \sqrt{\rho}/\rho]$. Therefore we conclude that for $\rho \geq 12$ the Hohmann transfer is definitely not optimal among the class of admissible impulse controllers. Therefore our theory shows that a circular orbit transfer from the vicinity of Earth to that of Venus or Mars is most efficiently effected by a Hohmann transfer. However for a transfer from the vicinity of Earth to Uranus the Hohmann transfer is not optimal.

We must now show that for small ratios $\rho > 1$ the Hohmann transfer is the optimal controller. For convenience in graphing we introduce new coordinates (ξ, η) by

$$\xi = l - f, \qquad \eta = l + f$$

so that the line of circular orbits $f = 0$ becomes the line $\eta = \xi$. The admissible controllers are represented by polygons with horizontal and vertical sides and with costs computed by corresponding changes in the functions

$$\sqrt{\frac{4k\xi}{(\xi + \eta)\eta}} \quad \text{and} \quad \sqrt{\frac{4k\eta}{(\xi + \eta)\xi}}\,.$$

Thus we can compute the cost of a polygon by evaluating the line integral

$$\int \sqrt{\frac{k\eta}{(\xi + \eta)^3\xi}}\,d\xi + \sqrt{\frac{k\xi}{(\xi + \eta)^3\eta}}\,d\eta.$$

Consider all admissible polygonal paths joining $\xi_0 = \eta_0 = l_0$ and $\xi_1 = \eta_1 = l_1 = \rho l_0$ with the additional condition that each such polygon remains always within the sector

$$0 < \tfrac{1}{2}\xi \leq \eta \leq 2\xi.$$

Any such polygon can be replaced by a polygon in $\xi \leq \eta \leq 2\xi$ without changing the cost. This follows from the symmetry of the cost functionals about the line $\eta = \xi$. Also any polygon can be improved (cost decreased) by eliminating any downwards movement in the sojourn from (ξ_0, η_0) to (ξ_1, η_1). This follows from the monotonicity of cost

$$\frac{\partial}{\partial\eta}\sqrt{\frac{\eta}{(\xi + \eta)^3\xi}} = \left[\frac{\eta}{(\xi + \eta)^3\xi}\right]^{-1/2} \frac{\xi - 2\eta}{2\xi(\xi + \eta)^4} < 0.$$

Any rise of the polygon above the level n_1 can be eliminated. This follows from a direct calculation that the base side of any (infinitesimally flat) rectangle costs less than the sum of the other three sides. Similarly a leftward horizontal movement can be eliminated to improve a control polygon. Therefore each polygon can be improved by replacing it by a step polygon always rising towards the right. Finally we show that each step can be filled in to reduce the cost further. We must compute the cost along the left vertical and top sides of a rectangle and show that this is less than the cost along the bottom side and right vertical side. To do this we verify

$$\oint \sqrt{\frac{k\eta}{(\xi + \eta)^3\xi}}\, d\xi + \sqrt{\frac{k\xi}{(\xi + \eta)^3\eta}}\, d\eta > 0$$

where the counterclockwise circuit lies within the specified angular sector. But Green's theorem shows that this line integral is equal to the double integral over the rectangle

$$\iint \left[\frac{\partial}{\partial \xi}\sqrt{\frac{k\xi}{(\xi + \eta)^3\eta}} - \frac{\partial}{\partial \eta}\sqrt{\frac{k\eta}{(\xi + \eta)^3\xi}}\right] d\xi\, d\eta.$$

The integrand has the same sign as

$$q(\lambda) = \lambda^3 - 2\lambda^2 + 2\lambda - 1 = (\lambda - 1)(\lambda^2 - \lambda + 1)$$

where $\lambda = \eta/\xi$ lies in the interval $1 < \lambda < 2$. An elementary investigation shows that $q(\lambda) > 0$ and hence each step of our controller can be filled to form a concave step polygon with smaller cost. Therefore the optimal controller, inside the sector $\frac{1}{2}\xi \leq \eta \leq 2\xi$, is the Hohmann transfer described by a polygon of two sides: a vertical side from (ξ_0, η_0) to (ξ_0, η_1) and a horizontal side from (ξ_0, η_1) to (ξ_1, η_1). It is clear that for $1 \leq \rho \leq 2$ the Hohmann transfer does correspond to a polygon that lies in the given sector.

To complete the problem we must show that controllers corresponding to polygons leading outside the sector always cost more than the Hohmann transfer when $\rho > 1$ is suitably restricted. Consider any polygon, corresponding to an admissible controller, leading from $\xi_0 = \eta_0 = l_0$ to the boundary of the sector. Using the rules for reducing the cost as described above we can show that the optimal path is either a horizontal segment, a vertical segment, or a vertical segment followed by a horizontal segment. We shall show that for an arbitrary point $(\bar{\xi}, \bar{\eta})$ in the sector $\xi \leq \eta \leq 2\xi$ the cost of the vertical path to the line $\eta = 2\xi$ is less than or equal to the cost of the horizontal segment to the line $\eta = 2\xi$. This observation, together with the earlier remarks, will prove that the optimal path from $\xi_0 = \eta_0 = l_0$ to the line $\eta = 2\xi$ is indeed the vertical segment.

We must show that

$$\left|\sqrt{\frac{4k}{\xi+\bar{\eta}}}\sqrt{\frac{\bar{\eta}}{\xi}}-\sqrt{\frac{4k}{\xi+2\xi}}\sqrt{2}\right|\le\left|\sqrt{\frac{4k}{\xi+\bar{\eta}}}\sqrt{\frac{\xi}{\bar{\eta}}}-\sqrt{\frac{4k}{3\bar{\eta}/2}}\frac{1}{\sqrt{2}}\right|$$

for $\xi \le \bar{\eta} \le 2\xi$. Let $\lambda = \bar{\eta}/\xi$ and this inequality reduces to

$$\sqrt{\frac{2}{3}}-\sqrt{\frac{\lambda}{1+\lambda}}\le\sqrt{\frac{1}{\lambda(1+\lambda)}}-\sqrt{\frac{1}{3\lambda}}.$$

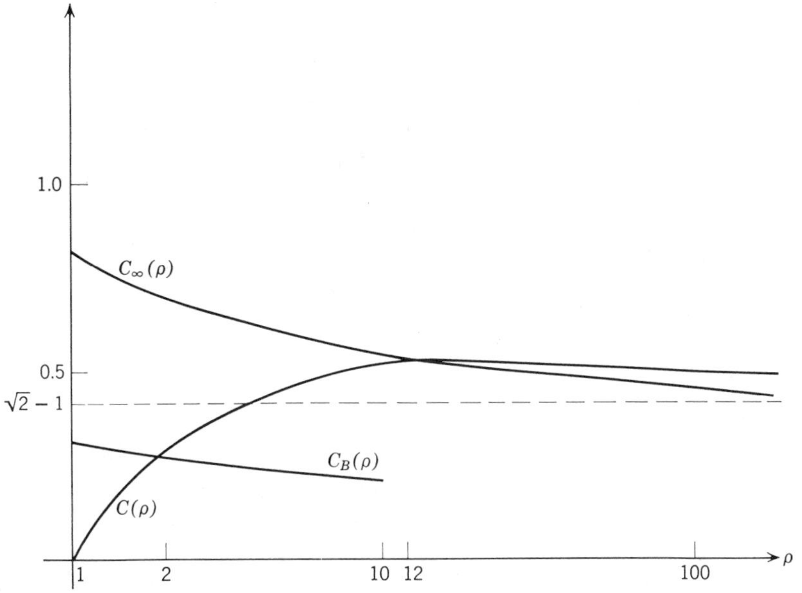

Figure 7.17 Cost of circular orbit transfer.
$C(\rho) = \sqrt{2\rho/1+\rho} - 1 + \sqrt{1/\rho} - \sqrt{2/\rho(\rho+1)}$ and $C_\infty(\rho) = (\sqrt{2}-1)(1+\sqrt{\rho}/\rho)$ with $\rho = l_1/l_0$ and $2k/l_0 = 1$.

Multiply by $\sqrt{\lambda}$ to obtain the equivalent inequality

$$\frac{\sqrt{2\lambda}+1}{\sqrt{3}}\le\frac{1+\lambda}{\sqrt{1+\lambda}}=\sqrt{1+\lambda}.$$

Square and simplify to obtain

$$2\sqrt{2\lambda}\le 2+\lambda$$

or

$$(\sqrt{\lambda}-\sqrt{2})^2\ge 0.$$

Therefore the vertical segment costs less than the horizontal segment to the line $\eta = 2\xi$.

It is now easy to compute the minimal cost of a controller that steers $\xi_0 = \eta_0 = l_0$ to the boundary of the basic sector and that continues onwards to the point $\xi_1 = \eta_1 = l_1 = \rho l_0$. This total cost must exceed (normalizing with $2k/l_0 = 1$)

$$C_B(\rho) = \left(\sqrt{\frac{1}{\rho}} + 1\right)\left(\sqrt{\frac{4}{3}} - 1\right).$$

Note $C_B(2) \geq 0.26$ and $C_B(\rho)$ is a decreasing function of ρ. We compute that

$$C(\rho) < C_B(\rho) \quad \text{for} \quad 1 \leq \rho \leq 1.8.$$

This inequality is our main result. It states that for impulsive control between two circular orbits, with diameter ratio $\rho \leq 1.8$, the Hohmann transfer is optimal even when compared to controllers which vary greatly over a global range. It is very likely that the Hohmann transfer is optimal for $\rho = 2$, and even somewhat beyond, but we have seen that the Hohmann transfer is not optimal for $\rho \geq 12$.

The comparison of the costs is summarized in Figure 7.17.

APPENDIX A

Steepest Descent and Computational Techniques for Optimal Control Problems

In many of the controller-optimization problems of the previous chapters it was possible to specify necessary and sufficient conditions. This was accomplished by showing that the general necessary condition, the maximal principle, was in some cases a sufficient condition. In many of the various problems it was also possible to answer the question of existence and uniqueness of optimal solutions. However, even with this knowledge the problems were not completely solved, that is, the optimal controller was in some cases not known as a function of the measured (sensed), initial data and the given terminal conditions. The problem that remained was essentially a two-point boundary-value problem: a problem of finding the solution of a $2n$ system of differential equations in which m of the boundary conditions are initial conditions and $2n - m$ are terminal conditions.

The steepest-descent techniques can be used at times to solve the two-point boundary-value problem that arises from application of the maximal principle, or they may be used without reference to this principle, to construct a sequence of control functions that has as its limit the optimal controller. The later approach, a so-called direct approach, can be used to find necessary conditions for optimal control because of the properties of the limit controller and, also, if the limit controller exists, can prove the existence of an optimal controller. We shall consider examples that display this constructive feature. The main reason for considering steepest-descent techniques or other constructive methods here is the ability of computer machines to approximately construct the sequence of control functions, or parameters that determine optimal-control functions, by a steepest-descent algorithm and that, after a limited amount of computing, have determined a control function close enough to the optimal control function to be used in the actual physical situation (see the discussion of feedback control at the beginning of Chapter 7).

In this appendix we consider the method of steepest descent for general optimization problems and show how it may be applied to the problem of selection of optimal controllers. The first section is an introduction to the technique of steepest descent for general optimization problems. The second section contains examples of application of the steepest-descent technique to controller optimization problems. The final section is a brief discussion of the literature associated with iterative (computational) methods and an associated bibliography.

A.1 THE METHOD OF STEEPEST DESCENT

We first consider the method of steepest descent for finite-dimensional spaces under the assumption that the finite-dimensional space is a Euclidian space E^m. Similar results can be obtained for other finite dimensional spaces if the space has defined on it an inner product $\langle \cdot, \cdot \rangle$ and is complete in terms of the metric $|\cdot| = \langle \cdot, \cdot \rangle^{1/2}$. A similar assumption is made for the infinite dimensional problems considered subsequently, namely, it is assumed that the underlying space is a Hilbert space $\mathcal{H}$. The norm for E^m with elements $u = (u_1, \ldots, u_m)$ is denoted by $\|u\| = (\Sigma\,(u_i)^2)^{1/2}$, and the inner product of u, and v in E^m by $\langle u, v \rangle = \sum_{i=1}^{m} u_i v_i$.

Let $C(u)$ be a real-valued function which is continuously differentiable on an open set $\mathcal{D}$ of E^m. For convenience we now consider the case where $m = 2$, and then $C = C(u_1, u_2)$ could be as shown in Figure A.1. The point $u^0 = (u_1{}^0, u_2{}^0)$ in $\mathcal{D}$ is such that the gradient $\partial C_0/\partial u \equiv (\partial C/\partial u)(u^0) = [(\partial C/\partial u_1)(u^0), (\partial C/\partial u_2)(u^0)]'$ is not the zero vector. Here, as before, prime denotes transpose of a vector or matrix. The direction of steepest descent at u^0 in E^m is the direction at u^0 for which the rate of change of C with respect to arc length is a minimum. Let S be the collection of all smooth curves in E^2 that pass through u^0. Let γ: $u_1 = u_1(s)$, $u_2 = u_2(s)$ be the parametric representation of a member of S where s denotes arc length measured from u^0. Since γ is a smooth curve

$$\left(\frac{du_1(s)}{ds}\right)^2 + \left(\frac{du_2(s)}{ds}\right)^2 = 1$$

for all s. Denote the direction cosines of the curve γ at u^0 by du_1/ds, du_2/ds.

For the continuously differentiable function $C(u)$ the rate of change of C with respect to arc length is

$$\frac{dC}{ds} = \frac{\partial C_0}{\partial u_1}\frac{du_1}{ds} + \frac{\partial C_0}{\partial u_2}\frac{du_2}{ds}$$

for any $\gamma \in S$. Thus the direction of steepest descent at u^0 is given by finding the direction cosines du_1/ds, du_2/ds that minimize dC/ds subject to the constraint

$$\left(\frac{du_1}{ds}\right)^2 + \left(\frac{du_2}{ds}\right)^2 = 1.$$

The minimizing values at u^0 are

$$\frac{du}{ds} = -\frac{1}{\|(\partial C/\partial u)(u^0)\|}\frac{\partial C}{\partial u}(u^0).$$

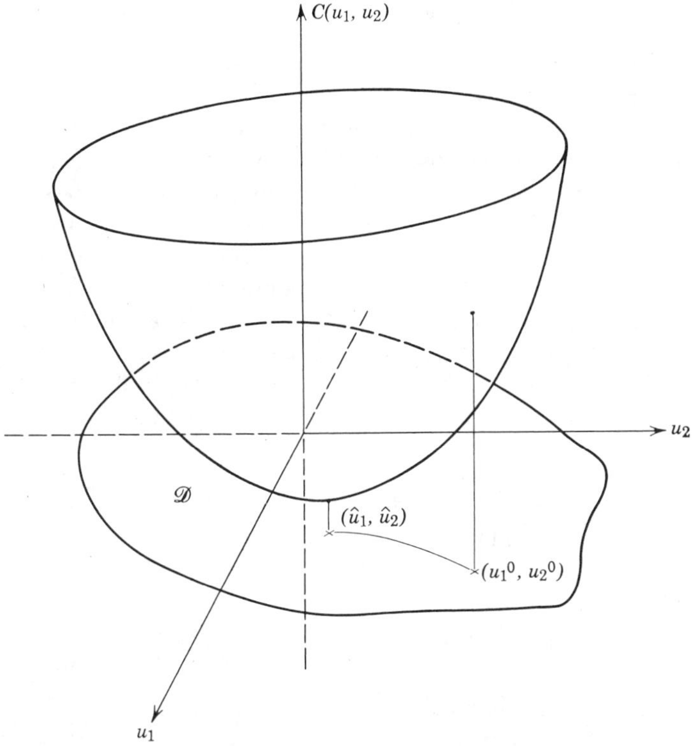

Figure A.1 Steepest descent paths in E^2.

Thus the direction of steepest descent at u^0 is in the direction of the negative gradient.

To determine a path of steepest descent consider the ordinary differential equation

$$\frac{du}{d\sigma} = -\frac{\partial C}{\partial u}(u), \qquad u(0) = (u_1^0, u_2^0), \qquad \sigma \geq 0,$$

which has a solution for σ sufficiently small. If the solution is assumed unique the resulting space curve $(u_1(\sigma), u_2(\sigma))$ defines a smooth path whose tangent lies in the direction of steepest descent for C at each point (u_1, u_2) along the path. The arc length along this curve is given by

$$s = \int_0^\sigma \left\| \frac{\partial C}{\partial u}(u(\tau)) \right\| d\tau \quad \text{and} \quad \frac{ds}{d\sigma} = \left\| \frac{\partial C}{\partial u}(u(\sigma)) \right\| > 0$$

($ds/d\sigma$ can be considered as velocity along the path). Along the space curve we compute

$$\frac{dC}{d\sigma} = \frac{dC}{ds}\frac{ds}{d\sigma} = -\left\| \frac{\partial C}{\partial u}(u(\sigma)) \right\| < 0, \quad \text{when} \quad \frac{\partial C}{\partial u} \neq 0$$

so that C is decreasing along the path of steepest descent, and as $\sigma \to +\infty$ a point of $\mathfrak{D}$ where C has a minimum may be approached. A theorem will be stated later that gives sufficient conditions for the existence of a unique local minimum and shows that the solution of the differential equation starting sufficiently near the minimum point goes to it as $\sigma \to \infty$ when certain local conditions are satisfied. We now consider the problem of a steepest descent calculation subject to side constraints, since such side constraints frequently occur in controller optimization problems.

Let $C(u)$ and $g_i(u)$, $i = 1, 2, \ldots r < m$ be of class C^1 in $\mathfrak{D} \subset E^m$, and suppose $\partial g_i/\partial u$ form a set of linearly independent vectors at each point of $\mathfrak{D}$. Let $u \in \mathfrak{D}$ be a point where gradient $C \neq 0$ but $g_i = 0$, $i = 1, 2, \ldots r$. The direction of steepest descent at u^0 is in the direction

$$-\frac{\partial C}{\partial u} + \sum_{k=1}^{m} \lambda_k \frac{\partial g_k}{\partial u}$$

where the λ_k are determined by the linear equation

$$G\lambda = y.$$

Here

$$y = \left(\left\langle \frac{\partial C}{\partial u}, \frac{\partial g_1}{\partial u} \right\rangle, \ldots, \left\langle \frac{\partial C}{\partial u}, \frac{\partial g_m}{\partial u} \right\rangle \right)'$$

and G is the Grammian matrix

$$\left\langle \frac{\partial g_i}{\partial u}, \frac{\partial g_j}{\partial u} \right\rangle \qquad i, j = 1, 2, \ldots, m,$$

which is positive definite by the above assumption.

Thus we consider solutions to the differential equation

$$\frac{du}{d\sigma} = -\frac{\partial C}{\partial u} + (G^{-1}y)' \frac{\partial g}{\partial u}, \qquad u^0 = u(0), \qquad \sigma \geq 0,$$

to generate paths of steepest descent.

If there are side constraints such as $g_j(u) \leq 0, j = r + l, \ldots,$ $r + l < m$—a case that occurs frequently in controller optimization problems—one can use a technique credited to Valentine to reduce the problem to one with equality constraints. This is accomplished by defining

$$\hat{g}_j(u_1, \ldots, u_m, u_{m+1}, \ldots, u_{m+l})$$
$$= g_j(u_1, \ldots, u_m) + u_{m+j-r})^2 = 0 \quad for \ j = r + l \ldots, r + l$$

and by defining $\hat{C}(u_1, \ldots, u_m, \ldots, u_{m+l}) = C(u_1, \ldots, u_m)$ and by considering the minimization of $\hat{C}$ in E^{m+l} subject to the constraints

$$g_i(u_1, \ldots, u_m) \quad = 0, \qquad i = 1, 2, \ldots, r$$
$$\hat{g}_j(u_1, \ldots, u_{m+l}) = 0, \qquad j = r + l \ldots, r + l$$

as above for the corresponding problem with equality constraints.

We next consider the problem of steepest descent in a function space and state results on convergence that apply also to the finite dimensional problems.

Before the method of steepest descent can be given for functionals defined on a function space, the concept of gradient must be extended (see also the exercises of Section 6.1). We now develop the necessary functional analysis for the extension. The results cited here are found mainly in Chapter 6 of Liusternik. The concept of arc length in a Banach space is developed first.

Let U be a real Banach space with norm $|\cdot|$, and suppose $u(t)$ is a function defined on the finite time interval $a \leq t \leq b$ with values in U. $u(t)$ is continuous on $[a, b]$ if it is continuous in the norm topology of U. The function $u(t)$ is continuously differentiable on $[a, b]$ if there is an element $u'(t) \in U$, such that

$$\lim_{\Delta t \to 0} \left| \frac{u(t + \Delta t) - u(t)}{\Delta t} - u'(t) \right| = 0$$

for every $t \in [a, b]$. Derivatives at the endpoints are taken in the one-sided sense. More general definitions are given in Hille and Phillips, Chapter 3.

Let $\pi_m: a \leq t_0 \leq t_1 \leq \cdots \leq t_m = b$ be a partition of the interval $[a, b]$, and define $S(u, \pi_m)$ by

$$S(u, \pi_m) = \sum_{i=1}^{m} |u(t_i) - u(t_{i-1})|.$$

The continuous curve in U defined by $u(t)$ is said to be rectifiable if L, defined by $L = \sup S(u, \pi_m)$, is finite. The supremum is taken over all finite partitions of $[a, b]$. If $u'(t)$ is also continuous and $|u'(t)| \neq 0$ for $t \in [a, b]$, then $u(t)$ is said to be a smooth curve. If $u(t)$ is smooth, then in a manner analogous to the case for E^n, we can show that the arc length s is given by

$$s = \int_0^t |u'(\sigma)|\, d\sigma.$$

From this we see that if the curve is parametrized by arc length, then $|u'(t)| = 1$, and conversely.

For example, if U is the space of square-integrable functions on $[0, 1]$, $L_2[0, 1]$, and if $u(s, \sigma)$ $0 \leq \sigma \leq 1$ is a smooth curve in $L_2[0, 1]$, then—taking s to be the arc length—we have

$$\int_0^1 \left(\frac{\partial u}{\partial s}(s, \sigma)\right)^2 d\sigma = 1, \quad \text{where} \quad \frac{\partial u}{\partial s}(s, \sigma) = u'(s).$$

This fact is often quoted when using the method of steepest descent in function space.

The concept of a Frechet derivative is essential for the method of steepest descent in function space. The definition is given now in sufficient generality for the problems to be considered here. For convenience, let $\mathcal{H}$ be a real Hilbert space with inner product $\langle \cdot, \cdot \rangle$ and let R denote the reals. Let C be a function C: $\mathcal{H} \to R$ and suppose $u_0, h \in \mathcal{H}$. The function C has a Frechet differential (or strong differential) $C'(u_0)h$, if there is a continuous linear functional $C'(u_0)$ on $\mathcal{H}$ such that

$$|C(u_0 + h) - C(u_0) - C'(u_0)h| = o(\|h\|)$$

as $\|h\| \to 0$, with $\|h\| = (\langle h, h \rangle)^{1/2}$. A function $g(h)$ is $o(\|h\|)$ as $\|h\| \to 0$ if $|g(h)|/\|h\| \to 0$ as $\|h\| \to 0$. The linear functional $C'(u_0)$ is called the Frechet derivative of C at u_0.

The expression

$$D\, C(u_0, h) = \frac{\partial C}{\partial \lambda}(u_0 + \lambda h)\bigg|_{\lambda=0} = \lim_{\lambda \to 0} \frac{C(u_0 + \lambda h) - C(u_0)}{\lambda},$$

if it exists, is called the direction derivative or weak differential of C at u_0. If the weak differential has certain properties, then $D\, C(u_0, h) = C'(u_0)h$, which gives a method for computing the Frechet derivative. The sufficient conditions on $D\, C(u_0, h)$ are formalized in the following theorem.

THEOREM 1. *If the direction derivative $D\, C(u, h)$ exists in $\|u - u_0\| \leq \alpha$, $\alpha > 0$, and if it is uniformly continuous in u and continuous in h, then the Frechet differential exists and $C'(u)h = D\, C(u, h)$.* (*The proof is given in Chapter* 6 *of Liusternik.*)

Higher-order Frechet derivatives are defined in a similar fashion. Let $\mathcal{H}^*$ denote the Banach space of continuous linear functionals on $\mathcal{H}$ with norm $|\cdot|_1$. If C has a Frechet derivative, then C is said to have a second

Frechet differential $C''(u_0)h$ if

$$|C'(u_0 + h) - C'(u_0) - C''(u_0)h|_1 = o(\|h\|)$$

as $\|h\| \to 0$. $C''(u_0)$ is a continuous linear operator from $\mathcal{H}$ into $\mathcal{H}^*$. An application of Theorem 1 shows that if the second direction derivative

$$\frac{\partial^2 C}{\partial \lambda_1 \, \partial \lambda_2}(u_0 + \lambda_1 h + \lambda_2 h_1)\Big|_{\substack{\lambda_1=0,\\ \lambda_2=0}} \qquad u_0,\ h,\ h_1 \in \mathcal{H}$$

is sufficiently continuous, then

$$(C''(u_0)h_1)h = \frac{\partial^2 C}{\partial \lambda_1 \, \partial \lambda_2}(u_0 + \lambda_1 h + \lambda_2 h_1)\Big|_{\substack{\lambda_1=0.\\ \lambda_2=0}}$$

The operator $C''(u)$ is called the second Frechet derivative.

Let C be as before, and suppose it has two continuous Frechet derivatives in $\mathcal{H}$. The first Frechet derivative is a linear functional on $\mathcal{H}$, and by the Riesz representation theorem we have

$$C'(u)h = \left\langle \frac{\partial C}{\partial u}(u), h \right\rangle, \qquad h \in \mathcal{H}$$

where $(\partial C/\partial u)(u)$ is a unique element in $\mathcal{H}$. The element $(\partial C/\partial u)(u)$ is called the gradient of C at u. This definition coincides with the finite-dimensional definition of a gradient if $\mathcal{H} = E^n$. Because $C''(u)h$ is a continuous linear functional on $\mathcal{H}$, the Riesz representation theorem states that

$$(C''(u)h)h = \langle H_C(u)h, h\rangle$$

where $H_C(u)$ is a continuous linear operator on $\mathcal{H}$. $H_C(u)$ is called the Hessian of C at u. For finite-dimensional spaces, $H_C(u)$ reduces to the symmetric matrix of second partial derivatives of C.

The underlying space for all of the previous considerations of this section was a Hilbert space $\mathcal{H}$. This requirement can be weakened slightly. Let U be a linear topological space with an inner product $\langle \cdot, \cdot \rangle$ defined on it and let $\bar{U}$ denote its completion with respect to the metric $d(u, y) = \|u - y\| = (\langle u - y, u - y \rangle)^{1/2}$. Let C be a real-valued function on U with a Frechet derivative $C'(u)$. By the Riesz representation theorem there is a unique element $C_u \in \bar{U}$ such that $C'(u)h = \langle C_u, h\rangle$, $h \in \bar{U}$. If $C_u \in U$, then C_u is called the gradient of C and $(\partial C/\partial u)(u) = C_u$. Since C would not necessarily have a gradient for each u, this motivates the choice of the basic space to insure that C always has a gradient. Thus, for example, instead of considering the Banach space of real continuous function on an interval I, it is more advantageous to consider its completion in the norm

$$\left(\int_I u^2 \, d\mu\right)^{1/2}$$

the Hilbert space of Lebesgue square-integrable functions with inner product $\int_I uy\, d\mu$.

The method of steepest descent is now given for functions defined on a Hilbert space $\mathcal{H}$. Let C be a real-valued function defined on $\mathcal{H}$, with one continuous Frechet derivative. Let $u_1 \in \mathcal{H}$ and let γ be a smooth curve in $\mathcal{H}$ passing through u_1. If the curve is parametrized by arc length, then $\|u'(s)\|^2 = 1$ and

$$\frac{dC}{ds} = \lim_{\Delta s \to 0} \frac{C(u(s + \Delta s)) - C(u(s))}{\Delta s} = \left\langle \frac{\partial C}{\partial u}(u(s)), u'(s) \right\rangle.$$

The direction of steepest descent is found by minimizing

$$\left\langle \frac{\partial C}{\partial u}(u_1), u'(0) \right\rangle$$

subject to

$$\|u'(0)\|^2 = 1.$$

The symbolism here coincides with that of the case for E^n, and hence all the results for E^n carry over to this case with the proper interpretation. The path of steepest descent is then found by solving the differential equation

$$\frac{du}{d\sigma} = -\frac{\partial C}{\partial u}(u), \qquad u(0) = u_1 \in \mathcal{H}$$

$$\sigma \geq 0.$$

The solution is a function with values in $\mathcal{H}$, and along this path $C(u(\sigma))$ is decreasing because

$$\frac{dC}{d\sigma} = \left\langle \frac{\partial C}{\partial u}, \frac{du}{d\sigma} \right\rangle = -\left\| \frac{\partial C}{\partial u}(u(\sigma)) \right\|^2 < 0,$$

if $\partial C/\partial u \neq 0$.

These techniques are formalized in the following theorems, credited to P. C. Rosenbloom. For these theorems we consider real-valued functions defined on a Hilbert space $\mathcal{H}$.

THEOREM 2. *Let C have two continuous Frechet derivatives on a convex domain D of $\mathcal{H}$, and suppose the sphere $S(u) = \{u \mid u \in \mathcal{H}, \|u - u_0\| \leq a/A\}$, where $a = \|(\partial C/\partial u)(u_0)\|$ and A is chosen later, is contained in D. Furthermore assume that*

$$\langle H_C(u)v, v\rangle \geq A\, \|v\|^2$$

for $u \in D$, $v \in \mathcal{H}$, $A > 0$ *and fixed, and that* $u(\sigma)$ *satisfies*

$$(*) \qquad \frac{du}{d\sigma}(\sigma) = -\frac{\partial C}{\partial u}(u(\sigma))$$

(*for* $\sigma \geq 0$, $u(0) = u_0 \in D$).

$H_C(u)$ *is the Hessian of* C *at* u. *Then*

a. $\lim_{\sigma\to\infty} u(\sigma) = u_\infty$ *exists in* $\mathcal{H}$,

b. $\lim_{\sigma\to\infty} C(u(\sigma)) = c$ *exists*,

c. $\|u(\sigma) - u_\infty\| \leq (a/A)\exp(-A\sigma)$

$$0 \leq C(u(\sigma)) - c \leq \frac{a^2}{2A}\exp(-2A\sigma),$$

and, for all $u \in D$,

$$C(u) \geq c + \frac{A}{2}\|u - u_\infty\|^2.$$

Thus if the hypotheses are satisfied the steepest descent solution converges to a minimizing element at an exponential rate. From part (a) of the conclusion we see that the minimizing element u_∞ is unique. It is also noted that the solution of (*) must exist for all nonnegative σ, a fact that is difficult to verify in certain cases. The paper of Rosenbloom indicates several sufficient conditions for the existence of a solution for all σ. One of the simplest conditions is that $\partial C/\partial u$ satisfy a Lipschitz condition in the sphere $S(u)$. A consequence of the proof is that

$$\lim_{\sigma\to\infty}\left\|\frac{\partial C}{\partial u}(u(\sigma))\right\| = 0.$$

The method of steepest descent via Theorem 2 can also be useful in solving certain isoperimetric problems. For simplicity, consider the problem of minimizing a function C on $\mathcal{H}$, subject to the side condition $g = 0$. The path of steepest descent is defined by

$$(**) \qquad \frac{du}{d\sigma} = -\frac{\partial C}{\partial u} + \lambda(u)\frac{\partial g}{\partial u}, \qquad \lambda(u) = \frac{\langle \partial C/\partial u, \partial g/\partial u\rangle}{\|\partial g/\partial u\|^2},$$

provided that $\left\|\dfrac{\partial g}{\partial u}\right\| \neq 0$. With this choice of path,

$$\frac{dC}{d\sigma} = -\left\{\frac{\|\partial C/\partial u\|\ \|\partial g/\partial u\| - \langle \partial C/\partial u, \partial g/\partial u\rangle}{\|\partial g/\partial u\|^2}\right\} < 0$$

$$\frac{dg(u)}{d\sigma} = 0.$$

The following theorem, attributed to Rosenbloom, gives sufficient conditions for the method of steepest descent to yield a unique solution of this problem.

THEOREM 3. *Let C and g have two continuous Frechet derivatives on a convex domain D of $\mathcal{H}$ and let $\mathcal{M}$ be the manifold defined by $g(u) = 0$, $u \in D$. Suppose that $\partial g/\partial u \neq 0$ on $\mathcal{M}$ and that*

$$\langle (H_C(u) - \lambda(u)\, H_g(u))v, v\rangle \leq A\, \|v\|^2, \qquad A > 0$$

for $u \in \mathcal{M}$, $v \in \mathcal{H}$, $\langle \partial g(u)/\partial u, v\rangle = 0$. Let $k(u)$ be the distance from u to the boundary of $\mathcal{M}$ and let E be the set of points $u_0 \in \mathcal{M}$ such that

$$a(u_0) = \left\| \frac{\partial C(u_0)}{\partial u} - \lambda(u_0)\frac{\partial g(u_0)}{\partial u} \right\| < Ak(u_0).$$

and such that the solution of (**) with $u(0) = u_0$ exists for all $\sigma \geq 0$. Then for all $u_0 \in E$ this soution $u(\sigma)$ has the properties expressed in the inequalities in the conclusion of Theorem 2. Furthermore, if u_0 and $\tilde{u}_0$ in E can be joined by an arc in E of class C^1, then $u_\infty = \tilde{u}_\infty$, and finally, if $c = \lim_{\sigma\to\infty} C(u(\sigma))$ for $u(0) = u_0 \in E$, then

$$C(u_0) \geq c + \frac{4A^2\delta^3}{27a}\left(1 + \frac{A\delta}{3a}\right), \qquad \delta = \|u_0 - u_\infty\|, \qquad a = a(u_0)\cdot$$

The theorem is also true in the case of several side conditions.

So far we have considered how a steepest-descent path can be constructed based on a knowledge of local information (gradient information) about the function (or functional) $C(u)$. With such gradient information one can implement a solution to the differential equation for the path of steepest descent on an analog computer in the finite-dimensional case. However, the most frequent usage of the steepest-descent technique, as a computational technique, is with a digital computer, where one considers changes in the numbers u (or function u) by means of an iteration technique based on the steepest-descent directions. That is, one considers an iteration procedure given by

$$u_{k+1} = u_k - \rho_k\, \nabla C(u_k) \quad \text{for} \quad \rho_k > 0$$

and $k = 0, 1, 2, \ldots$. So, the computer calculates an improvement in u at the $k + 1$th step based on the value of u and of the gradient ∇C at the kth step. Convergence results similar to those above for the continuous-variable case can be found in the papers of Goldstein.

This completes our introduction to steepest descent techniques and we now turn to their application in controller-optimization problems.

A.2 APPLICATION OF STEEPEST DESCENT TO CONTROLLER-OPTIMIZATION PROBLEMS AND FORMULATION OF COMPUTATIONAL ALGORITHMS

In this section we apply the steepest-descent techniques to controller optimization problems. First, in Examples 1 to 3, we illustrate how steepest descent can be considered a constructive approach to controller optimization problems yielding, for example, necessary properties of optimal controllers similar to those already derived from the maximal principle. We then consider, in Examples 4 to 7, how computational algorithms for optimal control may be derived. One approach is a direct approach, the other is to use the maximal principle to determine the form of the optimal controller, up to a set of parameters that are adjusted by steepest descent. Examples 4 to 7 are only carried to a point where a computer algorithm is in sight and hence are not complete.

***Example* 1.** Consider the first-order linear control process

$$\dot{x} = ax + u(t)$$

with scalar controller $u(t)$ and fundamental solution matrix $\Phi(t)^{-1} = e^{at}$ on $[0, T]$ with $x(0) = 0$, $\Phi(0) = 1$, and $T > 0$ fixed. The controller optimization problem we consider is that of steering $x(t)$ from $x(0) = x_0 = 0$ to $x = \Phi(T)^{-1}c$ for some constant c at time T with minimum cost

$$C(u) = \int_0^T (u(t))^2\, dt.$$

It is required, as in Chapter 3 for the equivalent problem, that $u \in L_2[0, T]$, where $L_2[0, T]$ denotes the Hilbert space of Lebesgue square-integrable functions on $[0, T]$. The problem is thus to minimize

$$\int_0^T u^2(t)\, dt$$

subject to

$$\int_0^T \Phi(t)\, u(t)\, dt = c,$$

where u is a scalar function in $L_2[0, T]$, $\Phi(t) \in L_2[0, T]$, and c is a given constant. If $c = 0$, $u = 0$ is the optimal solution. Note that $\|\Phi\| \neq 0$. In the notation of Theorems 2 and 3 we have

$$C(u) = \int_0^T u^2(t)\, dt$$

$$g(u) = \int_0^T \Phi(t)\, u(t)\, dt - c.$$

The gradients of C and g are now computed. The direction derivatives are computed as

$$\frac{\partial C}{\partial \lambda}(u + \lambda z)\bigg|_{\lambda=0} = \int_0^T 2u(t)\, z(t)\, dt$$

and

$$\frac{\partial g}{\partial \lambda}(u + \lambda z)\bigg|_{\lambda=0} = \int_0^T \Phi(t)\, z(t)\, dt, \qquad z \in L_2[0, T].$$

Because bounded linear functionals on $L_2[0, T]$ have the representation $\int_0^T \psi(t)\, z(t)\, dt$ for some fixed $\psi \in L_2[0, T]$, we see that

$$\frac{\partial C}{\partial u} = 2u$$

and

$$\frac{\partial g}{\partial u} = \Phi\cdot$$

The Hessians of C and g are found by examining the second direction derivatives

$$\frac{\partial^2 C}{\partial \lambda^2}(u + \lambda z)\bigg|_{\lambda=0} = \int_0^T 2z^2(t)\, dt$$

$$\frac{\partial^2 g}{\partial \lambda^2}(u + \lambda z)\bigg|_{\lambda=0} = 0.$$

The Hessian of C is the continuous linear operator

$$H_C(u)z = 2z, \qquad z \in L_2[0, T]$$

and the Hessian of g is the zero operator

$$H_g(u)z = 0, \qquad z \in L_2[0, T].$$

The hypotheses of Theorem 3 are now shown to hold. Since

$$\langle (H_C(u) - \lambda(u)\, H_g(u))v, v\rangle = \langle H_C(u)v, v\rangle = 2\, \|v\|^2$$

we have $A = 2$.

If u_0 and $\tilde{u}_0$ are in E, that is, if

$$\int_0^T \Phi(t)\, u_0(t)\, ds = c, \qquad \int_0^T \Phi(t)\, \tilde{u}_0(t)\, dt = c,$$

then the function $\hat{u} = \mu u_0 + (1 - \mu)\tilde{u}_0, 0 \leq \mu \leq 1$ is a C^1 arc in E joining u_0 and $\tilde{u}_0$. Consequently a solution will be unique if it exists.

The set $B = \{u \mid u \in L_2[0, T], \int_0^T \Phi(t)\, u(t)\, dt = c\}$ is a hyperplane, and consequently the condition

$$a(u_0) < Ak(u_0)$$

of Theorem 3 is always satisfied for any element of $\mathcal{M}$. $\mathcal{M}$ is given in Theorem 3. A function u_0 that satisfies $g(u_0) = 0$ can be found by considering the resulting overdetermined equation when step functions are used.

The differential equation that determines $u(\sigma)$ of Theorem 3 is given by

$$\frac{du}{d\sigma} = -2u + \lambda(u)\Phi, \qquad u(0) = u_0, \qquad \sigma \geq 0.$$

But since

$$\lambda(u) = \frac{\langle \partial C/\partial u, \partial g/\partial u \rangle}{\|\Phi\|^2} = \frac{2\int_0^T \Phi(t)\, u(t)\, dt}{\|\Phi\|^2} = \frac{2c}{\|\Phi\|^2},$$

we have that

$$\frac{du}{d\sigma}(t, \sigma) = -2u(t, \sigma) + \frac{2c}{\|\Phi\|^2}\Phi(t), \qquad \begin{array}{l} 0 \leq t \leq T \\ 0 \leq \sigma \\ u(t, 0) = u_0(t). \end{array}$$

The solution of this linear equation is given by

$$u(t, \sigma) = e^{-2\sigma} u_0(t) + \frac{c\,\Phi(t)}{\|\Phi\|^2}(1 - e^{-2\sigma}).$$

Thus, by Theorem 3, the unique minimizing element is given by

$$u_\infty(t) = \frac{c}{\|\Phi\|^2}\Phi(t),$$

and the value of $C(u_\infty)$ is c^2. This result can also be determined by using the necessary conditions from the maximal principle as in Chapter 3.

***Example* 2.** Consider the linear control process

$$\dot{x}^1 = ax^1 + u(t)$$

$$\dot{x}^2 = bx^2 + u(t)$$

with scalar controller $u(t)$ and $a \neq b$. The controller optimization problem is to steer $x^1(t)$ from $x^1(0) = 0$ to $x^1 = \Phi_1(T)^{-1}c$ for some constant c with minimum cost

$$C(u) = \int_0^T \Phi_2(t)\, u(t)\, dt + \int_0^T (u(t))^2\, dt$$

where the fundamental solution matrix

$$\psi(t) = \begin{bmatrix} \Phi_1(t)^{-1} & 0 \\ 0 & \Phi_2(t)^{-1} \end{bmatrix}$$

$$\psi(0) = \begin{bmatrix} 1 & 0 \\ 0 & 1 \end{bmatrix}$$

and $T > 0$ is fixed. Thus we are considering the problem of minimizing

$$\left(\int_0^T \Phi_2(t)\, u(t)\, dt\right)^2 + \int_0^T u^2(t)\, dt$$

subject to the constraint

$$\int_0^T \Phi_1(t)\, u(t)\, ds = c.$$

The functions $\Phi_2(t)$ and $\Phi_1(t)$ are elements of $L_2[0, T]$, $\Phi_1 \neq \Phi_2$, and $\|\Phi_1\| \neq 0$. Let $C(u) = \left(\int_0^T \Phi_2(t)\, u(t)\, dt\right)^2 + \int_0^T u^2(t)\, dt$ and $g(u) = \int_0^T \Phi_1(t)\, u(t)\, dt - c$. By computing the direction derivative of C and g we have that

$$\frac{\partial C}{\partial u} = 2\int_0^T \Phi_2(\tau)\, u(\tau)\, d\tau\, \Phi_2(t) + 2u(t)$$

and

$$\frac{\partial g}{\partial u} = \Phi_1(t).$$

A similar computation shows that

$$H_C(u)z = 2\int_0^T \Phi_2(\tau)\, z(\tau)\, d\tau\, \Phi_2(t) + 2z(t)$$

$$H_g(u)z = 0, \qquad z \in L_2[0, T].$$

Since

$$\langle (H_C(u) - \lambda(u)\, H_g(u))v, v\rangle = 2\left(\int_0^T \Phi_2(\tau)\, v(\tau)\, d\tau\right)^2 + 2\langle v, v\rangle \geq 2\langle v, v\rangle,$$

we conclude, by using Theorem 3, that the unique optimal solution is found by solving the differential equation

$$\frac{du}{d\sigma} = -\frac{\partial C}{\partial u} + \lambda(u)\frac{\partial g}{\partial u}, \qquad u(0) = u_0, \qquad \sigma \geq 0$$

$$\lambda(u) = \frac{2c + 2\langle \Phi_2, \Phi_1\rangle \int_0^T \Phi_2(\tau)\, u(\tau, \sigma)\, d\tau}{\|\Phi_1\|^2}.$$

If we write out this equation we have that

$$\frac{du}{d\sigma}(t, \sigma) = -2u(t, \sigma) - 2\int_0^T \Phi_2(\tau)\, u(\tau, \sigma)\, d\tau\, \Phi_2(t)$$

$$+ \left\{2\langle\Phi_2, \Phi_1\rangle \int_0^T \Phi_2(\tau)\, u(\tau, \sigma)\, d\tau + 2c\right\} \frac{\Phi_1(t)}{\|\Phi_1\|^2}$$

$$u(t, 0) = u_0(t), \qquad \sigma \geq 0.$$

This differential equation has a unique solution for all $\sigma \geq 0$ because it satisfies a uniform Lipschitz condition in $L_2[0, T]$. By using Theorem 3 we find a unique optimal solution $\hat{u}(t)$ for the problem. From the remark following Theorem 2 and the Schwarz inequality, we have

$$\hat{u}(t) = \left\{\frac{\langle\Phi_2, \Phi_1\rangle}{\langle\Phi_1, \Phi_1\rangle}\Phi_1(t) - \Phi_2(t)\right\}\int_0^T \Phi_2(\tau)\, \hat{u}(\tau)\, d\tau + \frac{c\Phi_1(t)}{\langle\Phi_1, \Phi_1\rangle}.$$

It remains to determine $\int_0^T \Phi_2(t)\, \hat{u}(t)\, dt$. The function

$$u(t, \beta) = \left\{\frac{\langle\Phi_2, \Phi_1\rangle}{\langle\Phi_1, \Phi_1\rangle}\Phi_1(t) - \Phi_2(t)\right\}\beta + \frac{c\,\Phi_1(t)}{\langle\Phi_1, \Phi_1\rangle}$$

defines an analytic family of curves in $L_2[0, T]$ which satisfy $g(u(t, \beta)) = c$ for all real β. There is a value of β, call it β_0, for which $\beta_0 = \langle\Phi_2, \hat{u}\rangle$. This value of β_0 is found by minimizing $C(u(t, \beta))$ with respect to β. Letting

$$u(t, \beta) = (\zeta\, \Phi_1(t) - \Phi_2(t))\beta + \hat{c}\, \Phi_1(t)$$

with

$$\zeta = \frac{\langle\Phi_1, \Phi_2\rangle}{\langle\Phi_1, \Phi_1\rangle}, \qquad \hat{c} = \frac{c}{\langle\Phi_1, \Phi_1\rangle}$$

we see that β_0 is given by

$$\beta_0 = \frac{-\hat{c}\langle\Phi_2, \Phi_1\rangle}{\zeta\langle\Phi_1, \Phi_2\rangle - \langle\Phi_2, \Phi_2\rangle} = \frac{c}{\|\Phi_2\|^2\, \|\Phi_1\|^2 - \langle\Phi_1, \Phi_2\rangle^2}.$$

The denominator of this expression is nonzero because Φ_2 and Φ_1 are not identical. The optimal solution is given as

$$\hat{u}(t) = \frac{-c}{G}\Phi_2(t) + \left\{\frac{\langle\Phi_2, \Phi_1\rangle + G}{G}\right\}\frac{c}{\|\Phi_1\|^2}\Phi_1(t),$$

where $\|\Phi_2\|^2\, \|\Phi_1\|^2 - \langle\Phi_2, \Phi_1\rangle^2 = G$.

***Example* 3.** A Generalization of Examples 1 and 2. We shall be concerned with the linear control process

$$(\mathfrak{L}) \qquad \dot{x} = A(t)x + B(t)u,$$

where $B(t)$ and $A(t)$ are continuous matrices of size $n \times m$ and $n \times n$ respectively on any finite interval of interest, $[t_0, T]$. We can, without loss of generality, take $t_0 = 0$. The state vector $x(t)$ is an n-vector with an initial value $x(0) = x_0$, and u is a bounded and measurable m-vector that comprises the control variable for the process. The cost functional corresponding to a given control is prescribed by

$$C(u) = g(x(T)) + \int_0^T u(s)'\, U(s)\, u(s)\, ds.$$

$g(x(T))$ is a given real, convex function on R^n, and $U(t)$ is a given continuous, symmetric, and positive definite matrix on $[0, T]$. The problem is to minimize $C(u)$ subject to the given differential equation and $u \in L_2[0, T]$. In certain fixed-endpoint problems further side conditions are also given. Chapter 3 contains a detailed treatment of this type of control problem. Several cases will be presented to illustrate how the method of steepest descent can be utilized to solve this type of problem.

Case 1. The problem is to minimize

$$C(u) = \int_0^T u(s)'\, U(s)\, u(s)\, ds, \qquad u \in L_2[0, T]$$

subject to the endpoint condition

$$x(T) = \hat{x}.$$

We can assume that $U(t) \equiv I$ for this problem because a standard change of control variables reduces the given problem to the case when $U = I$ that is, take as the inner product $\int u' U(t) v$. There is no immediate assurance that there is any solution of the given problem. To make the problem reasonable we assume that the system is controllable, that is, any given final state in R^n can be reached in time T with a bounded measurable controller u. A controllable process is characterized in the following manner: The process $\mathcal{L}$ is controllable on the interval $[0, T]$ if and only if the matrix

$$M(T) = \int_0^T \Phi(t)^{-1}\, B(t)\, B'(t)(\Phi(t)^{-1})'\, dt$$

has rank n. $\Phi(t)$ is the fundamental solution matrix of the equation $\dot{x} = A(t)x$ with $\Phi(0) = I$, the $n \times n$ identity matrix. A proof is given in Chapter 3. The variation of parameters solution for any bounded measurable u is given by

$$x(t) = \Phi(t)x_0 + \int_0^T \Phi(t, s)\, B(s)\, u(s)\, ds$$

where $\Phi(t, s) = \Phi(t)\, \Phi(s)^{-1}$, and thus the endpoint condition can be written as

$$\int_0^T \Phi(T, s)\, B(s)\, u(s)\, ds = \hat{x} - \Phi(T)x_0.$$

To solve this problem by the method of steepest descent we let $\mathcal{H}$ denote the direct sum Hilbert space $L^2[0, T] \oplus L^2[0, T] \oplus \cdots \oplus L^2[0, T]$ of m-tuples $u = (u_1, \ldots, u_m)'$ with inner product $\langle \cdot, \cdot \rangle_1$ defined by

$$\langle u, v \rangle_1 = \sum_1^m \langle u_i, v_i \rangle = \int_0^T u'v\, dt.$$

Let $\psi_k(s)$ denote the kth row of the matrix $\Phi(T, s)\, B(s)$ and α_k the kth component of $\hat{x} - \Phi(T)x_0$. The control problem is then to minimize

$$C(u) = \int_0^T u'u\, dt$$

subject to

$$g_k(u) = 0, \qquad k = 1, 2, \ldots, n$$

where $g_k(u) = \int_0^T \psi_k(t)\, u(t)\, dt - \alpha_k$. A computation shows that

$$\frac{\partial C}{\partial u} = 2u,$$

$$\frac{\partial g_k}{\partial u} = \psi_k', \qquad k = 1, \ldots, n,$$

$$H_C(u)z = 2z, \qquad z \in \mathcal{H}$$

$$H_g(u) = 0,$$

hence Theorems 2 and 3, applied to the case of several side conditions, will solve the problem, provided that the Grammian matrix $G = (\langle \psi_j, \psi_k \rangle_1)$ is nonsingular. A computation shows that

$$G = \Phi(T)\, M(T)\, \Phi(T)'$$

where $M(T)$ was defined above and consequently G is nonsingular. To solve the problem we consider the differential equation

$$\frac{du}{d\sigma} = -2u + \sum_{k=1}^n \lambda_k \psi_k', \qquad u(0) = u_0, \qquad \sigma \geq 0,$$

where $G\lambda = c$,

$$c = \begin{pmatrix} \left\langle \dfrac{\partial C}{\partial u}, \dfrac{\partial g_1}{\partial u} \right\rangle_1 \\ \cdot \\ \cdot \\ \cdot \\ \left\langle \dfrac{\partial C}{\partial u}, \dfrac{\partial g_n}{\partial u} \right\rangle_1 \end{pmatrix} = 2\alpha.$$

Upon rewriting we have that

$$\frac{du}{d\sigma}(t, \sigma) = -2u(t, \sigma) + 2B(t)' \Phi(T, t)' G^{-1}\alpha.$$

If we solve this differential equation and let $\sigma \to \infty$ we see that the optimal control is given by

$$u_\infty(t) = B(t)' \Phi(T, t)' G^{-1}\alpha = B(t)' \Phi(T, t)' G^{-1}(\hat{x} - \Phi(T)x_0).$$

This result agrees with results of Chapter 3 when the maximum principle is used.

Case 2. The problem is to minimize

$$C(u) = x_u(T)' G\, x_u(T) + \int_0^T u'u\, ds, \qquad u \in \mathcal{H},$$

where

$$x_u(T) = \Phi(T)x_0 + \int_0^T \Phi(T, s)\, B(s)\, u(s)\, ds$$

and $\mathcal{H}$ is the Hilbert space considered in Example 1. G is an $n \times n$ positive semidefinite constant symmetric matrix. The gradient of C is

$$\frac{\partial C}{\partial u} = 2B(t)' \Phi(T, t)' G\, x_u(T) + 2u,$$

and the Hessian is

$$H_C(u)z = 2z(s) + \left(\int_0^T \Phi(T, \tau)\, B(\tau)\, z(\tau)\, d\tau\right)' G\, \Phi(T, s)\, B(s), \qquad z \in \mathcal{H}.$$

Furthermore, $\langle H_C(u)z, z\rangle_1 \geq 2\langle z, z\rangle_1$; hence, by Theorem 2, we must solve the differential equation

$$\frac{du}{d\sigma}(t, \sigma) = -2u(t, \sigma)$$

$$- 2B'(t)\, \Phi(T,t)' G\left(\Phi(T)x_0 + \int_0^T \Phi(T, s)\, B(s)\, u(s, \sigma)\, ds\right) \quad u(t, 0) = 0,$$

Because the right-hand side satisfies a uniform Lipschitz condition, the solution will exist for all $\sigma \geq 0$. By Theorem 2 there is a unique optimal solution $\hat{u}$ given by

$$\hat{u}(t) = -B(t)' \Phi(T, t)' G\, x_{\hat{u}}(T).$$

On substituting and evaluating, we have that

$$\left(I + \int_0^T \Phi(T, s)\, B(s)\, B(s)' \Phi(T, s)'\, ds\, G\right) x_{\hat{u}}(T) = \Phi(T)x_0.$$

Since $I + \int_0^T \Phi(T, s)\, B(s)\, B(s)'\, \Phi(T, s)'\, ds\, G$ is nonsingular, we can solve for $x_{\hat{u}}(T)$ so that finally

$$\hat{u}(t) = -B(t)'\,\Phi(T, t)'\, G(I + \Phi(T)\, M(T)\, \Phi(T)'G)^{-1}\, \Phi(T)x_0.$$

The control $\hat{u}(t)$ is an open-loop control. A closed-loop or feedback control with time-varying gains can also be given for this problem. We attempt to find a matrix $E(t)$ such that $\hat{u}(t) = E(t)\, x(t)$, where $x(t)$ is the measured state at time t. The solution for time t is

$$x(t) = \Phi(t)x_0 + \int_0^t \Phi(t, s)\, B(s)\, \hat{u}(s)\, ds$$

$$= \Phi(t)x_0 - \int_0^t \Phi(t, s)\, B(s)\, B(s)'\, \Phi(T, s)'\, ds\, G\, x_{\hat{u}}(T).$$

We have, further,

$$x(t) = \Phi(t, T)\left(I + \int_t^T \Phi(T, s)\, B(s)\, B(s)'\, \Phi(T, s)'\, ds\, G\right) x_{\hat{u}}(T).$$

If we solve for $x_{\hat{u}}(T)$ and substitute we have $u(t) =$

$$-B'(t)\,\Phi'(T, t)\left(I + \int_t^T \Phi(T, s)\, B(s)\, B(s)'\, \Phi(T, s)'\, ds\, G\right)^{-1} \Phi(T, t)\, x(t).$$

The feedback matrix $E(t)$ is then given by

$$E(t) = -B(t)'\,\Phi(T, t)'\left(I + \int_t^T \Phi(T, s)\, B(s)\, B(s)'\, \Phi(T, s)'\, ds\, G\right)^{-1} \Phi(T, t),$$

and the feedback program defined by $E(t)$ is independent of the initial state x_0. A feedback program is also given for this problem in Chapter 3.

***Example* 4.** Consider the autonomous linear control process

$$(\mathfrak{L}) \qquad \dot{x} = Ax + b\, u(t)$$

where x, an n-vector, is the system state, u is a scalar control function subject to the restraint $|u(t)| \leq 1$ on $[0, T]$, and A and b are $n \times n$ and $n \times 1$ real constant matrices respectively. The results below can be readily extended to time-varying linear equations involving several control variables.

Consider the cost functional $C(u) = g(x(T)) = x(T)'\, \tilde{H}\, x(T)$, where $\tilde{H} = \tilde{H}' > 0$. Problems other than final-value problems can often be converted to final-value ones by adding a new coordinate to the system, as in Chapter 3.

Thus, for fixed time $T > 0$ we seek a scheme for computing the function $u^*(t)$, $t \in [0, T]$, so that the corresponding response endpoint $x_{u^*}(T)$ of the above system $\mathcal{L}$ is the point of the set of attainability $K(T)$ that is closest to the origin as measured by the distance function $x'\tilde{H}x$. Here $K(T)$ is the collection of endpoints of responses $x_u(t)$, which initiate at $x_0 = x(0)$ for all measurable control functions $u(t)$ on $[0, T]$, such that $|u(t)| \leq 1$. Properties of $K(T)$ as discussed in Chapter 2 will be freely used here.

The solution to this problem can be used in the solution of several other problems. For example, we can increase T, starting with $T = 0$, until the error function $\mathcal{E}(x) = x'\tilde{H}x$ is zero for some point $x_u(T)$ of the set of attainability $K(T)$, at which time a solution to the problem of time optimal control to the origin is obtained. If the problem is one of steering as close as possible to a convex target set G at a fixed time T, and if G can be approximated by a quadratic form—that is, G: $x'\tilde{H}x \leq c$, where $\tilde{H} = \tilde{H}' > 0$ and c is a constant—the problem is solved by finding the point of $K(T)$ where $x'\tilde{H}x$ is a minimum when $x'\tilde{H}x$ is the distance function. Time-optimal control to a convex target set G is handled in the same manner as the problem of time-optimal control to the origin: approximate G by $x'\tilde{H}x \leq c$ and then increase T starting at 0 until the error function $\mathcal{E}(x) = x'\tilde{H}x - c$ is zero for some point of the set of attainability $K(T)$.

We are concerned with finding $x_u{}^*(T)$—the point of $K(T)$ closest to the origin in terms of the distance function $x'\tilde{H}x$—and the controller $u^*(t)$, $0 \leq t \leq T$, which gave rise to $x_u{}^*(T)$. From Chapter 2 it is known that $K(T)$ is a compact, convex, subset of R^n for the above constant system $\mathcal{L}$ and $K(T)$ has an interior in R^n if and only if det $[b, Ab, \ldots, A^{n-1}b] \neq 0$. Thus, if $K(T)$ does not contain the origin there is a unique point x^* in $K(T)$ where $g(x) = x'\tilde{H}x$ achieves a minimum value. We will use the steepest-descent results to develop a computer algorithm for finding a controller u that steers to $x_{u*}(T)$.

Thus we seek the dependence of $u(t)$, $0 \leq t \leq T$, on another parameter σ so that corrected controls can be calculated as σ increases. We simply write $u(t, \sigma)$ to indicate this dependence, and identify σ with time on the computer machine.

If we discard the bound constraint on u, paths of steepest descent in $\mathcal{H}$ with norm $|\cdot| = \langle \cdot, \cdot \rangle^{1/2}$ can be determined by solving the equation

$$\frac{\partial u}{\partial \sigma}(t, \sigma) = -k \frac{\partial g}{\partial x}(x(T, \sigma))'h(t) \quad \text{where} \quad h(t) = e^{A(T-t)}b$$

and we have added a possible change in the computer time parameter σ by means of the positive constant $k > 0$.

We impose the constraint $|u(t, \sigma)| \leq 1$ by choosing the dependence of u on σ so that (for $\sigma \geq 0$)

$$(***) \qquad \frac{\partial u}{\partial \sigma}(t, \sigma) = -k \frac{\partial g}{\partial x}(x(T, \sigma))' \, h(t) \quad \text{if} \quad |u(t, \sigma)| < 1$$

or if

$$u(t, \sigma)\frac{\partial g}{\partial x}(x(T, \sigma))'\, h(t) > 0$$

$$\frac{\partial u}{\partial \sigma}(t, \sigma) = 0 \quad \text{otherwise on} \quad 0 \leq t \leq T.$$

It is assumed that the starting value $u(t, 0)$ satisfies $|u(t, 0)| \leq 1$.

With this choice of dependence of u on σ we have

$$\frac{dg}{d\sigma} = -k\int_{\mathfrak{J}}\left[\frac{\partial g}{\partial x}(x(T, \sigma))'\, h(t)\right]^2 dt$$

where

$$\mathfrak{J} = \left\{t \,\middle|\, \frac{\partial u}{\partial \sigma}(t, \sigma) \neq 0, \qquad t \in [0, T]\right\}.$$

Note $dg/d\sigma \leq 0$. Assuming $K(T)$ has an interior in R^n let us show formally that

$$\frac{dg}{d\sigma} < 0 \quad \text{except at} \quad x(T, \sigma) = x^*$$

for the above choice of dependence of u on σ. For the present, also, assume that $x = 0 \notin K(T)$.

Suppose $dg/d\sigma = 0$ at a point $x(T, \sigma)$ interior to $K(T)$. This can happen only if

$$\frac{\partial g}{\partial x}(x(T, \sigma))'\, h(t) = 0$$

or

$$u(t, \sigma) = -\operatorname{sgn}\left\{\frac{\partial g}{\partial x}(x(T, \sigma))'\, h(t)\right\}$$

for almost all $t \in [0, T]$ if $x = 0 \notin K(T)$. According to the results of Chapter 2 the response to the maximizing control

$$u(t) = \operatorname{sgn}\{\eta(t)b\},$$

[where $\eta(t) = ce^{-At}$ is a row vector with n components] is a point on $\partial K(T)$ for all $c \neq 0$. But the controller

$$u(t, \sigma) = -\operatorname{sgn}\left\{\frac{\partial g}{\partial x}(x(T, \sigma))'\, h(t)\right\}$$
$$= -\operatorname{sgn}\{2x(T, \sigma)'\tilde{H}e^{AT}e^{-At}b\}$$

is a controller of this class when $c = -2x(T, \sigma)'He^{AT}$. Thus there are no points interior to $K(T)$ where $dg/d\sigma = 0$, if $x = 0 \neq K(T)$.

Consider points of $\partial K(T)$ where $dg/d\sigma = 0$. Again this can only happen if

$$u(t, \sigma) = -\operatorname{sgn}\left\{\frac{\partial g}{\partial x}(x(T, \sigma))'\, h(t)\right\}$$

or

$$\frac{\partial g}{\partial x}(x(T, \sigma))'h(t) = 0$$

for almost all $t \in [0, T]$. This is a maximizing controller with endpoint on $\partial K(T)$, at which point $\eta(T)' = e^{-A'T}c'$ is an exterior normal of $K(T)$ according to results of Chapter 2.

Thus at $x(T, \sigma)$ on $\partial K(T)$ we have

$$\eta(T)' = e^{-A'T}c' = e^{-A'T}(-2x(T, \sigma)'\tilde{H}e^{AT})' = -2\tilde{H}\, x(T, \sigma)$$

an exterior normal to $K(T)$. This can only occur at the point $x(T, \sigma) = x^*$ where $\tilde{H}x$ is a normal vector to the surface $x'\tilde{H}x = x^{*\prime}\ \tilde{H}x^* = k^*$. A similar argument can be used in the case $x = 0 \in K(T)$ to obtain that x^* is the only point of $K(T)$ for which $\partial g/\partial\sigma = 0$. Thus the form of the optimum (limit) controller agrees with the results of Chapter 2.

We now consider how equation (***) can be approximately solved by a computer machine. One way to solve it is to replace the differential equation (***), depending on a continuous parameter σ, by the recurrence equation (iteration equation)

$$u(t, i + 1) = u(t, i) - \bar{k}_i \frac{\partial g}{\partial x}(x(T, i))'\, h(t) \quad \text{if} \quad |u(t, i)| < 1$$

or if
$$u(t, i)\frac{\partial g}{\partial x}(x(T, i))'\, h(t) > 0;$$

$$u(t, i + 1) = u(t, i) \quad \text{otherwise for} \quad 0 \leq t \leq T$$

[here $\bar{k}_i > 0$ is chosen such that $|u(t, i + 1)| \leq 1$] depending on a discrete parameter $i = 1, 2, \ldots$.

Thus we compute the sequence of corrected controls $u(t, i)$, $i = 1, 2, \ldots$ by the above scheme with the hope that the sequence is such that $g(x(T, i)) \to g(x^*)$ as i is increased. This computation is readily performed by a combination analog-digital computer and has been shown to be feasible [Ho; Gilbert].

A second method of approximation is to assume that the optimum controller can be approximated by step functions, that is, divide the interval from t_0 to T into a finite number of subintervals on each of which the controller is assumed to be a constant. The above computation is then performed to determine the best set of such constants. For simplicity assume the interval $[t_0, T]$ has been divided into ν similar subintervals each of length $(T - t_0)/\nu$. Let $u(0), u(1), \ldots, u(\nu - 1)$ be the set of constants to be determined, where $u(j)$ is the value of the approximating control function during the interval

$$t_0 + j\,\frac{T - t_0}{\nu} \le t \le (j + 1)\,\frac{T - t_0}{\nu} + t_0, \qquad j = 0, 1, 2, \ldots, \nu - 1.$$

Thus the differential equation for $u(t, \sigma)$, as a function of σ defined on the strip $0 = t_0 \le t \le T$, becomes the finite set of equations

$$\frac{\partial u}{\partial \sigma}(j, \sigma) = -k\,\frac{\partial g}{\partial x}(x(T, \sigma))'\,\hat{h}(j) \quad \text{if} \quad |u(j, \sigma)| < 1$$

or if

$$u(j, \sigma)\,\frac{\partial g}{\partial x}(x(T, \sigma))'\,\hat{h}(j) > 0;$$

$$\frac{du}{d\sigma}(j, \sigma) = 0 \quad \text{otherwise for} \quad j = 0, 1, 2, \ldots, \nu - 1.$$

Here

$$\begin{aligned} x(T, \sigma) &= e^{AT}x_0 + e^{AT}\int_0^T e^{-AT}b\,u(t, \sigma)\,dt \\ &= e^{AT}x_0 + \sum_{i=0}^{\nu-1}\left[\int_{iT/\nu}^{(i+1)T/\nu} e^{A(T-t)}\,b\,dt\right]u(i, \sigma) \\ &= e^{AT}x_0 + \sum_{i=0}^{\nu-1}\hat{h}(i)\,u(i, \sigma). \end{aligned}$$

The last equation defines $\hat{h}(i)$, which is some average weighting of the control during the various time intervals. We have taken $t_0 = 0$.

In a manner similar to the continuous-time case considered above, it can be shown that $dg/d\sigma \neq 0$ except at $x^* \in K_\nu(T)$, where $K_\nu(T)$ is the set of attainability when the controls are restricted to the class of bounded step functions on the above ν subintervals of $[t_0, T]$ and x^* is the optimum point of $K_\nu(T)$.

The above calculation is also easily performed by an analog-digital computer [Luh].

***Example* 5.** Consider the autonomous control process in R^n

$$(\mathcal{S}) \qquad \dot{x} = f(x, u) \qquad x(0) = x_0$$

with $f \in C^1$ in $R^n \times R^1$ and scalar controller $u(t)$ subject to the restraint $|u(t)| \le 1$. The cost functional of control

$$C(u) = g(x(T)),$$

where $T > 0$ is fixed and $g \in C^1$ in R^n.

Denote the solution of $\mathcal{S}$ corresponding to a controller $u(s)$, $0 \leq s \leq t$, with $x(0) = x_0$ by $x_u(t)$. We can also calculate $C(u) = g(x_u(T))$, since $x_u(T)$ is known to be a point of the set of attainability $K(T)$ as shown schematically in Figure A.2 for a two-dimensional problem. In computing an optimum controller we seek a path from the initially calculated point $x_u(T) \in K(T)$ that moves to values of $x \in K(T)$ where $g(x)$ is smaller than $g(x_u(T))$. In

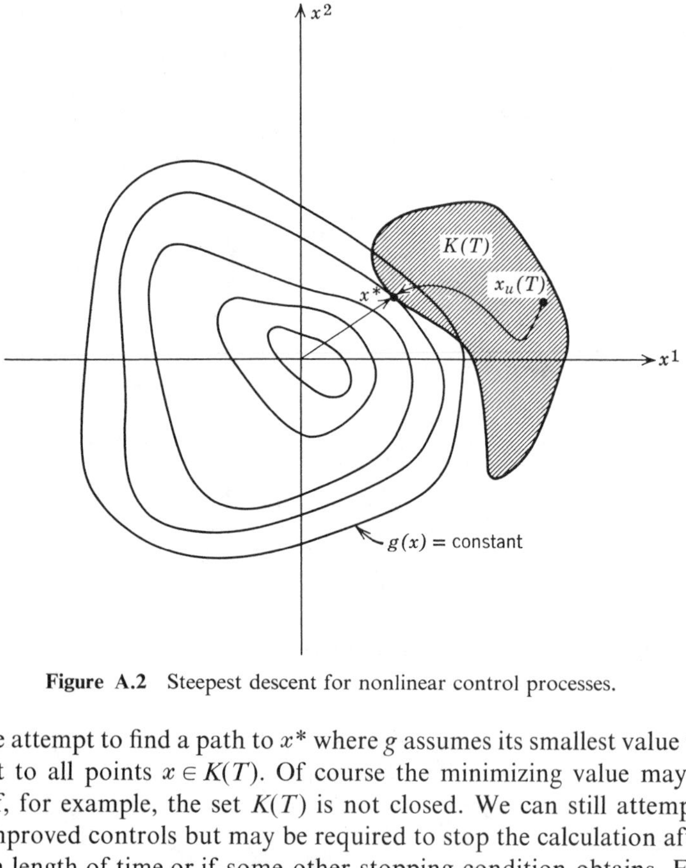

Figure A.2 Steepest descent for nonlinear control processes.

fact we attempt to find a path to x^* where g assumes its smallest value with respect to all points $x \in K(T)$. Of course the minimizing value may not exist if, for example, the set $K(T)$ is not closed. We can still attempt to find improved controls but may be required to stop the calculation after a certain length of time or if some other stopping condition obtains. For a discussion of cases in which the set $K(T)$ is closed see Chapters 4 to 6.

If u takes values in the Hilbert space $\mathcal{H}$ with norm $|\cdot| = \langle\cdot, \cdot\rangle^{1/2}$ and the restraint on u is discarded it can be shown that the gradient direction of steepest descent at a point $u_0 \in \mathcal{H}$ is defined by a path that is a solution of

$$\frac{du}{d\sigma} = -\frac{\partial g}{\partial x}(x_u(T))'\, h(t) \quad \text{starting at} \quad u_0,$$

where $h(t) = \Phi_u(T)\,\Phi_u(T)^{-1}\,B_u(t)$ and Φ_u is a solution of the linear variational equation

$$\dot{\Phi}_u(t) = \frac{\partial f}{\partial x}\,(x_u(t), u(t))\,\Phi_u(t)$$

with $\Phi_u(t_0) = I$ and $B_u(t) = (\partial f/\partial u)\,(x_u(t), u(t))$ on $[t_0, T]$.

To impose the constraint $|u| \leq 1$ pick

$$(\dagger) \qquad \frac{du}{d\sigma}(t, \sigma) = -\frac{\partial g}{\partial x}\,(x_u(T))'\,h(t) \quad \text{if} \quad |u(t, \sigma)| < 1$$

or if

$$u(t, \sigma)\,\frac{\partial g}{\partial x}\,(x_u(T))'\,h(t) > 0;$$

$$\frac{du}{d\sigma}(t, \sigma) = 0 \quad \text{otherwise on} \quad t_0 \leq t \leq T, \quad \text{for } \sigma \geq 0$$

where $|u(t, 0)| \leq 1$. Further we see that

$$\frac{dC}{d\sigma} = \frac{dg}{d\sigma} = \frac{\partial g}{\partial x}\,(x_u(T))'\,\frac{dx_u}{d\sigma}$$

$$= -\int_{\mathfrak{J}} \left[\frac{\partial g'}{\partial x}\,(x_u(T))\,h(t)\right]^2 dt \leq 0$$

where $\mathfrak{J} = \{t \mid \partial u/\partial \sigma \neq 0,\ t \in [t_0, T]\}$, so there is some hope that an approximate implementation of Eq. ($\dagger$) on a computer, as in the previous example, would lead to a convergent algorithm. See Kelly and Bryson for an indication of the computing efficiency of an algorithm so derived.

***Example* 6.** In this example and the next the results of Chapters 2, 3, 4, and 5 are applied to the control problem, in which case our knowledge of the optimum controller is usually complete when the initial values for the adjoint response are known. To solve for these initial values, functions are constructed such that they achieve an extreme value at the correct initial values for the adjoint response. The problem then reduces to the one considered in Section A.1 where the minimum (or maximum) of a function of a finite number of variables is to be obtained by the technique of a steepest-descent calculation. A time-optimal problem is now treated. Extensions of the results for computing time optimal control can be made to other cost functionals and nonlinear problems [Neustadt].

Consider the autonomous linear control process

$$(\mathfrak{L}) \qquad \dot{x} = Ax + b\,u(t)$$

with scalar control variable $u(t)$ satisfying $|u(t)| \leq 1$. Later more general linear systems are considered. Suppose the problem is one of steering to the origin in minimum time, that is, a problem of the type discussed in Chapter 2. In Chapter 2 the optimum control was found for each initial condition up to a set of constants, $\eta(t_0) = c$. Calculation of these constants is the objective in this example. In steering to the origin, the problem of time-optimal control is to find the smallest $t > 0$ such that the equation

$$x(t) = e^{At}x_0 + e^{At}\int_0^t e^{-At'}b\, u(t')\, dt' = 0$$

is satisfied for some $u(t')$, $0 \leq t' \leq t$ with $|u(t')| \leq 1$ at each point of the interval $[0, t]$. This requirement can be written as

$$-x_0 = \int_0^t e^{-At'}b\, u(t')\, dt'$$

for some admissible $u(t')$, $0 \leq t' \leq t$.

Let

$$C(t) = \left\{x \mid x = \int_0^t e^{-At'}b\, u(t')\, dt'; \qquad u(t') \text{ admissible}\right\}$$

= set of initial states from which the origin can be reached in time less than t. $C(t)$ is just a special case of the set of attainability as discussed in Chapter 2. In fact, $C(t) = K(-t)$ starting at $x_0 = 0$. Thus $C(t)$ is a closed, convex subset of R^n with

$$\partial C(t) = \left\{x \mid x = \int_0^t e^{-At'}b \operatorname{sgn}\{ce^{-At'}b\}\, dt', \qquad \|c\| = 1\right\}.$$

Moreover if $\zeta \in \partial C(t)$, c' is an exterior normal to $C(t)$ at ζ, and if the system $\mathfrak{L}$ is normal, $\partial C(t)$ contains no line segments. Note that $C(t_2) \supset C(t_1)$ for $t_2 \geq t_1$, and $C(t)$ grows continuously with t.

Assume that there is some control that steers $x(t)$ from x_0 to 0 in finite time; then there is a time-optimal control of the form

$$\operatorname{sgn}\{\eta(t)b\} = \operatorname{sgn}\{ce^{-At}b\}$$

that steers $x(t)$ from x_0 to 0, where c' is an exterior normal to $C(t^*)$ at $-x_0$. Denote the set of all such vectors c, with $\|c\| = 1$, at $-x_0$ on $\partial C(t)$ by Z. Then if $c \in Z$ the control

$$u(t) = \operatorname{sgn}\{ce^{-At}b\}$$

steers $x(t)$ from x_0 to 0 and is optimal.

Let

$$z(t, c) = \int_0^t e^{-At'}b \operatorname{sgn}\{ce^{-At'}b\}\, dt'.$$

Now $z(t, c)$ is on $\partial C(t)$ and therefore for the normal system

$$c\, z(t, c) > c\zeta \quad \text{all} \quad \zeta \in C(t), \qquad \zeta \neq z(t, c).$$

Note that

$$c\, z(t, c) = \int_0^t ce^{-At'}b \operatorname{sgn} \{ce^{-At'}\, b\}\, dt'$$

$$= \int_0^t |ce^{-At'}b|\, dt' > 0 \quad \text{for} \quad t > 0,$$

and therefore $c\, z(t, c)$ is a monotone increasing function of t, continuous in c for a normal system.

Consider (as Neustadt does) the function

$$\bar{f}(t, c, x_0) = c[z(t, c) + x_0],$$

which is continuous in t and c. For fixed (x_0, c) it is strictly increasing in t since

$$\bar{f}(t, c, x_0) = \int_0^t |ce^{-At'}b|\, dt' + cx_0$$

is strictly increasing in t. Consider only those c with $cx_0 = \bar{f}(0, c, x_0) < 0$. If c does not belong to Z, then—from the condition that $c\, z(t, c) > c\zeta$ all ζ in $C(t)$—it follows that $c\, z(t^*, c) > -cx_0$ or $\bar{f}(t^*, C, x_0) > 0$, where t^* is the optimum time. Hence, for some unique $0 < t < t^*$, we have

$$\bar{f}(t, c, x_0) = 0.$$

Denote this t by $T(c, x_0)$ so that for each choice of c with $cx_0 < 0$ we have $\bar{f}(T(c, x_0), c, x_0) = 0$.

Because $\bar{f}$ is continuous in its arguments, T is continuous in c. Hence if c is in Z, $T(c, x_0)$ takes on its maximum value. This is the function that is to be maximized in computing the required c.

To find the c that maximizes $T(c, x_0)$, make c depend on a continuous parameter σ by computing corrections in the c's by means of the (steepest ascent in E^n) differential equation

$$\frac{dc}{d\sigma} = k\frac{\partial T}{\partial c} \quad \text{where} \quad k > 0$$

and $\partial T/\partial c$ is the gradient vector at c.

To proceed as above we need to evaluate

$$\frac{\partial T}{\partial c} = -\frac{\partial \bar{f}/\partial c}{\partial \bar{f}/\partial T}.$$

From the defining equation for $\bar{f}(T, c, x_0)$ it is found that

$$\frac{\partial \bar{f}}{\partial T} = |ce^{-AT}b|$$

and

$$\frac{\partial \bar{f}}{\partial c_i} = x_0^{\,i} + \int_0^T \operatorname{sgn}\{ce^{-At'}b\}[e^{-At'}b]^i\,dt,$$

where $[e^{-At}b]^i$ is the ith component of the vector $e^{-At}b$ for the normal system.

Thus

$$\frac{\partial T}{\partial c} = -\frac{[x_0 + z(T, c)]'}{|ce^{-At}b|}.$$

If we choose $k = |[ce^{-At}b]|$, then we compute corrections to c from the equation

$$(\dagger\dagger) \qquad \frac{dc'}{d\sigma} = -[x_0 + z(T(c, x_0), c)].$$

Since the right side of this equation is continuous in c this equation has a solution. When $|ce^{-AT(c,\,x_0)}b| > 0$,

$$\frac{dT}{d\sigma} = \frac{\partial T}{\partial c}\frac{dc'}{d\sigma} = |ce^{-AT(c,x_0)}b|\,\frac{\partial T}{\partial c}\frac{\partial T'}{\partial c} \geq 0.$$

In this case, if $c \notin Z$, $\partial T/\partial c > 0$. When $|ce^{-AT(c,\,x_0)}b| = 0$, $dT/d\sigma$ is not defined, but $\partial \bar{f}/\partial c = [x_0 + z(T(c, x_0), c)]'$ is defined for fixed T and $(\partial \bar{f}/\partial \sigma)(T, c, x_0) = \partial \bar{f}/\partial c\ \partial c'/\partial \sigma = -\partial c/\partial \sigma\ \partial c'/\partial \sigma < 0$ if $c \notin Z$. Thus $\bar{f}$ monotonically decreases with σ, for fixed T, but, as we proved earlier, $\bar{f}$ increases with T and therefore T increases with σ for fixed $\bar{f}(T, \sigma, x_0) = 0$.

Note $T(c, x_0)$ is defined if and only if $cx_0 \leq 0$, and if the system is normal $cx_0 = 0$ implies $T(c, x_0) = 0$. Let D be the domain of definition of $T(c, x_0)$. If c is initially in D the solution $c(\sigma)$ of Eq. (††) never leaves D. In order for $c(\sigma)$ to leave D it is necessary for $c(\sigma_0)x_0 = 0$ for some σ_0, but then $T(c(\sigma_0), x_0) = 0$. This is impossible because $T(c(0), x_0) > 0$ and T increases with σ.

Thus it is clear that $T(c, x_0)$ vanishes on ∂D, takes on a positive maximum on the convex set $Z \subset D$, and has no other local maxima or minima in D. Thus when $c(\sigma)$ approaches a limit as $\sigma \to \infty$ this limit is in Z. Note that $\|c(\sigma)\| =$ constant because

$$\frac{d(cc')}{d\sigma} = 2c\frac{dc'}{d\sigma} = -2c[x_0 + z(T, c)] = 0.$$

To implement the solution to the Eq. (††) above, for making corrections to c, it is necessary to consider some approximations. This is true because it takes computer time to calculate the zero of the function $\bar{f}(T, c, x_0)$, which defines $T(c, x_0)$. Thus, to carry out the above schedule of computation a discrete version of Eq. (††) will be discussed; let

$$c^{(j+1)} = c^j + k\frac{\partial T}{\partial c}(c^{(j)}, x_0) \quad \text{for} \quad k > 0,$$

$j = 1, 2, \ldots,$ where $T(c^{(j)}, x_0)$ is determined by increasing t in the expression $\bar{f}(t, c^{(j)}, x_0) = c^{(j)}[x_0 + z(t, c^{(j)})]$ until its value is zero. This computation only requires integration, which is readily done by computers

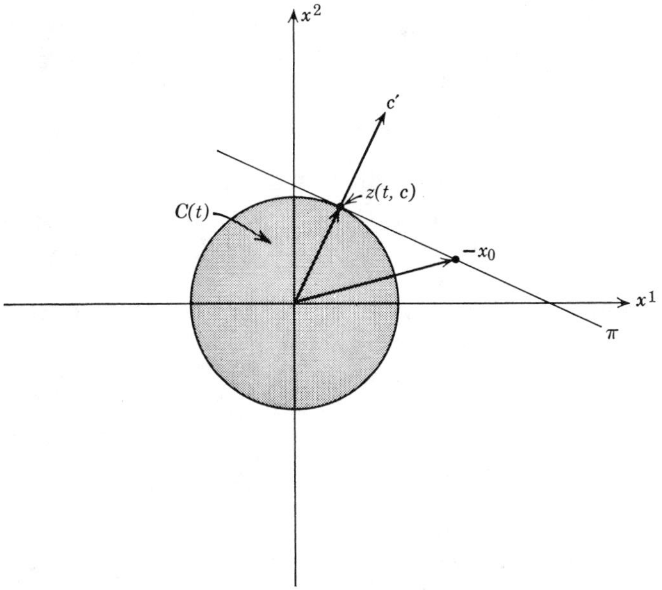

Figure A.3 Construction of steepest descent for time optimal controllers.

[Paiewonsky]. After $T(c^{(j)}, x_0)$ is found, $c^{(j+1)}$ is calculated by the above recurrence equation. Then the procedure is repeated by calculating $T(c^{(j+1)}, x_0)$ as above, and so on. $k > 0$ is a constant whose choice determines the convergence rate.

The method has a simple geometric interpretation. Consider, in Figure A.3, any point $-x_0$ and any nonzero vector c such that $cx_0 < 0$. To increase t until $\bar{f}(t, c, x_0) = c[x_0 + z(t, c)] = 0$ corresponds to carrying the hyperplane π: $c[-x + z(t, c)] = 0$ parallel to itself until it passes through $-x_0$.

A correction to c is then made in the direction $-x_0 - z(t, c) = v$, a vector orthogonal to c'. The iteration on this is then performed as indicated above.

Let us now generalize our results to the time-varying case with m control variables. Consider

$$(\mathcal{L})\quad \dot{x} = A(t)x + B(t)\,u(t) + v(t), \qquad |u^j(t)| \le 1, \qquad j = 1, 2, \dots, m$$

where as usual $A(t)$, $B(t)$ are continuous matrices of size $n \times n$ and $n \times m$ respectively. $v(t)$ is a known continuous vector function on the time interval of concern.

For this linear equation the response to any measurable control function $u(t)$ is

$$x(t) = \Phi(t)x_0 + \int_{t_0}^{t} \Phi(t)\,\Phi(t')^{-1}[B(t')\,u(t') + v(t')]\,dt',$$

where

$$\dot{\Phi}(t) = A(t)\,\Phi(t) \quad \text{and} \quad \Phi(t_0) = I.$$

If we rewrite the above equation we find

$$-x_0 - \int_{t_0}^{t} \Phi(t')^{-1}\,v(t')\,dt' + \Phi(t)^{-1}\,x(t) = \int_{t_0}^{t} \Phi(t')^{-1}\,B(t')\,u(t')\,dt'.$$

Suppose the problem is that of determining the allowable controller so that $x(t) = \xi(t)$ in minimum time, where $\xi(t)$ is the parametric representation of a continuous curve in R^n.

Let

$$\omega(t) = -\left[x_0 + \int_{t_0}^{t} \Phi(t')^{-1}\,v(t')\,dt' - \Phi(t)^{-1}\,\xi(t)\right].$$

Then the problem of time-optimal control is that of determining the allowable controller $u(t)$ and time t for which

$$\omega(t) = \int_{t_0}^{t} \Phi(t')^{-1}\,B(t')\,u(t')\,dt'$$

with t a minimum.

Since $\omega(t)$ is just a point in R^n for each t, we can consider the set $C(t)$ as above and the problem reduces to finding the smallest t for which $\omega(t) \in C(t)$.

Again consider maximizing controllers

$$u^j(t) = \operatorname{sgn}\{c\,\Phi(t)^{-1}\,B(t)\}_j, \qquad j = 1, 2, \dots, m$$

and assume that each component $\{c\,\Phi(t)^{-1}\,B(t)\}_j$ has no collection of zeros for $\|c\| = 1$, that is, the system is normal.

As usual let

$$z(t, c) = \int_{t_0}^{t} \Phi(t')^{-1}\,B(t')\operatorname{sgn}\{c\,\Phi(t')^{-1}\,B(t')\}\,dt'.$$

$z(t, c)$ is on $\partial C(t)$ and c' is an external normal of $C(t)$, the corresponding set of attainability, at $z(t, c)$.

Consider $\bar{f}(t, c, \omega(t)) = c[z(t, c) - \omega(t)]$. Geometrically the problem is as shown in Figure A.4, where we consider only discrete corrections to c.

Choose any $c^{(1)}$ with $c^{(1)}\,\omega(t_0) > 0$, $\|c^{(1)}\| = 1$ as the initial guess of c. Next increase t until $\bar{f}(t, c^{(1)}, \omega(t)) = 0$, which corresponds to carrying π parallel to itself until the plane π intercepts $\omega(t)$. Call this t, $t^{(1)}$. If

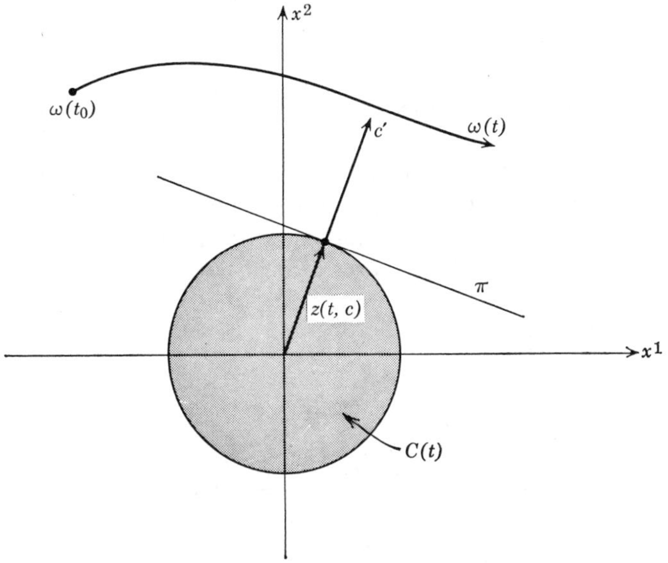

Figure A.4 Steepest descent with time varying target.

$\omega(t^{(1)}) \neq z(^{(1)}, c^{(1)})$ we must make a correction; for this purpose choose $c^{(2)} = c^{(1)} + kv^{(1)}$, where $v^{(1)}$ is a vector directed from the tip of $z(t^{(1)}, c^{(1)})$ to the tip of the vector $\omega(t^{(1)})$.

Thus choose

$$c^{(2)} = c^{(1)} + k[\omega(t^{(1)}) - z(t^{(1)}, c^{(1)})]$$

where $k > 0$. Again it can be shown that v is in the direction of gradient of T where $T > t_0$ is the smallest real root of the equation $\bar{f}(t, c, \omega(t)) = 0$.

If an optimum control exists, and if k is properly chosen, iterative corrections by the above scheme should lead to a $c \in Z$, call this c, c^*. Then the optimum control for steering to $\omega(t)$ is

$$u(t) = \text{sgn}\,[c^*\,\Phi(t)^{-1}\,B(t)]\ .$$

Solutions on analog and digital computer machines for this technique are reported by Paiewonsky. He also considers some techniques developed by N. N. Krasovski that are closely related to those given above.

***Example* 7.** Suppose T is the fixed time at which the problem is completed and consider the controlled process

$$(\mathcal{S}) \qquad \dot{x} = f(x, u), \qquad x(t_0) = x_0, \qquad u \in \Omega$$

with cost functional

$$C(u) = g(x(T)).$$

It is assumed that $f \in C^1$ in $R^n \times \Omega$ and $g \in C^1$ in R^n.

By the use of the maximal principle (Chapters 1 and 5) it is often possible to obtain u as a function of the state x and adjoint variable η by maximizing $H = \eta\dot{x} = \eta f(x, u)$ with respect to the explicit occurrence of u in H for $u \in \Omega$. We shall consider only those problems where such a unique u exists. In this case the problem of optimum control is reduced to finding the initial vector $\eta(t_0)$ so that an extremal curve satisfies the given boundary conditions and minimizes $C(u) = g(x(T))$.

Consider only maximal-solution curves, that is, curves satisfying the differential equation system

$$(\mathcal{S}_E) \qquad \dot{x} = \frac{\partial H(x, \eta, U)}{\partial \eta} = f(x, U(x, \eta)) = \tilde{f}(x, \eta)$$

$$\dot{\eta}' = -\frac{\partial H(x, \eta, U)}{\partial x} = -\frac{\partial f}{\partial x}(x, U(x, \eta))\eta'$$

$$= \tilde{g}(x, \eta)',$$

where $U = U(x, \eta)$ is defined from the condition that

$$H(x, \eta, U) = \operatorname*{Max}_{u \in \Omega} \{H(x, \eta, u)\} \quad \textit{for all} \quad x, \eta \in R^n.$$

Here $x(t_0) = x_0$ and $\eta(t_0)$ is to be determined to satisfy given boundary conditions (discussed below) and to provide a minimum for $g(x(T))$.

The problem has been imbedded in a family of curves depending on n parameters, $\eta(t_0)$. The object of the calculation methods is to determine $\eta(t_0)$ so that a curve of the family defined by $\mathcal{S}_E$ minimizes g.

For each $\eta(t_0) = c$ the solution of $\mathcal{S}_E$ is written as $\begin{pmatrix} x(t, c) \\ \eta(t, c)' \end{pmatrix}$, and it is assumed that such a solution can be calculated, given $u = U(x, \eta)$. In this case $g(x(T, c))$ is readily determined. We seek the dependence of c on a parameter σ so that $c(\sigma)$, as a function of σ, moves to values where $g(x(T, c(\sigma)))$ assumes smaller and smaller values as σ increases.

Let the dependence of c on σ be determined by the (steepest-descent) differential equation

$$(\dagger\dagger\dagger) \qquad \frac{dc}{d\sigma} = -k\frac{\partial g}{\partial c},$$

where $[\partial g/\partial c]'$ is the gradient vector and k = constant > 0. Then

$$\frac{dg}{d\sigma} = \frac{\partial g}{\partial c}\frac{\partial c'}{\partial \sigma} = -k\left[\frac{\partial q}{\partial c}\right]^2 \leq 0,$$

where $[\partial g/\partial c]^2$ is the inner product. When $\partial g/\partial c \neq 0$, $g(\sigma) = g(x(T, c(\sigma)))$ decreases as σ increases. Hence a point where $\partial g/\partial c = 0$ is approached as $\sigma \to \infty$.

To calculate $c(\sigma)$ by the above equation $(\dagger\dagger\dagger)$ it is necessary to know how changes in c effect g. For this purpose consider

$$\frac{\partial g}{\partial c} = \frac{\partial g'}{\partial x}\frac{\partial x}{\partial c}$$

where $x = x(T, c)$ and

$$\frac{\partial x}{\partial c'} = \begin{pmatrix} \dfrac{\partial x^1}{\partial c_1} & \cdots & \dfrac{\partial x^1}{\partial c_n} \\ \vdots & & \\ \dfrac{\partial x^n}{\partial c_1} & \cdots & \dfrac{\partial x^n}{\partial c_n} \end{pmatrix}$$

To evaluate this matrix of partial derivatives we are required to calculate the sensitivity of the solutions to variations in c. Since

$$\dot{x} = \tilde{f}(x, \eta)$$
$$\dot{\eta} = \tilde{g}(x, \eta)$$

the maximal solution pair $x(t, c)$, $\eta(t, c)$ must be such that

$$\dot{x}(t, c) = \tilde{f}(x(t, c), \eta(t, c))$$
$$\dot{\eta}(t, c) = \tilde{g}(x(t, c), \eta(t, c))$$

for $t_0 \leq t \leq T$.

Consider

$$\frac{\partial \dot{x}(t, c)}{\partial c_i} = \frac{\partial \tilde{f}}{\partial x}(x(t, c), \eta(t, c))\frac{\partial x}{\partial c_i}(t, c) + \frac{\partial \tilde{f}'}{\partial \eta}(x(t, c), \eta(t, c))\frac{\partial \eta'}{\partial c}(t, c)$$

and

$$\frac{\partial \dot{\eta}(t, c)'}{\partial c_i} = \frac{\partial \tilde{g}'}{\partial x}(x(t, c), \eta(t, c)) \frac{\partial x}{\partial c_i}(t, c) + \frac{\partial \tilde{g}}{\partial \eta}(x(T, c), \eta(t, c)) \frac{\partial \eta'}{\partial c_i}(t, c).$$

If we assume that the order of differentiation can be interchanged, we write

$$(\bar{\mathfrak{L}}) \qquad \begin{aligned} \frac{d(\partial x/\partial c_i)}{dt} &= \frac{\partial \tilde{f}}{\partial x}\frac{\partial x}{\partial c_i} + \frac{\partial \tilde{f}'}{\partial \eta}\frac{\partial \eta'}{\partial c_i} \\ \frac{d(\partial \eta'/\partial c_i)}{dt} &= \frac{\partial \tilde{g}'}{\partial x}\frac{\partial x}{\partial c_i} + \frac{\partial \tilde{g}}{\partial \eta}\frac{\partial \eta'}{\partial c_i}, \end{aligned}$$

where $\partial\tilde{f}/\partial x$, $\partial\tilde{f}/\partial\eta$, $\partial\tilde{g}/\partial x$, $\partial\tilde{g}/\partial\eta$ are evaluated along the extremal curve $x(t, c)$, $\eta(t, c)$; $t_0 \leq t \leq T$. Let

$$\psi(t) = \begin{pmatrix} \psi_1{}^1(t) & \psi_2{}^1(t) \\ \psi_1{}^2(t) & \psi_2{}^2(t) \end{pmatrix}$$

be a fundamental solution matrix for the linear system of equations $\bar{\mathfrak{L}}$ with $\psi(t_0) = I$. Then

$$\frac{\partial g}{\partial c} = \frac{\partial g'}{\partial x}\psi_2{}^1(T),$$

where, of course, $\psi_2{}^1(T)$ is dependent on the maximal curve $x(t, c)$, $\eta(t, c)$ so that $\psi_2{}^1(T)$ must be continuously reevaluated as σ increases. Thus the local sensitivity can be calculated from a variational approach.

Another technique often used on a computer machine is to approximately calculate $\partial g/\partial c$ by perturbation techniques (see, for example [Scharmack]).

Various applications of a computer algorithm derived from the steepest-descent equation (†††) to space-vehicle control problems [Scharmack] and even second-order schemes for such problems are available [Kelley-Bryson].

A.3 BIBLIOGRAPHY ON STEEPEST DESCENT AND COMPUTATIONAL TECHNIQUES FOR OPTIMAL-CONTROL PROBLEMS

In the previous two sections we have proposed the method of steepest descent as a constructive technique for optimal controllers. Many of the computer algorithms available for computing optimal controllers had their origins with the technique of steepest descent, that is, most computer algorithms, even those of Examples 5 and 6 above, are based on a modification of the steepest-descent calculation where the gradient is used in the appropriate space. The first modification is to introduce a change in time scale, as in Example 5, by means of a constant k. A further modification is to change the geometry in the function space (the second-order method of Newton–Raphson, which is accomplished by introducing a locally linear transformation). Other modifications have been made to increase the efficiency of the algorithm when applied to special classes of problems, and there are other computer algorithms for optimal controllers that bear little resemblance to the technique of steepest descent, for example, the technique of dynamic programming as applied to such problems [Bellman]. The general mathematical theory of steepest descent is discussed in papers of Goldstein, Kantorovich, and Rosenbloom.

There are many excellent examples of the application of steepest descent to control optimization problems in the literature. The papers of Bryson and Kelly contain several examples. Since neither of the authors of this book have done any serious calculations of optimal controllers by computer machines, we do not feel qualified to discuss the relative merits of one computer technique over another. We therefore complete this section by presenting a bibliography on steepest descent and other computational results. We have, however, divided the bibliography into a number of subsections to indicate what we feel to be the general area of contribution of any one reference.

Theory of Steepest Descent and Controller Optimization

Balakrishnan, A. V., "An Operator Theoretic Formulation of a Class of Control Problems and a Steepest Descent Method of Solution," *J. Soc. Ind. Appl. Math, Ser. A, Control,* **1,** No. 2, 109–127 (1963).

Curry, H. B., "The Method of Steepest Descent for Nonlinear Minimization Problems," *Quart. Appl. Math.*, **2,** No. 4 (Oct. 1944).

Denham, W. F., *Steepest-Ascent Solution of Optimal Programming Problems*, Ph.D. Thesis, Div. Engr. and Applied Physics, Harvard University (1963).

Goldstein, A. A., "Minimizing Functionals on Hilbert space," In A. V. Balakrishnan and L. W. Neustadt (eds.), *Computing methods in Optimization Problems* Academic Press Inc., New York, 1964, pp. 159–165. [Also in *J. Soc. Ind. Appl. Math. Ser. A, Control,* **4,** 81–89 (1966)].

Gollwitzer, H. E., *Application of the Method of Steepest Descent to Optimal Control Problems*, Thesis, University of Minnesota, 1965.

Hillsley, R. H., and H. M. Robins, "Steepest Ascent Trajectory Optimization Method Which Reduces Memory Requirements," in A. V. Balakrishnan and L. W. Neustadt (eds.), *Computing Methods in Optimization Problems*, Academic Press Inc., New York, 1964, pp. 107–133.

Kantorovich, L. V., and G. P. Akilov, *Functional Analysis in Normed Space*, Pergamon Press, New York, 1964.

Rosenbloom, P. C., "The Method of Steepest Descent," *Proceedings of Symposia in Applied Mathematics*, Vol. 6, pp. 127–176, The American Mathematical Society, 1956.

Tompkins, C. B., "Method of Steepest Descent," in E. F. Beckenbach (ed.), *Modern Mathematics for the Engineer*, McGraw-Hill Book Co., Inc., New York, 1956, Chapter 18.

Vachino, R. F., "Steepest Descent with Inequality Constraints on the Control Variables," *J. Soc. Ind. Appl. Math. Ser A, Control*, **4,** No. 1,245–261 (1966).

Vainberg, M. M., "On Convergence of the Method of Steepest Descent for Nonlinear Equations" *Am. Math. Soc. Transl.*, **1,** 1960.

Steepest Descent Applied to Controller Optimization Problems

Bryson, A. E., and W. F. Denham, "A Steepest Ascent Method for Solving Optimum Programming Problems," *J. Appl. Mech.*, **29,** 247–257 (1962).

Bryson, A. E., W. F. Denham, F. J. Caroll, and K. Mikamic, "Determination of the Lift or Drag Program that Minimizes Re-entry Heating with Acceleration or Range Constraints Using a Steepest Descent Computation Procedure," *Jour. of Aerospace Sci.*, **29,** No. 4, 420–430 (1962).

Denham, W. F., and A. E. Bryson, "Optimal programming with inequality constraints II: Solution by steepest ascent," *AIAA J.* **2,** 25–34 (1964).

Neustadt, L. W., "Minimum Effort Control Systems," *J. Soc. Ind. Appl. Math. Ser. A, Control*, **1,** No. 1, 16–31 (1962).

Neustadt, L. W., "Synthesizing Time Optimal Control Systems," *J. Math. Anal. Appl.*, **1,** No. 4, 484–493 (1960).

Computational Methods for Controller Optimization Problems

Aoki, M., "On a Successive Approximation Technique in Solving Some Control Problems," *Trans. ASME, Ser. D, J. Basic Eng.* **85,** 177–180 (1963).

Balakrishnan, A. V., and H. C. Hsieh, "Function Space Methods in Control System Optimization," *Proc. Optimum System Synthesis Conference*, Dayton, Ohio, September 1962, pp 10–40.

Bohn, E. V., "A Numerical Trajectory Optimization Method Suitable for a Computer of Limited Memory," *Proc. JACC* Seattle, Washington, 1966, 177–186.

Breakwell, J. V., J. L. Speyer, and A. E. Bryson, "Optimization and Control of Non-linear Systems Using Second Variation," *J. Soc. Ind. Appl. Math., Ser. A, Control*, **1,** No. 2, 193–223 (1963).

Canon, M. D., and J. H. Eaton, "A New Algorithm for a Class of Quadratic Programming Problems with Applications to Control," *J. Soc. Ind. Appl. Math, Ser. A, Control*, **4,** No. 1, 34–45 (1966).

Eaton, J. H., "An Iterative Solution to Time-Optimal Control," *J. Math. Anal. Appl.*, **5,** 324–344 (1962).

Eaton, J. H., "An On Line Solution to Sampled-Data Time Optimal Control," *J. Electron. Control*, **15,** No. 4, 333–341 (1963).

Eaton, J. H., "Improper Solutions under Existence Assumptions: an Example," *Tech. Note on NASA Grant NsG*-354, Ser. 5, Issue 16 University of California (1964).

Fadden, E. J., and E. G. Gilbert, "Computational Aspects of the Time-Optimal Control Problem," in A. V. Balakrishnan and L. W. Neustadt (eds.), *Computing Methods in Optimization Problems*, Academic Press Inc., New York, 1964, pp. 167–193.

Fancher, P. S., "Iterative Computation Procedures for an Optimum Control Problem," *IEEE Trans. Auto. Control* **AC-10,** 346–348 (1965).

Fletcher, R., and M. J. D. Powell, "A Rapidly Convergent Descent Method for Minimization," *Computer Journal,* **6,** No. 2, 163–168 (1963).

Fletcher, R., and C. M. Reeves, "Function Minimization by Conjugate Gradients," *Computer Journal,* **7,** No. 2, 149–154 (1964).

Forsythe, G. E., "Acceleration of the Optimum Gradient Method," (Abstract), *Bull. Am. Math. Soc.,* **57,** 304–305 (1951).

Forsythe, G. E., and T. S. Motzkin, "Asymptotic Properties of the Optimum Gradient Method," (Abstract), *Bull. Am. Math. Soc.,* **57,** 183 (1951).

Frank, M., and P. Wolfe, "An Algorithm for Quadratic Programming," *Naval Res. Logist. Quart.,* **3,** 95–110 (1956).

Gilbert, E. G., "An Iterative Procedure for Computing the Minimum of a Quadratic Form on a Convex Set." *J. Soc. Ind. Appl. Math., Ser. A, Control,* **4,** No. 1, 61–80 (1966).

Gottlieb, R. G., "Rapid Convergence to Optimum Solutions using Min-H Strategy," *Proc. JACC,* Seattle, Washington, 1966, pp. 167–174.

Goldstein, A. A., "Convex Programming and Optimal Control," *J. Soc. Ind. Appl. Math, Ser. A, Control,* **3,** No. 1, Ser., A, 147–151 (1965).

Halkin, H., "Method of convex ascent," in A. V. Balakrishnan and L. W. Neustadt (eds.), *Computing Methods in Optimization Problems,* Academic Press Inc., New York, 1964, pp 211–239.

Ho, Y. C., "A Successive Approximation Technique for Optimal Control Systems Subject to Input Saturation," *Trans. ASME, Ser. D. J. Basic Eng.,* **84,** No. 1, 33–40 (1962).

Ho, Y. C., "Computational Procedure for Optimal Control Problem with State Variable Constraints," *J. Math. Anal. Appl.,* **5,** 216–224 (1962).

Ho, Y. C., and P. B. Brentani, "On Computing Optimal Control with Inequality Constraints," *Soc. Ind. Appl. Math. Ser A, Control,* **1,** 319–348 (1963).

Ho, Y. C., and R. L. Kashyap, "A Class of Iterative Procedures for Linear Inequalities," *J. Soc. Ind. Appl. Math, Ser. A, Control,* **4,** No. 1, 112–115 (1966).

Isaacs, D., C. T. Leonodes, and R. A. Nieman, "On a Sequential Optimization Approach in Nonlinear Control," *Proc. JACC,* Seattle, Washington 158–166 pp. 1966.

Jurovics, A. S., and J. E. McIntyre, "The Adjoint Method and its Application to Trajectory Optimization," *ARS J.,* **32,** 135*s* (1962).

Kazda, L., "Control System Optimization Using Computers as Control System Elements," *Proc. of Computer in Control Systems Conference,* New York, 1958.

Kelley, H. J., "Methods of Gradients," in G. Leitman (ed.), *Optimization Techniques,*" Academic Press Inc., New York, 1962, Chapter 6.

Knapp, C. H., and P. A. Frost, "Determination of Optimal Control and Trajectories Using the Maximum Principle in Association with a Gradient Technique," *Proc. JACC,* Stanford, California, 1964, p. 222 ff.

Knudsen, H. K., "An Iterative Procedure for Computing Time-Optimal Controls," *IEEE Trans. Auto. Control,* **9,** 23–30 (1964).

Kopp, R. E., and R. McGill, "Several Trajectory Optimization Techniques," in A. V. Balakrishnan and L. W. Neustadt (eds.), *Computing Methods in Optimization Problems,* Academic Press Inc., New York, 1964, pp. 65–89.

Kulikowski, R., "Synthesis of a Class of Optimum Control Systems," *Bull. Acad. Polon. Sci., Ser. Sci. Tech.,* **7,** 663–671 (1959).

Luh, J. Y. S., "On a Computational Scheme for Time Optimal of Linear Discrete Systems," *IEEE Trans. Auto. Control,* **AC11,** No. 1, 145 (1966).

McGill, R., "Optimal Control, Inequality State Constraints, and the Generalized Newton-Raphson Algorithm," *J. Soc. Ind. Appl. Math. Ser. A, Control,* **3,** No. 2, 291–298 (1965).

McReynolds, S. R., *A Successive Sweep Method for Solving Optimal Programming Problems*, Ph.D. Dissertation, Div. of Engineering and Appl. Physics, Harvard University, 1966.

Meerov, M. V., and V. G. Fridmand, "Linear Programming in Hilbert Space for Optimizing a Class of Multivariable Systems," *Proc. Third IFAC Congress*, London, 1966.

Merriam, C. W., "An Algorithm for the Iterative Solution of a Class of Two Point Boundary Value Problems," *SIAM J. Control,* **2,** 1–10 (1964).

Mikami, T., "An Iterative Computing Method for Solving Time-Optimal Control Problems," *Proc. Third IFAC Congress, London*, 1966.

Mitter, S. K., "Successive Approximation Methods for the Solution of Optimal Control Problems," *Automatica*, **3.** 135–149, Pergamon Press (1966).

Moyer, H. G., and G. Purham, "Several trajectory optimization techniques, Part II," in A. V. Balakrishnan and L. W. Neustat (eds.), *Computing Methods in Optimization Problems*, Academic Press Inc., New York, 1964, pp. 91–105.

Neustadt, L. W., "A Synthesis Method for Optimal Controls," *Proc. of Optimum System Synthesis Conference*, Dayton, Ohio, September 1962, pp 373–382.

Noton, A. R., "Numerical computation of automatic control," *Proc. JACC*, Seattle, Washington, 1966, pp 193–204.

Paiewonsky, B., P. Woodrow, W. Brunner, and P. Halbert, "Synthesis of Optimal Controllers Using Hybrid Analog-digital Computers," in A. V. Balakrishnan and L. W. Neustadt (eds.), *Computing Methods in Optimization Problems*, Academic Press Inc., New York 1964, pp 285–304.

Plant, J. B., and M. Athans, "An Iterative Technique for the Computation of Time Optimal Controls," *Proc. Third IFAC Congress*, London, 1966.

Rosen, J. B., "Iterative Solution of Nonlinear Optimal Control Problems," *J. Soc. Ind. Appl. Math. Ser. A, Control,* **4,** No. 1, 223–244 (1966).

Rosen, J. B., "Optimal Control and Convex Programming," in J. Abadie (ed.), *Nonlinear Programming, A Course*, North Holland Publishing Company, Amsterdam, 1966.

Scharmack, D. K., "The Equivalent Minimization Problem and the Newton–Raphson Optimization Method," *Proc. of the Optimum System Synthesis Conference*, Dayton, Ohio, September 1962, pp 119–158.

Scheley, C. H., "Optimal Control Computation by the Newton–Raphson Method and the Riccati Transformation," *Proc. JACC*, Seattle, Washington, 1966, pp 186–192.

Shah, B. V., R. J. Buehler, and O. Kempthrone, "Some Algorithms for Minimizing a Function of Several Variables," *J. Soc. Ind. Appl. Math.*, **12,** 74 (1964).

Applications of Computational Techniques in Engineering Control Problems

Battin, R. H., "A Statistical Optimizating Navigation Procedure for Space Flight," *ARS J.*, **32,** 1681–1698 (1962).

Bellman, R., and S. Dreyfus, "An application of Dynamic Programming to the Determination of Optimal Satellite Trajectories," *J. Brit. Interplanet. Soc.*, No. 3–4, **17,** 78–83 (May–August 1958).

Kelley, H. J., "Successive Approximation Techniques for Trajectory Optimization," *Proc. of the IAS Symposium on Vehicles System Optimization*, Institute of Aerospace Sciences, New York, 1961, p. 10.

Landgraf, S. K., "Some Practical Application of Performance Optimization Techniques to High-Performance Aircraft," *J. Aircraft*, **2,** No. 2, 153–154 (1965).

Meditch, J., "Optimal Thrust Programming for Minimal Fuel Midcourse Guidance," *Proc. of Optimum Synthesis Conference*, Dayton, Ohio, 1962, pp. 55–68.

Melbourne, W. G., and C. G. Sauer, "Constant Attitude Thrust Program Optimization," *AIAA J.*, **3,** 8, 1428–1431 (1965).

Melbourne, W. G., and C. G. Sauer, Jr., "Optimum Thrust Program for Power Limited Propulsion Systems," *Astronaut. Acta* **8,** 1962.

Paiewonsky, B. H., "The Synthesis of Optimal Controller," *Proc. of the Optimum System Synthesis Conference*, Dayton, Ohio, September 1962, pp. 69–88.

Smith, F. T., "Optimization of Multistage Orbit Transfer Processes by Dynamic Programming," *ARS J.* **31,** pp. 1553–1559 (1961).

Spang, H. A., III, "Optimum Control of an Unknown Linear Plant Using Bayesian Estimation of the Error," *IEEE Trans. Auto. Control*, **AC-10,** No. 1 80–83 (1965).

Swerling, P., "A Proposed Stagewise Differential Condition Procedure for Satellite Tracking and Prediction," *J. Astro. Sci.*, **6,** 46–52 (1959).

Tsien, H. S., and R. C. Evans, "Optimum Thrust Programming for Sounding Rocket," *ARS J.*, **21,** 99–107 (1951).

Review Articles on Controller Optimization using Computational Techniques

Bell, D. J., "A Review of Flight Path Optimization . . . in the Period 1945–1960," *J. Roy. Aeron. Soc.*, **67,** 119 (1963).

Greenley, R. R., "Comments on 'The Adjoint Method and its Application to Trajectory Optimization'," *AIAA J.*, **1,** 1463 (1963).

Paiewonsky, B. H., "A Study of Time Optimal Control," in J. P. Lasalle and S. Lefshetz (eds.), *Proceedings of International Symposium on Nonlinear Differential Equations and Nonlinear Mechanics*, Academic Press Inc., New York, 1963, pp. 333–365.

Paiewonsky, B. H., "Optimal Control: A Review of Theory and Practice," *AIAA J.*, **3,** 11, 1985–2006 (1965).

Spang, H. A., III, "A Review of Minimization Techniques for Nonlinear Functions," *Soc. Ind. Appl. Math Rev.*, **4** (1962).

General References

Aris, R., *Discrete Dynamic Programming*, Blaisdell Publishing Co., Waltham, Mass., 1964.

Battin, R. H., *Astronautical Guidance*, McGraw-Hill Book Co., Inc., New York, 1964.

Bellman, R. E., *Dynamic Programming*, Princeton University Press, Princeton, N.J., 1957.

Bliss, G. A., *Lectures on Calculus of Variations*, University of Chicago Press, Chicago, 1946.

Bryson, A. E., W. F. Denham, and S. E. Dreyfus, "Optimal Programming Problems with Inequality Constraints I: Necessary Conditions for Extremal Solutions," *AIAA J.*, **1,** 2544–2550 (1963).

Chang, S. S. L., "General Theory of Optimal Processes," *J. Soc. Ind. Appl. Math. Ser. A, Control*, **4,** No. 1, 46–55 (1966).

Chang, S. S. L., *Synthesis of Optimal Control Systems*, McGraw-Hill Book Co., Inc., New York, 1961.

Cicala, P., *An Engineering Approach to the Calculus of Variations* (in English). Levrotto and Bella, Torino, Italy, 1957.

Coddington, E. A., and N. Levinson, *Theory of Ordinary Differential Equations*, McGraw-Hill Book Co., Inc., New York, 1955.

Courant, R., *Integral and Differential Calculus* Vol. 2, Academic Press Inc., New York, 1936.

Gelfand, I. M., and S. V. Fomin, *Calculus of Variations*, Prentice-Hall, Inc., Englewood Cliffs, New Jersey, 1963 (translated from Russian).

Hestenes, M. R., "Variational Theory and Optimal Control Theory," in A. V. Balakrishnan and L. W. Neustadt (eds.), *Computing Methods in Optimization Problems*, Academic Press Inc., New York, 1966, pp. 1–22.

Hille, E., and R. S. Phillips, "Functional Analysis and Semi-graphs," *AMS Colloquium Publications*, **31** (1957).

Kantorovich, L. V., "Functional Analysis and Applied Mathematics," *Usp. Mat. Nauk* **3**, 89–185 (1953); English translation, National Bureau of Standards.

Kelley, H. J., "Guidance Theory and Extremal Fields," *IRE Trans. Auto. Control*, 75-82 (1962).

Kuhn, H., and A. W. Tucker, *Nonlinear Programming*, Second Berkley Symposium of Math. Statistics and Probability, University of California Press, Berkley, 1951.

Leitmann, G., *Optimization Techniques*, Academic Press Inc., New York, 1962.

Liusternik, L. A., and V. J. Sobolev, *Elements of Functional Analysis*, Hindustan Publishing Corp., Delhi, India, 1961.

Merriam, C. W., III, *Optimization Theory and Design of Feedback Control Systems*, McGraw-Hill Book Co., Inc. New York, 1964.

Riesz, F., and B. Sz-Nagy, *Functional Analysis*, Friedrick Ungar Publishing Co., New York, 1955.

Saaty, T. L., and J. Bram, *Nonlinear Mathematics*, McGraw-Hill Book Co., Inc., New York, 1964.

Taylor, A. E., *Introduction to Functional Analysis*, John Wiley & Sons, Inc., New York, 1958.

Wilde, D. J., *Optimum Seeking Methods*, Prentice Hall, Inc., Englewood Cliffs, New Jersey, 1964.

Zoutendijk, G., "Nonlinear Programming: A Numerical Survey," *J. Soc. Ind. Appl. Math. Ser A, Control*, **3**, No. 1 (1966).

APPENDIX B

Bibliography on Optimal Processes Governed by Ordinary and Partial Functional-Differential Systems

In this appendix we describe recent studies of optimal processes governed by differential equations with delayed or retarded arguments, and by certain more complicated integral-differential functional systems. All these systems are deterministic and we do not survey the extensive literature on stochastic control processes. A special bibliography has been attached to this appendix for easy reference on the problems discussed here.

B.1 CONTROL PROCESSES DESCRIBED BY FUNCTIONAL-DIFFERENTIAL OR PARTIAL DIFFERENTIAL SYSTEMS, AND THE RELEVANCE OF FUNCTIONAL ANALYSIS

Consider the control of a linear oscillator with state $(x(t), \dot{x}(t))$, with a scalar controller $u(t)$, and suppose that the elastic restoring force has a delay or retardation of one second in its effect. Then the control equation of motion is

$$\frac{d^2\, x(t)}{dt^2} + x(t-1) = u(t),$$

or

$$\dot{x}(t) = y(t)$$
$$\dot{y}(t) = x(t-1) + u(t).$$

This is an example of a *differential-delay* or *differential-difference system.* If the state $(x(t), y(t))$ is prescribed on the unit interval $-1 \leq t \leq 0$, then it is uniquely determined on $t > 0$ once the control $u(t)$ is also given on

$t > 0$. Let us note the remarkable fact that the initial state (and all later states, once the problem is appropriately formulated) are functions defined on a real unit interval rather than points in a finite dimensional space R^n. Hence we are lead immediately to dynamical systems in infinite dimensional function spaces and the techniques of functional analysis play a basic role.

In order to describe a general ordinary functional-differential process we introduce the notation $x_t(\theta)$ denoting an n-vector function on $-1 \leq \theta \leq 0$ for each n-vector function $x(t)$. If $x(t)$ on $-1 \leq t \leq t_1$, for some $t_1 > 0$, is a function with values in R^n, then $x_t(\theta)$, for each $0 \leq t \leq t_1$, denotes the *unit segment of* $x(t)$ *ending at time* t, or

$$x_t(\theta) = x(t + \theta) \quad \text{on} \quad -1 \leq \theta \leq 0.$$

Clearly, if $x(t)$ is continuous, then so is $x_t(\theta)$; that is, $x_t(\theta)$ belongs to the function space $C([-1, 0], R^n)$ of continuous n-vector functions on $-1 \leq \theta \leq 0$. With the usual uniform norm $C([-1, 0], R^n)$ is a Banach space. Similarly, if $x(t)$ is integrable on finite intervals, then $x_t(\theta)$ belongs to $L_1((-1, 0), R^n)$.

An *ordinary functional-differential control process* is defined by an n-vector system

$$\dot{x}(t) = f(t, x_t, u_t, u(t))$$

where f is a continuous map from

$$R^1 \times C([-1, 0], R^n) \times L_1((-1, 0), R^m) \times R^m \quad \text{into} \quad R^n.$$

If the continuous initial state

$$x_0(\theta) = \phi(\theta) \quad \text{on} \quad -1 \leq \theta \leq 0$$

and the integrable controller

$$u(t) \quad \text{on} \quad -1 \leq t \leq t_1$$

are prescribed, then there is a uniquely determined response $x(t)$, provided that f satisfies certain smoothness and growth hypotheses. The optimal control problem requires that we choose $u(t)$ from some admissible class of controllers so that the response $x(t)$ minimizes some prescribed cost functional.

An important example is the linear autonomous differential-difference system

$$\dot{x}(t) = \sum_{k=1}^{p} A_k\, x(t - \tau_k) + B\, u(t)$$

with constant coefficient matrices A_k and B and constant delays $0 \leq \tau_k \leq 1$. This is the best-known type of functional-differential process. The obvious generalization to a continuum of delays leads to the *differential-renewal process*

$$\dot{x}(t) = \int_{-1}^{0} A(\theta)\, x(t + \theta)\, d\theta + B\, u(t).$$

Other generalizations permit the delays $\tau_k \geq 0$ to vary with time and to be unbounded.

In order to introduce the concept of controlled processes with *distributed parameters* we study the heat partial differential equation

$$\frac{\partial T}{\partial t} = \frac{\partial^2 T}{\partial y^2} + u(t, y).$$

Here $T(t, y)$ is the temperature on an infinite rod, $-\infty < y < \infty$, when the time is $t \geq 0$. The distributed controller is $u(t, y)$, which can be regarded as an adjustable heat source distributed on the rod. In integral form (which we adopt as the definition of the basic control process) we have

$$T(t, y) = \int_{-\infty}^{\infty} H(t, y - \xi)\, T(0, \xi)\, d\xi + \int_0^t \int_{-\infty}^{\infty} H(t - \tau, y - \xi)\, u(\tau, \xi)\, d\xi\, d\tau$$

where the heat kernel is

$$H(t, y) = \frac{1}{\sqrt{4\pi t}} e^{-y^2/4t} \quad \text{for} \quad t > 0.$$

The initial temperature $T(0, y)$ will be taken in the Banach space C_0 of real continuous functions that approach zero as $|y| \to \infty$ (with the uniform norm as usual). The control functions $u(t, y)$ will be continuous in $t \geq 0$, $-\infty < y < \infty$, and approach zero as $|y| \to \infty$, uniformly on each compact time interval in $t > 0$.

We can now define the state space $\mathfrak{X}$ to be the Banach space C_0 so that the state $x(t)$ is the function $T(t, \cdot)$. We take the control space $\mathfrak{U} = C_0$ and then each controller $u(t, y)$ defines a continuous map $t \to u(t)$ of R^1 into $\mathfrak{U}$ and every such map is defined by just one controller $u(t, y)$. With these notations we find that the heat diffusion process can be described by a semigroup $\Phi(t)$ of linear transformations of $\mathfrak{X}$ into itself, as specified by,

$$\Phi(t)\, x(o) \equiv \int_{-\infty}^{\infty} H(t, y - \xi)\, T(0, \xi)\, d\xi.$$

The semigroup $\Phi(t)$ is strongly continuous on $t \geq 0$, and furthermore $\Phi(t)x$ is continuous in (t, x). Thus we obtain a formula for the controlled temperature distribution

$$x(t) = \Phi(t)\, x(0) + \int_0^t \Phi(t - \tau)\, u(\tau)\, d\tau,$$

where the Riemann integral is computed for the continuous function $\Phi(t - \tau)\, u(\tau)$ (from $R^1 \to C_0$) for each fixed $t \geq 0$.

This variation of parameters formula agrees with the treatment of the heat control-process as an ordinary differential equation

$$\frac{dx(t)}{dt} = A\, x(t) + u(t),$$

where $x(t) = T(t, \cdot)$ is an element in the Banach space C_0, and A is the unbounded linear operator of second-differentiation (or the Laplacian in more space variables).

The above examples show that ordinary functional-differential processes, partial differential processes, and even partial differential processes with delays can be treated as ordinary dynamic systems in infinite dimensional spaces. If the basic processes are autonomous and linear, then the theory of semigroups provides a unifying approach; otherwise more general evolutionary systems must be considered.

The methods of functional analysis also apply to the classical optimization problem defined by ordinary differential systems in R^n:

$$\dot{x} = f(t, x, u)$$

with constraints $x(t) \subset \Lambda \subset R^n$ and restraints $u(t) \subset \Omega \subset R^m$. The cost functional $C(u)$ can be treated as a real value function defined on a control function space $\mathfrak{U}$. The optimal controller u^* then corresponds to a critical point of $C(u)$, that is, a point where a certain generalized gradient of $C(u)$ vanishes. The constraints and restraints merely limit the admissible variations of u to some subset $\mathfrak{V}$ (often submanifold) of $\mathfrak{U}$. Hence a necessary condition for the optimizing u^* is the vanishing of an appropriate projection of the gradient of $C(u)$ on $\mathfrak{V}$. This necessary condition can be expressed by Lagrange multipliers as a Kuhn-Tucker condition which becomes the Pontryagin maximal condition in the classical problem.

B.2 AN ABSTRACT MAXIMAL PRINCIPLE

An abstract or axiomatic treatment of the maximal principle can be phrased within general functional analysis. Let $\mathcal{F}$ be a linear topological space and $\mathcal{F}_1 \subset \mathcal{F}$ a *quasi-convex subset*; that is, each linear map of a finite dimensional simplex σ into $\mathcal{F}$, with vertices mapped into $\mathcal{F}_1$, can always be uniformly approximated by continuous maps of σ into $\mathcal{F}_1$. Let $\phi: \mathcal{F}_1 \to R^n : f \to x$ be a continuous map with a minimal value of x^1 at some point $f^* \in \mathcal{F}_1$. If ϕ is suitably differentiable (with a strong type of Gateaux differential) near f^*, the linearization $d\phi$ of ϕ maps the convex hull of $\mathcal{F}_1$ onto a convex set $\mathfrak{C}$ in R^n with $d\phi(f^*)$ lying on its boundary. The maximal principle, then, is just the assertion that there exists a hyperplane π in R^n separating $\mathfrak{C}$ from the ray parallel to the x^1-axis leading downward from $\phi(f^*)$.

We next show how the usual optimal control problem can be phrased in functional analytic terms. Consider the control problem in R^n

$$\text{(S)} \qquad \dot{x} = f(x, t, u),$$

with f in $\mathfrak{C}^1$ in R^{n+1+m}. The controllers $u(t)$ are measurable m-vector functions on a fixed finite interval $\mathfrak{J}: t_0 \leq t \leq T$, and we specify that $u(t)$ belongs to some function collection $\mathfrak{U}$. Often $\mathfrak{U}$ consists of all measurable functions $u(t) \subset \Omega \subset R^m$ on $t_0 \leq t \leq T$, where Ω is a fixed restraint set—but this need not be the prescription of $\mathfrak{U}$. We fix the initial state $x(t_0) = x_0$ in R^n. Let us assume that each admissible controller $u(t) \in \mathfrak{U}$ will yield a response $x(t)$ lying in a given compact constraint set $X \subset R^n$ and that

$$|f(x, t, u(t))| \leq m_u(t) \quad \text{for all} \quad x \in X, \qquad t \in \mathfrak{J},$$

where $m_u(t)$ is an integrable function on $\mathfrak{J}$, depending on the controller $u(t)$. Then each response $x(t)$ is defined on all $t_0 \leq t \leq T$, and we seek to minimize the response datum $x^1(T)$.

Let $\mathfrak{F}_1$ consist of all functions $\{f(x, t, u(t))\}$ for $u(t) \in \mathfrak{U}$, that is, we define functions of (x, t) by replacing u in $f(x, t, u)$ by a controller $u(t)$. We denote such a function by $f(x, t) = f(x, t, u(t)) \in \mathfrak{F}_1$. For each $f \in \mathfrak{F}_1$ we have a response $x(t)$ on $\mathfrak{J}$ with the endpoint $x(T) \in R^n$; hence we study the map

$$\phi: \mathfrak{F}_1 \to R^n : f \to x(T).$$

We embed $\mathfrak{F}_1$ in a linear topological space $\mathfrak{F}$ consisting of all n-vector functions $g(x, t)$ on $X \times \mathfrak{J}$, where

$$\begin{aligned} &g(x, t) \quad \text{is continuous in} \quad x \quad \text{for each} \quad t \in \mathfrak{J},\\ &g(x, t) \quad \text{is measurable in} \quad t \quad \text{for each} \quad x \in X,\\ &|g(x, t)| \leq m_g(t) \in L_1(\mathfrak{J}) \quad \text{for} \quad (x, t) \in X \times \mathfrak{J}, \end{aligned}$$

and the integrable function $m_g(t) \in L_1(\mathfrak{J})$ depends on the function g. The topology on $\mathfrak{F}$ is determined once a collection of neighborhoods of the origin $g_0(x, t) \equiv 0$ is specified. Many topologies have been suggested for $\mathfrak{F}$ and all are quite involved. A recent suggestion of L. Neustadt is to specify a neighborhood $N_{\epsilon,Y}$ of the origin for each positive $\epsilon > 0$ and for each auxiliary family Y of equicontinuous n-vector functions $y(t)$ on $\mathfrak{J}$, namely,

$$g \in N_{\epsilon,Y} \quad \text{in case} \quad \left| \int_{t_0}^{t} g(y(s), s)\, ds \right| < \epsilon, \quad \text{for all} \quad t \in \mathfrak{J}, y \in Y.$$

This prescription of neighborhoods $N_{\epsilon,Y}$ is sometimes called the *chattering topology* on $\mathfrak{F}$.

The derivation of the maximal principle for an optimal controller $u^*(t) \in \mathfrak{U}$, or for the corresponding element $f^* = f^*(x, t, u^*(t)) \in \mathfrak{F}_1$, now depends on proving that $\mathfrak{F}_1$ is quasi-convex in $\mathfrak{F}$ and that ϕ is differentiable (both requirements need hold only in a neighborhood of f^*). The most difficult part of the proof of the maximal principle, as expounded in Chapter 4, consists in verifying these two requirements in the case in which $\mathfrak{U}$ consists of all $u(t) \subset \Omega$.

In the full generality discussed here the maximal principle occurs in the integral form [involving the Hamiltonian function $\tilde{H}(\tilde{\eta}, \tilde{x}, u)$ of Theorem 5.2]

$$\int_{\mathfrak{J}} \tilde{H}(\tilde{\eta}^*(t), \tilde{x}^*(t), u^*(t))\, dt \geq \int_{\mathfrak{J}} \tilde{H}(\tilde{\eta}^*(t), \tilde{x}^*(t), u(t))\, dt,$$

where $u^*(t)$ is the optimal controller and $u(t)$ is an arbitrary controller in $\mathfrak{U}$. In the special case in which $\mathfrak{U}$ is specified by the restraint $u(t) \subset \Omega$ the integral form of the maximal principle immediately implies the pointwise formulation

$$\tilde{H}(\tilde{\eta}^*(t), \tilde{x}^*(t), u^*(t)) = \max_{u \in \Omega} \tilde{H}(\tilde{\eta}^*(t), \tilde{x}^*(t), u) \quad \text{a.e.}$$

The abstract treatment of the maximal principle unifies many classical optimization theories. The abstract approach, however, becomes truly significant when it is modified to deal with problems with bounded phase constraints or with minimax optimizations. Here the classical analysis in R^n becomes unwieldy and the description of the problem in terms of infinite dimensional function spaces is essential.

B.3 A BRIEF GUIDE TO THE BIBLIOGRAPHY

The items in the following outline bear numbers referring to the attached special bibliography. After the outline we present a few remarks on certain of these references.

1. Ordinary differential-functional processes.
   (a) Differential-difference equations
   [12; 13; 14; 15; 38; 39; 40; 41; 52; 53; 56; 57]
   (b) Special functional equations
   [16; 27; 33; 42; 61]
   (c) Examples and applications
   [12; 13; 40]
2. Partial differential processes.
   (a) Linear theory

(α) Existence of optimal controllers
[2; 3; 24; 28; 43; 59]
(β) Maximal principle, bang-bang control
[2; 3; 18; 26; 28; 58; 59]
(γ) Controllability and qualitative theory
[25; 28; 47]

(b) Nonlinear theory
(α) Existence of optimal controllers
[44]
(β) Maximal principle and further theory
[23; 24]

(c) Special examples and applications
[7; 8; 9; 10; 18; 19; 24; 26; 61]

3. Minimizing functionals in infinite-dimensional vector spaces.
(a) Generalized Kuhn-Tucker conditions
[17; 29; 30; 36; 37; 50; 51; 60]
(b) Applications to control theory
[11; 17; 29; 35; 36; 50; 51; 60]
(c) Steepest descent and computational methods
[32; 54; 55]
(d) Related topics in functional analysis
[5; 34; 58]

4. General theory of linear systems in infinite dimensions.
(a) Input-output transfer functions
[4; 46]
(b) Controllability, observability
[4; 25; 46; 47]
(c) Applications and examples
[4; 46; 47]

One of the earliest works on the optimal control of differential-difference equations is contained in [38; 56], where the maximal principle for nonlinear systems is proved. A systematic development of the control theory for differential-difference systems is given in [13] where the treatment closely parallels the present text. The most important parts of [13] are further developed in [15]. One of the main unsolved problems in the control of differential-difference systems is the feedback synthesis of the optimal controller. A start on this problem was made in [40].

In reference [33] the maximal principle is proved for nonlinear difference equations. This paper also contains the details of the topological arguments required for the transversality theorems of differential and difference optimal control processes. In [27] some further hereditary control processes are examined with respect to the maximal principle, bang-bang behavior, and controllability.

We remark that there seems to be little treatment of the optimal control problem for general functional-differential systems.

The existence theory for optimal controllers in partial differential processes has been attacked in many papers. However, the most general existence theorems, for linear evolutionary (parabolic and certain hyperbolic type) partial differential systems, are proved in [43]. A continuation of this work to nonlinear systems is given in [44].

The most thorough treatment of the maximal principle (with related topics of bang-bang synthesis of optimal control) for linear partial differential systems is given in [28]. A form of the maximal principle for nonlinear partial differential systems is offered in [23]. However, this topic still needs further clarification, especially as to how the various hypotheses relate to the standard theories of partial differential equations.

Some special linear partial differential systems, together with the maximal principle and optimal control synthesis, are presented in [3; 18; 59]. Several very interesting examples and applications are developed in these last references.

The original extension of the Kuhn-Tucker critical-point condition to infinite dimensional space was proved in [37]. The first application of these ideas to the proof of the maximal principle for bounded-phase coordinate problems appears in [17], and the most thorough treatment is [29; 36; 50; 51]. A further extension of the Kuhn-Tucker conditions to weakly continuous functionals is found in [30].

The minimizing of a functional on a Banach space by steepest-descent methods is reviewed in [32]. Here computational methods for the solution of the classical optimization problems are explored.

The final topic in the above outline refers to input-output transfer theory in general linear spaces. These questions of controllability, observability, and plant recognition are studied in [4; 46; 47].

B.4 BIBLIOGRAPHY ON FUNCTIONAL-DIFFERENTIAL CONTROL

[1] Arrow, K. J., L. Hurwicz, and H. Uzawa, "Constraint Qualifications in Maximization Problems," *Naval Res. Logist. Quart.*, **8**, 175–191 (1961).

[2] Balakrishnan, A. V., "Optimal Control Problems in Banach Spaces," *J. Soc. Ind. Appl. Math., Ser. Control*, **3**, 1, 152–180 (1965).

[3] ———, "Semigroup Theory and Control Theory," *Proc. IFIP Congress*, Tokyo, 1965.

[4] ———, "A Theory of Linear Systems of Non-Finite Dimension," to be published in the *Proceedings of the Symposium of System Theory*, Polytechnic Institute of Brooklyn, April, 1965.

[5] Berger, M., "Generalized differentiation and utility functionals for commodity spaces of arbitrary dimensions," to appear.

[6] Boltyanskii, V. G., R. V. Gamkrelidze, and L. S. Pontryagin, "The Theory of Optimal Processes I. The Maximum Principle," *Izv. Akad. Nauk SSSR, Ser. Mat.*, **24,** 3–42 (1960). English translation in *Am. Math. Soc. Trans.*, Ser. 2, **18,** 341–382 (1961).

[7] Butkovskii, A. G., "Optimum Processes in Systems with Distributed Parameters," *Avtomatika i. Telemekhanika*, **22,** 17–26 (1961).

[8] ———, "The Maximum Principle for Optimum Systems with Distributed Parameters," *Ibid.*, **22,** 1288–1301 (1961).

[9] Butkovskii, A. G., and A. Ya. Lerner, "Optimal Control of Systems with Distributed Parameters, *Avtomatika i Telemekhanika*, **21,** 682–691 (1960).

[10] ———, "Optimal Control Systems with Distributed Parameters," *Dokl. Akad. Nauk SSSR*, **134,** 778–781 (1960).

[11] Chang, S. S. L., "General Theory of Optimal Processes," *J. Soc. Ind. Appl. Math., Ser. A, Control*, **4,** 46–55 (1966).

[12] Chosky, N. H., "Time Lag Systems—A Bibliography," *IRE Trans. Auto. Control*, **AC5,** 66–70 (1960).

[13] Chyung, D. H., *Optimal Control Systems with Time Delays*, Ph.D. Thesis, University of Minnesota, 1965.

[14] Chyung, D. H., and E. Bruce Lee, "Linear Optimal Systems with Time Delays," *SIAM Journal on Control*, **4,** No. 3 (1966).

[15] ———, "Optimal Systems with Time Delay," *Proc. Third IFAC, Conf.*, London, 1966.

[16] Corduneanu, C., "Sur une Équation Intégrale de la Théorie du Réglage Automatique," *C.R. Acad. Sci.*, **256,** 3564–3567 (1963).

[17] Dubovitskiy, A. Ya., and A. A. Milyutin, "Extremum Problems in the Presence of Constraints," *Zh. Vychisl. Mat. i Mat. Fiz.*, **5,** 395–453 (1965).

[18] Egorov, A. I., "On Optimal Control of Processes in Distributed Objects," *J. Appl. Math. Mech.*, **27,** 1045–1058 (1963).

[19] ———, "On a Variational Problem in the Theory of Equations of Elliptic Type," *Sibirski Mat. J.*, **5,** 500–508 (1964).

[20] Egorov, Ju. V., "Certain Problems in the Theory of Optimal Control," *Dokl. Akad. Nauk SSSR*, **145,** 720–723 (1962).

[21] ———, "On Optimal Control of Processes in Distributed Objects," *Prikl. Matemat. Mekh.*, **27,** 688–696 (1963).

[22] ———, "Optimal Control in Banach Space," *Soviet Math. Dokl.*, **4,** 630–633 (1963).

[23] ———, "Necessary Conditions for Optimal Control in Banach Space, *Mat. Sbornik*, **64** (106), 79–101 (1964).

[24] ———, "Certain Problems in the Theory of Optimal Control," *Soviet Mathematics*, **3,** No. 4, 1080–1084 (July, 1962). (Russian Original Tom-165, Nos. 1–6.)

[25] Falb, Peter L., "Infinite Dimensional Control Problems I: On the Closure of the Set of Attainable States for Linear Systems," *J. Math. Anal. Appl.*, **9,** 12–22 (1964).

[26] Fattorini, H. O., "Time-Optimal Control of Solutions of Operational Differential Equations," *J. Soc. Ind. Appl. Math., Scr. A, Control*, **2,** 54–59 (1964).

[27] Friedman, A., "Optimal Control for Hereditary Processes," *Arch. Rat. Mech. Anal.*, **15,** 396–416 (1963).

[28] Friedman, A. "Optimal Control in Banach Spaces," to appear.

[29] Gamkrelidze, R. V., "On Some Extremal Problems in the Theory of Differential Equations with Applications to the Theory of Optimal Control," *J. Soc. Ind. Appl. Math. Ser. A, Control,* **3,** No. 1, 106–128 (1965).

[30] Gapushkin, V. F., "On Critical Points of Functionals in Banach Spaces," *Mat. Sb.,* **64** (106), 499–617 (1964).

[31] Girsanov, I. V., "Minimax Problems in the Theory of Diffusion Processes," *Dokl. SSSR,* **136,** No. 4 (1960).

[32] Gollwitzer, H. E., *Applications of the Method of Steepest Descent to Optimal Control Problems,* MS Thesis, University of Minnesota, September, 1965.

[33] Halkin, H. "A Maximum Principle of the Pontryagin Type for Systems Described by Nonlinear Difference Equations," *J. Soc. Ind. Appl. Math., Ser. A, Control,* **4,** No. 1, 90–111 (1966).

[34] ———, "Finite Convexity in Infinite Dimensional Spaces," to appear in *Proc. Colloquium on Convexity,* Copenhagen, 1965.

[35] ———, "An Abstract Framework for the Theory of Process Optimization," to appear in *Bull. Am. Math. Soc.*

[36] Halkin, H., and L. W. Neustadt, "General Necessary Conditions for Optimization Problems," *University of Southern California Report* 173 (1966).

[37] Hurwicz, L., *Programming in Linear Spaces, in Studies in Linear and Nonlinear Programming,* by K. J. Arrow, L. Hurwicz, and H. Uzawa, Stanford University Press, pp. 38–102, 1958.

[38] Kharatishvili, G. L., "The Maximum Principle in the Theory of Optimal Processes Involving Delay," *Soviet Mathematics,* **2,** No. 1, 28–32 (1961). (Russian Original Tom 136, Nos. 1–6.)

[39] Kramer, J., "On Control of Linear Systems with Time Lags," *Inform. Control,* **3,** No. 4 (1960).

[40] Krasovskii, N. N., "Optimal Processes in Systems with Time Lag," *Proc. Second IFAC Congress,* Basel, 1963.

[41] ———, "Analytic Construction of an Optimal Regulator in a System with Time Lags," *Prikl. Matemat. Makh.,* **26,** No. 1 (1962).

[42] Lee, E. B., "Recurrence Equations and the Control of their Evolution," *J. Math. Anal. Appl.,* **7,** 1, 118–126 (1963).

[43] Lions, J. L., "Sur quelques problems d'optimisation dans les équations d'evolution lineaires de type parabolique," in E. Caianiello (ed.), *Applications of Functional Analysis to Optimization,* Academic Press Inc., New York, 1966.

[44] ———, "Optimisation pour certaines classes d'équations d'evolution non lineaires," Proc. Symp. on Math. Theory of Control, USC, 1967.

[45] Lur'e, K. A., "The Mayer-Bolza Problem for Multiple Integrals and the Optimization of the Performance of Systems with Distributed Parameters," *Prikl. Matemat. Mekh.* **27,** No. 5, pp. 842–853, 1963.

[46] Markus, L., "Controllability and Observability," in E. Caianiello (ed.), *Applications of Functional Analysis to Optimization,* Academic Press Inc., New York, 1966.

[47] Miranker, W., "Approximate Controllability for Distributed Linear Systems," *J. Math. Anal. Appl.,* **10,** 378–387 (1965).

[48] Mischenko, E. F., and L. S. Pontryagin, "On a Statistical Problem of Automatic Control," *Izv. Akad. Nauk SSSR, Ser. Mat.,* **25,** 477 (1961).

[49] Neustadt, L. W., "Optimal Control Problems as Extremal Problems in a Banach Space," *Proc. of Poly. Inst. of Brooklyn Symposium on System Theory,* pp. 215–224, April, 1965.

[50] ———, "An Abstract Variational Theory with Applications to a Broad Class of Optimization Problems I: General Theory," *J. Soc. Ind. Appl. Math. Ser. A, Control*, **4** (1966). Available as report of the Electronic Sciences Laboratory, University of Southern California, Los Angeles, California.

[51] ———, "An Abstract Variational Theory with Applications to a Broad Class of Optimization Problems II: Applications," to appear. Available as report of the Electronic Sciences Laboratory, University of Southern California.

[52] Oguztoreli, M. N., "A Time Optimal Control Problem for Systems Described by Differential Difference Equations," *J. Soc. Ind. Appl. Math., Ser. A, Control*, **1,** 3, 290–310 (1963).

[53] Oziganova, I. A., "On the Theory of Optimal Control of Systems with Time Lag," *Tr. Sem. Teor. Diff. Urav. s Otklon. Argumentom, Univ. Druzby*, **2,** 136–145 (1963).

[54] Pishenichnii, B. N., "Numerical Methods in Solving Problems of Optimal Control," *N. M. and M. Ph.*, **4,** No. 2, 292–305 (1964).

[55] Pishenichnii, B. N., "Convex Programming in a Normed Space" (in Russian), *Kibernetika* (in Russian), **1,** No. 5, 46–54 (1965).

[56] Pontryagin, L. S., V. G. Boltyanskii, R. V. Gamkrelidze, and E. F. Mischenko, *The Mathematical Theory of Optimal Processes*, Interscience Publishers, New York, 1962.

[57] Popov, V. M., and A. Halanay, "A Problem in the Theory of Time Delay Optimum Systems," *Autom. Remote Control*, **24,** 129–131 (1963).

[58] Porter, W. A., "On the Optimal Control of Distributive Systems," Department of Electrical Engineering, The University of Michigan, Ann Arbor, Michigan, 1965.

[59] Russell, D. L., "Optimal Regulation of Linear Symmetric Hyperbolic Systems with Finite Dimensional Controls," *J. Soc. Ind. Appl. Math., Ser. A, Control*, **4,** No. 2, 276–284 (1966).

[60] ———, "The Kuhn-Tucker Conditions in Banach Space with an Application to Control Theory," Math. Research Center, University of Wisconsin.

[61] Wang, P. K. C., "Asymptotic Stability of a Time Delayed Diffusion System," *Trans. Am. Soc. Mech. Eng., Ser. E, J. Appl. Mech.*, **30E** 500–504 (1963).

[62] Wang, P. K. C., "Control of Distributive Parameter Systems," *Advances in Control Systems*, **1,** 75–171 (1964).

[63] Wang, P. K. C., and F. Tung, "Optimum Control of Distributed Parameter Systems," *Proc. Joint Automatic Control Conference*, pp. 16–31, 1963.

References

The following references, for which the full citation appears in the main bibliography, were useful to the authors in their preparation of the chapters of the text.

Chapter 1

Aris, R., *The Optimal Design of Chemical Reactors*
Bass, R., Equivalent Linearization—and Optimization of Control Systems
Bellman, R., I. Glicksberg, and O. Gross, *Some aspects of the Mathematical Theory of Control Processes*
Bliss, G., *Lectures on the Calculus of Variations*
Bushaw, D., Optimal Discontinuous Forcing Terms
Coddington, E., and N. Levinson, *Theory of Ordinary Differential Equations*
Dunford, N., and J. Schwartz, *Linear Operators* I
Gantmacher, F., *Theory of Matrices*
Graves, L., *Theory of Functions of Real Variables*
Lee, E. B., and L. Markus, Optimal Control of Nonlinear Processes
Leitmann, G., *Optimization Techniques with Applications to Aerospace Systems*
McShane, E., *Integration*
Pontryagin, L., V. Boltyanskii, R. Gamkrelidze, and E. Mischenko, *The Mathematical Theory of Optimal Processes*
Tsien, H., *Engineering Cybernetics*

Chapter 2

Antosiewicz, H., Linear Control Systems
Bellman, R., I. Glicksberg, and O. Gross, *Some Aspects of the Mathematical Theory of Control Processes*
Conti, R., Contributions to Linear Control Theory
Dunford, N., and J. Schwartz, *Linear Operators* I
Eggleston, H. G., *Convexity*

Filippov, A. F., On Certain Questions in the Theory of Optimal Control
Gamkrelidze, R. V., Theory of Optimal High Speed Processes in Linear Systems
Gilbert, E. G., Controllability and Observability in Multivariable Control Systems
Halkin, H., Some Further Generalizations of a Theorem of Lyapunov
Harvey, C., and E. B. Lee, On Necessary and Sufficient Conditions for Time Optimal Control of Linear Systems
Harvey, C., E. B. Lee, and L. Markus, On Time Optimal Control of Systems with Numerator Dynamics
Kalman, R. E., Mathematical Description of Linear Dynamical Systems
Kalman, R. E., Y. C. Ho, and K. S. Narendra, Controllability of Linear Dynamical Systems
Krasovski, N. N., On the Theory of Optimum Control
Laning, J. H. Jr., and R. H. Battin, *Random Processes in Automatic Control*
LaSalle, J. P., Time Optimal Control Systems
LaSalle, J. P., and S. Lefschetz, *Stability by Liapunov's Direct Method with Applications*
Liapunov, A. M., Sur les functions-vecteurs complètement additives
Neustadt, L. W., The Existence of Optimal Controls in the Absence of Convexity Conditions
Polya, G., and G. Szegö, *Aufgaben und Lehrsatze aus der Analysis*
Pontryagin, L., et al., *The Mathematical Theory of Optimal Processes*
Zadeh, L. A., and C. A. Desoer, *Linear System Theory: The State Space Approach*

Chapter 3

Bellman, R., et al., *Some Aspects of the Mathematical Theory of Control Processes*
Chang, A., An Optimal Regulator Problem
Kalman, R. E., Contributions to the Theory of Optimal Control
Lee, E. B., Design of Optimum Multivariable Control Systems
Lee, E. B., A Sufficient Condition in the Theory of Optimal Control
Letov, A. M., Analytical Controller Design, Parts I–III
Neustadt, L. W., The Existence of Optimal Controls in the Absence of Convexity Conditions
Neustadt, L. W., Time Optimal Control Systems with Position and Integral Limits

Chapter 4

Albrekht, E. G., On the Optimal Stabilization of Nonlinear Systems
Cesari, L., An Existence Theorem in Problems of Optimal Control
Chang, S. S. L., Optimal Control in Bounded Phase Space
Filippov, A. F., On Certain Questions in the Theory of Optimal Control
Gamkrelidze, R. V., On Sliding Optimal States
Graves, L. M., *Theory of Functions of a Real Variable*
Halkin, H., On the Necessary Condition for Optimal Control of Nonlinear Systems
Jones, G. S., Asymptotic Fixed-Point Theorems and Periodic Systems of Functional-Differential Equations
Lee, E. B., and L. Markus, Optimal Control of Nonlinear Processes
Lukes, D., *Optimal Control of Nonlinear Systems*
Neustadt, L. W., The Existence of Optimal Controls in the Absence of Convexity Conditions
Neustadt, L. W., A General Theory of Minimal-Fuel Space Trajectories

Neustadt, L. W., Optimization, a Moment Problem, and Nonlinear Programming
Pontryagin, L., et al., *The Mathematical Theory of Optimal Processes*
Roxin, E., The Existence of Optimal Controls
Schmaedeke, W. W., Optimal Control Theory for Nonlinear Vector Differential Equations Containing Measures
Warga, J., Relaxed Variational Problems
Ważewski, T., Sur les systems de commande

Chapter 5

Berkowitz, L., The Equivalence of Some Necessary Conditions for Optimal Control in Problems with Bounded State Variables
Boltyanskii, V. G., Sufficient Conditions for Optimality
Carathéodory, C., Variationrechnung und partielle Differentialgleichungen erster Ordnung
Cullum, J., Private communication on bounded phase problems
Falb, P. L., A Simple Local-Sufficiency Condition Based on the Second Variation
Hermes, H., The Equivalence and Approximation of Optimal Control Problems
Hestenes, M. R., On Variational Theory and Optimal Control Theory
Kalman, R. E., When Is a Linear Control System Optimal? Also numerous private discussions on the Hamilton-Jacobi theory
Lee, E. B., A Sufficient Condition in the Theory of Optimal Control
Lee, E. B., and L. Markus, On Necessary and Sufficient Conditions for Time Optimal Control of Second-order Nonlinear Systems
Neustadt, L. W., Optimization, a Moment Problem, and Nonlinear Programming
Pontryagin, L. S., et al., *The Mathematical Theory of Optimal Processes*
Russell, D. L., Penalty Functions and Bounded Phase Coordinate Control
Schmaedeke, W., and D. L. Russell, Time-Optimal Control with Amplitude and Rate Limited Controls

Chapter 6

Al'brekht, E., and N. N. Krasovskii, The Observability of a Nonlinear Controlled System in the Neighborhood of a Given Motion
Athans, M., and P. L. Falb, *Optimal Control: An Introduction to the Theory and Its Applications*
Gilchrist, J. D., *n*-Observability for Linear Systems
Halkin, H., On a Generalization of a Theorem of Lyapounov
Kalman, R. E., Liapunov Functions for the Problem of Luré in Automatic Control
Kirillova, F. M., On the Correct Statement of One Problem on Optimal Regulation
LaSalle, J. P., and S. Lefschetz, *Stability by Liapunov's Direct Method with Applications*
Lee, E. B., and L. Markus, Optimal Control of Nonlinear Processes
Lefschetz, S., *Stability of Nonlinear Control Systems*
Markus, L., Controllability for Nonlinear Processes
Markus, L., Controllability and Observability
Markus, L., Stability of the Optimal Control Problem
Markus, L., The Bang-Bang Principle
Markus, L., and H. Yamabe, Global Stability Criteria for Differential Systems
Meyer, K. R., On a System of Equations in Automatic Control Theory
Zadeh, L. A., and C. A. Desoer, *Linear System Theory: The State Space Approach*

Chapter 7

Athans, M., P. L. Falb, and R. T. Lacoss, On Optimal Control of Self-Adjoint Systems
Burmeister, H. L., Genäherte Bestimmung der Schaltkurve in seitoptimalen Regelkreisen mit nichtlinearen Strecke 2. Ordnung
Busemann, A., Minimal problem der Luft-und Raumfahrt
Ewing, G. M., and W. R. Haseltine, Optimal Programs for an Ascending Missile
Lawden, D. F., *Optimal Trajectories for Space Navigation*
Lee, E. B., Discussion of Satellite Attitude Control
Lee, E. B., and L. Markus, Synthesis of Optimal Control for Nonlinear Processes with One Degree of Freedom
Munick, H., *On Nonlinear Optimal Control Problems with Control Appearing Linearly*
Smith, F. B., Time-Optimal Control of Higher-Order Systems

Bibliography

Adorno, D. S., "Optimal Control of Certain Linear Systems with Quadratic Loss," *Inform. Control*, **5-1**, p. 1 (1962).

Aizerman, M. A., and F. R. Gantmakher, *Absolute Stability of Control Systems* (in Russian), 1963; English transl., Holden-Day, Inc., San Francisco, 1964.

Aizerman, M. A., and F. R. Gantmakher, "Some Aspects of the Discontinuities in a Nonlinear Automatic Control System with a Piecewise Linear Response in the Nonlinear Part," *Avtomatika i Telemekhanika*, **18,** No. 11 (1957).

Aizerman, M. A., *Lectures on the Theory of Automatic Regulation* (in Russian), *Gosudarstv. Izdat. Fiz-Mat. Lit.*, Moscow, 1958.

Al'brekht, E. G., "On the Optimal Stabilization of Non-linear Systems" (in Russian), *Prikl. Matemat. Mekh.*, **25,** 836–844 (1961).

Al'brekht, F., "On a Certain Problem in the Theory of Processes with an Optimal Speed of Response in Linear Systems" (in Russian), *Avtomatika i Telemekhanika*, **22,** 733–738 (1961).

Al'brekht, E. G., and N. N. Krasovskii, "The Observability of a Nonlinear Controlled System in the Neighborhood of a Given Motion" (in Russian), *Avtomatika i Telemekhanika*, **25,** No. 7, 1047–1057 (1964).

Andreyev, N. I., "Determination of an Optimal Dynamic System from the Criterion of a Functional of Partial Form," *Avtomatika i Telemekhanika*, **18,** No. 7 (1957).

Andreyev, N. I., "On the Theory of Determining an Optimal Dynamic System," *Avtomatika i Telemekhanika*, **19,** No. 12 (1958).

Andreyev, N. I., "A Method of Determining the Optimum Dynamic System from the Criterion of the Extreme of a Functional which is a Given Function of Several Other Functionals," *Proc. First IFAC Congress*, Moscow, 1960.

Andreyev, N. I., "Determining a Pulse System Optimum Weighting Function Which Ensures the Extreme of a Functional," *Avtomatika i Telemekhanika*, **21,** No. 4 (1960).

Antosiewicz, H. A., "Linear Control Systems," *Arch. Rat. Mech. Anal.*, **12,** 313–324 (1963).

Aoki, M., "Optimal Control Policies for Dynamical Systems whose Characteristics Change Randomly at Random Times," *Proc. Third IFAC Congress*, London, 1966.

Aoki, M., "Dynamic Programming Approach to a Final Value Control System with a

Random Variable Having an Unknown Distribution Function," *IRE, Trans. Auto. Control*, **5,** 270–283 (1960).

Aoki, M., "On Optimal and Sub-optimal Policies in the Choice of Control Forces for Final Value Systems," *IRE, Auto. Control* **5,** 171–178 (1960); **6,** 83 (1961).

Aoki, M., "Stochastic Time Optimal Control Systems," *AIEE,* **80,** II, 41–46 (1961).

Aoki, M., "Minimal Effort Control Systems with an Upper Bound of the Control Time, *IEEE Trans. Auto. Control,*" **AC-8,** No. 1, 60–61 (1963).

Aris, R., "Studies in Optimization—III. The Optimum Operating Conditions in Sequences of Stirred Tank Reactors," *Chem. Eng. Sci.*, **13,** 75–81 (1960).

Aris, R., "Studies in Optimization—II. Optimum Temperature Gradients in Tubular Reactors," *Chem. Eng. Sci.* **13,** 18–29 (1960).

Aris, R., *The Optimal Design of Chemical Reactors*, Academic Press, Inc., New York, 1961.

Aris, R., R. Bellman, and R. Kalaba, "Some Optimization Problems in Chemical Engineering," *Nucl. Sci. Eng.*, **6,** 486–493 (1959).

Ash, M., R. Bellman, and R. Kalaba, "On Control of Reactor Shut-Down Involving Minimum Xenon Poisoning." *Nucl. Sci. and Eng.*, **6,** 152–156 (1959).

Athans, M., "Minimum-Fuel Feedback Control Systems: Second Order Case," *IEEE Trans. Appl. Ind.* 8–17 (March 1963).

Athans, M., "Minimum Fuel Control of Second Order Systems with Real Poles," *Proc. JACC*, 232–240 (1963).

Athans, M., and P. L. Falb, "Time-Optimal Control for a Class of Nonlinear Systems," *IEEE Trans. Auto. Control*, **AC-8,** 379 (1963).

Athans, M., P. L. Falb, and R. T. Lacoss, "On Optimal Control of Self-Adjoint Systems, *IEEE Trans. Appl. Ind.*, **83,** 161–166 (1964).

Athans, M., and P. L. Falb, *Optimal Control: An Introduction to the Theory and Its Applications*, McGraw-Hill Book Company, Inc., New York, 1965.

Athans, M., P. L. Falb, and R. Lacoss, "Time, Fuel, Energy—Optimal Control of Nonlinear Norm Invariant Systems," *IEEE Trans. Autom. Control*, **AC-8,** 196–201 (1963).

Athanassiades, M., and O. J. M. Smith, "Theory and Design of High-Order Bang-Bang Control Systems." *IRE Trans. Auto. Control*, **6,** 125–134 (1961).

Athanassiades, M., "Optimal Control for Linear Time-Invariant Plants with Time, Fuel, and Energy Constraints," *IEEE Trans. Appl. Ind.*, 321–325 (January 1963).

Athanassiades, M., and P. Falb, "Time Optimal Control for Plants with Numerator Dynamics," *IRE Trans. Auto. Control.*, **7,** 47 (1962).

Averbuch, A. I., "Connection Between S. A. Tchaplygin's Theorem and the Theory of Optimal Processes" (in Russian), *Avtomatika i Telemekhanika*, **22,** 1309–1313 (1961).

Babister, A. W., "Determination of the Optimum Response of Linear Systems," *Quart. Mech. Appl. Math.*, **10,** 360–388, 502–512 (1957); **11,** 119–128 (1958).

Babunashvili, T. G., "The Synthesis of Linear Optimal Systems," *J. Soc. Ind. Appl. Math.*, Ser. A, *Control*, **2,** No. 2, 261–265 (1964).

Balakrishnan, A. V., "An Operator Theoretic Formulation of a Class of Control Problems and a Steepest Descent Method of Solution," *J. Soc. Ind. Appl. Math.*, Ser. A, *Control*, **1** (1963).

Balakrishnan, A. V., "Optimal Control Problems in Banach Spaces," *Soc. Ind. Appl. Math.*, Ser. **A,** *Control*, **3,** No. 1 (1965).

Balakrishnan, A. V., and H. C. Hsieh, "Function Space Methods in Control Systems

Optimization," *Proc. of the Optimum System Synthesis Conf.*, Dayton, Ohio. 1962, pp. 10–40.

Balakrishnan, A. V., and L. W. Neustadt (eds.) *Computing Methods in Optimization Problems*, Academic Press Inc., New York, 1964.

Barbashin, E. A., "On a Problem of the Theory of Dynamic Programming" (in Russian), *Prikl. Mathemat. Mekh.* **24,** 1002–1012 (1960).

Barbashin, E. A., "Estimating the Mean Square Deviation from a Given Trajectory" (in Russian), *Avtomatika i Telemekhanika*, **21,** N. 7, 941–950 (1960).

Bass, R. W., "Equivalent Linearization, Nonlinear Circuit Synthesis and the Stabilization and Optimization of Control Systems," *Proceedings of the 2nd Nonlinear Circuit Analysis Symposium*, Polytechnic Institute of Brooklyn, New York, 1956, pp. 163–198.

Bass, R. W., and I. Gura, "Canonical Forms for Controllable Systems with Application to Optimal Nonlinear Feedback," *Proc. Third IFAC Congress*, London, 1966.

Bellman, R., "Functional Equations in the Theory of Dynamic Programming. V: Positivity and Quasilinearity." *Proc. Nat. Acad. Sci. U.S.*, **41,** 743–746 (1955).

Bellman, R., "On the Application of Dynamic Programming to the Study of Control Processes." *Proceedings of the Symposium on Non-linear Circuit Analysis*, Polytechnic Institute of Brooklyn, 1956, 199–213.

Bellman, R., *Dynamic Programming*, Princeton University Press, Princeton, N.J., 1957.

Bellman, R., "Dynamic Programming, Terminal Control and General Nonlinear Control Processes," *Proc. Nat. Acad. Sci. U.S.*, **43,** 927–930 (1957).

Bellman, R., "Notes on Control Processes. I: On the Minimum of Maximum Deviation," *Quart. Appl. Math.*, **14,** 419–423 (1957).

Bellman, R. "A Markovian Decision Process," *J. Math. Mech.*, **6,** 679–684 (1957).

Bellman, R., "Some New Techniques in the Dynamic Programming Solution of Variational Problems." *Quart. Appl. Math.*, **16,** 295–303 (1958).

Bellman, R., *Adaptive Control Processes, a Guided Tour*, Princeton University Press, Princeton, N.J., 1961.

Bellman, R., *Mathematical Optimization Techniques*, University of California Press, Berkeley, 1963.

Bellman, R., I. Cherry, and G. M. Wing, "A Note on the Numerical Integration of a Class of Non-linear Hyperbolic Equations," *Quart. Appl. Math.*, **16,** 181–183 (1958).

Bellman, R., and S. Dreyfus, "An Application of Dynamic Programming to the Determination of Optimal Satellite Trajectories." *J. Brit. Interplanet. Soc.* **17,** 78–83 (1959).

Bellman, R., W. H. Fleming, and D. V. Widder, "Variational Problems with Constraints *Ann. Mat.*, **41,** 301–323 (1956).

Bellman, R., I. Glicksberg, and O. Gross, "On Some Variational Problems Occurring in the Theory of Dynamic Programming," *Rend. Circ. Mat. Palermo*, **3,** 1–35 (1954).

Bellman, R., I. Glicksberg, and O. Gross, "On the 'Bang-Bang' Control Problem, *Quart. Appl. Math.*, **14,** 11–18 (1956).

Bellman, R., I. Glicksberg, and O. Gross, "Some Non-Classical Problems in the Calculus of Variations," *Proc. Am. Math. Soc.*, **7,** 87–94 (1956).

Bellman, R., I. Glicksberg, and O. Gross, *Some Aspects of the Mathematical Theory of Control Processes*, Rand Corporation, Santa Monica, California, Report R-313, 1958.

Bellman, R., and R. Kalaba, "Dynamic Programming and Feedback Control," *Proc. IFAC Congress*, Moscow, 1960. (Butterworths & Co., Ltd., London, 1961), **1,** 460–464.

Benes, V. E., "A Nonlinear Integral Equation from the Theory of Servomechanisms," *The Bell System Technical Journal*, **40,** No. 5, 1309–1321 (1961).

Berkovitz, L. D., "An Optimum Thrust Control Problem," *J. Math. Anal. Appl.*, **3,** 122–132 (1961).

Berkovitz, L. D., "Variational Methods in Problems of Control and Programming," *J. Math. Anal. Appl.*, **3,** 145–169 (1961).

Berkovitz, L. D., "The Equivalence of Some Necessary Conditions for Optimal Control in Problems with Bounded State Variables," *J. Math. Anal. Appl.*, **10,** No. 2, pp. 275–283 (1965).

Bertram, J. E., and P. E. Sarachik, "On Optimal Computer Control," Proc. IFAC Congress, Moscow, 1960. (Butterworths & Co., Ltd., London, 1961), **1,** 419–422.

Bhavnani, K. H., "Optimization of Time-Dependent Systems by Dynamic Programming," *JACC*, 516–521 (1966).

Bilous, O., and N. R. Amundsen, "Optimum Temperature Gradients in Tubular Reactors, *Chem. Eng. Sci.*, **5,** 81–92, 115–126 (1956).

Birch, B. J., and R. Jackson, "The Behaviour of Linear Systems with Inputs Satisfying Certain Bounding Conditions," *J. Electron. Control*, **6,** 366–375 (1959).

Birkhoff, G., and S. MacLane, *A Survey of Modern Algebra*, The Macmillan Company New York, 1948.

Blackwell, D., "On the Functional Equation of Dynamic Programming," *J. Math. Anal. Appl.*, **2,** 273–276 (1961).

Blakemore, N., and R. Aris, "Studies in Optimization—V. The Bang-Bang Control of a Batch Reactor," *Chem. Eng. Sci.*, **17,** 591–598 (1962).

Blaquière, A., "Further Investigation into the Geometry of Optimal Processes," *J. Soc. Ind. Appl. Math., Ser. A, Control*, **4,** No. 1, 19–33 (1966).

Bliss, C., "Calculus of Variations," Mathematical Association of America, The Open Court Publishing Company, LaSalle, Ill., 1925.

Bliss, G. A., *Lectures on the Calculus of Variations*, Phoenix Science Series, The University of Chicago Press, Chicago, 1961.

Bogner, I., and L. F. Kazda, "An Investigation of the Switching Criteria for Higher Order Servomechanisms," *Trans. AIEE*, **73,** II, 118–127 (1954).

Bolza, O., *Lectures on the Calculus of Variations*, Chelsea Publishing Co., New York, 1904.

Boltyanski, V. G., "The Maximum Principle in the Theory of Optimal Processes" (in Russian), *Dokl. Akad. Nauk SSSR*, **119,** 1070–1073 (1958).

Boltyanski, V. G., "Optimal Processes with Parameters" (in Russian), *Dokl. Akad. Nauk Uzb. SSR* No. 10, 9–13 (1959).

Boltyanski, V. G., "Models of Linear Optimal High Speed Systems for Control Schemes" (in Russian), *Dokl. Akad. Nauk SSSR*, **139,** No. 2, 275–278 (1961).

Boltyanski, V. G., "Application of the Theory of Optimal Processes to Problems of Function Approximation" (in Russian), *Tr. Mat. Inst. Akad. Nauk SSSR*, **60,** 82–95 (1961).

Boltyanski, V. G., "Sufficient Conditions for Optimality" (in Russian), *Dokl. Akad. Nauk SSSR*, **140,** 994–997 (1961).

Boltyanski, V. G., R. V. Gamkrelidze, and L. S. Pontryagin, "On the Theory of Optimal Processes" (in Russian), *Dokl. Akad. Nauk SSSR*, **110,** 7–10 (1956).

Boltyanski, V. G., R. V. Gamkrelidze, and L. S. Pontryagin, "Theory of Optimal Processes. 1. The Maximum Principle" (in Russian), *Izv. Akad. Nauk SSSR, Ser. Mat.*, **24,** 3–42 (1960); Translated in *Am. Math. Soc. Transl.*, **18,** 341–382 (1961).

Boltyanski, V. G., R. V. Gamkrelidze, E. F. Mischenko, and L. S. Pontryagin, "The

Maximum Principle in the Theory of Optimal Processes of Control," *Proc. IFAC Congress*, Moscow, 1960. (Butterworths & Co., Ltd., London, 1961), **1,** 454–459.

Boltyanskii, V. R. Gamkrelidze, E. Mischenko, and L. Pontryagin, *The Mathematical Theory of Optimal Processes*, John Wiley & Sons, Inc., 1962.

Booton, R. C., "Optimum Design of Final-Value Control Systems," *Proceedings of the Symposium on Non-Linear Circuit Analysis*, Polytechnic Institute of Brooklyn, 1956, pp. 233–241.

Bor-Ramenski, A. E., and Sung Chien, "Optimum Servo Drive with Two Control Parameters" (in Russian), *Avtomatika i Telemekhanika*, **22,** 157–170 (1961).

Box, G. E. P., and K. B. Wilson, "On the Experimental Attainment of Opt. Cond.," *J. Roy. Stat. Soc.*, B, **13,** 1 (1951).

Breakwell, J. V., "The optimization of trajectories," *J. Soc. Ind. Appl. Math.*, **7,** 215–247 (1959).

Breakwell, J. V., "Approximations in Flight Optimization Techniques," *Aero/Space Eng.*, September 1961, **20,** No. 9, Part I, p.26.

Breakwell, J. V., "A Doubly Singular Problem in Optimal Interplanetary Guidance," *J. Soc. Ind. Appl. Math., Ser. A, Control*, **3,** No. 1, 1965.

Bridgland, T. F., Jr., "On the Existence of Optimal Feedback Controls;" *J. Soc. Ind. Appl. Math., Ser. A, Control*, **1,** No. 3, 261–274 (1963).

Bridgland, T. F. Jr., "On the Existence of Optimal Feedback Controls. II," *J. Soc. Ind. Appl. Math., Ser. A, Control*, **2,** No. 2, 137–150 (1964).

Brown, R. F., "A Calculation of Switching Functions as a Means of Minimizing Error in an On-Off Control System," *Proc. Inst. Elec. Eng.*, **107C,** 249–256 (1960).

Brunovsky, P., "On the Best Stabilizing Control under a Given Class of Perturbations," *Czech. Math. J.*, **15** (90), 329–369 (1965).

Bryson, A. E., W. F. Denham, and S. E. Dreyfus, "Optimal Programming Problems with Inequality Constraints I: Necessary Conditions for Extremal Solutions," *AIAA J.*, **1,** No. 11, 2544–2550 (1963).

Buckner, Hans, "A Formula for an Integral Occurring in the Theory of Linear Servomechanism and Control-Systems," *Quart. Appl. Math.*, **10,** 205–213 (1952–1953).

Burkov, V. N., "Application of Optimal Control Theory to the Problems of Resources Distribution," *Proc. Third IFAC Congress*, London, 1966.

Burmeister, H. L., "Zeitoptimal Übergangsvorgange mit beschränkter *n*-ter Ableitung," *Z. Messen, Steuern, Regeln*, **4,** 407–409 (1961).

Burmeister, H. L., "Genäherte Bestimmung der Schaltkurve in seitoptimalen Regelkreisen mit nichtlinearer Strecke 2. Ordnung," *Intern. Colloq.*, 113–117 (1963).

Burns, K. N., "The Transient Response of a Single Point Non-linear Servomechanism," *Proc. National Electronics Conference*, **8,** 22–30 (1952).

Busemann, A., "Minimal problem der Luft- und Raumfahrt," Zeitschr. fur Flugwissenschaften 13 (1965), pp 401–411.

Bushaw, D., "Optimal Discontinuous Forcing Terms," in S. Lefschetz (ed.), Vol. 4, *Contributions to the Theory of Non-linear Oscillations*, Princeton University Press, Princeton., N.J., 1958, pp. 29–52.

Bushaw, D., "Dynamical Polysystems and Optimization," *Contrib. Differential Eq.*, **2,** 351–365 (1963).

Butkovski, A. G., and A. Ya. Lerner, "The Optimal Control of Systems with Distributed Parameters" (in Russian), *Avtomatika i Telemekhanika*, **21,** 682–691 (1960).

Butkovski, A. G., "Optimum Processes in Systems with Distributed Parameters" (in Russian), *Avtomatika i Telemekhanika*, **22,** 17–26 (1961).

Butkovski, A. G., "Maximum Principle for Optimum Systems with Distributed Parameters" (in Russian), *Avtomatika i Telemekhanika*, **22,** 1288–1301 (1961).

Butkovski, A. G., "Some Approximate Methods for Solving Problems of Optimum Control of Distributed Parameter Systems" (in Russian), *Avtomatika i Telemekhanika*, **22,** 1565–1575 (1961).

Cadzow, J. A., "Optimal Control of a System Subject to Parameter Variation," *JACC*, 807–810 (1966).

Carathéodory, C., "Variationrechnung und partielle Differentialgleichungen erster Ordnung," B. G. Teubner Verlagsgesellschaft, Leipzig, 1935.

Cartaino, T. F., and S. E. Dreyfus, "Application of Dynamic Programming to the Airplane Time-to-Climb Problem," *Aeron. Eng. Rev.*, **16,** No. 6, 74–77 (1957).

Cesari, L., "Semicontinuita e Convessita nel Calcolo della Variazioni," *Estratto degli Annali della Scuola Normale Superiore di Pisa Science Fisiche e Matematiche*, Ser. III, **27,** Fasc. 4 (1964).

Cesari, L., "An Existence Theorem in Problems of Optimal Control," *Soc. Ind. Appl. Math., Ser. A, Control*, **3,** No. 1 (1965).

Cesari, L., "Existence Theorems for Optimal Solutions in Pontryagin and Lagrange Problems," *J. Soc. Ind. Appl. Math., Ser. A, Control*, **3,** No. 3, 475–498 (1965).

Chandaket, P., and C. T. Leondes, "Synthesis of Quasi-Stationary Optimum Nonlinear Control Systems." *Trans. AIEE, Electrical Engineers*, **80,** II, 313–325 (1961).

Chandaket, P., and C. T. Leondes, "Optimum Non-linear Bang-Bang Control Systems with Complex roots," *Trans. AIEE*, **80,** II, 82–102 (1961).

Chang, A., "An Optimal Regulator Problem," *J. Soc. Ind. Appl. Math., Ser. A, Control*, **2,** No. 2, 220–233 (1964).

Chang, Jen-Wei, "A problem of Synthesizing Optimal Systems by Means of the Maximum Principle" (in Russian), *Avtomatika i Telemekhanika*, **22,** 1302–1308 (1961).

Chang, Jen-Wei, "Synthesis of Relay Systems According to Minimum Integral Square Error" (in Russian), *Avtomatika i Telemekhanika*, **22,** 1601–1607 (1961).

Chang, Jen-Wei, "The Problem of Synthesizing an Optimal Controller in Systems with Time Delay," *Avtomatika i Telemekhanika*, **23,** No. 2 133–137 (1962).

Chang, S. S. L., "Optimum Switching Criteria for Higher Order Contactor Servos with Interrupted Circuits," *Trans. AIEE*, **74,** II, 273–276 (1955).

Chang, S. S. L., "An Airframe Pitch Linear Acceleration Controller," *Proceedings of the National Electronics Conference* **12,** 134–151 (1956).

Chang, S. S. L., "Digitized Maximum Principle," *Proc. IRE, Engineers*, **48,** 2030-2031 (1960).

Chang, S. S. L., "Computer Optimization of Nonlinear Control Systems by Means of Digitized Maximum Principle," *IRE Conv. Record, Part* 4, 654 (1961).

Chang, S. S. L., *Synthesis of Optimum Control Systems*, McGraw-Hill Book Company, Inc., New York, 1961, pp. 144–149.

Chang, S. S. L., "Optimal Control in Bounded Phase Space," *Automatica*, Vol. 1, Pergamon Press, New York, 1962, pp. 55–67.

Chang, S. S. L., "Minimum Time Control with Multiple Saturation Limits," *IEEE Trans. Auto. Control*, **AC-8,** No. 1, 65 (1963).

Chang, S. S. L., and F. J. Alexandro, Jr., "The Approximate Realization of Optimum Control Systems," *Proc. of the Optimum System Synthesis Conf.*, 269–322 (1963).

Chang, S. S. L., "General Theory of Optimal Processes," *J. Soc. Ind. Appl. Math. Ser. A, Control*, **4,** No. 1, 46–55 (1966).

Chernetskii, V. I., "An Interpolation Method for Analyzing Automatic Control System Accuracy Under Random Stimuli," *Avtomatika i Telemekhanika*, **21,** 481–488 (April 1960).

Chestnut, H., R. R. Duersch, and W. M. Gaines, "Automatic Opt. of a Poorly Defined Process," *JACC Conf.* 1962.

Chow, C. K., "An Optimum Character Recognition System Using Decision Functions," *IRE Trans. Electron. Computers*, **6** (1957).

Chu, Hsin, "A Remark on Complete Controllability," *J. Soc. Appl. Math., Control, A, Ser.* **3,** No. 3, 439–442 (1965.)

Chyung, D. H., and E. B. Lee, "Control of Linear Time Delay Systems with Essentially Linear Cost Functionals" in J. P. LaSalle (ed.), *Differential Equations and Dynamical Systems*, Academic Press Inc., New York, 1966.

Chyung, D. H., and E. B. Lee, "Optimal Systems with Time Delay, "*Third International Federation of Automatic Control Congress*, London (June 1966).

Chyung, D. H., "Discrete Linear Optimal System with Essentially Quadratic Cost Functional," *IEEE Trans. Auto. Control*, **11,** No. 3 (July 1966).

Chyung, D. H., "An Approximation to Bounded Phase Coordinate Problem for Linear Discrete Systems with Quadratic Cost Functionals," *Proc. JACC* (August 1966).

Chyung, D. H., and E. B. Lee, "Linear Optimal Systems with Time Delays," *J. Soc. Ind. Appl. Math. Ser. A*, **4,** No. 3, 548–575 (1966).

Chzhan, Sy-In, "On Sufficient Conditions for an Optimum," *Appl. Proc. Math. Mech.*, **25,** 1420–1423 (1961).

Chzhan Sy-In., "On the Theory of Optimum Control" (in Russian), *Prikl. Mathemat. Mekh.*, **25,** 413–479 (1961).

Chzhan, Sy-In., "On sufficient conditions for an optimum" (in Russian), *Prikl. Matemat. Mekh.*, **25,** 946–947 (1961).

Coales, J. F., and A. R. M. Noton, "An On-Off Servomechanism with Predicted Changeover," Proc. Inst. Elec. Eng., **103B,** 449–462 (1956).

Coddington, E. A., and N. L. Levinson, *Theory of Ordinary Differential Equations*, McGraw-Hill Book Co., Inc., New York, 1955.

Conti, R. "Contributions to Linear Control Theory," *J. Differential Equations*, **1,** No. 4, 427–445 (1965).

Cotter, J. E., "Evaluation of Optimal Control Strategies," *A.I. CH.E. J.*, 585–590, July 1964.

Coviello, G. J., "An Organization Approach to the Optimization of Multivariable Systems," *Proc. JACC* Session XVII, Paper 2 (1964).

Datko, R., "An Implicit Function Theorem with an Application to Control Theory," *Michigan Math. J.*, **11,** 345–351 (1964).

De Backer, W., "Synthesis of Optimal Control and Hybrid Computation," *Proc. Fourth AICA* Conf., Brighton (1964).

Desoer, C. A., "The Bang-Bang Servo Problem Treated by Variational Techniques," *Inform. Control*, **2,** 333–348 (1959).

Desoer, C. A., "Pontryagin's Maximum Principle and the Principle of Optimality," *J. Franklin Inst.*, **271,** 361–367 (1961); **272,** 313 (1961).

Desoer, C. A., and J. Wing, "An Optimal Strategy for a Saturating Sampled-Data System," *IRE, Trans. Auto. Control*, **6,** 5–15 (1961).

Desoer, C. A., and J. Wing, "A Minimal Time Discrete System," *IRE Trans., Auto. Control*, **6,** 111–125 (1961).

Desoer, C. A., E. Polak, and J. Wing, "Theory of Minimum Time Discrete Regulators, *Proc. IFAC, Basel*, 406/1–406/6 (1963).

Desoer, C. A., and J. Wing, "The Minimal Time Regulator Problem for Linear Sampled-Date Systems: General Theory," *J. Franklin Inst.*, **272,** 208–228 (1961).

Diesel, J. W., "Extended Switching Criterion for Second-Order Servos," *Trans. AIEE*, **76,** II, 388–393 (1957).

Doll, H. G., and T. M. Stout, "Design and Analogue Computer Analysis of an Optimum Third-Order Non-linear Servomechanism," *Trans. ASME*, **79,** 513–525 (1957).

Dommasch, D. O., "The Doliac Macro-Micro Control Logic, Its Synthesis, Evaluation, Potential and Problems," *Proc. Optimum System Synthesis Conf.*, pp. 241–268 (1963).

Douglas, J. M., "The Use of Optimizator Theory to Design Simple Multivariable Control Systems," *Proc. JACC*, 649–660 (1966).

Drenick, R. F., and L. Shaw, "Optimal Control of Linear Plants with Random Parameters," *IEEE Trans. Auto. Control*, **AC-9,** 236–244 (July 1964).

Dreyfus, S. E., "Dynamic Programming and the Calculus of Variations," *J. Math. Anal. Appl.*, **1,** 228–239 (1960).

Dreyfus, S. E., "Variational Problems with Inequality Constraints," *J. Math. Anal. Appl.*, **4,** 297–308 (1962).

Dreyfus, S. E., "Some Types of Optimal Control of Stochastic Systems," *J. Soc. Ind. Appl. Math., Ser. A, Control*, **2,** 120–134 (1964).

Dubovitskii, A. Y., and A. A. Milyutin, "Certain Optimality Problems for Linear Systems," *Automat. Remote Control*, **24,** 1471–1481 (1964).

Dunford, N., and J. Schwartz, *Linear Operators*, Interscience Publishers, Inc., New York, 1959.

Eckman, D. P., and I. Lefkowitz, "Optimizing Control of a Chemical Process," *Control Eng.* **4,** 197–204 (1957).

Edelbaum, T. N., "Optimum Low-Thrust Rendezvous and Station Keeping," *AIAA J.* **2,** No. 7, 1197–1201, 1964.

Edelbaum, T. N., "Optimum Low-Thrust Transfer Between Circular and Elliptic Orbits." Proc. Fourth U.S. National Congress of Applied Mechanics, 137–141 (1962).

Eggleston, H. G., *Convexity*, Cambridge University Press, London, 1958.

Egorov, A. I., "On Optimal Control of Processes in Distributed Objects," *Prikl. Matemat. Mekh.*, **27,** No. 4, 688–696 (1963).

Egorov, A. I., "Optimal Control Processes in Certain Systems with Distributed Parameters," *Automat. Remote Control*, **25,** 557–567 (1964).

Ellert, F. J., and C. W. Merriam, III, "Synthesis of Feedback Controls Using Optimization Theory—An Example, *IEEE Trans. Auto. Control*, **8,** No. 2, 89–103 (1963).

Ewing, G. M., and W. R. Haseltine, "Optimal Programs for an Ascending Missile," *Soc. Ind. Appl. Math., Ser. A, Control*, **2,** No. 1, 66–88 (1964).

Eykhoff, P., "Adaptive and Optimalizing Control Systems," *IRE Trans. Auto. Control* (Correspondence), **5,** 148–151 (June 1960).

Falb, P. L., "A Simple Local-Sufficiency Condition Based on the Second Variation," *IEEE Trans. Auto. Control*, 348–350 (July 1965).

Falb, P. L., and M. Athans, "A Direct Proof of the Criterion for Complete Controllability of Time-Invariant Linear Systems," *IEEE Trans. Auto. Control*, **9,** 189–190 (1964).

Faulders, C. R., "Optimum Thrust Programming of Electrically Powered Rockets in a Gravitational Field," *ARS J.*, **30,** No. 10 (October 1960).

Fel'dbaum, A. A., "The Simplest Relay Control Systems" (in Russian), *Avtomatika i Telemekhanika*, **10,** 249–266 (1949).

Fel'dbaum, A. A., "Optimal Processes in Automatic Control Systems" (in Russian), *Avtomatika i Telemekhanika*, **14,** 712–728 (1953).

Fel'dbaum, A. A., "An Approach to the Question of Synthesis of Optimal Systems of

Automatic Control" (in Russian), *Proc. Second All-Union Conf. on Automatic Control Theory*, 1953, Vol. 2, Izd-vo. Akad. Nauk, SSSR, Moscow, 1955, pp. 325–364.

Fel'dbaum, A. A., *Electrical Systems or Automatic Control* (in Russian), Oborongsiz, Moscow, 1954.

Fel'dbaum, A. A., "On the Synthesis of Optimal Systems with the Aid of Phase Space" (in Russian), *Avtomatika i Telemekhanika*, **16,** 129–149 (1955).

Fel'dbaum, A. A., "On the Question of Synthesizing Optimal Automatic Control Systems, *Trans. Second All-Union Conf. on Automatic Control Theory*, Vol. 2, Izd-vo. Akad. Nauk SSSR, Moscow, 1955.

Fel'dbaum, A. A., *Computers in Automatic Systems*, *Fizmatgiz* (1959).

Fel'dbaum, A. A., "Error Evaluation in Automatic Systems" (in Russian), *Gosudarstv. Izdat. Fiz.-Mat. Lit.* (1959).

Fel'dbaum, A. A., "Automatic Optimalizer," *Proc. JACC*, 718–728 (1960).

Fel'dbaum, A. A., "Dual Control Theory (4 parts)," *Avtomatika i Telemekhanika*, **21,** Nos. 9 and 11 (1960); **22,** Nos. 1 and 2 (1961).

Fel'dbaum, A. A., "On the Optimal Control of Markov Objects" *Automat. Remote Control*, **23,** 927–941 (1962).

Fickeisen, F. C., and T. M. Stout, "Analogue Methods for Optimum Servomechanism Design," Trans. AIEE, 244–250 (November 1952).

Filatov, A. N., "Problem of Speed of Response with an Arbitrary Number of Control Functions without Switching" (in Russian), *Avtomatika i Telemekhanika*, **22,** 834–837 (1961).

Filippov, A. F., "On Certain Questions in the Theory of Optimal Control," *Vestnik Moskov. Univ., Ser. Mat., Mekh., Abstr., Fiz., Khim.*, **2,** 25–32 (1959); English trans., *J. Soc. Ind. Appl. Math., Ser. A, Control*, **1,** 76–84 (1962).

Fimple, W. R., "Optimum Midcourse Plane Changes for Ballistic Interplanetary Trajectories," *AIAA J.*, **1,** No. 2, 430–434 (1963).

Fisher, E. E., "An Application of the Quadratic Penalty Function Criterion to the Determination of a Linear Control for a Flexible Vehicle," *AIAA J.* **3,** No. 7, 1262–1267 (1965).

Fitsner, L. N., "Two Types of Optimum-Extremal Systems" (in Russian), *Avtomatika i Telemakhanika*, **21,** 1115–1121 (1960).

Florentin, J. J., "Optimal Control of Continuous Time, Markov, Stochastic Systems," *J. Electron. Control*, **10,** 473–488 (1961).

Florentin, J. J., "Optimal, Probing, Adaptive, Control of a Simple Bayesian System," *J. Electron. Control*, **11,** 165–177 (August 1962).

Flügge-Lotz, I., *Discontinuous Automatic Control*, Princeton University Press, Princeton, N.J., 1953.

Flügge-Lotz, I., "Discontinuous Automatic Control," *Appl. Mech. Rev.*, **14,** 581–584 (1961).

Flügge-Lotz, I., and Mih Yin., "The Optimum Response of Second-Order Velocity-Controlled Systems with Contactor Control," *Trans. ASME, Ser. D, J. Basic Eng.*, **83,** 59–64 (1961).

Flügge-Lotz, I., "Synthesis of Third-order Contactor Control Systems," *Proc. First IFAC Congress*, 390–397 (1961).

Flügge-Lotz, I., and H. A. Titus, "The Optimum Response of Full Third-order Systems with Contactor Control," *J. Basic Eng.*, **84,** 554–558 (1962).

Flügge-Lotz, I., and H. Marbach, "The Optimal Control of Some Attitude Control Systems for Different Performance Criteria," *J. Basic Eng.*, **85,** 165–176 (1963).

Flügge-Lotz, I., and R. Marbach, "The Optimal Control of Some Attitude Control Systems for Different Performance Criteria," *Trans. ASME Ser. D, J. Basic Eng.*, **85,** 165–176 (1963).

Flügge-Lotz, I., and M. D. Maltz, "Attitude Stabilization Using a Contactor Control System with a Linear Switching Criterion," *Automatica*, **2,** 255–274 (1965).

Flügge-Lotz, I., and H. D. Marbach, "On the Minimum Effort Regulation of Stationary Linear Systems," *J. Franklin Inst.*, **279,** No. 4, 229–245 (1965).

Foy, W. H., "Fuel Minimization in Flight Vehicle Attitude Control," *IEEE Trans. Auto. Control*, **8,** 84–88 (1963).

Fraeys de Veubeke, B., "Méthodes variationnelles et performances optimales en aéronautique," *Bull. Soc. Math. Belg.* **8,** 136–157 (1956).

Freimer, M., "A Dynamic Programming Approach to Adaptive Control Processes," *IRE, Nat. Conv. Record*, **7,** Part 4, 12–17 (1959).

Fried, B. D., and J. M. Richardson, "Optimum Rocket Trajectories," *J. Appl. Phys.*, **27,** 955–961 (1956).

Friedman, A., "Optimal Control for Hereditary Processes," *Arch. Rat. Mech. Anal.*, **15,** 396–416 (1964).

Friedland, B., "A Minimum Response-Time Controller for Amplitude and Energy Constraints," *IRE Trans. Auto. Control*, **7,** 73–74 (January, 1962).

Friedland, B., "The Structure of Optimum Control Systems," *Trans. ASME. Ser. E, Basic Eng.* (March 1962).

Friedland, B., "The Design of Optimal Controllers for Linear Processes with Energy Constraints," *Melpar Technical Note*, 62/2 (March 1962).

Friedland, B., "Optimum Control of Descrete-Time Dynamic Processes," *Proc. Second IFAC, Congress, Basel*, 410/1-410/6 (1963).

Friedland, B., and P. E. Sarachik, "A Unified Approach to Suboptimum Control," *Proc. Third IFAC Congress*, London (1966).

Fukunaga, K., and T. Shibataki, "A Method for Optimizing Control, *J. Inst. Elec. Engrs., Japan*, **81,** No. 6 (1961).

Fuller, A. T., "Performance Criteria for Control Systems," *J. Electron. Control*, **7,** 456–462 (1959).

Fuller, A. T., "Relay Control Systems Optimized for Various Performance Criteria," *Proc. First IFAC Congress*, Moscow, 1960. Butterworth & Co., Ltd., London, 1961, Vol. 1, pp. 510–519.

Fuller, A. T., "Phase Space in the Theory of Optimum Control," *J. Electron. Control*, **8,** 381–400 (1960).

Fuller, A. T., "Optimization of Non-linear Control Systems with Transient Inputs." *J. Electron. Control*, **8,** 465–479 (1960).

Fuller, A. T., "Optimization of Non-linear Control Systems with Random Inputs," *J. Electron. Control*, **9,** 65–80 (1960).

Fuller, A. T., "Optimization of a Non-linear Control System with a Random Telegraph Signal Input," *J. Electron. Control*, **10,** 61–80 (1961).

Fuller, A. T., "Optimization of a Saturating Control System with Brownian Motion Input," *J. Electron. Control*, **10,** 157–164 (1961).

Fuller, A. T., "Bibliography of Optimum Non-linear Control of Determinate and Stochastic-Definite Systems," *J. Electron. Control* **13,** 589–611 (1962).

Gambill, R. A., "Generalized Curves and the Existence of Optimal Controls," *RIAS Techn. Rep.*, **63–2,** 1–26 (February 1963).

Gamkrelidze, R. V., "On the Theory of Optimum Processes," *Dokl. Akad. Nauk SSSR*, **116,** No. 1 (1957).

Gamkrelidze, R. V., "On the Theory of Optimal Processes in Linear Systems" (in Russian), *Dokl. Akad. Nauk SSSR*, **116,** 9–11 (1957).

Gamkrelidze, R. V., "On the General Theory of Optimal Processes" (in Russian), *Dokl. Akad. Nauk SSSR*, **123,** 223–226 (1958).

Gamkrelidze, R. V., "Theory of Optimal High Speed Processes in Linear Systems" (in Russian), *Izv. Akad. Nauk SSSR, Ser. Mat.*, **22,** 449–474 (1958).

Gamkrelidze, R. V., "Optimum High Speed Processes with Limitations on the Phase Coordinates" (in Russian), *Dokl. Akad. Nauk SSSR*, **125,** 475–478 (1959).

Gamkrelidze, R. V., "Optimal Control Processes with Restricted Phase Coordinates" in Russian), *Izv. Akad. Nauk SSSR, Ser. Mat.*, **24,** 315–356 (1960).

Gamkrelidze, R. V., "Optimal Sliding States," *Dokl. Akad. Nauk SSSR* (in Russian), **143,** 1243–1245 (1962).

Gamkrelidze, R. V., "On Sliding Optimal States," *Sov. Mat. Dokl.* **3,** 559 (1962).

Gamkrelidze, R. V., "On Some Extremal Problems in the Theory of Differential Equations with Applications to the Theory of Optimal Control," *J. Soc. Ind. Appl. Math., Ser. A, Control*, **3,** No. 11 (1965).

Garfinkel B., "A solution of the Goddard Problem" *J. Soc. Ind. Appl. Math. Ser. A, Control*, **1,** No. 3 (349–368) (1963).

Garrilovic, M., R. Petrovic, and D. Siljak, "Adjoint Method in Sensitivity Anal. of Optimal Systems," *J. Franklin Inst.*, **276,** 26–38 (July 1963).

Gantmacher, F. R., *Theory of Matrices*, Chelsea Publishing Co., N.Y., 1957.

Gelfand, I. M., and S. V. Fomin, *Calculus of Variations*, Prentice-Hall, Inc., Englewood Cliffs, N.J., 1963.

Gelb. A., *The Analysis and Design of Limit Cycling Adaptive Automatic Control Systems*, Sc. D. dissertation, Department of Aeronautics and Astronautics, M.I.T., Cambridge, September, 1961.

Ghonaimy, M. A. R., and B. Bernholtz, "On a Direct Method of Optimization: Linear Control Systems," *Proc. Third IFAC Congress, London* (1966).

Gibson, J. E., *Nonlinear Automatic Control*, McGraw-Hill Book Company, Inc., New York 1963.

Gilbert, E. G., "Controllability and Observability in Multivariable Control Systems," *J. Soc. Ind. Appl. Math., Ser. A, Control*, **1,** No. 2, 128–151 (1963).

Gilchrist, J. D., "n-Observability for Linear Systems," *IEEE Trans Auto. Control*, **11,** No. 3 (July 1966); and *Proc. JACC* (1966).

Gnoenski, L. S., "On the Accumulation of Disturbances in Nonstationary Linear Impulsive Systems" (in Russian), *Prikl. Matemat. Mekh.*, **23,** 1136–1141 (1959).

Gnoenski, L. S., "On the Accumulation of Disturbances in Linear Systems" (in Russian *Prikl. Matemat. Mekh.*, **25,** 319–331 (1961).

Gnoenski, L. S., "On a Method of Optimization of Servosystems" (in Russian), *Prikl. Matemat. Mekh.*, **25,** 948–953 (1961).

Gobetz, F. W., "Optimum Transfers Between Hyperbolic Asymptotes," *AIAA J.* **1,** No. 9, 2034–2041 (1963).

Gobetz, F. W., "Optimal Variable-Thrust Transfer of a Power-Limited Rocket Between Neighboring Circular Orbits," *AIAA J.* **2,** No. 2, 339–343 (1964).

Gollwitzer, H. E., "Applications of the Method of Steepest Descent to Optimal Control Problems," University of Minnesota, TR No. CS-104-65, September 1965.

Goddard, R. H., "A Method for Reaching Extreme Altitudes," *Smithsonian Collection* No. 2 (1919).

Gosienski, A., "Optimal Relay Servomechanisms of the Second Order," *Rozprany Elektrotech.* (*Poland*), **7,** 17–68 (1961).

Gould, L. A., and W. Kipiniak, "Dynamic Optimization and Control of a Stirred-tank Chemical Reactor," *Trans. AIEE*, **79,** I, 734–746 (1960).

Graham, Keith D., "Minimax Control of Large Launch Boosters," *Proc. JACC*, 359–373 (1966).

Graves, L. M., *Theory of Functions of Real Variables*, McGraw-Hill Book Co., Inc., New York, 1956.

Gröbner, W., "Steuerungsprobleme mit Optimalbedingung," *Math. Tech. Wirtschaft*, **8,** 62–64 (1961).

Groginsky, H. L., "On a Property of Optimal Controllers with Boundedness Constraint, *IRE Trans. Auto. Control*, **6,** 98–110 (1961).

Gumowski, I., and C. Mira, "Optimization by Means of the Hamilton-Jacobi Theory," *Third IFAC Congress, London* (1966).

Gurley, Joseph G., "Optimal-Thrust Trajectories in an Arbitrary Gravitational Field, *J. Soc. Ind. Appl. Math. Ser. A, Control*, **2,** No. 3, 423–432 (1965).

Gusev, L. A., "Determination of Periodic Behavior of Automatic Control Systems Containing a Nonlinear Element with Broken-Line Characteristic," *Avtomatika i Telemekhanika* (*USSR*) **19,** No. 10 (1958).

Hagin, E. J., and G. H. Fett, "Phase-Space Metrization for Relay control of Two Time-Constant Servomechanism," *Proc. National Electronics Conference* (*Chicago*), **13,** 537–548 (1957).

Halkin, H., *Trajectoires optimales*, Engineering Dissertation, Université de Liège, 1960.

Halkin, H., "Nondegenerate Variational Problems and the Principle of Optimal Evolution," *Proc. Symposium on Vehicle Systems Optimization, Inst. Aerospace Sci.*, New York, 43–44 (1962).

Halkin, H., "Liapounov's Theorem on the Range of a Vector Measure and Pontryagin's Maximum Principle," *Arch. Rat. Mech. Anal.*, **10,** 296–304 (1962).

Halkin, H., "On the Necessary Condition for Optimal Control of Nonlinear Systems," *J. Anal. Math.*, **12,** 1–82 (1963).

Halkin, H., "The Principle of Optimal Evolution," in J. P. LaSalle and S. Lefschetz (eds.), *Nonlinear Differential Equations and Nonlinear Mechanics*, Academic Press, Inc., New York, 1963.

Halkin, H., "Method of Convex Ascent," in A. V. Balakrishnan and Lucien W. Neustadt (eds.), *Computing Methods in Optimization Problems*, Academic Press Inc., New York, 1964, pp. 211–239.

Halkin, H., "Some Further Generalizations of a Theorem of Lyapounov," *Arch. Rat. Mech. Anal.*, **17,** No. 4, 272–277 (1964).

Halkin, H., "Topological Aspects of Optimal Control of Dynamical Polysystems," *Contrib. Differential Equations*, **3,** No. 4, 377–385 (1964).

Halkin, H., "A generalization of LaSalle's Bang-Bang Principle, *J. Soc. Ind. Appl. Math., Ser. A, Control*, **2,** No. 2, 199–202 (1965).

Halkin, H., "A Maximum Principle of the Pontryagin Type for Systems Described by Nonlinear Difference Equations," *J. Soc. Ind. Appl. Math., Ser. A, Control*, **4,** No. 1, 90–111 (1966).

Halkin, H., "On a Generalization of a Theorem of Lyapounov," *J. Math. Anal. Appl.*, to appear.

Halmos, Paul R., *Measure Theory*, D. Van Nostrand Co., Inc., Princeton, N.J.

Hammurabi, see: *The oldest code of laws in the world* by C. Johns, Edinburgh, 1903, sections 53–56.

Hartman, P., "A Lemma in the Theory of Structural Stability of Differential Equations," *Proc. Amer. Math Soc.*, **2,** 610–620 (1960).

Harvey, C. A., "Determining the Switching Criterion for Time-Optimal Control," *J. Math. Anal. Appl.*, **5,** No. 2, 245–257 (1962).

Harvey, C. A., "Modes of Finite Response Time Control," *J. Soc. Ind. Appl. Math., Control, Ser. A*, **2,** No. 1, 60–65 (1964).

Harvey, C. A., "Synthesis of Time-Optimal Control for Linear Processes," *J. Math. Anal. Appl.*, **10,** No. 2, 334–341 (1965).

Harvey, C. A., and E. B. Lee, "On Necessary and Sufficient Conditions for Time Optimal Control of Linear Systems," *J. Math. Anal. Appl.*, **5,** 258–268 (October 1962).

Harvey, C. A., E. B. Lee, and L. Markus, "On Time Optimal Control of Systems with Numerator Dynamics," Presented at ASD Symp. on Optimization, Dayton, Ohio, September 11, 1962.

Hayashi, C., and Sakawa, Y., "Optimum Switching Criteria for Higher-Order Contactor Servomechanisms," *J. Inst. Elec. Engrs, Japan*, **77,** 1601 (1957).

Hermes, H., and G. Haynes, "On the Non-Linear Control Problem with Control Appearing Linearly," *J. Soc. Ind. Appl. Math., Ser. A, Control*, **1,** No. 2, 85–108 (1963).

Hermes, H., "Controllability and the Singular Problem," *J. Soc. Ind. Appl. Math., Ser. A, Control*, **2,** No. 2, 241–260 (1964).

Hermes, H., "A Note on the Range of a Vector Measure; Application to the Theory of Optimal Control," *J. Math. Anal. Appl.*, **8,** No. 1, 78–83 (1964).

Hermes, H., "The Equivalence and Approximation of Optimal Control Problems," *J. Diff. Equations*, **1,** No. 4, 409–426 (1965).

Hestenes, M. R., "On Variational Theory and Optimal Control Theory," *J. Soc. Ind. Appl. Math., Ser. A, Control.* **3,** No. 1, 23–48 (1965).

Hestenes, M., *Calculus of Variations and Optimal Control Theory*, Wiley, 1966.

Hibbs, A., "Optimum Burning Program for Horizontal Flight," *J. ARS*, **22** (1952).

Higgins, T. J., "A Résumé of the Basic Literature of State-Space Techniques in Automatic Control Theory," *Proc. JACC* (1962).

Hinz, H. K., "Optimal Low-Thrust Near-Circular Orbit Transfer," *AIAA J.* **1,** 1367–1371 (1963).

Ho, Y. C., "Solution Space Approach to Optimal Control Problems," *Trans. ASME Ser. D*, **83,** 53–58 (1961).

Ho, Y. C., P. B. Brentani, "On Computing Optimal Control with Inequality Constraints," *J. Soc. Ind. Appl. Math., Control, Ser. A*, **1,** No. 3, 319–348 (1963).

Hopkin, A. M., "A Phase-Plane Approach to the Compensation of Saturating Servomechanisms," *Trans. AIEE*, **70,** 631–639 (1951).

Hopkin, A. M., and M. Iwana, "A Study of a Predictor-Type Airframe Controller," *Trans. AIEE*, **75,** II, 1–9 (1956).

Hopkin, A. M., and P. K. C. Wang, "Further Studies of Relay-Type Feedback Control Systems Designed for Random Inputs," *Proc. IFAC Congress, Moscow*, 1960, Butterworths & Co. Ltd., London, 1961, Vol. 1, pp. 369–389.

Hopkin, A. M., and P. K. C. Wang, "A Relay-Type Feedback Control System Designed for Random Input," *Trans. AIEE*, **78,** II, 228–233 (1959).

Hung, J. C., and S. S. L. Chang, "Switching Discontinuities in Phase Space," *IRE Nat. Con. Record*, pt. 4 22–26 (1957).

Isaacs, D., D. T. Leondes, and R. A. Niemann, "On a Sequential Optimization Approach in Nonlinear Control," *Proc. JACC*, 158–166 (1966).

Isaev, V. K., "L. S. Pontryagin's Maximum Principle and Optimal Programming of Rocket Thrust" (in Russian), *Avtomatika i Telemekhanika*, **22,** 986–1001 (1961).

Jackson, R., "The Design of Control Systems with Disturbances Satisfying Certain Bounding Conditions, with Application to Simple Level Control Systems," *Proc. IFAC Congress, Moscow*, 1960, Butterworths & Co. Ltd., London, 1961, Vol. 1, pp. 498–509.

Johnson, C. D., "Optimal Control with Chebyschev Minimax Performance Index," *Proc. JACC*, 345–358 (1966).

Johnson, C. D., and J. E. Gibson, "Singular Solutions in Problems of Optimal Control," *IEEE Trans. Auto. Control*, **8,** No. 1, 4–15 (January 1963).

Johnson, C. D., and J. E. Gibson, "Optimal Control with Quadratic Performance Index and Fixed Terminal Time," *IEEE Trans. Auto. Control*, **9,** No. 4, 355–360 (October 1964).

Jones, G. S., "Asymptotic Fixed-Point Theorems and Periodic Systems of Functional-Differential Equations," *Contrib. Diff. Equations*, **II,** 385–405 (1963).

Jordan, B. W., and E. Polak, "Theory of a Class of Discrete Optimal Control Systems," *J. Electron. Control*, **17,** 697–713 (1964).

Jordan, B. W., and E. Polak, "Optimal Control of Aperiodic Discrete-Time Systems," *J. Soc. Ind. Appl. Math., Ser. A, Control*, **2,** No. 3, 332–346 (1965).

Kahn, A., "An Analysis of Relay Servomechanisms," *AIEE Trans.* **68,** 1079–1088 (1949).

Kalaba, R., "On Non-linear Differential Equations, the Maximum Operation, and Monotone Convergence," *J. Math. Mech.*, **8,** 519–574 (1959).

Kalman, R. E., "Phase-Plane Analysis of Automatic Control Systems with Non-linear Gain Elements," *Trans. AIEE*, **73,** 383–390 (1954).

Kalman, R. E., "Analysis and Design Principles of Second and Higher Order Saturating Servomechanisms," *Trans. AIEE*, **74,** II, 294–310 (1955).

Kalman, R. E., "Optimal Non-linear Control of Saturating Systems by Intermittent Action," *IRE, WESCON Conv. Record*, Part 4, 130–135 (1957).

Kalman, R. E., "Contributions to the Theory of Optimal Control," *Bol. Soc. Mat. Mex.*, **5,** 102–119 (1960).

Kalman, R. E., "On the General Theory of Control Systems," *Proc. IFAC Congress*, Butterworths & Co., Ltd, London **4,** 2020–2030 (1961).

Kalman, R. E., "First Order Implications of the Calculus of Variations in Guidance and Control," *Proc. Optimum System Synthesis Conf.*, 365–372 (1962).

Kalman, R. E., "Liapunov Functions for the Problem of Luré in Automatic Control," *Proc. Nat. Acad. Sci.*, **49,** 201–205 (1963).

Kalman, R. E., "Mathematical Description of Linear Dynamical Systems," *J. Soc. Ind. Appl. Math., Ser. A, Control*, **1**, No. 2. 152–192 (1963).

Kalman, R. E., "When is a Linear Control System Optimal?" *Trans. ASME, Ser. D, J. Basic Eng.*, **86** (1964), to appear.

Kalman, R. E., and J. E., Bertram, "General Synthesis Procedure for Computer Control of Single and Multiloop Linear Systems," (An optimal sampled-data system) *Trans. AIEE*, **77,** II, 602–609 (1958).

Kalman, R. E., and J. E. Bertram, "Control System Analysis and Design via the Second Method of Lyapunov," *Trans ASME, Ser. D, J. Basic Eng.*, **82,** 371–393 (1960).

Kalman, R. E., Y. C. Ho, and K. S. Narendra, "Controllability of Linear Dynamical Systems," *Contrib. Diff. Equations*, **1,** No. 2, 189–213 (1963).

Kalman, R. E., and R. W. Koepcke, "Optimal Synthesis of Linear Sampling Control Systems Using Generalized Performance Indexes," *Trans. ASME*, **80,** 1820 (1958).

Kalman, R. E., and R. W. Koepcke, "The Role of Digital Computers in the Dynamic Optimization of Chemical Reactions," *Proc. Western Joint Computer Conference*, San Francisco (March 1959).

Kang, C. L., and G. H. Fett, "Metrization of Phase Space and Non-linear Servo Systems," *J. Appl. Phys.*, **24,** 96–97 (1953).

Katz, S., "Best Temperature Profiles in Plug-Flow Reactors: Methods of the Calculus of Variations." *Ann. N.Y. Acad. Sci.*, **84,** 441–478 (1960).

Kazda, L. F., "Errors in Relay Servo Systems," *Trans. AIEE*, **72,** II, 323–328 (1953).

Kelendzheridze, D. L., "On the Theory of Optimal Tracking," *Dokl. Adak. Nauk SSSR*, **138,** No. 3 (1961).

Kelendzheridze, D. L., "Theory of an Optimal Pursuit Strategy" (in Russian), *Dokl. Akad. Nauk SSSR*, **138,** 529–532 (1961).

Kelendzheridze, D. L., "On a Certain Problem of Optimal Tracking," *Avtomatika i Telemekhanika*, **23,** No. 9, 1008–1013 (September 1962).

Kelley, H. J., "Guidance Theory and Extremal Fields," *IRE Trans. Auto. Control*, **7,** No. 5, 75–81 (1962).

Kelley, H. J., "A Transformation Approach to Singular Subarcs in Optimal Trajectory and Control Problems," *J. Soc. Ind. Appl. Math., Ser. A, Control*, **2,** No. 2, 234–240 (1964).

Kelley, H. J., "An Ensemble Averaging Approach to Optimal Guidance Polynomial Approximations," *Proc. Third IFAC Congress, London* (1966).

Kelley, H. J., and J. C. Dunn, "An Optimal Guidance Approximation for Quasi-Circular Orbital Rendezvous," *Proc. Second IFAC Congress, Basle* (September 1963).

Kharatishvili, G. L., "The Maximum Principle in the Theory of Optimal Processes with Delay" (in Russian), *Dokl. Akad. Nauk SSSR*, **136,** No. 1, 39–42 (1961).

Kipiniak, W., *Dynamic Optimization and Control: A Variational Approach*, MIT Press and John Wiley & Sons, Inc., New York, 1961.

Kirillova, F. M., "On the Correct Statement of One Problem on Optimal Regulation," *Izv. Vysshikh Ucheby a Zavednii, Matem.* **4**(5), 113–126 (1958).

Kirillova, F. M., "A Limiting Process in the Solution of an Optimal Control Problem" (in Russian), *Prikl. Matemat. Mekh.*, **24,** 277–282 (1960).

Kirillova, F. M., "Contribution to the Problem on the Analytical Construction of Regulators" (in Russian), *Prikl. Matemat. Mekh.*, **25,** 433–439 (1961).

Kirillova, L. S., "An Existence Theorem in Terminal Control Problems," *Automat. Remote Control*, **24,** 1071–1074 (1964).

Kleinman, D. L., *Fuel Optimal Control of Second and Third Order Systems with Different Time Constraints*, M.S. Thesis, Massachusetts Institute of Technology, Cambridge, Massachusetts, June 1963.

Koivuniemi, A. J., "Parameter Optimization in Systems Subject to Worst (Bounded) Disturbance," *Proc. JACC*, 668–674 (1966).

Kolmogorov, A. N., E. F. Mischenko, and L. S. Pontryagin, "On A Certain Probability Problem in Optimal Control," *Dokl. Akad. Nauk SSSR*, **145,** 993–995 (1962); *Soviet Math. Dokl.*, **3,** pp. 1143–1145 (1962).

Kolosov, G. E., and R. L. Stratonovič, "A Problem of Synthesis of an Optimal Regulator by Methods of Dynamic Programming," *Avtomatika i Telemekhanika*, **24,** 1165–1173 (1963).

Komarnitskaya, O. I., "Stability of Nonlinear Automatic Control Systems," *Prikl. Matemat. Mekh.*, **23,** No. 3 (1959); *J. Appl. Math. Mech.*, **23,** 716–729 (1959).

Kopp, R. E., "Pontryagin Maximum Principle," in G. Leitmann (ed.), *Optimization Techniques*, Academic Press Inc., New York, 1962.

Kosmodemyanski, A. B., "Extremal Problems for Particles of Variable Mass," *Dokl. Akad. Nauk SSSR*, **53** (1946). English translation in *Compt. Rend. Acad. Sci. URSS*, **53,** 17–19 (1946).

Kranc, G. M., and P. E. Sarachik, "An Application of Functional Analysis to the Optimal Control Problem," *ASME Paper* 62, *JACC*-4 (1962).

Kranc, G. M., and P. E. Sarachik, "On Optimal Control of Systems with Multi-Norm Constraints," *JACC Paper* No. I-4 (1963).

Krasovski, A. A., "Problems of Continuous Systems Theory of Extreme Control of Industrial Processes, *Proc. IFAC Congress, Basel*, 530 (1963).

Krasovski, N. N., "On the Theory of Optimum Control" (in Russian), *Avtomatika i Telemekhanika*, **18,** 960–970 (1957).

Krasovski, N. N., "On a Certain Problem of Optimum Control" (in Russian), *Prikl. Matemat. Mekh*, **21,** 670–677 (1957).

Krasovski, N. N., "A Contribution to the Theory of Optimum Regulation of Nonlinear second order systems," *Dokl. Akad. Nauk SSSR*, **126**, 267–270 (1959).

Krasovski, N. N., "On a Problem of Optimal Control of Non-linear Systems" (in Russian), *Prikl. Matemat. Mekh.* **23,** 209–229 (1959).

Krasovski, N. N., "Sufficient Conditions for Optimization" (in Russian), *Prikl. Matemat. Mekh,.* **23,** 592–594 (1959).

Krasovski, N. N., "On the Theory of Optimum Control" (in Russian), *Prikl. Matemat. Mekh.*, **23,** 625–639 (1959).

Krasovski, N. N., "On Optimum Control in the Presence of Random Disturbances" (in Russian), *Prikl. Matemat. Mekh.*, **24,** 64–79 (1960).

Krasovski, N. N., "Approximate Computation of Optimal Control by the Direct Method" (in Russian), *Prikl. Matemat. Mekh.*, **24,** 271–276 (1960).

Krasovski, N. N., "Choice of Parameters of the Optimum Stable System," *Proc. IFAC Congress, Moscow*, 1960, Butterworths & Co., Ltd., London, 1961, Vol. 1, pp. 465–468.

Krasovski, N. N., "On a Method of Constructing Optimal Trajectories" (in Russian), *Mat. Sb.*, **53**(95), pp. 195–206 (1961).

Krasovski, N. N., "On the Analytical Design of an Optimal Controller in Systems with Time Delay," *Prikl. Matemat. Mekh.*, **26,** No. 1 (1962).

Krasovski, N. N., "On the Theory of Controllability and Observability of Linear Dynamic Systems," *Prikl. Matemat. Mekh.*, **28,** No. 1, 3–14 (1964).

Krasovski, N. N., and A. M. Letov," The Theory of Analytic Design of Controllers" (in Russian), *Avtomatika i Telemekhanika*, **23,** No. 6, 713–720 (June 1962).

Krasovski, N. N., and E. A. Lidski, "Analytical Design of Controllers in Systems with Random Attributes" (in Russian), *Avtomatika i Telemekhanika*, **22,** 1145–1150, 1273–1278, 1425–1431 (1961).

Krasovski, N. N., and E. A. Lidski, "Analytical Design of Controllers in Stochastic Systems with Velocity-Limited Controlling Action" (in Russian), *Prikl. Matemat. Mekh.*, **25,** 420–432 (1961).

Kreindler, E., "Contributions to the Theory of Time-Optimal Control," *J. Franklin Inst.*, **275,** No. 4, 314–344 (April 1963).

Kreindler, E., and P. E. Sarachik, "On the Concepts of Controllability and Observability of Linear Systems; *IEEE Trans. Auto. Control*, **9,** No. 2, 129–136 (April 1964).

Krotov, V. F., "Methods for Solving Variational Problems on the Basis of the Sufficient

Conditions for an Absolute Minimum. I.," *Automat. Remote Control*, **23,** 1473–1484 (1963).

Krotov, V. F., "Methods for Solving Variational Problems. II. Sliding Regimes, *Automat. Remote Control*, **24,** 539–553 (1963).

Krug, E. K., and O. M. Mirina, "On Optimal Transients in an Automatic Control System with Limited Position of the Regulating Device" (in Russian), *Avtomatika i Telemekhanika*, **19,** 10–25 (1958).

Kuba, R. F., and L. F. Kazda, "A Phase-Space Method for the Synthesis of Non-linear Servomechanisms," *Trans. AIEE*, **75,** II, 282–290 (1956).

Kuba, R. I., and L. F. Kazda, "The Design and Performance of a Model Second Order Non-linear Servomechanism," *IRE Trans. Auto. Control*, 43–48 (July 1958).

Kulikowski, R., "On Optimum Control with Constraints," *Bull. Acad. Polon. Sci., Ser. Sci. Tech.*, **7,** 385–394 (1959).

Kulikowski, R., "Concerning the Synthesis of the Optimum Non-linear Control Systems," *Bull. Acad. Polon. Sci., Ser. Sci. Tech.*, **7,** 391–399 (1959).

Kulikowski, R., "Synthesis of a Class of Optimum Control Systems," *Bull. Acad. Polon. Sci., Ser. Sci. Tech.*, **7,** 663–671 (1959).

Kulikowski, R., "On the Synthesis of Adaptive Systems," *Bull. Acad. Polon. Sci. Ser. Sci. Tech.*, **7,** 697–707 (1959).

Kulikowski, R., "Optimizing Processes and Synthesis of Optimizing Automatic Control Systems with Non-linear Invariable Elements," *Proc. IFAC Congress, Moscow*, 1960, Butterworths & Co., Ltd., London, 1961, Vol. 1, pp. 469–476.

Kulikowski, R., "Concerning a Class of Optimum Control Systems," *Bull. Acad. Polon. Sci., Ser. Sci. Tech.*, **8,** 595–600 (1968).

Kulikowski, R., "On the Synthesis of Optimum Sampled Data Control Systems," *Bull. Acad. Polon. Sci., Ser. Sci. Tech.*, **8,** 673–679 (1960).

Kulikowski, R., "On Optimalization of Time-Varying, Inertial, and Non-linear Control Systems," *Bull. Acad. Polon. Sci., Ser. Sci. Tech.*, **9,** 477–486 (1961).

Kulikowski, R., "Optimalization of Non-linear Random Control Processes," *IFAC Congress, Basel*, 283/1-283/7 (1963).

Kumar, K. S. P., and R. Sridhar, "On the Identification of Control Systems by the Quasi-Linearization Method," *IEEE Trans. Auto. Control*, **9,** No. 2, 151–154 (April 1964).

Kumar, T. K., *On Optimal Programs in Dynamic Economic Policy Models*, Department of Economics, Iowa State University, May 10, 1965.

Kundren, H. K., "Maximum Effort Control of an Oscillatory Element," *IRE, WESCON Conv. Record*, **3,** Part 4, 116–124 (1959).

Kuntsevich, V. M., and Yu. V. Krementulo, "Invariance of Sampled-date and Adaptive Sampled-date Systems," *Proc. IFAC Congress, Basel*, 531/1-513/8 (1963).

Kurzweil, Ya., "The Analytical Design of Control Systems" (in Russian), *Avtomatika i Telemekhanika*, **22,** 688–695 (1961).

Kurzweil, F., "Dynamic Synthesis of Higher-Order Optimum Saturating Systems," *Trans. ASME, Ser. D, J. Basic Eng.*, **83,** 45–52 (1961).

Kurzweil, J., and Z. Vorel, "On Linear Control Systems," *Bull. Inst. Politeh. Iasi, S. N.* **6** (X) 13–20 (1960).

Kushner, H. J., "Optimal Stochastic Control," *IRE Trans. Auto. Control*, **7,** No. 5, 120–122 (October 1962).

Kushner, H. J., "On the Optimum Location of Observations for Linear control systems with unknown initial state," *IRE Trans. Auto. Control*, **9,** 144–150 (1963).

Kushner, H. J., "Near Optimal Control in the Presence of Small Stochastic Perturbations," *Proc. JACC*, Session XIV, Paper 4 (1964).

Kushner, H. J., "On the Dynamical Equations of Conditional Probability Density Functions with Applications to Optimum Stochastic Control Theory," *J. Math. Anal. Appl.*, **8,** 332–344 (April 1964).

Kushner, H. J., "A Time-Domain Successive Approximation Method for Some Linear Optimal Control Systems," *IEEE Trans. Auto. Control*, **9,** No. 3, 294–295 (July 1964).

Kushner, H. J., "On the Existence of Optimal Stochastic Controls," *J. Soc. Ind. Appl. Math., Ser. A, Control*, **3,** No. 3, 463–474 (1965).

Kushner, H. J., "Sufficient Conditions for the Optimality of a Stochastic Control," *J. Soc. Ind. Appl. Math., Ser. A, Control*, **3,** No. 3, 499–508 (1965).

Kushner, H. J., "On the Stochastic Maximum Principle with 'Average' Constraints," *J. Math. Anal. Appl.*, **12,** No. 1, 13–26 (1965).

La Barriere, R. Pallu de, "Duality in Dynamic Optimization," *J. Soc. Ind. Appl. Math., Ser. A, Control*, **4,** No. 1, 159–163 (1966).

Lakshmikantham, V., and N. Onuchi, "On the Comparison Between the Solutions of Ordinary Differential Systems," *Sep. Bol. Soc. Mat. S. Paulo*, **15,** Fasc. 1–2 (1960).

Laning, Jr., J. H., and R. H. Battin, *Random Processes in Automatic Control*, McGraw-Hill Book Co., Inc., New York, 1956.

Larichev, O. I., "Time Optimum Control of Systems Connected by Limitation," *Proc. Third IFAC Congress*, London, 1966.

Larsen, A., "Optimal Control and the Calculus of Variations," Ser. 60, Issue 462, Electronics Res. Lab., Univ. of Cal., 1962.

LaSalle, J. P., "Basic Principle of the 'Bang-Bang' Servo," *Bull. Am. Math. Soc.*, **60,** 154 (1954).

LaSalle, J. P., "Time Optimal Control Systems," *Proc. Nat. Acad. Sci. U.S.*, **45,** 573–577 (1959).

LaSalle, J. P., "The Time Optimal Control Problem," *Contrib. Theory Non-linear Oscillations* **5,** 1–24 (1959).

LaSalle, J. P., "The Time Optimal Control Problem," *Ann. Math.* **45,** 1 (1960).

LaSalle, J. P., "Time Optimal Control," *Bol. Soc. Mat. Mex.*, **5,** 120–124 (1960).

LaSalle, J. P., "The 'Bang-Bang' Principle," *Proc. IFAC Congress, Moscow*, 1960, Butterworths & Co. Ltd., London, 1961, Vol. 1, pp. 493–497.

LaSalle, J. P., "Complete Stability of a Nonlinear Control System," *Proc. Nat. Acad. Sci.*, **48,** 600–603 (1962).

LaSalle, J. P., "Automation and Control in the Soviet Union," *Mathematical Optimization Techniques*, University of California Press, Berkeley, Calif., 1963, pp. 303–308.

LaSalle, J. P., and S. Lefschetz, *Stability by Liapunov's Direct Method with Applications*, Academic Press Inc., 1961.

LaSalle, J. P., and S. Lefschetz, "Recent Soviet Contributions to Ordinary Differential Equations and Nonlinear Mechanics," *J. Math. Anal. Appl.*, **2,** 467–499 (1961).

LaSalle, J. P., and R. J. Rath, "Eventual Stability," *Proc. IFAC Congress, Basel*, 415/1–415/4 (1963).

Lawden, D. F., "Optimal Escape from a Circular Orbit," *Astronaut. Acta.* **41,** Fasc. 3 (1958).

Lawden, D. F., "Optimal Programme for Correctional Maneuvers," *Astronaut. Acta*, **6,** Fasc. 4 (1960).

Lawden, D. F., "Optimal Intermediate-Thrust Arcs in a Gravitational Field," *Astronaut. Acta*, **8,** 106–123 (1962).

Lawden, D. F., *Optimal Trajectories for Space Navigation*, Butterworths & Co. Ltd., London, 1963.

Lee, E. B., "Mathematical Aspects of the Synthesis of Linear Minimum Response Time Controllers," *IRE Trans. Auto. Control*, **5,** 283–290 (1960).

Lee, E. B., "Design of Optimum Multivariable Control Systems, *Trans. ASME*, **83,** 85–90 (1961).

Lee, E. B., "On the Time-Optimal Regulation of Plants with Numerator Dynamics," *IRE, Trans. Auto. Control*, **6,** 351–352 (1961).

Lee, E. B., "Discussion of Satellite Attitude Control," *ARS J.* (June 1962).

Lee, E. B., "A Sufficient Condition in the Theory of Optimal Control," *J. Soc. Ind. Appl. Math., Ser. A, Control*, **1,** No. 3, 241–245 (1963).

Lee, E. B., "Recurrence Equations and the Control of their Evolution," *J. Math. Anal. Appl.*, **7,** No. 1, 118–126 (August 1963).

Lee, E. B., "An Approximation to Linear Bounded Phase Coordinate Control Problems," *J. Math. Anal. Appl.*, **13,** No. 3, 550–564 (March 1966).

Lee, E. B., and L. Markus, "Optimal Control of Nonlinear Processes," *Arch. Rat. Mech. Anal.*, **8,** 36–58 (1961).

Lee, E. B., and L. Markus, "Synthesis of Optimal Control for Nonlinear Processes with One Degree of Freedom," *Symp. Nonlinear Vibrations.*, Kiev (September 1961).

Lee, E. B., and L. Markus "On Necessary and Sufficient Conditions for Time Optimal Control of Second-order Non-linear Systems, *Proc. IFAC Congress, Basel*, 416 (1963).

Lee, Socker, *Optimization of Servomechanisms Having Velocity Saturation*, Thesis, University of Minnesota, 1958.

Leonardo da Vinci, in C. Singer et al. (eds.), *A history of technology III*, Oxford University Press, London (1957), pp. 445–450 and p. 330.

Lefschetz, S., *Differential Equations: Geometric Theory*, Interscience Publishers, Inc., New York, 1957.

Lefschetz, S., "On Indirect Automatic Controls," *Intern. Union Theoret. Appl. Mech., Symp. Nonlinear Vibrations, Kiev* (September 1961).

Lefschetz, S., "Some Mathematical Considerations on Nonlinear Automatic Controls," *Contrib. Diff. Equations*, **1,** 1–28 (1963).

Lefschetz, S. "Liapunov Stability and Controls," *J. Soc. Ind. Appl. Math., Ser. A, Control*, **3,** No. 1 (1965).

Lefschetz, S., *Stability of Nonlinear Control Systems*, Academic Press, Inc., New York, 1965.

Leitmann, G., "On a Class of Variational Problems in Rocket Flight," *J. Aerospace Sci.*, **26,** 586–591 (1959).

Leitmann, G., "An Elementary Derivation of the Optimum Control Conditions," *Proc. Twelfth Int. Astronaut. Congr., Washington, D.C.* (1961).

Leitmann, G. (ed.), *Optimization Techniques with Applications to Aerospace Systems*, Academic Press Inc., New York, 1962.

Leitmann, G., "Variational Problems with Bounded Control Variables," in George Leitmann (ed.), *Optimization Techniques*, Academic Press Inc., New York, 1962.

Leitmann, G., "Some Geometrical Aspects of Optimal Processes," *J. Soc. Ind. Appl. Math., Ser. A, Control*, **3,** No. 1 (1965).

LeMay, J. L., "Recoverable and Reachable Zones for Control Systems with Linear

Plants and Bounded Controller Outputs" *Proc. Fifth Joint Automatic Control Conference, Stanford, California, June* 24–26 (1964).

Lerner, A. Ya., "Improvement of the Dynamical Properties of Automatic Compensators by Means of Non-linear Coupling," Parts I and II (in Russian), *Avtomatika i Telemekhanika*, **13,** 134–144, 429–444 (1952).

Lerner, A. Ya., "On the Limit of Time-Response in Automatic Control Systems" (in Russian), *Avtomatika i Telemekhanika*, **15,** 461–477 (1954).

Lerner, A. Ya., "The Construction of High-Speed Systems of Automatic Control with Limitations on the Phase Coordinates of the Plant" (in Russian), *Proc. Second All-Union Conference on the Theory of Automatic Control*, 1953, Akad. Nauk SSSR, Moscow, 1953, pp 305–324.

Lerner, A. Ya., and A. G. Butkovski, "On Optimal Control," *Proc. IFAC Congress, Moscow*, 1960, Butterworths & Co. Ltd., London, Vol. 1, pp. 544–546.

Letov, A. M., *Stability of Nonlinear Control Systems* (in Russian), State Publishers, Moscow, 1955.

Letov, A. M., "Analytical Controller Design," Parts I–III (in Russian), *Avtomatika i Telemekhanika*, **21,** 436–441, 561–568, 661–665 (1960).

Letov, A. M., "The Analytical Design of Control Systems" (in Russian), *Avtomatika i Telemekhanika*, **22,** 425–435 (1961).

Levine, W. S., and M. Athans, "On the Optimal Error Regulation of a String of Moving Vehicles," *Proc. JACC*, 661–667 (1966).

Levinson, N., "Minimax, Liapunov and 'Bang-Bang'," *J. Diff. Equations*, **2,** No. 2 218–241 (April 1966).

Lewis, D. C., and P. Mendelson, "Contributions to the Theory of Optimal Control. A General Procedure for the Computation of Switching Manifolds," *Trans. Am. Math. Soc.*, **110,** No. 2, 232–244 (February 1964).

Lewis, J. B., "The Use of Nonlinear Feedback to Improve the Transient Response of a Servomechanism," *Trans. AIEE*, **71,** pt. 11, 449–453 (1952).

Li, Y. T., "Optimalizing Systems for Process Control," *Instruments*, **25,** No. 1, 72–77; No. 2, 190–193; No. 3, 324–359 (1962).

Liapunov, A., "Sur les functions-vecteurs complètement additives" *Bull. Acad. Sci. USSR*, Vol. 4, 465–478 (1960).

Liapunov, A. M., "Problème général de la stabilité du mouvement," *Ann. of Math. Study 17*, Princeton (1947).

Litovchenko, I. A., "On a Problem of Optimum Control" (in Russian), *Avtomatika i Telemekhanika*, **21,** 1122–1133 (1960).

Litovchenko, I. A., "On an Isoperimetric Problem of Analytical Design" (in Russian), *Avtomatika i Telemekhanika*, **22,** 1553–1559 (1961).

Lubbock, J. K., "The Optimization of a Class of Non-linear Filters," *Proc. Inst. Elec. Eng., P.C.*, **107,** 60–74 (1960).

Lukes, D., "Application of Pontryagin's Maximum Principle in Determining the Optimal Control of a Variable Mass Vehicle," *ARS Paper*, No. 1927–61 (August 1961).

Lukes, D., *Optimal Control of Nonlinear Systems*, Ph.D. Thesis, University of Minnesota, 1966.

Lur'e, A. I., *Some Nonlinear Problems in the Theory of Automatic Control* (in Russian), State Publishers, Moscow, 1951.

Lur'e, K. A., "The Mayer-Bolza Problem for Multiple Integrals and the Optimization of the Performance of Systems with Distributed Parameters," *Prikl. Matemat. Mekh.*, **27,** No. 5, 842–853 (1963).

Lur'e, K. A., "Optimal Problems for Distributed Systems," *Proc. Third IFAC Congress*, London (1966).

MacDonald, D. C., "Non-linear Techniques for Improving Servo Performance," *Proc. Nat. Electron. Conf. Chicago*, **6**, 400–421 (1950).

MacDonald, D. C., "Multiple Mode Operation of Servos," *Rev. Sci. Instr.*, **23** (1952).

Magnin, J. P., and J. R. Burnett, "A Comparison of a Contactor Servomechanism with an Average-Power Constrained Linear Servomechanism," *Proc. Nat. Electron. Conf., Chicago*, **11**, 974–986 (1955).

Mangasarian, O. L., "Sufficient Conditions for the Optimal Control of Nonlinear Systems," *J. Soc. Ind. Appl. Math., Ser. A, Control*, **4**, No. 1, 139–152 (1966).

Markus, L., "Global Structure of Ordinary Differential Equations in the Plane," *Trans. Am. Math. Sol.*, **76**, 127–148 (1954).

Markus. L., "Controllability for Nonlinear Processes," *J. Soc. Ind. Appl. Math., Ser. A, Control*, **3**, 78–90 (1965).

Markus, L., "Stability of the Optimal Control Problem," *Proc. I.B.M. Symp.*, to appear.

Markus, L., *Controllability and Observability, Applications of Functional Analysis to Optimization*, in Academic Press, Inc., to appear.

Markus, L., "The Bang-Bang Principle," Lecture Series in Differential Equations, *AFOSR Report* (1965), to appear.

Markus, L., and E. B. Lee, "On the Existence of Optimal Controls," *Trans. ASME, Ser. D*, **84**, 1–8 (1962).

Markus, L. and E. B. Lee, "Synthesis of Optimal Control for Nonlinear Processes with One Degree of Freedom," *Symp. Nonlinear Vibrations, Kiev* **3**, pp. 200–218 (1963).

Markus, L., and E. B. Lee, "On Necessary and Sufficient Conditions for Time Optimal Control of Second-Order Nonlinear Systems," *Proc. IFAC Congress, Basel*, 416, 1–6 (1963).

Markus, L., and N. Wagner, "On Stability Theorems for Nonlinear Servomechanisms," *J. Math. Mech.* **6**, 393–400 (1957).

Markus, L., and H. Yamabe, "Global Stability Criteria for Differential Systems," *Osaka Math. J.* **12**, 305–318 (1960).

Mathews, K. C., and R. C. Boe, "The Application of Non-linear Techniques to Servomechanisms," *Proc. Nat. Electron. Conf., Chicago*, **8**, 10–21 (1952).

Mathews, M. V., and C. W. Steeg, "Final Value Control System Synthesis," *IRE Trans. Auto. Control*, **2**, 6–16 (February 1957).

Matystin, V. D., and V. A. Ryapolov, Use of Integral-Square Estimate to Determine the Optimal Parameters of an Autopilot with Rate Feedback," *Avtomatika i Telemekhanika* **20**, No. 4 (1959).

Maxwell, J. C., "On Governors," *Proc. Roy. Soc. London*, **16**, 270–283 (1868).

McCann, M. J., "Introduction to Variational Methods for Optimal Control," *Trans. Soc. Instr. Tech.*, **13**, 232–237 (1961).

McGill, R., "Optimal Control, Inequality State Constraints, and the Generalized Newton-Raphson Algorithm," *J. Soc. Ind. Appl. Math., Ser. A, Control*, **3**, No. 2, 291–298 (1965).

McShane, E. J., "On Multipliers for Lagrange Problems," *Am. J. Math.*, **61**, 809–819 (1939).

McShane, E. J., *Integration*, Princeton University Press, Princeton, N.J., 1944.

Meditch, J. S., "Optimal Thrust Programming for Minimal Fuel Midcourse Guidance," *Proc. Optimum System Synthesis Conf.*, 55–68 (1962).

Meditch, J. S., "On Minimal Fuel Satellite Attitude Control," *Proc. Fourth JACC*, 558–564 (June 1963).

Meditch, J. S., "Synthesis of a Class of Linear Feedback Minimum Energy Controls," *IEEE Trans. Auto. Control*, **8,** No. 4, 376–378 (1963).

Meditch, J. S., and L. W. Neustadt, "An Application of Optimal Control to Midcourse Guidance," *Proc. IFAC Conference, Basle* (1964).

Meditch, J. S., "A Class of Suboptimal Linear Controls," *Proc. JACC*, 776–782 (1966).

Melbourne, W. G., "Three Dimensional Optimum Thrust Trajectories for Power-Limited Propulsion Systems," *ARS* J. **31,** 1723–1728 (1961).

Melbourne, W. G., and C. G. Sauer, Jr., "Optimum Thrust Programs for Power-Limited Propulsion Systems," *Astronaut. Acta*, **8,** 205–227 (1962).

Melbourne, W. G., and C. G. Sauer, Jr., "Optimum Interplanetary Rendezvous with Power-Limited Vehicles," *AIAA J.* **1,** 54–60 (1963).

Merriam, C. W., "A Class of Optimum Control Systems," *J. Franklin Inst.*, **267,** 267–282 (1959).

Merriam, C. W., "Use of a Mathematical Error Criterion in the Design of Adaptive Control Systems," *Trans. AIEE*, **78,** II, 506–512 (1959).

Merriam, C. W., "An Optimization Theory for Feedback Control System Design," *Inform. Control*, **3,** 32–59 (1960).

Merriam, C. W., *Optimization Theory and the Design of Feedback Control Systems*, McGraw-Hill Book Co., Inc., New York, 1964.

Mesarovic, M. D., "On the Existence and Uniqueness of the Optimal Multivariable System Synthesis," *IRE Intern. Conv. Record*, Part 4, 10–14 (1960).

Meyer, K. R., "On a System of Equations in Automatic Control Theory," *Contrib. Diff. Equations*, **3,** pp. 163–173 (1964).

Miele, A., "An Extension of the Theory of the Optimum Burning Program for the Level Flight of a Rocket-Powered Aircraft," *J. Aeronaut. Sci.*, **24,** No. 12, 874–884 (1957).

Miele, A., "Generalized Variational Approach to the Optimum Thrust Programming for the Vertical Flight of a Rocket, Part I," *Z. Flugwiss*, **6,** 69–77 (1958).

Miele, A., and C. R. Cavoti, "Generalized Variational Approach to the Optimum Thrust Programming for the Vertical Flight of a Rocket, Part II," *Z. Flugwiss*, **6,** pp. 102–109 (1958).

Mikami, K., C. T. Battle, and R. S. Goodell, "Sortie Boost and Glide Trajectories Determined for Specified Mission Requirements by the Steepest-Ascent Technique," *Proc. Optimum System Synthesis Conf.*, 189–240 (1962).

Mikhailov, F. A., "Integral Indicators of Automatic Control Systems Quality," in V. V. Solodovnikov (ed.), *Automatic Control Fundamentals*, Mashgiz, 1954.

Mikhailov, F. A., "Integral Exponents of the Quality of the Automatic Regulation System," in V. V. Solodovnikvo (ed.), *Principles of Automatic Regulation*, Mashgiz, 1954.

Mikhailov, F. A., "On Boundary Values of Quadratic Estimations of the Quality and their Application in the Selection of Parameters in Automatic Regulation Systems," *Proc. Second All-Union Conference on the Theory of Automatic Regulation*, Vol. 2, Akad. Nauk SSSR, Moscow, 1955.

Minorsky, N., "Control Problems," *J. Franklin Inst.*, **232,** 451–487, 519–551 (1941).

Mischenko, E. F., and L. S. Pontryagin, One Statistical Problem of Optimal Control (in Russian), *Dokl. Akak. Nauk SSSR*, **128,** 890–892 (1959).

Mischenko, E. F., and L. S. Pontryagin, "On a Statistical Problem of Automatic Control" (in Russian), *Izv. Akad. Nauk SSSR, Ser. Mat.*, **25,** 477–498 (1961).

Mischenko, E. F., "On a Certain Problem for Parabolic Differential Equations Connected with Optimal Pursuit," *J. Soc. Ind. Appl. Math., Ser. A, Control*, Vol. 3, No. 1, 1965.

Mitsumaki, T., "Optimum Non-linear Control of Retarded Systems," *Bull. JSME*, **2,** 348–355 (1959).

Mufti, I. H., "Comments on An optimal strategy for a saturating sampled-data system." *IRE Trans. Auto. Control*, **8,** 350–351 (1961).

Munick, H., "Optimum Orbital Transfer Using N-Impulses," *ARS J.* **32,** No. 9, 1347–1350 (September 1962).

Munick, H., *On Nonlinear Optimal Control Problems with Control Appearing Linearly* Ph.D. Thesis, Adelphi University, June 1965.

Munick, H., R. McGill, and G. E. Taylor, "Analytic Solutions to Several Optimum Orbit Transfer Problems," *J. Astronaut. Sci.*, **7,** No. 4 (1960).

Nagata, A., S. Kodama, and S. Kumagai, "Time-Optimal Discrete Control System with Bounded State Variable," *IEEE Trans. Auto Control*, **10,** No. 2, April 1965, pp. 155–164.

Nahi, N. E., "Optimum Control of Linear Systems with a 'Modified' Energy Constraint," *IEEE Trans. Auto. Control*, **9,** 137–143 (April 1964).

Nahi, N. E., "On the Design of Time Optimal Systems Via the Second Method of Liapunov," *IEEE Trans. Auto. Control*, **9,** No. 3, 274–275 (July 1964).

Neimark, Yu. I., "Certain Numerical Methods for Determining Periodical Motions of Automatic Control Systems," *Avtomatika i Telemekhanika*, **22,** No. 1, 47–56 (January 1961).

Neiswander, R. S., "An Experimental Treatment of Non-linear Servomechanisms," *Trans. AIEE*, **75,** II, 308–316 (1956).

Neiswander, R. S., and R. H. MacNeal "Optimization of Nonlinear Control Systems by Means of Non-linear Feedbacks," *Trans. AIEE*, **72,** II, 262–272 (1953).

Nelson, W. T., "Pulse-Width Relay Control in Sampling Systems," *Trans. ASME, Ser, D, J. Basic Eng.* **83,** p. 15 (1961).

Nesbit, R. A., "The Problem of Optimal Mode Switching," *Proc. Optimum System Synthesis Conf.*, 41–54 (1962).

Nesbit, R. A., "The Use of the Technique of Linear Bounds for Applying the Direct Method of Liapunov to a Class of Non-linear and Time-varying Systems," *Proc. IFAC, Basel*, 420 (1963).

Nesline, F. W., "Optimum Response of Discontinuous Feedback Control System," *Trans. AIEE*, **77,** II, 651–658 (1958).

Neustadt, L. W., "Synthesizing Time Optimal Control Systems," *J. Math. Anal. Appl.*, **1,** 484–493 (1960).

Neustadt, L. W., "Russian Contributions to Control Theory," *IRE, Trans. Auto. Control*, **6,** 348 (1961).

Neustadt, L. W., "Time Optimal Control Systems with Position and Integral Limits," *J. Math. Anal. Appl.*, **3** (1961).

Neustadt, L. W., "Minimum Effort Control Systems," *J. Soc. Ind. Appl. Math., Ser. A, Control*, **1,** No. 1, 16–31 (1962).

Neustadt, L. W., "A Synthesis Method for Optimal Controls," *Proc. Optimum System Synthesis Conf.*, 373–382 (1962).

Neustadt, L. W., "The Existence of Optimal Controls in the Absence of Convexity Conditions, *Jour. Math. Anal. Appl.*, **7,** 110–117 (August 1963).

Neustadt, L. W., "On Synthesizing Optimal Controls," *Proc. Second IFAC Congress, Basle*, Paper 421 (1963).

Neustadt, L. W., "Optimization, A Moment Problem, and Nonlinear Programming," *J. Soc. Ind. Appl. Math., Ser. A, Control*, **2,** No. 1, 33–53 (1964).

Neustadt, L. W., "A General Theory of Minimum-Fuel Space Trajectories," *J. Soc. Ind. Appl. Math., Ser. A, Control*, **3,** No. 2, 317–356 (1965).

Nixon, Floyd E., *Principles of Automatic Controls*, Prentice–Hall, Inc., Englewood Cliffs, N.J., 1953.

Noton, A. R. M., P. Dyer, and C. A. Markland, "Numerical Computation of Optimal Control," *Proc. JACC*, 193–204 (1966).

Novoseltsev, V. N., "Optimum Process in Second-Order Pulse-Relay System," *Avtomatika i Telemekhanika*, **21,** No. 5 (May 1960).

Novoseltsev, V. N., "Optimal Control in a Second-Order Pulse-Relay System with Random Disturbances" (in Russian), *Avtomatika i Telemekhanika*, **22,** 865–875 (1961).

O'Donnell, J. J., "Bounds on Limit Cycles in Two-Dimensional Bang-Bang Control Systems with an Almost Time Optimal Switching Curve," *Proc. JACC* (1964).

Oǧuztöreli, M. N., "On a Time Optimal Control Problem," *J. Soc. Ind. Appl. Math. Ser. A, Control,* **1** (1963).

Oǧuztöreli, M. N., "A Time Optimal Control Problem for Systems Described by Differential-Difference Equations," *J. Soc. Ind. Appl. Math., Ser. A, Control,* **1,** No. 3, 290–310 (1963).

Oǧuztöreli, M. N., "Optimal Pursuit Strategy Processes with Retarded Control Systems," *J. Soc. Ind. Appl. Math., Ser. A, Control,* **2,** No. 1, 89–105 (1964).

Okamura, K., "Some Mathematical Theory of the Penalty Method for Solving Optimum Control Problems," *J. Soc. Ind. Appl. Math., Ser. A, Control,* **2,** No. 3, 317–331 (1965).

Okhotsimski, D. E., and T. M. Ennev, "Some Variational Problems Connected with the Launching of an Artificial Satellite of the Earth" (in Russian), *Usp. Fiz. Nauk*, **63,** 5–32 (1957).

Oldenburger, R., *Optimal Control*, Holt, Rinehart, Winston, 1966.

Orford, R. J., "Optimal Stochastic Control Systems," *J. Math. Anal. Appl.*, **6,** 419–429 (1963).

Osborn, H., "The Problem of Continuous Programs." *Pacific J. Math.*, **6,** 721–731 (1956).

Osborn, H., "On the Foundations of Dynamic Programming," *J. Math. Mech.*, **8,** 867–872 (1959).

Paiewonsky, B. H., "The Synthesis of Optimal Controllers," *Proc. Optimum System Synthesis Conf.*, 69–98 (1962).

Paiewonsky, B. H., and P. J. Woodrow, "The Synthesis of Optimal Controls for a Class of Rocket Steering Problems," *AIAA Summer Meeting* (1963).

Paiewonsky, B. H., P. J. Woodrow, W. Brunner, and P. Halbert, "Synthesis of Optimal Controllers Using Hybrid Analog-Digital Computers," in A. V. Balakrishnan and L. W. Neustadt (eds.), *Computing Methods in Optimization Problems*, Academic Press Inc., New York, 1964, pp. 285–303.

Paiewonsky, S. H., "Time-Optimal Control of Linear Systems with Bounded Control," in *International Symposium on Nonlinear Differential Equations and Nonlinear Mechanics*, Academic Press Inc., New York, 1963.

Pallu de la Barrière, R., *Optimal Control Theory*, Saunders Co., 1967.

Pavlov, A. A., "Synthesis of Certain Optimal Relay Systems by Phase Space Methods" (in Russian), *Izv. Akad. Nauk SSSR, Energ. Avtomat.*, No. 6, 118–126 (1959).

Pavlov, A. A., "On the Optimal Law of Control for a Relay System of the Third Order" (in Russian), *Izv. Akad. Nauk SSSR, Energ. Avtomat.*, No. 4, 102–108 (1960).

Pearson, A. E., "Aspects of Adaptive Optimal Steady State Control," *Proc. JACC*, 783–799 (1966).

Pearson, J. D., "Note on a Solution Due to A. T. Fuller," *J. Electron. Control*, **10,** 323–324 (1961).

Pearson, J. D., "Approximation Methods in Optimal Control," *J. Electron. Control*, **13,** 453–469 (November 1962).

Pearson, J. D., "Reciprocity and Duality in Control Programming Problems," *J. Math. Anal. Appl.*, **10,** 388–408 (1965).

Pechorina, N. N., "The Stability of Pulse-Width Modulated Automatic Control Systems," *Izv. Akad. Nauk SSSR, Energ. Avtomat.*, No. 2 (1960); STL Translation: STL-TR-61-6110-3.

Pervozvansky, A. A., "Uniformly Optimal Control Systems," *Proc. Third IFAC Congress, London* (1966).

Petras, S., "Certain Problems of Optimum Control with Incomplete Information," *Proc. Third IFAC Congress, London* (1966).

Pfeiffer, P. E., "The Maximum Response Ratio of Linear Systems," *Trans. AIEE*, **73,** II, 480–484 (1954).

Petrov, B. N., C. M. Ulanov, and S. V. Emelyanov, "Optimization and Invariance in Control Systems with Constant and Variable Structure," *Proc. IFAC Congress, Basel*, 532 (1963).

Pittel, B. G., "Some Problems of Optimum Control. I," *Automat. Remote Control*, **24,** 1078–1091 (1964).

Pocard, M., "Optimal Control of an Industrial Plant," *Proc. Third IFAC Congress, London* (1966).

Polya, G., and G. Szego, *Aufgaben und Lehrsatze aus der Analysis*, J. Springer, Verlag, Berlin, 1925; Dover Publications, Inc., New York, 1945.

Pontryagin, L. S., "On Some Differential Games," *J. Soc. Ind. Appl. Math. Ser. A, Control*, **3,** No. 1, 49–52 (1965).

Pontryagin, L. S., Optimal Control Processes (in Russian), *Usp. Mat. Nauk*, **14,** 3–20 (1959); translated in *Am. Math. Soc. Translations*, **18,** 321–339 (1961).

Pontryagin, L. S., V. G. Boltyanski, R. V. Gamkrelidze, and E. F. Mischenko, *The Mathematical Theory of Optimal Processes*, Interscience Publishers, Inc., New York, 1962.

Potter, J. E., "Matrix Quadratic Solutions," *J. Soc. Ind. App. Math.*, **14,** No. 3, 496–502 (1966).

Popov, E. P., "Approximate Methods of Studying Non-Linear Oscillations in Automatic Control Systems," *International Union of Theoretical and Applied Mechanics, Kiev*, 1–39 (September 1961).

Popov, E. P., *The Dynamics of Automatic Control Systems*, Addison-Wesley Publishing Co., Reading, Mass., 1962.

Popov, V. M., "Absolute Stability of Nonlinear Systems of Automatic Control," *Automat. Remote Control*, **22,** 857–875 (1961).

Popkov, Yu S., "Structural Design of Statistically Optimum Systems of Optimalizing Control," *Proc. Third IFAC Congress, London* (1966).

Pugachev, V. S., "On a Possible General Solution of the Problem of Determining Optimum Dynamic Systems," *Automat. Remote Control*, **17,** 585–589 (1956).

Pugachev, V. S., "The Use of Canonical Expansions of Random Functions in Determining an Optimum Linear System," *Automat. Remote Control*, **17,** 489–499 (1956); translation pp. 545–556.

Pugachev, V. S., *Theory of Random Functions and Its Application to Problems of*

Automatic Control (1957 Russian book), English translation of first Russian edition, WADD Translation No. F-TS-9546/V, No. 59.

Pugachev, V. S., "A Method for Solving the Basic Integral Equation of Statistical Theory of Optimum Systems in Finite Form," *Prikl. Mathemat. Mekh.*, **23,** 3–14 (1959); English translation pp. 1–16.

Pugachev, V. S., "The Method of Determining Optimum Systems Using General Bayes Criteria," *IRE Trans. Auto. Control*, **7,** 491–505 (December 1960).

Pyatnitskii, G. I., "The Effect of Stationary Random Processes on an Automatic Control System Containing Essentially Nonlinear Elements," *Avtomatika i Telemekhanika*, **21,** No. 4, 474–480 (April 1960).

Radner, R., "Paths of Economic Growth that are Optimal with Regard to the Final States, A Turnpike Theorem," *Rev. Econ. Studies*, **28** (2), No. 76, 98–104 (February 1961).

Razumikhin, B. S., "Mechanical Model and Method of Solution of the Resources Distribution Problem," *Proc. Third IFAC Congress, London* (1966).

Reid, W. T., "Ordinary Linear Differential Operators of Minimum Norm," *Duke Math. J.* **29,** 591–600 (1962).

Rekasius, Z. V., "A General Performance Index for Analytical Design of Control Systems," *IRE Trans. Auto. Control*, **6,** No. 2, 217–222 (1961).

Reynolds, P. A., and E. C. Rynaski, "Application of Optimal Linear Control Theory to the Design of Aerospace Vehicle Control Systems," *Proc. Optimum System Synthesis Conf.*, 159–188 (1962).

Riabov, Iu. A., "The Application of the Small-Parameter Method to an Investigation of Automatic Control Systems with Delay," *Avtomatika i Telemekhanika*, **21,** No. 6 (June 1960).

Rishel, R. W., "An Extended Pontryagin Principle for Control Systems Whose Control Laws Contain Measures," *J. Soc. Ind. Appl. Math. Ser. A, Control*, **3,** No. 2, pp. 191–205 (1965).

Roberts, J. D., "Extreme Regulation with Self-Adjustment of the Perturbation Amplitude," *Proc. Third IFAC, Congress London* (1966).

Roitenberg, Ya. N., "On the Accumulation of Disturbances in Transient Linear Impulsive Systems" (in Russian), *Prikl. Matemat. Mekh.*, **22,** 534–536 (1958).

Roitenberg, Ya. N., "On Some Indirect Methods of Obtaining Information on the Position of a Controlled System in the Phase Space" (in Russian), *Prikl. Matemat. Mekh.*, **25,** 440–444 (1961).

Roitenberg, Ya. N., "The Determination of the Position of a Nonlinear Controlled System in Phase Space," *Soviet Math.* **3,** No. 3, 888–892 (1962).

Rose, N. J., "Optimum Switching Criteria for Discontinuous Automatic Controls," *IRE Nat. Conv. Record*, pt. 4, 61–67 (1956).

Rosenbloom, A., "Final Value Systems with Total Effort Constraints," *Proc. IFAC Congress, Moscow*, 1960, Butterworth & Co. Ltd., London, 1961, Vol. 1, pp. 535–544.

Rosenman, E. A., "On the Limit Response of Servo Systems with Power, Torque and Velocity Limitations" (in Russian), *Avtomatika i Telemekhanika*, **19,** 633–653 (1958).

Ross, R., "Minimality for Problems in Vertical and Horizontal Rocket Flight," *ARS J.* **28,** No. 1, 55–56 (1958).

Ross, S. and G. Leitmann, "Low Acceleration Trajectory Optimization in a Strong Central Force Field," *Proc. IAS Symposium on Vehicle Systems Optimization* (November 1961).

Roxin, E., "Reachable Zones in Autonomous Differential Systems," *Bol. Soc. Mat. Max.*, **5,** 125–135 (1960).

Roxin, E., "The Existence of Optimal Controls," *Mich. Math. J.* **9,** 109–119 (1962).

Roxin, E., "A Geometric Interpretation of Pontryagin's Maximum Principle," in J. P. LaSalle and S. Lefschetz (eds.), *Nonlinear Differential Equations and Nonlinear Mechanics*, Academic Press Inc., New York, 1963.

Roxin, E., "Axiomatic Foundation of the Theory of Control Systems," *Proc. IFAC, Basel* (1963).

Roxin, E., "On Generalized Dynamical Systems Defined by Contingent Equations," *J. Diff. Equations*, **1,** No. 2, 188–205 (1965).

Roxin, E., "On Stability in Control Systems," *J. Soc. Ind. Appl. Math. Control, Ser. A*, **3,** No. 3, 357–372 (1965).

Roxin, E., "Stability in General Control Systems," *J. Diff. Equations*, **1,** No. 2, 115–150 (1965).

Roxin, E. and V. Spinadel, "Reachable Zones in Autonomous Differential Systems," *Contrib. Diff. Equations*, **1,** No. 3, 275–315 (1961).

Rozeman, E. A., "Optimal Transient Response in Power-Constrained Systems" (in Russian), *Avtomatika i Telemekhanika*, **18,** 497–513 (1957).

Rozenvasser, E. N., "On the Accurate Determination of Periodic Regimes in Sectionally Linear Automatic Control Systems," *Avtomatika i Telemekhanika*, **21,** No. 9 (1960).

Rozonoer, L. I., "Necessary Conditions of Optimality" (in Russian), *Dokl. Akad. Nauk SSSR*, **127,** 520–523 (1959).

Rozonoer, L. I., "The Maximum Principle of L. S. Pontryagin in the Theory of Optimum Systems," Parts I–III (in Russian), *Avtomatika i Telemekhanika*, **20,** 1320–1334, 1441–1458, 1561–1578 (1959).

Rozonoer, L. I."On the Variational Methods of Analysis of Automatic Control Systems Performance" *Proc. IFAC Congress, Moscow*, 1960, Butterworth & Co. Ltd., London, 1961, Vol. 1, pp. 477–480.

Russell, D. L., "Penalty Functions and Bounded Phase Coordinate Control," *J. Soc. Ind. Appl. Math., Ser. A, Control*, **2,** No. 3 (1965).

Rybashov, M. V., and E. E. Dudnikov, "Solution of Linear and Non-linear Programming Problems by Analogue Computers," *Proc. Third IFAC Congress, London* (1966).

Saltzer, C., and C. W. Fetheroff, "A Direct Variational Method for the Calculation of Optimum Thrust Programs for Power-Limited Interplanetary Flight," *Astronaut. Acta*, **7,** 8–20 (1961).

Salukvadze, M. E., "Analytic Design of Regulators (Constant Disturbances)" (in Russian), *Avtomatika i Telemekhanika*, **22,** 1279–1287 (1961).

Sarachik, P. E., and C. M. Kranc, "On Optimal Control of Systems with Multi-norm Constraints," *Proc. IFAC Congress, Basel*, 423 (1963).

Saridis, G. N., and Z. V. Rekasius, "Design of Approximately Optimal Feedback Controllers for Systems with Bounded State Variables," *Proc. JACC*, 70–75 (1966).

Scharmack, D. K., "The Equivalent Minimization Problem and the Newton-Raphson Optimization Method," *Proc. Optimum System Synthesis Conf.*, 119–158 (1962).

Schmaedeke, W. W., "Optimalizing Techniques for Injection Guidance" (with G. Swanlund), Presented at Amer. Rocket Society Guid., Control and Nar. Conf., Stanford, California, Aug. 7–9, 1961. Published in R. Robertson and J. Farrior (eds), *Guidance and Control*, Academic Press Inc., New York, 1961, Vol. 8, pp. 31–54.

Schmaedeke, W. W., "Optimal Control Theory for Nonlinear Vector Differential

Equations Containing Measures," *J. Soc. Ind. Appl. Math., Ser. A, Control* **3,** No. 2, 231–280 (1965).

Schmaedeke, W. W., "Time-Optimal Control with Amplitude and Rate Limited Controls" (with D. L. Russell) *J. Soc. Ind. Appl. Math., Ser. A, Control,* **2,** No. 3 (1965).

Schmaedeke, W. W., "Existence Theory of Optimal Controls," in C. T. Leondes (ed), *Advances in Control Systems: Theory and Applications,* Vol. 3, Academic Press Inc., New York, 1965.

Schmidt, S. F., "The Analysis and Design of Continuous and Sampled Data Feedback Control Systems with a Saturation Type Non-linearity," *NASA Tech. Note* D-20 (1959).

Shimanov, S. N., "On Almost-Periodic Oscillations in Non-Linear Systems with Lag," *Dokl. Akad. Nauk* **125,** No. 6, 1203–1206 (April 1959).

Shinbrot, M., "Optimization of Time-Varying Linear Systems with Nonstationary Inputs," *Trans. ASME,* **80,** 457–462 (1958).

Shinohara, Y., and J. Valat, "On the Junction Points of Optimal Trajectories," *Proc. Third IFAC Congress, London* (1966).

Sibuya, Y., K. Ogata, and T. Sekiguchi, "Optimal Control of a Linear Random Parameter System with Quadratic Performance Index," *Proc. JACC,* 457–462 (1966).

Siebenthal, C. D., and R. Aris, "Studies in Optimization—VII. The Application of Pontryagin's Methods to the Control of Batch and Tubular Reactors," *Chem. Eng. Sci.* **19,** 747–761 (1964).

Siebenthal, C. D., and R. Aris, "Studies in Optimization—VI. The Application of Pontryagin's Methods to the Control of a Stirred Reactor," *Chem. Eng. Sci.* **19,** 729–746 (1964).

Silva, L. M., "Predictor Servomechanisms," *IRE, Trans. Circuit Theory,* **1,** 56–69 (1954).

Silva, L. M., "Predictor Control Optimalizes Control System Performance," *Trans. ASME,* **77,** 1317 (1955).

Sivan, R., "Necessary and Sufficient Conditions for an Optimal Control to be Linear," *JACC Preprints,* 297–304 (1964).

Smith, E. S., *Automatic Control Engineering,* McGraw-Hill Book Co. Inc., New York, 1944.

Smith, F. B., "Time-Optimal Control of Higher-Order Systems," *IRE, Trans. Auto. Control,* **6,** 16–21 (1961).

Smith, F. B., and J. A. Lovingood, "An Application of Time-Optimal Control Theory to Launch Vehicle Regulation," *Proc. Optimum System Synthesis Conf.,* 99–118 (1962).

Smith, O. J. M., *Feedback Control Systems,* McGraw-Hill Book Co. Inc., New York, 1958 pp. 563–564.

Snow, D. R., *Reachable Regions and Optimal Controls,* Ph.D. Dissertation, Department of Mathematics, Stanford University, October 1964.

Snow, D. R., "Singular Optimal Controls for a Class of Minimum Effort Problems," *J. Soc. Ind. Appl. Math., Ser. A, Control,* **2,** No. 2, 203–219 (1964).

Solodovnikov, V. V., *Introduction to the Statistical Dynamics of Automatic Control Systems,* Dover Publications, Inc., New York, 1960.

Sonneborn, L. M., and F. S. Van Vleck, "The Bang-Bang Principle for Linear Control Systems," *J. Soc. Ind. Appl. Math., Ser. A, Control,* **2,** No. 2, 151–159 (1964).

Stakhovskii, R. I., "On the Comparison of Some Seeking Methods for the Automatic Optimizer," *Reports in the Conference on Theory and Application of Discrete Systems,* Moscow, 1958.

Stakhovskii, R. I., L. N. Fitsner, and A. B. Shubin, "Automatic Optimizers and Their Use for Solving Variational Problems and For Automatic Synthesis," *Proc. First IFAC Congress, Moscow* (1960).

Steeg, C. W., "A Time-Domain Synthesis for Optimum Extrapolators," *Trans. IRE, Auto. Control,* 32–41 (November 1957).

Stewart, E. C., and W. P. Kavanaugh, "Optimal Control of Saturating Systems with Stochastic Inputs," *Proc. Third IFAC Congress, London* (1966).

Stiles, J. A., *Time Optimal Control of a Two Variable System*, Ph.D. dissertation, 1964, Cambridge University.

Stout, T. M., "Effects of Friction in an Optimum Relay Servomechanism," *Trans. AIEE,* **72,** II, 329–336 (1953).

Stout, T. M., "Switching Errors in an Optimum Relay Servomechanism," *Proc. Nat. Electron. Conf.*, Chicago, **9,** 188–198 (1953).

Stratonovich, R. L., "On the Theory of Optimal Control: Sufficient Coordinates," *Avtomatika i Telemekhanika*, **23,** No. 7, 910–917 (1962).

Stratonovich, R. L., "Most Recent Development of Dynamic Programming Techniques and Their Application to Optimal Systems Designs," *Proc. Second IFAC Congress, Basel* 1963, Butterworth & Co. Ltd., London, 1964.

Stromer, P. R., "Adaptive or Self-Optimizing Control Systems–A Bibliography" *IRE Trans. Aut. Control,* 65–68 (May 1959).

Stubberud, A. R., "A Controllability Criterion for a Class of Linear Systems," *Wescon Paper* 12.1, 1963.

Sun-Jian, and Hang King-ching, "Analysis and Synthesis of Time-optimal Control Systems," *Second IFAC Congress, Basle* (September 1963).

Sun-Tsyan, "Synthesis of the Control Unit of a Servo System Which is Optimum for High Speed" (in Russian), *Avtomatika i Telemekhanika*, **20,** 273–288 (1959).

Sun-Tsyan, "Optimal Control of a Non-linear System" (in Russian), *Avtomatika i Telemekhanika*, **21,** 3–14 (1960).

Sun-Tsyan, "Synthesis of Optimal Control Systems on the Basis of Isochrones" (in Russian), *Izv. Akad. Nauk SSSR, Energ. Avtomat.* No. 5, 96–103 (1960).

Sutherland, J. W., and E. V. Bohn, "A Numerical Trajectory Optimization Method Suitable for a Computer of Limited Memory," *Proc. JACC,* 177–185 (1966).

Swinnerton-Dyer, H. P. F. "On an Extremal Problem," *Proc. London Math. Soc.*, **7,** 568–583 (1957).

Syui-Yan, T., and Lui-Vy, T., "Auto-Oscillations in a Single-Loop Automatic Control System Containing Two Symmetric Relays," *Avtomatika i Telemekhanika*, **20,** No. 1 (January 1959).

Szegö, G. P., "On a New Partial Differential Equation of the Stability Analysis of Time Invariant Control Systems," *J. Soc. Ind. Appl. Math. Ser. A, Control,* **1,** 63–75 (1962).

Szegö, G. P., "On the Absolute Stability of Sampled-Data Control Systems," *Proc. Nat. Acad. of Sci., U.S.*, **50,** 558–560 (1963).

Szegö, G. P., "A Contribution to Liapunov's Second Method: Nonlinear Autonomous Systems," *Proc. of the International Symposium on Nonlinear Differential Equations and Nonlinear Mechanics*, Academic Press, Inc. New York, 1963, pp. 421–430.

Szegö, G. P., "On the Application of Zubov's Method of Constructing Liapunov Functions for Nonlinear Control Systems," *Trans. ASME, Ser. D, J. Basic Eng.* **85,** No. 2, 137–142, 1963.

Szegö, G. P., "New Methods for Constructing Liapunov Functions for Time-invariant Control Systems," *Proc. IFAC Congress, Basel*, 600 (1963).

Szegö, G. P. and G. R. Geiss, "A Remark on 'A New Partial Differential Equation For the Stability Analysis of Time Invariant Control Systems'," *J. Soc. Ind. Appl. Math., Ser. A, Control*, **1,** No. 3, 369–376 (1963).

Takamatsu, T. and E. Nakanishi and I. Hashimoto, "Optimal Control and its Sensitivity for a Multi-Stage Process," *Proc. Third IFAC Congress, London* (1966).

Takhovskiy, R. I., "Multichannel Automatic Optimizer for Solving Variational Problems," *Automatics and Telemechanics*, **20,** No. 11 (1959).

Tinbergen, J., "Optimum Savings and Utility Maximization Over Time," *Econometrica*, **28,** No. 2, 481–489 (April 1960).

Ting, L., "Optimum Orbital Transfer by Several Impulses," *Astronaut. Acta*, **6,** Fasc. 5 (1960).

Tokumaru, H. and N. Saito, "On the Absolute Stability of Automatic Control System with Many Nonlinear Characteristics," *Mem. Fac. Eng. Kyoto Univ.*, **27,** Part 3 (1965).

Tou, J. R., *Modern Control Theory*, McGraw-Hill Book Co., Inc., 1964.

Traksal, Dzh., *Synthesis of Automatic Control Systems*, Mashgiz, 1959.

Troitski, V. A., "The Mayer-Bolza Problem of the Calculus of Variations and the Theory of Optimum Systems" (in Russian), *Prikl. Matemat. Mekh.*, **25,** 994–1010 (1961).

Troitski, V. A., "Variational Problems in Control Process Optimization" *Prikl. Matemat. Mekh.*, **26,** No. 1 (1962).

Truxal, J. G., *Automatic Feedback Control System Synthesis*, McGraw-Hill Book Co., Inc., New York, 1955.

Truxal, J. G., "Computers in Automatic Control Systems," *Proc. IRE*, **49,** 308 (1961).

Truxal, J. G., "Feedback Design and Optimal Control Theory," *Proc. Optimum System Synthesis Conf., Dayton, Ohio*, 1–9 (1962).

Tsien, H. S., and R. C. Evans, "Optimum Thrust Programming for a Sounding Rocket," *J. ARS*, **21,** 99–107 (1951).

Tsien, H. S., *Engineering Cybernetics*, McGraw-Hill Book Co., Inc., New York, 1954.

Tsypkin, Ya. Z., *Theory of Relay-Systems of Automatic Control* (in Russian), Gostechnizdat, Moscow, 1955.

Tsypkin, Ya. Z., "On Optimal Processes in Pulsed Automatic Systems," *Proc. Acad. Sci., USSR*, **134,** No. 2 (Sept. 1960); STL Translation: STL-TR-61-5110-12.

Tsypkin, Ya. Z., "Fundamentals of the Theory of Non-linear Pulse Control Systems," *Proc. IFAC Congress, Basel*, 537 (1963).

Ulanov, G. M., "The Analysis of Processes of Autom. Regulation at the Presence of Reactions, Restricted According to the Modulus," in V.V. Solodovnikov (ed.) *The Principles of Automatic Regulation*, Mashgiz, 1954.

Uzgiris, S. C., and A. F. D'Souza, "Optimal Control of Distributed Parameter Systems with Nonlinear Boundary Conditions," *Proc. JACC*, 675–683 (1966).

Van Eijk, C. J., and J. Sandee, Quantitative Determination of an Optimum Economic Policy," *Econometrica*, **27,** No. 1, 1–13 (January 1959).

Van Gelder, A., J. Dunn, and J. Mendelsohn, "The Final Value Optimal Stochastic Control Problem with Bounded Controller," *Proc. JACC*, 441–449 (1966).

Van Slyke, R., and R. Wets, "Programming Under Uncertainty and Stochastic Optimal Control," *J. Soc. Ind. Appl. Math., Ser. A, Control*, **4,** No. 1, 179–193 (1966).

Vichnevetsky, R., "The Non-linear Optimization of Servomechanisms" (in French), *Acad. Roy. Belg., Bull. Classe Sci.*, **44,** 493–502 (1958).

Vyshregradskii, J., "Sur la théorié géneral de régulateure," *Compt. Rend. Acad. Sci, Paris*, **83,** 318–331 (1876).

Wang, P. K. C., "Comments on 'Mathematical Aspects of the Synthesis of Linear Minimum Response-Time Controllers'," *IRE, Trans. Auto. Control.* **6,** 349–350 (1961).

Warga, J., "Relaxed Variational Problems," *J. Math. Anal. Appl.*, **4,** 111–128 (1962).

Warga, J., "Necessary Conditions for Minimum in Relaxed Variational Problems," *J. Math. Anal. Appl.*, **4,** 129–145 (1962).

Watson, J. W., *Preliminary Investigation: Optimum Re-entry Trajectories with Independent Lift and Drag Control.* Preliminary report prepared on Air Force Contract AF33(657)-11477 Leondes, University of California, Los Angeles, August 1963.

Ważewski, T. "Sur les systèms de commande non lineaires dont le contre de maine de commande n'est pas forcément convexe" *Bull. Acad. Polon. Sci., Ser. Sci. Math. Astron. Phys*, **10,** 17–21 (1962).

Weidemann, Henry L., *Optimum Re-entry Trajectories Using Dynamic Programming*, M.S. Thesis, Department of Engineering, University of California, Los Angeles, July 1963.

Weiss, L., "On a Question Related to the Control of Linear Systems," *IEEE Trans. Auto. Control*, **9,** No. 2 (1964).

Weiss, L., "The Concepts of Differential Controllability and Differential Observability," *J. Math. Anal. Appl.*, **10,** No. 2, 442–449 (1965).

Weiss, L., and R. E. Kalman, "Contributions to Linear System Theory," *Int. J. Eng., Sci.*, **3,** 141–171 (1965).

West, J. C., and I. R. Dalton, "The Step-Function Response of an R.P.C. Servomechanism Possessing Torque Limitation." *Proc. Inst. Elec. Eng.*, **101,** II, 166–177 (1954).

West, J. C., J. L. Douce, and R. Naylor, "The Effects of the Addition of some Nonlinear Elements on the Transient Performance of a Simple R.P.C. System Possessing Torque Limitation," *Proc. Inst. Elec. Eng.*, **101,** II, 156–165, 173–177 (1954).

West, J. C., and P. N. Nikiforuk, "The Frequency Response of a Servomechanism Designed for Optimum Transient Response," *Trans. AIEE*, **75,** II, 234–239 (1956).

West, J. C., and Somerville, M. J., "Integral Control with Torque Limitation," *Proc. Inst. Elec. Eng.*, **103C,** 407–419 (1956).

Westcott, J. H., "Design of Multivariable Optimum Filters," *Trans. ASME*, **80,** 463–467 (1958).

Westcott, J. H., J. J. Florentin, and J. D. Pearson, "Approximation Methods in Optimal and Adaptive Control," *Proc. Second IFAC Congress, Basel*, 1963, to be published by Butterworths & Co. Ltd, London.

Westcott, J. H., D. Q. Mayne, G. F. Bryant, and S. K. Mitter, "Optimal Techniques for On-Line Control," *Proc. Third IFAC Congress, London* (1966).

Whitback, Richard F., "The Application of Linear Optimal Control to Systems Containing Uncertain Parameters," *Proc. JACC*, 463–474 (1966).

Williams, A. J., "Combined thyratron and tachometer Speed Control of Small Motors," *Trans. AIEE*, **57,** 565–568 (1938).

Wilts, C. H., *Principles of Feedback Control*, Addison-Wesley Publishing Company, Reading, Massachusetts, 1960.

Witsenhausen, H. S., "Hybrid Techniques applied to Optimization Problems," *Spring Joint Computer Conference* (1962).

Wonham, W. M., "Note on a Problem in Optimal Nonlinear Control," *J. Elec. Control*, **15,** 59–62 (1963).

Wonham, W. M., and C. D. Johnson, "Optimal Bang-Bang Control with Quadratic Performance Index," *Trans. ASME Ser. D*, **86,** No. 1, 107–115 (March 1964).

Xirokostas, D. A., and J. O. Henderson, "Extremum Control of Dynamic Systems in the Presence of Random Disturbances and Noise," *Third IFAC Congress*, London (1966).

Yaari, M. E., "On the Existence of an Optimal Plan in a Continuous Time Allocation Process," *Econometrica*, **3**, No. 4, 576–590 (October 1964).

Yakubovich, V. A., "On Asymptotic Stability in the Large of the Unperturbed Motion of the Equations of a Nonlinear Automatic Control System," *Vest. Len. Univ.*, **19**, 172–176 (1957).

Yakubovich, V. A., "The Stability on the Whole of the Undisturbed Motion for Equations of Indirect Automatic Regulation," *Vestnik LGU* No. 19 (1957).

Yakubovich, V. A., "On Nonlinear Differential Equations of Automatic Control Systems," *Vest. Len. Univ., Ser. Mat., Mekh., Astron.*, 120–153 (1960).

Yakubovich, V. A., "The Solution of Certain Matrix Inequalities in Automatic Control Theory," *Dokl. Akad. Nauk SSSR*, **143**, 1304–07 (1962).

Yeh, H. H., "Design of Optimum Control for a Class of Distributed Parameter Systems," *Proc. JACC*, 684–693 (1966).

Yemelyanov, S. V., and A. I. Fedotova, "The Design of Optimal Automatic Control Systems of the Second Order Using Limiting Values of the Elements of the Control Circuit," *Avtomatika i Telemekhanika*, **21**, No. 12 (1960).

Zadeh, L. A., "Optimum Nonlinear Filters," *J. Appl. Phys.*, **24**, 396–404 (April 1953).

Zadeh, L. A. "Optimality and Non-Scalar-Valued Performance Criteria," *IEEE Trans. Auto. Control*, **8**, No. 1, 53–60 (January 1963).

Zadeh, L. A., and C. A. Desoer, *Linear System Theory: The State Space Approach*, McGraw-Hill Book Company, Inc., New York, 1963.

Zadeh, L. A., and B. H. Whalen, "On Optimal Control and Linear Programming," *IRE Trans. Auto. Control*, **7**, No. 4, 45 (July 1962).

Zubov, V. I., "Mathematical Methods of Investigating Automatic Regulation Systems" (1959); English Trans. AEC-tr-4494.

Zubov, V. I., "The Theory of Analytical Design of Controllers," *Avtomatika i Telemekhanika*, **24**, No. 8, 1037–1041 (1963).

Index